MICROCOMPUTER ELECTRONICS

Other books by the same author:

ELECTRONIC CIRCUIT BEHAVIOR

A quantitative yet highly simplified approach to Basic Electronics.

ELECTRONICS POCKET HANDBOOK

Most-needed formulas, charts, and component data,
unit conversions, and glossary of terms—all in a convenient pocket size.

22 MICROCOMPUTER PROJECTS TO BUILD, USE, AND LEARN

Learn microcomputers by building your own games and instruments from readily available microchips.

Cover Photo

Most semiconductors are created from quartz rock (not sand as is commonly believed), which is melted in furnaces at 2000° centigrade to produce molten silicon. After further processing, a silicon seed crystal is lowered into the molten material and the resulting crystal-growing process reproduces the atomic structure of the seed, producing a silicon ingot. The ingots are sliced into wafers about the thickness of a business card, and later thousands of chips will be diced from these wafers. The microcircuitry on each chip is created through a variety of photographic and chemical processes. Today tens of thousands of transistors are formed on a single microchip to make a microcomputer. In a few years it is projected that over 10 million transistors will crowd on a single chip. This creation of one of man's greatest inventions from one of nature's most basic elements epitomizes man's control of, and dependence upon, his environment.

MICROCOMPUTER ELECTRONICS

A Practical Approach to Hardware, Software, Troubleshooting, and Interfacing

DANIEL L. METZGER

PRENTICE HALL

Englewood Cliffs, New Jersey 07632

Library of Congress Cataloging-in-Publication Data

Metzger, Daniel L.
Microcomputer electronics.

Bibliography: p.
1. Microcomputers—Maintenance and repair.
2. Microprocessors—Maintenance and repair. 3. Computer interfaces—Maintenance and repair. I. Title.
TK7887.M43 1989 621.391'6 87-19352
ISBN 0-13-579871-X

Editorial/production supervision
and interior design: *Theresa A. Soler/Anne Kenney*
Cover design: *Photo Plus Art*
Manufacturing buyer: *Pete Havens*

Printed in the United States of America

10 9 8 7 6 5 4 3 2 1

ISBN 0-13-579871-X

Prentice-Hall International (UK) Limited, *London*
Prentice-Hall of Australia Pty. Limited, *Sydney*
Prentice-Hall Canada Inc., *Toronto*
Prentice-Hall Hispanoamericana, S.A., *Mexico*
Prentice-Hall of India Private Limited, *New Delhi*
Prentice-Hall of Japan, Inc., *Tokyo*
Simon & Schuster Asia Pte. Ltd., *Singapore*
Editora Prentice-Hall do Brasil, Ltda., *Rio de Janeiro*

Dedication

When I was a child I brought her a fistfull of dandelions from the edge of her own yard, thinking it an elegant gift of flowers. It seems that we are seldom able to repay those who have been our greatest benefactors in any coin that is not already their own. We can, however, pass on the gifts we have received to those who follow us. If, as its author truly hopes, this book shall be judged by the reader to have a value greater than the purchase price, think of it as the passing on, in a different coinage, of the gifts that were originally given to me by my Mother

Catharine Kroeger Metzger

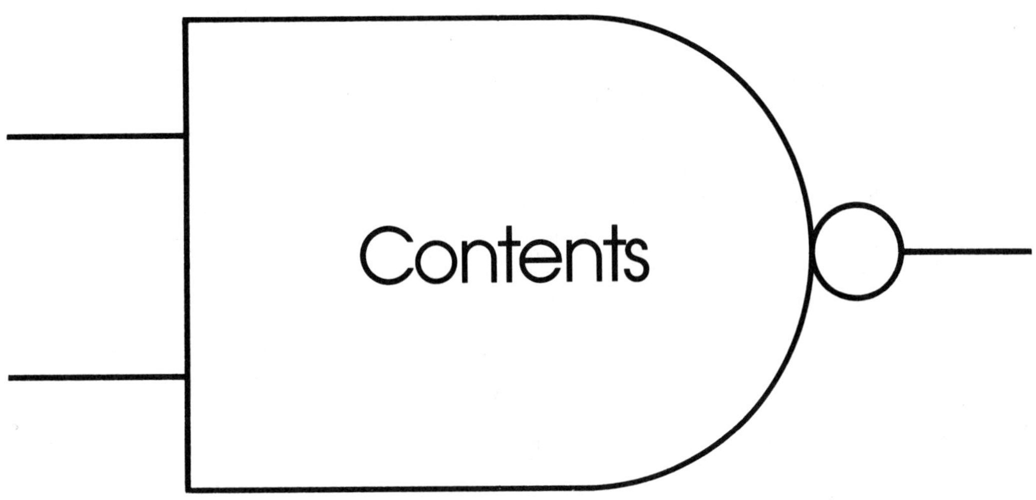
Contents

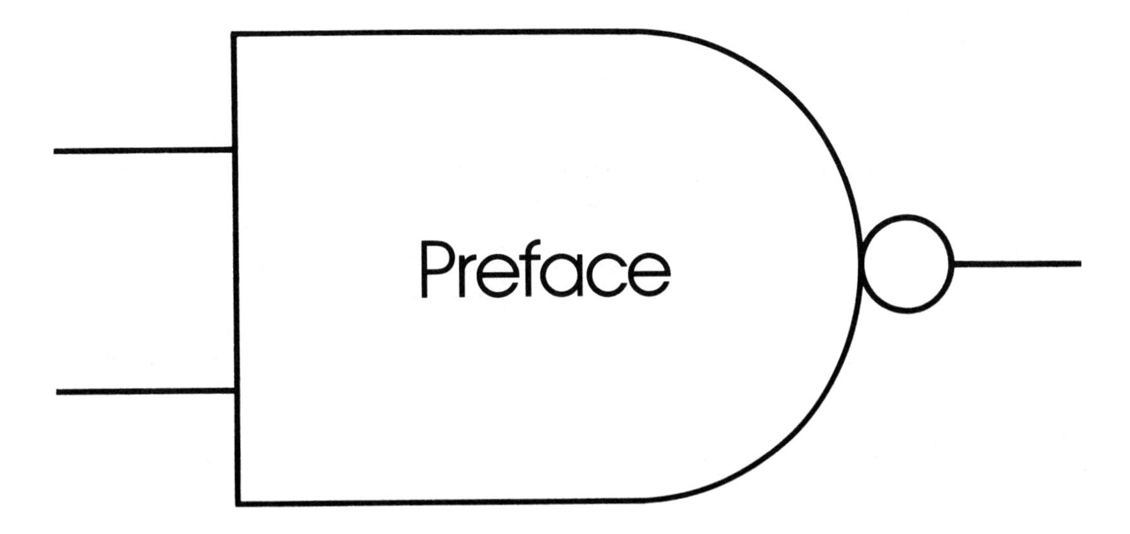

This text has been written to overcome two obstacles that have long plagued students of electronics in their efforts to learn microcomputers. These are *system complexity* and the *dominance of programming*.

Complexity. Many of us remember fondly the "one transistor" or "one IC" project that gave us our first foothold in understanding conventional electronics. But with microcomputer ICs, it has seemed that one has to *know* almost everything about them before one can *do* anything with them. There seemed to be no easy place to start. Well, this book has the one-chip microcomputer project, and it moves from there to two chips, and then three . . . always slowly, one step at a time, with real, buildable hardware applications and reviews along the way to help you maintain a firm footing as you advance toward the level of system complexity that is common in today's industry.

Programming has tended to dominate microcomputer instruction, so that many microcomputer courses are really machine-language programming courses. One can understand the reluctance of instructors to risk microcomputer systems costing hundreds of dollars in student lab projects, but one can also understand the impatience of the technicians to get their hands off the keyboard and on to the IC chips. These concerns are now obsolete. The prices of microcomputer chips have dropped to only a few dollars, and this book will show you how to wire them and program them for working projects that are no more expensive than were the vacuum-tube projects of a few decades ago. It is no longer necessary to keep hands off the chips for fear of damaging something expensive.

Why the 6802? If you accept the proposition that it is time for student electronics technicians to get their hands on the microprocessor chips, then it will be necessary to choose one chip to learn first. Hypothetical or generic processors only introduce an unnecessary extra step into the process. Tackling two or three processors at once in order to compare them is a sure-fire recipe for confusion, because there are numerous differences among them, not only in their hardware and programming but also in the definitions of the words used to explain them. The assertion is often made that all processors are basically similar and that if you learn one it is easy to learn another—but my experience and that of my students is that this is simply not true.

The 6802 is featured in the first three of the four parts of this book for the very good reason that it is the easiest to learn and simplest for building small experimental systems. This is not an opinion but a demonstrable fact. Other processors may be newer, more computationally powerful, and more popular, but if the objective is simplicity, no other processor has a simpler instruction set, simpler clock timing, simpler memory interfacing, and fewer machine registers.

The 6802 is functionally identical to the 6800, so any of the several 6800-based training systems can be used to run the programs and interfacing projects in this book. The last quarter of the book introduces three of the newer and more computationally powerful processors, including the 16/32-bit 68000 family. The approach here is still "hands on the chips."

The Structure of the Text

This book begins with Chapter Ø, which is a review of digital electronics, emphasizing those concepts that are most important to a study of microcomputers. The book is then divided into four parts, each with a specific objective.

Part I Elementary Microcomputer Concepts

The first part of the text is designed to answer three fundamental questions about microcomputers:

1. How does a computer use electrical signals to execute a single program instruction, such as ADD? This is the subject of Chapters 1 and 2.
2. How are instructions grouped to form a program, and how can the computer be made to execute a series of instructions repeatedly until the desired result is obtained? Chapters 3 and 4 explain memory accessing and program branching.
3. How can the computer be made to accept inputs and deliver outputs that are useful to the human operator? Chapters 5 and 6 explain elementary I/O interfaces.

Once you understand these three key concepts, computers will lose their aura of mystery, and you will be in a strong position to continue learning the more advanced topics, and the hardware and software details presented in the rest of the text.

Part II Software, Subroutines, and Troubleshooting

The second part of the text contains six "bread-and-butter" chapters, full of information that is used every day by microcomputer engineers and technicians. Indexed addressing (Chapter 7) is the source of the computer's power, because it allows the computer to solve a problem for a thousand different conditions, whereas the programmer only has to solve it once.

Software development requires more effort than any other task in microcomputing. Chapters 8 and 10 explain program instructions and present some aids for making the programming task more efficient. Subroutines and interrupts (Chapter 9) are universally used to ease the programming burden and expand the hardware capabilities of computer systems.

Troubleshooting microcomputer systems requires tools and procedures that are often very different from those used in analog electronic systems. Chapters 11 and 12 are devoted entirely to microcomputer troubleshooting.

Part III Microcomputer Interfacing

The third quarter of the text explains a variety of techniques for connecting microcomputers to the real world. Some of the examples presented include:

- Analog inputs and digital recording in Chapter 13.
- Parallel interfacing, including stepper motors and voice synthesis, in Chapters 14 and 15.
- The RS-232 serial interface, telephone data communication, and disk-drive systems, in Chapters 16 and 17.
- Video displays and graphics techniques, in Chapter 18.

Part IV Some Advanced Microprocessors

The last part of the text introduces three more-modern microprocessors, each very popular among computer engineers. With the solid background obtained by studying a simpler microcomputer and its interfacing, you will be ready to understand these more advanced machines and to make comparisons among them. The new processors, presented in two chapters each, are:

- The 6809, which is an advanced version of the 6800/6802. It retains nearly all of the features of the 6800 and adds an impressive set of new registers, instructions, and features. It is chosen to represent a modern 8-bit microprocessor.
- The Zilog Z80, which is an improvement on the Intel 8080—the machine that started the microprocessor revolution. It contrasts sharply with the 6800 and 6809 in many ways.
- The 68000, representing the new generation of 16- and 32-bit microprocessors. The hardware circuits presented use the 68008 because it requires only one chip each for RAM and ROM. (The 68000 would require two each.) The Intel 8086 processor is described briefly for comparison.

Each part of the text contains six chapters. Each chapter contains six informational sections plus a seventh section presenting a hands-on project illustrating the material of that chapter. There are 10 review questions at the end of each even-numbered section, with answers at the ends of the chapters. Each chapter concludes with a number of questions, problems, and project suggestions.

Throughout the book the unslashed zero is used for the normal decimal number system. The slashed zero (Ø) is used when the number system is binary, hexadecimal, or decimal with a starting count of Ø rather than 1. A computer BLOCK typeface has been used in the text to refer to computer instruction codes and computer machine registers.

The index is placed at the back of the book to make it easy to find, not because you are to save looking at it until last. Thumb through it now. Use it to help you find the answers to your questions.

Acknowledgments

Several persons have contributed to the quality of this book, and my gratitude for their efforts must not be left unexpressed. For proofreading the manuscript and making valuable suggestions for improving its currency, accuracy, and clarity:

- David Clay, Vice President of Engineering, TL Industries, Northwood, Ohio
- Elmer Poe, Professor of Electronics, Eastern Kentucky University
- Chuck Kelly, Chief Engineer, Prodev, Inc., Monroe, Michigan
- Larry Dorazio, Engineer, Hilgraeve, Inc., Monroe, Michigan
- Les Nelson, Student, Microprocessors I, Monroe, Michigan

For contributing their Microprocessor II term projects for use in the book, thanks to my former students:

- Bill Harrison, now System Manager for Autotote, Ltd., Toledo, Ohio—68008 "Chicken Picker"
- Kimberly Kelly, now Systems Repair Technician with Unisys Corp., Holland, Ohio—Z80 Morse Code Sender
- Richard Stone, now a technician with the Hydromatic division of General Motors in Toledo; and Darrin Gaynier, now at Le Tournau College, Longview, Texas—6809 MEM Game with Sound

In a special category, thanks to my good friend and colleague at Monroe County Community College, Tim Maloney, for generously making time—in the midst of updating his own excellent texts in electronics—to do a detailed proofreading of mine.

Personal Notes

Users of this book may contact the author with comments, suggestions, or problems at Post Office Box 466, Temperance, Michigan, 48182. Please enclose a stamped self-addressed envelope if a reply is desired. Although my time is limited, I will endeavor to respond if a brief answer is possible.

Finally, if a masculine pronoun is noticed here and there in the text it is attributable entirely to my abhorrence of the awkward he-or-she construction, and not at all to any discriminatory intent. My own feeling, which seems to be nearly unanimous among my colleagues, is that the recent influx of women to the electronics professions is an entirely welcome phenomenon.

DANIEL L. METZGER
Temperance, Michigan

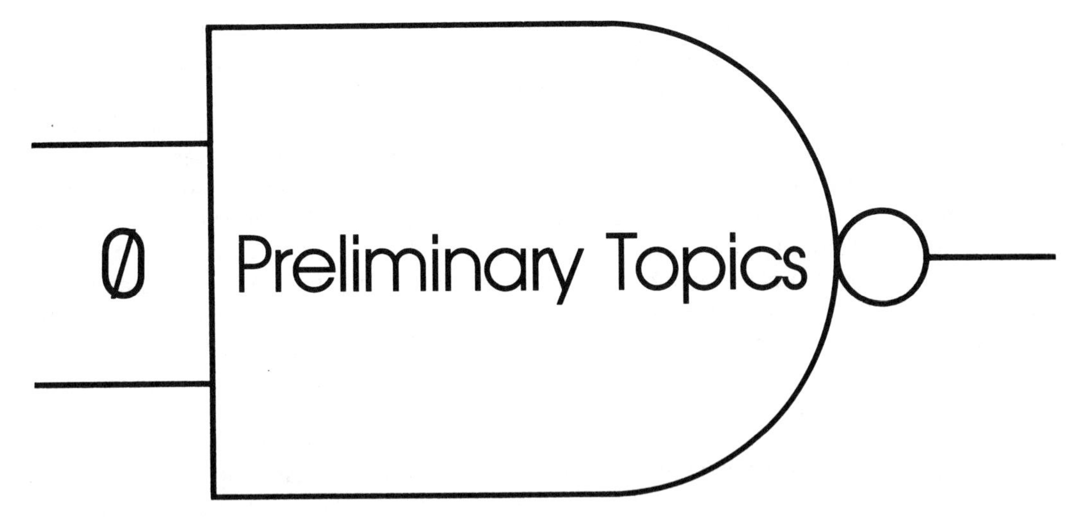

Chapter Overview

This chapter presents an overview of computers (Sections 1 and 2), a review of digital logic (Sections 3 through 7), and computer memory chips (Sections 8 and 9). If you have already had a course in digital electronics and logic, you may still wish to review Sections 3 through 7 to fix in your mind the topics that you will need in your study of microcomputers. If you have not had a digital logic course, it would be a good idea to obtain a book on this topic, since Sections 3 through 7 of this chapter are intended primarily as a review.

0.1 WHAT GOES ON INSIDE A COMPUTER?

Analog and digital: Computers are of two fundamentally different types. In an electronic *analog* computer, electrical quantities (such as current, capacitance, and voltage) are used to represent physical quantities (such as force, mass, and velocity). In designing an elevator system, an engineer might set up an electric circuit in which current acting on a capacitor produces a voltage change, analogous to the way force acting on a mass produces a velocity change. Since the two systems obey analogous sets of equations, results observed in the electrical system can be interpreted in terms of results to be expected in the mechanical system. Changes in current and capacitance are much easier to implement than changes in force and mass in a real elevator, however, and there is much less potential for destruction. Analog computers have been used to simulate rockets, automotive suspension systems, and nuclear power plants, to name a few examples.

It is a feature common to all analog systems that the parameter values may change smoothly and continuously between some maximum and minimum limits, and these values all have different meanings. For example, if 4.00 V represents 4.00 m/s velocity, then 3.99 V represents 3.99 m/s. By contrast, the voltages in a *digital computer* change by discrete "jumps" between certain allowed values. For example, a digital microcomputer treats any voltage between 2.0 V and 5.0 V as the same value (called a *high* level) and any voltage between 0 and 0.8 V as a second value (called a *low* level). Voltages between 0.8 V and 2.0 V are forbidden, and there are no other values besides the *high* and *low* levels. Differing values are represented by various combinations of *high* and *low* levels on a large number of lines.

The digital advantage is superior accuracy and reliability. Poor component tolerances may cause an analog computer to mistake 3.99 V for 4.00 V, but a digital computer almost never mistakes a *low* level for a *high* level. The digital disadvantage used to be complexity—10 000 vacuum tubes or transistors for a digital computer versus 100 tubes or transistors for an analog computer. In 1955 analog computers were far more common than digital computers. Today 10 000 transistors are sold in a single integrated circuit package for less than a dollar, and analog computers have been forced into the background. In fact, the single word *computer* is understood to mean *electronic digital computer* unless another type is specified.

Digital computers operate by switching sets of voltages at a very fast rate. The simpler microcomputers have only about 250 different voltage sets, which appear as *high* and *low* levels on eight lines, called a data bus. The voltage sets by themselves have no meaning. Rather, meaning is assigned to groups of voltage sets by computer programmers and designers. As an analogy, we might say that a computer is like a typewriter. A typewriter is a machine for making very regularized marks on paper. There are about 90 different marks. The marks by themselves have no meaning. Consider, for example,

M

A person skilled in writing can begin to arrange the marks in groups which do have meaning:

MAN

And as the skill develops, truly useful things can be achieved by arranging these marks:

NO MAN IS AN ISLAND

Unless we can tap the skills of someone who knows how to write, a typewriter is just a useless piece of hardware. Unless we tap the skills of someone who knows how to program, a computer is just the same. Just as the typist who advances is likely to be one who acquires some skill in writing, the computer technician who advances is likely to be one who acquires some skill in programming. Elementary programs to check out a piece of computer hardware may consist of only five or ten voltage sets (we call them *words*), but complete programs commonly run to 10 000 or 100 000 words.

The computer's forte is *speed*. The slow ones do a million operations per second. The fast ones do a hundred times that. To gain some appreciation of what this means, imagine a store clerk who takes 1 year to calculate your change if you give him a dollar for a 68¢ purchase. That's about how slow a human seems compared to a computer.

0.2 A BRIEF HISTORY OF COMPUTING

Mechanical digital computers are usually traced back to the ancient oriental abacus, but these are really calculators, not computers. A computer must be capable of executing a sequence of calculations unaided and of modifying that sequence based on intermediate results. A mechanical "analytical engine" to do this was designed by Charles Babbage for the British Post Office in 1833. The plans proved too ambitious for the resources available at the time and the machine was never completed, but Babbage is still generally credited with the invention of the digital computer. An electromechanical digital computer was finally completed in 1944 for IBM at Harvard University under the direction of Howard Aiken. It used relays, shafts, and gears, and could perform about ten additions per second.

The first full-scale electronic digital computer was completed in 1946 at the University of Pennsylvania by John Mauchly and J. Presper Eckert. It occupied an area about the size of a modern ranch home, contained 18 000 vacuum tubes, and required a special maintenance staff just for replacing burned-out tubes. It could perform about 5 000 additions per second and was used to calculate ballistics tables for the U.S. Army. It was called ENIAC, for Electronic Numerical Integrator And Calculator. Unlike the Babbage and Aiken machines, which stored programs in the same way they stored data, the ENIAC was fixed-programmed by its circuit wiring.

The first commercial vacuum-tube computer was sold by Remington Rand in 1951. It was called UNIVAC, for Universal Automatic Computer. In the mid-1950s the IBM model 650 made the computer an established part of many large business operations.

The first transistorized computer was introduced in 1954, and within a few years vacuum-tube computers were obsolete. Then in 1964, IBM introduced its *System 360* computers using integrated circuits, and in a few more years discrete-transistor computers were themselves obsolete.

Minicomputers provided hands-on experience with computers for thousands of engineers and technicians in the late 1960s and early 1970s. These machines, epitomized by Digital Equipment Corporation's PDP-8, eventually brought the price of a computer down tenfold, to less than $10 000.

Microprocessors, which integrated the entire computer—except for memory and input/output (I/O) functions—on one chip, began to appear in primitive forms in 1971. Many of the processors that are still used today had been introduced by 1976. Microcomputer chips, in which even the memory and I/O functions are included, appeared in the late 1970s. Prices for these chips have fallen rapidly, from over $300 at their introduction to generally less than $10 today. Such prices have permitted the incorporation of microcomputers in everything from automobiles to washing machines.

Today the distinctions are arbitrary and blurred, but we may think of large, or *mainframe*, computers as costing over $100 000 and requiring a special area and staff for their operation. *Minicomputers* may be thought of as costing $10 000 to $100 000 and might require a corner of a room and only occasional attention from the staff. A *microcomputer* (the complete system—not just the chips) will cost less than $10 000 and will be turned off and left unused when the individual is finished with it.

The future of computing is a compelling subject for speculation, and much is being written about machine intelligence and vast extensions of the capabilities of the human brain. Such predictions should be offered cautiously by those who are at the leading edge of the new technology. Those who have not yet mastered the basics of elementary computer systems hardly have grounds to form an opinion. Therefore, we defer a discussion of this topic until the end of the text.

REVIEW OF SECTIONS 0.1 AND 0.2

Answers appear at the end of Chapter 0.

1. Systems in which voltages can change continuously between two limiting values are called ________.
2. Systems in which voltages change by discrete jumps from one "allowed" level to another are called ________.
3. Which type of computer has inherently simpler circuitry?
4. Digital computers operate by ________ voltage sets at a very fast rate.
5. Meaning is assigned to groups of digital voltage sets by ________.
6. Charles Babbage designed the first digital computer in (1833, 1876, 1908, or 1936) ________.
7. The PDP-8 was a popular ________ of about 1970.
8. An IC that contains all the circuitry for a computer except for memory and input/output is called a ________.
9. A large computer, costing $250 000, would be called a ________.
10. Which is more powerful, a computer or a human brain?

0.3 DIGITAL LOGIC—RULES AND STANDARDS

Voltage levels: Nearly all microcomputer chips are designed to operate at standard TTL voltage levels because of the widespread use of this family of logic components.

- The main supply, designated V_{CC}, is $+5.0$ V, generally with a tolerance of $\pm 5\%$. Ground (0 V) is often designated V_{SS}.

- The input pins are generally guaranteed to treat any voltage from 0 up to +0.8 V as a *low* logic level and any voltage from +2.0 V up to +5.0 V as a *high* logic level. The specification sheets list V_{IL} = +0.8 V MAX, V_{IH} = +2.0 V MIN, where the subscripts I, L, and H stand for *input*, logic *low*, and logic *high*, respectively. Voltages between +0.8 and +2.0 might be interpreted as either a *low* or a *high* and are called *forbidden* levels.
- The output pins are generally guaranteed to hold the outputs to less than +0.4 V for a *low* level and more than +2.4 V for a *high* level. The spec sheets list V_{OL} = +0.4 V MAX, V_{OH} = +2.4 V MIN. Thus there is a 0.4-V margin of safety between what the inputs require and what the outputs deliver.
- In positive-true logic a *high* level is designated by the symbol 1, or by the words *true* or *asserted*. A *low* level is then designated by the symbol Ø, or by the words *false* or *negated*.

The current convention universally used by IC and microprocessor manufacturers is that current passes from positive to negative through a resistive load. Many technicians prefer to visualize current as electrons flowing from negative to positive. Since we cannot actually see the electrons, it makes no difference in practice which view we take. The only requirement is that we be consistent. In this text we will adopt the positive-to-negative convention to be consistent with manufacturers' data sheets.

Another convention used in the data sheets is that current into an IC pin is listed as positive and current out of an IC is listed as negative. This is true regardless of whether the pin is an input or an output. Figure Ø.1 illustrates the convention for one TTL gate driving another with a *low* output (a) and for a TTL gate driving a transistor LED driver with a *high* output (b).

Ø.4 LOGIC GATES

Logic gates are often used to interface (connect together) various parts of a computer system. There are six basic types of gates. You may already be familiar with them, but we summarize them here.

The inverter is also called a NOT gate. It has one input and one output. It simply inverts the state of a logic signal; a 1 at the input becomes a Ø at the output. If we call the input signal by the variable name *A* (*A* may equal 1 or *A* may equal Ø), then the output signal is $\overline{A}$ (read *A*-not). Some logic diagrams and text material may substitute the notations *A**, or *A*′, or /*A* for $\overline{A}$ because of typographical limitations. Figure Ø.2 shows the logic symbols for an inverter. There are two logic-symbol systems currently in use; the MIL-STD (military standard) system, which has predominated through the 1970s and early 1980s, and the IEEE system (Standard 91-1984), which began gaining ground, especially with computerized drafting systems, in the mid-1980s. Both symbols are shown. The small circle (or triangle) at the right stands for logic inversion.

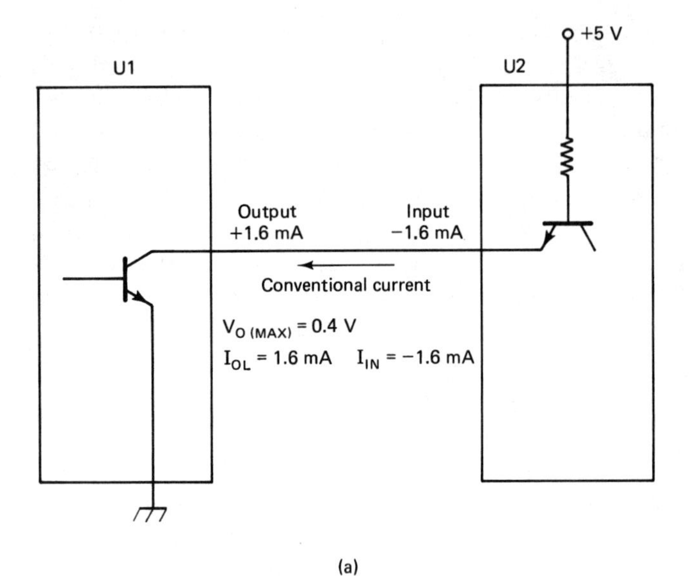

Figure 0.1 (a) A *low* TTL output level sinks current supplied by the TTL input being driven. (b) A *high* TTL output level is capable of supplying current to devices which require it.

A logic AND gate produces a *true* (logic 1) output only when all its inputs are *true*. Otherwise, its output will be *false* (logic 0). An AND gate may have any number of inputs, but gates with 2, 3, or 4 inputs are most common. Figure 0.3 shows a two-input AND gate with a *truth table* listing the logic state of the output for all four possible input conditions. Note that the operation of *anding* A with B is indicated by writing

$$A \cdot B$$

It is perhaps unfortunate that the AND sign is the same as the *multiply* sign in ordinary algebra, but the context usually makes the meaning of the sign clear. As in algebra, the dot is sometimes omitted, so the simple juxtaposition *AB* is taken to mean *A anded* with *B*. Other writers may use the symbol $\wedge$ to indicate the logic AND operation, since it eliminates ambiguity. Hence $A \cdot B^* = A\overline{B} = A \wedge B'$.

The NAND gate is the same as the AND except that the output is inverted. Figure Ø.4 shows a 3-input NAND gate and its truth table.

OR gates produce a *true* output if one or more of their several inputs is true. **NOR gates** are the same except that the output is inverted. The logic OR operation is indicated by the plus sign (+), borrowed from arithmetic, or, by those who wish to avoid ambiguity, the sign V.

A special type of OR gate is the **exclusive-OR**, also called an XOR gate. It usually has only two inputs. It produces a true output when one or the other input is true but not when both are true. The XOR operation is indicated by the sign $\oplus$ or, alternatively, the sign $\forall$. Figure Ø.5 shows the three OR-type gates with their truth tables.

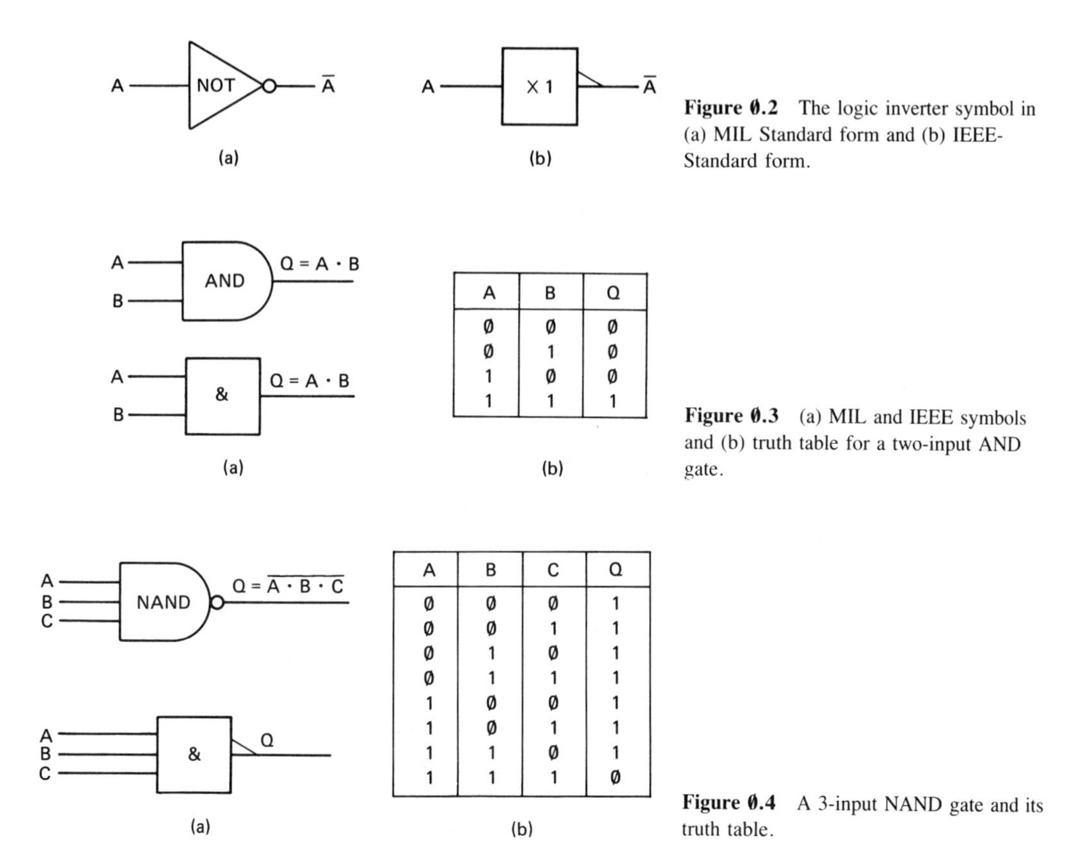

Figure Ø.2 The logic inverter symbol in (a) MIL Standard form and (b) IEEE-Standard form.

A	B	Q
Ø	Ø	Ø
Ø	1	Ø
1	Ø	Ø
1	1	1

Figure Ø.3 (a) MIL and IEEE symbols and (b) truth table for a two-input AND gate.

A	B	C	Q
Ø	Ø	Ø	1
Ø	Ø	1	1
Ø	1	Ø	1
Ø	1	1	1
1	Ø	Ø	1
1	Ø	1	1
1	1	Ø	1
1	1	1	Ø

Figure Ø.4 A 3-input NAND gate and its truth table.

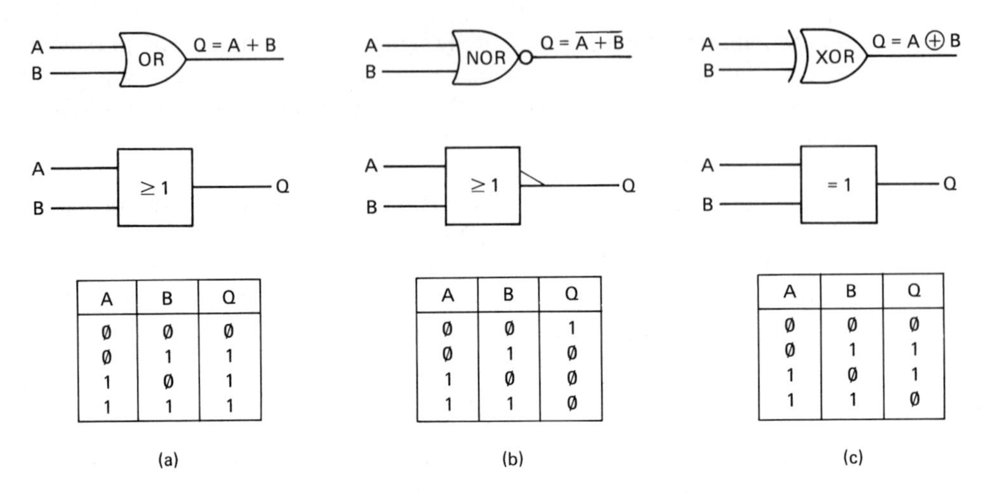

A	B	Q
0	0	0
0	1	1
1	0	1
1	1	1

A	B	Q
0	0	1
0	1	0
1	0	0
1	1	0

A	B	Q
0	0	0
0	1	1
1	0	1
1	1	0

Figure 0.5 (a) A 2-input OR gate, (b) NOR gate, and (c) exclusive-OR gate, with their respective truth tables.

Negative-true logic: If a line is marked HALT we know that a logic *high* (about +3 or +4 V) will appear on it if the condition called HALT is true. We say that HALT is positive-true or *high* active.

If a line is marked $\overline{\text{HALT}}$, we know that a logic *high* will be present if the condition HALT is *not* true. Another way of saying this is that a logic *low* (about 0 V) will be present if HALT is true. In this case we say that $\overline{\text{HALT}}$ is negative-true. Positive-true logic is thought of as the standard, but it is common to find many parts of a microcomputer system working in negative-true logic. Be on the alert for overbars, asterisks, or prime marks on labels and the little invert circles or triangles at the IC pins, which denote negative-true logic.

You must also be careful about using the symbols 1 and 0 to refer to logic levels. In positive-true TTL logic, 1 means approximately +4 V and 0 means 0 V, but in negative-true logic

$$\overline{\text{HALT}} = 0$$

means that the *low*-active HALT line is *not* asserted; in other words it is at the inactive or *high* (+4-V) state. To avoid confusion it is a good idea to use the letters H and L to specify *high* and *low* logic levels, and to avoid the use of 1 and 0 except where strictly positive-true logic is being employed.

A positive-true AND gate becomes a negative-true OR gate. This fact is often used to implement a logic OR function with the common 74LS00 NAND gate, as shown in Figure 0.6. An intuitive appreciation of this can be gained from the following equivalent logic statements about a coffee-vending machine (35¢ per cup).

1. If QUARTER *AND* DIME, then COFFEE (positive-logic AND).
2. If $\overline{\text{QUARTER}}$ *OR* $\overline{\text{DIME}}$, then $\overline{\text{COFFEE}}$ (negative-logic OR).

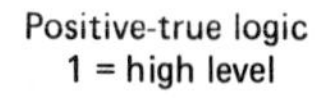
Positive-true logic
1 = high level

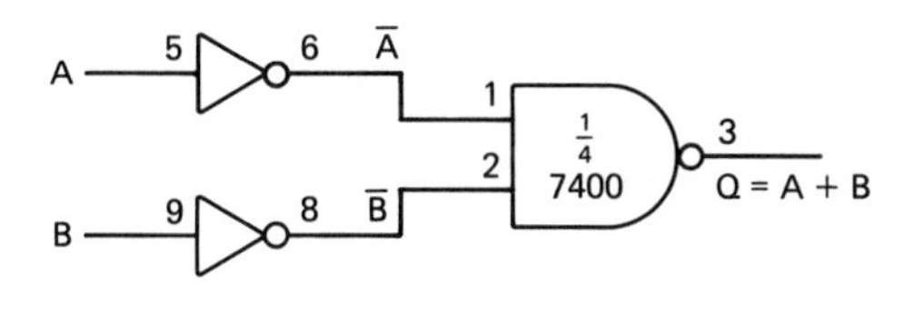

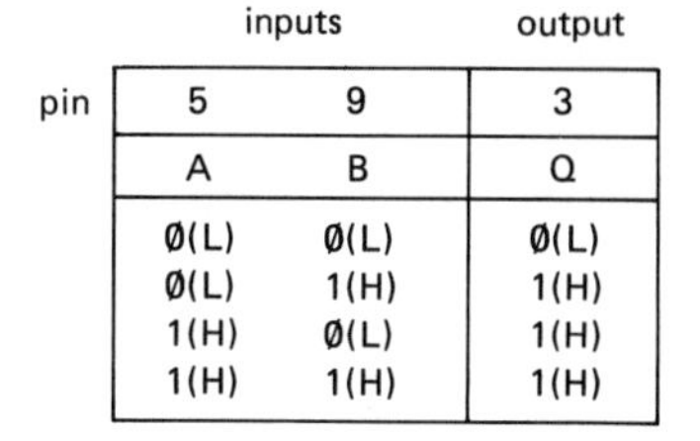
OR function performed

pin	inputs		output
pin	5	9	3
	A	B	Q
	Ø(L)	Ø(L)	Ø(L)
	Ø(L)	1(H)	1(H)
	1(H)	Ø(L)	1(H)
	1(H)	1(H)	1(H)

Negative-true logic
1 = low level

NOR function performed

pin	inputs		output
pin	1	2	3
	Ā	B̄	Q
	Ø(H)	Ø(H)	1(L)
	Ø(H)	1(L)	Ø(H)
	1(L)	Ø(H)	Ø(H)
	1(L)	1(L)	Ø(H)

Figure Ø.6 A NAND gate, when fed with negative-true inputs, performs an OR function.

The second statement can be read, "If the quarter is missing OR the dime is missing, then you get no coffee. Study the truth table to convince yourself that the function is NAND if pins 1 and 2 of the 7400 are regarded as the inputs but OR if the inputs are inverted before being applied to the 7400 (*A* and *B* regarded as the inputs).

Any logic function can be implemented using combinations of NAND gates. Figure Ø.7 shows three examples that are often used to take advantage of the common 74LS00 IC.

Open-collector outputs are available in many logic ICs. For example, a 74LS03 is a 74LS00 with open-collector outputs. The open-collector output simply omits the pull-up transistor Q_2, which is used in normal, or *totem-pole*, outputs [see Figure Ø.8(a) and (b)]. Open-collector outputs can drive LEDs directly without wasting current [Figure Ø.8(c)]. They can also be used in a *wired-OR* configuration shown in Figure Ø.8(d), in which the outputs of several gates are tied together.

Programmable Array Logic (PAL) is a recent development which allows a large number of logic ICs to be replaced by a single chip. A typical PAL might have 12 inputs and 10 outputs in a 24-pin package. Each output can be programmed to be *true* for any desired logical combination of the inputs, as expressed by AND, OR, and NOT conditions. Programming is done by blowing microscopic fuse links inside the chip with a special programming fixture. Some PALs use charge storage in MOS switching transistors instead of fuse links, and these can be erased under a strong ultraviolet light and reprogrammed.

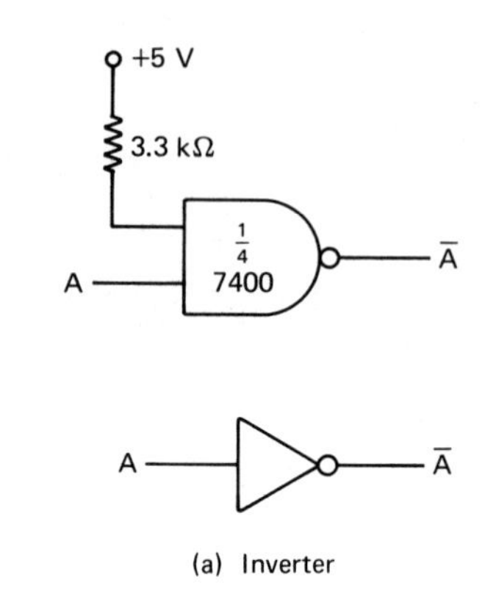

(a) Inverter

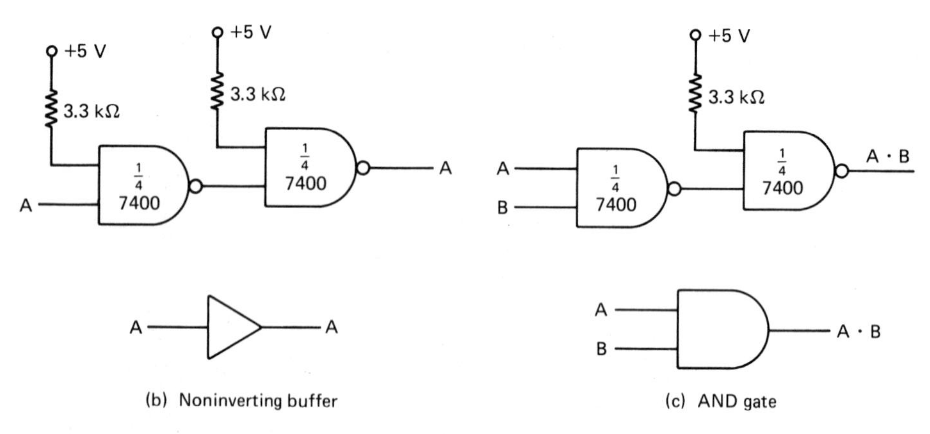

(b) Noninverting buffer

(c) AND gate

Figure 0.7 A standard 7400 NAND gate can be pressed into service as an inverter, noninverting buffer, or AND gate.

REVIEW OF SECTIONS 0.3 AND 0.4

Answers appear at the end of Chapter 0.

11. Which of these *do not* meet the specs of TTL and microprocessor system voltages?
(a) $V_{CC} = +4.8$ V **(b)** $V_{IH} = +2.1$ V
(c) $V_{IL} = +0.6$ V **(d)** $V_{OL} = +0.6$ V
(e) $V_{OH} = +4.5$ V

12. IC and microprocessor data sheets use the convention that current passes through a load from ________ to ________.

13. What is the algebraic sign given to these currents on an IC spec sheet?
(a) Current into an output pin.
(b) Current into an input pin.
(c) Current out of an output pin.

14. If the input of an inverter is logic 1, the output is ________.

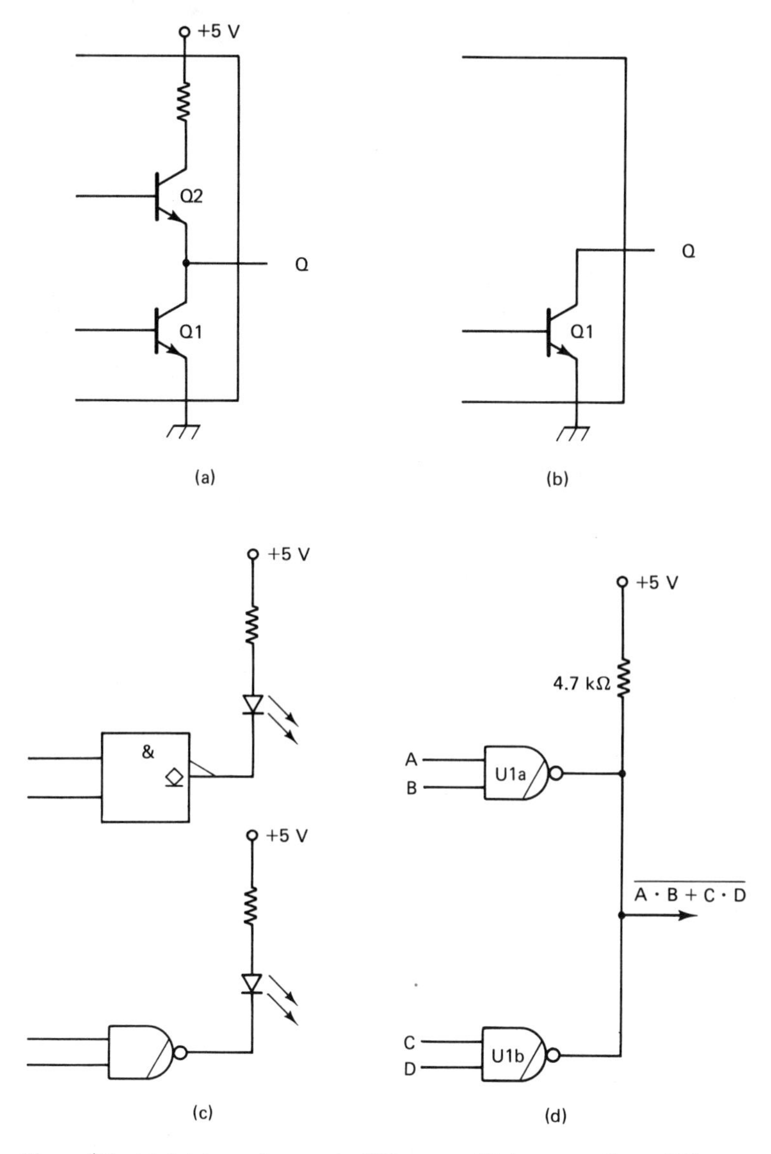

Figure 0.8 (a) A totem-pole or *active* TTL output. (b) An open-collector TTL output. (c) IEEE and MIL symbols for an open-collector NAND gate. (d) Wired-OR, in which U1a OR U1b can pull the output *low*.

15. If all the inputs of a positive-true NAND gate are *high*, the output is ________.

16. Spell out in words how each of these logic expressions is to be read aloud:
(a) $\overline{A}$ **(b)** $A \cdot B$ **(c)** $A + B$ **(d)** $\overline{A \cdot B}$

17. An exclusive-OR gate has input $A = 1$ and input $B = 1$. The output is ________.

18. A positive-logic NAND gate is a negative-logic ________.

19. A ________ ________ output pulls *high* for a positive-true logic 1, but a ________ ________ output floats unless pulled *high* externally.

20. In the wired-OR circuit of Figure 0.8(d), using positive-true logic, $A = 1$, $B = 0$, $C = 1$, and $D = 1$. The output is ________.

0.5 FLIP-FLOPS

A flip-flop is an electronic memory element. If its output is set to a logic 1, it will stay there until an appropriate input signal is applied to change the output to logic-0. The switching is accomplished with no moving parts and takes much less than a microsecond. Three different types of flip-flops are commonly used in microcomputer systems.

The *RS* flip-flop is also called a set-reset flip-flop. It has two inputs, S and R. A *true* input at S causes the output Q to become *true*. Output Q remains true until a *true* input at R causes it to *reset* to the opposite (false) state. If both inputs are made *true* together, the output state is unpredictable, and this is called a *forbidden* condition. *RS* flip-flops are commonly made by cross-connecting two NAND gates, as shown in Figure 0.9(a). The inputs are negative-true, since the resistors pull the inputs to the *high* state and the action of the input switch is to force either $\overline{S}$ or $\overline{R}$ *low*.

This circuit functions as a switch-bounce eliminator. When a toggle switch is actuated, its contacts scrape intermittently for a few milliseconds as they leave the first pole, spend a few milliseconds in transition, and then bounce on the second pole for a while until making firm contact. Computer circuits are so fast that they would count each bounce as a switch actuation if the switch were connected directly to a counting circuit. The debounce circuit's output switches when the moving contact first touches the new terminal. Transition time and bounces into the space between

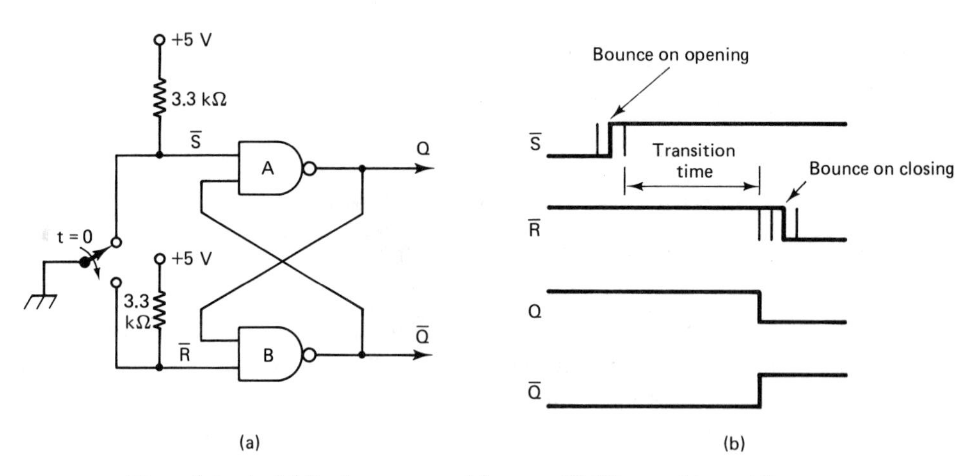

Figure 0.9 An *RS* flip-flop constructed from two NAND gates (a) can deliver a single-switching output from an input which chatters as the switch contacts bounce.

the terminals have no effect, since both inputs are *high* (not asserted) during those times. The forbidden state (both inputs *low*) is never applied by the switch.

The D latch is a flip-flop so-named because it functions as a *data* latch. It has a *data* input and a *gate* input. The output Q assumes the state of the data input D when the gate input G becomes true. The output follows the state of D until G becomes false. Thus the D latch holds the data that is at D when G transitions from true to false. Figure Ø.10(a) shows a common type-D latch. Note that the gating input G may also be called E (enable). These latches are sometimes called *transparent* because, as long as G is asserted (true) the D input "sees" straight through to the Q output.

Figure Ø.10(b) shows an edge-triggered type-D flip-flop. The triangle at the triggering input indicates that the data is latched at the instant when the trigger voltage transitions from *low* to *high* (the positive *edge*). The trigger input may be called T or C or CK (clock).

The data-latch concept is important to microcomputer operation because it allows the processor to direct data from a single group of signal lines to a number of different

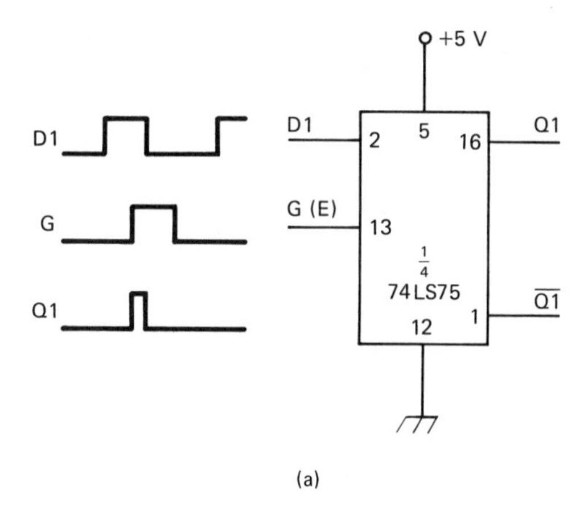

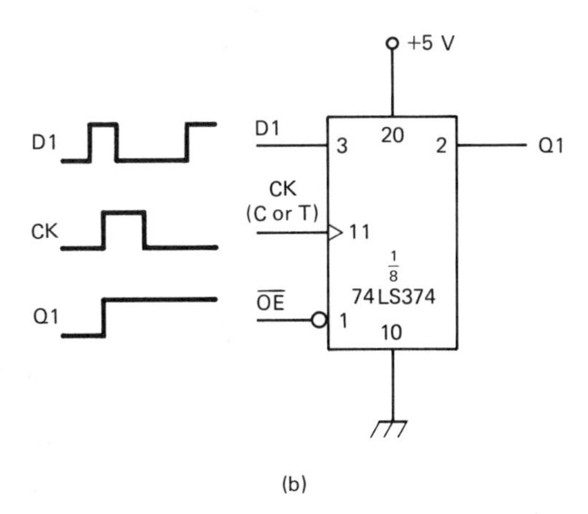

Figure Ø.10 (a) A type-D transparent latch. (b) A positive-edge triggered type-D flip-flop.

devices. A single data output bus from the microcomputer is connected to the data-input buses of several devices, but they all ignore the data until the processor triggers one of their enable inputs. The data intended for that device is then latched and held while the processor "talks" to the other devices in their turns.

***JK* flip-flops** are remarkably versatile. They have a triggering input and two control inputs, *J* and *K*. The letters *J* and *K* do not stand for any words; they are just consecutive letters of the alphabet. *J* corresponds vaguely to *set* and *K* vaguely to *clear*. There are four possible combinations of logic inputs to *J* and *K*, and these control four modes of operation of the flip-flop.

J	*K*	Function
Ø	Ø	Ignore trigger input
Ø	1	Clear *Q* (to Ø) upon trigger input
1	Ø	Set *Q* (to 1) upon trigger input
1	1	Toggle (change) *Q* upon trigger input

Figure Ø.11 shows a typical *JK* flip-flop with its truth table. The *low*-active (negative-true) clear input overrides the other inputs and holds the Q_1 output *low* as long as it is *low*. The *X* entries in the truth table indicate "don't care—no effect." The ↓ entries in the table indicate a negative transition—logic 1 to logic Ø. The entries $Q_\emptyset$ refer to the state of *Q before* the application of the trigger signal.

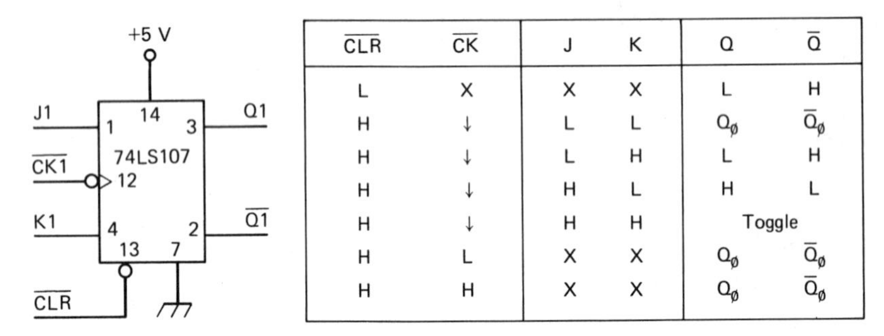

$\overline{CLR}$	$\overline{CK}$	J	K	Q	$\overline{Q}$
L	X	X	X	L	H
H	↓	L	L	$Q_\emptyset$	$\overline{Q}_\emptyset$
H	↓	L	H	L	H
H	↓	H	L	H	L
H	↓	H	H	Toggle	
H	L	X	X	$Q_\emptyset$	$\overline{Q}_\emptyset$
H	H	X	X	$Q_\emptyset$	$\overline{Q}_\emptyset$

↓ High to low transition
X Don't care; H or L
$Q_\emptyset$ State of Q before clock transition

Figure Ø.11 A *JK* flip-flop can be made to do nothing, *clear*, *set*, or *toggle* in response to a trigger input.

Ø.6 SHIFT REGISTERS

A register is a device for storing and manipulating groups of binary digits. Flip-flops are ideally suited for building registers.

A shift register allows the binary digit in each flip-flop to be moved to the adjacent flip-flop (let us say the one to the right) under control of a *shift* pulse. Figure

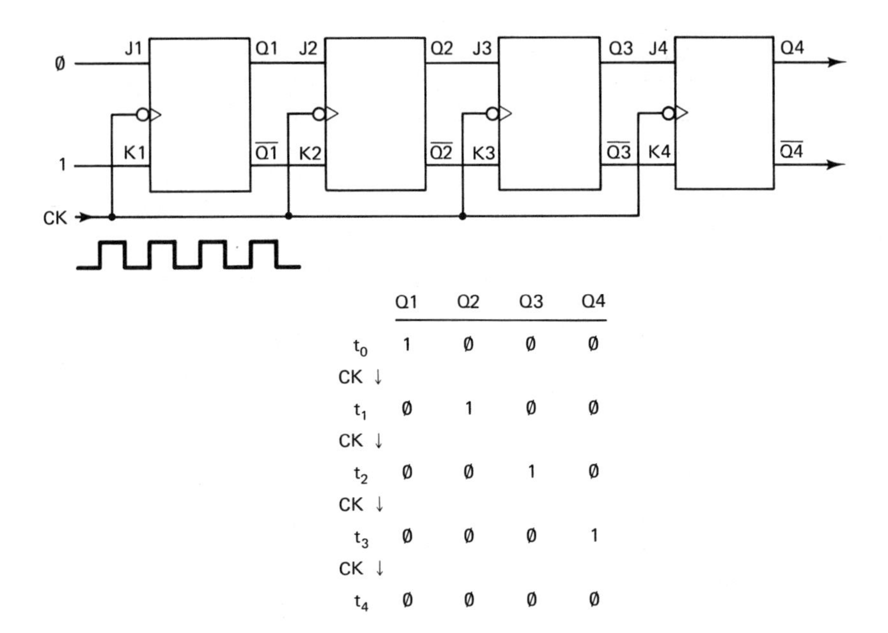

	Q1	Q2	Q3	Q4
t_0	1	Ø	Ø	Ø
CK ↓				
t_1	Ø	1	Ø	Ø
CK ↓				
t_2	Ø	Ø	1	Ø
CK ↓				
t_3	Ø	Ø	Ø	1
CK ↓				
t_4	Ø	Ø	Ø	Ø

Figure Ø.12 A shift register composed of four *JK* flip-flops.

Ø.12 shows a 4-bit shift register using JK flip-flops. At each negative edge of the clock the flip-flops respond by placing their *Q* outputs at the previous state of their *J* inputs. Study the truth table given in Figure Ø.11 to see why this is so.

A ring counter is a shift register with its outputs connected back to its inputs. If there are *n* flip-flops, the same patterns of 1s and Øs will repeat after *n* clock pulses.

IC shift registers, such as the 74LS194, may have parallel inputs for loading all four flip-flops at once and left/right shift capability.

The race problem. In Figure Ø.13 someone has attempted to build a shift register using *transparent* latches rather than edge-triggered flip-flops. This is not

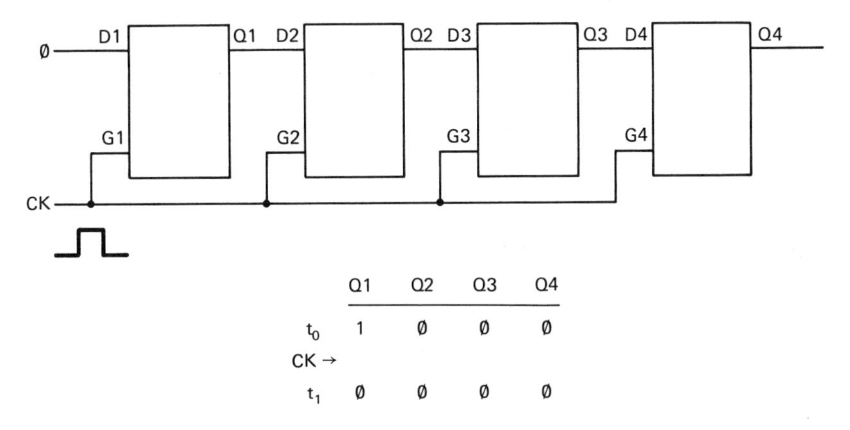

	Q1	Q2	Q3	Q4
t_0	1	Ø	Ø	Ø
CK →				
t_1	Ø	Ø	Ø	Ø

Figure Ø.13 An unsuccessful attempt to build a shift register with type-D flip-flops. Master-slave flip-flops are used to solve the race problem which defeats this circuit.

likely to work reliably because Q_1 may go *low* while G2 is still true, causing Q_2 to clear by the end of the first clock pulse. We might try reducing the width of the clock pulse so that CK is back to zero by the time Q_1 has had a chance to go *low*. However, the pulse width required might vary from something like 5 ns to 50 ns, depending on the individual flip-flop, so this is not a practical solution.

The edge-triggered *JK* flip-flops of Figure 0.12 are immune to this *race problem* because of their internal *master/slave* circuitry. The J, K, and CK inputs are connected to a *master* flip-flop, which responds to data present while CK is *high*. The outputs of the master are not connected to Q and $\overline{Q}$ but to the inputs of a *slave* flip-flop, which responds when CK transitions *high* to *low*. Thus a master/slave flip-flop responds to data present during the positive duration of CK, but the response is delayed until the negative duration of CK.

REVIEW OF SECTIONS 0.5 AND 0.6

21. A cross-connected pair of NAND gates forms a ________.

22. In MIL standard logic symbols, signal inversion is represented by a ________, and edge-triggering is indicated by a ________.

23. A device that catches the data at its input upon receiving a trigger signal and holds that data at the output until the next trigger is called a ________.

24. What function will a *JK* flip-flop implement upon trigger with $J = 1$ and $K = 0$? ________.

25. Latches that use level-sensitive *enable* inputs rather than edge-triggered clock inputs are called ________.

26. A series of flip-flops used to store and manipulate binary data is called a ________.

27. How many flip-flops are required to build a ring counter that will return to its original state after ten clock pulses? ________.

28. In a shift register using *JK* flip-flops Q_1 and Q_2 are connected, respectively, to ________ and ________.

29. What problem is eliminated by using edge-triggered master/slave flip-flops?

30. What does the symbol X mean in digital logic tables?

0.7 BINARY COUNTERS

Binary numbers: The contents of a register often represent numerical data. Our familiar decimal number system has 10 different symbols (0, 1, 2, 3, 4, 5, 6, 7, 8, and 9), so after we have counted from zero through nine we make a group of 10. We indicate this by writing a 1 in the next column and indicating that we have a group of one 10 and no units:

Tens	Units
1	0

If there are more than nine groups of 10 we must form a group of 100 and move to a third column. This is all very familiar from early elementary school.

A flip-flop has two different states, represented by the symbols 0 and 1, so after it has counted from 0 to 1 it is necessary to record a group of 2 in a second flip-flop and indicate this in a second column:

Twos	Units	
1	0	= two

If there is more than one group of 2 we must form a group of 4 and move to a third column. If there is more than one group of 4, we move to a fourth column and set up a group of 8. Binary numbers take up a lot more columns than decimal numbers and, although humans may find this awkward, computers don't seem to mind it a bit.

Here is the binary counting series from 0 to 15. Notice that the *first* count is 0, the second is 1, . . ., and the *sixteenth* is 15. This offset of 1 is common in computer systems.

8s	4s	2s	1s	Decimal
0	0	0	0	0
0	0	0	1	1
0	0	1	0	2
0	0	1	1	3
0	1	0	0	4
0	1	0	1	5
0	1	1	0	6
0	1	1	1	7
1	0	0	0	8
1	0	0	1	9
1	0	1	0	10
1	0	1	1	11
1	1	0	0	12
1	1	0	1	13
1	1	1	0	14
1	1	1	1	15

In a computer system these binary digits would be carried on a set of four signal lines called a 4-bit bus.

A binary counter that implements the preceding counting series electronically is shown in Figure 0.14. Each time the input clock makes a negative transition, FF_0 toggles (changes its Q_0 output state). On every second toggle Q_0 will make a *high-*

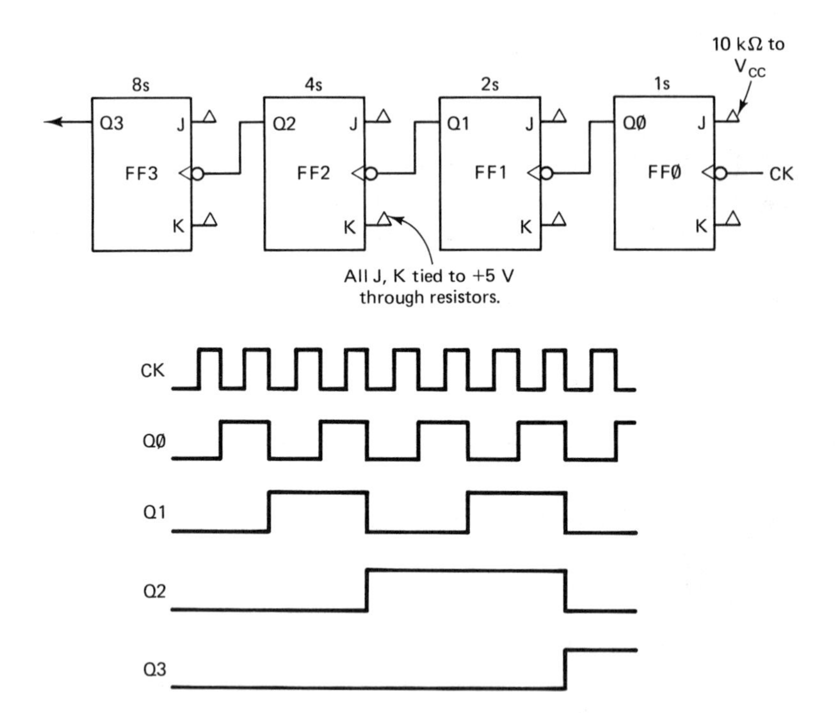

Figure Ø.14 A four-stage binary counter counts zero through 15, decimal. The frequency of the Q3 output is one-sixteenth that of the clock input.

to-*low* transition, causing FF_1 to toggle. Thus FF_2 will toggle on every fourth clock pulse, and FF_3 will change states on every eighth clock cycle. This pattern is the same as the Ø-to-1 transition pattern in the 1s, 2s, 4s, and 8s columns, respectively, of the binary counting sequence. The flip-flops are drawn with the signal propagating from right to left. This is backward from the usual practice of showing electronic signal flow from left to right, but it corresponds to the mathematical convention of placing the more significant digits toward the left.

Frequency division: Notice from Figure Ø.14 that each flip-flop divides the frequency of the input signal by two. The frequency at the output of the fourth flip-flop (Q_3) is $\frac{1}{16}$ of the original clock frequency. As an example of the usefulness of this frequency division, the 6802 microprocessor, which we will meet in the next chapter, is designed to operate with a clock of 1 MHz. A 1-MHz crystal for generating this signal would typically cost $3.00, but a thinner 4-MHz crystal might be obtained for $2.00. Two flip-flops are thus incorporated within the *6802* to allow us to take advantage of the lower-priced crystal. The cost of adding two flip-flops to a chip containing over 6000 integrated components is hardly a penny, but the savings if 10 million of these chips are sold become quite impressive.

Synchronous counters: The counter of Figure Ø.14 obtains the clock input for each flip-flop from the previous flip-flop. The delay time from input to output may be as much as 40 ns for each TTL flip-flop. For a 16-bit counter the total delay, as

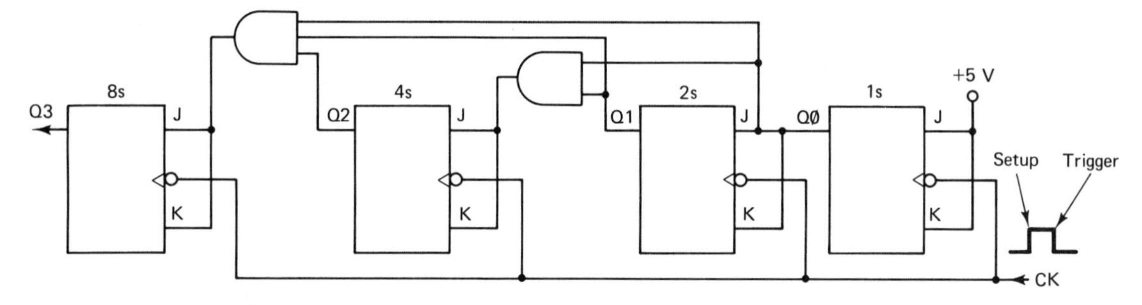

Figure Ø.15 A four-stage synchronous binary counter. Synchronous counters apply the clock signal to all flip-flops simultaneously for speed. The counter of Figure Ø.14 is asynchronous; it is slower because the clock signal ripples through the delay of each flip-flop.

the signal ripples from input to output, could be 16 × (40 ns) or 640 ns, making the counter unacceptably slow.

Ripple or asynchronous counters have generally been replaced by synchronous counters in computer systems. Figure Ø.15 shows a 4-bit synchronous counter. Notice how the clock signal is applied simultaneously to all the flip-flops. The logic gates determine the conditions upon which each flip-flop toggles.

Odd-modulus counters: Binary counters count to (or divide by) 2, 4, 8, 16, 32, and so on. By clever gating and routing of the signals to *J* and *K* of the respective flip-flops, counters can be devised that skip certain parts of the binary-count sequence, thus making a different number of counts before recycling. The number of counts made is called the *modulus* of the counter. A count of 10 is often desired, and Figure Ø.16 shows a synchronous *mod*-10 counter. Since this chapter is intended primarily as a review, we will not pursue the analysis or design of odd-modulus counters. Such details can be found in most books on digital circuits.

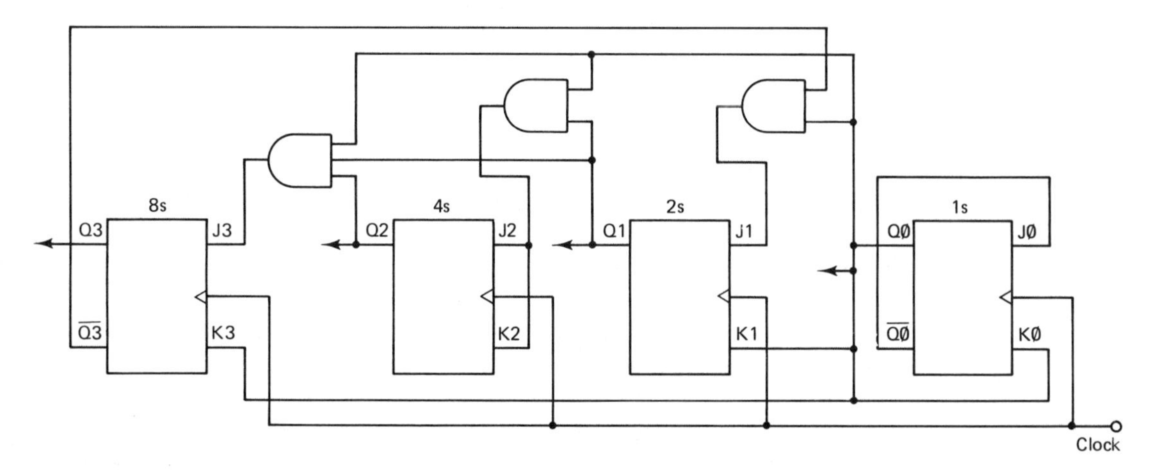

Figure Ø.16 A synchronous modulus-ten counter.

Integrated-circuit counters are available in binary and in a variety of odd-modulus types. Some can be precleared to all zeros by a separate pulse. Some can be preset to a value loaded by special parallel inputs. Some have a programmable modulus; they count to a preloaded value and then automatically return to zero.

0.8 MEMORY

Microprocessors spend most of their time reading data from memory. We will first sort out the various types of memories that are available and then show how the binary data in a memory can be accessed.

Read-Only Memories are called ROMs (pronounced RAHMZ). All ROMs have two features in common.

1. Data can be read from them quickly and easily, but writing new data to them is either slow, difficult, or impossible, depending on the type.
2. They are *nonvolatile*, which means that if the power to them is shut off, their data will remain intact, ready for use at the next power-up.

ROMs containing 16 000 to 512 000 bits are readily available. Here are the common types of ROMs, in order of increasing ease of writing to them.

- Mask-programmed ROMs have a fixed program placed in them during the photomask portion of their manufacture. This program data can never be changed. The cost of producing the special mask for the desired data is on the order of $1000, so mask-programmed ROMs are generally used for production quantities of 1 000 or more.
- Field-programmable ROMs, or PROMs, are programmed "in the field" rather than at the factory, but once in place the data likewise can never be changed. The programming is done in a special fixture, which blows microscopic fuse links at selected bit locations inside the IC.
- EPROMs are Erasable Programmable Read-Only Memories. They are programmed by applying an overvoltage (typically +25 V) at selected bit locations, which stores a charge behind the near-perfect insulating layer of an MOS transistor. This charge can be retained for years. EPROMs are programmed ("burned," in the vernacular) in special fixtures, which are often computer-driven. They are erased by shining a strong ultraviolet light through a window in the top of the package, which exposes the chip.
- EEPROMs, or E^2PROMs, are electrically erasable and reprogrammable in the system of which they are part. They do not need to be removed and reprogrammed in a special fixture. Erasure time is generally several milliseconds (compared to a read time of less than a microsecond), and the whole memory or large blocks of it may be erased at once. Older EEPROMs require special erase and reprogramming voltages and waveforms from the system of which they are a part. E^2PROMs are currently more expensive than EPROMs.

Read/write memories can be read from and written to equally as fast. The chief representatives of this type are the static RAM[1] and the dynamic RAM.

- Static RAMs use banks of flip-flops to store binary data. RAMs are *volatile*—if their power is turned off, the data stored in them will be lost. The name *static* means that as long as the power is maintained, the memories hold the data by maintaining unchanging, or dc, voltages and currents. To overcome the volatility problem, manufacturers are incorporating rechargeable batteries on their circuit boards or even within the IC packages to supply the RAMs with standby power if the main power fails. These are often called NVRAMs. Static RAMs containing 64 000 bits are available for less than $4.00, and 262 000-bit units can be obtained for a somewhat higher price.
- Dynamic RAMs, or DRAMs, use integrated capacitors to hold a charge for a few milliseconds. Each capacitor represents 1 bit of information and is charged or discharged to store a 1 or 0 binary output. DRAMs are considerably less expensive than static RAMs and can store more bits per chip, but their data must be refreshed approximately every 4 ms by reading groups of bits and writing back the same data that was read. The main processor can be programmed to do this and still have time for other functions if the processor has on-board DRAM-refresh capabilities. For larger systems a special RAM-refresh processor is generally used. Dynamic RAM chips containing 262 000 bits are available for less than $3.00, with 1-megabit chips available at a higher price.

Memory organization: Memory chips contain address pins and data pins. Different data is called up by applying different binary combinations to the address pins. The number of addresses available is 2^a where a is the number of address lines. For example, a memory with address lines A0 through A9 would contain 2^{10}, or 1024, different data words.

The number of data lines equals the number of bits stored at each memory address, which is the number of bits per word.

Figure 0.17 shows the pin diagram of a small MOS RAM. We can see by the pinout that there are 2^{10}, or 1024, different addresses in the memory. Each address stores a word consisting of 4 binary digits, or *bits*. The memory is being read and data is output to lines D0 through D3 if pin 8 (chip select, low-active) is a *low* level and pin 14 (write-enable, low–active) is a *high* level. With pin 8 *high* the data lines are disconnected from the memory output drivers so these lines can be controlled by another part.

Writing to the RAM is accomplished by applying the desired address and data bits to the *A* and *D* lines, holding $\overline{\text{CS}}$ *low* and causing $\overline{\text{WE}}$ to go *low*.

[1]The name RAM (Random-Access Memory) was given in the early years of IC memory technology when one of the RAM's big advantages was thought to be the fact that every data word in the memory was accessible almost immediately. Whereas some other memories of the time had to sort through data serially to get at the desired word, RAMs could access data at random. Subsequent developments have made the acronym RAM meaningless. All the various ROMs discussed previously are also random access. However, the name has stuck and the word RAM is today understood to mean "read/writable memory."

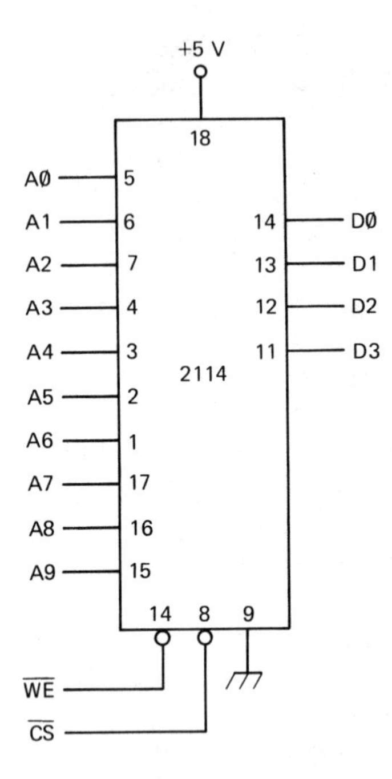

Figure Ø.17 A 1024 × 4 RAM chip.

REVIEW OF SECTIONS Ø.7 AND Ø.8

31. The binary number 1Ø11Ø is equivalent to decimal ________.

32. A binary counter using four flip-flops will divide the input frequency by ________.

33. A 10-stage ripple counter has a maximum clock frequency of 1 MHz. We might raise this to 5 MHz by using a ________ counter.

34. A mod-five binary counter counts Ø through ________.

35. Mask-programmed ROMs can be erased by ________.

36. EPROMs can be erased by ________.

37. Is a PROM a random-access memory?

38. A static RAM is volatile. This means that the data in it will be lost if ________.

39. A RAM with address lines AØ–A1Ø and data lines DØ–D7 stores ________bits of data.

40. The type of memory that must be refreshed every few milliseconds is the ________ ________.

Ø.9 A RAM TEST CIRCUIT

The Motorola 6810 is a 128 word × 8-bit-per-word read/write memory which was designed some years ago for use with the 6800 family of microprocessors. Its capacity is quite meager by today's standards, but it is ideal for its simplicity, and we will see it again in later chapters. You can familiarize yourself with RAM operation by building the circuit of Figure 0.18 on a breadboard and storing data words at a few of the address locations. You do not need to use switches as shown in the figure. You can simply plug wires from the address and data lines to the breadboard *ground* and V_{CC} lines for 0 and 1 inputs. You may wish to note the following points:

1. The data outputs are random when the power is first turned on. Try switching a few address inputs to read some other addresses. Turn the power off and then on again to see if the bits come up the same.
2. The data outputs are inactive or floating unless all of the CS (chip-select) lines are in their indicated states. Switch CS0 to ground and remove the LED from D0. Measure with a voltmeter that D0 can be pulled *high* or *low* with a 10-kΩ resistor to V_{CC} or ground.

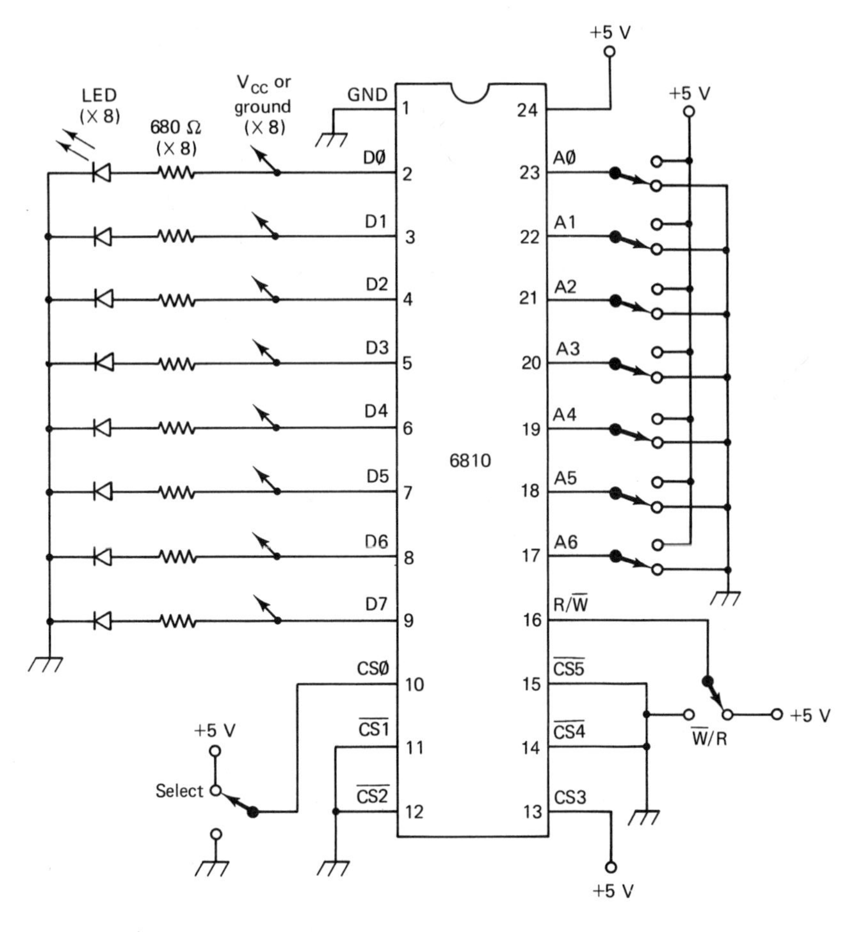

Figure 0.18 A 128 × 8 RAM with circuitry for loading and reading data in binary.

3. Switch the $R/\overline{W}$ line *low* (all CS lines active again), and apply a pattern of *highs* and *lows* to the data bus by plugging wires to V_{CC} or ground. Now hold CS0 *low* while you remove the data wires and return $R/\overline{W}$ to the *high* state. Then reselect the chip by switching CS0 *high*. Note that the LEDs read the binary pattern just stored. Turn off the power and see what happens to your data. Store your phone number in seven successive addresses and have someone else read it out by switching address lines. Be sure to deselect the chip while changing any data or address inputs.

Answers to Chapter 0 Review Questions

1. Analog **2.** Digital **3.** Analog **4.** Switching
5. Computer programmers and designers. **6.** 1833 **7.** Minicomputer
8. Microprocessor **9.** Mainframe
10. We're not ready to form an opinion on that question yet.
11. Only (d) is out of spec. **12.** Positive, negative
13. Positive, positive, negative **14.** Logic 0
15. *Low* **16.** *A*-not; *A* and *B*; *A* or *B*; *A* and *B*, not
17. 0 **18.** NOR gate **19.** Totem pole, open-collector
20. 0 **21.** *RS* flip-flop or bounce eliminator
22. Circle, triangle **23.** Latch **24.** Set *Q* output to 1.
25. Transparent **26.** Register **27.** 10 **28.** J2 and K2
29. The race problem **30.** Don't care; *high* or *low* the same.
31. 22 **32.** 16 **33.** Synchronous **34.** 4
35. No means whatsoever **36.** Ultraviolet light
37. Yes **38.** The power is interrupted. **39.** 16 384
40. Dynamic RAM

CHAPTER 0 QUESTIONS AND PROBLEMS

Digits after the decimal point refer to section numbers in Chapter 0.

Basic Level

1.1 Tell whether each of these is basically digital or basically analog:
(a) A game of checkers **(b)** A mercury thermometer **(c)** A standard-transmission automobile clutch **(d)** An automobile gearshift lever

2.1 How does a microcomputer interpret these voltage levels?
(a) +0.1 V **(b)** +1.1 V **(c)** +2.1 V **(d)** +3.1 V

3.2 In what year was the first digital computer designed? In what year was the first electronic digital computer completed?

4.2 What minicomputer of the early 1970s brought the price of a computer below $10 000?

5.3 What is the voltage at a microprocessor pin labeled V_{SS}?

6.3 Do IC data sheets use electron-flow or conventional-current direction?

7.4 How many rows of the truth table for a four-input NOR gate will show logic 0 as the output?

8.4 In negative-true logic, what voltage level is associated with the binary symbol 1?

9.4 In Figure 0.8(d) the output line has a stray capacitance to ground of 50 pF. What is the time constant of the pull-up from *low* to *high* output?

10.5 What would be the output levels at Q and $\overline{Q}$, respectively, if both $\overline{S}$ and $\overline{R}$ in the circuit of Figure 0.9(a) were held *low*?

11.5 Define the word toggle.

12.6 When does a 74LS107 flip-flop accept input data? When does it switch its outputs in response to this data?

13.6 The shift register of Figure 0.12 is made into a ring counter by connecting Q4 to J1 and $\overline{Q4}$ to K1. Data 1101 is loaded in. What will be the data at Q1 through Q4 after three clock pulses?

14.7 Give the value of each of the five digits of the binary number 10001.

15.7 What is the advantage of synchronous counters over ripple counters?

16.7 A three-stage binary counter is fed with a 1-MHz square wave. What is the output frequency?

17.8 How many address lines are required by an 8 K-word memory?

18.8 What would be a more accurately descriptive name for the devices we call RAMs?

Advanced Level

19.1 Why have analog computers, which were more common in the 1950s and 1960s, been largely replaced by digital computers in the 1970s and 1980s?

20.1 Why does a digital computer so rarely make a mistake?

21.1 As the answer to a problem, a computer outputs the following set of voltages

+4 V 0 V +4 V +4 V 0 V 0 V 0 V +4 V

Where do we go to find the meaning of this "answer"?

22.2 Distinguish among mainframes, minicomputers, and microcomputers.

23.3 The output of IC1 drives the input of IC2. I_O for IC1 is +0.5 mA. Draw the circuit with a milliammeter connected and marked for proper polarity to measure I_O, and give the value of I_{IN} for both ICs.

24.3 Write two mathematical expressions for noise margin, one for logic *low* levels and one for logic *high*. Substitute the numbers given in the text to prove that in both cases there is a 0.4-V margin between input voltage requirements and guaranteed output drive.

25.4 Make a table showing all the symbols currently used to express each of the following concepts in logic: NOT, AND, OR, XOR.

26.4 Draw a logic diagram of a 4-input OR gate constructed entirely of 7400 2-input NAND gates.

27.4 Draft a complete truth table for the circuit of Figure Ø.8(d).

28.5 Explain the difference between a *D* flip-flop and a latch.

29.5 Explain what voltages must be applied to make Q1 of a 74LS107 flip-flop go to a *high* level, regardless of its present state.

30.6 Differentiate among a shift register, a ring counter, and a binary counter.

31.6 Draw a diagram showing the two separate flip-flops contained in a *JK* flip-flop. Show the logic sense and connections of the clock inputs.

32.7 For the mod-16 counter of Figure Ø.15, list the states of the *J* and *K* inputs to each flip-flop when the counter contains data 9. Then list whether each flip-flop will set, clear, hold, or toggle on the next clock. Finally, list the new binary state of each flip-flop after the clock.

33.7 Repeat Problem 32.7 for the mod-10 counter of Fig. Ø.16.

34.8 What is the exact number of bits that can be stored in a ROM having address lines AØ through A13 and data lines DØ through D7?

35.8 Define these memory-related words: volatile; static; dynamic; EEPROM.

PART I ELEMENTARY MICROCOMPUTER CONCEPTS

1 Machine Cycle and Bus Concepts

1.1 COMPUTER ORGANIZATION

Section Overview

The basic computer consists of a main processor, a memory, an input device, and an output device, interconnected by buses. A bus is a group of signal lines acting together. The processor sends signals to the memory via an *address bus*, which accesses a succession of address locations. Data from these locations are sent back to the processor via the *data bus*. These data constitute the program that the processor follows.

Input and output devices are also connected on the same data bus, and the processor occasionally receives data from or sends data to them. Several devices can be connected in parallel on the data bus because the processor (with the help of external logic) generates control signals that permit only one device to apply data to the bus at a time.

A computer works by reading data from a series of memory locations and reacting to each piece of data according to a strict set of rules. The *data* constitutes a *program* and must be carefully structured to produce the desired responses from the computer. The *rules* are the computer's *instruction set* and are different for each type of machine. The instruction set for the 6800 and 6802 microprocessors is summarized on the Programmer's Reference Card (the last two pages of this text). Have a look at it now. It will be a good measure of your progress to see it become more understandable and useful, chapter by chapter.

Parallel bus structure: Figure 1.1 illustrates the main operational concepts for most 8-bit microcomputers. The MPU is the Main Processing Unit or, simply, the processor. It generates a series of binary numbers with its *program counter*, which is a 16-bit internal binary counter. This succession of logic-signal sets appears on the 16 lines of the processor's *address bus*, which the memory decodes to call up a succession of data words. The memory places these data words on the 8-line data bus. The data words are then read and acted upon by the processor.

Occasionally the processor's interpretation of the data it has been reading may indicate that it is being instructed to do something other than simply read the data from the next program-memory address. Here are some of the possibilities.

- The processor may write new data to a selected memory address. In this case the address bus outputs this address and the processor asserts (activates) the *write* line to tell the memory to store the new data.
- The processor may read data being input from a peripheral device, such as a keyboard. In this case the *select memory* line will be deactivated and the *select input* line will be asserted while the processor reads the data applied to the bus by the input device.
- The processor may send data to an output device, such as a CRT (cathode-ray tube) monitor. Now the *select output* line will be asserted and the *memory* and *input* devices will be deselected.

In each case the processor will assure, via the *select* lines, that only one of the three external devices is active at a time. Also, if the processor is writing data to the bus,

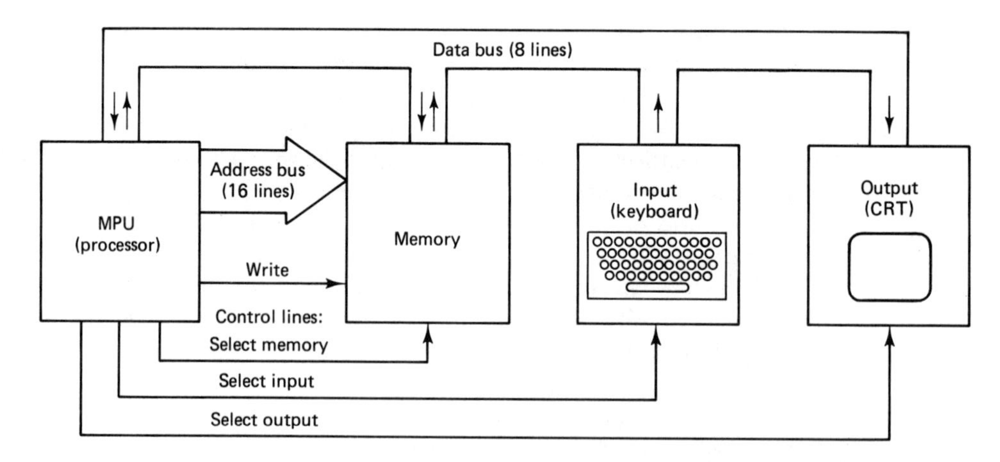

Figure 1.1 Computer organization. Several devices are connected in parallel on the data bus, but only one is allowed to apply data to the bus at a time.

the other device on the bus will be directed to pick up that data. And, if the other device is applying data to the bus, the processor will be reading it. Thus many devices can be wired in parallel on the single data bus, but there is never a *fight for the bus* because only one device is allowed to apply data to it at a time.

Once the processor has completed one of these special tasks, the processor's address output lines resume the task of calling up successive memory addresses, and the processor continues reading and interpreting program words until another special operation is called for.

1.2 THE SYSTEM CLOCK AND 6802 PIN FUNCTIONS

Section Overview

The microprocessor *clock* signal specifies the time available for data to be transferred from the selected sending device to the receiving device.

In addition to the obvious pin requirements for V_{CC}, ground, address bus, data bus, and clock, the following pins are common among microprocessors.

- ***Reset* input to start the main operating program over from the beginning.**
- ***Read/write-not* output to specify whether the processor is receiving or sending data on the data bus.**
- ***Interrupt* inputs to force the processor to set aside its regular program and go to an interrupt-service program.**

The various parts of a computer system are generally kept in step with a system clock. The clock signal is generated by an oscillator running at a frequency that is typically between 0.5 and 15 MHz. While all microprocessor clocks have essentially the same function, some have multiple outputs and rather complex timing relationships. One of the reasons that we have elected to use the 6802 in our more specific discussions of microprocessor characteristics is the fact that it has the simplest clock and timing characteristics of any microprocessor on the market. We will encounter more sophisticated clock timing in Part IV of this text.

Figure 1.2 shows the 6802 clock signal, which is labeled E (enable) in Motorola literature. Essentially, the first (*low*) half of the cycle is devoted to address and device selection; data is transferred on the data bus during the second (*high*) half of the E cycle. The E signal is output via pin 37 of the 6802. The falling edge of E identifies the safest data-valid point, and is commonly used for latching data.

The oscillator circuitry for generating the E signal is included in the 6802 chip. The only external components required are two capacitors and a frequency-determining crystal, as shown in Figure 1.3 at pins 39 and 38. The frequency of E is one-fourth of the crystal frequency. We will use a 2-MHz crystal for a 0.5-MHz clock frequency in most of the microprocessor circuit examples in this book. This is slower than the standard 1-MHz clock, but it will make the timing and wiring less critical.

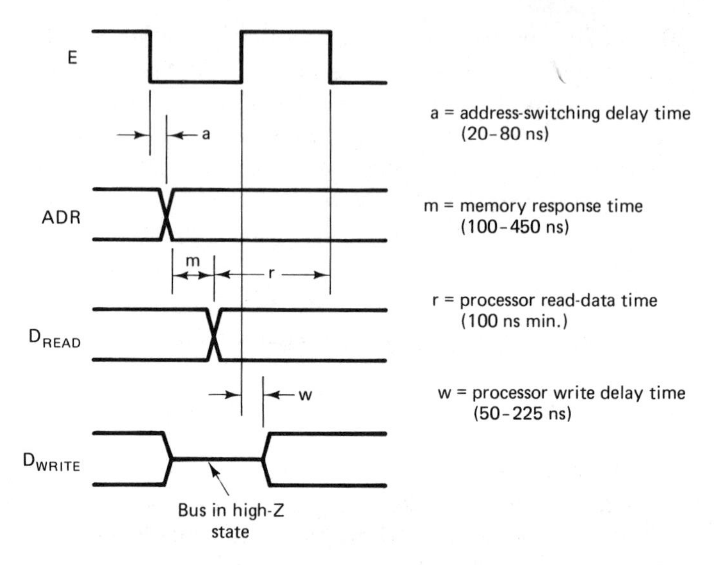

Figure 1.2 Clock timing for the 6802 microprocessor. The address is switched during the first (*low*) half of the *E* cycle, and data is read or written in the second (*high*) half.

An external square-wave oscillator can be used to produce the *E* signal for the 6802. The square wave must be TTL level (0 to +4 V) and must be connected to EXTAL (external/crystal) pin 39. Pin 38 should be grounded, and the crystal and capacitors should, of course, be removed. The *E* frequency will be one-fourth of the EXTAL square-wave frequency and may be made to vary from 0.1 to 1.0 MHz. The 68B02 is a fast version, which can operate with *E* up to 2.0 MHz.

Pin functions for the 6802 microprocessor are labeled in Figure 1.3. The following pins of the 6802 are found in some form among microprocessors generally.

- Reset pin 40 stops the program when it is pulled *low*. When it is returned *high* the processor starts the main operating program over from the beginning.
- Read/Write-not pin 34 is an output that is connected to the $R/\overline{W}$ input of a RAM or other read/writable device to direct whether data is being read from it or written to it. Some processors use two output pins to assert *read* and *write* conditions separately.
- V_{SS} is simply the power-supply and signal ground. The 6802 has two grounds, pins 21 and 1.
- The power supply, +5.0 V (±5%), is connected to pin 8. Typical current demand for the 6802 is 120 mA.
- Bus available, pin 7, is an output that goes *high* to indicate that the processor is not using the address and data buses. It is used to permit another device (such as a second processor) to assume control of the buses.
- Nonmaskable Interrupt (pin 6) and Interrupt Request (pin 4) are *low*-active inputs that can cause the processor to leave its normal program and execute

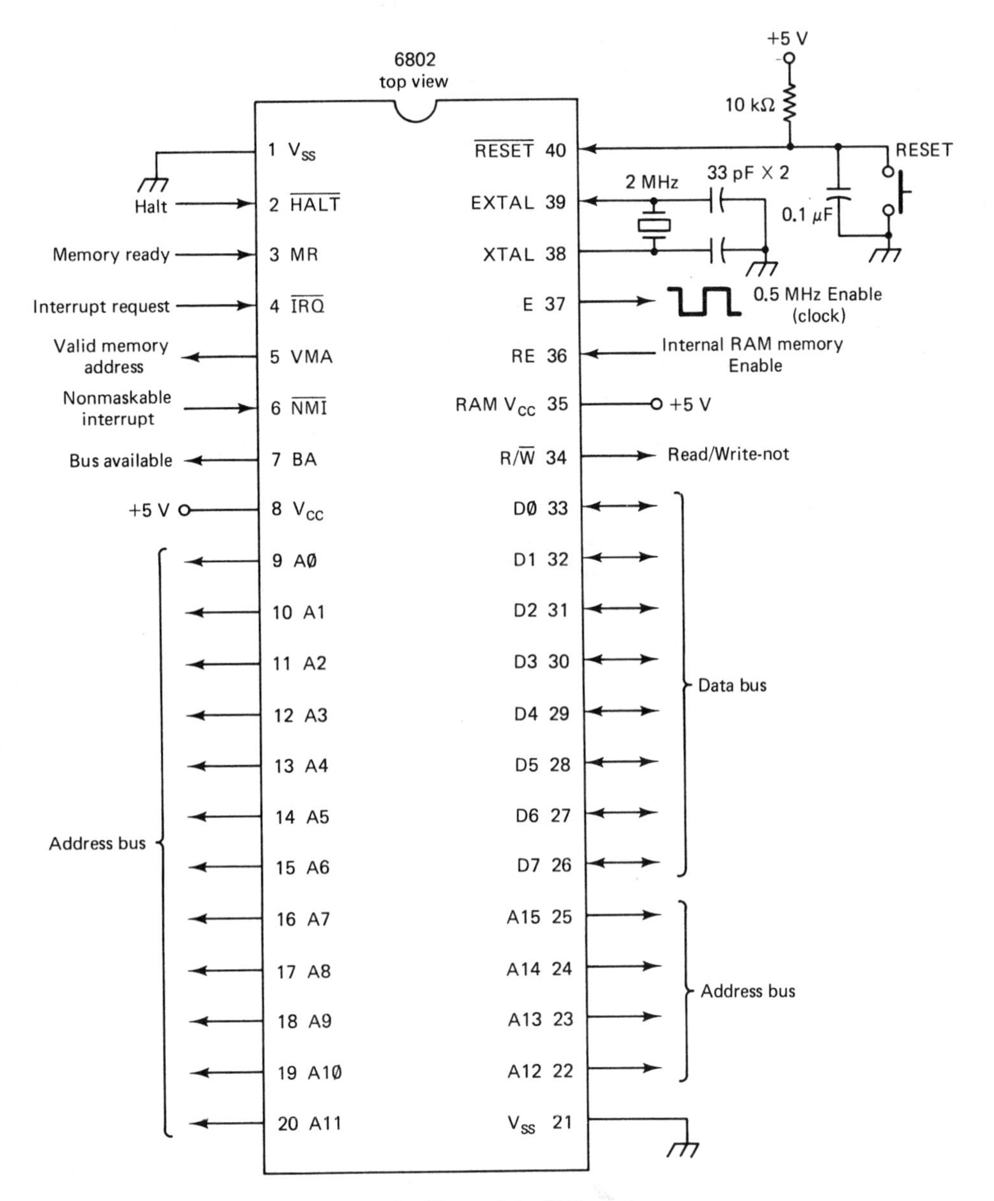

Figure 1.3 Pinout of the 6802 microprocessor.

one of several special *interrupt* programs. Interrupting is common in microcomputer systems. For example, a desktop micro may use its normal program to keep the display appearing on the CRT. Pressing a key may generate an interrupt, causing a short *keyread* program to be executed, followed in a millisecond or so by a return to the *display* program.

- The 16-bit address bus occupies pins 9 through 20 and 22 through 25. They are always outputs, and, for the 6802, they are not tristate. This means that they are always active and do not have a floating or high-Z state. The address bus *is* tristate in most other microprocessors.

- The 8-bit data bus occupies pins 26 through 33. These lines are outputs when the R/$\overline{\text{W}}$ pin 34 is *low* and inputs or floating when R/$\overline{\text{W}}$ is *high*.

The following pin functions of the 6802 are commonly, but not necessarily, found in other microprocessors.

- The $\overline{\text{HALT}}$ input, pin 2, stops the program execution when it is pulled *low*. With some external circuitry it can be used to single-step the processor, making it execute one instruction at a time.
- The Memory Ready input (MR, pin 3) can be pulled *low* by a slow memory to give it more time to apply data to the bus. Normally, this line is tied to V_{CC}.
- The Valid Memory Address output (VMA, pin 5) goes *low* during E-clock cycles when the processor is performing an operation that does not involve any memory address. Recall that the address bus is not tristate and so is always calling up *some* memory address. The system logic is generally wired to disable all memory devices when VMA is *low*.
- A RAM memory of 128 words is integrated right on the 6802 chip. RAM enable (RE, pin 36) enables this internal memory when *high* and disables it when *low*. The first 32 words of this RAM are powered from RAM V_{CC}, pin 35, independently of the V_{CC} at pin 8. With pins 35 and 36 powered from a battery supply delivering only 8 mA, critical data can be saved even with the processor turned off.

REVIEW OF SECTIONS 1.1 AND 1.2

Answers appear at the end of Chapter 1.

1. The microprocessor selects different words from a memory chip by signals sent out on the ________ ________.
2. The memory word selected is sent back to the processor via the ________ ________.
3. How many devices can be connected to the microprocessor's data bus? (1, 2, 3, or many?)
4. Does the processor always receive signals, always send signals, or both send and receive signals via its address bus?
5. Addresses are switched during the ________ half of the 6802 clock. Data is transferred during the ________ half. (Answer *high* or *low*.)
6. For the 6802, a 4-MHz crystal produces a clock frequency of ________.
7. The 6802 ________ bus is tristate, but the ________ bus is not.
8. The $\overline{\text{IRQ}}$ and $\overline{\text{NMI}}$ lines can ________ the processor's normal program and transfer it to a short special program.

9. The ________ line of the 6802 goes *low* to indicate that no memory address should be called up because the processor is handling data internally.

10. The 6802 has a ________-bit data bus and a ________-bit address bus.

1.3 THE ADDRESS BUS AND BINARY NUMBERS

Section Overview

A 16-bit address bus can access 2^{16}, or 65 536, different addresses. This is referred to as a 64K memory area.

The voltage patterns on the address bus represent binary numbers. Decimal numbers use 10 symbols and place values that increase by a factor of 10 in each column. Binary numbers use two symbols (Ø for approximately 0 V and 1 for approximately +4 V) and place values that increase by a factor of 2 in each column; hence binary 1Ø1 equals 5 in the decimal system:

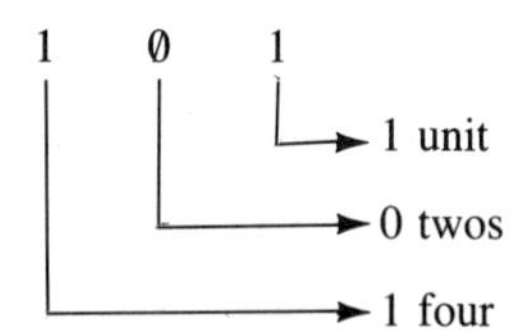

Most 8-bit microprocessors have a 16-bit address bus. The lines are labeled AØ through A15. Note that the first line is AØ; the second is A1, and so on, until the sixteenth, which is A15.

The number of binary combinations possible on 16 lines is 2^{16}, or 65 536. The lowest, middle, and highest addresses are listed next in binary, along with the place value of each binary digit.

32 768	16 384	8 192	4 096	2 048	1 024	512	256	128	64	32	16	8	4	2	1
Ø	Ø	Ø	Ø	Ø	Ø	Ø	Ø	Ø	Ø	Ø	Ø	Ø	Ø	Ø	Ø
1	Ø	Ø	Ø	Ø	Ø	Ø	Ø	Ø	Ø	Ø	Ø	Ø	Ø	Ø	Ø
1	1	1	1	1	1	1	1	1	1	1	1	1	1	1	1

If you take the trouble to add the place values for each 1 in the bottom row, you will see that the maximum count is 65 535. This means that the microprocesor can address, at a maximum, 65 536 different memory words.

K in computers. It has become common to refer to 2^{10}, or 1024, memory words as 1K. The value 1024 is approximated by 1000, and K is used because of the metric multiplier k, for *kilo*, as in 1 kW. Be sure not to confuse the metric prefix k with the computer term K, however. Using this terminology, the 16-bit address bus can access 64K of memory.

Counting in binary requires that the lowest-order digit (or 1's bit) be toggled (changed) at each count. Each higher-order digit changes if the digit below it changed from a logic 1 to a Ø. Here is the binary counting sequence from 0 to 15. We use only the four least significant binary digits to simplify the example, but the techniques can be extended to 8 or 16 digits without any conceptual differences.

A3	A2	A1	AØ	Decimal	Hex
Ø	Ø	Ø	Ø	0	Ø
Ø	Ø	Ø	1	1	1
Ø	Ø	1	Ø	2	2
Ø	Ø	1	1	3	3
Ø	1	Ø	Ø	4	4
Ø	1	Ø	1	5	5
Ø	1	1	Ø	6	6
Ø	1	1	1	7	7
1	Ø	Ø	Ø	8	8
1	Ø	Ø	1	9	9
1	Ø	1	Ø	10	A
1	Ø	1	1	11	B
1	1	Ø	Ø	12	C
1	1	Ø	1	13	D
1	1	1	Ø	14	E
1	1	1	1	15	F

Binary counting has the following properties:

- The lowest bit (AØ) changes at every count.
- AØ is Ø for all even numbers and is 1 for all odd numbers.
- The 4-through-7 sequence is a repeat of the 0-through-3 sequence with A2 *high*. Similarly, 8 through 15 repeats the 0-through-7 sequence with A3 *high*, and 16 through 31 repeats the 0-through-15 sequence with A4 *high*.
- The first count with A2 *high* has A1 and AØ *low*. When A3 first goes *high*, lower digits A2, A1, and AØ go *low*. At the count of 16, A4 goes *high* for the first time, with all lower digits going *low*.

Binary-to-decimal conversion is straightforward. Just add up the place values of all the 1 digits, ignoring all the Ø digits. This process is illustrated for an 8-bit binary number.

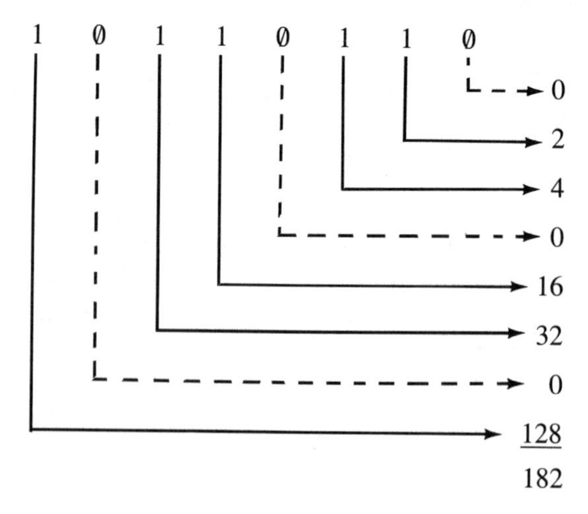

Decimal-to-binary conversion is tedious but not difficult. It is done by successive subtractions of the highest possible powers of 2. A table of powers of 2 (binary place values) is a helpful aid. As an example, we find the binary expression of decimal 147. First we write out the place values. Then we write in a 1 where subtraction is possible and a Ø where it is not.

128	64	32	16	8	4	2	1
1	**Ø**	**Ø**	**1**	**Ø**	**Ø**	**1**	**1**
147	19	19	19	3	3	3	1
−128	−64	−32	−16	−8	−4	−2	−1
19			3			1	0

Fortunately, these conversions are not often required. Where they are, low-cost calculators are available that perform them automatically.

1.4 HEXADECIMAL REPRESENTATION OF BINARY

Section Overview

Hexadecimal numbers use 16 different symbols and a place value that increases by a factor of 16 in each column, moving left. The symbols used, equivalent to decimal 0 through 15, are Ø, 1, 2, 3, 4, 5, 6, 7, 8, 9, A, B, C, D, E, F. The place values are *units* in the first column, *groups of* 16 in the second, groups of 16^2, or 256, in the third, groups of 16^3 or 4096 in the fourth, and so on.

Microcomputer programmers use hexadecimal numbers because they are not as confusing to humans as the long strings of 1s and Øs of binary, and they can be obtained immediately from the binary numbers used by the machine. Each group of four binary digits is simply replaced by a single *hex* digit. The chart on page 34 shows this correspondence.

Microcomputers are constrained, by the nature of their circuitry, to dealing with numbers in binary. The human operator, however, finds it tedious and error-prone to copy and check long strings of Øs and 1s. Decimal numbers are not often used to represent computer data because conversion to the machine's binary is time consuming, and there exists a far superior alternative.

Hexadecimal numbers are based on groups of 16. In decimal we count objects as units until we have a group of 10. Once we have accumulated 10 groups of 10, we place them in a third column called *hundreds*.

In binary we put groups of 2 in the second column, groups of 4 in the third, groups of 8 in the fourth, groups of 16 in the fifth, and so on.

In hexadecimal we put units in the first column, groups of 16 in the second column, groups of 16^2, or 256, in the third column, and groups of 16^3, or 4096, in the fourth column. This requires that we have 16 symbols, so we can count units from 0 to 15. We use the capital letters A through F to represent the numbers that, in decimal, we call 10 through 15.

The numbers from 0 to hex 2F (decimal 47) are as follows:

00	10	20
01	11	21
02	12	22
03	13	23
04	14	24
05	15	25
06	16	26
07	17	27
08	18	28
09	19	29
0A	1A	2A
0B	1B	2B
0C	1C	2C
0D	1D	2D
0E	1E	2E
0F	1F	2F

Some of these numbers look like regular decimal numbers, although they must not be interpreted as such. For example, 29 in hex means 32 plus 9, or 41, in decimal. To keep this clear it has become common practice to preceed hex numbers with a dollar sign where confusion with decimal numbers is possible; thus $29 = 41. (Some computer programs use a capital *H*, after the digits, as in 29H, to denote hexadecimal.) Verbally, you should cultivate the habit of saying "hex-two-nine" rather than "twenty-nine," which implies decimal.

Here are some examples to help you gain familiarity with hex numbers.

1. $A0 = 10 groups of 16 plus 0, or 160_{10}
2. $B5 = 11 groups of 16 plus 5, or 181
3. $12C = 256 + 2 (16) + 12 = 300
4. $F0D = 15 (256) + 13 = 3 853
5. $148E = 4096 + 4(256) + 8(16) + 14 = 5 262
6. $FFFF = 65 535, the highest count on our microprocessor's address bus

Binary numbers can be converted to hex, and vice-versa, by a process that is immediate and almost effortless. The binary number is simply written in groups of four digits, starting at the right and filling in leading zeros at the left as necessary. Each group of four binary digits is then replaced with one corresponding hex digit. No additions, subtractions, or carries between digits are required. Here are some examples.

1. 1001 0011 = $93
2. 1111 0001 = $F1
3. 100 1110 = $4E
4. 1011 1101 0111 1100 = $BD7C

Most of the time, these values do not need to be converted to decimal. The computer works in binary, the programmer works in hex, the two are easily convertible, and there's the end of it.

Soon you will develop a great facility for converting between groups of four binary digits and single hex digits. The digits 0 through 9 are familiar and generally give no trouble. Binary 1010 is easy to remember as $A, the first unfamiliar digit. Binary 1111 will stick in your memory as $F, the highest hex digit. If you can also remember that binary 1100 is $C, you can get along by mentally counting on your fingers up or down from these reference points to fill in the binary digits for $B, $D, and $E.

REVIEW OF SECTIONS 1.3 AND 1.4

Answers appear at the end of Chapter 1.

11. The Intel 8086 microprocessor has 20 address lines. How many different addresses can it access?
12. The type 2732 memory chip has 12 address lines. How many K-words does it contain?
13. What is the decimal value of binary 0111 0110?
14. What is the binary representation for decimal 173?
15. How many binary bits are needed to count up to 1 billion (10^9)?
16. Why are computer programs not listed in binary?
17. Why are computer programs listed in hex rather than in decimal?
18. What is the hex expression for binary 1010 1011?
19. What is the binary representation of $8B?
20. Is $A3D an even or odd number? *Hint:* Convert to binary and check bit 0.

1.5 BITS, BYTES, NIBBLES, AND WORDS

Section Overview

A *byte* is a fixed unit of data equal to 8 bits. It can be expressed in two hex digits.

A *word* is a variable amount of data—anything from 1 bit to 64 bits or more. Word length depends on the device using the data, how it uses the data, and, sometimes, on the judgment of the individual interpreting this use.

A bit is a Binary digIT, the amount of data that can be represented by a single flip-flop or a single digital line at one instant. It is the smallest unit of data.

A byte is a standard unit of digital data consisting of 8 bits. These bits are usually labeled D0 (place value 1) through D7 (place value 128). One byte can represent one of 256 different values—in straight binary, the values 0 through 255.

A nibble is, of course, half a byte. This is 4 bits—the amount of data represented by one hex digit.

A word is an amount of data that is handled or treated as a unit. Thus the number of bits in a word depends on the piece of hardware being discussed. Sometimes it also depends on the scruples of the advertising department that is trying to sell the hardware. Here's what we mean.

A memory chip that stores 4 bits at each address inarguably treats 4 bits as a unit and has a 4-bit word length. The 8085 and 6802 microprocessors have 8-bit data buses, and nearly all their instructions deal with 1 byte of data at a time. Nobody claims that these are anything but 8-bit microprocessors.

The 8088 microprocessor, however, has an 8-bit data bus, even though most of its instructions deal with 16 bits of data at a time. (Each instruction generally requires two accesses of the data bus to transfer the 16 bits of data to or from the processor.) Is this a 16-bit processor or an 8-bit processor? The argument is not yet settled, but the point to remember is that word length is variable and sometimes dependent on the definitions and perceptions of an individual.

More Definitions: Here are some other computer-related definitions:

- *Hardware*. The physical equipment making up a computer system; for example, IC chips, circuit cards, wiring, printers, display CRTs
- *Software*. Computer programs.
- *Firmware*. Computer programs that reside in the machine (usually in ROM) and do not need to be reloaded each time the computer is powered up.
- *Documentation*. Manuals, notes, diagrams, and other information intended to help users understand software or hardware.

1.6 MICROPROCESSOR COMPARISONS

Section Overview

The first microprocessor to attain lasting success was Intel's 8080. Intel's improved version of this processor is the 8085. Zilog markets a more extensively improved version called the Z80.

Motorola's first processor was the 6800, which is essentially our 6802 without the on-board clock or RAM. Motorola's enhanced version is the 6809. MOS Technology developed, and Rockwell now markets, a simpler enhanced version called the 6502.

All these are 8-bit microprocessors.

Although we will concentrate our more specific discussions on the 6802 processor for the first three-fourths of this book, we will occasionally pause to survey the characteristics of some other popular microprocessors.

Engineers are often chauvinistic about microprocessors, ascribing all virtues to the one they use and heaping scorn on the ones with which they have less experience. However, there are several processors that are selling well in a competitive market, so it is obvious that no one chip is clearly superior to all others in all applications. Indeed, a better case can be made that there are several excellent or even superior processors that are not selling well simply because they arrived too late to gain a following among the engineers, who had already committed themselves to one or another of the better-known lines of components.

Broadly speaking, the most successful 8-bit microprocessors are the descendents of the Motorola 6800 (1974) and the Intel 8080 (1972). All are 8-bit machines with 16-bit address buses. We look first at the Motorola line.

- The 6800 is reputed to have been named for the number of integrated components it contains. It is identical to the 6802, which we have elected to treat in detail, except that it contains no on-board RAM, requires an external two-phase clock generator, and has an address bus that can go tristate.
- The 6502 (1975) was designed by a group of 6800 veterans and is very similar to that processor. It has an on-board clock circuit but no RAM. Its most striking differences are in addressing modes and programming flexibility. It was designed at MOS Technology (Commodore) and is also made by Rockwell.
- The 6809 is a greatly enhanced version of the 6800 and was introduced by Motorola in 1978. It has a more powerful and complex instruction set, including several instructions for 16-bit data and a multiply instruction. The 6809 has an internal clock but no RAM. The 6809E requires an external clock and adds some extra control functions.
- The 6805 is a family of single-chip microcomputers by Motorola. They contain 1K to 4K of program memory (ROM), 64 to 112 bytes of RAM, a clock, and 16 to 32 input/output lines—all in a single package of 28 or 40 pins. In terms of processing power and sophistication, they are similar to—but a small step down from—the 6800. Since the program memory is masked ROM, they must be ordered in large quantities with a fixed program. Their address and data buses are not brought out to the package pins. They are, therefore, not suitable for classroom experimentation.
- The 6801 is a more advanced single-chip microcomputer from Motorola. It has 31 input/output lines and 2K of masked ROM. However, the I/O lines can be programmed as address and data lines to access additional external memory. The 6803 is a 6801 without the on-board ROM. These computers use the same basic instruction set and therefore run the same programs as the 6800 and 6802. Several enhanced instructions, such as *multiply* and *add 16-bit data* have been included.

The 8080 and its descendents differ from the 6800 line more in instructions and programming than in hardware. There is a remarkable parallelism between the histories of the two groups.

- The 8080 preceded all of the other microprocessors mentioned here, and quickly gained a large following. Many argue that it represents the standard by which all others should be measured. Its instruction set is in some respects more powerful but certainly less straightforward than that of the 6800. It requires a special external clock-generator chip and three power supplies (+5 V, −5 V, and +12 V).
- The Z80 was designed by a group of 8080 veterans and released by Zilog in 1976. It follows the 8080 instruction set but adds several powerful new instructions and many new machine registers for ease of handling multiple tasks. The Z80 requires a simple single-phase TTL clock generator and requires only the standard +5-V power supply.
- The 8085 (1976) is Intel's upgraded version of the 8080. It contains an on-board clock and needs only the standard +5-V supply. Serial (single-line) input and output pins are included. The data bus lines are multiplexed with the low-order 8 address lines. This means that the lines occupy the same pins, which output address bits during the first part of the machine cycle and transfer data bits during the second part. Memory and peripheral chips must either be Intel parts, which also have multiplexed buses, or they must be interfaced through an external address-latch chip.
- Intel markets a family of single-chip microcomputers with on-board clock, I/O pins, and ROM. All of them follow the basic 8080 instruction set or a limited version of it. The 8051 has 4K of ROM and 32 I/O lines and is capable of expansion by using some of these lines to access external memory. The 8031 is a ROMless version of the 8051. The 8048 offers 1K of ROM, 16 I/O lines, and full access to the 8-bit data bus. The 8035 is a ROMless version that can interface to external memory.

In addition to these 8-bit machines, Motorola offers the 68000 and Intel the 8086 in the 16-bit field. Several processors with 32-bit data buses are now available. We will discuss some of these chips in detail in Part IV of the text.

REVIEW OF SECTIONS 1.5 AND 1.6

21. How many hex digits does it take to represent a 6802 address?

22. A memory chip contains 16K bits and has address lines A0 through A10. What is its word length?

23. Does a microprocessor with an 8-bit data bus necessarily have an 8-bit word length? (Yes, no, or arguable.)

24. Name two components that are likely to be included on board in a single-chip microcomputer but not in a microprocessor.

25. A ROM chip contains a program that tells a personal computer how to read its keyboard and display on its CRT. Is this hardware, software, or firmware?
26. The Z80 is an enhancement of what original chip?
27. The 6809 is an enhancement of what original processor?
28. What company originated the 8085 processor?
29. If a processor uses the same pins for address and data, switching them on alternate halves of the clock cycle, we say that the buses are ________.
30. Which was the only popular microprocessor chip that required more than one power supply?

1.7 CHECKING OUT THE 6802 CLOCK AND PROGRAM COUNTER

Section Overview

A simple setup is described for the 6802 microprocessor to permit observation of several clock and address-line features on an oscilloscope.

At this point it would be a good idea to firm up the things you have learned about microprocessors by actually wiring and checking out a microprocessor's clock and address bus. Figure 1.4 shows a wiring diagram for such a test on a 6802 processor.

The crystal may be of any frequency between 1 and 4 MHz. The 2-MHz value shown has the advantage of producing a 2-μs clock period, so each half of the clock cycle occupies exactly 1 cm on an oscilloscope set to sweep at 1 μs/cm.

The resistor and capacitor values are not at all critical and may vary ±50% with no noticeable effect. This is part of the digital advantage, as you may recall from the first section of Chapter 0. The capacitor from V_{CC} to ground is there to suppress noise, which digital circuits tend to impose on the supply line. It is interesting to observe its effectiveness by removing it while monitoring V_{CC} with a fast oscilloscope (10 MHz or more).

The unused inputs $\overline{\text{HALT}}$, MR, $\overline{\text{IRQ}}$, and $\overline{\text{NMI}}$ are tied to a logic *high* level. We will not be using the 128-byte on-board RAM until later chapters, but we may as well power it up and enable it (pins 35 and 36) right away.

The data bus is wired to produce a continuous binary 0001 0000, which is hex 10. You can check in the 6802 programmer's reference (back flyleaf) that this is the command SBA, which takes 2 machine cycles (~) and tells the processor to subtract the contents of internal register B from the contents of internal register A and store the results in internal register A.

Breadboard-wiring tips for microcomputer projects are given in Appendix A. Read them carefully now, and refer to them again whenever you undertake a microcomputer-wiring project. Wiring errors in microcomputer systems are notoriously hard to find, so it is worth the extra effort to avoid making them in the first place. After wiring the circuit, have a friend check it for errors. Then set a regulated supply

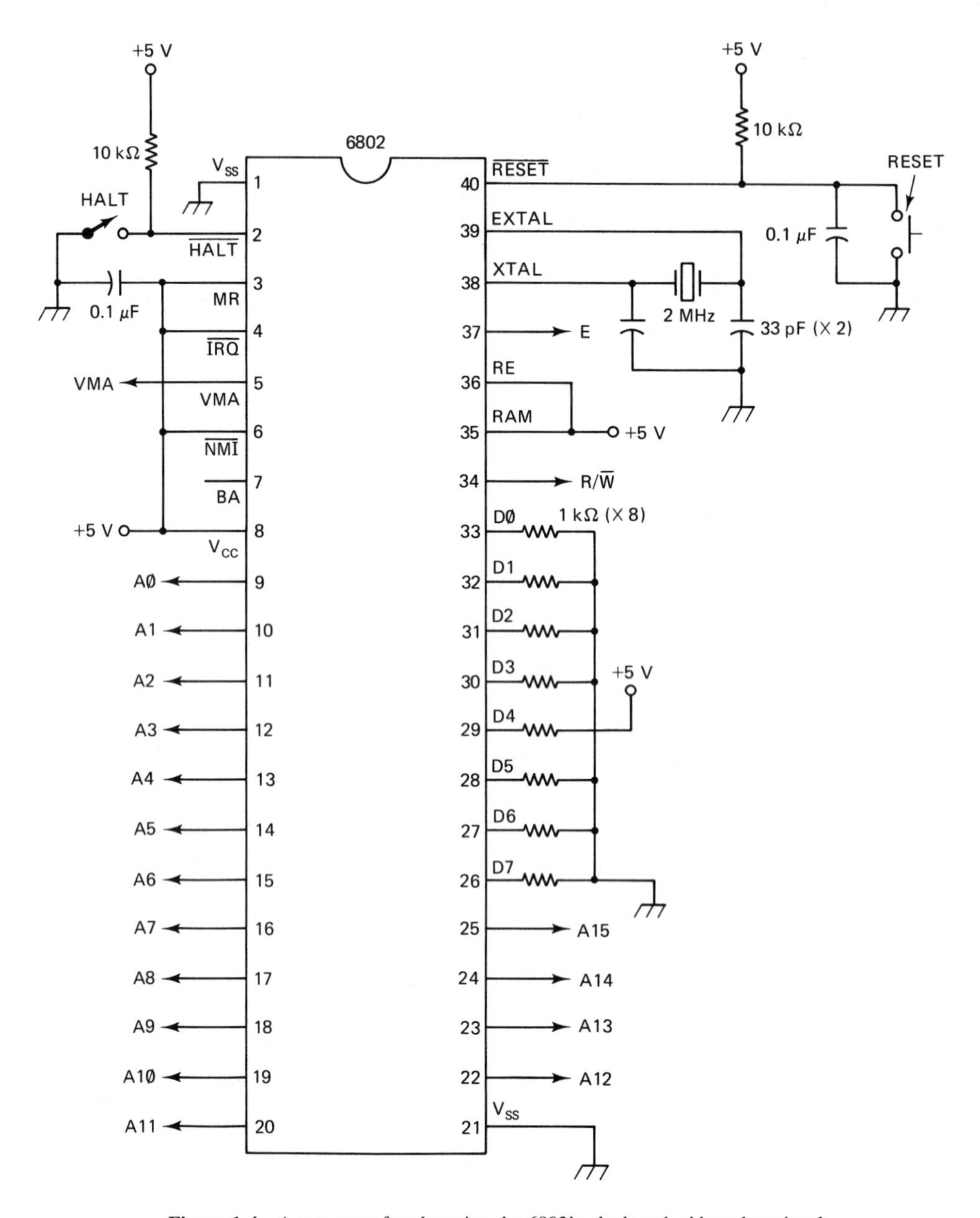

Figure 1.4 A test setup for observing the 6802's clock and address-bus signals.

to 5.0 V, ±0.1 V *before connecting it* to the circuit V_{CC}. Use an oscilloscope on dc coupling and 1 μs/cm sweep to measure the *high* and *low* levels of the *E*-clock. Measure the period T of one cycle. Observe A1 and A0, and compare their periods. Does A1 switch on the rise or fall of A0? Similarly, observe and measure A2 and A1, and explain what the address bus is doing.

Answers to Chapter 1 Review Questions

1. Address bus **2.** Data bus **3.** Many
4. Always send **5.** Low, high **6.** 1 MHz
7. Data, address **8.** Interrupt **9.** VMA
10. 8, 16 **11.** 1 048 576 **12.** 4K
13. 118 **14.** 1010 1101 **15.** 30
16. The binary digit strings would be too long and confusing.
17. Hex is much easier to convert to binary than is decimal.
18. $AB **19.** 1000 1011 **20.** odd **21.** four
22. 8 bits **23.** Arguable **24.** ROM program memory and input/output lines **25.** Firmware **26.** 8080
27. 6800 **28.** Intel **29.** Multiplexed
30. 8080

CHAPTER 1 QUESTIONS AND PROBLEMS

Digits after the decimal point refer to section numbers in Chapter 1.

Basic Level

1.1 List the four main parts of a computer divided conceptually by function.

2.1 What is the most common operation that a computer performs on its address and data buses?

3.2 If a 6802 microprocessor uses a surplus TV "color burst" crystal of 3.579545 MHz, what will be the period of its clock signal?

4.2 Do the address pins of the 6802 ever assume a high-*Z* state? What about the data pins?

5.2 At what levels should the following 6802 inputs be held for normal processor operation?

$$\overline{\text{MR}}, \quad \overline{\text{IRQ}}, \quad \text{and} \quad \overline{\text{NMI}}$$

6.2 Which pin(s) of the 6802 must be held *high* to enable the internal RAM?

7.3 What is the highest number that can be represented with 7 binary digits?

8.3 Exactly how many words can be stored in an 8K memory?

9.4 How many hex digits does it take to represent an address for a computer with a 1M-byte addressing range?

10.4 Convert binary 1100 1010 1011 1101 to hex.

11.5 Define the terms bit, byte, nibble, and word.

12.5 What is the difference between software and documentation?

13.6 How many address lines and how many data lines do the 6502 microprocessors have?

14.6 Why is the 6805 single-chip microcomputer not suitable for school lab experiments?

Advanced Level

15.1 List three things that the processor can do via its data bus, other than read data from its program memory.

16.1 What causes the processor to do one of the things listed above instead of reading program memory?

17.1 The *MPU*, the *memory*, and the *input* interface can all apply data to the data bus. How do we prevent one of them pulling a line *high* while another one is pulling it *low*?

18.2 Describe the timing of the 6802 address-line transitions with respect to the *E* clock. Do the same for the data transitions on a processor write to memory.

19.2 Describe the use of the VMA pin of the 6802 processor.

20.3 Why do computers use binary instead of the more familiar decimal arithmetic?

21.3 Add binary 10111 to 10001 and convert the result to decimal. Now convert each of the original numbers to decimal, add them, and show that additions in binary and in decimal produce the same result. Display the process clearly on your paper.

22.4 Make a table showing all three number-system representations for each of the three given numbers.

decimal 5280   hex 1A B7   binary 1010 0110 1110

23.4 Add $5E to $64 in hex, carrying groups of 16 from column 1 to column 2. Now change the two original numbers to decimal, add, and change back to hex to show that the hex addition was correct. Show the carry and conversion steps clearly on your paper.

24.5 What is the argument that says that the 8088 is a 16-bit microprocessor? What is the argument that says it is an 8-bit microprocessor?

25.6 Make two "family tree" charts, showing the original 8-bit Intel microprocessor and its descendents and the first Motorola microprocessor and its descendents.

26.6 Explain the concept of multiplexed buses.

2 Data Types and Instruction Execution

2.1 THE DATA BUS

Section Overview

The microprocessor's data lines are bidirectional. When the processor *writes* data to another device, its data lines force the bus lines *high* or *low*. When the processor *reads* data from another device, its data lines float (i.e., they pull neither *high* nor *low*. The other device then outputs data to the bus and the processor reads this data.

A microprocessor address bus generally consists of 16 output-only lines.[1] Addresses are represented by programmers as 4 hex digits. A microprocessor data bus generally consists of 8 bidirectional lines, whose contents are represented by 2 hex digits.

A bidirectional line can function as an input or an output. Figure 2.1(a) shows how a single line, D0, can both apply data to and receive data from a type-D latch. (It can't do both at once, though, you understand.)

[1]In referring to microprocessors ''generally'' we include the common types with 8-bit data buses, but exclude 4-bit controller chips and the newer 16-bit processors. We will meet the latter in Chapter 23.

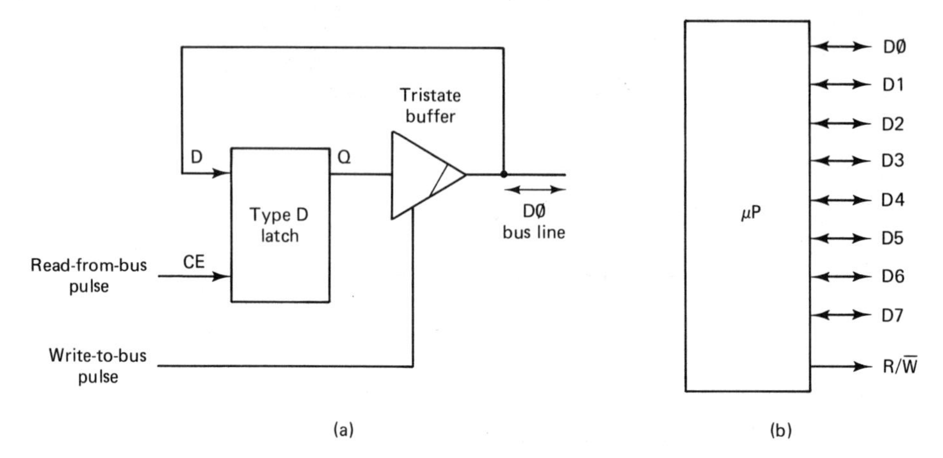

Figure 2.1 (a) A tristate data line presents a high impedance unless it is writing. (b) The processor data lines are tristate so they can either read or write on the bus.

The *D* input monitors the data-bus line continuously, presenting a high impedance. The data is not stored in the latch unless a *read-bus* pulse pulls chip-enable (CE) *high*. The output data *Q* is connected to the data-bus line through a tristate buffer, which is normally in the high-Z state. The data at *Q* is applied to the bus only when a *write-to-bus* pulse is present. *Read-bus* and *write-to-bus* pulses are never applied simultaneously.

In the case of the 6802 processor, the *read* and *write* pulses are generated internally and output via a single read/write-not (R/$\overline{\text{W}}$) line. A processor *read* of the bus occurs every time the processor accesses its program memory, and when it obtains data from its data memory or an input device. As an example, the instruction LDAA (Load Accumulator A) causes the processor to LOAD (read) data from the data bus to its accumulator A. A processor *write* to the bus is caused by such instructions as STAA (Store Accumulator A). The 6802, like all microprocessors, has within it a set of eight tristate buffers, which apply data to the bus only during a processor-write cycle.

2.2 DATA TYPES

Section Overview

The address bus carries bits that represent only addresses where information is to be stored or retrieved. The data bus carries bits that may have any of three distinctly different functions:

1. ***Instruction codes* tell the processor what operation to perform—for example, ADD or STORE.**
2. ***Addresses* where the next operation is to be performed are passed via the data bus. A 16-bit address can be obtained by two *reads* of the 8-**

bit data bus, and may later be output via the address bus to access that address.

3. ***User data* may have many different functions, but its interpretation is always determined by the programmer, not the processor.**

The signals output on the address bus always represent an address between $0000 and $FFFF.[2] The data bus, however, carries signals that may be interpreted in different ways at different times in the program. We can divide these signals into three basic data types.

The data bus can carry instruction codes. These will be sent to the microprocessor's internal *instruction decoder*, which will cause the microprocessor to execute one or another of the operations in its *instruction set*. We saw in Section 1.7 that the code $10 caused the 6802 to subtract the contents of register B from the contents of register A. Other instruction codes cause additions, logical AND operations, binary shift operations, and storing of data from the processor to its memory, to name a few of the possibilities.

The 8-bit data bus can carry 256 different binary patterns, so there can be that number of instructions for an 8-bit microprocessor. The 6802 processor actually uses only 197 of these as its instruction set. The instruction set is, more than anything else, the thing that distinguishes one microprocessor from another, and manufacturers are forever arguing that their instruction set is ''more powerful'' than those of their competitors.

The data bus can carry addresses. If, for example, we want to make the microprocessor store a data byte to memory location $12AB, we will have to read that address into the processor from the program memory via the data bus. Since an address consists of 16 bits and the data bus can handle only 8 bits at a time, it will be necessary to store the address $12AB in memory as two bytes at two adjacent memory locations. The 6802 stores the high-order byte (most-significant byte, or $12 in this example) first and then the low-order byte ($AB) at the next-higher memory location. Other processors, such as the 6502 and Z80, store the low-order byte first.

Thus, address bytes can be of two types:

1. Address high-order byte
2. Address low-order byte

These bytes will be concatenated (placed together) on the address bus to form a single 16-bit address.

For completeness, we must mention that address information can have a third form, called *relative* or *displacement*. In this form, the byte $12 would tell the processor to add $12 to its current program-counter value, thus skipping over 18 bytes. We will not encounter this data type until Chapter 4.

The data bus can carry user data. This data can have an immense variety of subtypes. They are similar only in the processor's impartial and indifferent treatment

[2]Other types may appear on the address bus during $\overline{\text{VMA}}$ cycles, but these bits is not generally usable.

of them all.[3] By contrast, instruction codes and address data types cause the processor to take specific actions, which depend on the circuit design of the particular microprocessor chip. For example, the 6802 may, at a certain point in its program, recognize the byte $4A as an instruction code telling it to decrement (decrease by one) the contents of its register A. At another time $4A may be interpreted as the high-order half of address $4A30. At another time it may be the low-order half of address $FE4A, and yet again it may tell the processor to skip its program counter from $F800 to $F84A. In each of these cases the processor has built-in logic that determines the effects of the byte $4A.

If the data bus interprets $4A as user data, however, the processor will simply store binary 0100 1010 somewhere. Just where it may be stored and for what it may be used are matters that we will address shortly; the point for the moment is that the ultimate effect of user data is entirely the responsibility of the user—that is, the programmer.

REVIEW OF SECTIONS 2.1 AND 2.2

Answers appear at the end of Chapter 2.

1. The microprocessor data bus can both read from and write to the bus. It is, therefore, a ________ bus.
2. When the microprocessor is not applying data to the data bus, its internal bus buffers are in the ________ state.
3. A microprocessor write to memory is called a ________ (load or store).
4. The list of operations that a computer can perform (such as add, shift, increment, and load) constitutes its ________ ________.
5. Name two general data types to which the computer hardware is designed to respond.
6. The data type of the bits appearing on the address bus is always ________.
7. How can a memory with 8-bit words store a 16-bit address?
8. Data that the computer handles without making specific hardware response to it is called ________ ________.
9. What is the data type of the data $4A?
10. What is the function of bits whose data type is ''user''?

2.3 STANDARD USER CODES

Section Overview

Three examples of standardized user codes are

1. **Straight binary numbers, in which 8 bits can represent the counting numbers from 0 to 255.**

[3]A few special instructions, such as *Decimal Adjust*, do make assumptions about the meaning of user data.

2. **Binary-coded decimal, in which 8 bits can represent the decimal numbers from 0 to 99.**
3. **ASCII, in which each byte represents one alphanumeric character, punctuation mark, or special symbol.**

Sometimes user data has a very widely recognized meaning. However, if the processor hardware treats this data the same, essentially without regard to its various binary patterns, we call it *user data*, or *user code*.

Binary numbers, for example, are a form of user code understood and utilized by everyone who uses microprocessors. But they represent a user code, not a machine-recognized data type. There is nothing "five-ish" about binary 0000 0101. The processor doesn't do or have five of anything when it loads $05 into its register A. Microprocessor users have simply agreed that a 1 in the D0 position shall represent *one*, a 1 in the D1 position shall represent *two*, D2 = 1 shall represent *four*, and so on, so they agree that 0000 0101 means five. If the processor loads $12 rather than $05, nothing different in kind happens inside it. No different hardware is activated, except that two different bits of A are *high*.

Binary-coded decimal is another example of a widely recognized user code. In this code each group of 4 binary digits represents 1 decimal digit. Thus binary 1001 simply means 9. The bit combinations whose straight binary values exceed 9 are not permitted in BCD code. Thus the binary digits

1100

are not a valid BCD code, since they equal $C, or 12 in decimal, and 12 requires 2 decimal digits for its expression.

BCD digits are generally *packed*, two to a byte. The most-significant nibble is understood to mean *group of 10*. Thus

0110 1000

denotes 6 groups of 10 plus 8 units, or

68

Four-digit decimal numbers can be packed into 2 bytes of binary data, 6 decimal digits, into 3 bytes, and so on.

If you have been thinking about binary versus BCD codes, you may question how the computer knows, upon encountering binary

0110 1000

whether this means $68 (6 groups of 16, plus 8, which equals 104 in decimal) or decimal 68. The answer is that the computer doesn't know, and it doesn't care. If 0110 1000 is a user code, the computer simply manipulates the bits according to the rules of its instruction set. The meaning of those bits is assigned by the programmer, not the computer, and it is the programmer's responsibility to stick to the rules of the game he has chosen.

Hex	ASCII	Hex	ASCII	Hex	ASCII	Hex	ASCII
ØØ	Null	2Ø	Space; blank	4Ø	@	6Ø	`
Ø1	Start heading	21	!	41	A	61	a
Ø2	Start text	22	" quote	42	B	62	b
Ø3	End text	23	#	43	C	63	c
Ø4	End transm'	24	$	44	D	64	d
Ø5	Enquiry	25	%	45	E	65	e
Ø6	Acknowledge	26	&	46	F	66	f
Ø7	Bell	27	' apost'	47	G	67	g
Ø8	Backspace	28	(	48	H	68	h
Ø9	Horiz tab	29	)	49	I	69	i
ØA	Line feed	2A	*	4A	J	6A	j
ØB	Vert tab	2B	+	4B	K	6B	k
ØC	Form feed	2C	, comma	4C	L	6C	l
ØD	Carriage ret'n	2D	- hyph	4D	M	6D	m
ØE	Shift out	2E	. period	4E	N	6E	n
ØF	Shift in	2F	/	4F	O	6F	o
1Ø	Data link esc.	3Ø	Ø	5Ø	P	7Ø	p
11	Dev ctr'l 1	31	1	51	Q	71	q
12	Dev ctr'l 2	32	2	52	R	72	r
13	Dev ctr'l 3	33	3	53	S	73	s
14	Dev ctr'l 4	34	4	54	T	74	t
15	Negative ack	35	5	55	U	75	u
16	Sync idle	36	6	56	V	76	v
17	End tr block	37	7	57	W	77	w
18	Cancel	38	8	58	X	78	x
19	End medium	39	9	59	Y	79	y
1A	Substitute	3A	: colon	5A	Z	7A	z
1B	Escape	3B	; semicolon	5B	[	7B	{
1C	File separate	3C	<	5C	\	7C	¦
1D	Group sep.	3D	=	5D	]	7D	}
1E	Record sep.	3E	>	5E	∧	7E	≈
1F	Unit sep.	3F	?	5F	– dash	7F	Delete

Figure 2.2 The American Standard Code for Information Interchange (ASCII—pronounced ASK-ee.)

ASCII is a widely accepted user code for handling alphanumeric characters via keyboards, printers, and CRT displays. The letters stand for American Standard Code for Information Interchange. (The last two letters are often read as the Roman numeral two—this is an error.) ASCII represents each number, letter, and punctuation mark with 1 byte of data. Thus 256 different characters could be represented, although only 128 characters are commonly used. Figure 2.2 gives the ASCII character set.

The ASCII code is quite arbitrary. There is no reason why *A* is $41 rather than $51. Other alphanumeric codes have been used, notably *Baudot*, which used only 5 bits per character, and EBCDIC (Extended Binary Coded Decimal Interchange Code), which is similar to ASCII.

2.4 AD HOC USER CODES

Section Overview

User data does not have to represent numbers or alphanumeric characters. Codes can be made up to allow binary data to represent almost any parameter, quality, or concept. In this section, we give an example of a code made up to light a seven-segment display with any desired hex character.

The Latin phrase *ad hoc* (literally *to this*) is widely used to refer to committees and devices that are brought into being for some special one-time purpose (as opposed to general-purpose or widely used). Many microcomputer data codes fall in this category. They are made up by a system designer to solve a problem that is unique to the system being designed. To illustrate this data type, we will consider the case of a common-anode seven-segment LED to be used in displaying the hex digits Ø through F. We wish to use *software decoding* to eliminate the expense of a seven-segment decoder IC and to provide the flexibility to display such characters as L and – at will.

Figure 2.3(a) shows the hardware connection of the LED segments to the data output. This connection is quite arbitrary. (We could have connected the data lines to the LED segments in any of 8!, or 40 320, ways.) A low level on a data line lights its corresponding LED segment. Figure 2.3(b) shows the binary bit patterns required to display the digits 0 through F on the LEDs. We have thus generated an ad hoc code for lighting the LEDs, in which, for example, $42 represents the digit d.

When the microprocessor outputs $42 it neither knows nor cares whether that binary pattern represents the decimal number 66, the ASCII character B, or the hex digit d. It is up to the programmer to write code that causes these various user data types to be sent to the proper places at the proper times to achieve the desired effects.

Ad hoc, or special, user codes have been developed to control machine-tool and robot-arm position, color spectra on video games, air/fuel mixture in automobile engines, and player positions on a chess board, to name a few examples.

Figure 2.4 gives a summary of microcomputer data types. Some of the subtypes are still unfamiliar, but we will explain them presently. Figure 2.5 shows how the

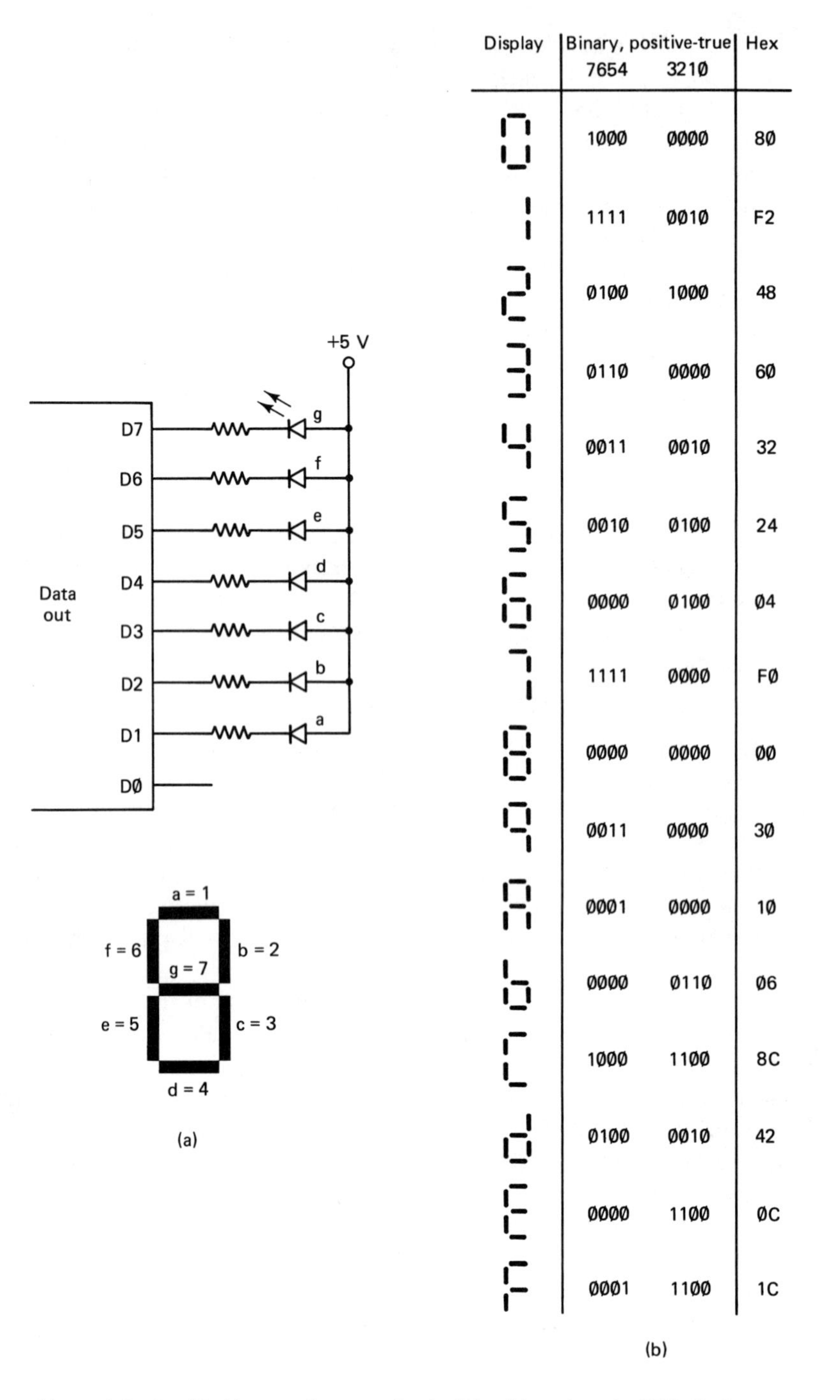

Display	Binary, positive-true 7654	321Ø	Hex
0	1ØØØ	ØØØØ	8Ø
1	1111	ØØ1Ø	F2
2	Ø1ØØ	1ØØØ	48
3	Ø11Ø	ØØØØ	6Ø
4	ØØ11	ØØ1Ø	32
5	ØØ1Ø	Ø1ØØ	24
6	ØØØØ	Ø1ØØ	Ø4
7	1111	ØØØØ	FØ
8	ØØØØ	ØØØØ	ØØ
9	ØØ11	ØØØØ	3Ø
A	ØØØ1	ØØØØ	1Ø
b	ØØØØ	Ø11Ø	Ø6
C	1ØØØ	11ØØ	8C
d	Ø1ØØ	ØØ1Ø	42
E	ØØØØ	11ØØ	ØC
F	ØØØ1	11ØØ	1C

Figure 2.3 Specific binary codes output by the IC at (a) produce the LED displays listed at (b).

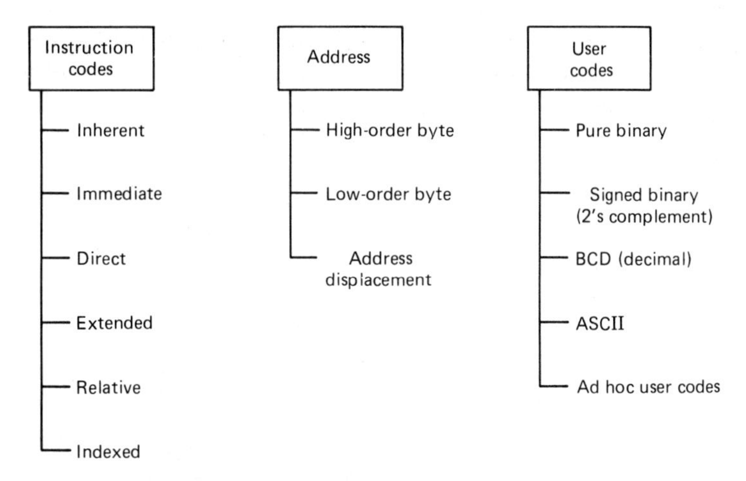

Figure 2.4 A summary of computer data types. The various subtypes of instruction codes will be sorted out in subsequent chapters.

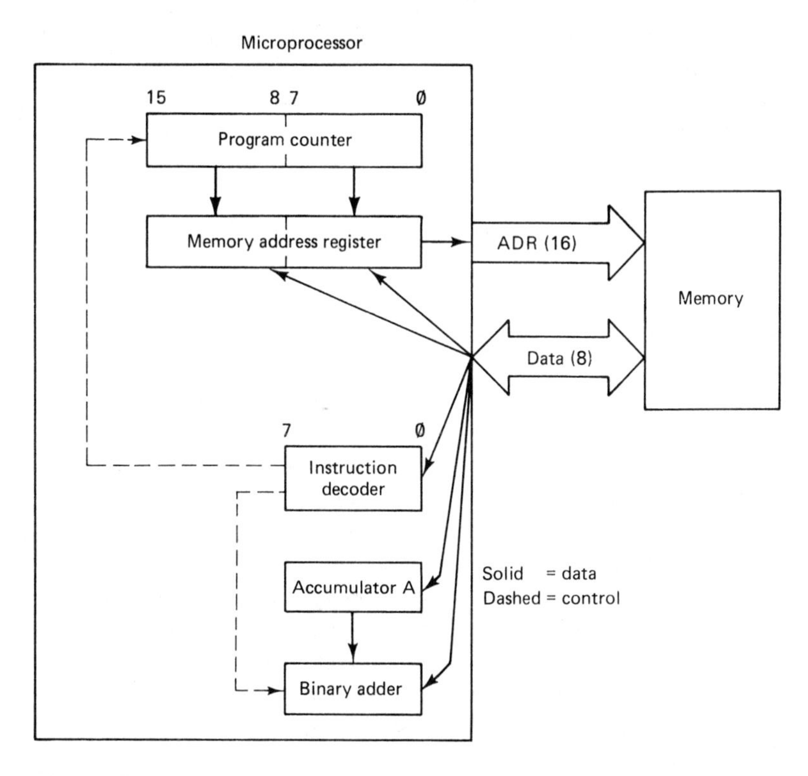

Figure 2.5 Some of the microprocessor's internal registers. The instruction decoder determines where to send the data read from memory.

data bus can be used to send the three main data types to different registers within the microprocessor. Address bytes are sent to the two halves of the memory-address register, which can output them to the address bus. Instructions are sent to the instruction decoder, and user data is sent to internal registers such as the accumulator A and binary adder.

REVIEW OF SECTIONS 2.3 AND 2.4

Answers appear at the end of Chapter 2.

11. The number 19 in binary is 0001 0011. What does the microcomputer have 19 of when this user data is read into it?
12. What is the decimal value of the binary number 1000 1100?
13. What is the decimal value of the binary-coded-decimal number 1000 1100?
14. What is the decimal value of the binary-coded-decimal number 0010 0011?
15. Write the hex digits to encode the word *MICRO* in ASCII.
16. Using the hardware of Figure 2.3, what user code would display the capital letter *H*?
17. Interpreted as instruction codes, what is the difference in the 6802's response to the byte $F6 and the byte $B7? (Refer to the programmer's aid at the end of the book.)
18. Interpreted as user data, what is the difference in response to data $F6 and data $B7?
19. How does the computer recognize the difference between an ASCII code and a binary number?
20. An ad hoc code is one that is ________.

2.5 INSTRUCTION EXECUTION SEQUENCE

Section Overview

A microprocessor interprets each byte it receives as an instruction code, address, or user code, depending on its position relative to the last instruction code read. Each instruction byte may be followed by 1 or 2 bytes having data-type *address* or *user code*. The instruction byte itself informs the processor of the data types of the bytes between the last instruction-code byte and the next instruction-code byte.

Inherent-mode instruction bytes are followed by another instruction-code byte. Extended-mode instruction bytes have 2 address bytes following them and then another instruction byte.

You must have noticed that we've been building suspense by emphasizing that the various data types are not inherently distinguishable. Data $4A could be an instruction code, a part of an address, an ASCII character, a binary number, or some ad hoc code. Very well then—how does the microcomputer tell one from the other?

The first step: When the RESET button is pushed and released, the processor is sent to the first memory address in its main program and it reads the byte that it finds there. The processor *assumes* that this byte is an instruction code. You may have miscalculated and placed an address byte at that location. Maybe you *really* miscalculated and there is no memory at all at the starting address—just random noise on a floating bus. Nevertheless, the processor is going to call up the starting address, read the data bus, and treat whatever it reads as an instruction code.

Interpreting the instruction: Each instruction code tells the processor three things:

1. What operation is to be performed.
2. How many data bytes after the instruction-code byte are still part of this total instruction (with the 6802, this may be 0, 1, or 2 additional bytes).
3. The data type of those next 1 or 2 bytes, if present.

For example, if the first byte read is $1B (instruction mnemonic ABA), the processor hardware is activated to add the contents of internal register B to those of internal register A, and then to store the result in A.

It is also instructed that this operation can be completed without reading any more information into the processor, so the next byte in the program sequence is to be interpreted as another instruction code. We say that this instruction is inherently complete with the instruction-code byte—it is an *inherent-mode* instruction.

In some microcomputer literature such instructions are called *implied* because the operand is implied by the instruction. (The operand is the data to be operated upon.) In the instruction ABA, the operand data to be added to A is contained in B, and no external data source is required. The A and B internal registers are called *accumulators*, and they serve as central data-handling stations for the 6802.

Extended-mode instructions: Suppose that the second byte in your program (the one after $1B) is $B6. We know that this is an instruction code (also called an op-code) because the first instruction $1B was complete in 1 byte. The instruction $B6 (mnemonic LDAA) tells the processor three things:

1. Load accumulator A with new data.
2. The next 2 bytes in the program are still a part of the LDAA instruction.
3. These 2 bytes are the high-order and low-order bytes, respectively, of the address where the data to be loaded resides.

This instruction and the inherent-mode instruction preceeding it are depicted in Figure 2.6. A number of implications of this instruction need to be examined.

- The instruction LDAA is not inherently complete. The command *load accumulator* A leaves unanswered the question, With what data?

- The 2 bytes following the LDAA op-code ($B6) are *address* data type, not because of any properties of themselves, but because they follow an op-code that instructs the processor to treat them as such.
- The operand—that is, the data to be loaded into A—is contained at the memory address obtained by concatenating this instruction's second byte (high-order half) and third byte (low-order half).
- The fourth byte in the program is another op-code, again not by its own properties but because it follows the last of the 3 bytes comprising the previous instruction.

Instructions in which the op-code is followed by 2 address bytes are called *extended mode* because they extend the processor's access to all 65 536 possible memory locations. Some manufacturers refer to this as the *absolute* addressing mode, because the address of the operand is specified in absolute terms—that is, without reliance on anything other than the 2 operand-address bytes.

Program execution. If anything about microcomputers is awe-inspiring it is the fact that components selling for $5.00 can routinely operate at 1 million clock cycles per second for days going into years without ever once getting out of step. Each op-code defines the boundaries of its instruction and the data types of any following operand bytes, leaving the assumption that the next byte is another op-code. Programs tens of thousands of words long are executed repeatedly, with each instruction defining the position of the next instruction.

If in Figure 2.6, for example, the processor just once landed at address F8Ø2 while expecting an op-code, it would execute instruction F8, which is an Exclusive-OR of accumulator B, extended mode—an entirely unintended operation. It would probably never get back on the right track. If it happened to pick up the data at address F8Ø3 as an op-code (a WAIT instruction), it would simply stop. Amazingly,

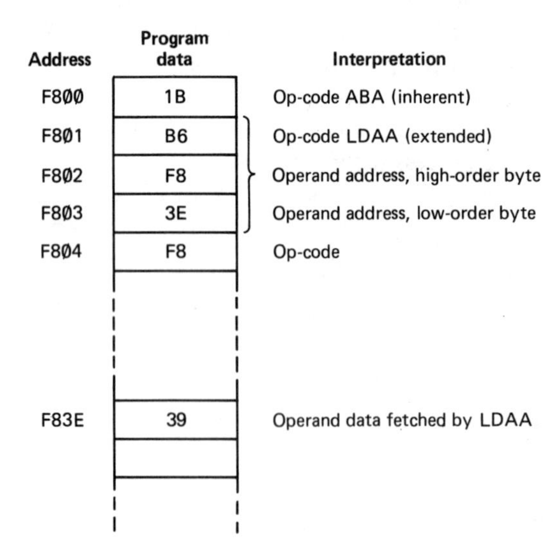

Figure 2.6 Inherent-mode instructions are complete in 1 byte. Extended-mode instructions have 2 operand-address bytes following the op-code.

through billions of instruction executions, this just doesn't happen,[4] largely because of the *digital advantage* discussed in Section 0.1.

2.6 CYCLE-BY-CYCLE OPERATION

Section Overview

Each 6802 instruction requires a specified number of *E*-clock cycles to complete. The data on the address bus, data bus, R/$\overline{W}$ line, and VMA line can be determined for each of these cycles by consulting the Instruction Cycle-by-Cycle Execution chart in Appendix B.

Simple instructions take only two cycles:

1. **Read instruction-code byte.**
2. **Perform indicated operation.**

More complex instructions take more clock cycles:

- **Read instruction-code byte (1 cycle).**
- **Read operand-address (1 or 2 cycles).**
- **Perform indicated operations (1 to 6 cycles).**

6802 instructions take anywhere from 2 to 12 clock cycles to execute. The exact number for each instruction is given in the programmer's reference at the end of the book. A detailed description of what happens during each cycle appears in Appendix B. These aids have been placed at the end of the book to make them easy to find. You will need to refer to them often, so be sure to spend the necessary time to become familiar with them.

In the programmer's reference, the *inherent-mode* column is shared with *immediate-mode* instructions to conserve space. (Immediate-mode instructions are explained in Section 3.3.) The inherent instructions are identifiable because they are logically complete without an operand. ASLA, for example, shifts the bits of accumulator A left and is inherent mode. LDAA, however, does not specify inherently *what* data is to be loaded and is immediate mode in this column. Also note that only inherent instructions are single byte (1 in the # column.)

An inherent instruction starts, as do all instructions, with a read of the op-code. Let us take the example of Figure 2.6, an ABA instruction at address $F800.

1. On the *low* half of the first *E*-clock cycle, the processor's address bus outputs $F800. By the *high* half of this cycle the memory has had sufficient time to

[4]In electrically noisy environments, such as factories or automobiles, the processor does occasionally get out of step. Much engineering effort is spent on noise suppression, as well as *fail-soft* programming techniques for recovering from this condition.

place the data from this address on the data bus, and the processor reads data $1B, interpreting it as the inherent-mode ABA instruction.

2. During the second clock cycle the processor internally adds the contents of its A and B accumulators. Externally, the address bus outputs $F801, which is the address of the *next* instruction. This early call-up of the next instruction sometimes shortens execution time.

Most of the 8-bit inherent instructions operate in this general manner, as shown in Appendix B.

Extended-mode instructions all take three program bytes, one for the op-code and two for the address. It therefore takes three machine cycles just to read in the instruction and its operand address. It generally takes additional cycles to complete the instruction operations. Here is a cycle-by-cycle description of the LDAA instruction from Figure 2.6.

1. The processor's address lines output $F801 on the first (*low*) half of *E* and the data lines read data $B6 from the program memory on the second half. It interprets this as the extended-mode LDAA instruction.
2. The processor outputs address $F802 and reads data $F8. Interpreting this as the high-order half of the operand address, it loads it into the top byte of the memory-address register.
3. The processor outputs address $F803, reads data $3E, and loads this into the low-order half of the memory address register.
4. The processor outputs the address $F83E on the first half of *E* and reads the operand data ($39) from that address on the second half of *E*.

The next machine cycle is another op-code read. You may notice that most of the other extended-mode instructions in Appendix B are more complex than LDAA, but we will take them up as they arise.

REVIEW OF SECTIONS 2.5 AND 2.6

21. What must be the data type of the first byte in any program?
22. In addition to specifying the operations to be performed, what information does the op-code contain?
23. How many bytes does an inherent-mode instruction take?
24. How many bytes does an extended-mode instruction take?
25. Could the STAB instruction be inherent mode, extended mode, or either?
26. What is the data type of the second and third bytes of an extended-mode instruction?
27. Write the hex digits for an extended-mode instruction to clear memory location $029A.

28. How much time will it take to execute the ABA and LDAA instructions of Figure 2.6 if each E cycle takes 2 μs?

29. What data is contained in accumulator A at the end of this four-byte program segment?

Address	Program Data
FA00	B6
FA01	FA
FA02	93
FA03	4A
•	•
•	•
•	•
FA93	50

30. Use Appendix B to find the logic state of the 6802 $R/\overline{W}$ line during the fourth cycle of an extended-mode STAB instruction.

2.7 TRACING A ONE-INSTRUCTION PROGRAM

Section Overview

A standard oscilloscope can be used to observe signals only if they are repetitive. Normal microcomputer programs do not produce such signals, but the processor can be made to execute *test loops*, consisting of only a few instructions endlessly repeated, to allow oscilloscope observation of the signals involved in program execution.

Working with microcomputers consists in large part of arranging strings of instructions and op-codes and examining lists of hex digits—in other words, programming. We will get to this point soon enough, but first we will examine how the computer, as an electronic switching device, executes a single instruction.

Real microcomputer programs consist of hundreds or thousands of instructions, which are executed in a continuously changing variety of sequences. The voltage patterns on the address and data lines do not repeat within ten to one hundred clock cycles and cannot, in general, be viewed on an oscilloscope. Digital storage oscilloscopes and logic analyzers are used in place of conventional oscilloscopes to capture and display a short sequence of the processor's activity.

A short program consisting of only a few instructions executed repetitively can be viewed on an oscilloscope, however. Such a program loop won't accomplish any of the things that computers are famous for, but it will allow us to examine the details of the processor's response to an instruction. For our first program we will examine the 6802 as it executes repetitively a single extended-mode *jump* instruction.

The jump instruction (JMP, op code $7E) simply reloads the program counter with the address obtained by concatenating the two operand bytes following the op-code. For example, the following program segment would cause the processor to skip over the bytes from $FD56 through FD7B and jump to the op-code at $FD7C.

Address	Program Data	
FD53	7E	Jump
FD54	FD	to address
FD55	7C	$FD7C
FD56	B6	
FD57	00	
FD58	24	
|		Skipped
|		
|		
|		
FD7B	0F	
FD7C	1B	
		Add B to A

Our program will start at address $7E7E and will consist of a **JMP** instruction with two operand bytes telling the processor to jump right back to address $7E7E. Here is the program listing:

7E7E	7E	JMP op-code
7E7F	7E	Operand high byte
7E80	7E	Operand low byte

Notice that all 3 data bytes read by the data bus during this program are $7E. This presents us with an irresistible temptation to cheat on the hardware. We won't use a memory at all to supply the program data. Since it's all the same, we will just wire the data bus for binary 0111 1110, or $7E. The processor will read the desired data on each cycle and never know that it isn't actually coming from the three memory locations it is calling up.

Tracing the loop: Figure 2.7 shows the diagram for a 6802 processor, hard-wired to execute the **JMP** instruction repetitively. The eight 1-kΩ resistors can easily pull the data lines *high* or *low* when they are acting as inputs but will not force them one way or the other if a processor *write* outputs data to the bus. The address lines are not connected to anything, since we are executing this first "program" without really calling up a memory, but we will use the 'scope to read the address bits and thus follow the processor through the steps of the program's execution.

1. Wire the circuit as shown in Figure 2.7. Apply V_{CC} and *reset* the processor. Check the *E* clock on the scope, and measure its period.
2. Measure the period of the waveform on line A7. How many clock cycles is it? How many clock cycles does the **JMP** instruction take, according to the programmer's reference?
3. Trigger the scope from line A7, negative slope, and observe the logic levels on each of the address and data lines for the first three machine cycles (6 μs) of the scope display. (If you have a single-trace scope it will be necessary to

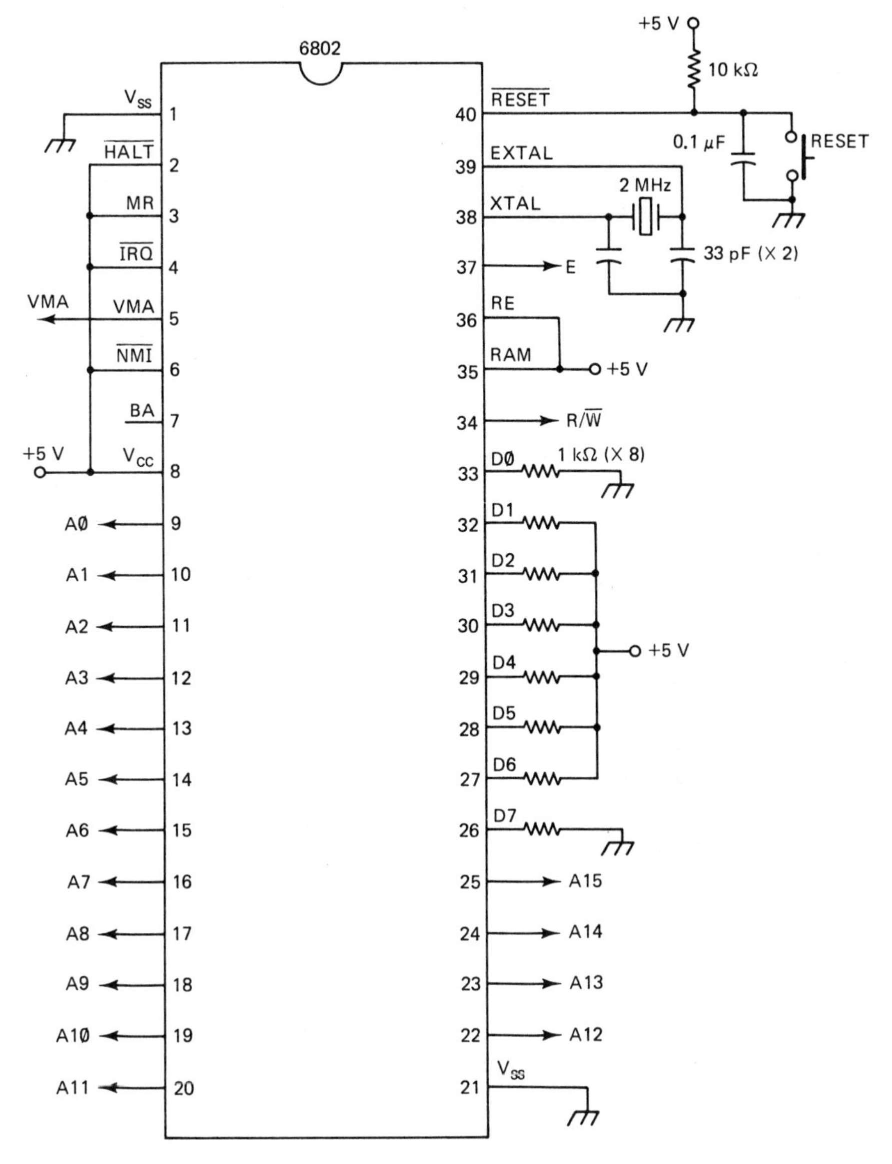

Figure 2.7 Hardware to make the 6802 execute a single-instruction 3-cycle test loop.

use external trigger to keep the negative transition of A7 as the zero-time reference.) Check the binary levels you observe against the following table.

Cycle	Binary Address				Binary Data		Hex Address	Hex Data	Meaning
	15 14 13 12	*11 1Ø 9 8*	*7 6 5 4*	*3 2 1 Ø*	*7 6 5 4*	*3 2 1 Ø*			
1	Ø 1 1 1	1 1 1 Ø	Ø 1 1 1	1 1 1 Ø	Ø 1 1 1	1 1 1 Ø	7E7E	7E	JMP
2	Ø 1 1 1	1 1 1 Ø	Ø 1 1 1	1 1 1 1	Ø 1 1 1	1 1 1 Ø	7E7F	7E	MSB
3	Ø 1 1 1	1 1 1 Ø	1 Ø Ø Ø	Ø Ø Ø Ø	Ø 1 1 1	1 1 1 Ø	7E8Ø	7E	LSB

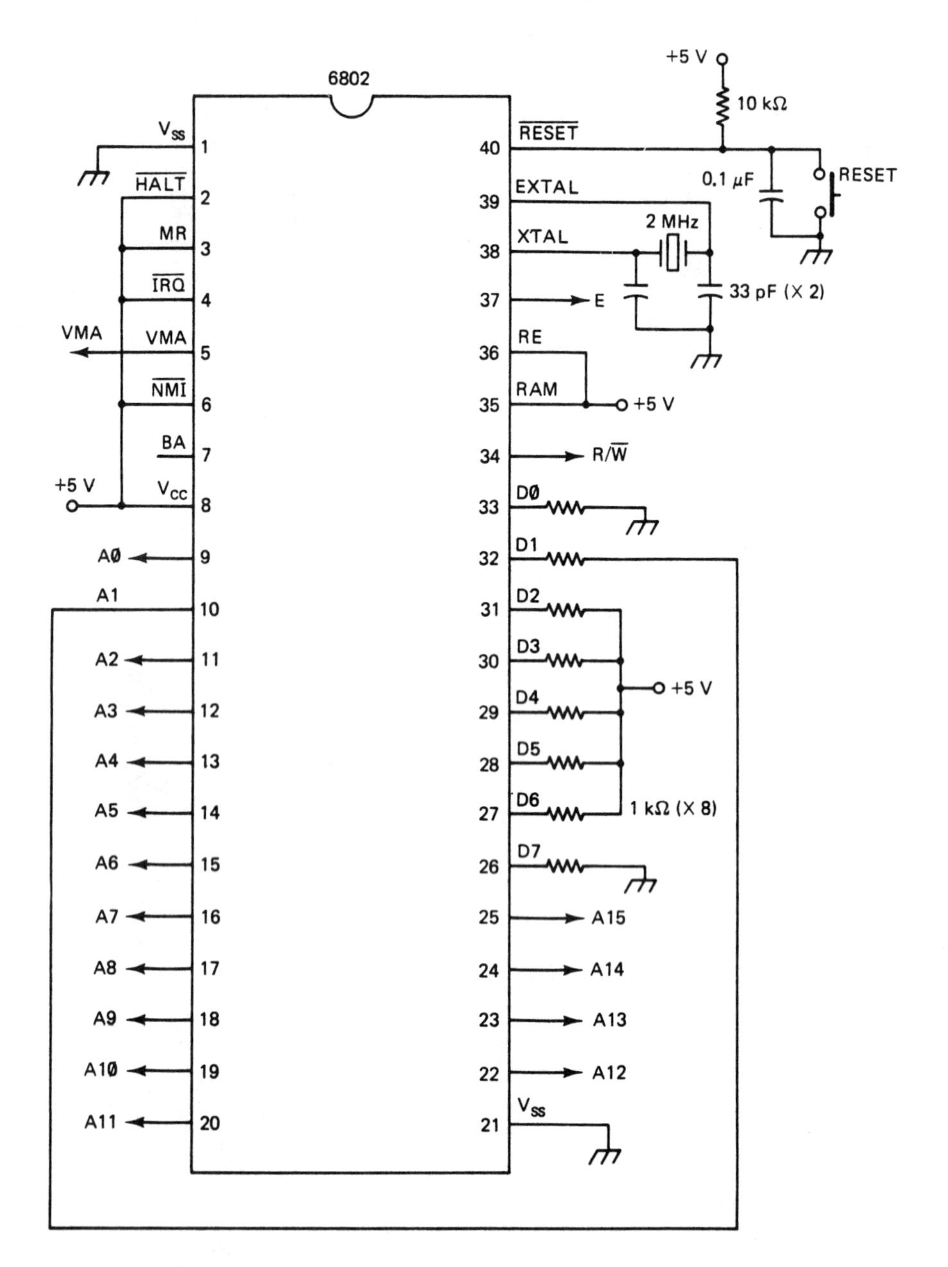

Figure 2.8 Hardware setup to make the 6802 execute a two-instruction, 9-cycle loop. Line D1 changes the op code from 7E (Jump) to 7C (Increment data).

4. Consult the cycle-by-cycle specifications for the extended-mode JMP instruction in Appendix B. Write three sentences explaining in your own words what the processor is doing during each of the three machine cycles.
5. Modify the hardware wiring as shown in Figure 2.8. Address line A1 is now used to modify the data byte read by the processor after each two addresses.

Op codes $7E (Jump) and $7C (Increment data at operand address) will be executed alternately. The increment instruction takes 6 machine cycles. Trigger the scope on A7, negative slope, and display the 9 cycles of the loop, recording the binary bits and converting them to hex, as in the previous table. Explain what the processor is doing on each cycle, referring to Appendix B to interpret the INC, Extended, cycles. Demonstrate that the data read was actually incremented, and that the operand addresses read were actually the addresses operated upon for both instructions. Point out and explain the cycles in which VMA or R/$\overline{W}$ were *low*.

Answers to Chapter 2 Review Questions

1. Bidirectional **2.** High-Z **3.** Store
4. Instruction set **5.** Instruction codes and addresses
6. Address **7.** The high-order 8 bits are stored at one address and the low-order 8 bits are stored at the following address. **8.** User data **9.** There is no way to tell. It could be instruction, address, or user. **10.** Their function is determined entirely by the user; that is, the programmer. **11.** Nothing. The value 19 is a user interpretation. The machine is indifferent to it.
12. 140 **13.** No value. This is invalid as BCD data.
14. 23 **15.** 4D 49 43 52 4F **16.** 0001 0010, or $12 **17.** F6 causes the processor to load its *B* register from memory. B7 causes it to store its *A* register in memory.
18. No difference **19.** It can't. This is the programmer's responsibility.
20. Designed for a specific and limited purpose. **21.** Instruction code
22. The data types of the next 1, 2, or 3 program bytes.
23. 1 **24.** 3 **25.** Extended
26. Address **27.** 7F 02 9A **28.** 12 μs
29. $4F **30.** *High* voltage level.

CHAPTER 2 QUESTIONS AND PROBLEMS

Digits after the decimal point refer to section numbers in Chapter 2.

Basic Level

1.1 Are the 6802 address lines input-only, output-only, or bidirectional? Answer the same question for the 6802 data lines.

2.1 Which instruction causes R/$\overline{W}$ to go *low*; LOAD or STORE?

3.2 List the three general data types handled by a computer.

4.2 Which of these data types is least dependent for its effect on the particular microprocessor being used?

5.3 Name three standard user codes, and give the accepted meaning of binary 0101 1001 in each.

6.3 Interpret this hex data as ASCII code:

332E3134

7.4 Refer to Figure 2.3 and tell what character will be displayed on the LED if data $16 is output.

8.4 In Figure 2.5 there are five lines showing data flow from the data bus to internal machine registers. From the top line to the bottom, list the data types sent on each line.

9.5 What is the data type of the first program byte that the microprocessor reads after the RESET line is returned *high*?

10.5 Four extended-mode instructions appear at addresses $E000 through $E00B. Which addresses contain instruction codes?

11.6 Use Appendix B to determine the number of clock cycles required for a 6802 COM, extended instruction. Also tell on which of these cycles $R/\overline{W}$ and VMA go *low*.

12.6 Use Appendix B to determine what data is on the data bus during the second cycle of an inherent INCA instruction.

Advanced Level

13.1 Describe the direction of data transfer during the 6802 LOAD and STORE operations, respectively.

14.1 Explain how the 6802 data lines can read *high* or *low* levels from the data bus but can also output *high* and *low* levels to the bus. Why doesn't the output override the input?

15.2 Make a chart showing the three types of data that may appear on the microprocessor data bus. Break the second type into three subtypes, and leave room to break the third type into several subtypes later.

16.2 Explain how a 16-bit address can be stored in memory and sent to the processor, when the memory words and the data bus are only 8 bits wide.

17.3 How can we tell whether the user data 0001 0000 represents binary (16) or BCD (10)?

18.4 Give two reasons why software decoding, as shown in Figure 2.3, is preferred over hardware decoders, such as the 7447 IC.

19.4 Redraw the data bus and LEDs of Figure 2.3(a) with segment **g** on D0, in order, up to segment **a** on D6. Now determine what hex data will produce a display of 4. Leave D7 at 0.

20.5 Explain the *inherent-* and *extended* instruction modes by listing the number of bytes and the data type of those bytes for each.

21.5 Examine the 6802 programmer's reference chart at the back of the book to determine which binary bits the processor decodes to determine the instruction mode. You will find that these bits remain the same for all *extended-mode* instructions but change to a different pattern, which remains the same for all inherent-mode instructions. Note that the inherent-mode instructions list 1 byte as their size. Other instructions sharing the same column but listing 2 or 3 bytes are not inherent mode.

22.6 List in order the hex program bytes required to perform the following operations:

```
LOAD ACCUMULATOR B WITH THE BYTE FROM ADDRESS $1234.
ADD TO THIS THE BYTE FROM ADDRESS $5678.
INCREASE THE SUM BY 1.
STORE THE RESULT AT ADDRESS $9ABC.
```

23.6 Interpret the following program. The first byte is an instruction code.

```
4F
4C
F6
1B
4A
1B
B7
01
16
```

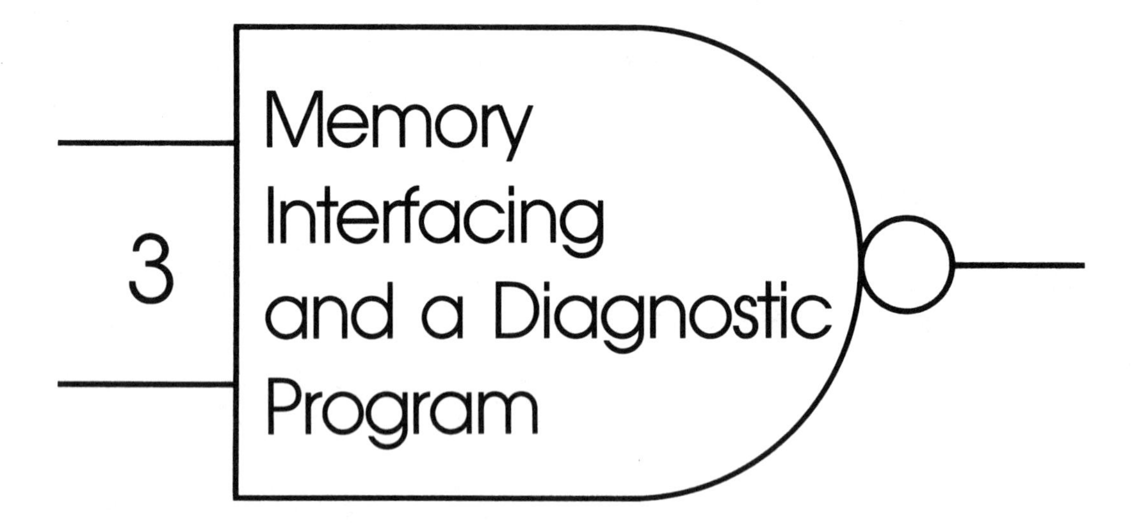

3.1 MEMORY MAPPING AND ADDRESS DECODING

Section Overview

In designing a microcomputer system, a few high-order address lines generally drive logic that selects one of several devices (ROMs, RAMs, and various I/O chips) with which the processor is going to communicate. Three high-order lines could be decoded to select among 2^3, or 8, different devices.

Low-order address lines (starting from AØ) are used to select the particular address accessed within a selected device. If there are lines not used for device select or address select, the device can be addressed identically with these lines in either binary state. Four unused lines means 2^4, or 16, redundant address ranges for the device. Larger systems use complete address decoding to avoid redundant address ranges.

A memory map shows graphically which addresses can be used to access each device in the system.

Eight-bit microprocessors generally have a 16-bit address bus, indicating that they can access 65 536 data locations. Commonly available memory chips have from 7 to 15 address lines, indicating that they have from 128 to 32 768 data locations each.

Generally, several memory chips must be connected to the processor to provide both ROM (permanent) and RAM (writable) memory functions and to provide the memory needed in large systems. How are these chips connected, and how does the microprocessor select one of them without selecting them all?

Address select and device select: Let us say that we are building a microcomputer system in which the largest memory chip is a 2716 EPROM with 11 address lines (A0 through A10) and 2048 bytes of data. Let us assume further that the complete system consists of the processor surrounded by four peripheral chips: The EPROM, a 1K RAM, a data-input chip, and a data-output chip. We could divide the 13 address lines of the processor shown in Figure 3.1 into two groups: 11 to select the specific address in a memory chip and 2 to select (by four possible bit combinations) which peripheral chip is being activated. We might assume that the processor is a type 6504, which has only 13 address lines and is available in a 28-pin IC pack; or we may assume that we are using a processor with the standard 16-line address bus but have chosen to leave the high three lines at 000 to simplify the system design.

Address decoding: In Figure 3.1, address lines A11 and A12 are each inverted, and the four resulting lines are used to drive four AND gates to select one of the four peripheral chips

$$\overline{A12} \cdot \overline{A11} = \text{RAM}$$

$$\overline{A12} \cdot A11 = \text{INPUT}$$

$$A12 \cdot \overline{A11} = \text{OUTPUT}$$

$$A12 \cdot A11 = \text{ROM}$$

Assuming that the unused lines A13, A14, and A15 are left at binary 0, the addresses that will activate the ROM are

A	15	14	13	12	11	10	9	8	7	6	5	4	3	2	1	0
	0	0	0	1	1	*A*	*A*	*A*	*A*	*A*	*A*	*A*	*A*	*A*	*A*	*A*

where the bits marked *A* select the particular address in the EPROM. If we first set all the *A* bits to 0 and then to 1, we can determine the range of microprocessor addresses that will access data bytes in the EPROM.

```
0001   1000   0000   0000 = $1800
0001   1111   1111   1111 = $1FFF
```

Partial decoding: If we repeat this process for the RAM, we will notice that address line A10 is not used for chip select *or* address select, because the 1K RAM uses only address lines 0 through 9 to select 1024 address locations.

A	15	14	13	12	11	10	9	8	7	6	5	4	3	2	1	0
	0	0	0	0	0	X	*A*	*A*	*A*	*A*	*A*	*A*	*A*	*A*	*A*	*A*

If we let the *don't care* bit (*X*) be a 0, the range of RAM addresses is

```
        X
0000   0000   0000   0000 = $0000
0000   0011   1111   1111 = $03FF
```

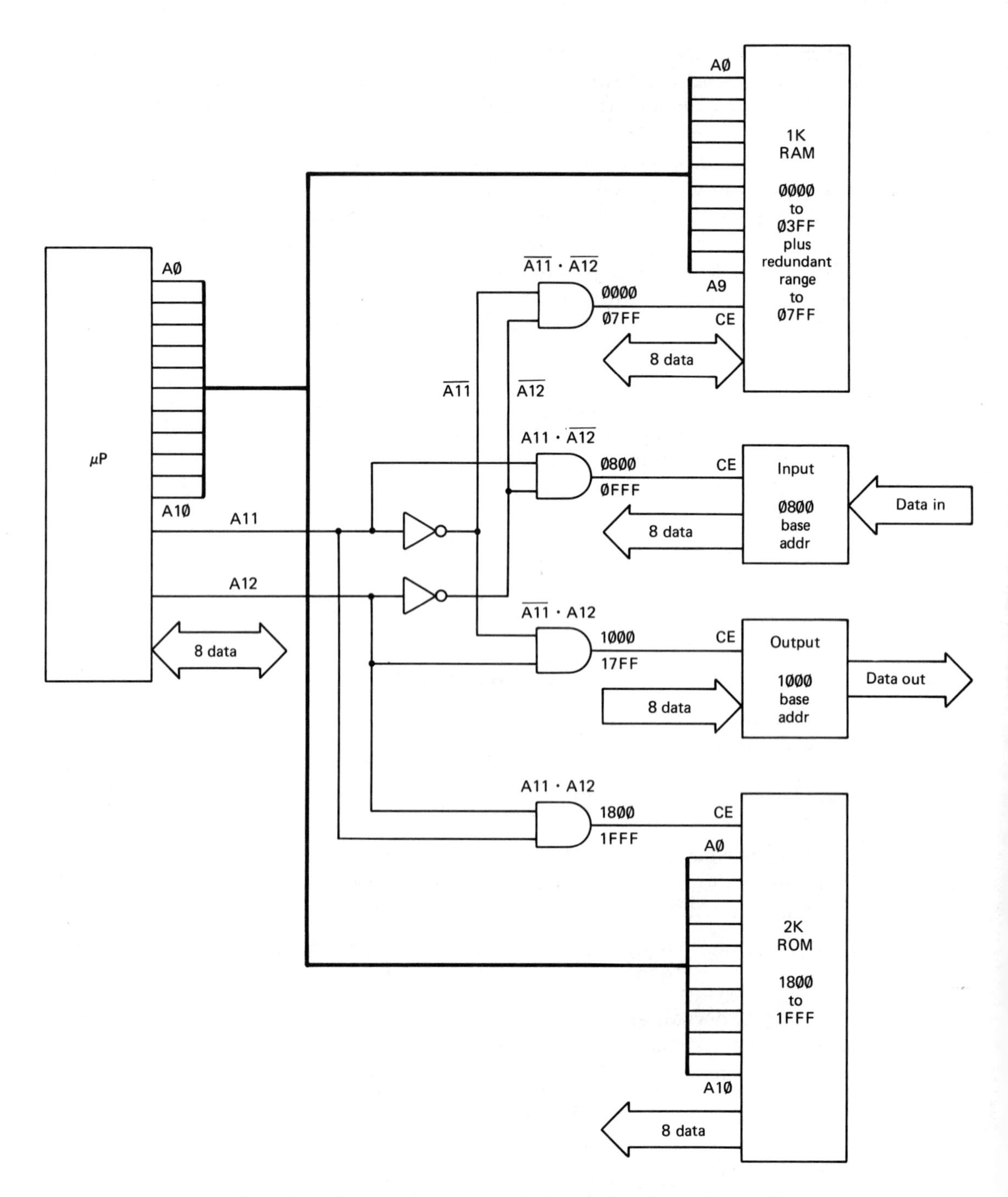

Figure 3.1 Memory-mapped device selection. Address lines A11 and A12 are used to select one of four peripheral chips.

However, we could also let the *X* bit (A10) be a logic 1 with no effect on the RAM data being accessed, since A10 drives neither the RAM chip-select logic nor the RAM address lines. The range of RAM addresses is now

```
         X
0000   0100   0000   0000 = $0400
0000   0111   1111   1111 = $07FF
```

Thus there are two ranges of addresses that access the same RAM data. If the processor reads data ($AB) at address $0012, it will read the very same byte ($AB) at address $0412 in the system of Figure 3.1. We say that the RAM addresses are redundant in two places.

A memory map is a chart showing the memory type and use for each address range. Figure 3.2 is a memory map for the system of Figure 3.1. The memory map shows only 8K of address space, because address lines A13, A14, and A15 have been specified as all logic 0s. If these lines were used, the memory map would cover a full 64K, up to address $FFFF.

The memory map for the system of Figure 3.1 shows that the 1K RAM appears from $0000 to $03FF and, identically, from $0400 to $07FF. We will refer to the lower range of addresses as the *base* range, and the upper range as a redundant range of the base range.

Redundant address ranges occur when the address lines are only partially decoded. In Figure 3.1, for example, line A10 is not used in address range $0000 through $07FF. It does not select a chip, nor does it select an address within a chip. In this range, A10 can assume either a 1 or a 0 value with no effect on the system.

We could fully decode the address range $0000 through $07FF by using 3-input AND gates instead of the 2-input gates of Figure 3.1. Then we could put two 1K RAMs in the memory space presently occupied by the single 1K RAM's two redundant ranges. This is illustrated in Figure 3.3.

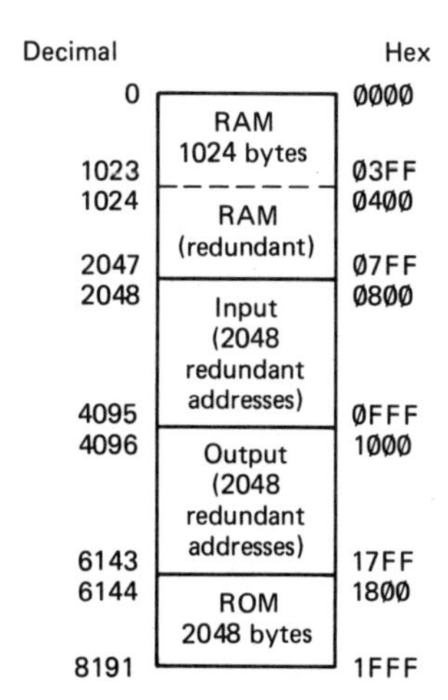

Figure 3.2 A memory map for the system of Figure 3.1. Partial address decoding results in wasted memory space through redundant ranges.

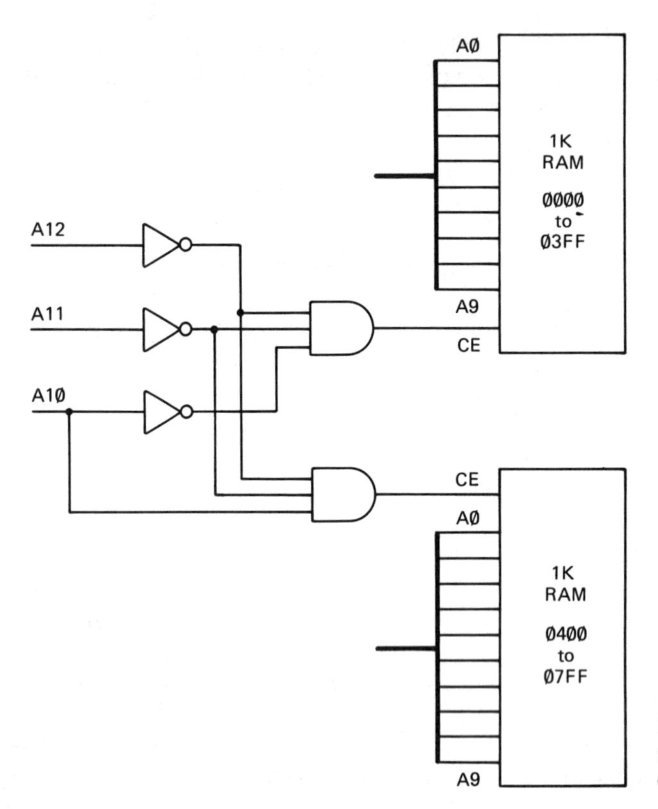

Figure 3.3 An extra input to the chip-select logic gates allows line A1Ø to enable one or the other of two RAMs.

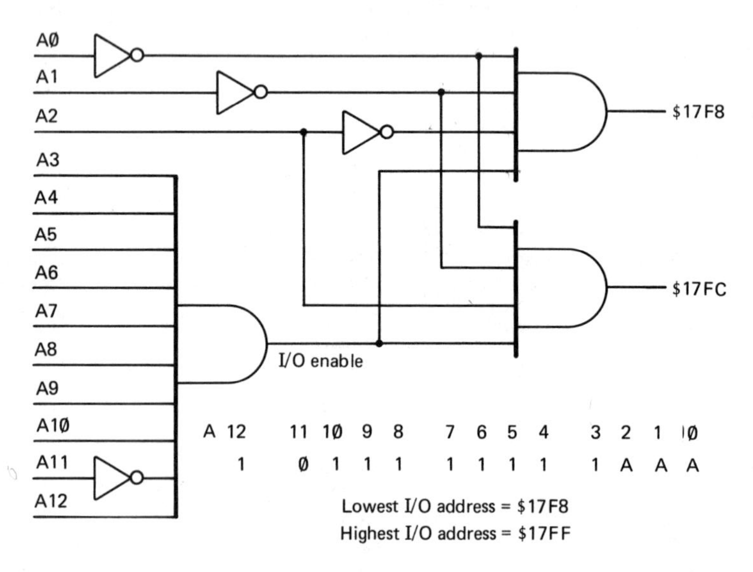

Figure 3.4 Unique address decoding for 8 I/O devices. Final logic is shown for only two of the possible 8 devices.

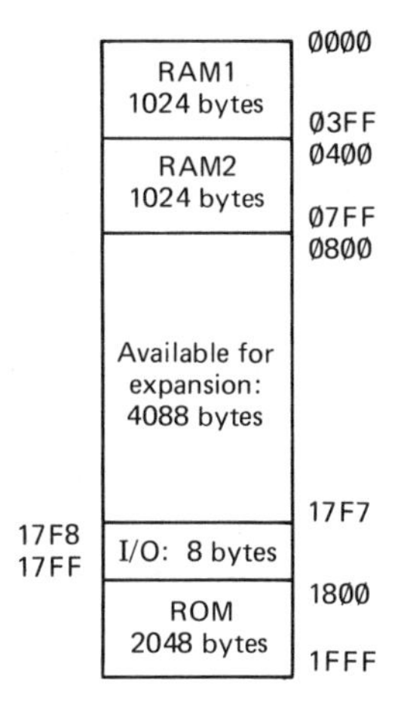

Figure 3.5 Memory map for a fully decoded microcomputer system with 13 address lines. Redundancy is eliminated so more memory and I/O devices can be accessed.

The input and output devices of Figure 3.1 actually require only one memory location each. The address lines A0 through A10 are not used in selecting them, so each is redundant in 2^{11}, or 2048, places. This waste of memory space can be avoided by providing full decoding of the address lines so that each I/O device has one unique address. Figure 3.4 shows one full-decoding scheme. The top ten address lines are used to select a base address of $17F8 for the I/O devices. The low three address lines can then be decoded by 4-input gates to select any of 2^3, or 8, I/O devices. The new memory map, using the additional decoding logic of Figures 3.3 and 3.4, is shown in Figure 3.5.

3.2 ZERO-PAGE ADDRESSING

Section Overview

Microprocessors use a variety of ways to specify the addresses at which they access data. These are called *addressing modes*. Zero-page addressing is like extended-mode addressing except that the high-order byte is *assumed* to be $00, and only the low-order byte needs to be given. Motorola calls this fast addressing mode *direct* in its 6800/6802 literature.

An address on a 16-bit bus is represented by four hex digits. The two most-significant digits are often referred to as the *page* number of an address, with the two least-significant digits denoting one of 256 words on that page. A word is the contents of a single address and, for an 8-bit processor, it is 8 bits *wide*. There are, of course, 256 pages and 256 × 256, or 65 536, words in the entire address range.

Extended-mode instructions require two address bytes following the op-code; one to specify the page (high-order half of address) and another to specify the word on the page (low-order half of address) of the operand. The processor requires three machine cycles just to read the op-code and operand address in extended-mode addressing.

The zero-page addressing mode assumes that the page number of the operand's address is $00, so only the word on page $00 needs to be specified. Zero-page instructions are thus only 2 bytes long and require only two machine cycles to read the op-code and operand address. A system using a microprocessor with zero-page addressing will generally have RAM on page $00 of its memory map. This RAM will be used as a "scratch pad" for quick storage and retrieval of data during computations. The 6500 and 6800 family processors use zero-page addressing. The Z80, 8080, and 8085 processors do not.

Motorola uses the term *direct addressing* to refer to zero-page addressing. This is most unfortunate because Intel uses the term *direct addressing* to refer to the addressing mode that we have been calling *extended*. In discussing systems to be implemented with the 6802 processor, we will use their term, *direct*, for zero-page addressing.

REVIEW OF SECTIONS 3.1 AND 3.2

Answers appear at the end of Chapter 3.

1. A microprocessor uses all 16 of its address lines. A ROM memory is selected under the following condition:

$$\text{ROM} = \overline{\text{A15}} \cdot \text{A14} \cdot \text{A13} \cdot \overline{\text{A12}} \cdot \text{A11}$$

   What are the highest and lowest hex addresses that will select the ROM?
2. For the system of Figure 3.1, what device is selected at binary address 0001 1010 0111 1001?
3. A data byte is stored in a 2K ROM at binary address 011 0010 1110. The ROM is then inserted into the system of Figure 3.1. At what hex address is that byte accessed by the processor?
4. A microprocessor has 16 address lines. A 4K-byte ROM chip is selected when the three highest-order address bits are all at positive-logic 1. In how many places does the ROM appear redundantly?
5. A microprocessor with a 16-bit address bus selects a memory chip with the following logic:

$$\text{CS} = \text{A15} \cdot \overline{\text{A14}} \cdot \overline{\text{A13}} \cdot \text{A12}$$

   What is the range of addresses (in hex) to which the chip will respond?
6. A microprocessor has 16 address lines. One page (256 bytes) is to be reserved for I/O devices. Which address lines should be involved in the ANDing for the I/O enable signal? *Hint:* See Figure 3.4.
7. How many 8K ROMs will fit into the memory map of a computer with a 16-bit address bus?
8. In a 6802 microcomputer system, where should a RAM be placed in the memory map for the fastest access?
9. What is the generally used term for the addressing mode that Motorola calls *direct addressing?*

10. Use the programmer's reference to find how many 6802 machine cycles are used for a zero-page LDAA versus an extended LDAA instruction.

3.3 IMMEDIATE-MODE INSTRUCTIONS

Section Overview

When an instruction needs to access a fixed piece of data, the *immediate* addressing mode is used. The data follows the instruction immediately in the sequence of program-memory locations.

We now have four addressing modes for the 6802:

1. **Inherent: No memory access except for the instruction code itself; no operand bytes.**
2. **Extended: Two bytes following op-code byte give operand address.**
3. **Direct, or zero-page: 1 byte following op-code byte gives low-order half of operand address; high-order half is ØØ.**
4. **Immediate: Data itself follows op-code byte.**

When operand data is a constant, it is unnecessary and inconvenient to have the processor access this data at a RAM location. RAM is volatile, and the supposedly constant data could easily be altered or lost. The most sensible place to store constant data is in ROM, *immediately* following the program instruction that requires the data.

When the processor reads an instruction of the immediate mode, it interprets the next byte accessed in the program-counter sequence not as an instruction, not as a part of an address, but as user data.

Immediate-mode instructions involving an 8-bit register are 2 bytes long—the first byte is the instruction and the second is the operand data. The third byte is then another instruction. An example is:

```
86    7F    4C
```

The first byte, 86, is the immediate-mode LOAD A instruction. The second byte, 7F, is then loaded into 8-bit register A. The next byte, 4C, is interpreted as the *increment register* A instruction.

Immediate-mode instructions involving a 16-bit register are 3 bytes long. As an example, the 4 bytes

```
CE 1A 2B Ø8
```

load the 16-bit *X* register with data $1A2B and then increment that register (instruction $Ø8). The programmer's reference card at the back of this text will help you to interpret these instruction codes.

A review of addressing modes will be helpful at this point because we have now seen four of them, and they do tend to become confusing.

- Inherent-mode instructions operate on internal machine registers and are inherently complete in one byte (the operation-code byte). In general, they do not access memory for their operands.[1] In 6800 code all these instructions have the first hex digit 0, 1, 3, 4, or 5. Examples are as follows:

TAB:	16	Transfer contents of internal 8-bit accumulator A to accumulator B.
DEX:	09	Decrement (subtract 1 from) internal 16-bit register X.

- Extended-mode instructions use 2 bytes following the operation code to point to one of 65 536 addresses. The processor then suspends its normal program-counter sequence of accessing memory and accesses the operand address pointed to. These instructions begin with hex digits 7, B, or F. If the register involved is 16 bits long, the operand consists of a high-order byte located at the address pointed to and a low-order byte located at the next sequential memory address. Examples are as follows:

LDAA:	B6 0B 12	Load accumulator A with the 8-bit data word found at address $0B12.
STX:	FF 01 23	Store the high-order 8 bits of register X at address $0123. Store the low-order 8 bits at address $0124.

- Direct-mode instructions are like extended-mode, but the high-order byte of the address is assumed to be 00. Thus only 2 bytes are needed to complete the instruction, 1 for the op-code and 1 for the low-order byte of the operand address. These instructions all begin with $9 or $D in 6800 machine code. Examples are as follows:

ANDB:	D4 A5	Perform a logic AND of each bit of accumulator B with the corresponding bit of the data found at memory address $00A5. Store the result in B.
CPX:	9C 12	Compare the 16 bits of register *X* with the 16 bits found at address $0012 (high-order half) and $0013 (low-order half). Chapters 4 and 8 detail what may be done with the results of this comparison.

- Immediate-mode instructions have the actual operand data in the byte following the op-code byte. If the data is 16 bits long, 2 bytes are needed following the op-code. These instructions begin with $8 or $C. Examples are as follows:

ADDA:	8B A0	Add the value 160 (hex A0) to the contents of accumulator A.
LDX:	CE 00 00	Load the 16-bit X register with all zeros (It would be nice if there were an inherent-mode CLX instruction to do this, but there isn't.)

A graphic summary of how a 6802 processor executes a series of instructions is given in Figure 3.6. The program consists of one instruction each in the *immediate*, *inherent*, *direct*, and *extended* modes. It is actually a short loop which endlessly repeats, alternately writing $A6 and $59 into memory location 0017. The COMA

[1]Exceptions are the instructions involving the stack. See Chapter 9.

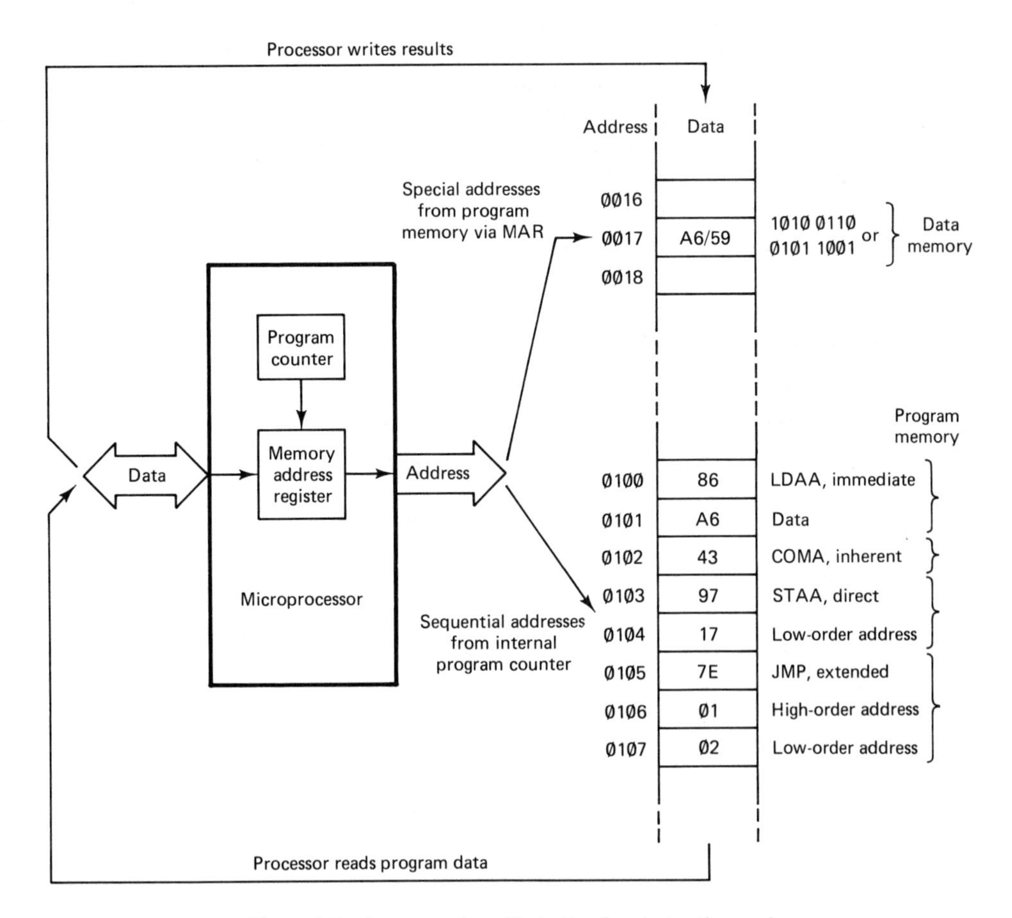

Figure 3.6 A program loop illustrating four instruction modes.

causes the data in accumulator A to be *complemented*; all binary 1s change to Øs and Øs change to 1s, causing the change from \$A6 to \$59 and back.

Once started in this program loop, the microprocessor will repeat it endlessly, many millions of times, interpreting each data type in the program memory as indicated: instruction, data, address low-order, address high-order. It is interesting to contemplate what would happen if the processor ever once got out of step, interpreting the data at address \$Ø1Ø4 as an operation code, for example. But of course, in smoothly functioning systems, it never does.

3.4 ASSEMBLY LANGUAGE

Section Overview

Microcomputer programs are written in *assembly language*, in which 3- or 4-letter mnemonics stand for the actual hex or binary digits of machine

op-codes. *Labels* to the left of certain mnemonics give easy-to-remember names to the addresses of these op-codes. *Operands*, to the right of the op-code mnemonics, refer to the operand addresses, or immediate operand data if preceded by the # sign. Operands can be listed by letter names, as decimal numbers, or as hex numbers if preceded by the $ sign.

Assembly language is referred to as *source code*. Source code is incomprehensible to humans by itself, so it must always be accompanied by English-language *comments*, which explain the overall purpose and strategy of the program in performing its task for the user. Programmers who write uncommented source code are frequently assassinated by those who have to use and modify their programs.

The microprocessor actually deals with a program as a rapidly switching series of voltages on 8 lines (the data bus) coming from memory. True machine language, with +4 V and 0 V represented by the symbols 1 and 0, respectively, looks like the column at the left:

Binary	Hex
1000 0110	86
1010 0110	A6
0100 0011	43
1001 0111	97
0001 0111	17
0111 1110	7E
0000 0001	01
0000 0010	02

The program is the four-instruction loop from Figure 3.6. We have placed the hex representation of the program at the right of the binary, and this could also be called machine language.

Machine language is almost impossible for humans to decipher, so programs are written, debugged, and modified in a human-recognizable form called *assembly language* and then translated, or *assembled*, into machine code.

Assembly language is different for each computer or microprocessor, but the format and underlying ideas are the same for all assembly languages. By learning 6800 assembly language you will be putting yourself in a position to "pick up" the languages for other processors easily.

The general rules for writing assembly language are simple:

1. Each line contains one complete instruction. The lines are broken into three columns. Every line must contain an entry in the middle column. The left and right columns contain entries only where needed.
2. The middle column contains a mnemonic representing the operation. Mnemonics for 6800 language consist of three or four letters. Examples are LDAB for *Load Accumulator* B and INC for *Increment*.

3. The right column contains an indication of the operand, unless the instruction is inherent mode, in which case no operand need be expressed.

a. The operand is presumed to be expressed in decimal unless it is preceded by the sign $, meaning *hex*. Thus

12 = twelve, but
$12 = eighteen

b. The operand is assumed to be an address unless it is preceded by the symbol #, which means *immediate-mode data*. Thus the operand

$12 means *the data at address eighteen*, but
#$12 means *the data itself is eighteen*

c. The operand may be an algebraic-type variable name that has been previously defined. Thus if CENT has been defined as 100, the operand CENT would equal address 100 (or address $0064) and #CENT would equal immediate data 100 (or data $64.)

4. The left column may contain a *label*, which would be an algebraic-type variable name. This variable would have the value of the address of the op-code for the line on which it appears. We say that the label *points to* the instruction with which it is associated. Labels are generally used only when we need to jump to a particular instruction rather than following the normal program-counter sequence.

Here is the 6800 assembly language for the four-instruction program loop diagrammed in Figure 3.6.

Label	Op-code	Operand
	LDAA	#$A6
LOOP	COMA	
	STAA	$17
	JMP	LOOP

Source code and object code. Assembly-language lists like the preceding one are referred to as *source code* because they are what programmers actually write, and they are the origin of the machine language that the computer finally runs. The machine language itself (usually expressed in hex digits) is then called *object code*. The process of converting from source code to object code is called *assembling* the program. Assembly can be done by hand for short programs written for microprocessors with straightforward instruction codes,[2] but universal practice in industry is to turn this tedious and error-prone task over to a computer program called, quite naturally, an assembler.

Here are the source code and the object code for the four-instruction loop in standard assembly-language form. Notice that the program bytes are expressed in

[2]The Motorola 6800/6801/6802 and the Rockwell 6502 are the only popular microprocessors on which hand assembly is practical.

hexadecimal, with the op-code and operand bytes for each instruction all on one line. Also note that the addresses are given only for the op-code bytes.

Object Code		Source Code		
Address	Program	Label	Op-code	Operand
0100	86 A6		LDAA	#$A6
0102	43	LOOP	COMA	
0103	97 17		STAA	$17
0105	7E 01 02		JMP	LOOP

The label LOOP has the value $0102 and is used as the operand in the fourth instruction to jump the program counter to the address that restarts the loop. Be sure you understand the difference between the operand #$A6, in which the data is hex A6, and the operand $17, in which the data is to be stored at address $17.

Comment your source code: The unforgivable sin among programmers is writing uncommented source code. Real-life programs are written to accomplish a specific useful task. Source code alone gives almost no clues to relate the program instructions to that task. Troubleshooting or modifying the code to change the performance of the task is impossible without an understanding of *how* the instructions accomplish the task. We will have much more to say about this in later chapters, where our programs will be directed toward useful outcomes. Even with this example program, however, we ought to start on the right foot by providing comments to explain what the program does and how.

Here is an example of how *not* to comment source code.

Source Code			Poor Comments
Label	Op-code	Operand	
	LDAA	#$A6	Load accumulator *A* with $A6.
LOOP	COMA		Complement accumulator *A*.
	STAA	$17	Store contents of *A* at address $0017.
	JMP	LOOP	Jump back to start of loop.

Comments like this are worse than useless. They tell nothing that is not already evident from the source code, and they insult the reader by implying that he doesn't know the assembly-language mnemonics. Here are a few rules for writing useful comments:

1. Good comments are not written line by line. They refer to groups of five to ten lines as functional units.
2. Good comments don't tell *what* happens; they tell *why* it happens.
3. Good comments don't talk about the machine instructions; they talk about the task that is being accomplished for the user.

4. Good comments never restate the obvious. They tell something that isn't already apparent.
5. Translations of the instruction mnemonics are not good comments.

Here is another try at commenting the source code.

Source Code			Good Comments
Labels	Op-code	Operand	
	LDAA	#$A6	Endless test loop demonstrating
LOOP	COMA		four instruction modes;
	STAA	$17	alternately writes $A6 and
	JMP	LOOP	$59 to RAM at address $0017.

REVIEW OF SECTIONS 3.3 AND 3.4

Answers appear at the end of Chapter 3.

11. How many bytes are required for a complete 6802 immediate-mode instruction?
12. Which addressing mode is capable of accessing directly any of the possible 65 536 memory addresses?
13. Which 6800 addressing mode requires no operand bytes?
14. Which addressing mode has operand bytes which are data rather than addresses?
15. Assembly language is called *source code*; machine language is called ______.
16. The symbol # preceding a number means the number is ______.
17. In assembly language, the column to the right of the instruction-code mnemonic contains ______.
18. The value assigned to a label is the ______ of its instruction.
19. Comments should relate the program instructions to the ______.
20. Why is "Complement Accumulator A" not a good comment for the instruction mnemonic COMA?

3.5 PROCESSOR-CONTROL LINES

Section Overview

The address of the first instruction executed by the 6802 is stored at $FFFE (high-order half) and $FFFF (low-order half). This *reset vector* is read when the $\overline{\text{RESET}}$ input is brought *low* and then returned *high*.

The 6802's VMA output pin stays *high* on cycles in which memory is being accessed and goes *low* on cycles when no valid memory address is on the bus.

The R/$\overline{W}$ line goes *low* only on *E*-clock cycles when the processor is writing data on the data bus.

To have the 6802 microprocessor run a program that we have stored in memory, three processor-control lines must be connected and used properly.

The reset vector: The $\overline{RESET}$ input (pin 40) of the 6802 is held *high* by a resistor to V_{CC} when the processor is running. A *reset* switch pulls this line *low* momentarily to start the processor at the beginning of your program. A capacitor is placed across the switch so the RESET-pin voltage rises smoothly when the switch is opened, even though the contacts may bounce. To find the beginning of your program, the 6802 reads its *reset vector*, which is an address stored at address $FFFE (high-order byte) and $FFFF (low-order byte). It then loads the program counter with this vector address and starts running your program from there.

Going back to the example program loop of Figure 3.6, the program starts at address $0100. It is up to the programmer, then, to provide a ROM that will be accessed by a processor read of address $FFFE and to store data $01 at that address. Likewise, data 00 (the low-order address byte) must be stored at address $FFFF. When the $\overline{RESET}$ line is brought *low* the 6802 will output address $FFFE continuously, but VMA will be *low* so no read will take place. When $\overline{RESET}$ goes *high* the processor will load the high-order half of its program counter with data 01 from address $FFFE. Then it will output address $FFFF and fill the low-order byte of its program counter with data 00. It will then immediately read address $0100 and *assume* that the byte it finds there is an instruction code—the first instruction of your program.

The VMA line: The processor spends a good portion of its time operating on its internal registers, and during those times memory ought not to be accessed at all. However, it is not possible for a digital bus to have "nothing" on it. A 16-bit address bus is always calling up one of its 65 536 addresses. Some of those addresses will contain RAM data, which we would rather not have disturbed. Some will access output devices that we would rather not have activated.

To avoid address call-ups when the processor is not, in fact, accessing memory, the VMA (Valid Memory Address) line is included in the AND logic, which selects a memory or I/O device. VMA goes *low* during machine cycles when no memory address is being called up.

The R/$\overline{W}$ line: A RAM chip can be read from or written to at the same address. To select which of these two functions to implement, RAMs and other readable/writable devices must have a read/write control input, which is driven by the processor's READ/$\overline{WRITE}$ output line. This line remains *high* for all cycles that read program or data memory and goes *low* only when the processor intends to write data to a memory or output device. STAB and INC (extended mode) are examples of instructions that cause R/$\overline{W}$ to go *low*.

When R/$\overline{W}$ is *low* the processor's data lines go to an active-output condition during the second half of the *E*-clock cycle; that is, they apply binary 1s or 0s to each line. When R/$\overline{W}$ is *high* the processor's data lines are in a high-impedance state, leaving the bus free for data to be applied by some other device.

3.6 A DIAGNOSTIC PROGRAM

Section Overview

A first diagnostic program for a microcomputer system should cause all the address, data, and control lines to switch *high* and *low* in an endless loop. The loop should take a minimum number of machine cycles so that voltage patterns can be observed on a regular oscilloscope.

Such a simple diagnostic program will verify that the processor and ROM are functioning. Further diagnostics can be written later to test other system components.

Real-life microcomputer programs are normally quite long, 1000 to 50 000 bytes in typical cases. However, electronics technicians responsible for installing, testing, and troubleshooting microcomputer hardware will usually be required to write only short programs, a few dozen or so words in length. These are diagnostic programs, and they do nothing more than repeatedly operate some portion of the system hardware so that failures can be spotted.

A 6802 diagnostic program can be written in three instructions. This program will not be particularly impressive to the nontechnical system user, but it will allow the technician to determine that the processor is reading from and writing to its memory, and that the address, data, and control lines are capable of driving their rated loads. List 3.1 contains the program object code and source code with comments.

```
                  0850 *-----------------------------------------------------------
                  0950 *
                  0980 * LIST 3-1    6802 DIAGNOSTIC LOOP
                  0985 * SWITCHES ALL ADR, DATA, R/W, VMA IN 9-CYCLE LOOP.
                  0986 * RUNS ON FIG 3.7 OR 5.7 HARDWARE; 2 JUL '87 * D. METZGER
                  0990 *
                  1000 * <<<<<<<SOURCE CODE>>>>>>>>>>>>>>>>>>>>>>>>>>>>
                  1010 *-----------------------------------------------------------
                  1020 *LABEL   OP   OPERAND         COMMENTS
                  1030 *        CODE
                  1040 *-----------------------------------------------------------
                  1050          .OR  $0182    DEFINES ORIGIN OR START ADR OF PROGRAM.
                  1055          .TA  $4180    RAM AREA IN HOST COMPUTER TO HOLD FILE.
                  1058 *
0182- CE FF 00    1060          LDX  #$FF00   LOOP WRITES DATA $FF TO
0185- FF FF FF    1070 LOOP     STX  $FFFF     ADDRESS $FFFF, THEN DATA $00
0188- 7E 01 85    1080          JMP  LOOP      TO ADDRESS $0000.
                  1082 *
                  1084          .OR  $FFFE    RESET VECTOR TO OBTAIN STARTING ADDRESS
                  1094          .TA  $47FE     OF PROGRAM. TARGET ADDRESS IS FREE RAM
FFFE- 01 82       1096          .HS  0182      IN HOST COMPUTER.  HEX STRING IS ADR.
                  1098 *
                  1110          .EN           DEFINES END OF PROGRAM.
```

The program consists of 9 bytes, the last 6 forming a loop that is repeated endlessly. The loop takes 9 machine cycles, so it can be displayed in its entirety on the 10 horizontal divisions of an oscilloscope. Appendix B, "Cycle-by-Cycle Operations for the 6800," is used to predict the state of the processor address, data, and control lines during each cycle of the loop. Here is a cycle-by-cycle summary of processor activity for the 9-cycle loop:

1. Processor reads op-code FF (STX, extended) on data bus from address $0185. R/$\overline{W}$ and VMA lines both *high*.
2. Processor reads most-significant operand-address byte FF on data bus from address 0186. R/$\overline{W}$ and VMA lines both *high*.
3. Processor reads least-significant operand-address byte FF on data bus from address 0187. R/$\overline{W}$ and VMA lines both *high*.
4. Processor floats data bus while address bus outputs operand address FFFF. Processor internally prepares to output X-register bytes via data bus. R/$\overline{W}$ *high*; VMA *low*.
5. Processor outputs most-significant X-register byte FF on data bus to operand address FFFF. R/$\overline{W}$ *low*; VMA *high*.
6. Processor outputs least-significant X-register byte 00 via data bus to operand address plus one (0000). R/$\overline{W}$ *low*; VMA *high*.
7. Processor reads op-code 7E (JMP, extended) on data bus from address $0188. R/$\overline{W}$ and VMA lines both *high*.
8. Processor reads most-significant operand-address byte 01 via data bus from address $0189. R/$\overline{W}$ and VMA lines both *high*.
9. Processor reads least-significant operand-address byte 85 via data bus from address $018A. R/$\overline{W}$ and VMA lines both *high*. Program counter loaded with 0185.

This loop causes all the data lines, all the address lines, the R/$\overline{W}$ line, and the VMA line to switch in a pattern that is predictable and observable on the oscilloscope.

REVIEW OF SECTIONS 3.5 AND 3.6

21. A 6802 program begins at address F820. What bytes must be stored and at what addresses to start the processor at this address upon RESET?
22. When does the 6802 read from address $FFFF?
23. What signal does the 6802 use in order not to call up any of its 65 536 possible addresses?
24. What is the state of the processor's data bus when R/$\overline{W}$ is *high*?
25. Which of the following instructions cause R/$\overline{W}$ to go *low*?
    (a) ASL, extended **(b)** ASLA **(c)** INX
    (d) JMP, extended **(e)** ORAB, direct **(f)** STAB, direct

Here is the source code for a continuous-loop program that will be used in Questions 26 through 30.

```
            LDX     #00
AGAIN       STX     $3FFF
            INX
            JMP     AGAIN
```

26. How many machine cycles does each pass through the loop take?
27. Considering the read of the STX op code to be cycle 1, on what cycle(s) does R/$\overline{W}$ go *low*?
28. On what cycle(s) does VMA go *low*?
29. What will be the binary levels on the data bus during the seventh cycle of the loop?
30. One of the data lines is observed to have a voltage of 1.3 V during the first half of cycle 5. (It goes to +4 V during the second half.) Does the invalid logic level indicate a problem?

3.7 RUNNING DIAGNOSTICS ON A 6802 SYSTEM

Section Overview

This section describes a simple hardware setup that can be used to run the diagnostic program of Section 3.6 as well as other programs. It consists of two chips—a 6802 processor and a 6810 RAM. The program is entered into the RAM in binary via address and data toggle switches, which can be isolated from the processor with 74LS244 tristate buffer ICs.

System timing, logic levels, and bus loading can be observed with an oscilloscope on this elementary microcomputer system.

To run a program, a microprocessor must be connected to a memory chip, from which it may read the successive bytes of the program. Figure 3.7 shows a 6802 processor connected to run a program stored in a 6810 RAM memory.

The eight data lines are each connected between the processor and the memory, so the processor may read data applied by the RAM, or the RAM may receive data applied by the processor. R_2 has a low-enough value to allow the R/$\overline{W}$ output of the processor (pin 34) to control the *read* or *write* state of the RAM via its R/$\overline{W}$ input, pin 16.

The seven address lines of the 6810 are driven by the corresponding pins (A0 through A6) of the processor. The 6810 has 2^7, or 128, bytes of addressable memory. This is a pitifully small memory by today's microcomputer standards, but it will be quite adequate for us to demonstrate most of the features of the microprocessor. Also, note that we will be using a RAM as program memory. Common practice is to store basic operating programs in ROM so they will not be lost if power is interrupted. We have chosen to use a RAM for our demonstration system because it allows the program to be changed quickly without a special programmer.

Address decoding for the 6810 is provided through an internal 6-input AND gate, 4 of whose inputs are inverted. Figure 3.8 represents this function.

CS3 is used to ensure that the memory is called up only when the processor is outputting a valid memory address, and CS0 is used to access the memory only at addresses for which bit A8 is *high*. This leaves lines A7 and A9 through A15 (8 address lines) unconnected, so the 128 RAM bytes are mirrored at 2^8, or 256, places.

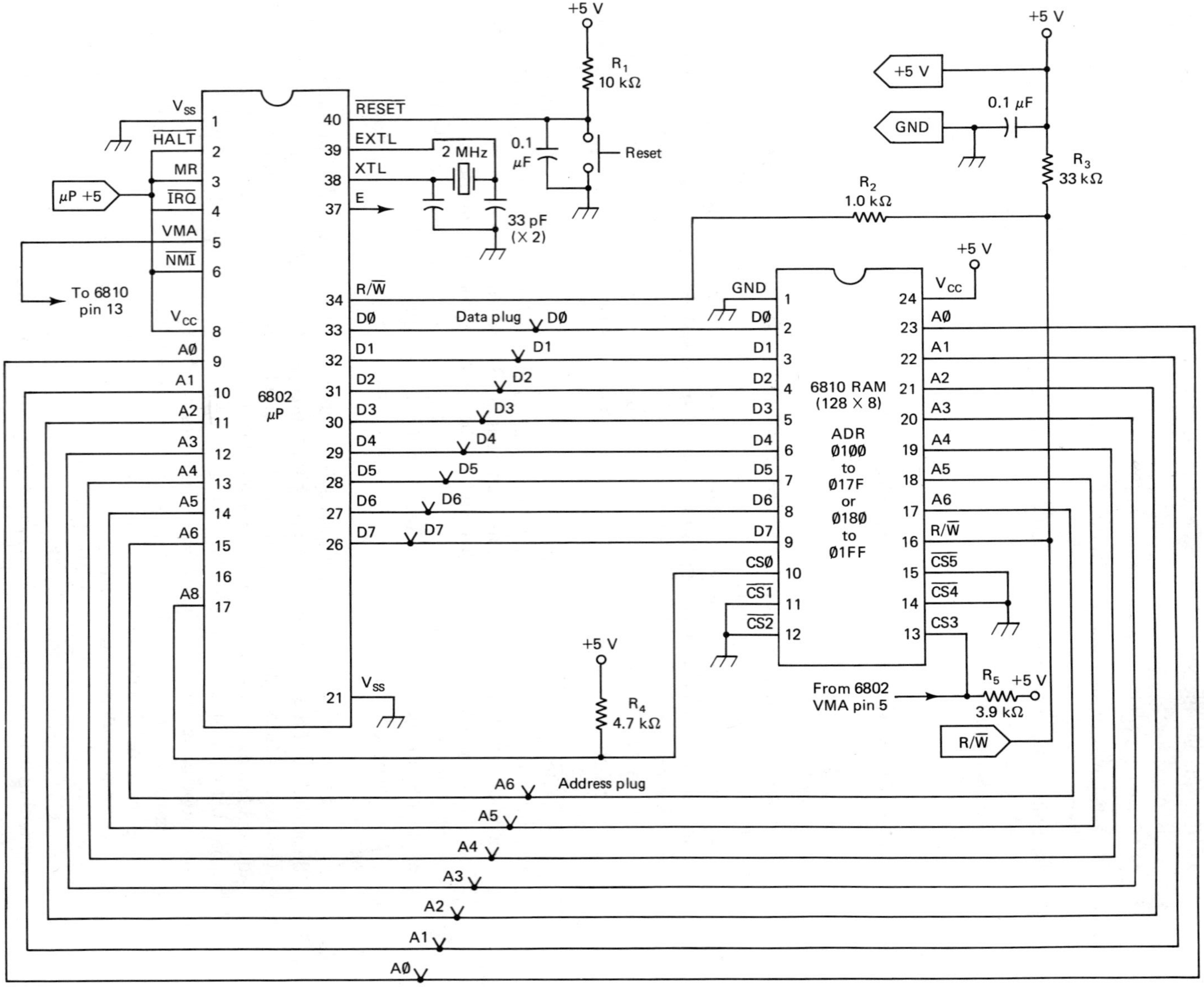

Figure 3.7 An elementary microcomputer system suitable for breadboarding and laboratory experimentation.

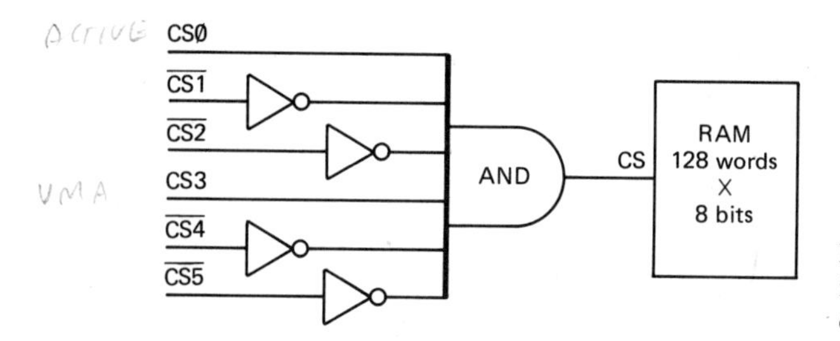

Figure 3.8 The 6810 RAM has an internal 6-input AND gate for address decoding.

The remaining 4 chip-select lines of the 6810 have not been used to decode the address more completely because they are all *low*-active. Consider that the RAM must contain the reset victor as well as the program, and the reset vector is accessed at $FFFF (all binary *highs*). Therefore, the chip-select for the 6810 cannot include any *low*-active inputs.

Memory map: Selection of the RAM in the system of Figure 3.7 requires only that address A8 be *high*. The lowest base address in the 6810 is then

ØØØØ ØØØ1 ØØØØ ØØØØ = $Ø1ØØ

At the highest RAM address, AØ through A6 will also be *high*, so the highest base address is

ØØØØ ØØØ1 Ø111 1111 = $Ø17F

The first redundant address range for the 6810 would have the unused bit A7 *high*, producing a range from

ØØØØ ØØØ1 1ØØØ ØØØØ = $Ø18Ø to

ØØØØ ØØØ1 1111 1111 = $Ø1FF

The second redundant range would have A7 *low*, A8 *high* (as always), and A9 *high*, giving a range from

ØØØØ ØØ11 ØØØØ ØØØØ = $Ø3ØØ to

ØØØØ ØØ11 Ø111 1111 = $Ø37F

Figure 3.9 shows the pattern of redundant addresses that is repeated until the end of the memory. We will use the second range for programs and the highest range for the reset vector. This will permit the programs developed for the 6810-based system of Figure 3.7 to run on the 6116-based system of Figure 5.7, which you may prefer to wire in a more permanent form.

Entering the program: We have indicated that the 6802 will run a diagnostic program contained in a 6810 RAM, but we have not yet explained how the program bytes will get into the RAM. They will be entered by the control circuit of Figure 3.10. It will enable us to read the data at any address in the RAM and to enter new data as desired.

In the PROGRAM/VERIFY mode, power is removed from the processor, and the RAM address lines are driven through noninverting buffers by a bank of seven toggle switches. Addresses and RAM data are read out in binary on discrete LEDs.

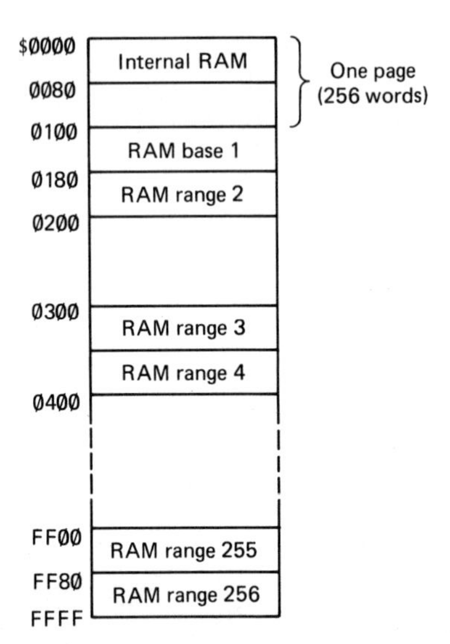

Figure 3.9 The system of Figure 3.7 uses incomplete address decoding, so the RAM appears at 256 redundant ranges in memory.

Pushing the WRITE button places the RAM in a writable condition and simultaneously drives the data bus from a bank of 8 toggle switches via another pack of noninverting buffers. Thus the data from the 8 toggle switches is programmed into the address selected by the 7 toggle switches, and the 8 LEDs indicate the new data.

In the RUN modes the address buffer is placed in a high-impedance state, and power is applied to the processor so that it drives the address inputs of the RAM. In the RUN-PROTECT mode the 470-Ω resistor holds the RAM's $R/\overline{W}$ line *high* even in the face of a pull-low from the processor through the 1.0-kΩ R_2. Thus the RAM data is protected from accidental writes by the processor. In the RUN-WRITEABLE mode the $R/\overline{W}$ line is allowed to be driven by the processor.

The hardware setup for running the diagnostic program is shown in the photograph of Figure 3.11. It consists of the processor-and-memory system of Figure 3.7 and the control unit of Figure 3.10. The processor circuit is set up on a solderless breadboard to facilitate modifications and additions. The control unit is built with perf-board and solder construction, since it can be used for many future projects without modification. Plugs for attaching the data and address-bus lines to the breadboard are made from halves of wirewrap IC sockets. The plugs will fit in the breadboard alongside the processor address and data pins.

The control unit is more time-consuming to build than the processor circuit, so it is recommended that it be constructed in permanent form before the 6802/6810 circuit is breadboarded. The processor circuit will work with the control unit unplugged so one control unit can service two or three 6802/6810 systems. It is best to *reset* the processor once in the *run*(*protect*) position before switching to run(*writable*) to avoid spurious writes over your program by an uncontrolled processor.

Here is a procedure for running the diagnostic program:

1. Set up and interconnect the circuits of Figures 3.7 and 3.10.
2. Set the FUNCTION switch to PROGRAM/VERIFY. Set the address switches to

   000 0010

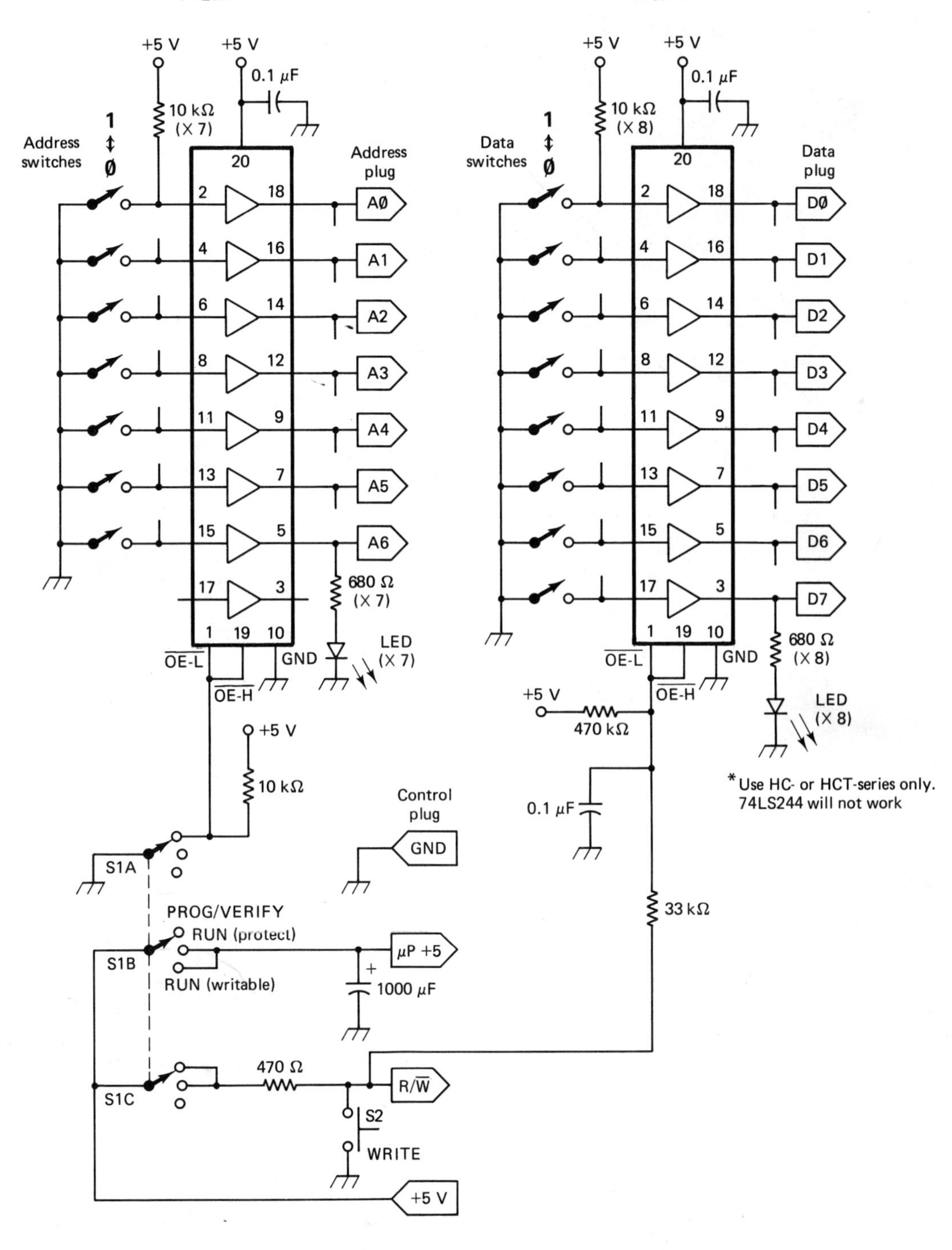

Figure 3.10 The control unit for the system of Figure 3.7 is used for programming and readout of the 6810 memory.

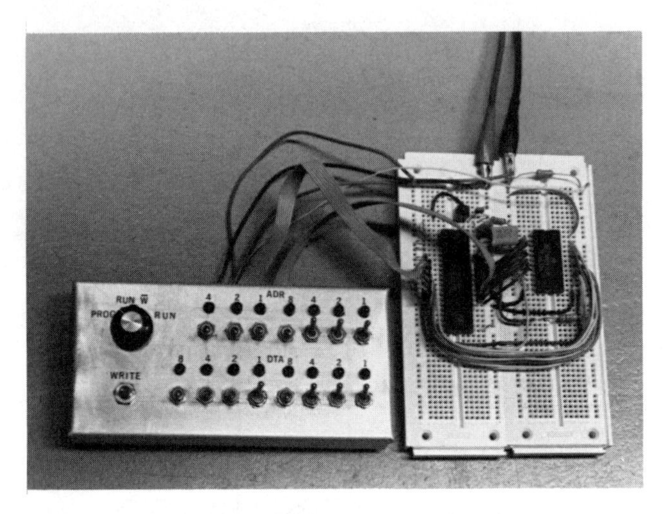

Figure 3.11 The microprocessor system of Figure 3.7 together with the control unit of Figure 3.10.

This address is accessed when the processor outputs

0000 0001 1000 0010

or $0182, since A8 = 1 is a required condition to access the RAM.

3. Set the data switches to the first byte of the program, as listed on page 81. This is $CE, or in binary

   1100 1110

4. Press the *write* button, and verify that the data LEDs show $CE when the button is released.
5. Change the address switches to 000 0011 (16-bit hex address 0103) and the data switches to the next program byte, and continue loading until the end of the program.
6. Set the address switches to the address that will be called up when the processor outputs $FFFE. This is 111 1110 in 7-bit binary. Load data 01, which is the most-significant byte of the program starting address.
7. Set the address to 111 1111 and load data 82, the least-significant byte of the starting address.
8. Ground the 6802 $\overline{\text{RESET}}$ line, pin 40, and switch to the RUN/WRITEABLE function. Unground pin 40.
9. Set an oscilloscope to 2 V/div, 2 μs/div, and observe the *E*-clock, the address lines, and the data lines, and verify that all lines are switching to valid logic levels (high ≥ 2.4 V, low ≤ 0.4 V).
10. Trigger the scope on the positive transition of the R/$\overline{\text{W}}$ line. This is the start of cycle 7 of the loop, which is the read of the JMP instruction $7E.
11. Predict the logic level of the following lines in the given cycles from the cycle-by-cycle operation given on page 000, and verify each with the scope:
    a. A0, cycles 2 and 3   **b.** D3, cycles 2 and 3

c. VMA, cycles 4 and 5 **d.** $R/\overline{W}$, cycles 6 and 7
e. D7, cycles 6 and 7 **f.** A8, cycles 5 and 6

12. Note that the test program destroys the second RESET vector byte by writing to $FFFF. Reprogram this byte, remove the address- and data bus plugs, and run the diagnostic loop again. The 6802 data sheets indicate that the address, data, VMA, and $R/\overline{W}$ outputs should be able to drive a standard TTL load and still pull down to a *low* of no more than +0.4 V. Power up a 7404 TTL chip and connect one of its inputs to selected lines to verify this.

13. The specs also indicate that a 330-pF load capacitance will typically cause a delay of 250 ns on data and $R/\overline{W}$ lines and 330 ns on address and VMA lines. Use a 330-pF capacitor to ground at selected pins to verify this.

14. Program and test the three-instruction loop in the program given before Question 26 on page 82.

Answers to Chapter 3 Review Questions

1. $6800, $6FFF **2.** The ROM **3.** $1B2E **4.** 2

5. $9000 through $9FFF **6.** A15 through A8

7. 8 **8.** Page 0 **9.** Zero-page addressing

10. 3 cycles versus 4 **11.** 2 or 3 **12.** Extended

13. Inherent **14.** Immediate **15.** Object code

16. Data for an immediate-mode instruction. **17.** Operands

18. Address **19.** Task that the program will accomplish.

20. Because it simply translates the mnemonic, which any programmer can do without the comment. Also, it comments only a single line; comments should refer to groups of lines.

21. $F8 at address $FFFE, $20 at address $FFFF.

22. When the RESET input goes *high*. **23.** VMA line *low*.

24. High-impedance or read state **25.** (a) and (f) only

26. 13 cycles **27.** cycles 5 and 6 **28.** Cycles 4, 9, and 10

29. 0000 1000, or $08

30. No. The processor applies data to the bus only during the second half-cycle. Nothing applies data during the first half, so the float to an invalid level is normal.

CHAPTER 3 QUESTIONS AND PROBLEMS

Digits after the decimal point refer to section numbers in Chapter 3.

Basic Level

1.1 A 6802 system uses a number of 2764 EPROMs and 6264 RAMs. These are both 8K devices. How many address lines are taken for address selection by the memories, and how many are left for device selection? How many 8K devices can be selected in this system?

2.1 For the system described in Question 1, write the decoding logic expression for selection of a ROM in the third-highest 8K block of memory, and give the range of addresses (in hex) to which this chip will respond.

3.1 In Question 2 the ROM has a particular data byte stored at its binary address 11Ø11 11Ø1 Ø111. At what hex address will this byte be accessed in the 6802 system?

4.1 In the system of Figure 3.1 a RAM byte is stored at address $2A5. At what other address could it be read besides $Ø2A5? Refer to Figure 3.2.

5.2 What are the lowest and highest addresses (in hex) that can be accessed by the 6802's direct addressing mode?

6.2 How many "pages" of address space are possible in a 6802 system?

7.3 Examine the 6802 programmer's reference at the back of the book to determine which binary bits stay the same for all immediate-mode instructions. Note that the first-column instructions having 1 byte are inherent, not immediate, mode.

8.3 List the 6802 immediate-mode instructions that require 3 bytes. Explain why these few require 3 bytes instead of 2.

9.4 In standard assembly-language format, what are the names of the three columns of source code?

10.4 When is an entry required in the first source-code column?

11.4 When is an entry *not* required in the last source code column?

12.5 When does the 6802 read addresses $FFFE and $FFFF?

13.5 When does the 6802 assert data on the data bus: during the first half of the *E* cycle, during the second half, or during the entire *E* cycle?

14.6 The diagnostic program of List 3.1 is run on a 1-MHz 6802 system. An oscilloscope is set to trigger on the negative transition of VMA, with its sweep speed at 1 μs/division. During which division(s) of the sweep will all of the data lines be pulled *low*?

15.6 In the setup of question 14, during which sweep division will all of the address lines be pulled *low*?

Advanced Level

16.1 A 6810 RAM (128 bytes × 8 bits/byte) is to be accessed at base address $12ØØ in the system of Figure 3.1. Draw the necessary decoding logic. Refer to Figure 3.4 for guidance. The 6810 pinout is given in Figure Ø.18.

17.1 A 6802 system has three 8K RAMs at the low end of the memory map and four 8K ROMs at the high end. The remaining 8K block is to be broken into eight 1K areas. Draw the memory map and the decoding logic for the lowest of these 1K areas.

18.2 Use the programmer's reference chart at the back of the book to make a list of all 6802 instructions that are available in both extended and direct modes (on the left) and those that are available in the extended mode only (on the right).

Address	Data	Address	Data
F8C0	86	F8C5	3C
F8C1	7E	F8C6	7E
F8C2	4C	F8C7	F8
F8C3	B7	F8C8	C6
F8C4	00		

19.3 Here is a list of memory bytes comprising a program routine. The first byte is an op-code. Interpret the program with the help of the dissassembly chart in Appendix C. Give the mnemonic and instruction mode of all op-codes, the decimal equivalent of any numeric data, and the concatenated hex address of any address bytes.

20.3 In Problem 19, trace what the 6802 would do if it somehow landed at address $F8C1 while expecting an op-code.

21.4 List three features of good source-code comments.

22.4 Here is a program in 6802 source code. Comment the program and assemble it into object code starting at address $C000.

```
START   LDAA    $C01A
        ADDA    #$10
        STAA    $C01A
DONE    JMP     DONE
```

23.4 What do you think of the comment "Write data from accumulator A to address $C01A" for the third line of the preceding program?

24.5 Consult the 6802 cycle-by-cycle operation chart (Appendix B) to determine what appears on the address bus during the third cycle of an INX instruction (increment the 16-bit X-register, inherent mode.) What prevents the 6802 from accessing a memory that happens to have this address?

25.5 Explain why the highest block of memory in a 6802 system should be a ROM and not a RAM.

26.6 When running the diagnostic program of List 3.1, stray bus capacitance usually causes the data lines to hold their cycle-3 data during VMA-not cycle 4. How could you demonstrate that the data bus really is floating?

27.7 Write a program for the experimental system of Figure 3.7 that will add two immediate data bytes and store the result at address $0120. Show source code comments and object code. Start the program at address $0100. Test the program by running it with the hardware of Figure 3.10 and then reading the result at address $0120.

28.7 Someone proposes to eliminate some of the redundant addressing of the 6810 RAM in Figure 3.7 by driving the inputs CS1, CS2, CS4, and CS5 from processor pins A9, A10, A11, and A12, respectively. What is the flaw in this proposal?

4 Branch Instructions and Relative Addressing

4.1 RELATIVE-MODE ADDRESSING

Section Overview

All the 6800/6802 *branch* instructions are of the *relative-addressing* mode. Following the op-code byte they have a single operand byte, which specifies the number of address bytes to be skipped over if the branch is taken. The address of the next instruction is thus given relative to the address of the current instruction.

Instructions are most often followed in program-counter sequence. The program counter calls up the 1, 2, or 3 bytes of an instruction, executes it, and then goes on to fetch the next instruction at the next higher address. Jump instructions, as you know, can move the program counter to an entirely different area of memory.

Relative-mode instructions are like the *jump* instruction in that they can interrupt the normal program-counter sequence and cause a series of instructions at a new section of memory to be executed. They are different from the jump instruction in

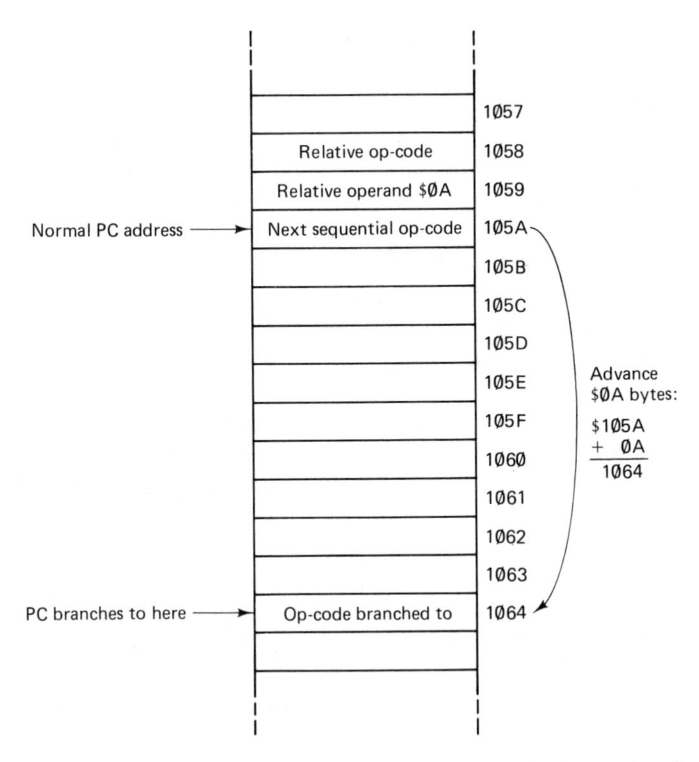

Figure 4.1 Relative addressing causes the program counter to skip the number of bytes given in the operand.

that the operand is not a 2-byte address but rather a 1-byte address *displacement*. Thus if a relative-mode instruction has an operand of $ØA, the program counter is told to skip ahead 10 bytes from the address where it would normally be after finishing the relative-mode instruction. This is the address of the next op-code, to which the program counter normally increments. Figure 4.1 illustrates this. Note that all 6802 relative-mode instructions are 2-byte instructions.

Relative addressing permits a routine to be located (and relocated) anywhere in memory without modification to the object code bytes. The 14 bytes illustrated in Figure 4.1, for example, span addresses $1Ø57 through $1Ø64. They could be moved, for instance, to the space from $3267 through $3274, and the routine would still function the same—the operand $ØA would still cause a 10-byte jump relative to the normal program-counter location. We say that the routine is *position-independent*, or *relocatable*.

Relative jumps of the program counter are called *branches*. All the 6802 instructions that use the relative mode are called *branch* instructions, and their source-code mnemonics all begin with the letter B.

4.2 NEGATIVE NUMBERS IN BINARY

Section Overview

The operand byte of a relative-mode instruction is interpreted as a signed binary number. In signed binary a byte having the most significant bit (b7) = 1 is negative. The negative numbers extend from $FF (which is −1) to $8Ø (which is −128). The hex values $ØØ through $7F have bit 7 = Ø and are simply +0 through +127 in decimal.

The negative numbers are said to be in *2's complement* form because of the technique used in obtaining them.

Relative-mode instructions can cause the program counter to be displaced forward or backward from the address of the op-code following the branch. Backward displacements require negative operands, so we must now learn how to express negative numbers in binary and hex arithmetic.

The odometer example: Consider a new car, fresh off the assembly line. Its odometer reads 00000. If we drive it forward for 3 miles, the odometer will read 00001, 00002, and finally 00003. But if, instead, we drove it backward for 3 miles, the odometer would read, successively, 99999, 99998, and 99997. Therefore, negative three (−3) is expressed by a mechanical counter as 99997.

In binary and hex, negative numbers are normally expressed using the same *count backward* idea.

Decimal	Binary	Hex
0	ØØØØ ØØØØ	ØØ
−1	1111 1111	FF
−2	1111 111Ø	FE
−3	1111 11Ø1	FD
−4	1111 11ØØ	FC
−5	1111 1Ø11	FB
|	|	|
|	|	|
|	|	|
−16	1111 ØØØØ	FØ
|	|	|
|	|	|
|	|	|
−128	1ØØØ ØØØØ	8Ø

Two's complement: There is a fairly simple procedure for changing a positive binary number to its negative binary representation. The following two steps are required:

1. Complement the binary number, that is, change all 1s to Øs and all Øs to 1s. The COM instruction in 6802 assembly language does this.
2. Add 1 to the complemented result.

The NEG instruction in 6802 language performs both of these operations. Here is an example of the procedure for obtaining the 2's complement representation of −5.

Binary positive	0000 0101 = $05
Complement	1111 1010
Add 1	1111 1011 = $FB

A 2's complement negative number can be converted back to a binary positive number simply by reversing the procedure: Subtract 1 and complement the resulting binary number.

Signed or unsigned? The binary number following a relative-mode instruction is always interpreted as a 2's complement address displacement. But in the case of user data how does one know, upon encountering the hex number $FB, for example, whether it represents −5 (its 2's complement value) or simply 251 (which is 15 groups of 16, plus 11) as we would have assumed before we started talking about negative numbers? The answer is that the programmer decides before the routine is written whether the numbers encountered are to be treated as unsigned (pure binary) or as signed (2's complement) numbers. The processor is designed to handle either without any changes in its hardware or setup. The programmer structures the routines and selects from the available instructions to treat the numbers as signed or unsigned.

If 2's complement signed-number arithmetic is being used, the most significant binary digit is interpreted as the sign bit. If it is 1, the number is negative and must be interpreted as a 2's complement. If it is 0, the number is positive and is to be interpreted as straight binary. The largest negative number in an 8-bit register is then

1000 0000 = $80 = −128

The largest positive 8-bit number (when signed-number arithmetic is being used) is

0111 1111 = $7F = +127

It may be helpful in spotting errors to note that even numbers remain even and odd numbers remain odd, regardless of whether a number is expressed in decimal, hex, pure binary, or 2's complement form.

REVIEW OF SECTIONS 4.1 AND 4.2

Answers appear at the end of Chapter 4.

1. A relative-mode instruction op-code is stored at address $FC2D. What is the address of the next op-code in the normal program-counter sequence?
2. A relative-mode instruction at address $FC2D has operand $1C. What will be the address of the next instruction after the relative jump is made?
3. The routine of Question 2 is relocated so the relative-mode op-code appears at $432D. What is the new next-instruction address?
4. A relative-mode instruction is located at address $016B. Its operand byte is $FE. What is the next-instruction address after the relative jump?
5. What types of 6802 instructions use relative addressing?

6. If 2's complement arithmetic is being used, what is the decimal value of binary 1011 0111?
7. What is the decimal value of the binary number in Question 6 if straight binary is being used?
8. What is the largest negative number that can be held in a 16-bit register using 2's complement arithmetic? (Answer in decimal.)
9. What is the hex expression for decimal −29, assuming 2's complement arithmetic?
10. A relative-mode instruction at address $C02D has operand $96. To what address will it jump the program counter?

4.3 BRANCH INSTRUCTIONS AND THE CONDITION CODE REGISTER

Section Overview

Branch instructions are the microcomputer's decision makers. They cause the program counter to be displaced from its normal *next-op-code* position only if a specified condition is true. Otherwise the program counter simply goes on to the next sequential op-code as usual.

The *conditions* are the result of previous instructions, which affect the various *flag bits* of the processor's condition-code register. Flag bits are the outputs of certain flip-flops within the processor which are set or cleared based on the result of certain preceding instructions. For example, the instruction CLRA would cause the Z flag to become true (Z = 1) because it produces a *zero* result. The Z = 1 condition would remain until another instruction (for example, LDAB #$2B) produced a nonzero result.

The BEQ instruction causes the program counter to be displaced only if Z = 1 (the *equal-to-zero* condition).

Conditional branching: All but one of the relative-mode instructions used by the 6802 are conditional-branch instructions. They will displace the program counter by the amount specified in the operand only if a certain condition (specified by the instruction) is true. If the specified condition is not true, the relative branching will not take place and the program counter will fetch the op-code that falls in normal sequence after the branch operand. In this case the branch instruction is passed through with no effect other than the use of four cycles of machine time.

The condition-code register: The conditions specified by the various branch instructions are held in an 8-bit register internal to the 6802. This is variously called the condition-code register (CCR) or the processor-status register (P, or SR). The bits of the CCR are not related as are the bits of most other registers. Bit 0 can never overflow into or be shifted into bit 1, for example. Rather, the bits are set or cleared independently by certain occurrences within the processor, and are used as "flags" to mark the most recent occurrence of each condition. For the moment we will concern ourselves only with three of the CCR flags:

1. The Z flag is bit 2 of the CCR. It is *set* to binary 1 by most instructions when they produce a zero result. As soon as one of these instructions produces a nonzero result, the Z flag is *cleared* back to binary Ø. Instructions that affect the Z flag have a letter *z* in the right column of the programmer's reference at the back of the book. In general, the Z flag is affected by all instructions except branches, jumps, and those involving the S register. We will not be concerned with S-register instructions until Chapter 9.
2. The C (carry) flag is bit Ø of the condition-code register. It is *set* when an add, shift, or rotate instruction causes a result greater than 255 in an 8-bit register. In many ways it is like a ninth bit, because it is *set* by an overflow from the eighth bit (bit 7). The C flag is also *set* when a subtract operation causes a borrow from bit 8. The C flag is *cleared* to Ø whenever an instruction is executed that affects C but fails to *set* it.
3. The N (negative) flag is bit 3 of the CCR. In general it follows the state of bit 7 for most 8-bit data operations. For the LDX, STX, and CPX instructions it follows the state of bit 15 of the 16-bit X register. N is *cleared* by any instruction that affects N but fails to *set* it. Refer to the programmer's reference at the back of the book for the *n*, which indicates instructions that affect the N flag.

Branch instructions never affect the condition-code register, but they check the CCR to see what state the specified flag was left in by the last instruction that did affect that flag. The program counter is then displaced by the operand amount (2's complement value) if the specified flag condition is met.

For example, the BEQ instruction checks for a Z flag equal to 1 (last result equal to 0). If Z is Ø the program proceeds to the next sequential instruction following the BEQ. If Z is 1, the program counter is displaced by the 2's complement value of the BEQ operand.

The BNE instruction checks for the opposite condition: The branch is ignored if Z = 1 (0 result) and the branch is taken if Z = Ø (result not equal to 0).

Similarly, the BCS instruction causes the branch to be taken (program counter to be displaced) if the carry flag is *set*; that is, C = 1. The companion BCC causes the branch to be taken if C = Ø.

The BMI (branch if minus) instruction causes a branch if flag N = 1, indicating a negative number if signed binary is being used. BPL causes branching if N = Ø, indicating a *plus* result from a previous instruction. If unsigned binary is being used by the programmer the branch simply depends on the most significant bit of the result from a previous instruction.

4.4 A TIME-DELAY LOOP

Section Overview

This section shows how the DECREMENT and BRANCH IF NOT EQUAL instructions can be used to keep the processor in a repetitive loop

for a given length of time. Such *delay loops* are commonly used for audio tone generation and switch-bounce elimination, among other applications.

If a computer can be called "smart" in any sense of the word, it is due to the branch instructions. They allow a decision to be made as to which of two program routines the computer will execute. The source of the computer's power is its ability to execute hundreds or millions of times a routine that the programmer has to write only once. Now we will see how a branch instruction lets the computer decide when to stop this repeated execution of a routine and go on to another part of the program.

Decrementing the contents of a memory location requires an extended-mode instruction occupying six machine cycles. A branch instruction to test when to stop decrementing takes another four machine cycles. On a 2-μs clock, the two instructions take 2 μs (6 + 4) or 20 μs. Let us say that we wish to have the computer delay in a *decrement-memory* loop for 1 ms and then leave the loop. The number of passes through the loop required is

$$n = \frac{1000\ \mu\text{s/delay}}{20\ \mu\text{s/pass}} = 50\ \text{passes/delay}$$

It will be necessary to preload the memory location with decimal 50 ($32), decrement, test for zero, and branch back to the decrement instruction until a zero result is obtained. List 4.1 shows the routine.

```
                   1000 * LIST 4-1      TIME-DELAY LOOP:
                   1010 *               50 LOOPS X 20 US = 1 MS DELAY.
                   1020 * PROGRAM SEGMENT. NOT TO RUN. 2 JUL '87 METZGER
                   1030 *-----------------------------------------------
                   1040             .OR  $0180     PROGRAM STARTING ADDRESS.
                   1050             .TA  $4180     FREE RAM AREA FOR ASSEMBLER.
0180- 86 32        1060 START       LDAA ##$32     INITIALIZE MEMORY FOR 50-COUNT
0182- B7 01 30     1070             STAA $0130      DECREMENT.
0185- 7A 01 30     1080 LOOP        DEC  $0130     BRANCH BACK AND DECREMENT
0188- 26 FB        1090             BNE  LOOP       FOR 50 LOOPS.
018A- 01           1100 OUT         NOP            YOU'RE DONE NOW.
                   1110             .EN
```

It is common to say that the program "falls through" the loop when the contents of the memory location reach 00, after 50 decrements. It may have been easier to decrement accumulator *A* itself rather than using a memory location (arbitrarily selected as $0130). However, decrementing a memory location places the decrementing counts on the data bus, whereas decrementing an accumulator keeps this activity inside the processor. If a logic analyzer is used to trace this routine, the former method makes a more interesting program readout.

REVIEW OF SECTIONS 4.3 AND 4.4

Answers appear at the end of Chapter 4.

11. Give two names for the 6802 register, which contains the flag bits sensed by branch instructions.

12. What is the status of the C flag after execution of

```
LDAA   #91
SUBA   #$83
```

13. Which of these instructions does *not* affect the C flag (use the programmer's reference to find out): ABA, ASRB, NEG, ORAB, TSTA?
14. Give the status of the N, Z, and C flags after the following program sequence is executed:

```
CLRA
CLRB
COMA
ABA
```

15. Give the mnemonic and hex op-code for the instruction that causes a branch if the N flag equals 1.
16. Will the N flag be *set* to 1 if the positive decimal value 209 is loaded into accumulator B?

Questions 17 through 20 refer to the following program:

```
        CLRA
BACK    INCA
        CMPA    #$9A
        BEQ     OUT
        JMP     BACK
OUT     NOP
```

17. Which instructions comprise the loop?
18. How long will it take to go from the CLRA instruction to the NOP instruction if the clock period is 2 μs?
19. If the CLRA instruction is at address $Ø129, what is the address labeled OUT?
20. What must be the operand of the BEQ instruction (in object code, hex form)?

4.5 FLOW CHARTS AND AN AUDIO TEST TONE

Section Overview

A pair of delay loops can be written to hold one of the processor's higher-order address lines *low* for 1 ms and then *high* for 1 ms, producing a square wave with a 2-ms period (or 500-Hz frequency) on that address line. This tone can be detected with a simple earpiece, and its presence permits the diagnosis that the processor and the memory chip containing the delay-loop program are functioning.

Programs consisting of more than one routine are easier to understand if a *flowchart* is drawn to show how the routines are related.

The most useful troubleshooting techniques are often those that require the least sophisticated test equipment. The diagnostic of Section 3.7 required a triggered-sweep oscilloscope. Now we are ready to write a diagnostic program that will check out the processor and its memory with just an earphone from a transistor radio or cassette player. The routine will produce a distinctive audio tone.

The test tone will be produced by a pair of delay loops, each similar to the one on page 98. Upon exiting the first loop the program will jump to the second, and upon exiting the second it will jump back to the first. Each loop will delay for about one 1 ms, so the pair will be executed at a rate given by

$$f = \frac{1}{T} = \frac{1}{2\text{ ms}} = 500\text{ Hz}$$

The loops will be positioned in RAM memory so that one loop keeps address line A6 *low* and the other keeps line A6 *high*. Line A6 will thus carry a 500-Hz square wave. It must be emphasized that this is a trick to obtain an audio test signal without using an output latch. Microcomputers do not normally produce output signals on their address lines. Figure 4.2 shows the memory map.

Flowcharting is a visual way of presenting a plan of a computer program. Boxes are used to represent program operations, and diamonds indicate decision-making or branching functions. Figure 4.3 shows a flowchart for the audio-test-tone program. Compare the flowchart with the program listing (List 4.2), which follows. After a few minutes' study you should see that the flowchart can be taken in *at a glance*, whereas the program listing must still be followed *line by line* to be deciphered. This ability to display the relationships between various parts of a program is the great advantage of flowcharts over program listings.

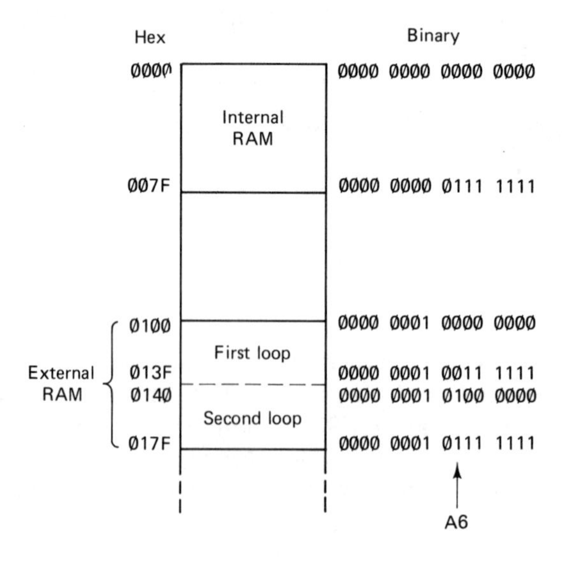

Figure 4.2 Address line A6 is *low* when the first half of the RAM memory is being accessed and *high* during accessing of the second half.

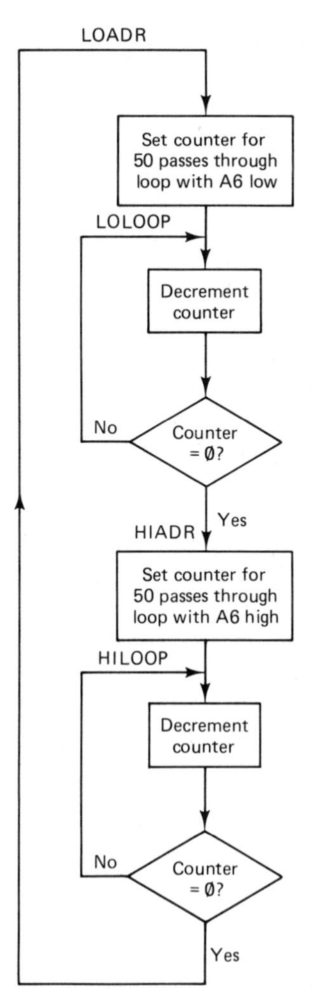

Figure 4.3 A flowchart facilitates visualization of the program as a whole.

Notice that the boxes and diamonds contain functional descriptions of machine operations but do not use specific instruction-code mnemonics. Thus a flowchart should be machine-independent, that is, applicable to a program written for a 6802, Z80, or other processor. Notice, however, that the *labels* on the flowchart are the same as the *labels* on the program listing. This permits interpreting the program listing in terms of the conceptual picture provided by the flowchart.

```
                1000 * LIST 4-2    TEST TONE GENERATOR FOR 6802, LINE A6
                1010 *             SWITCHES AT 500-HZ RATE.  FIG 3.7 OR
                1020 *             5.7 HARDWARE. RUNS, 2 JUL '87. METZGER
                1030 *
                1040        .OR  $0190  ORIGIN OF PROGRAM SEGMENT.
                1050        .TA  $4190  (TARGET AREA OF HOST COMPUTER)
0190- 86 32     1060 LOADR  LDAA #$32   INITIALIZE MEMORY FOR
0192- B7 01 B0  1070        STAA $01B0   50 PASSES THROUGH 10-CYCLE
0195- 7A 01 B0  1080 LOLOOP DEC  $01B0   DELAY LOOP; KEEPS LINE
0198- 26 FB     1090        BNE  LOLOOP  A6 LOW FOR 1 MS.
019A- 7E 01 D0  1100        JMP  HIADR
```

```
                   1110 *
                   1120              .OR  $01D0    JUMP PROGRAM TO NEW MEMORY AREA.
                   1130              .TA  $41D0    (ASSEMBLING COMPUTER RAM AREA)
01D0- B7 01 F0  1140 HIADR   STAA $01F0    IDENTICAL DELAY LOOP
01D3- 7A 01 F0  1150 HILOOP  DEC  $01F0     KEEPS LINE A6 HIGH
01D6- 26 FB     1160         BNE  HILOOP    FOR 1 MS, GIVING
01D8- 7E 01 90  1170         JMP  LOADR     500-HZ TONE.
                   1180 *
                   1190              .OR  $01FE    RESET VECTOR, REDUNDANT AT $FFFE.
                   1200              .TA  $41FE    (ASSEMBLER SAVES TO THIS ADR.)
01FE- 01 90     1210         .HS  0190     VALUE STORED AT RESET VECTOR.
                   1220              .EN           END PROGRAM.
```

This program may be entered and tested using the system of Figure 3.7 and the control unit of Figure 3.10. The only addition necessary is a high-impedance earphone, in series with a 1-kΩ resistor, from address line A6 (pin 15 of the 6802) to ground. An earphone is used rather than a speaker because it has a high impedance (typically 1 kΩ), whereas the low impedance of a speaker would short out the address signal.

Notice that, in repeating the loop program at a higher address, the operand of the BNE instruction did not have to be changed. This is because the relative addressing mode makes the object code position-independent. The operand of the STAA instruction did need to be changed because the addressing mode there was *extended*, not relative, and the code is not position-independent.

Take a few moments also to contemplate the fact that the processor is executing 100 passes through a two-instruction loop for *each cycle* of a 500-Hz audio waveform—and this is a relatively slow computer operating at half its rated speed.

4.6 NESTED DELAY LOOPS—A LAMP FLASHER

Section Overview

Microcomputers are so fast that counting loops with an 8-bit register produces maximum delays of only a few milliseconds. Nested delay loops place the first loop inside a second loop, so that an inner 256-loop routine can be required to run 256 times before the outer loop is exited. Two-level nesting gives delays on the order of a second. Four-level nesting can give delays of several hours. This section presents a two-level nest which causes an LED to flash at an observable rate.

A delay loop that operates by decrementing an 8-bit register has a rather limited range. Even if each pass through the loop takes 20 μs (which is a relatively long time for a computer), the maximum delay is 256 loops × 20 μs/loop, or about 5 ms. Longer delays can be produced by *nesting* a first delay loop inside a second, as illustrated in Figure 4.4.

This nesting causes the entire first delay routine to be executed up to 256 times, as given by the count initialized in the second RAM buffer. For a 20-μs basic inner-loop time, the maximum delay for two levels of nesting is (20 μs/loop × 256 loops/inner routine) (256 inner routines/outer routine) = 1.3 s.

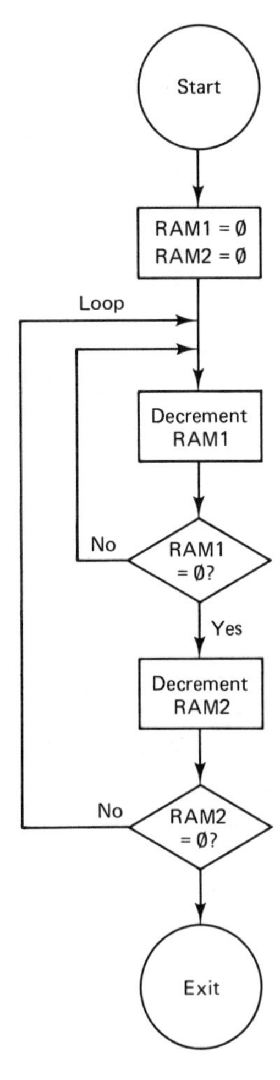

Figure 4.4 Nested loops with 8-bit registers give 256 × 256, or 65 536, passes through the inner loop.

List 4.3 uses two nested loops to flash an LED on the 6802 address line A6 at a 1.3-s rate. The system hardware is the same as that given in Figures 3.7 and 3.10, with the addition of a resistor-and-LED indicator, as shown in Figure 4.5.

```
                 1000 * LIST 4-3    LAMP-FLASHER TEST FOR 6802
                 1010 *  SWITCHES LINE A6 AT ABOUT 1.3 HZ
                 1015 * RUNS ON FIG 3.7 OR 5.7 HARDWARE, 1 JUL '87. METZGER
                 1020 *
                 1030        .OR  $0190    FIRST LOOP START.
                 1040        .TA  $4190    RAM STORAGE AREA IN HOST.
01B0-            1050 LORAM1 .EQ  $01B0    DEFINE RAM BUFFER VARIABLES.
01B1-            1060 LORAM2 .EQ  $01B1    "1" IS INNER LOOP,
01F0-            1070 HIRAM1 .EQ  $01F0    "2" IS OUTER LOOP.
01F1-            1080 HIRAM2 .EQ  $01F1    HI & LO REFER TO LEVEL OF A6.
                 1090 *
0190- 7A 01 B0   1100 LOADR   DEC  LORAM1   INNER LOOP COUNTS LORAM1
0193- 26 FB      1110         BNE  LOADR    FROM FF TO 00 IN 5 MS.
0195- 7A 01 B1   1120         DEC  LORAM2  OUTER LOOP COUNTS 256
0198- 26 F6      1130         BNE  LOADR    PASSES THROUGH INNER LOOP,
019A- 7E 01 D0   1140         JMP  HIADR    A6 HELD LOW, LAMP ON.
```

```
                  1145 *
                  1150          .OR   $01D0    SECOND LOOP START ADR.
                  1160          .TA   $41D0    TARGET ADDRESS IN HOST.
01D0- 7A 01 F0 1170 HIADR   DEC   HIRAM1   A6 HELD HIGH, LAMP
01D3- 26 FB     1180          BNE   HIADR     OFF, 1.3 SECONDS.
01D5- 7A 01 F1 1190          DEC   HIRAM2   INNER LOOP AND OUTER LOOP
01D8- 26 F6     1200          BNE   HIADR     LEAVE RAM BUFFERS AT 00,
01DA- 7E 01 90 1210          JMP   LOADR     FOR 256 COUNTS NEXT TIME.
                  1220          .OR   $01FE
                  1230          .TA   $41FE
01FE- 01 90     1240          .HS   0190     RESET VECTOR.
```

Some notes on this program might be worth observing:

1. The memory locations referenced by the program are given descriptive names at the head of the source-code list. The symbol .EQ, for EQUALS, is used for this. Motorola assemblers use the code RMB, for *Reserve Memory Byte*, to assign a series of addresses to variable names as they are entered in source code. The source code is then easier to read for sense because the operands are meaningful words or acronyms rather than abstract-looking addresses. This becomes increasingly important as longer programs are developed, in which various operands refer to dozens of memory locations.
2. No RAM initialization is included in the program because as each delay loop is exited, its RAM buffer is left at 00. The next entry to that loop will then decrement that RAM buffer to $FF, FE, and so on, back to 00. The first *lamp on-off* cycle will consist of random delays, depending on the power-up states of the RAM buffers, but subsequent delays will be 1.3 s *on*, 1.3 s *off.*
3. The comments deserve special attention. Notice these features, and try to emulate them.
   a. They refer to user-observable functions (lamp *on*, lamp *off*).
   b. They refer to groups of lines or whole routines, not to single lines.
   c. They *do not* simply translate the source-code mnemonics.
   d. Although the HIADR routine is essentially identical to the LOADR routine, the comments for HIADR do not simply repeat those for LOADR; they are used to add more information, which is applicable to both routines.

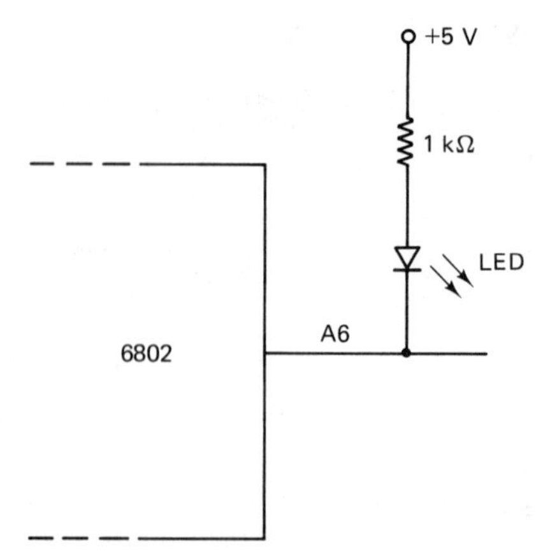

Figure 4.5 An address-line indicator for the test program of Section 4.6.

REVIEW OF SECTIONS 4.5 AND 4.6

21. What sort of instruction is implied by the diamond shape in a flowchart?
22. A microprocessor is caught in an endless loop, which spans all addresses from \$D3A9 to \$D429. Which binary address lines remain *high*, which remain *low*, and which change states during this loop? Express your answer as in the following example, which shows that A15 through A12 stay *high*, A11 through A8 stay *low*, and A7 through AØ switch:

    1111 ØØØØ XXXX XXXX

23. The following delay loop is to be used as one-half of a 440-Hz test-tone generator. What should be the hex operand of the LDAA, immediate, instruction if the *E*-clock period is 1 μs?

```
        LDAA   #XX
LOOP    DECA
        BEQ    OUT
        JMP    LOOP
```

24. The 6810 RAM has only 7 address lines, so binary 111 1111, or \$7F, is its highest address. How, then, can the TEST-TONE program (page 101) access a memory location such as \$Ø15A, which is higher than \$7F?
25. The LOADR routine on page 103 consists of five instructions. Why were only two of them counted in determining the loop time?
26. If the basic loop of a delay routine takes 20 μs, how many levels of nesting will be needed to produce a 24-hour delay, assuming all 8-bit registers?
27. Referring to the LAMP-FLASHER program of page 103, what is the 2s-complement value of \$F6, the operand of the second BNE instruction?
28. Continuing Question 27, algebraically add the operand value to the address of the next instruction in the program sequence. What is the resulting hex address and the name given to it?
29. An 8-bit register contains data \$ØØ. What does it contain after being decremented once?
30. In writing test programs designed to switch address line A6, what is the maximum number of bytes that can be allotted to each half of the program?

4.7 VARIABLE LOOPS—EMERGENCY SIRENS

Section Overview

This section presents two programs, which produce two different *siren* sounds when an earpiece is connected to address line A6. The purpose is to demonstrate an important feature of microcomputer systems: System function can be changed solely by changing software. No hardware changes

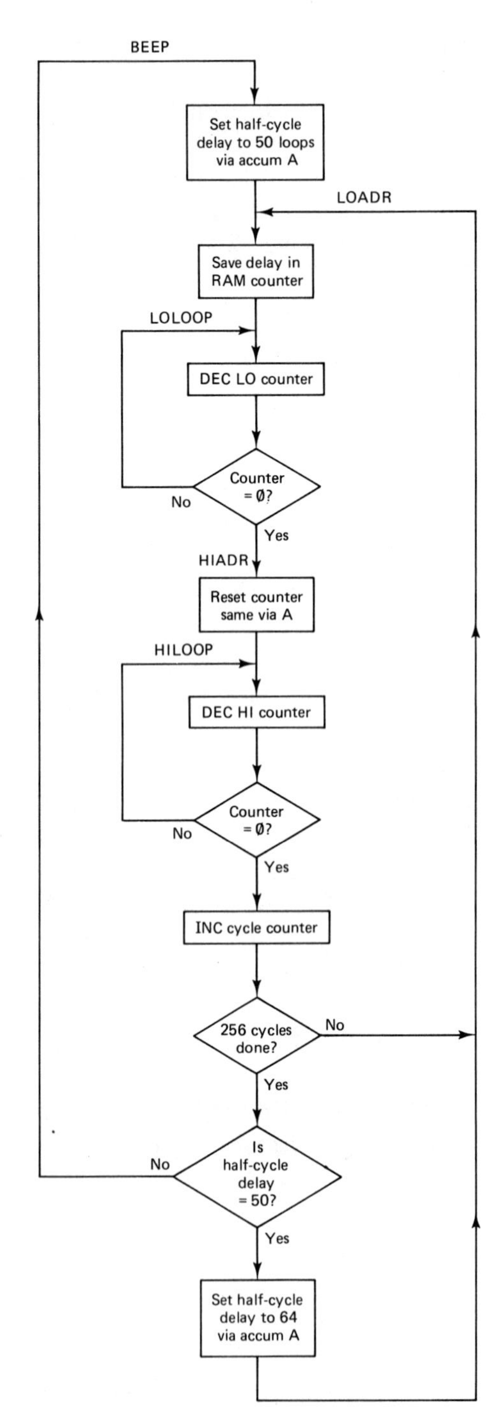

Figure 4.6 Flowchart for an Ambulance Siren program. The tone is switched after 256 cycles at the previous tone.

are necessary. Again, note that contriving to generate output signals on an address line is a trick to keep the hardware simple. Real microcomputers use output latches, which are the subject of Chapter 5.

Microprocessor control systems have come in like a tidal wave over the last decade. Few areas have been left untouched—one almost can't buy a toaster that doesn't have a microprocessor in it. Part of this phenomenon may stem from sales and promotional considerations, but often there is a solid technical reason for using a microprocessor to do a job that could have been done just as well by a few analog or TTL chips. That reason is usually *reprogrammability*—the ability to make a functional change in a system without any changes in the hardware, wiring, schematic diagrams, or manufacturing process. Production managers (who may have seen thousands of circuit boards scrapped because of a simple wiring-change requirement) are *very* interested in systems that can be modified by software changes instead of hardware changes.

This section presents an example of how an ambulance siren (beep-boop-beep-boop) can be changed to a police siren (the familiar "whooping" sound) simply by making program changes. We will be cheating a little, because we will be stealing the audio output signal from address line A6. (Real microcomputer systems use special output devices. We will meet these in the next chapter.) Also, we will store the operating program in RAM for ease of data entry and modification in the lab. Real-world operating programs for small systems are kept in ROM so they will not be lost when the power is shut down.

The Ambulance Siren program of List 4.4 is similar to the tone generator given on page 101. The delay per audio cycle is changed from $32 loops to $40 loops via accumulator *B* after each 256 cycles of the tone. A flowchart for this program appears in Figure 4.6. The hardware is, again, that of Figure 3.7.

```
                1000 * LIST 4-4     AMBULANCE SIREN: TWO TONES ALTERNATE.
                1005 * RUNS ON FIG 3.7 OR 5.7 HARDWARE; 1 JUL '87. METZGER
                1010 *
                1020         .OR  $0190
                1030         .TA  $4190
01B0-           1040 LODELY  .EQ  $01B0     DELAY-LOOPS RAM BUFFER WITH A6 LO.
01F0-           1050 HIDELY  .EQ  $01F0     RAM BUFFER FOR DELAY WITH A6 HI.
01F2-           1060 CYCLCT  .EQ  $01F2     CYCLE COUNTER (256 CY AT EACH TONE)
                1070 *
0190- 86 32     1080 BEEP    LDAA #$32      SET 50 LOOPS FOR HIGH TONE.
0192- B7 01 B0  1090 LOADR   STAA LODELY    COUNT DOWN TO 00 IN RAM
0195- 7A 01 B0  1100 LOLOOP  DEC  LODELY     ADR WITH A6 LOW.
0198- 26 FB     1110         BNE  LOLOOP    REMAIN IN LOOP ABOUT 1 MS.
019A- 20 34     1120         BRA  HIADR     GO TO LOOP WITH A6 HIGH.
                1125 *
                1130         .OR  $01D0
                1140         .TA  $41D0
01D0- B7 01 F0  1150 HIADR   STAA HIDELY    AUDIO TONE 400 - 500 HZ.
01D3- 7A 01 F0  1160 HILOOP  DEC  HIDELY     COUNTING DOWN RAM $01F0
01D6- 26 FB     1170         BNE  HILOOP     AT ADR $01DX KEEPS A6 HI.
                1180 *
01D8- 7C 01 F2  1190         INC  CYCLCT    COUNT OFF 256 CYCLES
01DB- 26 B5     1200         BNE  LOADR      OF A6 HIGH (OR LOW).
01DD- 81 32     1210         CMPA #$32      IF DELAY IS $32 LOOPS,
01DF- 26 AF     1220         BNE  BEEP       MAKE IT $40 LOOPS.
```

```
01E1- 86 40     1230          LDAA #$40
01E3- 20 AD     1240          BRA   LOADR     MAKE 256 MORE CYCLES.
                1250 *
                1260          .OR   $01FE
                1270          .TA   $41FE
01FE- 01 90     1280          .HS   0190      RESET VECTOR
```

Note that we have made the program position-independent by substituting the *branch always* instruction for the *jump* instruction. Also notice the introduction of the *compare* instruction CMPA. This instruction sets the Z, C, and N flags as if a subtraction (*accumulator* − *operand*) had been done, but the contents of the accumulator are not actually changed. This instruction allows us to test A for a certain value and branch to one routine or another depending on the contents of A.

The Police Siren program is also based on the *test-tone generator* of page 101. However, after each cycle is completed, the delay counter is initialized to a value 1 less than the previous value. Thus the frequency keeps increasing as the delay keeps decreasing from a maximum of $98 loops to a minimum of $20 loops. When the minimum delay is reached the BEQ instruction sends the program back to START, with the delay-per-half-cycle again set to produce the lowest frequency. A total of 120 cycles of increasing frequency are made before each restart.

Figure 4.7 gives a flowchart for the Police Siren program. The flowchart has been simplified by omitting the details of the *high* and *low* delay loops. This is common practice where such details are likely to be already familiar to the reader.

```
                 1000 * LIST 4-5      POLICE SIREN: TONE SWEEPS UP
                 1010 *               274 TO 1302 HZ IN 0.26 SEC.
                 1015 * RUNS ON FIG 3.7 OR 5.7 HARDWARE, 1 JUL '87. METZGER
                 1020 *
                 1030          .OR   $0190    FINAL SYSTEM START ADDRESS.
                 1040          .TA   $4190    ASSEMBLING COMPUTER STORAGE.
                 1050 *
0190- 86 98      1060 START    LDAA #$98      START WITH 1/2-CYCLE DELAY
0192- 16         1070 LOADR    TAB             OF 152 LOOPS X 12 US PER LOOP,
0193- 5A         1080 LOLOOP   DECB            OR 274 HZ.
0194- 26 FD      1090          BNE   LOLOOP   COUNT LOOPS IN ACCUM B,
0196- 7E 01 D0   1100          JMP   HIADR    KEEP A6 LOW.
                 1110 *
                 1120          .OR   $01D0    THIS ADDRESS RANGE KEEPS
                 1130          .TA   $41D0     ADDRESS LINE A6 HIGH.
01D0- 16         1140 HIADR    TAB            LOOP FOR A6 HIGH SAME AS
01D1- 5A         1150 HILOOP   DECB            LOLOOP; MAKES SECOND
01D2- 26 FD      1160          BNE   HILOOP    HALF OF AUDIO CYCLE.
                 1170 *
01D4- 4A         1180          DECA           SPEED UP EACH CYCLE
01D5- 81 20      1190          CMPA #$20       UNTIL HALF-CYCLE DELAY IS
01D7- 27 B7      1200          BEQ   START       32 LOOPS X 12 US OR 1302 HZ.
01D9- 7E 01 92   1205          JMP   LOADR    MAKE ANOTHER LO/HI CYCLE.
                 1210 *
                 1220 * TIME PER SWEEP = NO. CYCLES X AVG CYCLE TIME
                 1230 *    = (152 - 32) X (3.6 MS + .77 MS)/2 = 0.26 SEC.
                 1240 *
                 1250          .OR   $01FE
                 1260          .TA   $41FE
01FE- 01 90      1270          .HS   0190
                 1280          .EN
```

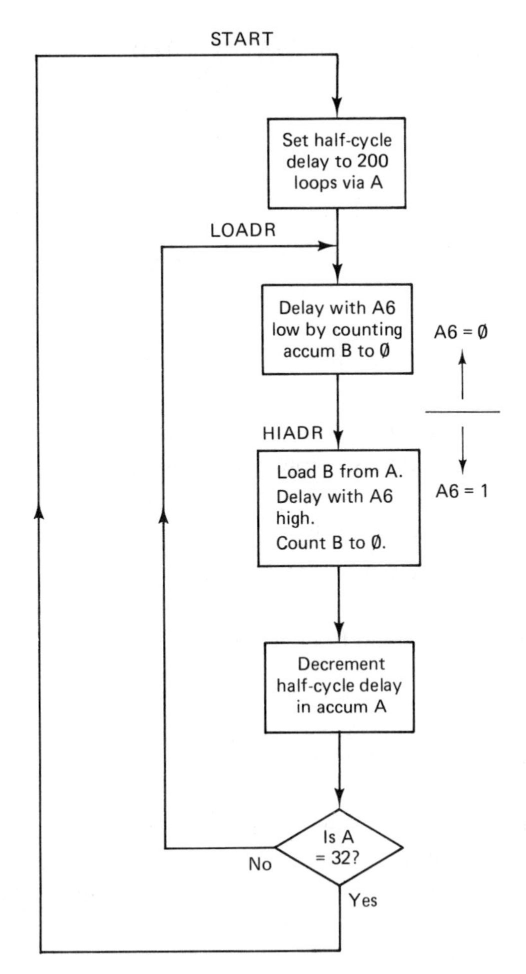

Figure 4.7 The Police Siren program provides an increasing pitch by decreasing the delay after each cycle.

Answers to Chapter 4 Review Questions

1. $FC2F **2.** $FC4B **3.** $434B **4.** $Ø16B
5. Branch instructions only. **6.** −73 **7.** 183
8. −32768 **9.** $E3 **10.** $BFC5
11. Condition-code register (CCR), processor status register (*P* or SR)
12. $C = 1$
13. ORAB **14.** $N = 1, Z = Ø, C = Ø$ **15.** BMI, $2B **16.** Yes
17. INCA, CMPA, BEQ, and JMP
18. 3.39 ms **19.** $Ø132 **20.** $Ø3 **21.** A branch instruction.
22. 11Ø1 ØXXX XXXX XXXX
23. $7E **24.** $Ø15A is binary ØØØØ ØØØ1 Ø1Ø1 1Ø1Ø. The binary 1 from line A8 is used for chip select of the 6810, not for address select within the chip.
25. The last three are executed once for every 256 executions of the first two instructions. **26.** Five (four misses it by about ½%).
27. −10 **28.** $Ø19Ø, LOADR **29.** $FF **30.** 2^6, or 64 bytes

CHAPTER 4 QUESTIONS AND PROBLEMS

Digits after the decimal point refer to section numbers in Chapter 4.

Basic Level

1.1 The relative-mode instruction $20 (branch always) appears at address $015B, followed by operand data $07. What is the address of the next instruction to be executed?

2.1 In Figure 4.1, what operand to the relative-mode instruction would branch the program counter to address $10A0?

3.2 Is $A7 a positive or negative number?

4.2 What is the decimal value of the 2s complement number $C9?

5.2 What hex operand to the relative-mode branch-always op-code will halt the processor by returning it endlessly to that same branch op-code?

6.3 Give three letter names variously used for the register that holds the 6802's flag bits.

7.3 What are the states of the C, Z, and N flags after the execution of the following instructions?

```
LDAA    #$FF
ADDA    #1
```

8.3 Continuing Question 7, assume that the X register contains $1234 and an INX instruction is executed. What are the new states of C, Z, and N?

9.3 In Figure 4.1, what is the address of the instruction that will be executed after the relative instruction if the branch is not taken?

10.4 In the program of List 4.1, what data should be stored at address $0130 to maximize the delay time, and what will this time be on a 1-MHz clock?

11.4 The Delay routine of List 4.1 is to be lengthened by the insertion of three NOP instructions between the DEC and BNE LOOP instructions. What is the new maximum delay time and the new hex operand for the BNE instruction?

12.5 The Test-tone program of List 4.2 is to be run on a system with a 0.500-MHz *E* clock. Calculate the exact audio frequency that it generates and compare this with the approximate value of 500 Hz given in the text. Don't forget the extra cycles for the JMP and STAA instructions.

13.5 In flowcharting, what geometric shapes are associated with (a) decision making and (b) program operations?

14.6 With reference to the lamp-flasher program of List 4.3:

a. What is the 2's complement value of operand $F6 (line 1130)?

b. What hex address does the processor access if the branch on line 1130 is *not* taken?

c. What address does the processor reach if this branch *is* taken?

d. Subtract the hex numbers of answers (b) and (c). Compare to the value obtained in (a).

Advanced Level

15.1 Use the programmer's reference card to identify the two 6802 instruction mnemonics that begin with the letter *B* but are *not* branch instructions. Use the index to find out what these instructions do.

16.1 Explain what is meant by *position-independent* programming, and tell how it is achieved.

17.2 Form the binary 2's complements of the decimal numbers −4 and −8. Add these two numbers in binary. Compare the binary result with the 2's complement of −12. Make a statement on the significance of this exercise.

18.3 Check the 6802 programmer's reference and then explain under what conditions the INCA instruction will set the C, Z, and/or N flags.

19.3 Write a program to add 10 to the binary data at address $Ø145 and store the result back to address $Ø145 if the result does not exceed decimal 255.

20.4 Write a time-delay routine that decrements the X register instead of an 8-bit memory location. Include comments and object code. What is the maximum delay obtainable if the frequency of *E* is 1 MHz?

21.5 For the test-tone program of List 4.2, make a list of the hex data on the address bus for each of the 10 cycles of the LOLOOP, and show that A6 is *low* on each of them.

22.5 Why should flowcharts generally not contain references to specific machine registers and instructions?

23.5 What technique can be used to correlate a program's source code to the flowchart so that points in one can be quickly located in the other?

24.6 Calculate the exact time of the nested LOADR loop in the Lamp-flasher program of List 4.3. Include the two instructions that are executed 256 times and the one that is executed once. By what percent does the exact time differ from an estimate obtained by calculating $256^2 \times$ *inner-loop instruction time*? Use a 0.500-MHz clock rate.

25.6 Write a three-level nested loop patterned after the LOADR loop of List 4.3. Include comments and object code. Use brackets to identify which instructions comprise the inner, middle, and outer loops. Estimate the maximum delay obtainable on a 0.500-MHz clock.

26.6 In List 4.3, why did we bother to equate four names with the four RAM addresses? Wouldn't it have been faster to simply write the instructions with numerical addresses as the operands—for example, DEC $Ø1BØ?

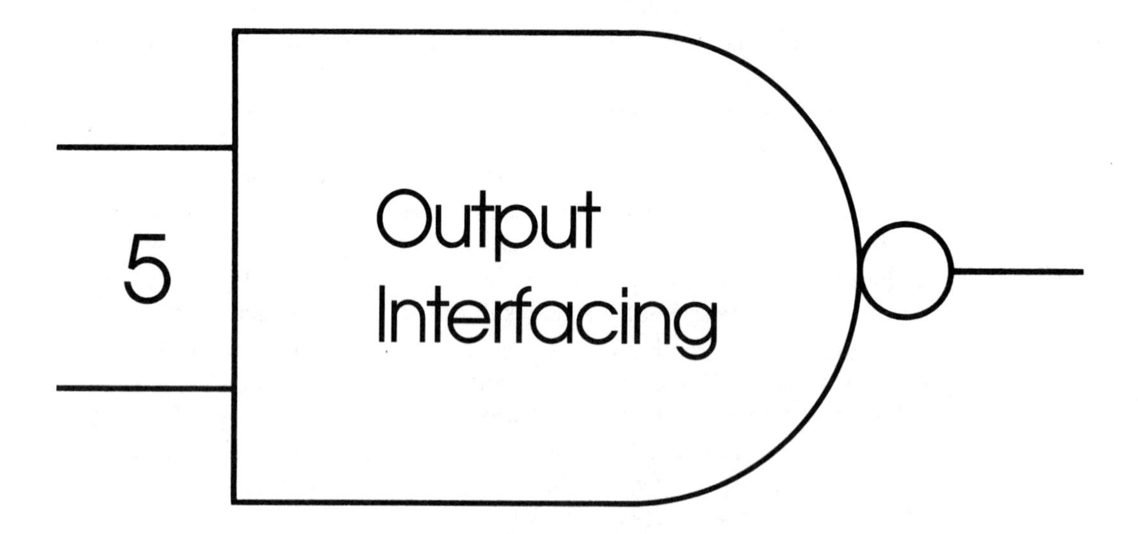

5.1 OUTPUT LATCHES

Section Overview

An output latch picks up data that the processor has applied to the data bus and holds that data at the output until the processor directs another byte of data to it.

The processor directs data to a latch by writing to a memory address, which activates the latch's chip-select logic. Latches may drive display LEDs, relays, and stepper motors, to name a few applications.

During the course of executing a program, millions of bits flit by on the data bus as the processor uses it to read instructions and operands, and to write data back to memory. A very few of these bits may be output data, destined to activate some output device, such as an LED display, a printer, or a control motor. The function of an output latch is to catch and hold these output data bits, delivering them to the output until another set of output bits is received.

Memory-mapped I/O: Intel and Zilog microprocessors have special input and output instructions, which activate a separate *I/O-enable* pin on the chip. This pin then causes the output latch to capture the data on the bus at that instant and hold it at the output. Motorola and Rockwell processors use memory-mapped I/O exclusively.

This means that the output latch is assigned an address in memory, and data is output by simply writing to that address using the STAA or STAB instructions. There are no special output instructions and no special processor pins to enable I/O devices.

An output latch is simply a set of type-*D* flip-flops, with the *D* inputs connected to the processor data bus and the *G* inputs connected to the output chip-enable logic. (See Section Ø.5 for a review of *D* flip-flops.) The 74LS75 provides a half-byte (4-bit-wide) set of flip-flops in a 16-pin package. The 74LS373 provides a full 8 bits in a 20-pin package. These chips also provide a buffering function for the processor. They can drive several standard TTL loads at their output sides but present only about one-fourth a standard TTL load at their input (data bus) sides.

5.2 DEVICE-SELECT LOGIC FOR WRITABLE DEVICES

Section Overview

The addresses to which a memory or I/O device responds are determined by device-select logic. This consists of AND gates driven by the processor's higher-order address lines or inversions thereof. The 6802's *E*-clock should be included in the ANDing for RAMs and latches to keep the devices from being called up accidentally during the first half of the cycle, when address and data have not settled.

In our simple experimental microcomputer system, we have been using address lines AØ through A6 to select memory locations in the 6810 RAM and have been requiring A8 = 1 for 6810 chip select. Leaving lines A9 through A15 uncommitted, there is one combination of A7 and A8 available for I/O device select. Figure 5.1 shows the logic and memory map.

A15-A9	A8	A7	A6-AØ
0000 000	Ø	Ø	Address select
Uncommitted	Ø	1	Available
Uncommitted	1	Ø	Address select
Uncommitted	1	1	Address select

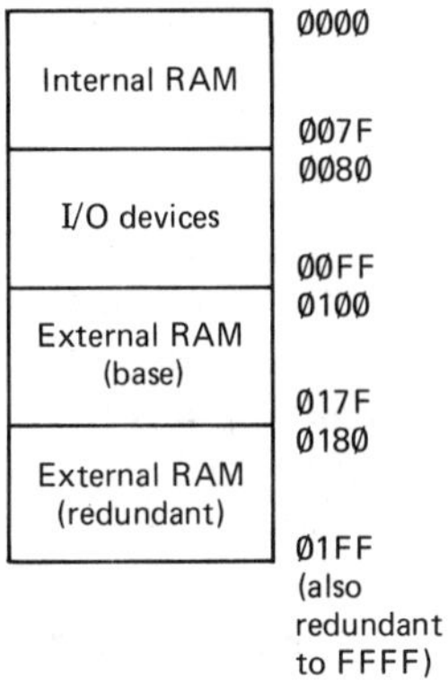

Memory map	Address range
Internal RAM	ØØØØ – ØØ7F
I/O devices	ØØ8Ø – ØØFF
External RAM (base)	Ø1ØØ – Ø17F
External RAM (redundant)	Ø18Ø – Ø1FF (also redundant to FFFF)

A8 = 1 selects external RAM.

A7 AND A8 = Ø selects internal RAM (A15-A9 also Ø).

A7 AND $\overline{A8}$ is available to select another device.

Figure 5.1 Decoding logic and memory map for the experimental microcomputer of Figure 3.7. An output latch can be called up at address $ØØ8Ø.

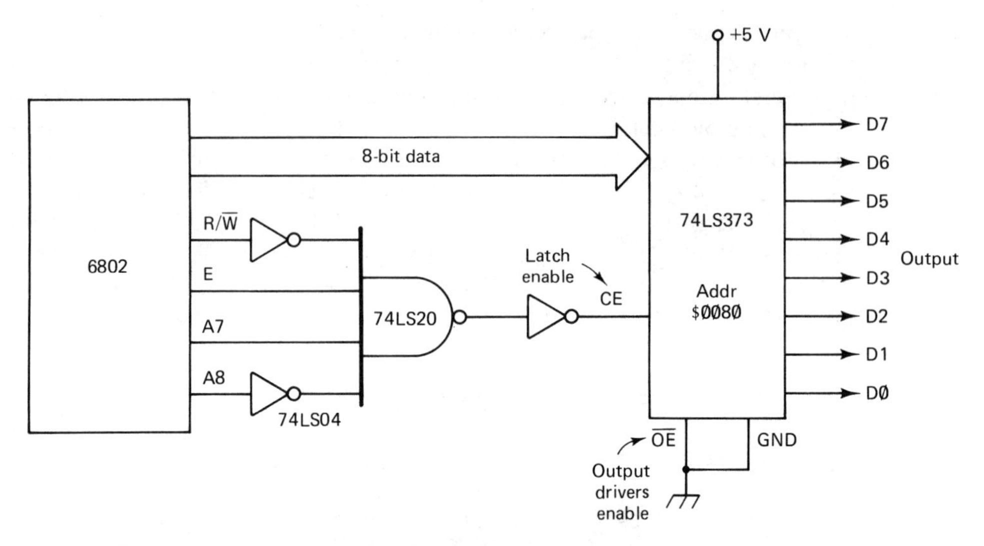

Figure 5.2 An output latch samples the data bus when its address is written to and holds that data at the output until it is written to again.

The output latch can be selected at any address between \$0080 and \$00FF. Since there is only one latch, it is redundant at 128 places, even disregarding the unused addresses above \$01FF produced by the address lines A15 through A9. This is wasteful of memory space, and is generally frowned upon in commercial applications. We will wink at the practice in this experimental system since it keeps the hardware simple. Figure 5.2 shows the additional connections that will be needed to add an 8-bit output latch to our experimental system of Fig. 3.7.

The R/$\overline{W}$ line is inverted and ANDed in the latch-select logic to ensure that an attempt to read the latch address does not inadvertently write to the latch. VMA is not included in the AND logic because the condition VMA = 0 AND R/$\overline{W}$ = 0 should never be encountered.[1] The cycle-by-cycle summary (Appendix B) shows that only the TST instruction causes this condition, and TST involves a read of the selected address. An attempt to read a latch will produce invalid data, since the data-bus side of the latch is an input-only connection. Thus TST \$0080 is invalid.

The *E*-clock should be ANDed into the select logic for any writable device to prevent false writes during the first part of the clock cycle, when address lines are in transition. It is possible that during this time a valid latch address could inadvertently be called up if R/$\overline{W}$ falls quickly while the address lines are still in transition. The data on the bus at this time might then be falsely written to the latch. The *E* signal does not rise until the address lines and R/$\overline{W}$ have had time to stabilize, so false writes are prevented by ANDing *E* into the latch select. Figure 5.3 illustrates the write-cycle timing relationships.

ROMs as address decoders: In large systems it is often desirable to divide the memory map among different devices in unequal amounts and at other than "round-number" boundaries. This is easily accomplished by using a ROM as a

[1]The fact that we ran out of inputs on the 74LS20 may also have had something to do with it.

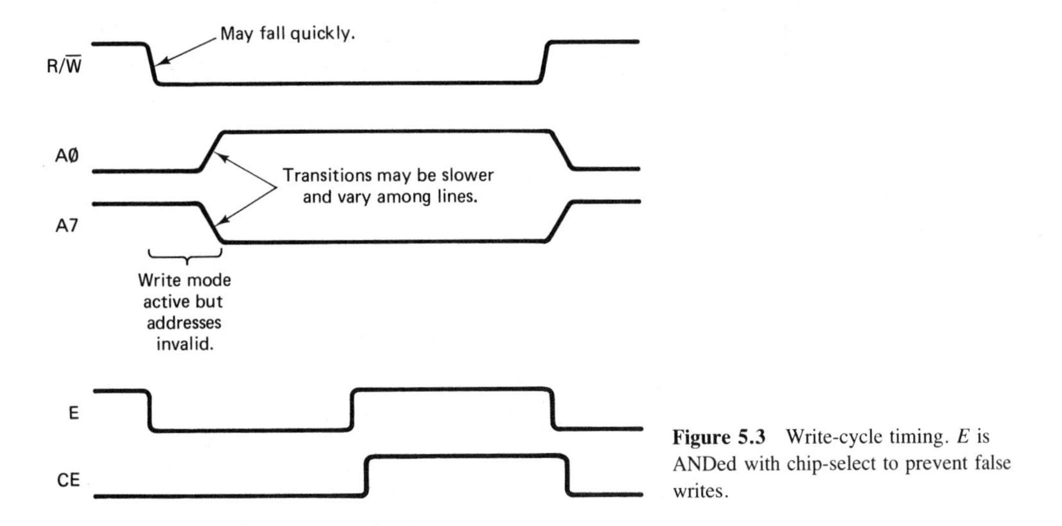

Figure 5.3 Write-cycle timing. *E* is ANDed with chip-select to prevent false writes.

device decoder. The binary data programmed into each ROM word always causes just 1 of the output lines to activate an I/O or memory device. Assuming all high-active chip-enable lines and a 4-bit ROM word, each word would consist of three binary Øs and one binary 1. The device selected at each address then depends entirely on the programming of the ROM. Figure 5.4 shows an example in which a 74S287 field-programmable ROM decodes a 16k segment of the memory map among three memory chips and one I/O chip. Note that each ROM is given the space required by its program—not necessarily the maximum addressable space it contains. EPROMs

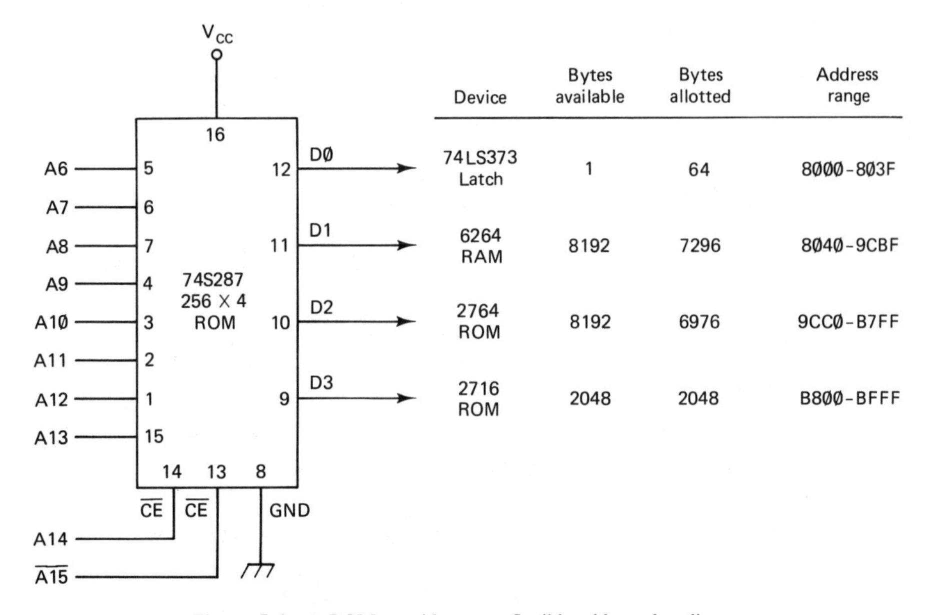

Device	Bytes available	Bytes allotted	Address range
74LS373 Latch	1	64	8ØØØ-8Ø3F
6264 RAM	8192	7296	8Ø4Ø-9CBF
2764 ROM	8192	6976	9CCØ-B7FF
2716 ROM	2048	2048	B8ØØ-BFFF

Figure 5.4 A ROM provides more-flexible address decoding.

are not generally used as address decoders because they are too slow. EPROM response times are on the order of 250 ns, compared to 35 ns for FPROMs.

REVIEW OF SECTIONS 5.1 AND 5.2

Answers appear at the end of Chapter 5.

1. If output devices are assigned an address in memory and are accessed by simply writing to that address, the I/O scheme is called ________.
2. To pick out and hold the data words destined for outputting we use a ________.

For Questions 3 through 6: A microprocessor system uses three 2K ROMs, a 2K RAM, and two output latches and has 16 address lines.

3. What is the minimum number of memory-cell-select lines required?
4. What is the minimum number of chip-select address lines required?
5. How many redundant address ranges does each memory chip have?
6. How many redundant addresses are there for each latch?
7. In a 6802 system, how can false writes by prevented while the address lines are in transition?
8. Why is only partial decoding sometimes used for I/O devices?
9. An 8K RAM is decoded with the logic

$$CE = \overline{A15} \cdot A14 \cdot \overline{A13} \cdot E$$

   To what range of addresses does it respond?

10. How can odd and differing amounts of address space be allocated to a number of memory and I/O devices?

5.3 BUS TIMING CONSIDERATIONS

Section Overview

During each cycle that a memory address is read, enough time must be available for

1. **The processor to apply the address signals to the bus,**
2. **The chip-select logic to access the memory,**
3. **The memory to apply its data to the data bus, and**
4. **The processor to read the data from the data bus.**

The sum of these times must not exceed the clock-cycle time.

A microprocessor uses the data bus to read data from or write data to external devices. In each case it is essential that the system clock allow enough time for the data transfer to be completed before dismissing the current address selected and going on to the next address. On the other hand, it is often desirable to run the clock as fast as possible to maximize the processing power of the system. Bus-timing calculations are necessary to determine the maximum permissible clock rate.

Read-cycle timing is the simpler of the two. Let us say that we are addressing a ROM. The timing problem, simply stated, is this: Once a new address is applied by the processor, the ROM requires a certain response time to get its data asserted on the data bus. (This is the data-access time, called t_{DDR}—time, data-delay read.) The processor then requires a certain amount of time to read this data (called t_{DSR} —time, data-setup read.) The clock cycle must be long enough to maintain each address for this total time. Figure 5.5 shows read-cycle timing for the 6802.

Note that the processor is guaranteed to hold the old address valid for about 20 ns after the fall of $E(t_{AH})$, but that under worst-case capacitive bus loading the new address is not guaranteed to be valid until 250 ns after the fall of E (t_{AD}). Thus the usable cycle time is decreased by the possible address-delay time minus the address-hold time, or 230 ns under worst-case conditions. Also be aware that EPROM delay times are much longer than LS TTL delay times, so the use of EPROMs as address decoders may introduce timing problems.

In some simple systems E is ANDed into all chip-select logic, even though it is necessary to do this only for writable devices. In such cases the chip-select timing starts from the rise of E, rather than from the address transition. This practice places unnecessarily stringent requirements on the EPROMs, which are usually slower than

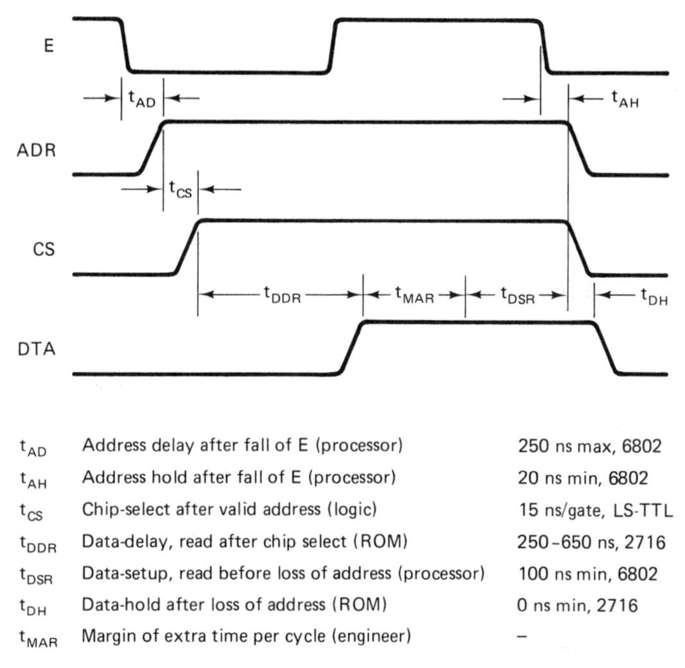

Figure 5.5 Read-cycle timing for a 6802 system. The memory device must apply the data with enough time left for the processor to read it before the cycle is over.

RAMs and latches, and it should be avoided in the interests of system performance and economy. The timing relationship, referring to Figure 5.5, is then:

$$\tfrac{1}{2}t_E = t_{CS} + t_{DDR} + t_{MAR} + t_{DSR} - t_{AH}$$

Example 5.1

Chip-select logic in a 6802 system decodes from *address*, R/$\overline{\text{W}}$, and VMA lines and uses three LS TTL parts in cascade. A 450-ns ROM is used. What is the maximum clock speed based on ROM-read considerations if a 150-ns margin of safety is required?

Solution

$$\begin{aligned} t_E &= t_{AD} + t_{CS} + t_{DDR} + t_{MAR} + t_{DSR} - t_{AH} \\ &= 350 + 3(15) + 450 + 150 + 100 - 20 \\ &= 1075 \text{ ns} \end{aligned}$$

$$f_E = \frac{1}{t_E} = \frac{1}{1075} \approx 0.93 \text{ MHz}$$

5.4 WRITE-CYCLE TIMING

Section Overview

During each cycle that a memory or latch is written to, the *E* clock must remain *high* long enough for

1. **The processor to write data to the bus, and**
2. **The memory or latch to read the data from the bus.**

The *E* signal generally remains *low* for an equal time. The minimum *E* period determines the maximum clock frequency.

Calculated design limits should be adhered to even in the face of experimental evidence that the system functions perfectly if they are exceeded because of the possibility of changes in the production process of the components.

On a write cycle the processor applies data to the data bus after a delay time t_{DDW} following the *rise* of the *E* clock. It is essential that there be enough time remaining in the cycle for the external device to read this data. Figure 5.6 shows the timing relationships. Chip-select time is not a factor, unless it becomes longer than the processor's write-data delay time. This situation would be likely only if a slow ROM were used for decoding. Notice that the 20-ns guaranteed address-hold time at the end of the cycle is usable on the write cycle. The somewhat longer data-hold time at the end of the write cycle should not be used, because there is no guarantee that the address will be stable during this entire time.

In the first half of the clock cycle, after the address lines have selected the

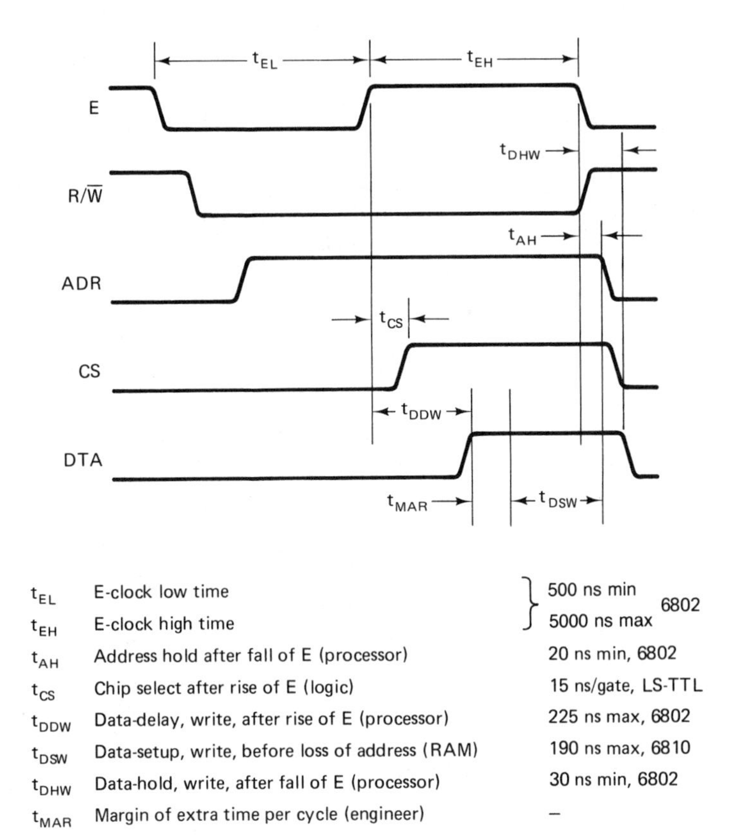

t_{EL}	E-clock low time	500 ns min	6802
t_{EH}	E-clock high time	5000 ns max	
t_{AH}	Address hold after fall of E (processor)	20 ns min, 6802	
t_{CS}	Chip select after rise of E (logic)	15 ns/gate, LS-TTL	
t_{DDW}	Data-delay, write, after rise of E (processor)	225 ns max, 6802	
t_{DSW}	Data-setup, write, before loss of address (RAM)	190 ns max, 6810	
t_{DHW}	Data-hold, write, after fall of E (processor)	30 ns min, 6802	
t_{MAR}	Margin of extra time per cycle (engineer)	–	

Figure 5.6 Write-cycle timing for a 6802 system. The processor must apply the data soon enough to give the memory device time to read it (t_{DSW}) before the cycle is over.

device to be written to but before the second half of the cycle, when the processor has not yet asserted the data being written on the bus, no device is applying data to the bus. During this time the data lines may hold their previous data due to stray capacitance. If there are TTL inputs on the bus they will tend to pull the voltage levels up. Invalid logic levels (not valid Ø or valid 1 level) are common (and harmless) during this time.

Worst-case timing: Repeated measurements of physical microprocessor systems may show that the actual address delay times (t_{AD}) and write-data delay times (t_{DDW}) are always much shorter than the "maximum" values specified. Indeed, it may be observed that the calculated maximum clock speed of 1.0 MHz can be pushed to 1.3 MHz with the system still functioning perfectly. One might even take the trouble to test a number of chips from a number of different vendors at the lowest specified supply voltages and widest specified extremes of hot and cold temperatures and find that the system continues to function at 1.3 MHz.

Nevertheless, it is a *very bad* idea to install a system that runs beyond the limits calculated from published specifications. Manufacturers are always striving to reduce the cost of producing the items they sell. Indeed, larger companies have whole engineering departments devoted to this effort. In the future a new production process may be developed that increases address and data delay times, but cuts production cost by 20%. As long as the new delay times do not exceed the published maximums, the manufacturers would certainly implement the new process. Very soon chips made under the old process (which all had shorter delay times) would be unavailable, and systems that relied on these better-than-spec components would be impossible to maintain for lack of spare parts.

Example 5.2

Chip-select logic for a 6810 RAM in a 6802 system decodes from E, R/$\overline{\text{W}}$ VMA, and address lines, using three 74LS TTL parts in cascade. What is the maximum clock speed on write cycles if a 100-ns safety margin is included?

Solution

$$
\begin{aligned}
t_{EH} &= t_{DDW} + t_{MAR} + t_{DSW} - t_{AH} \\
&= 225 + 100 + 190 - 20 \\
&= 495 \text{ ns} \\
t_E &= t_{EH} + t_{EL} \approx 990 \text{ ns} \\
f_E &= \frac{1}{t_E} = \frac{1}{990 \text{ ns}} = 1.01 \text{ MHz}
\end{aligned}
$$

REVIEW OF SECTIONS 5.3 AND 5.4

Answers appear at the end of Chapter 5.

11. Which device determines the read-cycle data-setup time?
12. A 6802-based system uses 250-ns ROMs. Address decoding is handled by another 250-ns ROM, with some inputs fed through one LS TTL gate. What is the safety time margin per cycle if the clock speed is 0.8 MHz?
13. The 6802 system in Example 5.1 is changed to AND the E signal into the chip-select logic. What is the new maximum clock speed?
14. Name a disadvantage of EPROMs as address decoders in place of logic elements.
15. Why is the time required for the chip-select logic to access the device being written to usually not a factor in write-cycle timing?
16. What is the maximum clock speed on read cycles for a 68B02 ($t_{AD} \leqslant 160$ ns, $t_{DSR} \leqslant 60$ ns, $t_{AH} \geqslant 20$ ns) with a 250-ns ROM decoded by two 74LS TTL gates cascaded? Allow a 50-ns time margin.
17. Invalid logic levels on the data bus are normal during what time period?
18. A 68B02 operates at 2 MHz. It has $t_{DDW} \leqslant 160$ ns and $t_{AH} \geqslant 20$ ns, and it feeds a RAM with $t_{DSW} \leqslant 90$ ns. What time margin remains on each write cycle?
19. In Example 5.2 the system is altered so that chip select is done with a 250-ns EPROM fed through a single LS TTL gate at some inputs. What is the new maximum clock speed?

20. Which should be used in determining maximum system speed: measured time delays or manufacturer's specified worst-case time delays?

5.5 DECIMAL ARITHMETIC FOR MICROPROCESSORS

Section Overview

In this section we leave our considerations of hardware to discuss a software topic that will be used shortly in a computer-game project. We will explain how the DAA (Decimal Adjust A) instruction is used to permit the 6802 to add two-digit decimal numbers in its *A* accumulator. You may wish to review Section 2.3, which introduced the binary-coded decimal (BCD) system.

In many microprocessors an 8-bit data word can be used to represent pure binary data (0 through 255) or binary-coded decimal data (0 through 99). In BDC each 4-bit nibble represents one decimal digit. Only the nibbles corresponding to valid decimal digits (0 through 9) are allowed. Binary nibbles corresponding to the hex digits A through F are considered invalid.

Perhaps it will be worthwhile to repeat at this point that the interpretation of any particular byte of data is a function of the program sequence and structure and is not discernable from any characteristic of the data itself. Thus if the binary pattern stored at address $F850 is

0010 0111

we have no way of knowing how to interpret that data, unless we analyze the program of which it is a part to see how the processor treats it when the program calls up address $F850. Some of the possibilities are

0010 0111 = $27 = op-code BEQ
0010 0111 = $27 = data (2 × 16) + 7 = 39
0010 0111 = 27 = BCD data 27
0010 0111 = address, least-significant byte = $XX27
0010 0111 = address, most-significant byte = $27XX
0010 0111 = user's binary code pulling 4 output lines high, 4 output lines low.

BCD addition is facilitated in the 6802 processor in the *A* accumulator by the Decimal Adjust A (DAA) instruction. The effect of this instruction is to add 6 (binary 0110) to either nibble of accumulator A if a preceding addition of data to A has caused that nibble to contain an invalid BCD digit (i.e., $A through F). The following examples show how this procedure produces correct results for BCD addition.

Example 5.3

Show how the 6802 would add decimals 9 + 1 = 10.

Solution

Decimal	BCD	BCD Adjust
09	0000 1001	ADDA
+ 1	+0000 0001	#01
10	0000 1010	Low nibble > 9
	+0000 0110	DAA adds 06
	0001 0000	

Example 5.4

Show how the 6802 would add decimals 95 + 8.

Solution

Decimal	BCD	BCD Adjust
95	1001 0101	ADDA
+ 8	+0000 1000	#08
103	1001 1101	Low nibble > 9
	+0000 0110	DAA adds 06
	1010 0011	High nibble > 9
	+0110 0000	DAA adds 60
	1 0000 0011	C flag holds third digit

The processor logic recognizes when an addition has produced a nibble result greater than 9 and sets the H flag, which is bit 5 of the condition-code register, to keep track of the fact. The H stands for half-carry and refers to the detection of a carry halfway in the register (low-order BCD digit to high-order BCD digit.)

Notice that the DAA instruction works for additions to the A accumulator only:

- DAA does not work for additions to the B accumulator.
- DAA does not work for subtractions.
- DAA does not work for increments (INCA). Instead, use ADDA #01.

5.6 A REACTION TIMER

Section Overview

In this section we show how a microprocessor can be programmed to measure human reaction time. The program uses nested delay loops and BCD addition. The hardware uses an output data latch. Of course, this same task could be accomplished with discrete counters and logic gates, but the exercise illustrates microprocessor arithmetic and data-output techniques, and we will see in the next section that the microprocessor has an added dimension of versatility.

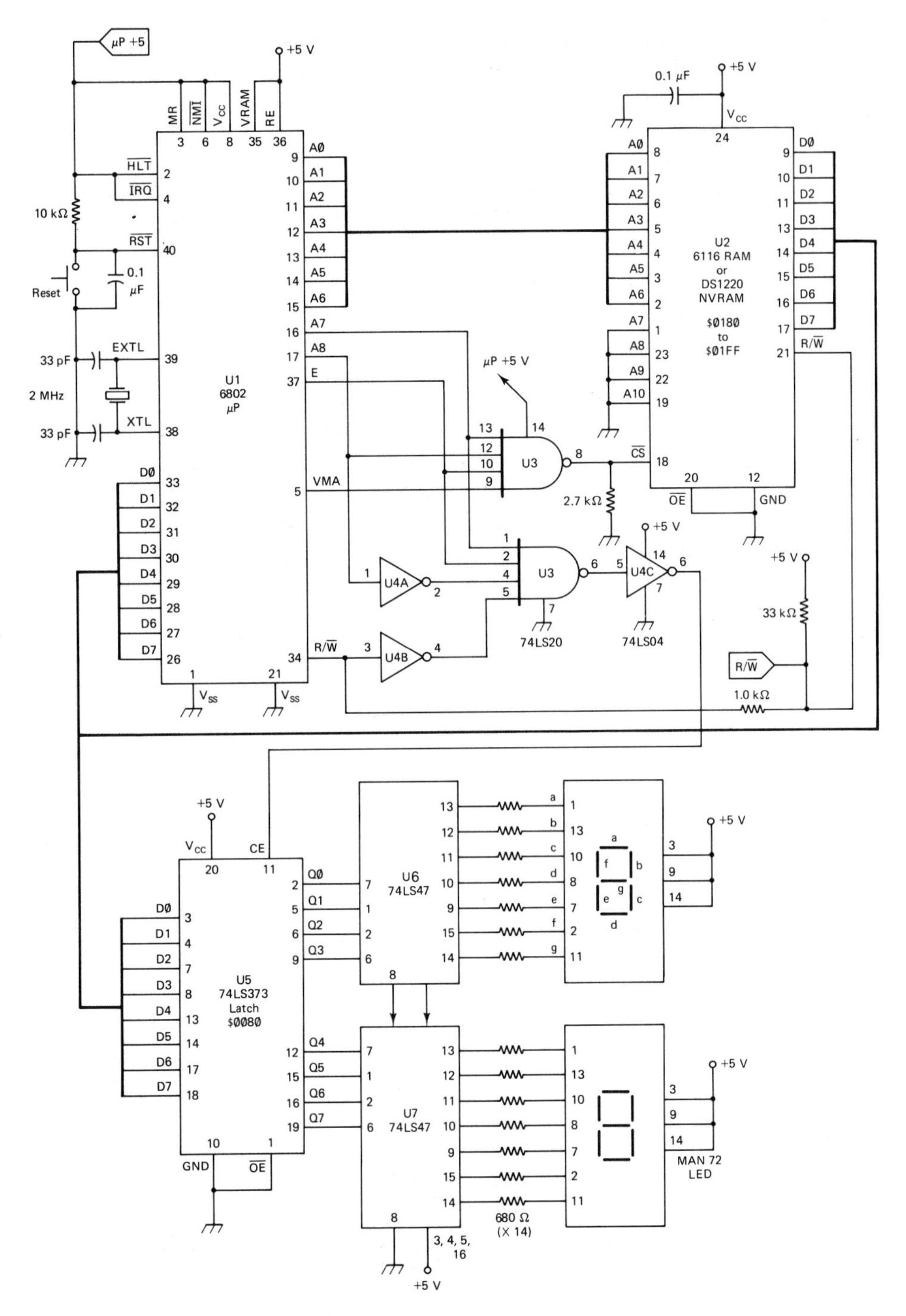

Figure 5.7 Hardware for the Reaction Timer game. The control unit of Figure 3.10 or the advanced unit of Figure 5.9 may be used to enter the program into RAM, or the EPROM-memory configuration of Figure 5.11 may be used.

The hardware diagram for the reaction timer is given in Figure 5.7. In this diagram we begin to use two simplifying techniques, which make large microcomputer systems easier to draw:

1. The address and data bus lines are not drawn individually. Rather, they are represented by a single heavy line and broken into individual lines only at the IC pins they service. This minimizes the confusion of having dozens of lines cross the diagram page.
2. The pins are placed around each IC for ease of drawing and interpreting their functions rather than counterclockwise in numerical order as they appear on the actual chips.

We also abandon the 6810 RAM in favor of the 6116 for program storage at this point. The 6810 may still be used if the pin connections of Figure 3.7 are observed, but the 6116 RAM offers greater flexibility because several alternative types of memory are available with essentially the same pinout. Although the 6116 contains 2048 bytes we will ground the highest four address lines so that only the first 128 bytes will be available, making the addressing the same as for the 6810.

Reaction times are to be counted in BCD and displayed in tenths and hundredths of a second on two 7-segment LEDs driven from the two BCD nibbles by a pair of 74LS47 decoders. A 74LS373 latch stores the nibbles during the display time. Figure 5.8 shows a flowchart of the program.

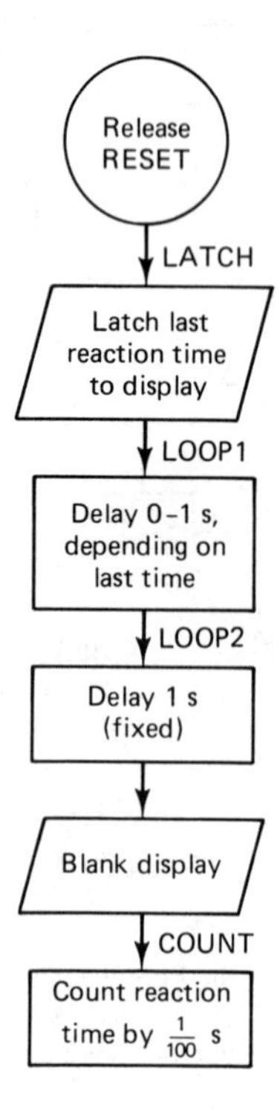

Figure 5.8 The flowchart for the Reaction Timer program.

The processor's RESET button serves as an input to stop the timer. When RESET is released the last reaction time will be displayed for a period of 1 to 2 s. Then the display will go dark, and the timer will begin counting by hundredths of a second. The operator is to press RESET as soon as possible upon seeing the display go dark. The display time is made variable (based on the low-order digit of the last reaction time) to prevent the operator from being able to anticipate the darkening of the display.

List 5.1 shows the program listing.

```
                         1000 * LIST 5-1        REACTION TIMER FOR FIG.5-7 HARDWARE
                         1005 * WAIT FOR DISPLAY TO GO DARK, THEN HIT RESET AS QUICKLY
                         1006 * AS POSSIBLE. TIME IN 1/100-SEC IS DISPLAYED.
                         1007 *  RUNS ON FIG. 5.7 HARDWARE, 2 JUL '87.  D. METZGER.
                         1010 *
                         1020        .OR   $0190    PROGRAM ORIGIN
                         1030        .TA   $4190    HOST ASSEMBLER TARGET ADR.
0010-                    1040 TIME   .EQ   $0010    RAM BUFFERS DEFINED:
0011-                    1050 DELY1  .EQ   $0011
0012-                    1060 DELY2  .EQ   $0012
0080-                    1070 LATCH  .EQ   $0080    OUTPUT LATCH BASE ADR.
                         1080 *
0190- D6 10              1090        LDAB TIME      PICK UP LAST TIME
0192- D7 80              1100        STAB LATCH      AND OUTPUT TO DISPLAY.
0194- 58                 1110        ASLB           PUT LOW-ORDER DIGIT OF
0195- 58                 1120        ASLB            LAST TIME INTO HIGH-ORDER
0196- 58                 1130        ASLB            NIBBLE OF ACCUM B.
0197- 58                 1140        ASLB
                         1150 *
0198- 7A 00 11           1160 LOOP1  DEC  DELY1     VARIABLE DELAY          6 CY
019B- 26 FB              1170        BNE  LOOP1      0 TO 1.3 SECONDS       4 CY
019D- 5A                 1180        DECB            ON A 0.5-MHZ CLOCK.
019E- 26 F8              1190        BNE  LOOP1
01A0- 7A 00 11           1200 LOOP2  DEC  DELY1     FIXED DELAY; 1.3 SECONDS
01A3- 26 FB              1210        BNE  LOOP2      AFTER FIRST RUN.
01A5- 7A 00 12           1220        DEC  DELY2
01A8- 26 F6              1230        BNE  LOOP2
01AA- 86 FF              1240        LDAA #$FF      BLANK DISPLAY
01AC- 97 80              1250        STAA LATCH
                         1260 *
01AE- 4F                 1270        CLRA           START WITH 00/100 SEC.
01AF- CE 02 6F           1280 COUNT  LDX  #$026F    WAIT FOR 10 MILLISEC.
01B2- 09                 1290 WAIT   DEX                                    4 CY
01B3- 26 FD              1300        BNE  WAIT      (10000)/(2 X 8)         4 CY
01B5- 8B 01              1310        ADDA #1             = 623 LOOPS.
01B7- 19                 1320        DAA            THEN COUNT TIME AND
01B8- 97 10              1330        STAA TIME       SAVE IN RAM.
01BA- 7E 01 AF           1340        JMP  COUNT
                         1350 *
                         1360        .OR  $01FE     RESET VECTOR
                         1370        .TA  $41FE
01FE- 01 90              1380        .HS  0190
                         1390        .EN
```

Program Commentary: The BCD reaction time, stored in RAM location TIME just before the RESET was pushed, is latched to the 74LS373 when RESET is released, and the time is displayed during the variable delay until the display goes out.

The least-significant BCD nibble from the last run will be quite random, so it is shifted up to the most-significant nibble, where it determines the length of the nested delay, LOOP1. A fixed delay, LOOP2, then follows to ensure that the previous reaction time will be displayed long enough to be observed.

After a delay of 1.3 to 2.6 s, the displays are blanked, signaling the user to hit the timer button (RESET). This will normally take a few tenths of a second, and the final timing loop, WAIT, delays for 0.01 s (10 000 μs) and then counts up in BCD via accumulator A. Each time the WAIT loop is exited, 1 is added to the BCD time and the loop is restarted for another 0.01-s wait.

The WAIT loop consists of two 4-cycle instructions, DEX and BNE, for a time-per-pass of 16 μs on a 2-μs clock. The count, decimal adjust, and loop-restart instructions take 14 cycles, or 28 μs, off the 10 000-μs time required for each count. The number of passes through the WAIT loop required to give 0.01 s per count is then

$$(10\ 000 - 28)\ \mu s \div 16\ \mu s/\text{loop} = 623\ \text{loops}$$

This is more than can be counted off in an 8-bit register, so we have taken the liberty of using the 6802's internal 16-bit X register to count off the 623 loops. (In hex, this is $Ø26F loops, the value loaded into X at the start of each count routine.) The X register was intended for quite another purpose, as we shall see in Chapter 7, but there is no harm in making its acquaintance a little early.

Wire the system of Figure 5.7 on a solderless breadboard and enter the reaction-timer program with the control unit of Figure 3.10 or Figure 5.9. The latter programming fixture allows data to be entered quickly from a hex keypad, rather than laboriously in binary. It also permits hex display of data at any address in program memory.

The 6116 RAM costs only about $3.00, but it can be frustrating to use it as the program memory in a breadboard system because even a millisecond power interruption will scramble the data you have so painstakingly entered. A battery-backed NOVRAM (nonvolatile RAM), such as the DS1220 from Dallas Semiconductor, can be substituted for the 6116 at about three or four times the cost. It will allow the power to be turned off, and even allow the memory chip to be removed and carried about, without loss of data. Figure 5.11 shows how an EPROM can be substituted for the RAM to make your program completely invulnerable to processor overwrites. Construction details for an EPROM programmer are given in Appendix D, if you do not already have one available.

1. Review and follow the wiring hints given in Appendix A, page 000. This system is about twice as complex as those from previous chapters, and the chances for making an error are great.
2. Enter a few program bytes, and then go back and read them out to be sure the RAM is taking data.
3. Run the program and check the E, R/$\overline{W}$, VMA, data, and address lines to be sure that none are stuck at V_{CC} or ground.
4. Use a logic analyzer to trace the program in operation. The axiom in microcomputer development work is that three times out of four the problem is in the software, not the hardware.

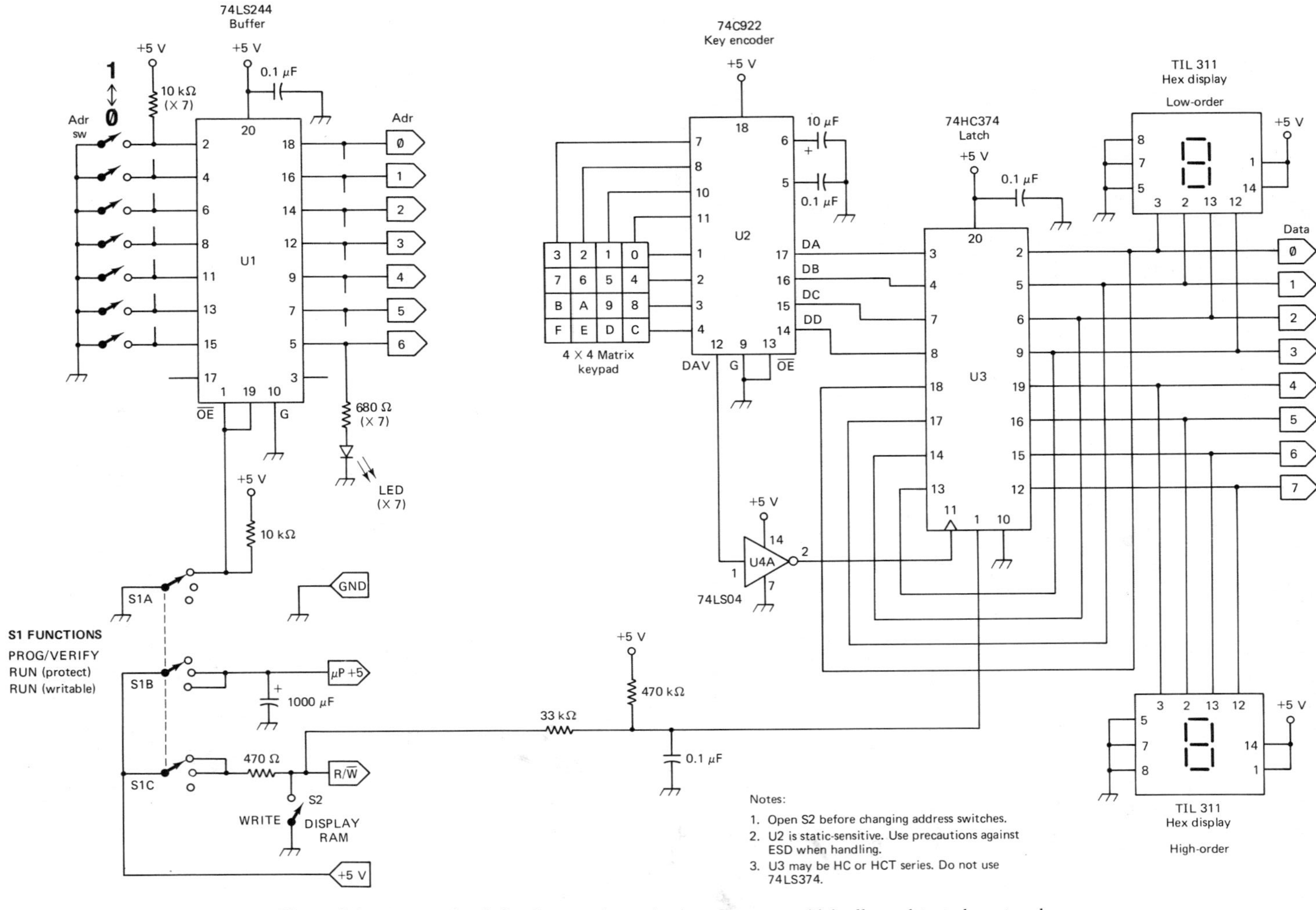

Figure 5.9 A control unit for the experimental microcomputer, which allows data to be entered and read in hex rather than in binary.

REVIEW OF SECTIONS 5.5 AND 5.6

21. Register A contains binary 1000 0101 and B contains binary 0011 0111. What binary data will register A contain after executing the two instructions ABA and DAA?
22. In Question 21, what is the state of the C flag, and what is the value represented thereby?
23. Write the logic expression for the enabling of the 74LS373 in Figure 5.7.
24. What base address and machine-cycle conditions are implied by the expression of the previous question?
25. When data is output via a latch, how long after it is written will it remain at the output?
26. Identify any writes to the 6116 program RAM in the reaction-timer program. What position of the control unit should be used to run the program?
27. Calculate the delay time for the inner loop (decrementing DELY1) of LOOP1 in the reaction timer program. The clock speed is 0.5 MHz.
28. Calculate the entire nested LOOP1 delay time if the previous time displayed was 37. f_E = 0.5 MHz.
29. In the reaction-timer program, what delay would be produced in the WAIT loop if X had been loaded with its maximum value of $FFFF? f_E = 0.5 MHz.
30. We wish to make the LOOP2 delay 2.6 s by inserting some NOP instructions between the DEC and BNE instructions. How many NOPs are needed?

5.7 SOFTWARE-DECODED OUTPUT

Section Overview

This section presents new software to show how the reaction time display game can be converted to a two-player reaction time challenge game. It also shows how programming can be used to eliminate two LED decoder chips from the system hardware.

System updating with software and cost saving by replacing hardware with software are two big reasons for the success of microcomputers.

The reaction timer of Figure 5.7 uses a pair of 7447 ICs to decode the BCD nibbles from the latch by having them turn on appropriate LEDs of the 7-segment displays. This is called *hardware decoding*, and it is relatively expensive.

Software decoding uses the outputs from the latch to drive the 7-segment-display cathodes directly. A user code of binary digits is established to form the required characters on the display. The decoder chips are eliminated, and the restriction to digits 0 through 9 plus *blank*, which they imposed, is removed. Any pattern conceivable on a 7-segment display can be produced. Figure 5.10 shows some ex-

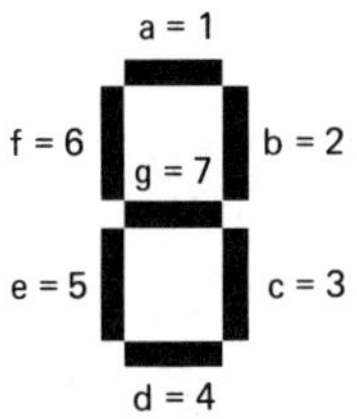

a = 1
f = 6
b = 2
g = 7
e = 5
c = 3
d = 4

Display	Binary 7654	Binary 3210	Hex
H	0001	0010	12
I	1111	0010	F2
L	1000	1110	8E
O	1000	0000	80

Figure 5.10 Software decoding for a 7-segment LED.

amples. This technique was exhibited previously in Section 2.4 (see Figure 2.3). The only penalty it imposes is to require slightly more complicated software.

Strobing, also called *multiplexing*, is a technique for driving more than one display device with a single output and decoder. This is accomplished by switching among the displays at a rate greater than 60 Hz. The displays actually flash *on* one at a time, but the flicker rate is so high that the eye perceives them as all being lit simultaneously. The speed of the computer is, of course, more than adequate to change the output pattern as each display is turned on, so each one shows a different character.

A reaction-time challenge game for two players is shown in Figure 5.11 (hardware) and 5.12 (program flowchart). The system uses two LED displays, which are software decoded and strobed to produce one of two messages: HI or LO.

To play the game the first player holds the first RESET button down and then releases it. The displays, lit from the previous run, will go dark. The program is now in the WAIT/COUNT loop, counting 3-ms intervals until the second player pushes the second RESET button. This stops the counting process, with the number of 3-ms intervals stored as a pure binary number in a RAM location called NEWTIM.

The DISPLAY portion of the flowchart is entered when the second player releases the pushbutton. The letters HI appear if this reaction time was longer than that recorded in the last pair of runs. The display reads LO for a shorter reaction time than the previous one. The program automatically records this new reaction time as the "previous" time, storing it in RAM OLDTIM. NEWTIM is set to zero as a flag that the next run is to be a timing run (WAIT loops), not a display run. To help yourself to obtain a more thorough understanding of the program, try to implement these modifications:

1. Change the displays to read YA and NO instead of HI and LO.
2. Add a routine to cause a display of TI in case of a tie.

List 5.2 gives the program listing for the Reaction-time Challenge Game.

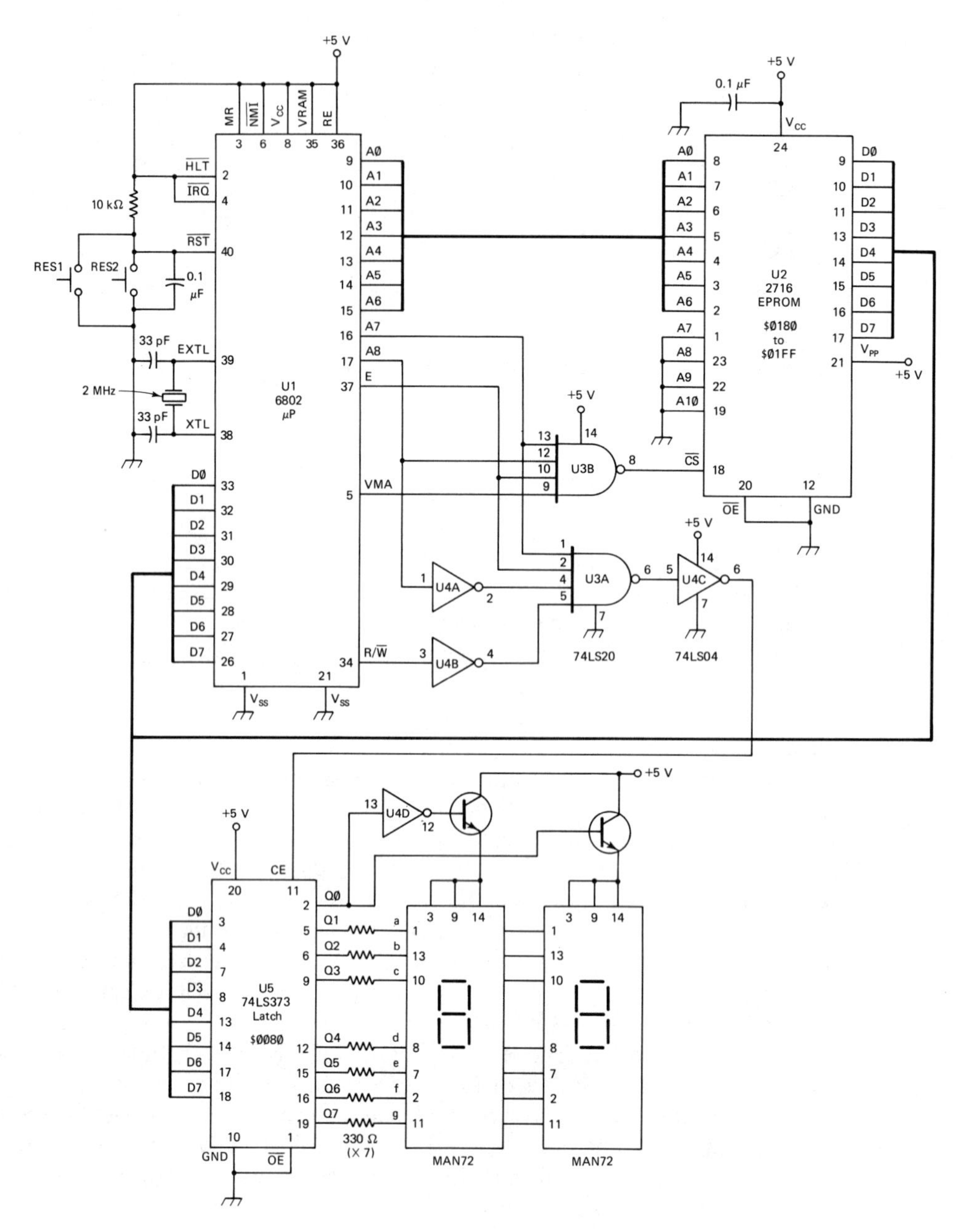

Figure 5.11 Hardware diagram for the Reaction Time Challenge game, with software-decoded output display reading HI or LO. The RAM configuration of Figure 5.7 may be substituted.

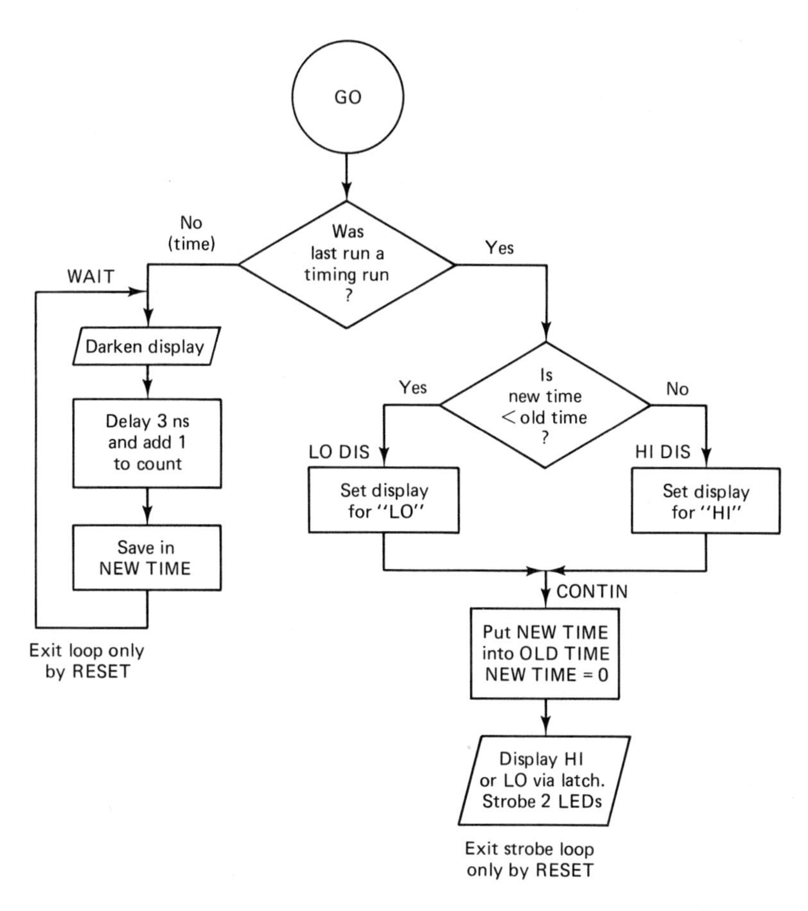

Figure 5.12 Flowchart for the Reaction Time Challenge program.

```
                    0900 * LIST 5-2 RUNS ON FIG. 5-11 SYSTEM, 7 JUL '87. METZGER
                    1000 * REACTION-TIME CHALLENGE GAME FOR 2 PLAYERS.
                    1002 *  PLAYER 1 HOLDS, THEN RELEASES BUTTON, CAUSING DISPLAY
                    1004 *  TO DARKEN.  PLAYER 2 THEN QUICKLY PUSHES THEN RELEASES
                    1006 *  HIS BUTTON. "LO" SHOWS PLAYER 2 WAS FASTER THAN LAST
                    1008 *  RUN.   "HI" SHOWS PLAYER 2 WAS SLOWER THAN LAST RUN.
                    1010 *
                    1020         .OR  $0190
                    1030         .TA  $4190
0010-               1040 NEWTIM .EQ  $0010    DEFINE RAM BUFFER AREAS:
0011-               1050 OLDTIM .EQ  $0011
0012-               1060 TWOBYT .EQ  $0012    (2 BYTES FOR X REGISTER)
0080-               1070 LATCH  .EQ  $0080
                    1080 *
0190- 96 10         1090         LDAA NEWTIM   IF NEWTIM NOT = 0, LAST
0192- 26 0D         1100         BNE  DISPLY    RUN WAS A TIMING, SO DISPLAY.
0194- C6 FF         1104         LDAB #$FF     BLANK DISPLAY.
0196- D7 80         1106         STAB LATCH
0198- 5A            1110 WAIT    DECB          WAIT FOR 6 X 256 US, OR
0199- 26 FD         1120         BNE  WAIT      3 MS; THEN COUNT UP NEWTIM.
019B- 7C 00 10      1130         INC  NEWTIM
```

```
019E- 7E 01 98 1140        JMP  WAIT
               1150 *
01A1- 91 11    1160 DISPLY CMPA OLDTIM   IF OLDTIM LARGER, C FLAG IS
01A3- 25 08    1170        BCS  LODIS     SET (BORROW); DISPLAY "LO".
01A5- CE 12 9F 1180 HIDIS  LDX  #$129F   IF NEWTIM LARGER, DISPLAY "HI".
01A8- DF 12    1190        STX  TWOBYT   USE 16-BIT X REGISTER TO
01AA- 7E 01 B2 1200        JMP  CONTIN    LOAD BIT PATTERNS
01AD- CE 8E 81 1210 LODIS  LDX  #$8E81    TO DISPLAY "HI" OR "LO".
01B0- DF 12    1220        STX  TWOBYT
01B2- 97 11    1230 CONTIN STAA OLDTIM   NEWTIM BECOMES OLDTIM;
01B4- 7F 00 10 1240        CLR  NEWTIM    NEWTIM = 0 SIGNALS NEXT
01B7- 96 12    1250        LDAA TWOBYT    RUN IS A TIMING.  FIRST
01B9- D6 13    1260        LDAB TWOBYT+1 BYTE FLASHES H OR L;
01BB- 97 80    1270 LOOP   STAA LATCH     SECOND BYTE FLASHES I OR O,
01BD- 01       1280        NOP            HAS BIT 0 = 1 TO TURN ON
01BE- D7 80    1290        STAB LATCH     RIGHT LED.
01C0- 7E 01 BB 1300        JMP  LOOP     ALTERNATE LEFT/RT DISPLAY.
               1310        .OR  $01FE
               1320        .TA  $41FE
01FE- 01 90    1330        .HS  0190     RESET VECTOR AT TOP SYSTEM ADR.
               1340        .EN
```

Answers to Chapter 5 Review Questions

1. Memory mapped **2.** Latch **3.** 11 (A0 - A10)
4. 3, since $2^3 = 8$, but $2^2 = 4$, and 6 chip-enable lines are needed.
5. 4 for memories **6.** 8192 for latches **7.** AND the *E*-clock in with the chip-select logic. **8.** A large number of gate inputs and inverters are needed for full decoding. **9.** \$4000 through \$5FFF
10. Use a ROM for address decoding.
11. The processor **12.** 405 ns **13.** 0.69 MHz
14. Longer delay times, hence slower clock speeds. **15.** Chip-select starts upon the rise of *E*, but so does data assertion by the processor, and this write-data delay (t_{DDW}) usually takes longer than chip select. **16.** 1.89 MHz
17. The first half of a write cycle. **18.** 20 ns **19.** 1.40 MHz
20. Specified worst-case **21.** 0010 0010 **22.** C = 1, value = 100
23. $CE = A7 \cdot \overline{A8} \cdot \overline{W} \cdot E$ **24.** This identifies the last half of a write cycle at base address \$0080.
25. Forever, unless new data is written or the power is removed.
26. There are none. Use the RUN/protect mode to avoid accidentally erasing the program. **27.** 5.12 ms **28.** 0.57 s **29.** 1.05 s **30.** 5 NOPs.

CHAPTER 5 QUESTIONS AND PROBLEMS

Digits after the decimal point refer to section numbers in Chapter 5.

Basic Level

1.1 What 6802 instruction(s) are used to output data via a latch?

2.1 What are the input and output sides of such a latch to connect to?

3.2 State whether the latch of Figure 5.2 will output data from each of these instructions:

a. LDAA $0080   c. CLR $0080
b. STAB $00C7   d. STX $007F

4.2 Refer to Figure 5.4 and tell what data words should be programmed into what raw ROM addresses to enable the 8K ROM from addresses $9CC0 through $B7FF.

5.3 A 6802 system is to operate at 1.0 MHz. It uses two cascaded LS-TTL gates to select the ROM. Allowing a 100-ns safety margin, what is the data-delay time of the slowest ROM that could be used? Refer to Figure 5.5.

6.3 In Problem 5.3, the decoding logic is changed so that E is ANDed with the address lines for chip select of the ROM. Now what is the slowest acceptable ROM response time?

7.4 Refer to Figure 5.6 and determine the longest acceptable data setup time for a RAM in a 1.00-MHz 6802 system. Allow a 100-ns safety margin.

8.4 What is the state of the data bus lines on a 6802 write cycle during the *low* half of the E-clock?

9.5 Accumulator A contains $17 and B contains $59. Write a 6802 routine to add these two numbers as decimal 17 + 59, and identify where the result, decimal 76, will be found.

10.5 Examine the 6802 Programmer's Reference and list all the instructions that can alter the H flag. Does DAA clear H if it has been set?

11.6 Refer to the Reaction Timer program of List 5.1 and calculate the LOOP1 delay time if the previous delay stored in TIME was 0.30 s.

12.6 In the WAIT loop of the Reaction Timer program, X assumes values that are in the range of the RAM and latch addresses. These values are placed on the address bus. What prevents false data from being written to each of these devices at those times?

13.7 In the circuit of Figure 5.7 the LED segment resistors are 680 Ω. Why were they changed to 330 Ω in the circuit of Figure 5.12?

Advanced Level

14.1 Describe the function of an output latch in a microcomputer system.

15.2 There is one instruction for which VMA and R/$\overline{W}$ are *low* simultaneously. Identify this instruction and explain what would happen if it were executed at the address of the latch in Figure 5.2. (Refer to Appendix B). Would it help to include VMA in the chip-select logic? Explain.

16.2 Add the necessary logic to the circuit of Figure 5.2 to enable two separate latches: one at addresses from $0080 through $00BF, and a second from $00C0 through $00FF.

17.3 What is likely to happen in a system in which the ROM responds so slowly that t_{MAR} shrinks to less than zero? (Refer to Figure 5.5.)

18.4 Refer to Figure 5.6 and explain what might happen on a write cycle if a RAM's chip-select logic did not include E as one of its AND inputs.

19.4 Laboratory tests show that a 6802 system designed for a 0.8-MHz clock runs perfectly at 1.2 MHz, even at maximum and minimum supply voltages and ambient temperatures. Would it be safe to increase the clock speed to 1.0 MHz in production units offered for sale? Explain.

20.5 Write a 6802 program to convert any hex number from $1 to $63 to a decimal number from 1 to 99. Do this by setting up a loop that subtracts 1 from the hex number while adding 1 to the decimal number, until the subtraction produces a zero result. Show a flowchart, source code, and comments.

21.6 Refer to the Reaction Timer program of List 5.1 and explain why the four ASLB instructions are needed to produce an essentially random time delay for the next game.

22.6 Show how a 74LS30 (8-input NAND) could be connected to make the RAM in Figure 5.7 respond only to the address range $F800 through $FFFF, eliminating all redundant addresses.

23.7 Explain what it means to *strobe* a display.

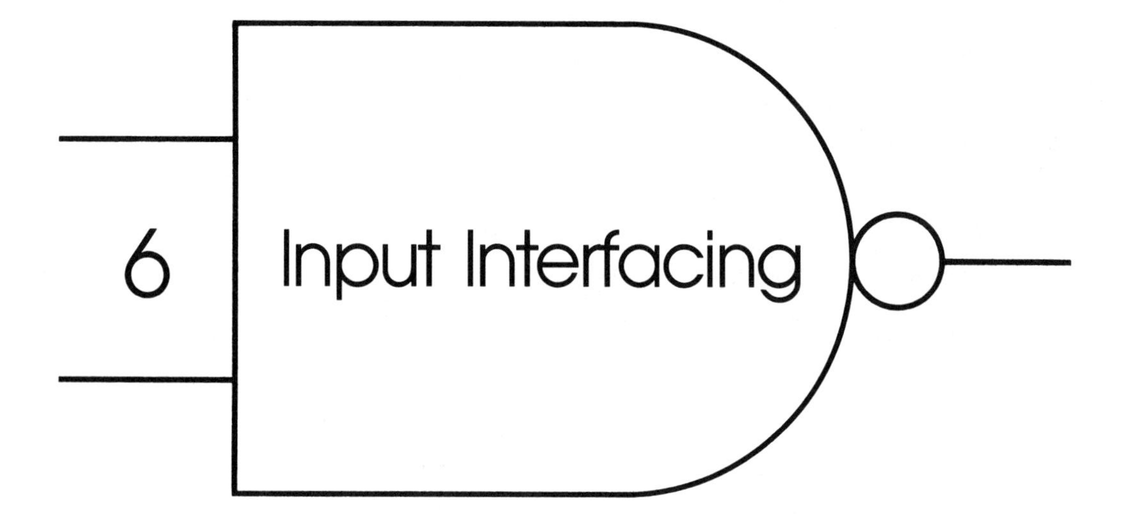

6.1 INPUT BUFFERS AND ADDRESS DECODERS

Section Overview

This section explains how tristate input-buffer chips are used to connect data from the outside world to the microprocessor data bus for a read-data cycle.

Since the number of devices that our microprocessor may call upon is increasing, we also introduce here the address-decoder IC. This chip contains in one package all the AND logic needed to activate any one of eight peripheral devices by decoding the processor's address lines.

When a microcomputer executes an instruction to input data from the outside world, it reads its data bus. At the instant of the read all memory devices must have their outputs disabled (high-impedance state), and a special *input buffer* must be enabled. The input buffer is a tristate device whose inputs are driven by binary data from the outside world and whose outputs are connected to the microcomputer system data bus, in parallel with the corresponding data lines of the processor, memories, and output latches. Figure 6.1 illustrates the input interface.

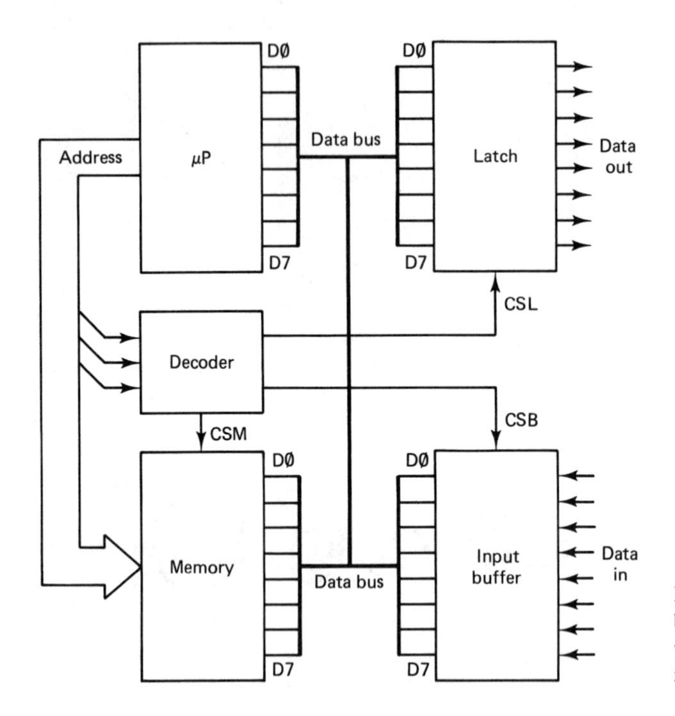

Figure 6.1 The tristate outputs of the buffer are paralleled on the data bus, along with the memory input/output lines and the inputs to the latch.

Of course, only one of the devices paralleled on the data bus may assert an active output at any given time. Decoder circuitry must be provided by which the processor can activate any one of its peripheral devices: memory, output, or input.

Address decoders are MSI (medium-scale integration) ICs that contain in one package all the AND gates and inverters needed to enable a number of different peripheral devices, each at its own block of addresses in the memory map. Figure 6.2 shows the 74LS138 address decoder. It has three inputs and 2^3, or 8, tristate outputs (low active) for enabling various memory and I/O devices. It also has three chip-select or *gate* inputs which must all be asserted to cause an active output. As the figure illustrates, the decoder can be used for producing large divisions or small subdivisions in the memory map. It is limited to producing equal divisions, however, unless additional decoding is used.

Example 6.1

Devise a simple address-decoding scheme to place eight 2K memory chips in the low half of the memory map, but do not use page zero ($ØØØØ through $ØØFF), as this is required for other devices.

Solution

The 2-K blocks require that lines AØ through A1Ø be used for address selection within each chip, since $2^{11} = 2048$. The next three lines (A11, A12, and A13) can then be used to distinguish among the eight chips, since $2^3 = 8$. Line A14 is used to select the zero-page device (A14 = Ø) or the eight 2K memory chips (A14 = 1).

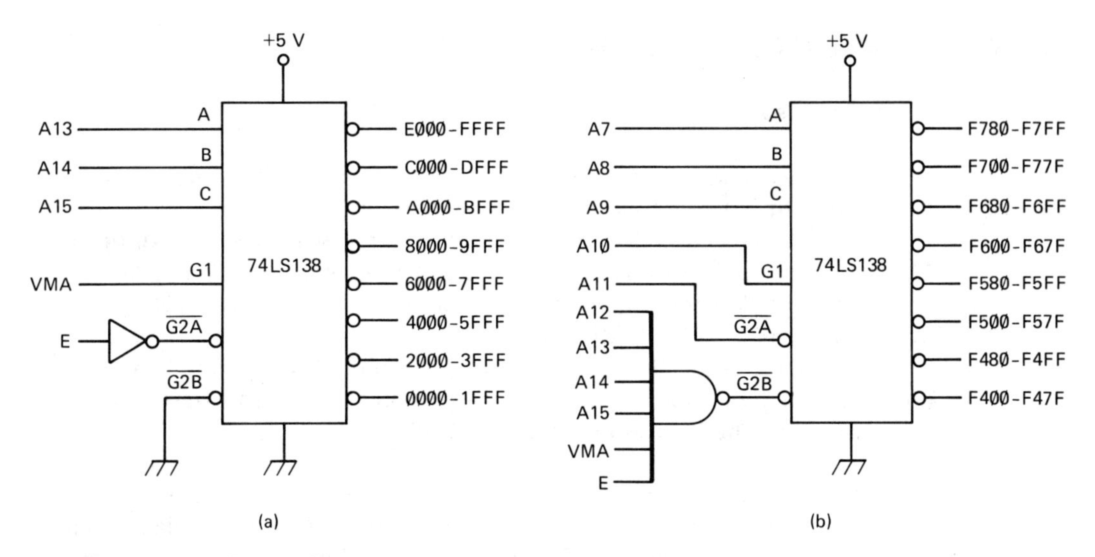

Figure 6.2 (a) The 74LS138 address-decoder breaks the entire 64K of memory into eight blocks of 8K each. (b) With a little additional logic it breaks a 1K segment near the end of the memory map into eight blocks of 128 bytes each.

A 15	14	13	12	*A* 11	10	9	8	*A* 7	6	5	4	*A* 3	2	1	0	
0	0	0	0	0	0	0	0	*A*	*A*	*A*	*A*	*A*	*A*	*A*	*A*	Zero-page device
0	1	*X*	*X*	*X*	*A*	*A*	*A*	*A*	*A*	*A*	*A*	*A*	*A*	*A*	*A*	Eight 2K blocks

The base addresses for the eight devices are found by letting all *A* bits equal 0 and counting the three *X* bits from 000 to 111. Figure 6.3 shows the logic diagram.

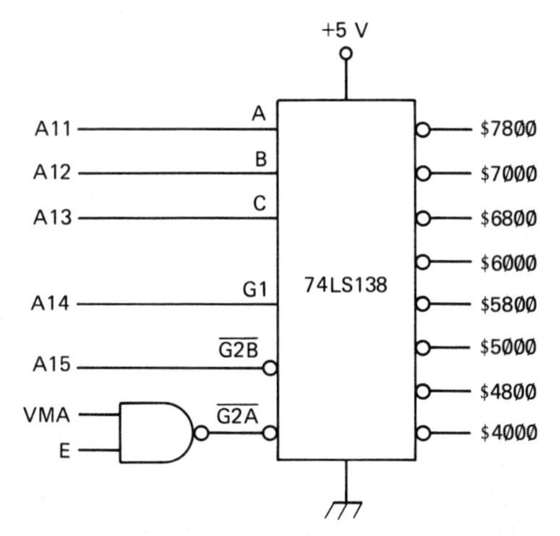

Figure 6.3 Solution for Example 6.1.

6.2 MICROPROCESSOR AND TTL COMPATABILITY

Section Overview

Microprocessor input pins generally present a high input impedance. An open input is likely to switch on noise pickup.

LS-TTL parts are often used with microprocessor systems. An open LS-TTL input floats *high* and *outputs* about 0.3 mA when pulled to ground.

A microprocessor output can typically drive 5 LS-TTL inputs or 10 microprocessor or CMOS inputs.

TTL outputs can drive microprocessor inputs directly, but TTL outputs may not pull high enough to drive CMOS inputs. An external pull-up resistor may be helpful in driving CMOS.

Modern microprocessors and their related components operate from +5-V supplies; they recognize voltages above +2.0 V as logic *high* and voltages below +0.4 V as logic *low* state, much as TTL chips do. Many chips used for microprocessor control and I/O are TTL devices, whereas the microprocessors themselves are usually NMOS devices. Generally, these devices can be interfaced with no problems, but you should be aware of some potential problem areas.

Current considerations: Although TTL and microprocessor voltages are about the same for logic-0 and logic-1 conditions, their current requirements are quite different and may vary from one type to the next. This sometimes necessitates a close reading of the IC spec sheets. The basic difference is that TTL uses *current-sourcing inputs*, whereas microprocessor components use *high-impedance* inputs.[1]

A simplified schematic of a TTL output driving a current-sourcing TTL input is shown in Figure 6.4(a). R_B sources a current that will

1. Pass through the collector of Q_3 to turn the base of Q_4 *on* if the INPUT is held *high* or open-circuited, or
2. Be diverted through the emitter of Q_3 to ground, turning Q_3 *on* and Q_4 *off* if the INPUT is held *low* by grounding it or turning *on* Q_1.

There are three things to remember about a TTL-type gate:

1. A TTL input sources current. The driving gate must sink this current to ground to establish a logic-0 (low-voltage) condition. A typical LS-TTL gate output can sink up to 8 mA.
2. An open TTL input will assume a logic 1 (*high*) level. Open inputs are not recommended because of the possibility of noise pickup, but note that even

[1]We will speak in this text of *conventional current*, which is assumed to pass from positive to negative through a resistive load. Many technicians prefer to visualize current as electron flow, which passes in the opposite direction. Since we cannot actually see the flow it makes no difference in practice which view we take. It is only essential that we stick consistently to one point of view. The manufacturers of ICs, without exception, have adopted the positive-to-negative convention.

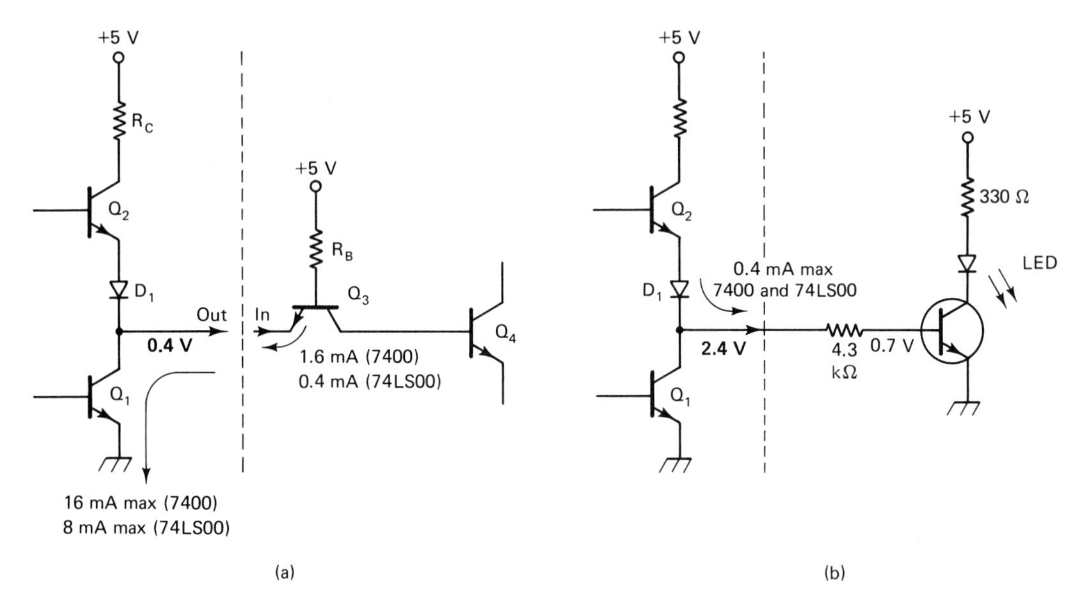

Figure 6.4 (a) TTL output and input internal circuitry. (b) A TTL-to-transistor driver circuit.

without actively pulling the input *high*, the TTL gate will generally assume a logic-1 input.

3. TTL outputs are capable of sourcing current to loads that require it, as shown in Figure 6.4(b). However, the output voltage cannot generally be guaranteed to pull higher than +2.4 V.

Today many popular TTL interface chips are available in high-power CMOS versions (e.g., 74HC244), which have essentially the output drive capability of TTL but a high-impedance input.

Microprocessor inputs are passive and high impedance. They source no current and require drive currents of only a few microamperes. Microprocessor outputs can sink current to pull an output down to logic Ø—but only about 1.6 mA in most cases. Microprocessor outputs can be counted upon to source only about 0.2 mA while pulling up to +2.4 V. This is illustrated in Figure 6.5(a).

The greatest potential for trouble arises when a microprocessor output is used to drive several TTL inputs, as shown in Figure 6.5(b). One standard TTL input or four LS-TTL inputs may source all the current that a microprocessor can guarantee to sink at logic-Ø level.

Microprocessor and CMOS inputs typically present about a 10-pF capacitive load. Microprocessor output timing is generally specified with a bus load of about 100 pF per line. Therefore, about 10 microprocessor or CMOS devices can be driven at full rated speed by a microprocessor. Manufacturers' data sheets should, of course, be consulted to obtain more accurate loading determinations.

Driving CMOS logic, such as the CD4000 series, presents a special problem,

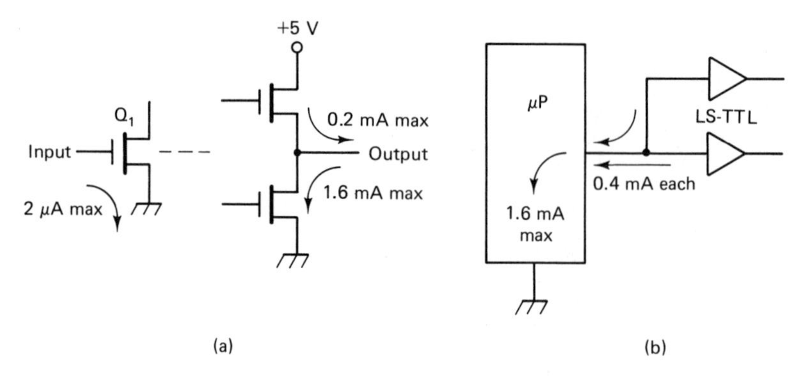

Figure 6.5 (a) Microprocessor and CMOS inputs require virtually no input current. (b) A microprocessor can sink current for a maximum of 4 LS-TTL inputs.

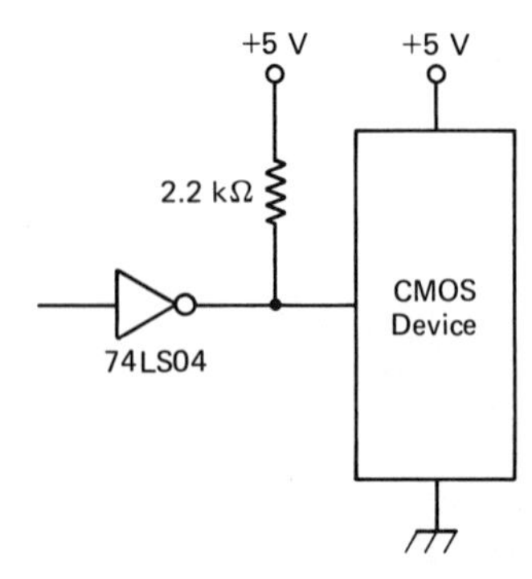

Figure 6.6 Pull-up resistors may be required to permit TTL devices to drive CMOS inputs.

since CMOS inputs may require a voltage of 0.7 V_{CC} (or 3.5 V on a 5-V supply) to guarantee a logic-1 input. Standard TTL and microprocessor outputs cannot be relied upon to pull that high, so external pullup resistors may be required, as shown in Figure 6.6.

REVIEW OF SECTIONS 6.1 AND 6.2

Answers appear at the end of Chapter 6.

1. How many devices may be asserting data on the data bus during a read cycle? During a write cycle?
2. What causes an input buffer to assert data on the data bus?
3. A 74LS138 decoder is driven by address lines A9, A10, and A11. How many words are there in each block of the memory map?
4. The 74LS154 decoder has 4 inputs and 16 outputs. It is driven by the most-significant 4 of the 16 address lines. What is the base address of the highest block in the memory map?
5. A 74LS04 inverter drives a second such inverter with a *low* level at the second

input. What are the magnitude and direction of conventional current between the two gates?

6. Answer Question 5 again, this time for a *high* level at the second input.
7. If microprocessor and CMOS components require virtually no input current, what limits the number of such inputs that can be paralleled on a bus?
8. What precaution should be taken if TTL outputs are used to drive CMOS inputs?
9. One of the inputs to a 74LS244 buffer falls off, leaving the input pin open. When the processor reads the buffer, will this bit read a *high* level, a *low* level, or random noise?
10. The 6802 processor uses a LOAD instruction both to read data from memory and to input data from an input buffer. What then distinguishes these two operations?

6.3 INPUT-SIGNAL CONDITIONING

Section Overview

The most reliable switch connection for a TTL input buffer is a normally open contact from input to ground with a pullup resistor of about 10 kΩ to V_{CC}. A key press produces a *low* level, but this can be inverted with the processor's bit-complement instruction.

Tests may show that the pull-up resistor can be eliminated without high-frequency noise pickup, but it is low-frequency and single-shot noise to which an open TTL input is vulnerable.

In spite of a variety of sophisticated alternatives, the most common microprocessor input device is still a pair of mechanical contacts—a switch. Figure 6.7 shows three ways of wiring the buffer input to the switch.

The first, Figure 6.7(a), while quite straightforward, is not recommended because the resistor R_1 would have to be quite low (680 Ω for LS-TTL, 160 Ω for standard TTL) to pull the input down to the middle of the "guaranteed logic 0"

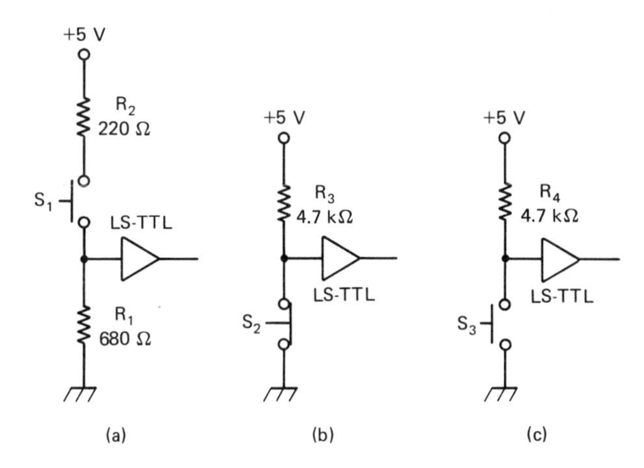

Figure 6.7 Three switch-input connections for microprocessor systems. Figure 6.7 (c) is preferred.

zone, which is about +0.2 V. This would result in excessive supply current when the switch was depressed. Also, the quiescent input level would be only 0.2 V from a possible *high*-level input signal, inviting noise pickup. The circuit may have more merit for driving CMOS inputs, when logic *low* is typically above +2.0 V, and high Z_{in} permits higher values of R_1, but still neither side of the input switch is grounded; one grounded input leg would be desirable for shielding, measurement, and common-switch-pole wiring.

The input circuit of Figure 6.7(b) works quite well but requires a normally closed switch to produce a logic-*high* input in response to a press of the button. Such switches are less common than the normally open variety. Also, a physical jolt to the switch could cause its contacts to part for a millisecond or so—more than long enough for the computer to read a false switch depression.

An inverted input is produced by the circuit of Figure 6.7(c); a press of the button produces a logic *low* input. However this is of little consequence because the signal can be reinverted with the software instruction COM, which changes all binary 1s to Øs, and vice versa. This circuit provides a common ground line for all inputs and produces a solid Ø V for a *low* level, with little possibility of logic-*high*-level noise across the shorted switch. The *high* level is near +5 V, allowing a 2.6-V margin before noise could produce a false logic-*low* input. It also uses the more-common switch form: normally open.

An open TTL input generally assumes a *high* level, but it is very unwise to depend on this and omit the pull-up resistor R_4 in Figure 6.7(c). Stray capacitive coupling of only few picofarads to a switching line can produce false *lows* on an open TTL input. This trouble is often particularly difficult to find, since a high-frequency signal will generally *not* produce false triggering, but a low-frequency or single-shot signal will. Figure 6.8 illustrates the time-constant "spikes" responsible for this fact. An external pull-up resistor, even of such a relatively high value as 100 kΩ, will eliminate false triggering.

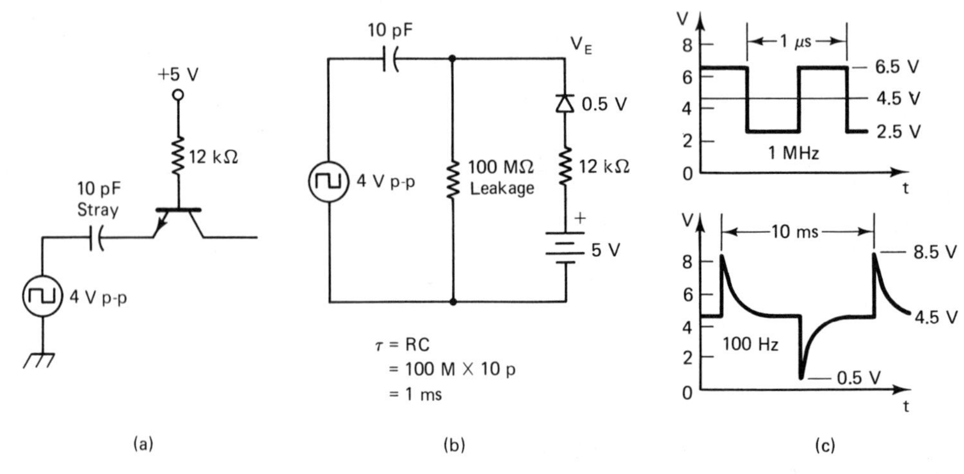

Figure 6.8 (a) An open TTL input, and (b) its equivalent circuit. At (c) a high-frequency stray signal does not cause false logic-Ø levels, but a low-frequency signal of the same amplitude does.

An open CMOS or microprocessor input, because of its very high impedance, will pick up noise even from the 60-Hz ac line. An external pull-up (or pull-down) resistor is absolutely necessary on any otherwise-open high-impedance inputs. This also provides added protection against static-discharge damage to the chip.

6.4 SOFTWARE SWITCH DEBOUNCING

Section Overview

If a switch's contacts are liable to bounce for a maximum of 20 ms, the microprocessor can be programmed to require two identical input reads separated by a 50-ms delay before accepting the switch input. This eliminates the need for hardware switch debounce circuitry.

The problem of switch-contact bounce was discussed in Section Ø.5 (see Figure Ø.9). The hardware debounce solution proposed there required an SPDT switch and an *RS* flip-flop for each line to be debounced. With the availability of a microprocessor, the simple input of Figure 6.7(c) can be used, with its more readily available SPST normally open switch. The debounce can be taken care of by programming. Figure 6.9 gives a flowchart for the software debounce routine.

The debounce routine reads a byte from 8 switches, such as the one in Figure 6.7(c), through 8 input buffers. It then delays for 0.05 s and reads the buffer again. If the two reads, 50 ms apart, are identical, it can be assumed that the inputs are valid (as opposed to noise) and that the switches have stopped bouncing. If the two reads are different, no input data is accepted, and the processor starts the *read* routine from the beginning.

Once valid data is accepted the processor inverts the logic sense of the data, so a depressed switch produces a binary 1. It then waits until all keys are released (all binary 1s at the buffer inputs) before leaving the routine.

Of course, the processor may enter the routine with no switches pressed, in which case the processor will remain trapped in the first Read loop until a key is pressed.

REVIEW OF SECTIONS 6.3 and 6.4

Answers appear at the end of Chapter 6.

11. An open-circuited TTL input will generally assume what logic level?
12. An open microprocessor input pin will generally assume what logic level?
13. A normally open switch to ground produces a positive-logic Ø for a press of the switch. How can this be changed to a logic 1 in the microprocessor?
14. Is an open TTL input more likely to trigger on a high- or low-frequency signal coupled in by stray capacitance?

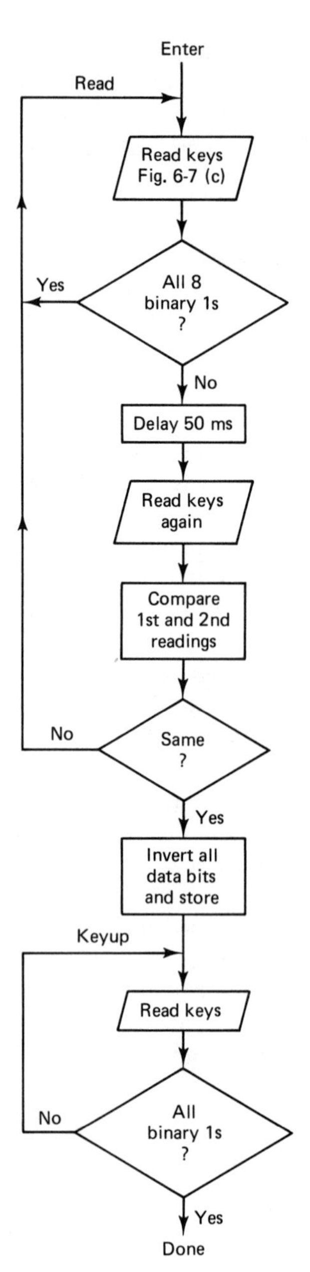

Figure 6.9 A software switch-debounce routine.

15. Why is it a good idea to use a pull-up resistor to +5 V on a TTL input but a poor idea to use a pull-down resistor to 0 V?

16. Give two advantages of software switch debouncing over hardware debouncing.

17. Why would a delay of 50 μs not be adequate in the flowchart of Figure 6.9?

18. What is the purpose of the third key-read block in the flowchart of Figure 6.9?
19. Could the delay required in Figure 6.9 be achieved in an 8-bit register on a 2-μs clock without nested loops? If yes, how many machine cycles per loop would be required?
20. If the 50-ms delay uses the loop

```
LOOP    DEX
        BNE    LOOP
```

what hex value should initially be loaded into X? Assume a 2-μs clock.

6.5 LOGIC INSTRUCTION APPLICATIONS

Section Overview

The 6802 has 67 different instruction types. With their various addressing modes, the total number of instructions is 197. A large part of learning to use a microprocessor consists of learning the instruction set and its applications. This section is devoted to applications of the 6802 logic instructions AND, OR, and EXCLUSIVE OR. Section 6.6 explains the SHIFT and ROTATE instructions. Chapters 8 and 10 contain a more complete catalog of 6802 instruction specifications.

The 6802, like most other processors, has a set of instructions for implementing binary logic. Often, the function of these instructions is obvious from the corresponding hardware element. Such is the case with the COM (complement) instruction, which simply inverts the logic sense of 8 binary digits. Other logic instructions have special uses, which we now enumerate.

AND—bit masking: The logic AND instructions implement, 8 bits at a time, the truth table

$$\begin{aligned} 0 \cdot 0 &= 0 \\ 0 \cdot 1 &= 0 \\ 1 \cdot 0 &= 0 \\ 1 \cdot 1 &= 1 \end{aligned}$$

One of the input bit sets always comes from an accumulator. The other set may be immediate data or data accessed at a memory location (which location *may*, in fact, be an input buffer). The resulting bit set is always stored in an accumulator.

The AND instruction is used for *bit masking*. This is the process of reducing all unwanted bits in a byte to logic 0 so the desired bit or bits can be examined. The technique is to AND the byte in question with an immediate byte, which contains binary 0s in all unwanted bits and binary 1s in bit positions to be retained.

The instruction ANDA #$0F, for example, masks off the most-significant nibble of accumulator A while preserving the least-significant nibble. For example,

		Hex
Original A	1001 1011	9B
Operand byte	0000 1111	0F
	0000 1011	0B

Example 6.2

Write the 6802 source code for a routine that will hang up until the data bit D3 at input-buffer address $2000 becomes 1. The other input bits may be assuming either binary state all the while.

Solution

```
WAIT    LDAA    $2000    Get input byte.
        ANDA    #$08     Mask off all but bit 3.
        BEQ     WAIT     If result = 0, bit 3 was zero.
        NOP              Otherwise bit 3 = 1, and exit.
```

The OR instructions can be used to *set* selected bits to logic 1, as the AND instructions cleared them to logic 0.

Example 6.3

Accumulator A contains data $A2. What does it contain after executing the instruction

```
ORAA #$C6
```

Solution

```
A2 = 1010 0010
C6 = 1100 0110
     1110 0110 = $E6
```

The Exclusive OR (EOR) instructions can be used to perform bit-by-bit comparisons between two bytes. Positions containing identical bits will produce logic-0 result bits. Positions containing a 1 in one byte and a 0 in the other will produce logic-1 results. Example 6.4 in the next section demonstrates the usefulness of this instruction.

6.6 SHIFT AND ROTATE INSTRUCTIONS

Section Overview

Shifting a binary byte 1 bit right divides the value by 2. Shifting it left multiplies the value by 2, providing that the 8-bit capacity of the register is not exceeded.

An *arithmetic* shift retains the most-significant (sign) bit.

The ROTATE instructions circulate 9 bits (1 byte plus carry flag), restoring the original conditions after nine *rotates*. A program routine is presented using *rotate* and *branch if carry clear* to compare 2 bytes of data, bit by bit, and count the number of bits that are different.

The Logic Shift Right (LSR) operation can be carried out on accumulators A or B or on a RAM memory location. When done on a RAM location, the processor actually reads the RAM byte, shifts the bits internally, and then writes the shifted data back to the RAM. Attempts to shift data in a nonreadable device, such as an output latch, would therefore be futile.

A logic shift simply causes the 8 data bits of the operand to be moved 1 bit position, as shown in Figure 6.10. The end bit shifts into the carry flag, where it can be tested with the BCS or BCC instructions. Zero bits shift into the other end, so after eight shifts the operand will be sure to contain all Ø bits.

Notice that if the byte to be shifted is considered to be a pure (unsigned) binary number, an LSR divides the number by 2, with the C flag representing a value of one-half.

The Arithmetic Shift Right (ASR) is for use with signed binary numbers. Bit 7, which is merely a positive/negative indicator, remains in place instead of having a Ø shifted in. A shift therefore divides positive or negative numbers by 2. Figure 6.11 shows an example.

The Arithmetic Shift Left (ASL) is identical with a logic shift left. Indeed, some processors such as the 6809 reference a single object-code instruction under two source-code mnemonics: ASL and LSL.

When used as an LSL for pure binary numbers, the ASL instruction is straightforward. When used as an ASL for signed numbers, the value of the positive or negative number is multiplied by 2. The maximum range for an 8-bit register is −128 to +127, and invalid results will be obtained if the multiply-by-two would produce a value outside this range. Figure 6.12 shows two examples of the ASL.

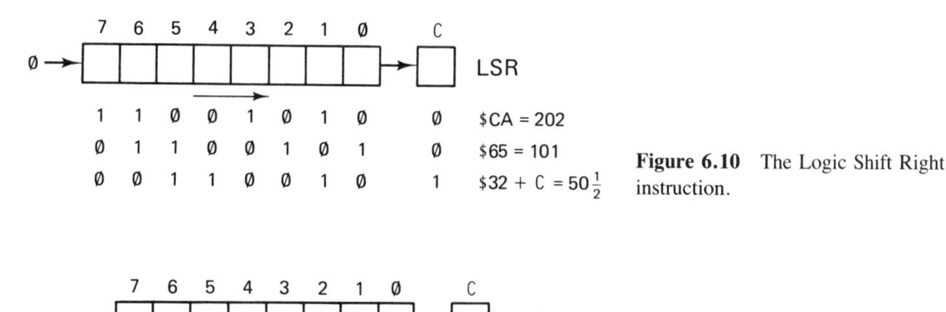

Figure 6.10 The Logic Shift Right instruction.

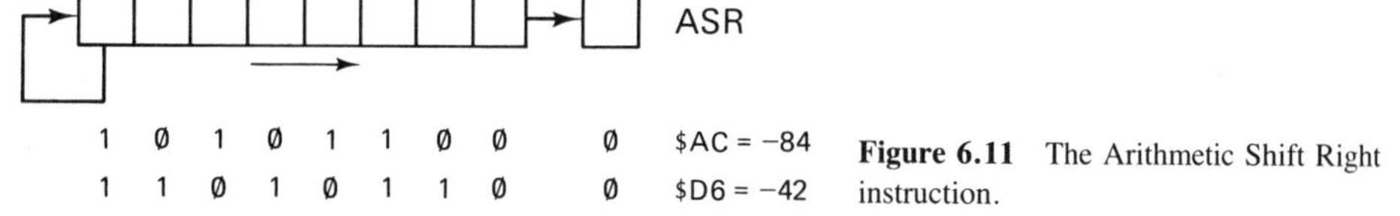

Figure 6.11 The Arithmetic Shift Right instruction.

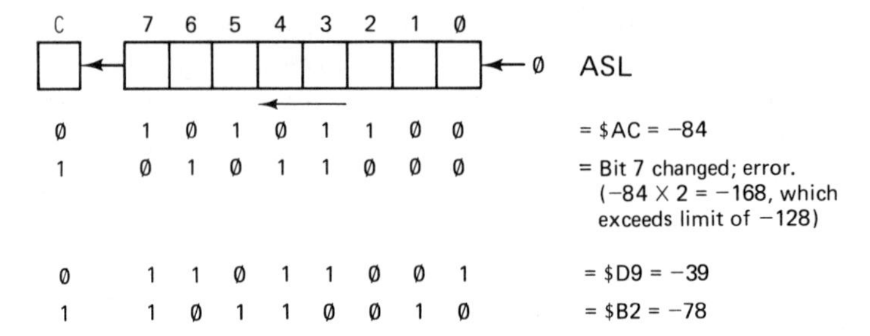

Figure 6.12 Two examples of the ASL instruction. The first is invalid in signed binary, since it would produce a result more negative than −128.

The first produces invalid results. The second produces results within the allowed range.

When using shift instructions to multiply it is necessary that the software routine check the status of bit 7 before and after the ASL. If bit 7 changes, there has been an overflow, and the result must either be discarded or corrected. Since bit 7 is retained in the carry flag, this check can be made by simply checking that the C and N flags are identical.

The 6802 provides a special flag in the condition-code register, the V (overflow) flag, which assumes a true (1) value when a shift instruction produces an overflow in signed-number arithmetic. The branch instructions BVS and BVC sense the V flag status.

The Rotate instructions are similar to the logic shifts, except that the bits shifted out of the carry are shifted back into the other side of the register. Nine *rotate*

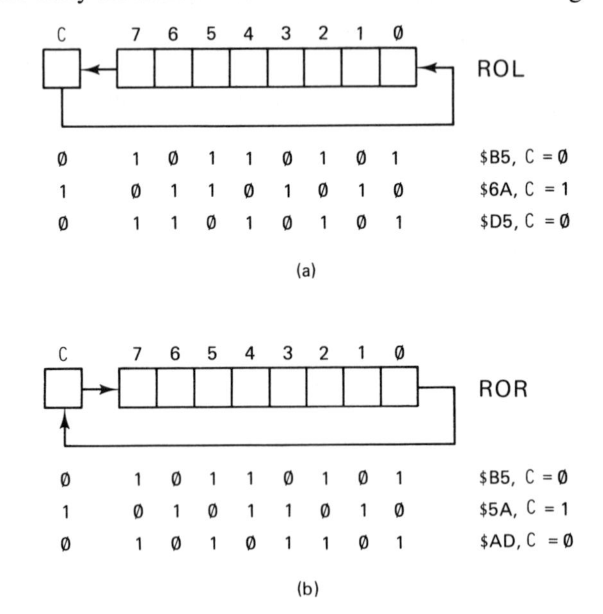

Figure 6.13 (a) The Rotate Left instruction, and (b) the Rotate Right instruction.

instructions will shift the 8 register bits and the carry in a full circle, restoring them all to their original positions. Rotate left (ROL) and rotate right (ROR) can be applied to the A or B internal registers or to an external RAM byte. Figure 6.13 diagrams the *rotate* instructions.

Example 6.4

Write the 6802 source code for a routine that will determine how many of the bits in two data bytes are different.

Solution

Assume that the first byte is in accumulator A and the second byte is in memory at address $0200. The two bytes will have an exclusive-OR operation applied, leaving binary 1s in accumulator A, where the two corresponding bits were not both 0 or both 1. Accumulator A will then be shifted until empty of 1s. Each time a 1 is shifted into the carry flag, accumulator B will be incremented. When A is empty, B will contain the number of differing bits.

```
        EORA   $0200   Set 1s where A and memory differ.
        CLRB           Start count at zero.
LOOP    LSRA           Shift a bit into C flag.
        BCC    SKIP    Don't count if bits same,
        INCB           Count if bits differ.
SKIP    TSTA           More 1s in register A?
        BNE    LOOP    Yes: go get 'em.
DONE    NOP            Else exit.
```

REVIEW OF SECTIONS 6.5 and 6.6

21. Give the instruction and operand that will mask to 0s the most-significant 3 bits of the data in accumulator A.
22. Accumulator A contains data $5A. Memory location $0031 contains data $E9. What does accumulator A contain after executing the following instruction?

    ANDA $31

23. What is the functional difference between the following instructions?

    LDAA #$FF and ORAA #$FF

24. Accumulator B contains data $91. What does it contain after executing the following instruction?

    ORAB #$B4

25. Accumulator A contains data $B7. What does it contain after executing the following instruction?

    EORA #$6D

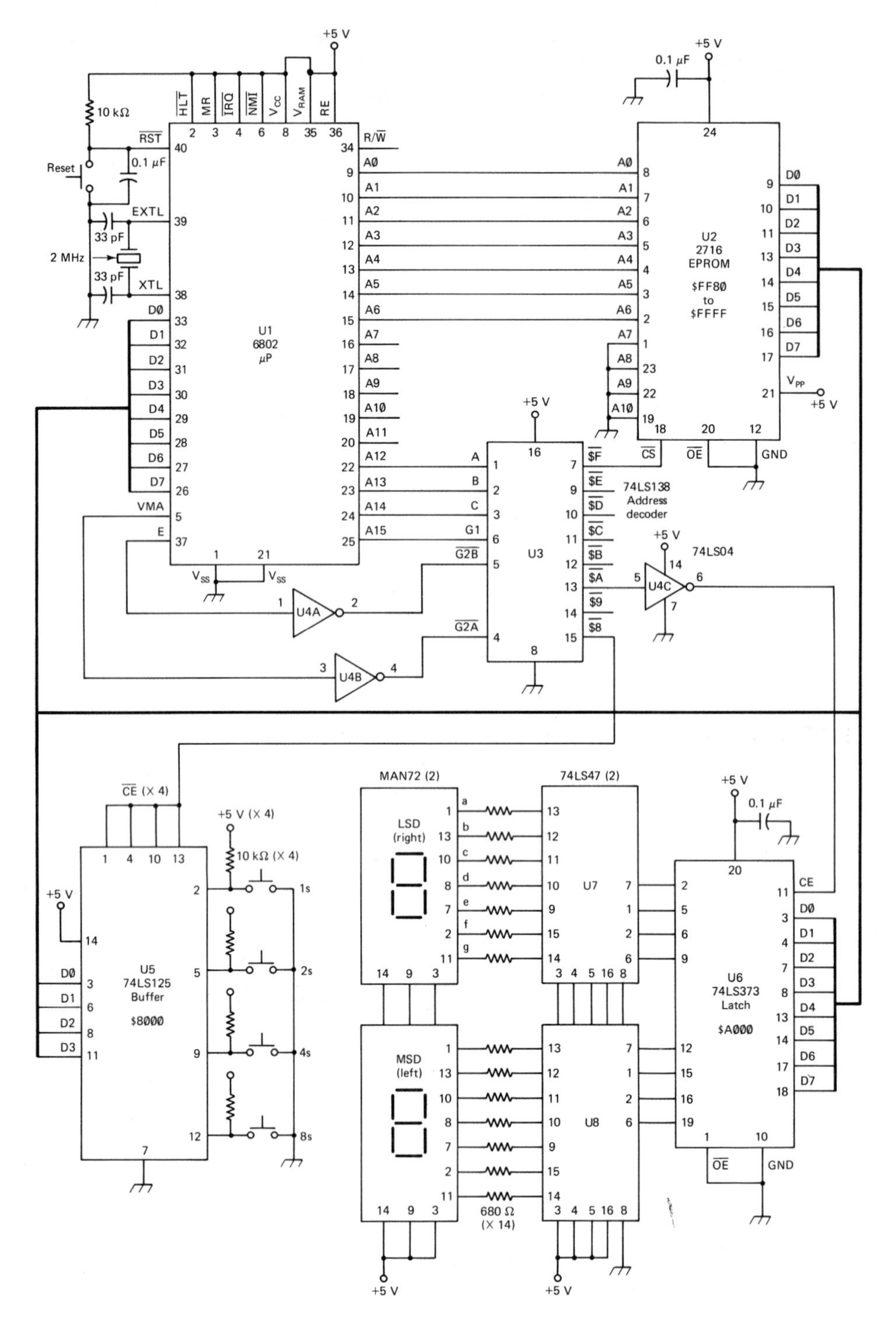

Figure 6.14 Hardware diagram for the decimal input-and-display system. See Figure 7.6 for a RAM configuration if an EPROM programmer is not available.

26. Accumulator A contains data $57. What does it contain after two executions of the LSRA instruction?
27. Accumulator B contains data $B9 (decimal −71). Give the hex contents and decimal equivalent after two executions of the ASRB instruction.
28. Accumulator A contains $41. An ASLA instruction is executed. Is the result valid in signed binary arithmetic? How does the 6802 indicate this?
29. RAM memory location $0035 contains data $C3. The carry flag is clear. What does this RAM location contain after four executions of the instruction:

ROR $0035

30. What CCR flag is set if two identical bytes of data are exclusive-ORed?

6.7 A BCD INPUT AND DISPLAY

Section Overview

This section describes a system that inputs BCD digits from four pushbuttons and packs the last two digits received into an output byte, which is displayed on two 7-segment LEDs. Software switch debouncing is used to screen out spurious inputs. Bit masking and shifting are used to allow two 4-bit inputs to fill a single 8-bit output.

The system hardware is diagrammed in Figure 6.14. The four switches should ideally be lever switches. Pulling a lever down inputs a binary 0. In the rest position the input nibble is $F, which is invalid in BCD and is ignored.

The program is flowcharted in Figure 6.15. The input routine tests the four input bits, accepting only valid BCD combinations, which remain stable for 0.5 second. The existing least-significant digit is then shifted to the most-significant position while the new input digit is ORed into the empty least-significant position. The two BCD digits are then displayed via the output latch and decoders. The program listing is given in List 6.1.

```
                  1000 * LIST 6-1    BCD INPUT AND DISPLAY; FIG 6-14 HARDWARE
                  1004 *  SUCCESSIVE 4-BIT INPUTS PRODUCE 2-DIGIT DECIMAL DISPLAY.
                  1006 * RUNS, 7 JULY '87.   D. METZGER
                  1010 *
                  1020          .OR  $FF81    BASE ADR OF 6810 PROGRAM RAM
                  1030          .TA  $4081     HIGHEST MIRROR IS $FF80.
8000-             1040 INBUF    .EQ  $8000    INPUT BUFFER BASE ADDRESS.
A000-             1050 LATCH    .EQ  $A000    OUTPUT LATCH BASE (74LS138 DECODED)
0001-             1060 DIGIT    .EQ  $0001    BUFFER SAVES KEY VALUE.
                  1070 *
FF81- 5F          1080          CLRB          B HOLDS DATA TO OUTPUT.
FF82- B6 80 00    1090 KEYIN    LDAA INBUF    INPUT BYTE;
FF85- 84 0F       1100          ANDA #$0F      MASK HIGH-ORDER NIBBLE TO 0.
FF87- 81 0A       1110          CMPA #$0A     ACCEPT ONLY IF KEY 0 THRU 9.
FF89- 24 F7       1120          BCC  KEYIN
FF8B- 97 01       1130          STAA DIGIT    SAVE KEY VALUE.
FF8D- CE 70 00    1140          LDX  #$7000   DELAY FOR 1/2 SEC.
```

```
FF90- 09        1150 DELAY   DEX
FF91- 26 FD     1160         BNE   DELAY
FF93- B6 80 00  1170         LDAA  INBUF     CHECK KEY AGAIN
FF96- 84 0F     1180         ANDA  #$0F       FOR DEBOUNCE.
FF98- 91 01     1190         CMPA  DIGIT
FF9A- 26 E6     1200         BNE   KEYIN     REJECT IF NOT SAME AFTER .5 S.
FF9C- B6 80 00  1210 KEYUP   LDAA  INBUF     WAIT UNTIL ALL 4
FF9F- 84 0F     1220         ANDA  #$0F       KEYS ARE UP.
FFA1- 81 0F     1230         CMPA  #$0F
FFA3- 26 F7     1240         BNE   KEYUP
FFA5- 58        1250         ASLB            SHIFT LAST DIGIT UP
FFA6- 58        1260         ASLB             TO MOST-SIGNIFICENT.
FFA7- 58        1270         ASLB             0000 FILLS LOW-ORDER
FFA8- 58        1280         ASLB             NIBBLE.
FFA9- DA 01     1290         ORAB  DIGIT     COMBINE NEW LOW- , OLD
FFAB- F7 A0 00  1300         STAB  LATCH      HIGH-ORDER NIBBLES.
FFAE- 7E FF 82  1310         JMP   KEYIN     OUTPUT TWO BCD DIGITS.
                1320         .OR   $FFFE
                1330         .TA   $40FE
FFFE- FF 81     1340         .HS   FF81      RESET VECTOR
                1350         .EN
```

Answers to Chapter 6 Review Questions

1. One; one **2.** A read of the address decoded to enable the buffer.
3. 512 **4.** $F000
5. 0.4 mA from the driven gate into the driving gate.
6. Essentially zero current. **7.** Input capacitance
8. Each line should have a pullup resistor to V_{CC}. **9.** A *high* level, generally.
10. The operand address. **11.** A high level, but not reliably so.
12. Alternating *high* and *low* from noise pickup, often 60 Hz.
13. Use the COM (bitwise complement) instruction. **14.** Low frequency.
15. Pull-up to only +2.5 V is adequate, but pull-down must be to 0.4 V or lower.
16. (a) Two logic gates are eliminated, (b) the more common SPST, normally open switch can be used. **17.** Switch bounce may continue for several milliseconds.
18. To keep the processor from accepting a second input until after the key is released. **19.** Not reasonably. 196 μs or 98 machine cycles per loop would be required on a 2-μs clock. **20.** $C35
21. ANDA #$1F **22.** $48 **23.** No difference
24. $B5 **25.** $DA **26.** $15 **27.** $EE, −18
28. No; V flag is set. **29.** $6C **30.** Z flag

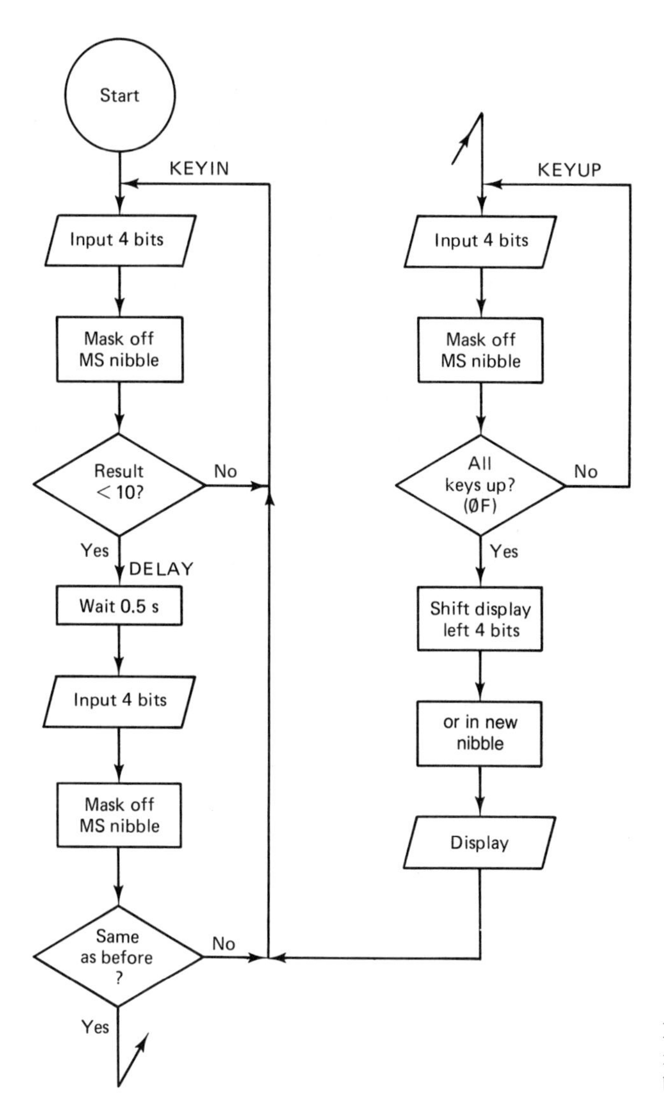

Figure 6.15 Flowchart for the decimal input-and-display program, demonstrating bit masking and shift operations.

CHAPTER 6 QUESTIONS AND PROBLEMS

Digits after the decimal point refer to section numbers in Chapter 6.

Basic Level

1.1 For how long is the output of a data buffer asserted on the data bus? For how long is the output of a data latch asserted on the output bus?

2.1 What 6802 instructions are normally used to input data from a buffer?

3.2 An LED is connected in series with a 680-Ω resistor to ground. Can the output from a 74LS373 latch fully light this LED by pulling toward V_{CC}? If the series is connected to V_{CC} instead of ground, can the IC safely light the LED by pulling toward ground?

4.2 How many LS-TTL inputs can be driven reliably from one microprocessor output?

5.3 Which of the three input-switch configurations shown in Figure 6.7 is generally preferred?

6.3 How can the levels from an input buffer be inverted without changes or additions to system hardware?

7.4 How many input keys can be debounced with the single software debounce routine of Figure 6.9?

8.4 What is the purpose of the final loop in the key-read flowchart of Figure 6.9?

9.5 What instruction and operand will mask the high-order nibble of a byte in accumulator B to zero, leaving the low-order nibble unchanged?

10.5 Assume that accumulator A contains data $19. What does it contain after executing EORA #$3C?

11.6 How can an ASL be accomplished in which 1 bits fill in from the right instead of 0 bits? (Be careful not to change the C or Z flags.)

12.6 The data in RAM location $0123 is $B9. What is the value after two LSR $0123 instructions are executed?

13.7 What instruction in the BCD Display program of List 6.1 makes sure that the display will be 00 when the program is first entered?

14.7 What instruction in the BCD display program makes sure that the data on open lines D4 through D7 is not accepted on a read of the first input buffer?

Advanced Level

15.1 Draw a logic diagram similar to that of Figure 6.2(b), showing how a 74LS138 can be connected to an address bus to provide eight blocks of 16 words each, from address $C580 through $C5FF.

16.1 A 74LS138 decoder has $\overline{G2B}$ fed from $\overline{A15 \cdot A14 \cdot A13 \cdot E \cdot VMA}$. A12 feeds $\overline{G2A}$ and A11 feeds G1. A10, 9, and 8 feed C, B, and A, respectively. Draw the logic diagram and determine the addresses to which the third lowest output (pin 13) responds.

17.2 Redraw the circuit of Figure 6.4(b), showing how a pullup resistor may be added to increase the transistor's base-drive current. Show calculations for the lowest permissible value of this resistor staying within the 8-mA sinking limitation, and predict the new base drive current. Use 0.7 V for V_{BE} and 0.1 V for $V_{CE(sat)}$.

18.2 How long will it take for the maximum pull-up current of a CMOS output to raise the output voltage from 0 to 3 V on a line with a shunt capacitance of 100 pF? How does this compare with the pull-down time? Can you suggest a method to equalize the two times? Refer to Figure 6.5(a).

19.3 Give two reasons why pull-down to ground is preferable to pull-up to V_{CC} when switches are used to input data to TTL buffers.

20.3 If an open TTL input assumes a logic-1 state, why is it necessary to pull unused inputs *high* with a resistor to V_{CC}?

21.4 Write the 6802 source code (and comments, of course) for the debounce routine of Figure 6.9. Use nested delay loops for the delays. Assume a 1-MHz clock.

22.5 Write the 6802 source code for a routine that will continue endlessly to read an input buffer at address $2040 until a data byte is applied for which the high-order nibble is equal to $A and the low-order nibble is *not* equal to $B.

23.6 Describe the difference between the ASL and LSL instructions.

24.6 Write a 6802 routine that will count the number of logic-1 bits in accumulator A and leave the total as a binary number in accumulator B. The contents of A must be returned to their original value.

25.7 In the BCD display program of List 6.1, explain how program lines 1110 and 1120 reject any nondecimal digit values.

PART II SOFTWARE, SUBROUTINES, AND TROUBLESHOOTING

7 Indexed Addressing and Data Tables

7.1 THE INDEXING CONCEPT

Section Overview

All microprocessors have at least one internal address-pointer register. This allows a fixed program routine in ROM to access data from a variety of addresses rather than from a fixed address.

In the 6800 family of microprocessors, this register is called the index register, X, and it contains the 16 bits of the operand address for an indexed-mode instruction.

A *table* is a block of memory containing similar-type data. Indexed addressing can be used to access all the bytes in a table by incrementing X to point to the next consecutive byte before each pass through a loop.

A frequent requirement of a microcomputer system is the storing of a large number of data words of a similar kind. As an example, let us say that the cargo hold of an aircraft is to be loaded with parcels until the maximum safe take-off weight is reached. The parcels are weighed as they are loaded, and the weights are stored in memory. The total weight in the hold can be determined at any time by having

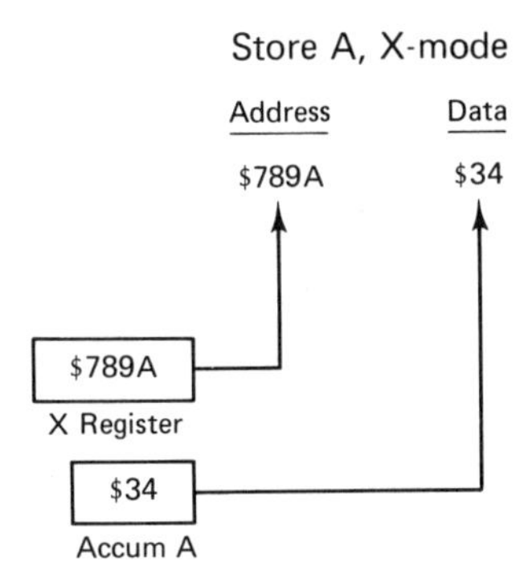

Figure 7.1 In an instruction using indexed addressing the X register contains the address of the operand.

the computer add all the data words. If a parcel is removed, the data word specifying its weight can be removed from memory.

Data tables: We would find it convenient to store the parcel-weight data in a group of consecutive RAM bytes, which we call a *data table*.

The index register: We can write numerous bytes of data into sequential data-table locations with a single program routine using *indexed addressing*. This is an addressing mode in which the address of the operand is obtained from an internal microprocessor register. In the 6802 this operand-address register is the 16-bit X *register*. We sometimes say that the X register *points to* the data to be operated on. Figure 7.1 illustrates how a *Store Accumulator* A instruction is executed in the *X*-indexed mode.

The assembly-language form for the indexed-addressing mode is illustrated here for the 6802 *Store Accumulator* A instruction:

Label	Mnemonic	Operand
	STAA	00,X

The operand field does not contain an operand at all but instead contains the *offset* and a comma, followed by a letter *X* to indicate *X*-indexed mode. The offset is often zero, in which case we can simply enter 0 and it will be of no more concern.

To access an entire table of data, the index register is incremented (to point to the next table address) before each pass through a loop. Here is a routine to clear a block of data (store data $00) extending from address $4000 through address $5FFF.

```
        LDX   #$4000   Start of block.
CLEAN   CLR   0,X      Clear a byte and
        INX            move on to the next.
        CPX   #$6000   Finished last byte?
        BNE   CLEAN    No: loop back for another.
```

The LDX, *immediate*, instruction points X to the beginning address of the data block at the start of the CLEAN routine. The CPX compares the latest value in *X* to the address just beyond the block area. When this address is reached, the CLEAN routine is finished and the program "falls through" the BNE instruction and proceeds to the next program segment.

A nonzero offset gives an 8-bit binary value that is added to the 16-bit binary value in the X register to obtain the final operand address (called the effective address, or EA). In the example of Figure 7.1, where X contains $789A, if the instruction given were

```
STAA    02,X
```

the address where data $34 is stored would be

EA = X + offset = $789A + $02 = **$789C**

The offset feature is useful when each entry to the data table includes several elements. In the aircraft cargo-hold example, each parcel may require three bytes:

Byte 1: Weight in pounds and ounces
Byte 2: Destination; one of 256 airports
Byte 3: Postage paid; 0 through $2.55, for the moment.

The X register would be loaded with the base address for the parcel in question:

```
LDX     #$789A
```

The weight would be stored by using zero offset:

```
LDAA    WEIGHT
STAA    00,X
```

The destination code would be stored at the next sequential address by using an offset of 1:

```
LDAA    DESTIN
STAA    01,X
```

And the postage would use the same *X*-register base address, with an offset of 2:

```
LDAA    POST
STAA    02,X
```

Note that the offset is an unsigned binary number. $FF means +255, not −1. Also remember that 6802 indexed-mode instructions are always 2 bytes long; the 00 offset must appear in the object code, even if the offset feature is not used. Some assemblers assume a zero offset if none is given, but for others the operand must be listed as 0,X.

7.2 BUILDING AND READING DATA TABLES

Section Overview

When table data is accessed sequentially, the end of the table can be recognized by

1. **Comparing the value in address-pointer X to the address of the end of the table, or**
2. **Reserving a special 8-bit data value (such as $FF) as an end-of-table flag and comparing each data byte processed to this value.**

The rate at which data is accessed can be controlled by

1. **Letting the read or write loop run as fast as the processor clock allows, or**
2. **Inserting a fixed delay loop between successive data accesses, or**
3. **In the case of data transfers from an input buffer to a RAM data area, requiring a special *flag* byte to be input before each valid data byte is accepted.**

Other methods of data-transfer control are discussed in Chapters 9 and 14.

A microprocessor can access data in a table at a rate of perhaps 100 000 bytes per second. This is often very much faster than the data rate of the system input or output. There are many techniques for slowing the accessing of a table down to a manageable rate. We examine two of them in this section. Others will be seen in Chapter 9 on interrupt processing.

Since memory space for the table is limited, it will be necessary somehow to ensure that data storage does not continue indefinitely. Thus we will see two techniques for identifying the end of the table, where storage must stop.

Building a table: Let us assume that we have a scale that outputs a data byte corresponding to parcel weight as soon as the weight settles. We would like to store one weight in a table location and not store again until the scale outputs 00 (indicating the package has been removed from the scale) and then outputs another nonzero value (indicating that another package has settled on the scale).

The length of the table is to be limited to 100 bytes. It will start at RAM address $0100. Figure 7.2 gives the BUILDTABLE program flowchart. The source-code listing is given in List 7.1.

```
                 0900 * LIST 7-1.  PROG SEGMENT; NOT TO RUN.  6 JUL '87  METZGER
                 1000 * BUILDTABLE ROUTIINE; INPUTS FROM SCALE; 100 BYTES MAX
                 1010 *
                 1020        .OR  $FF81
                 1030        .TA  $4081
4000-            1040 INBUF  .EQ  $4000
                 1050 *
FF81- CE 01 00   1060        LDX  #$0100   START OF TABLE.
```

```
FF84- B6 40 00 1070 NOWT    LDAA INBUF    LOOK FOR NONZERO WEIGHT.
FF87- 27 FB    1080         BEQ  NOWT
FF89- A7 00    1090         STAA 00,X     STORE IN TABLE.
FF8B- 08       1100         INX           POINT TO NEXT IN TABLE.
FF8C- 8C 01 64 1110         CPX  #$0164   IF END OF TABLE SPACE;
FF8F- 27 07    1120         BEQ  ENDTAB    GO TO END ROUTINE (NOT GIVEN).
FF91- B6 40 00 1130 HOLD    LDAA INBUF    WAIT UNTIL PARCEL IS
FF94- 26 FB    1140         BNE  HOLD      REMOVED (WEIGHT = 0).
FF96- 20 EC    1150         BRA  NOWT     LOOK FOR NEXT WEIGHT.
FF98- 01       1160 ENDTAB  NOP
               1170         .EN
```

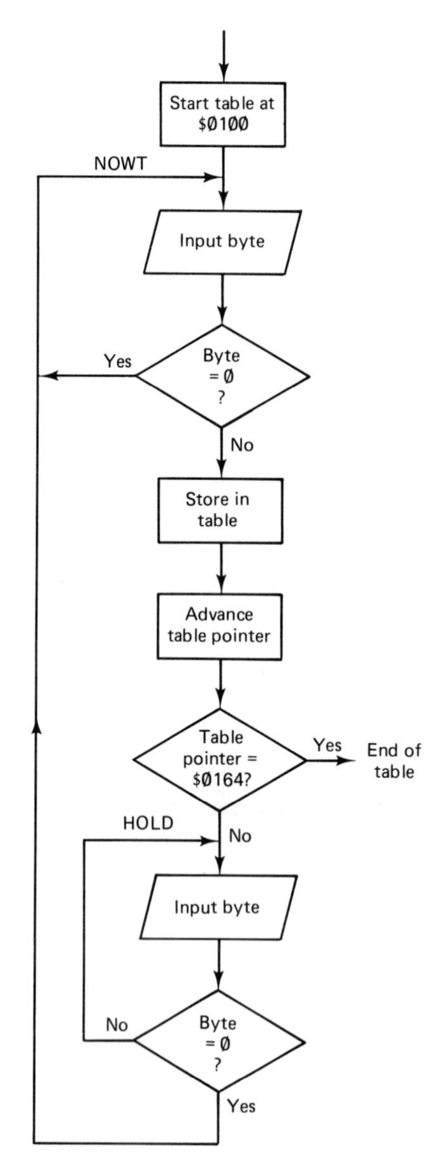

Figure 7.2 Flowchart for a routine to build a 100-byte table from data provided by a digital scale.

This routine slows down the table-access rate by requiring a code byte (00) to appear between each valid data byte. The code byte only appears as fast as the operator places new parcels on the scale. The end of table is specified as an absolute memory address and is recognized with the immediate-mode *Compare X* instruction.

Reading a table can be quite similar to building a table. However, this time we will demonstrate new techniques for slowing down the data-access rate and an alternative method for recognizing the end of table.

Let us assume that we wish to output data bytes from a table at a rate of 1 byte/ms. The table starts at address $0100 and may be of any length. The end of the table is *flagged* by a code byte, $FF. Of course, this means that the value $FF is forbidden as a valid data byte, and the BUILDTABLE routine must ensure that only data values $00 through $FE are allowed. Upon reading the end-of-table flag, the READTABLE routine is to return to its beginning for another in an endless series of table reads. The system operates on a 1-MHz clock.

Figure 7.3 gives the READTABLE flowchart. The program listing is given in List 7.2.

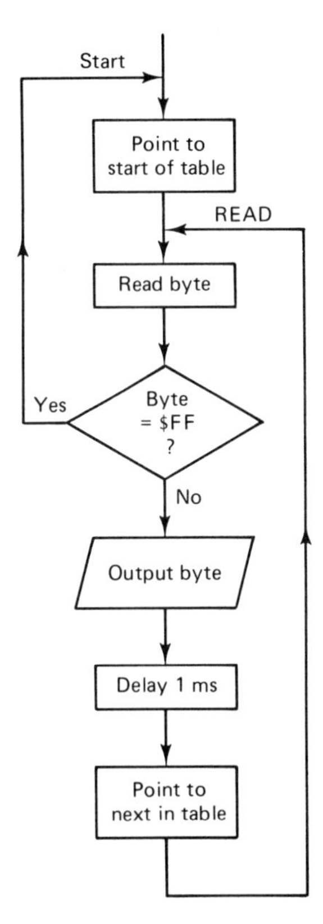

Figure 7.3 Flowchart for a routine to output 1 byte from a table every millisecond.

```
                1000 * LIST 7-2. PROG SEGMENT; NOT TO RUN. METZGER, 4 DEC '87
                1010 * READTABLE; OUTPUT ONE BYTE EVERY MILLISECOND.
                1020 *  STOP ON READING END-OF-TABLE FLAG $FF.
                1030 *
                1040           .OR   $FF81
                1050           .TA   $4081
6000-           1060 LATCH     .EQ   $6000
                1070 *
FF81- CE 01 00  1080 START     LDX   #$0100   3  SET START OF TABLE.
FF84- A6 00     1090 READ      LDAA  00,X     5  GET DATA FROM TABLE.
FF86- 81 FF     1100           CMPA  #$FF     2  IF DATA IS END FLAG
FF88- 27 F7     1110           BEQ   START    4   GO BACK TO START OF TABLE.
FF8A- B7 60 00  1120           STAA  LATCH    5  OUTPUT DATA BYTE.
FF8D- 08        1130           INX            4  POINT TO NEXT BYTE IN TABLE.
FF8E- C6 A2     1140           LDAB  #$A2     2  DELAY $A2 OR 162 LOOPS:
FF90- 5A        1150 DELAY     DECB           2  : 6 US PER LOOP; 1000 - 29 US DELY
FF91- 26 FD     1160           BNE   DELAY    4  : NEEDED; 971/6 = 162 LOOPS.
FF93- 20 EF     1170           BRA   READ     4  START OVER AFTER 1-MS DELAY.
                1180           .EN
```

REVIEW OF SECTIONS 7.1 AND 7.2

Answers appear at the end of Chapter 7.

1. In the *indexed* instruction mode, what is the data type contained in the X register? (Review Section 2.2.)
2. Give the complete assembly-language instruction you would use to establish the start of a table at address $1234.
3. Give the assembly-language instruction you would use to move to the next sequential byte in a table.
4. If the X register contains $1235, what address is accessed by the instruction LDAA 01, X?
5. Here is a program segment to stop a table-access loop. The table starts at address $2400 and is 200 bytes long. What should be the value of XXXX?

```
INX
CPX     #$XXXX
BNE     LOOP
```

6. In the BUILDTABLE routine (List 7.1), what is the highest table address to be loaded with data?
7. In the BUILDTABLE routine, what command makes sure that the table is not filled with zeros when the scale is empty?
8. In the BUILDTABLE routine, what command imposes the requirement that the first package be removed before the second data byte can be stored in the table?
9. In the READTABLE routine (List 7.2), what would happen if the table contained a weight of 15 lb 15 oz ($FF)?
10. If the delay required in the READTABLE routine were ½ s rather than 1 ms, would you use nested loops or *X*-register decrementing for the delay? Why?

7.3 MOVING A DATA BLOCK: THE STACK POINTER

Section Overview

The *stack pointer*, S, is another 16-bit address pointer register in the 6802. The *push* instructions store data to the address pointed to by S and then automatically decrement S. The *pull* instructions first increment S and then load data from the address pointed to by S.

A block-move routine can be written using X to point to the source-data addresses and S to point to destination-data addresses.

To move a block of data from one area of memory to another (for example, from ROM to RAM), we need two indexable address-pointer registers. Many microprocessors have two (or more) index registers, but the 6802 has only one—the X register. The 6802 does, however, have a second 16-bit address-pointer register. It is called the *stack pointer*.

The stack pointer register, S, is intended to control the *stack* area of RAM, which we will cover in Chapter 9. However, as long as we take care not to interfere with its stack-handling functions, we can use it as a second index register.

The stack pointer is initialized by a load-immediate command, much like the X register:

```
LDS    #$1234
```

Data is stored from accumulator A to the address pointed to by the S register with the *Push A* instruction:

PSHA (data from A stored to address $1234)

The PSHA instruction is inherent mode and requires only a single byte. It performs two distinct functions:

1. It stores the data from A to the address pointed to by register S.
2. It then automatically decrements the contents of S, so S now points to the next *lower* address.

The single stack instruction thus performs the function of two index instructions:

```
PSHA   ↔   STAA 00,X
           DEX
```

The *Push-on-stack* instruction is a *postdecrement* operation. Data is saved starting at the initialized address and moving down to consecutively lower addresses.

Data is read from addresses pointed to by the S register with the *Pull A* (PULA) instruction. The PULA instruction is a *preincrement* operation. The stack pointer is incremented first, and then the address pointed to is read and loaded into register A. PULA performs two operations similar to *increment* and *load* with the X register:

```
PULA   ↔   INX
           LDAA 00,X
```

Thus PULA reads data from the initialized address *plus one*, upward toward higher addresses.

The contents of the B register can be stored and loaded with PSHB and PULB instructions, just like the A register.

Moving a block of data: Let us say that a microcomputer operating program exists in ROM at the 2K block from $F800 to $FFFF. We would like to move it to the 2K block of RAM at addresses $C000 to $C7FF. We may then modify the program in RAM and finally burn it into a new ROM at the higher addresses. List 7.3 is a program to move the data. Notice that the end of the block must be recognized with a CPX instruction. There is no corresponding CPS instruction to allow branching based on the value in *S*.

```
                1000 * LIST 7-3: MOVE DATA; 1 K-BYTE; USES S REGISTER.
                1005 * TESTED ON DEVELOPMENT SYSTEM 6 JUL '87; METZGER.
                1010 *
                1020          .OR  $4000
                1030 *
4000- CE FF FF  1040          LDX  #$FFFF   END OF BLOCK TO READ.
4003- 8E C7 FF  1050          LDS  #$C7FF   END OF BLOCK TO STORE.
4006- A6 00     1060 NEXT     LDAA 00,X     GET DATA.
4008- 36        1070          PSHA          STORE, AND DECREMENT S.
4009- 09        1080          DEX           MOVE TOWARD START OF BLOCK.
400A- 8C FB FF  1090          CPX  #$F7FF   JUST PASSED START?
400D- 26 F7     1100          BNE  NEXT      NO: THEN KEEP GOING.
                1110          .EN
```

7.4 THE PROGRAMMER'S MODEL

Section Overview

A diagram of a microcomputer's internal machine registers showing their word lengths and basic functions is called a *programmer's model*. With due caution these models can be useful in comparing different microprocessor types.

We have now been introduced to all the accessible internal rgisters of the 6802 and are ready to draw a diagram of them, called a *programmer's model*. This diagram has two uses:

1. It serves as a reminder to the programmer of the available registers in the processor. Programming possibilities that might otherwise be elusive are made easier to recognize.

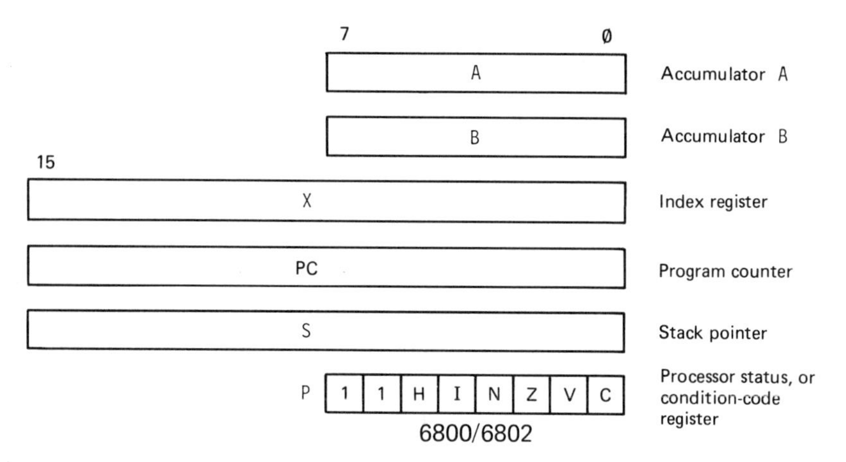

Figure 7.4 A programmer's model for the 6800 / 6802. All machine registers accessible to the programmer are shown.

2. It provides a basis for comparisons among microprocessors. This will become apparent in the last quarter of this text, when we examine a number of other processors. Generally, when a new microprocessor is brought out, the first thing that a computerist will say is, "Let's see the programmer's model."

Figure 7.4 gives the programmer's model for the 6802. This model is the same for the original 6800 and for the single-chip 6801 microcomputer.

Without denying their usefulness, you should guard against the oversimplification that programmer's models alone are a sufficient basis for comparing the "power" of various microprocessors. Hardware features, such as instruction-execution, data-bus and address-bus width, and multiplexing requirements, are as important as the number of accumulators a machine has, but these features are not specified by the programmer's model.

Even more important are software considerations such as a complete and consistent instruction set that is easy to learn and free from annoying "holes," that is, instructions that seem perfectly logical and possible—but are inexplicably missing from the actual machine instruction set.

REVIEW QUESTIONS FOR SECTION 7.3 AND 7.4

Answers appear at the end of Chapter 7.

11. What is the bit length of the 6802 stack pointer register?
12. The S register contains $1234. What address does PSHA store to?
13. The S register contains $1234. What address does PULA read from?
14. The stack pointer can be used to store data ________ (forward or backward) through a table using the ________ (PSH or PUL) instruction.

15. Could the S register be used to count the loops for a long time-delay routine? Check the DES instruction specifications in the programmer's reference and explain your answer.
16. The MOVE DATA program (page 164) uses a DEX instruction but no corresponding DES instruction. Why?
17. Use the programmer's reference at the back of the book to calculate how long the MOVE DATA routine will take to move 2K of data. Assume a 1-MHz clock.
18. The 6802 P register contains data $C5. What is the status of the N, Z, and C flags?
19. Excluding the program counter, what is the total number of bytes in the internal registers of the 6802?
20. What instruction would load the 6802 program counter with data $789A?

7.5 VIRTUAL INDEX REGISTERS

Section Overview

A single X register can be used to index through two data tables. The trick is to reserve 4 bytes of RAM for two *X*-buffers. When the first table is to be accessed, X is loaded from RAM buffer 1, used to load or store data, incremented to ready it for the next access of table 1, and then stored back in RAM buffer 1.

Now X is loaded from RAM buffer 2, used in indexed addressing for table 2, incremented or decremented as required for the next access of table 2, and then stored back in RAM buffer 2.

Data-block moves are so common, especially in video display systems, where screen displays are changed rapidly, that it is worthwhile to see one more way to implement them with the 6802.

Let us say that the S register cannot be freed from its stacking functions (see Chapter 9). We will have to make a single X register count through two data blocks. We can do this by establishing two RAM buffers to hold the contents of *X* as it points to the source block and the destination block, respectively. The X register can thus perform double duty as an address pointer to two blocks. List 7.4 gives the MOVE DATA program of page 164, using RAM buffers to implement multiple virtual X registers in place of using the stack pointer.

```
               0900 * LIST 7-4. RUNS ON DEVELOPMENT SYST,. 6 JUL '87. METZGER
               1000 * MOVE DATA 2; 1 K-BYTE; USES VIRTUAL X REGISTERS.
               1010 *
               1020        .OR  $4000
               1030 *
4000- CE FC 00 1040        LDX  #$FC00  START OF SOURCE TABLE.
4003- DF 10    1050        STX  $10     SOURCE BUFFER; XSH = 0010; XSL = 0011.
4005- CE C4 00 1060        LDX  #$C400  START OF DESTINATION TABLE.
4008- DF 12    1070        STX  $12     DESTIN. BUFR; XDH = $0012; XDL = $0013.
```

```
400A- DE 10     1080 NEXT   LDX   $10      POINT TO SOURCE.
400C- A6 00     1090        LDAA  00,X     GET DATA FROM SOURCE TABLE.
400E- 08        1100        INX            POINT TO NEXT IN SOURCE TABLE.
400F- DF 10     1110        STX   $10      SAVE SOURCE POINTER.
4011- DE 12     1120        LDX   $12      POINT TO DESTINATION.
4013- A7 00     1130        STAA  00,X     STORE DATA TO DESTINATION TABLE.
4015- 08        1140        INX            NEXT IN DESTINATION TABLE.
4016- DF 12     1150        STX   $12      SAVE DESTINATION POINTER.
4018- 8C C8 00  1160        CPX   #$C800   PASSED END OF TABLE.
401B- 26 ED     1170        BNE   NEXT      NO: GET ANOTHER BYTE.
401D- 20 FE     1175 HALT   BRA   HALT
                1180        .EN
```

The loop in this MOVE routine takes just twice as many instructions as the move in the first routine using the stack pointer. Still, if the S register cannot be spared and there is no need for speed, the virtual-index-register approach may be more desirable.

7.6 INDIRECT ADDRESSING

Section Overview

Sequential access of data in a table requires incrementing or decrementing the index register between accessing. However, random access of data in a table is often required. For example, it may be desired to have each key of an alphanumeric keyboard access a specific byte in a table. The keys are pressed not sequentially but randomly.

In indirect addressing the final, or *effective, address* (EA) is first computed and stored in a pair of RAM bytes called the *indirect address.*

Indirect addressing is a mode in which the address of the operand data is contained in another area of the computer's memory, invariably RAM. The address of the operand data is called the *effective address* (EA). The RAM locations that contain the effective address are the *indirect address*. The advantage of indirect addressing is that the effective address is pointed to by bytes in RAM, which can be changed very easily with simple *load* and *store* operations. Whereas indexed addressing permits data in a table to be accessed sequentially, indirect addressing permits this data to be accessed at random, anywhere in the table. Figure 7.5 illustrates indirect addressing.

The 6802 processor does not have an indirect addressing mode, but the function can be synthesized using the indexed mode and a RAM buffer area. Here is the procedure.

1. Store the desired effective address in two consecutive RAM bytes, most-significant byte first, least-significant byte second:

```
STAA    HIBYTE
STAB    LOBYTE
```

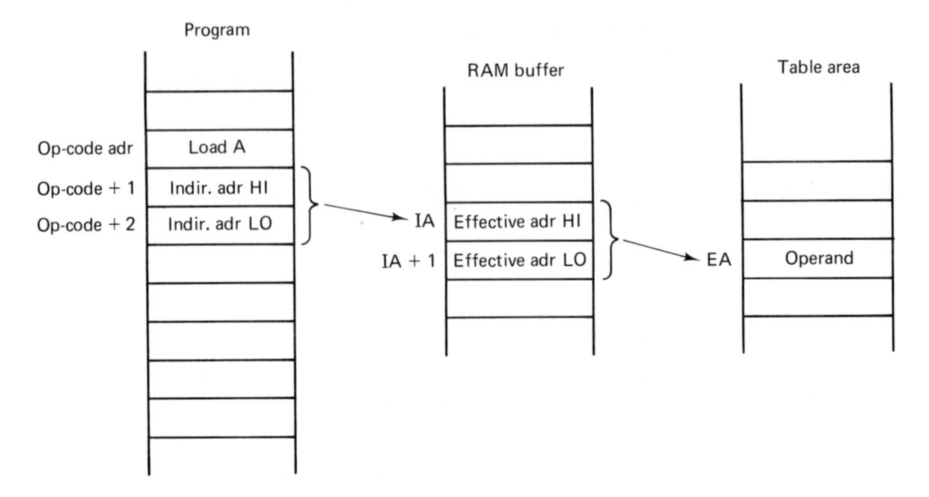

Figure 7.5 Indirect addressing sends the processor to a pair of addresses in RAM, which in turn contain the effective address, where the operand is stored. The program can be in ROM, but the effective address is stored in RAM and can be modified as the program runs. The 6802 cannot do this directly it must load X from the RAM buffer and then use indexed addressing.

2. Load the X register from these two locations:

```
LDX     HIBYTE
```

The next (low-order) byte will be loaded automatically into the low-order eight bits of X.

3. Use the indexed-addressing mode to access the data at the effective address:

```
LDAA     00,X
```

Software decoding: As an example of indirect addressing and random table access, let us say that the data table given in Figure 2.3(b) exists in ROM from addresses \$D000 to \$D00F. Refer back to the figure to see how the bits of each data byte are chosen to light the LEDs of a 7-segment display to form the hex characters 0 through F, respectively.

Now let us assume that an input buffer at address \$2000 receives data \$00 through \$0F and that a latch at \$3000 is to display the digit specified by the input data. Figure 2.3(a) shows the latch-to-LED connection. List 7.5 shows the program routine to read the input data, fetch the appropriate data bits from the table, and output the bits to the LEDs via the latch.

```
0900 * LIST 7-5. PROGRAM SEGMENT, NOT TO RUN. JUL '87, METZGER
1000 * SOFTWARE DECODING: INDIRECT ADDRESSING
1010 *                      LIGHTS 7-SEGMENT DISPLAYS.
1020 *
1030        .OR   $4000
1040 *
```

```
4000- 86 D0      1050     LDAA #$D0    SET INDIRECT ADDRESS HIGH BYTE TO
4002- 97 10      1060     STAA $0010    PAGE $D0.
4004- B6 20 00   1070     LDAA $2000   GET INPUT DATA FROM BUFFER.
4007- 97 11      1080     STAA $0011   POINT TO APPROPRIATE WORD IN TABLE.
4009- DE 10      1090     LDX  $0010   PICK UP PAGE & WORD IN X-POINTER.
400B- A6 00      1100     LDAA 00,X    GET WORD (BIT PATTERN) FROM TABLE.
400D- B7 30 00   1110     STAA $3000   LIGHT LEDS VIA OUTPUT LATCH.
                 1120     .EN
```

REVIEW QUESTIONS FOR SECTIONS 7.5 AND 7.6

21. In the MOVE DATA 2 program (page 166), what is held at addresses $0012 and $0013?

22. In the MOVE DATA 2 routine, what address is the last byte of data stored to?

23. Use the programmer's reference card at the back of the book to calculate how long it takes the MOVE DATA 2 routine to move 2K bytes. Assume a 1-MHz clock.

24. The *X* register contains data $ABCD. Where does data $CD reside after execution of the instruction

STX $53

25. Would it be possible to use virtual index registers to access three data tables, for example, the minuend, subtrahend, and result, in a series of subtraction problems? If so, what changes would be required?

26. In using indirect addressing, the address where the operand data resides is called the ________.

27. In synthesizing indirect addressing for the 6802, the indirect address is in ____ (ROM, RAM, or a machine register).

28. In the SOFTWARE DECODING routine, where are the high-order-half and low-order-half addresses of the indirect address?

29. In the SOFTWARE DECODING routine, the least-significant byte of the effective address is originally obtained from ________.

30. What word, frequently used in microcomputing, has these two definitions?
- An area of RAM used for temporary storage of data.
- A hardware device for driving or isolating digital signals.

7.7 MEM—A MEMORY CHALLENGE GAME

Section Overview

Here is a simple game that illustrates the use of the X register for building and reading a table. The table contains a record of the sequence in which 4 buttons are pushed by the first player—15 pushes of the buttons are stored. The second player then attempts to repeat the exact same sequence of 15 button entries.

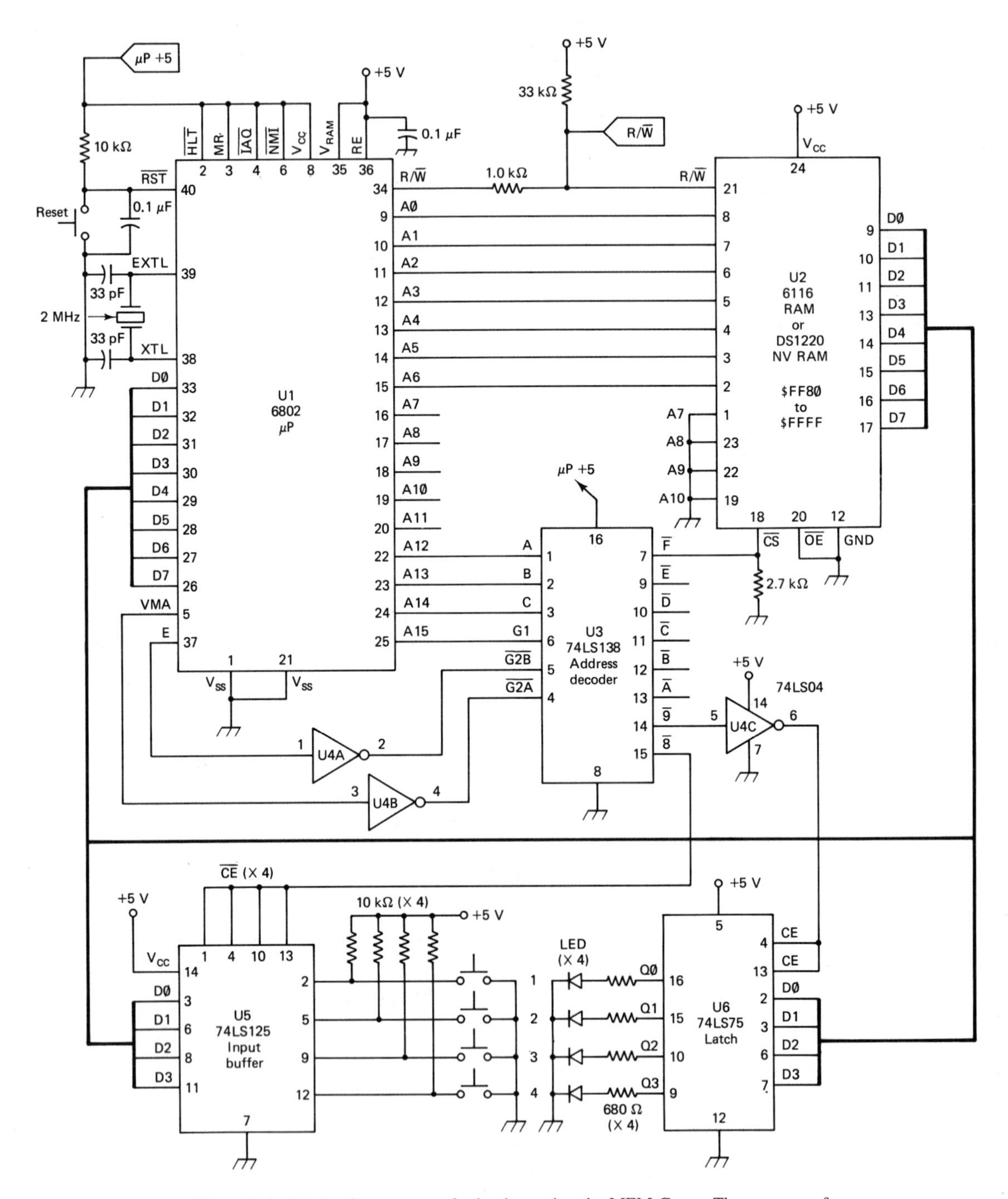

Figure 7.6 The hardware system for implementing the MEM Game. The program of page 171 can be loaded into the RAM using the circuit of Figure 3.10 or Figure 5.9. See Figure 6.14 for a ROM configuration. Start the table at $0000 if a ROM is used.

Figure 7.6 shows the hardware for the MEM game. The table data stored is one of four binary combinations:

Button 1	0000 0001	$01
Button 2	0000 0010	$02
Button 3	0000 0100	$04
Button 4	0000 1000	$08

Building the table: As the first player presses a button, the corresponding LED lights, and another entry will not be accepted for 1.4 s. The switch-input byte is complemented and the high-order nibble is masked off, as discussed in Sections 6.3 and 6.5, respectively. When 15 entries have been made, the data pointer is at $FFDF. This is recognized by the CPX instruction, and the low-order nibble of X ($F) is stored in the low-order 4 bits of the output latch. This lights all four LEDs, signaling the first player that data entry is complete.

Reading the table: The program is now altered slightly; the *store input data to table* routine is replaced with a *compare input data to table data* routine. If the comparison fails, the low-order nibble of the data pointer is stored to the output latch. Since this pointer (the X register) increments each time a correct comparison is made, it contains, in binary, the number of correct repetitions from the original sequence. The four LEDs thus display the second player's score in binary.

Figure 7.7 gives the program flowchart for the MEM game. The program listing follows (List 7.6). After you build and test the game you may want to modify the program to avoid the need for changing it between the LOAD and PLAY portions of the game. This will make the program twice as long.

LIST 7-6

```
                 0900 *  -----TESTED AND RUNS, 7 JULY '87, BY D. METZGER-----
                 1000 * MEM GAME FOR BREADBOARD COMPUTER.  SAVE "LOAD" ROUTINE
                 1010 *   IN RAM; MAKE 15 KEY INPUTS; THEN CHANGE 4 BYTES TO
                 1020 *  CONVERT TO "PLAY" ROUTINE, AND REPEAT KEY SEQUENCE.
                 1030 *  SEE FIG 7-6 FOR HARDWARE.
                 1040 *
                 1050           .OR  $FF81
                 1060           .TA  $4081
8000-            1070 INPUT   .EQ  $8000
9000-            1080 OUTPUT  .EQ  $9000
                 1090 *
FF81- CE FF D0   1100 LOAD    LDX  #$FFD0   START OF 15-BYTE TABLE.
FF84- B6 80 00   1110 KEY     LDAA INPUT    GET KEY (0 = KEY PRESSED)
FF87- 43         1120         COMA          CONVERT TO "1 = PRESSED."
FF88- B7 90 00   1130         STAA OUTPUT   LIGHT LED BY PRESSED KEY.
FF8B- 84 0F      1140         ANDA #$0F     MASK OFF HIGH-ORDER 4 BITS.
FF8D- 27 F5      1150         BEQ  KEY      LOOK AGAIN IF NO KEY.
                 1160 *                    (REPLACE SECTION BELOW FOR "PLAY")
FF8F- A7 00      1170         STAA 00,X     STORE KEY VALUES IN TABLE.
FF91- 01         1180         NOP           LEAVE 2-BYTE SPACE FOR "PLAY"
FF92- 01         1190         NOP            BRANCH INSTRUCTION.
                 1200 *
FF93- C6 C0      1210         LDAB #$C0     SET OUTER-LOOP DELAY COUNTER.
FF95- 7C FF E0   1220 DELAY   INC  $FFE0    INNER-LOOP COUNTER.
FF98- 26 FB      1230         BNE  DELAY     256 LOOPS X 10 CYCLES X 2 US/CY
FF9A- 5A         1240         DECB            TIMES 192 OUTER LOOPS
FF9B- 26 F8      1250         BNE  DELAY       IS ABOUT 1 SECOND.
FF9D- 08         1260         INX           POINT TO NEXT TABLE ADDRESS.
```

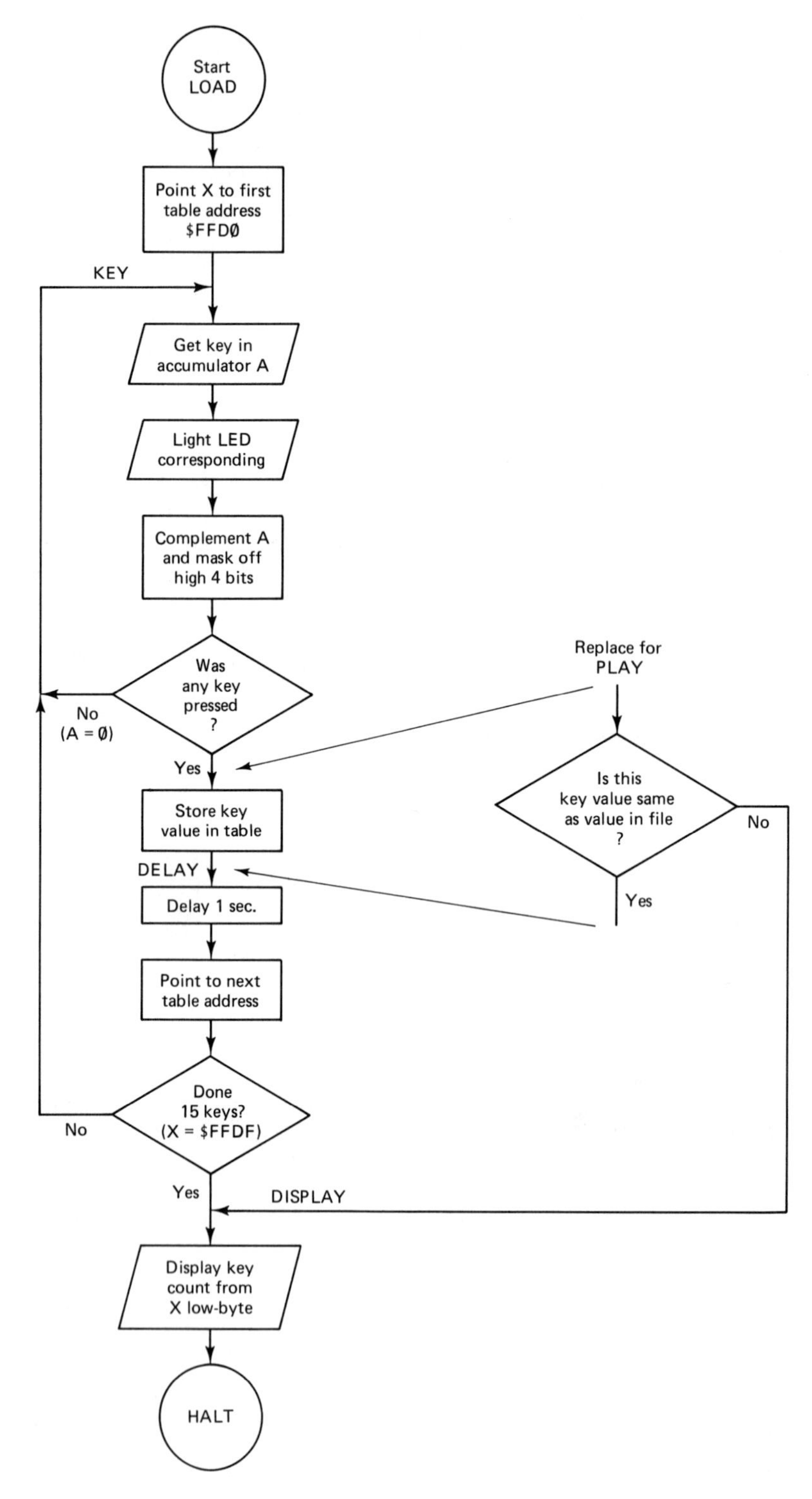

Figure 7.7 Flowchart for the MEM Game. First a table is created by a series of 15 button entries. Then an attempt is made to repeat the same series.

```
FF9E- 8C FF DF 1270         CPX   #$FFDF   15 ADDRESSES FILLED?
FFA1- 26 E1    1280         BNE   KEY       NO: GET NEXT KEY.
FFA3- FF 90 00 1290 DISPLY  STX   OUTPUT   LIGHT 4 LEDS AFTER 15 KEYS. SHOW
               1300 *                       SCORE AFTER ERROR IN "PLAY".
FFA6- 20 FE    1310 HALT    BRA   HALT
               1320 *
               1330         .OR   $FF8F
               1340         .TA   $408F
               1350 * REPLACE LINES 1170, 1180, & 1190 WITH 1370 & 1380
               1360 *  TO CHANGE FROM "ENTER" TO "REPEAT" SEQUENCE.
FF8F- A1 00    1370         CMPA 00,X      IS KEY VALUE SAME AS TABLE VALUE?
FF91- 26 10    1380         BNE   DISPLY    NO: LO BYTE OF X TO LATCH LEDS.
               1390         .OR   $FFFE
               1400         .TA   $40FE
FFFE- FF 81    1410         .HS   FF81
               1420         .EN
```

Answers to Chapter 7 Review Questions

1. Address **2.** LDX #$1234 **3.** INX **4.** Address $1236
5. $24C8 **6.** $0163 **7.** BEQ NOWT **8.** BNE HOLD
9. The routine would not output the data $FF but would return to the start of the table. **10.** Nested loops, because X must be preserved as the data pointer.
11. 16 bits **12.** $1234 **13.** $1235 **14.** Backward; PSH
15. No, because DES affects no flags in the CCR, so the zero count could not be recognized.
16. The PSHA instruction automatically decrements the S register.
17. 20 μs/loop × 2048 loops = 41 ms. **18.** N = 0, Z = 1, C = 1
19. 7 **20.** JMP $789A
21. The current destination address, high- and low-order bytes **22.** $C7FF
23. 44 μs/loop × 2048 loops = 90 ms **24.** At address $0054 **25.** Yes. It requires only 6 bytes of RAM for three X buffers. **26.** Effective address
27. RAM **28.** High-order byte is at $0010, low-order byte is at $0011.
29. The input data **30.** Buffer

CHAPTER 7 QUESTIONS AND PROBLEMS

Digits after the decimal point refer to section numbers in Chapter 7.

Basic Level

1.1 Write a 6802 mnemonic and operand that will load accumulator A with the data stored in a memory location whose address is contained in the X register.

2.1 Write a 6802 mnemonic and operand that will set the X register to point to address $1234.

3.1 If X contains $1234, what address is written to by the instruction STAA $FF,X?

4.2 What modification could be made to the BUILDTABLE routine of List 7.1 to extend the length of the table to 200 bytes?

5.2 What limits the length of the table in READTABLE routine of List 7.2?

6.3 If the S register contains $89AB and the A accumulator contains $1F, what data is stored to what location by the instruction PSHA? Answer again for PULA.

7.3 List all six 6802 instructions that operate directly on the stack pointer, and indicate which flag bits are affected by each one.

8.4 Use the index of this book to find the programmer's model of the Intel 8085 microprocessor. Which machine has the greater number of bytes of programmable registers, the 6802 or the 8085?

9.5 If the X register contains $A1B2, what data is stored to what addresses by the instruction STX $0123?

10.5 Calculate the time required for the routine of List 7.4 to move 8K-words of data. Assume a 1-MHz clock.

11.6 Indexed addressing is useful for accessing tables sequentially. Indirect addressing is useful for accessing tables ________.

12.6 In the program of List 7.5, what would be the address of the table data accessed if the data input via the buffer was $08?

Advanced Level

13.1 In the CLEARBLOCK routine presented on page 157, why is the operand of the CPX instruction $6000 when the last address to be cleared is $5FFF?

14.2 Explain how the data-access rate is slowed down in the BUILDTABLE routine of List 7.1 and in the READTABLE routine of List 7.2.

15.3 Rewrite the MOVEDATA routine of List 7.3 to transfer data from $F800 through $FFFF to $C000 through $C7FF by working from low to high addresses.

16.4 Check out the TAP and TPA instructions in the 6802 programmer's reference, and then write a routine that will set the H flag of the CCR without disturbing any of the other flag bits.

17.5 A table of data is stored in the 2048 bytes from $C800 through $CFFF. We wish to reverse the order of the data, so that the same 2048 bytes are stored from $CFFF through $C800. Write a flowchart and the source code for a 6802 program to accomplish this. Use virtual X registers.

18.6 Can an indirect address be stored in RAM, ROM, or either? Explain why.

19.6 Modify the routine of List 7.5 to accept only digits $00 through $0F from the input buffer and to return to the read-buffer instruction if any other data is received.

20.7 Examine the system of Figure 7.6 and list all the separate ranges of addresses to which the 6116 responds.

21.7 Modify the program of List 7.6 so that a separate REPEAT sequence follows the ENTER sequence, with no need to change any part of the program.

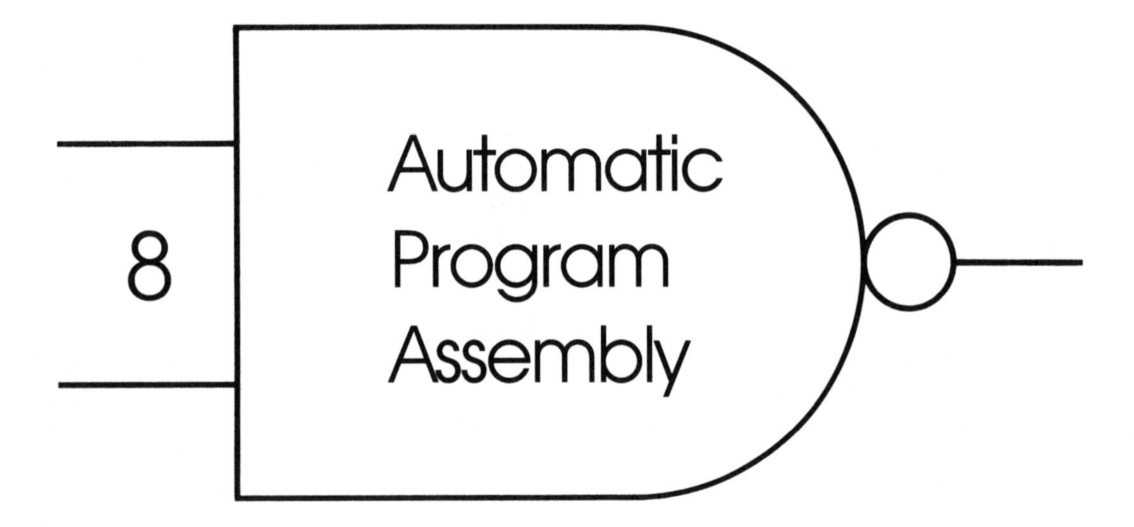

8.1 ASSEMBLER TYPES AND FUNCTIONS

Section Overview

Assemblers are computer programs that translate mnemonic instruction statements into binary machine code. The mnemonic statements are called *source code*. The hex representations of the binary machine language are called *object code*.

Source code consists of five columns:

- Line numbers
- Address labels
- Op-code mnemonics
- Operand values or names
- Comments

Mnemonic statements, such as

```
LDAA     BUFR
```

are called *assembly language*, or *source code*. This source code is unintelligible to a computer and must be changed into a string of binary bytes before it can be executed as machine code. We represent this binary machine code by hex digits, such as

```
B6 20 00
```

which we call *object code*.

Assembly: The process of translating assembly language into object code is called assembling a program. The 6800/6801/6802/6805 family is unusual in that assembly of short programs can reasonably be accomplished by hand. The only other widely used processor on which this is possible is the 6502 which we will cover briefly in Section 20.7. These processors have instruction sets that are so straightforward that all the equivalencies between instruction mnemonics and hex op-codes can be printed on a single card, called the programmer's reference. The first byte of each instruction embodies nothing but an op-code. It completely defines the instruction type and addressing mode, and contains no operand data. Any subsequent bytes are solely operands and do not modify the operation specified by the first byte. Most other microprocessors have instructions in which bits referring to the operand appear in the first byte, or bits referring to the op code appear in a later byte of the machine-language instruction. This means that the object code cannot simply be looked up; the machine code must first be built up bit by bit, and then translated into the hex digits of object code.

Computerized assembly: Although it is a good educational experience to assemble a few programs by hand, professional programmers always use a computer program to do this. These programs are called *assemblers* if the *target* processor (the one on which the machine language is going to run) is the same as the processor in the computer that runs the assembler program (the *host* computer). If the target processor and the host computer's processor are different, the program is called a *cross-assembler*.

The programs in this book have been assembled on cross-assemblers from S-C Software (P.O. Box 280300, Dallas, TX 75228), which run on the Apple II personal computers (6502 host processor).

Assembler formats vary slightly, but they all include the columns listed in the following example:

Object Code		**Source Code**				
Address	Instruction	Line No.	Label	Op.	Operand	Comment
0123	7E 01 23	1000	START	JMP	START	JUMP UP AND DOWN

The *line number* is simply a reference, which is useful when it becomes necessary to examine or modify specific parts of a large assembly-language program. The other columns should already be familiar to you from earlier chapters.

contains no operand data. This makes ________ ________ of 6800 programs reasonable.

3. A cross-assembler is used when the processors in the ________ computer and the ________ computer are not the same.
4. An assembler uses ________ code to generate ________ code.
5. The assembler can be directed to assign a numeric value to a mnemonic name with the pseudo-op ________.
6. Is the number 1Ø1, appearing in a source program, interpreted as binary, decimal, or hex?
7. What is the meaning of the number assigned by the assembler to the name appearing in the label field of a line of source code?
8. Misspellings and other forms that the assembler is unable to interpret will cause the message ________ ________.
9. If your assembly-language program contains the instruction

   ```
   BEQ NEXT
   ```

   but you have failed to assign an address value to the variable name NEXT, the message ________ will appear.
10. What, if anything, is wrong with this instruction:

   ```
   LDAB #291
   ```

THE 6800 INSTRUCTION SET

The next four sections begin a catalog of 6800/6802 instructions, with details of their assembly-language and machine-language forms and notes on their applications where this is not immediately obvious. Since the complete 6800 instruction set is a lot to digest, we will take a break from it in Chapter 9 and complete the catalog in Chapter 10.

To help you remember the proper spellings, note that all 6800 assembler mnemonics contain three letters, except that mnemonics referencing *accumulators* A *or* B *only* contain four letters.

8.3 DATA-MOVE INSTRUCTIONS FOR THE 6800

Section Overview

LOAD instructions copy data into a 6802 machine register. STORE instructions copy data from a machine register to a memory location. CLEAR instructions can be used to place data *zero* in a memory location or in certain machine registers.

8.2 ASSEMBLER DIRECTIVES AND FEATURES

Section Overview

Assembler directives are statements that do not generate machine code. They may define address or data names or cause lists of data to be generated.

Most assemblers interpret numbers as decimals unless a sign indicating hex is included.

Labels are names given to the address of the op-code for the line where they appear. Operands in assembly language can be decimal or hex values, address labels, data names, or algebraic expressions.

Directives, also called *pseudo-ops*, appear in the op-code column of the source code, but they are not assembly mnemonics for the target processor. Rather, they are directions, supplying information that is not contained in the source program. Here are the most commonly used assembler directives, as used by assemblers from Motorola (left column) and S-C Software (next column).

	Motorola		S-C Software	
	ORG $1234		.OR $1234	The program is to start at address $1234 (address of first op-code).
BUFR	EQU $2ØØØ	BUFR	.EQ $2ØØØ	The label BUFR is to be interpreted by the assembler as $2ØØØ.
	FCB $56, $78		.HS 5678	Form constant bytes (Motorola) or hex string (S-C Software): Insert the two bytes $56 and $78 in the object code list.
	END		.EN	This is the end of the source-code listing.

There are many other assembler directives, but the four listed should give you an idea of their character. It would do little good to go into detail about assembler directives because they vary so much among the different software suppliers. Of course, every assembler package comes with a manual, which should contain complete specifications and examples of use for each directive.

Number-system options: Most assemblers allow constant address and data to be specified in a variety of number systems, as signaled by a pre- or postcharacter.

- **Decimal:** Digits or groups of digits containing only the characters 0 through 9 are assumed to be in decimal, unless an assembler directive has specified another number system. Note that object-code lists are in hex, however.
- **Hexadecimal numbers** may be specified by a prefix $ or, with some assemblers, a suffix H:

$21 = 21H = 33

Some assemblers will not accept the characters A through F as the first digit of a hex number. Thus, $C7 may be listed as 0C7H, but not as C7H.

- **Binary numbers** may be specified with some assemblers by using the prefix % or the suffix B.

```
%100001 = 100001B = $21 = 33
```

Operands: To avoid the distraction of having to remember large numbers of buffer memory addresses, output-latch addresses, and input-buffer addresses, we assign mnemonic expressions to these numbers when writing the source code. The definition can be assigned by an EQUALS directive or by using the operand as a label.

```
         .OR     $0100
LATCH    .EQ     $3000
START    STAA    LATCH
HALT     JMP     HALT
```

In the preceding examples the assembler will assign the address $3000 as the operand of the STAA op-code because of the .EQ directive. It will assign the address $0100 to the label START because of the .OR directive; this means that the op-code for STAA will appear at address $0100. The STAA instruction and its two operand bytes occupy addresses $0100, $0101, and $0102. The JMP op-code thus is placed at $0103, and the assembler assigns this address to the label HALT.

Operand expressions. Operands may contain certain algebraic expressions as well as mnemonic labels. For example

```
STX     LATCH-1
```

(where LATCH = $3000) will store the most-significant byte of X at $2FFF and the least-significant byte at $3000. The line of source code

```
LDX     #HALT - START + 3
```

will assemble an operand to the LDX command that equals the length of the program in bytes, assuming that the first op-code is at START and the last is a JMP at address HALT.

Beware of the difference between immediate-mode and extended-mode instructions, especially when using operand labels. In the preceding example, LDX #START loads the *value* $0100 into the *X* register. LDX START loads the two bytes from *addresses* $0100 and $0101 into the X register.

Error indication is provided with most assemblers. However, the errors that can be detected by the assembler are those of format and syntax only. There is no guarantee that a program that assembles smoothly will even run on the target processor, much less perform the functions intended by the writer. Only human troubleshooters can isolate and correct errors in program logic.

Here are some of the more common errors that can be caught by the assembler program:

- **Multiple definition:** For example, the label LOOP cannot be used at two places in the program. Two different addresses obviously cannot have the same name.[1]
- **Undefined label:** This often happens through inadvertent changes of spelling or abbreviation. For example, if we define an input buffer

```
BUFFER    .EQ    $3000
```

and then later try to access it by

```
LDAA    BUFR
```

the assembler will have no address to assign to BUFR.

- **Illegal operand:** For the 6802 processor, extended-mode instructions must b followed by an operand in the range 0 through 65 535. Relative-mode instruc tions must be followed by an operand in the range −128 through +127 Indexed-mode instructions must have an operand in the form nn, X where n is a value from 0 to +255. Most assemblers will catch and report on missin or illegal operands.
- **Reserved words or characters:** Most assemblers will not accept instructic mnemonics as labels or operands. For example, it is easy to write

```
BMI NEG
```

when testing for a negative number, but NEG is itself an instruction mnemon and cannot be used as an operand name. Most assemblers will also not pern labels or operands to begin with a nonalphabetic character or to contain certa reserved characters, such as ? and = signs.

- **Syntax errors:** The assembler will generally catch and report on such err as ORA for ORAA and ANDAA for ANDA. It will also catch the use of lett *l* and *O* for numbers 1 and 0 and the appearance of the numbers A through in any number not preceded by a hex ($) sign. In most assemblers such err will cause the message SYNTAX ERROR to be printed.

REVIEW OF SECTIONS 8.1 and 8.2

Answers appear at the end of Chapter 8.

1. Does object code consist of mnemonics, decimal numbers, or hex digits?
2. The first byte of each 6800 instruction completely defines the operation ;

[1]Some assemblers have a *local label* feature that allows a name to be used over, and to be interpr as a different address, but not within the same area of the program.

Most of these instructions *set* the Z flag if the value loaded or stored is zero and set the N flag to the value of the most-significant bit of the data loaded or stored. The programmer's reference should be consulted for more details of effects on condition-code flags.

LDAA Load Accumulator A

Operation. 8-bit operand data is copied into accumulator A.

Addressing modes, machine cycles (~), example forms, and comments

Immediate	2~	86 FF	LDAA #$FF	Operand data is $FF.
Direct[1]	3~	96 0E	LDAA $0E	Operand is at address $000E.
Extended	4~	B6 12 34	LDAA $1234	Operand is at address $1234.
Indexed	5~	A6 00	LDAA 0,X	Operand is at address contained in 16-bit machine register X.

Flags. Negative flag set to 1 if value loaded is greater than $7F (bit 7 = 1). Zero flag set if value loaded is zero. Overflow flag (V) always cleared to 0.

LDAB Load Accumulator B

Operation. 8-bit operand data is copied into accumulator B.

Addressing modes, machine cycles (~), example forms, and comments

Immediate	2~	C6 20	LDAB #32	Operand is decimal 32, hex 20.
Direct[1]	3~	D6 20	LDAB 32	Operand is at address $0020.
Extended	4~	F6 01 00	LDAB 256	Operand is at address 256 decimal, 0100 hex.
Indexed	5~	E6 01	LDAB 1,X	Operand is at address contained in *X* register, plus 1.

[1]Most assembler programs will automatically use direct mode if the operand address is less than 256 ($00FF or lower) and extended mode if the address is 256 or higher. Some assemblers allow a special character to be added to the mnemonic to force extended mode even when direct mode is possible.

Flags. Same as LDAA. N = bit 7, Z = 1 if operand = 0, V cleared always.

STAA Store Accumulator A

Operation. 8-bit data from accumulator A is copied to operand address

Addressing modes, machine cycles (~), example forms, and comments

Direct	4~	97 80	STAA $80	*A* is stored to address $0080.
Extended	5~	B7 02 AB	STAA $02AB	*A* is stored to address $02AB.
Indexed	6~	A7 FF	STAA $FF,X	*A* is stored to address *X* + 255 (decimal). This is maximum offset.

Flags. N = bit 7 of data in A, Z = 1 if A = 0, V cleared always.

STAB Store Accumulator B

Operation. 8-bit data from accumulator B is copied to operand address.

Addressing modes, machine cycles (~), example forms, and comments

Direct	4~	D7 20	STAB RAM1	Value $20 previously assigned to expression RAM1.
Extended	5~	F7 C0 01	STAB LATCH	Previously LATCH EQU $C001.
Indexed	6~	E7 0A	STAB 10,X	Offset to X address is decimal 10.

Flags. N = b_7, Z = 1 if B = 0, V cleared, others unchanged.

LDX Load X Register

Operation. 16-bit operand data is copied into X register.

Addressing modes, machine cycles (~), example forms, and comments

Immediate	3~	CE 47 80	LDX #$4780	Data $4780 placed in *X*.
Direct	4~	DE 30	LDX $30	8 bits from address $0030 placed in most-significant byte of *X*. 8 bits from $0031 placed in least-significant byte of *X*.
Extended	5~	FE FA 30	LDX $FA30	($FA30) → X_{HI}, ($FA31) → X_{LO}[2].
Indexed	6~	EE 00	LDX 0,X	New value in *X* obtained from address pointed to by old value in *X* (call it *M*) and next address (*M* + 1).

[2]Parentheses around a value indicate *data found at this address*.

Flags. N = X_{15} (most-significant bit of X), Z = 1 if X = 0, V cleared.

STX Store X Register

Operation. 16 bits from X register are copied to two successive operand addresses.

Addressing modes, machine cycles (~), example forms, and comments

Direct	5~	DF 7F	STX 127	X_{HI} → address \$007F; X_{LO} → \$0080.
Extended	6~	FF 1B 58	STX 7000	*X* stored to addresses \$1B58, 1B59.
Indexed	7~	EF 00	STX 0,X	Value of *X* stored to address contained in *X* and next address. If *X* = \$1234, then data \$12 → address \$1234 and data \$34 → address \$1235.

Flags: N = X_{15}, Z = 1 if X = 0 (Z = 0 if X ≠ 0), V cleared.

CLEAR INSTRUCTIONS

CLRA Clear Accumulator A

This instruction is identical to LDAA #0, but it takes 1 less byte of program memory.

CLRB Clear Accumulator B

CLR Clear Operand Address

This instruction allows data \$00 to be stored to any memory address without disturbing either accumulator. For any other data, use LDAA #DATA then STAA ADDR.

Operation. 8-bit data is cleared to \$00.

Addressing modes, machine cycles (~), example forms, and comments

Inherent	2~	4F	CLRA	No operand required.
Inherent	2~	5F	CLRB	No operand required.
Extended[3]	6~	7F C0 00	CLR \$C000	Data \$00 stored to address \$C000.
Indexed	7~	6F 10	CLR \$10,X	00 → (*X* + 16).

[3]There is no direct-mode clear instruction in the 6800 set.

Flags. N = 0, Z = 1, V = 0, C = 0.

Notice that there are no 6800 instructions such as CLX or CLS to clear the X register and stack pointer. Use LDX #0 and LDS #0 for these functions.

8.4 LOGIC INSTRUCTIONS FOR THE 6800

Section Overview

The 6800/6802 logic instructions operate on individual bits within 8-bit registers. They include AND, OR, and EOR (exclusive OR) of accu-

mulator bits with corresponding bits of an operand byte, the resulting bits being stored in the accumulator. BIT is a dummy AND, which affects flag bits but not the accumulator.

COM complements the bits of an accumulator or operand. NEG performs a 2's complement on the entire 8-bit operand.

ANDA AND Accumulator A

ANDB AND Accumulator B

Operation. Each bit of the operand is logic-ANDed with the corresponding bit of the accumulator. The result of the AND is loaded into the accumulator. The operand data is left unchanged.

Addressing modes, machine cycles (~), example forms, and comments

Immediate	2~	84 0F	ANDA #$0F	Masks high 4 bits of *A* to 0000.
Immediate	2~	C4 08	ANDB #8	Masks all but bit B_3 to 0.
Direct	3~	94 3A	ANDA $3A	Masks *A* with bits from address $003A.
Direct	3~	D4 3A	ANDB MASK	Operand address MASK defined previously as $3A.
Extended	4~	B4 F7 10	ANDA BITSEL	BITSEL defined previously as $F710.
Extended	4~	F4 F7 12	ANDB BITSEL+2	Assembler adds 2 to BITSEL value.
Indexed	5~	A4 00	ANDA 0,X	Mask bits are at address contained in *X*.
Indexed	5~	E4 A0	ANDB 160,X	Operand is at address *X* + 160.

Flags. N = bit 7 of result, Z = 1 if result = 0, V cleared.

BITA and BITB

BITA and BITB have the same forms and functions as the ANDA and ANDB instructions, except that the accumulator values remain unchanged. BIT is, in effect, a dummy AND instruction used to set condition-code flags without altering any data.

	Immediate	Direct	Extended	Indexed
BITA	85 (2~)	95 (3~)	B5 (4~)	A5 (5~)
BITB	C5 (2~)	D5 (3~)	F5 (4~)	E5 (5~)

This instruction is named BIT because it is useful in testing the state of a given bit of operand data. Let us say that we wish to search a block of bytes and count how many have a logic 1 in the bit-5 position. Here's how it can be done.

```
        LDAA    #$20       A is mask, with 1 in bit 5 only.
NEXT    BITA    0,X        Result = 0 unless bit 5 = 1.
        BEQ     SKIP       No count if bit 5 = 0.
        INC     COUNT      Count if bit 5 = 1.
SKIP    INX                Point to next in block.
        CPX     #ENDBLK    End of block?
        BNE     NEXT       No: then check next byte.
```

ORAA OR Accumulator A

ORAB OR Accumulator B

Operation. Each bit of the operand byte is logic-ORed with the corresponding bit of the accumulator. The result is loaded into the accumulator.

Addressing modes, machine cycles (~), example forms, and comments

Immediate	2~	8A 0F	ORAA #$0F	High nibble retained, low nibble set to all 1s.
Immediate	2~	CA 08	ORAB #8	Bit 3 set to 1, others retained.
Direct	3~	9A A0	ORAA 160	*A* ORed with bits from $00A0 (decimal 160).
Direct	3~	DA A0	ORAB BITS	Name BITS previously set to $A0.
Extended	4~	BA 47 30	ORAA $4730	If accumulator = $49
Extended	4~	FA 47 30	ORAB DATA	and operand = $63
Indexed	5~	AA 00	ORAA 0,X	result in accumulator
Indexed	5~	EA 01	ORAB 1,X	will be $6B[4].

[4]

	Binary	Hex
Original accum.	0100 1001	49
Operand byte	0110 0011	63
New accum.	0110 1011	6B

Flags. N equals bit 7 of the new accumulator data, Z = 1 if the new accumulator data = 0, V is always cleared.

EORA Exclusive-OR Accumulator A

EORB Exclusive-OR Accumulator B

Operation. The 8 bits of the operand are exclusive-ORed with the corresponding bits of the accumulator. For each pair of bits the result is 1 if one *or* the other bit is 1 but not if both are 1s. The resulting bits are loaded into the accumulator.

Addressing modes, machine cycles (~), and object codes

	Immediate	Direct	Extended	Indexed
		plus 1-byte operand	plus 2-byte operand	plus 3-byte operand
EORA	88 (2~)	98 (3~)	B8 (4~)	A8 (5~)
EORB	C8 (2~)	D8 (3~)	F8 (4~)	E8 (5~)

Flags. N = bit 7 of result, Z = 1 if result = 0, V cleared.

The exclusive-OR Truth table:					
	Original accumulator bit	0	1	0	1
	Operand bit	0	0	1	1
	New accumulator	0	1	1	0

If an accumulator containing $49 is EORed With $63:

	Binary	Hex
Original accumulator	0100 1001	49
Operand byte	0110 0011	63
New accumulator	0010 1010	2A

COMA Complement Accumulator A

COMB Complement Accumulator B

COM Complement Data at Operand Address

Operation. Each of the 8 bits of the operand is individually complemented; that is, the 1s are changed to 0s and the 0s are changed to 1s.

Addressing modes, machine cycles (~), example forms, and comments

Inherent	2~	43	COMA		Data $7F in either accumulator
Inherent	2~	53	COMB		becomes $80.
Extended	6~	73 00 5A	COM	$5A	No zero-page instruction.
Indexed	7~	63 00	COM	0,X	*X* register contains operand address.

Flags. N = 1 if bit 7 of result = 1, Z = 1 if result = 0, V = 0 always, C = 1 always.

NEGA Negate Accumulator A

NEGB Negate Accumulator B

NEG Negate Data at Operand Address

Operation. The 8-bit operand is 2's complemented, that is, each bit is complemented, and binary 0000 0001 is added to the result.

Addressing modes, machine cycles (~), example forms, and comments

Inherent	2~	40	NEGA	Data $C1 in A or B would
Inherent	2~	50	NEGB	become $3F.
Extended	6~	70 01 A5	NEG SUM	Try subtracting 00 − C1 in
Indexed	7~	60 01	NEG 01,X	binary; result = 3F.

Flags. N = bit 7, Z = 1 if result = 0, V = 1 if result ≥ $80, C = 1 if result = 0.

Note: Extended and Indexed operands for COM and NEG must access only RAM addresses, since both *read* and *write* are performed by these instructions.

REVIEW OF SECTIONS 8.3 AND 8.4

Answers appear at the end of Chapter 8.

11. Identify any differences in the way the 6800 LDAA and STAA instructions affect the CCR flags.
12. Why is the immediate-mode LDAB a 2-byte instruction, but LDX, immediate, is a 3-byte instruction?
13. Referring to Question 12, why is it that in the extended mode, both LDAB and LDX are 3-byte instructions?
14. What is the smallest decimal operand that will cause N = 1 when the instruction LDX #OPERAND is executed?
15. Other than the byte and cycle times, are there any differences in the effect of CLRA versus LDAA #0?
16. What instruction and operand would you use to mask all but bit 5 of accumulator A to zero?
17. If A contains $9B, what will it contain after executing BITA #$37?
18. What values will be possible in A after executing ORAA #$01?
19. When an EOR is done with A and an operand byte, the number of 1 bits in the result equals the number of bits of A and the operand that were ________.
20. The instructions

```
COMB
ADDB #1
```

are combined in the single instruction ________.

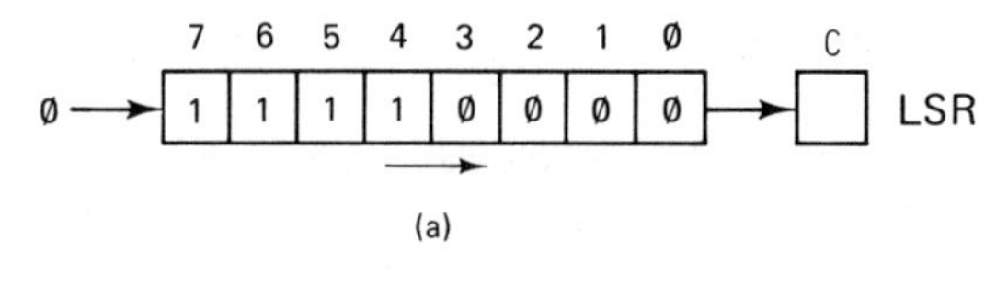

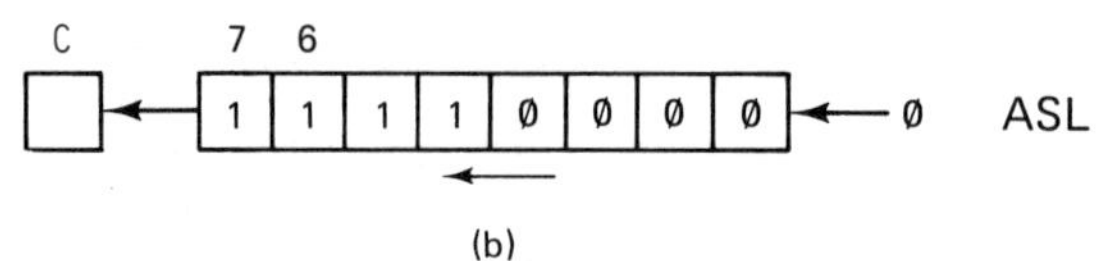

Figure 8.1 (a) The Logic Shift Right instruction. (b) The Arithmetic Shift Left, which is also a logic shift left.

8.5 SHIFT AND ROTATE INSTRUCTIONS

Section Overview

Logic shift instructions move all the data bits in a register to the adjacent bit position, right or left. In the 6800/6802, the last bit shifts into the C flag as Øs shift into the other end of the register.

In rotate instructions the C bit shifts back around into the other end of the register, so that the original register contents are restored after nine shifts.

LSRA Logic Shift Right, Accumulator A

LSRB Logic Shift Right, Accumulator B

LSR Logic Shift Right, Data at Operand Address

Operand address must be both readable and writable.

Operation. Each bit of the 8-bit operand is transferred to the next-lower bit position. The lowest bit (b_0) is transferred into the *carry* flag, and a Ø bit is moved into the highest bit, b_7. Figure 8.1 diagrams the operation.

Addressing modes, machine cycles (~), example forms, and comments

Inherent	2~	44	LSRA	Data $5Ø would become $28, C = Ø.
Inherent	2~	54	LSRB	
Extended	6~	74 47 A3	LSR BITS	Eight shifts clears operand to $ØØ.
Indexed	7~	64 01	LSR 1,X	

Flags. N = Ø always, Z = 1 if result = ØØ, C = 1 if previous bit Ø = 1, V = C.

The logic shift is so called because it treats all the bits equally, without regard for arithmetic place value. If a register containing the value 6 is shifted right, the result is 6 ÷ 2, or 3:

$$6 = 0000\ 0110$$
$$\text{LSR} \rightarrow 0000\ 0011 = 3$$

However, logic shifts do not produce proper division on 2's complement signed numbers.

$$-6 = 1111\ 1010$$
$$\text{LSR} = 0111\ 1101 = 125$$

ASLA Arithmetic Shift Left, Accumulator A

ASLB Arithmetic Shift Left, Accumulator B

ASL Arithmetic Shift Left, Data at Operand Address

Operand address must be both readable and writable.

Operation. Each bit of the operand is moved to the next higher bit position. The highest-order bit is moved into the C flag. A logic 0 is moved into the bit-0 position.

Addressing modes, machine cycles (~), example forms, and comments

Inherent	2~	48	ASLA	Data $F0 would become $E0,
Inherent	2~	58	ASLB	C = 1.
Extended	6~	78 01 5D	ASL $15D	Data +3 becomes +6; data −3
Indexed	7~	68 00	ASL 0,X	becomes −6.

Flags. N = bit 7, Z = 1 if result = 0, C = previous bit 7, V = N $\oplus$ C (V = 1 if N = 1, or C = 1, but not if both equal 1).

The shift left effectively multiplies the operand value by two. The shift left is called *arithmetic* because this multiplication is valid for any numbers, provided that the result does not exceed the capacity of an 8-bit register, which is −128 to +127. Here are two examples showing $6 \times 2 = 12$ and $-6 \times 2 = -12$:

$$6 = 0000\ 0110$$
$$\text{ASL} \leftarrow 0000\ 1100 = 12$$
$$-6 = 1111\ 1010$$
$$\text{ASL} \leftarrow 1111\ 0100 = -12$$

The V flag is *set* to 1 when an ASL produces a result less than −127. It is expected that a routine using ASL to multiply by two would include a check of the V flag and an error-correction procedure if V = 1. Figure 8.2 shows how valid ASL operations produce identical values in bit C and bit 7 (C exclusive-OR N equals 0), but invalid ASLs produce differing C and N bits (C $\oplus$ N = 1).

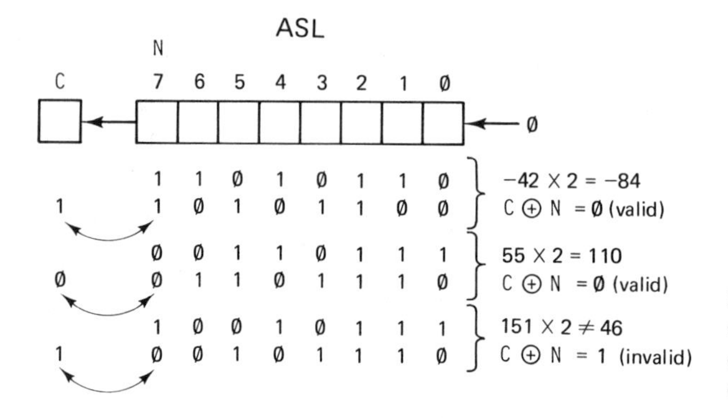

Figure 8.2. The C and N flags are the same after an ASL that stays within the capacity of the 8-bit register. An overflow condition is flagged by V = 1.

The logic shift left would be identical with ASL, so there is no need for a separate instruction. Some assembler programs will accept the mnemonic LSL, treating it as ASL.

ASRA Arithmetic Shift Right, Accumulator A

ASRB Arithmetic Shift Right, Accumulator B

ASR Arithmetic Shift Right, Data at Operand Address

Operation. Each bit of the operand is shifted to the next-lower bit position, except that the sign bit (b_7) is retained. Bit Ø is shifted to the C flag. Figure 8.3 illustrates the operation as a signed number divide-by-2.

Addressing modes, machine cycles (~), example forms, and comments

Inherent	2~	47	ASRA	Data +6 becomes +3;
Inherent	2~	57	ASRB	data −6 becomes −3;
Extended	6~	77 01 5D	ASR RAM4	data +7 becomes +3
Indexed	7~	67 00	ASR 00,X	with C = 1 indicating ½.

Flags. N = 1 if result is negative (bit 7 = 1); Z = 1 if result = 0; C = 1 if previous bit 7 = 1, indicating that the result should be augmented by ½; V = N + C, although V = 1 does not indicate a range overflow for ASR as it did for ASL.

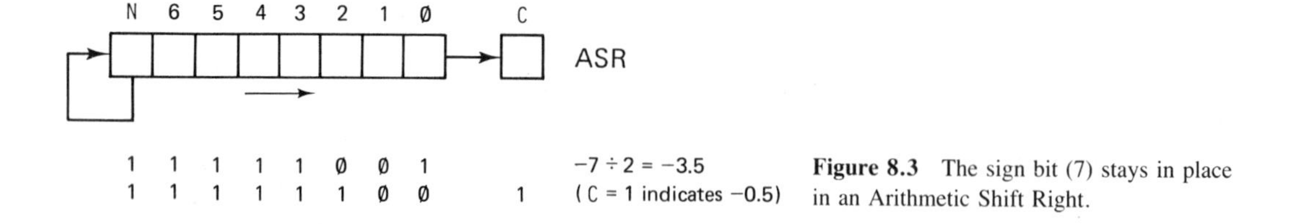

Figure 8.3 The sign bit (7) stays in place in an Arithmetic Shift Right.

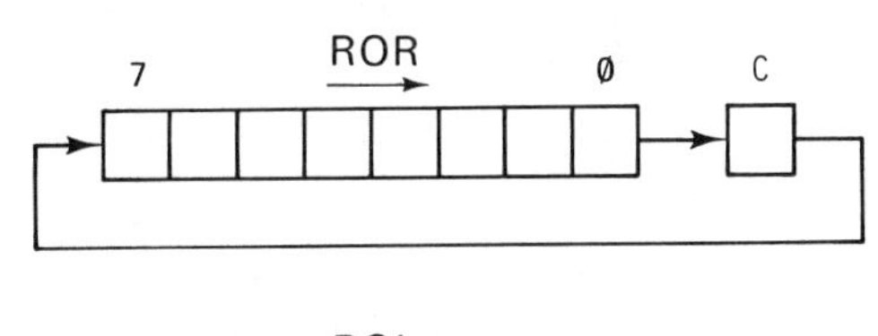

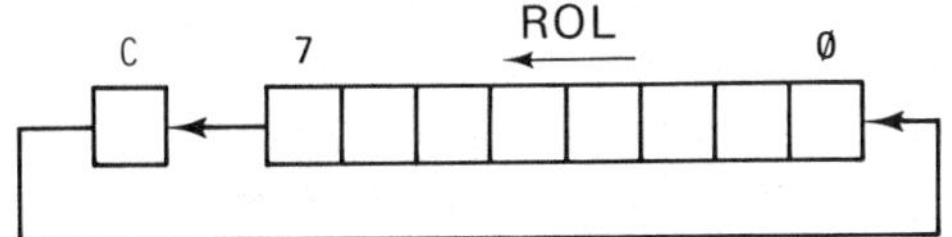

Figure 8.4 The Rotate Right and Rotate Left instructions.

RORA, RORB, ROR Rotate Right, A, B, or Operand

The 8 bits of the operand are shifted to the next-lower bit position. The C bit enters the bit-7 position and the C flag is filled by bit Ø.

ROLA, ROLB, ROL Rotate Left, A, B, or Operand

Eight bits are rotated left through the carry flag. Figure 8.4 illustrates the ROR and ROL instruction operation.

Addressing modes, machine cycles (~), and object codes

	Inherent (*A* accumulator)	Inherent (*B* accumulator)	Extended (plus 2-byte operand)	Indexed (plus 1-byte offset)
ROR	46 (2~)	56 (2~)	76 (6~)	66 (7~)
ROL	49 (2~)	59 (2~)	79 (6~)	69 (7~)

Flags. N = 1 if result bit 7 = 1, Z = 1 if result = Ø; ROR: C = 1 if previous bit Ø = 1; ROL: C = 1 if previous bit 7 = 1; V = N XOR C.

The rotate instruction is useful for examining each bit of a byte of data. Here is a routine that counts the number of 1 bits in accumulator A. The result is stored in RAM location COUNT, and A is returned to its original value.

```
* COUNT NUMBER OF 1 BITS IN A, STORE IN RAM "COUNT"

        CLR     COUNT     Zero RAM bit counter and C flag.
        LDAB    #9        Set up for 9 shifts.
ROT     RORA              If a 1 shifts into the
        BCC     SKIP      carry flag,
        INC     COUNT     count another bit.
SKIP    DECB              Nine counts done?
        BNE     ROT       No: shift another bit.
```

8.6 BRANCH INSTRUCTIONS

Section Overview

None of the branch instructions affects any of the CCR flag bits, but they do cause the program to be displaced by the amount specified in the operand if a given set of CCR bits is present.

In writing and interpreting branching routines, it is helpful to remember that the *accumulator* is the first data referenced and operand is the second in all cases where both are involved. Thus

• SUBA NUMBR	means	Accumulator *A* minus NUMBR.
• BHI	used after a subtraction means	Branch if accumulator was higher than the operand.
• BLE	used after a subtraction means	Branch if accumulator was less than or equal to the operand.

BHI and BLS refer to unsigned binary subtractions. BGT, BGE, BLT, and BLE refer to 2's complement signed-binary subtractions.

All branch instructions consist of 2 bytes (op-code plus 2's complement branch displacement), and all take four machine cycles.

BCC Branch on Carry Clear; Op-code $24

Operation. The branch is taken if the last operation affecting the C flag leaves C = 0. If the branch is *not* taken, the program continues with the instruction following the BCC in the program sequence. If the branch *is* taken the program fetches its next instruction from an address calculated by adding the "following instruction" address and the operand displacement byte.

Example 8.1 Branch Not Taken

```
0123   0D         LOOP    SEC            Deliberately, C = 1
0124   24 03              BCC    NEVER   so branch never taken;
0126   B6 C0 20           LDAA   KEY     and load always done.
0129   7E 01 23   NEVER   JMP    LOOP    Then loop is repeated.
```

Example 8.2 Branch Always Taken

```
0123   0C         LOOP    CLC            Deliberately, C = 0
0124   24 03              BCC    EVERY   so branch 3 bytes ahead
0126   B6 C0 20           LDAA   KEY     always, and skip load.
0129   7E 01 23   EVERY   JMP    LOOP    Then loop is repeated.
```

Assembler programs allow the branch operand to be expressed as a variable name (NEVER and EVERY) or as a numerical address ($0129 in the preceding example) when the source code is written. The assembler will then calculate the address displacement from the "following" op-code and list it as the branch operand in the object code.

BCS Branch on Carry Set; Op-code $25

Operation. If C = 0 the instruction following in normal program sequence is executed. If C = 1 the 2's complement operand of the BCS instruction is added to the normal address to obtain the branch address.

BEQ Branch on result Equal to zero; Op-code $27

Operation. The branch-operand displacement is applied if the last instruction affecting the Z flag left Z = 1 (zero result). Otherwise the normal program sequence is followed.

BNE Branch on result Not Equal to zero; Op-code $26

Operation. The branch is taken if Z = 0 (nonzero result.) If Z = 1 the program "falls through" the branch to the next sequential instruction.

BMI Branch on Minus result (N = 1); Op-code $2B

Operation. The branch is taken if the last instruction affecting the N flag left N = 1. In 2's complement binary, N = 1 indicates a negative (minus) result. However, since N simply duplicates the status of bit 7 for most instructions, the BMI can be used to test the status of this bit. In straight binary, for example, $b_7 = 1$ indicates a number greater than 127. The BMI could be used to sort data into greater than 127 and 127 or less.

BPL Branch on Plus result (N = 0); Op-code $2A

Operation. The branch is taken if N = 0, indicating a positive result in signed binary.

Notice that LDX, STX, and CPX set the N flag to the status of bit 15 of X, permitting BMI and BPL to distinguish between $X \geq 32\,768$ and $X < 32\,768$. INX does not affect the N flag, but a STX to a dummy address can be used to set $N = X_{15}$.

BVS Branch on Overflow Set (V = 1); Op-code $29

Operation. The branch is taken if the V flag is set. This flag is set when an 8-bit addition or subtraction causes a carry or borrow across bits 7 and 6. In signed-number arithmetic, bit 7 is the sign bit and does not have a place value. A carry or borrow on it is therefore illegal, and V = 1 indicates an overflow of the allowable −128 to +127 range for signed numbers.

The V flag is also used to indicate a number of special conditions resulting from operations other than *add* and *subtract*. These uses are listed in the *notes* at the end of the programmer's reference and are explained with the instructions involved. (See, for instance, the use of V with the ASL instruction on page 189.)

BVC Branch on Overflow Clear (V = 0); Op-code $28

Operation. This instruction is the complement of BVS. The branch is taken if V = 0. The normal program sequence is followed if V = 1.

BRA Branch always; Op-code $20

Operation. This is an unconditional branch. The 2's complement operand is always added to the address of the instruction following the BRA to obtain the address of the next instruction to be executed. The instruction following the BRA instruction in the program list is not executed (unless the operand of the BRA is a $00.) No flags are checked.

This instruction is a relative-mode JMP, except that it can only access addresses in a range from −127 to +128 around the current address. It is useful as a replacement for the JMP in routines that are recopied into various areas of memory, because the operand does not have to be changed each time the routine is moved.

BSR Branch always to Subroutine; Op-code $8D

Subroutines are the major topic of the next chapter, so we will defer an extended discussion of this instruction until then. For the moment, suffice it to say that there is an extended-mode JSR (jump to subroutine) instruction, and BSR is a relative-mode version of the same.

BHI Branch if accumulator Higher than operand; Op-code $22

Operation. This instruction is meant to be used after subtract or compare instructions CBA, CMP, SBA, or SUB in unsigned (straight binary) arithmetic. The branch is taken if the comparison showed the accumulator value to be larger than the operand value. If the accumulator was equal to or smaller than the operand in the subtract operation, the branch is not taken, and the following instruction is executed in normal program sequence.

With respect to the CCR flags, the BHI branch is taken if C and Z = 0. If C = 0, there was no borrow, so the operand was not larger than the accumulator, and if Z = 0, the operand and the accumulator were not the same. The conclusion is that the accumulator was larger than the operand.

BLS Branch if accumulator Lower or Same as operand; Op-code $23

Operation. This operation is the complement of BHI and is likewise meant for use after one of the instructions CBA, CMP, SBA, or SUB in unsigned arithmetic.

The Boolean logic is that the branch is taken if C or Z = 1. If C = 1, a borrow was necessary, so the accumulator was smaller than the operand. If Z = 1, the two were the same.

BGT Branch if accumulator Greater Than operand; Op-code $2E

Operation. This instruction is intended for use in 2's complement signed-number arithmetic after one of the instructions CBA, CMP, SBA, or SUB. It causes the branch to be taken if the algebraic value of the accumulator was greater than the algebraic value of the operand. Note that algebraic -20 is greater than -21.

The Boolean test for BGT is that the branch is taken if Z is zero and N and V are identical, that is,

$$Z + (N \oplus V) = 0$$

BLT Branch if accumulator Less than operand; Op-code $2D

Operation. This is another signed-number instruction for use after CBA, CMP, SBA, or SUB instructions. The branch condition is that N or V must be true but not both true, that is,

$$N \oplus V = 1$$

BGE Branch if accumulator Greater than or Equal to operand; Op-code $2C

Operation. BGE is the complementary instruction to BLT. The branch condition is that N and V must both be 1 or both be 0, that is,

$$N \oplus V = 0$$

REVIEW OF SECTIONS 8.5 and 8.6

21. Accumulator A contains decimal 124. What does it contain after two LSRA instructions are executed?
22. Answer Question 21 if A contains -124 (decimal).
23. If A contains -124 (decimal), what will it contain after two ASRA instructions are executed?
24. If A contains -124 and a single ASLA instruction is executed, what will be the status of the V flag?
25. Accumulator B contains decimal 112. How many RORB instructions must be executed to cause the C flag to be set?
26. A branch-instruction op-code resides at address $530B. The byte at $530C is $12. What is the address of the next instruction if the branch is taken?
27. Answer Question 26 if the branch is not taken.

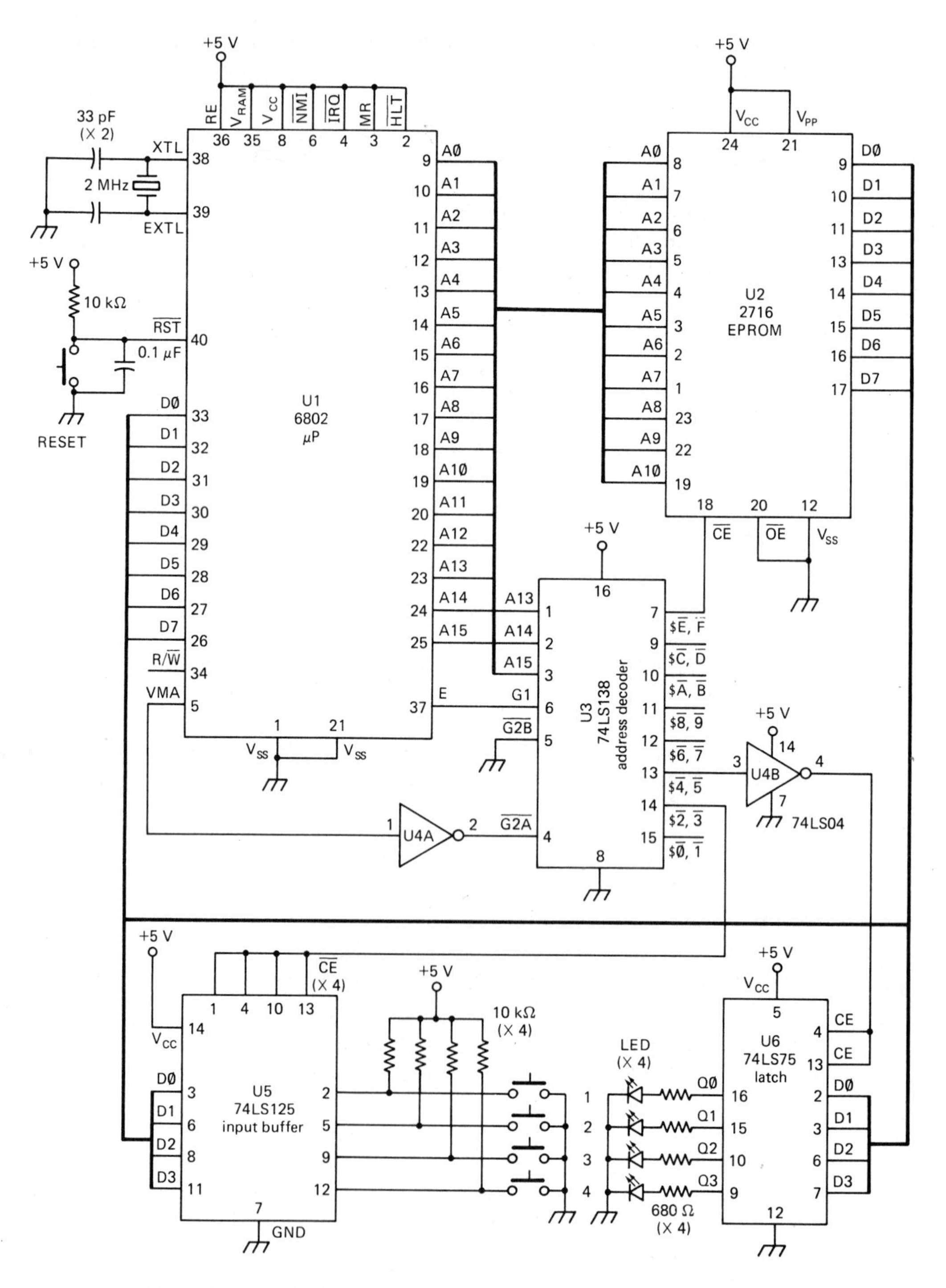

Figure 8.5 The MEM Game hardware of Figure 7.6 with an EPROM replacing the RAM for permanent program storage.

28. Answer Question 26 if the byte at $530C is $E4 and the branch is taken.
29. Accumulator A contains decimal 212 and accumulator B is loaded with unsigned data from an input buffer. What are the appropriate compare and branch instructions to cause a branch if the input data is 212 or less?
30. What are the appropriate compare and branch instructions to take a branch if accumulator B contains data that is more negative than −47?

8.7 AN EPROM-BASED SYSTEM

Section Overview

The usual procedure for the development of a microcomputer-based product includes computer-terminal entry of source code, automatic assembly of the object code, and programming of an EPROM with this code. The EPROM is then inserted into a test system for checkout and troubleshooting of the system hardware and software.

Figure 8.5 shows how a 2716 EPROM can be interfaced to the MEM game system of the previous chapter. The EPROM program is nonvolatile, so the system power can be turned off without loss of program data, and there is no danger of an accidental write destroying the program data. The complete program listing for the EPROM-based system is given in List 8.1. EPROM programmers for most popular desktop computers are available, but if you do not have one you might consider building the stand-alone programmer of Appendix D.

MEM Game Version One: Program Listing

```
                 1000 * ---------------------------------------------------------
                 1010 ***********      MEM GAME -- VERSION ONE     ************
                 1020 *
                 1030 * INPUT A SERIES OF 16 BUTTON PUSHES AND HAVE AN OPPONENT
                 1040 *  TRY TO REPEAT THE SAME SERIES.  SCORE OF CORRECT MATCHES
                 1050 *  IS DISPLAYED IN BINARY AFTER AN ERROR IS MADE.  YOU MAY
                 1060 *  TRY AGAIN TO REPEAT THE SERIES AFTER AN ERROR OR CORRECT
                 1070 *  RUN.  PUSH RESET TO LOAD A NEW SERIES.
                 1080 *
                 1090        .OR  $FF80     PROGRAM ORIGIN
                 1100        .TA  $4780     ASSEMBLER WORKING RAM ADDRESS
0030-            1110 COUNT  .EQ  $30       DELAY-LOOP COUNTER
2000-            1120 INPUT  .EQ  $2000     BUFFER ADR (74125)
4000-            1130 OUTPUT .EQ  $4000     LATCH ADR (7475)
                 1140 *
FF80- CE 00 00   1150        LDX  #0000     X REGISTER COUNTS 16 KEY INPUTS AND
                 1160 *                      POINTS TO FILE SAVING KEY VALUES.
FF83- B6 20 00   1170 LOAD   LDAA INPUT     GET KEY VALUE (1, 2, 4, OR 8).
FF86- 43         1180        COMA           INVERT SO "PRESSED" = LOGIC 1.
FF87- B7 40 00   1190        STAA OUTPUT    LIGHT LED BY PRESSED KEY.
FF8A- 84 0F      1200        ANDA #$0F      MASK OFF HI NIBBLE.
FF8C- 26 03      1210        BNE  GOOD      SCAN FOR KEY IN AGAIN IF
FF8E- 7E FF 83   1220        JMP  LOAD       NO KEY IN SENSED.
FF91- A7 00      1230 GOOD   STAA 00,X      STORE KEY VALUE IN FILE.
FF93- 5F         1240        CLRB           DELAY 1.2 SEC  WITH
FF94- 7F 00 30   1250        CLR  COUNT      NESTED LOOPS; ACC B GOES
```

```
FF97- 5C        1260 INNER  INCB           00 TO $FF IN INNER LOOP.
FF98- 27 03     1270        BEQ  OUTER
FF9A- 7E FF 97  1280        JMP  INNER
FF9D- 7C 00 30  1290 OUTER  INC  COUNT    RAM "COUNT" GOES 00 TO $FF
FFA0- 27 03     1300        BEQ  DONE      IN OUTER LOOP.
FFA2- 7E FF 97  1310        JMP  INNER
FFA5- 08        1320 DONE   INX           ADVANCE TO NEXT FILE SLOT.
FFA6- 8C 00 10  1330        CPX  #$0010   LAST FILE SLOT?
FFA9- 27 03     1340        BEQ  FLASH
FFAB- 7E FF 83  1350        JMP  LOAD      NO: GET ANOTHER KEY IN.
FFAE- 86 0F     1360 FLASH  LDAA #$0F      YES: LIGHT ALL 4 LEDS.
FFB0- B7 40 00  1370        STAA OUTPUT
FFB3- 08        1380 DELAY  INX           DELAY 1.4 SEC BY
FFB4- 27 03     1390        BEQ  DARK      COUNTING X TO $FFFF.
FFB6- 7E FF B3  1400        JMP  DELAY
FFB9- 7F 40 00  1410 DARK   CLR  OUTPUT   DARKEN ALL LEDS.
                1420 *
FFBC- B6 20 00  1430 PLAY   LDAA INPUT    GET KEY IN.
FFBF- 43        1440        COMA
FFC0- 84 0F     1450        ANDA #$0F     MASK OFF HI NIBBLE.
FFC2- 26 03     1460        BNE  VALID    IF NO KEY IN THEN
FFC4- 7E FF BC  1470        JMP  PLAY      LOOK AGAIN.
FFC7- B7 40 00  1480 VALID  STAA OUTPUT   LIGHT LED BY KEY PRESSED.
FFCA- A1 00     1490        CMPA 00,X     IS KEY VALUE SAME AS STORED
FFCC- 27 03     1500        BEQ  MORE      IN FILE BY "LOAD" ROUTINE?
FFCE- 7E FF F2  1510        JMP  DISPLY   NO: DISPLAY SCORE
FFD1- 5F        1520 MORE   CLRB          YES: DELAY 1.2 SEC WITH
FFD2- 7F 00 30  1530        CLR  COUNT     NESTED LOOPS WHILE
FFD5- 5C        1540 INNR   INCB           DISPLAYING KEY LED.
FFD6- 27 03     1550        BEQ  OUTR
FFD8- 7E FF D5  1560        JMP  INNR
FFDB- 7C 00 30  1570 OUTR   INC  COUNT
FFDE- 27 03     1580        BEQ  FINISH
FFE0- 7E FF D5  1590        JMP  INNR
FFE3- 7F 40 00  1600 FINISH CLR  OUTPUT
FFE6- 08        1610        INX           ADVANCE TO NEXT FILE ADR.
FFE7- 8C 00 10  1620        CPX  #$0010   LAST FILE LOCATION (16TH)?
FFEA- 27 03     1630        BEQ  PERFCT
FFEC- 7E FF BC  1640        JMP  PLAY      NO: GET NEXT KEY.
FFEF- 7E FF AE  1650 PERFCT JMP  FLASH     YES: ALL 16 RIGHT; FLASH & PLAY.
FFF2- FF 40 00  1660 DISPLY STX  OUTPUT   SCORE IN LO BYTE OF X STORED LAST AT
FFF5- 08        1670 DELY   INX            ADR $4001 (STILL LATCH ADR).
FFF6- 27 03     1680        BEQ  AGAIN    DISPLAY SCORE FOR 1.4 SEC
FFF8- 7E FF F5  1690        JMP  DELY
FFFB- 7E FF BC  1700 AGAIN  JMP  PLAY      AND PLAY AGAIN.
                1710 *
                1720        .OR  $FFFE    RESET VECTOR AT $FFFE, FFFF
                1730        .TA  $47FE     POINTS TO PROGRAM START
FFFE- FF 80     1740        .HS  FF80      ADR $FF80.
                1750 * 6802 ADR $FF80= APPLE TA $4780 & 2716 ADR $0780
                1760        .EN           END LISTING
```

Answers to Chapter 8 Review Questions

1. Hex digits **2.** Hand assembly **3.** Host; target

4. Source; object **5.** EQU (or .EQ) **6.** Decimal

7. It is the address of the op-code for the instruction on that line.

8. SYNTAX ERROR **9.** UNDEFINED LABEL ERROR

10. ILLEGAL OPERAND. Accumulator B can hold no value greater than decimal 255. **11.** There are no differences.

12. It takes 2 operand bytes to fill X, but only one to fill B.

13. It takes 2 bytes to specify a complete 6800 address. In LDAB the operand resides at the specified address. In LDX the operand resides at the specified address (high-order byte) and at the next sequential address (low-order byte).

14. 32 768 **15.** CLRA sets C = 0; LDAA #0 does not affect the C flag.
16. ANDA #$20 **17.** $9B **18.** All odd values, 1 through 255
19. Different **20.** NEGB **21.** Decimal 31
22. +33 decimal (not a valid divide-by-4).
23. −31 decimal (a valid divide-by-4)
24. V = 1, indicating overflow of −128 negative limit. **25.** Five
26. $531F **27.** $530D **28.** $52F1 **29.** CBA, BLS
30. CMPB #−47, BLT. (CMPB #−48, BLE would also work.)

CHAPTER 8 QUESTIONS AND PROBLEMS

Digits after the decimal point refer to section numbers in Chapter 8.

Basic Level

1.1 Define the terms machine code, object code, source code, and assembly language.

2.1 Define the terms assembler, host processor, target processor, and cross-assembler.

3.2 How can you tell your assembler that you want to use the name LATCH as an operand instead of using the address of the latch, which is $9000?

4.2 What hex value will an assembler assign to an operand listed as 10110011?

5.2 When a label is placed at the left of an instruction mnemonic, what value does the assembler give to this label?

6.2 What, if anything, is wrong with this instruction?

```
LDX     #89AB
```

7.3 Examine the eight op-codes for LDAA and LDAB in their binary forms. Which bits are common to all eight load-accumulator instructions? Which bit distinguishes between accumulators A and B?

8.3 If the C flag has been set, can A be stored without changing this flag? Can A be cleared without changing the C flag?

9.4 If B contains $12, what will be the contents of B, N, V, and Z after executing BITB #$C5?

10.4 What single instruction will set bit 0 of accumulator A to a binary 1 without disturbing any of the other bits?

11.5 Is ASL the same as LSL? Is ASR the same as LSR?

12.5 How many RORA instructions must be executed to return the contents of A to the original value?

13.6 In the instruction BGT, is the branch taken if the *accumulator* is the greater or if the *operand* is the greater?

14.6 Assume that A = $B5 and SUBA #$9F is executed. Show the states of Z, N, and V and show whether or not the following "branch taken" statement is true.

$$Z + (N \oplus V) = 1$$

Advanced Level

15.1 List and explain the reason for each of the seven columns in an assembler printout.

16.2 List some types of errors that an assembler can generally catch and some types of errors that it cannot catch.

17.3 A block of memory from address $C000 through address $DFFF is to be cleared to all $00 bytes. Is it faster to do this with a routine using a CLR 0,X instruction, or is it faster to use LDS #0 and then STS 0,X to clear 2 bytes at a time? Show both routines to support your answer.

18.4 Explain the difference between the complement instructions and the negate instructions of the 6802.

19.5 Write the 2's complement binary for −117. Shift the binary digits as they would be shifted by two ASR instructions, and demonstrate that the result is equivalent to a divide-by-4.

20.6 List in the left-hand column the 6802 branch instructions that apply exclusively to pure binary (unsigned) numbers. List in the right-hand column those that apply exclusively to 2's complement signed numbers.

21.6 Write two diagnostic routines to test the 6802's BLT instruction. In one loop the branch should be taken repeatedly, and in the other the branch should never be taken. Both routines should place the 6802 in an endless loop with a minimum number of machine cycles.

22.7 Draft a complete flowchart for the MEM game (List 8.1). Use the labels from the source code on the flowchart.

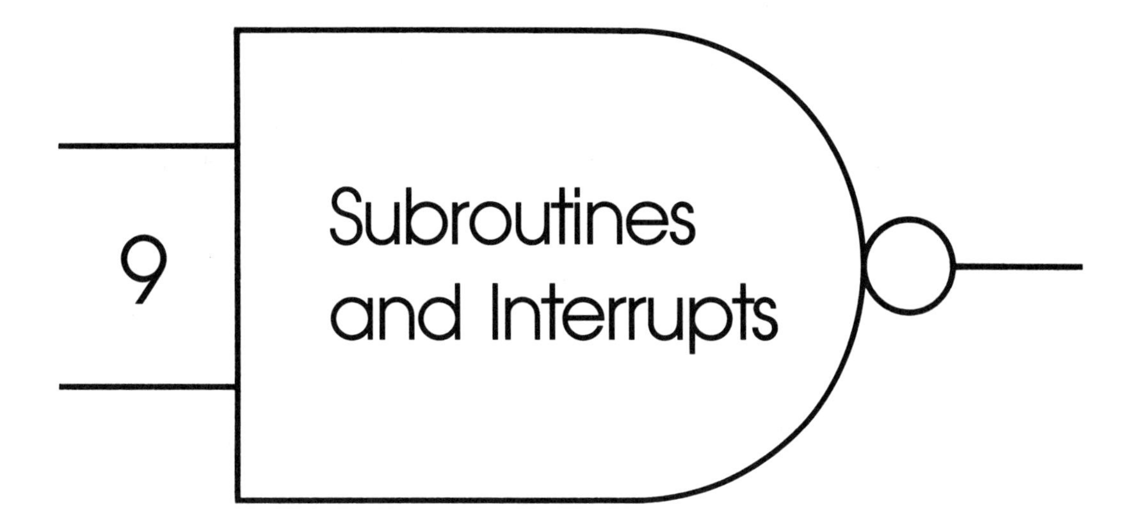

9.1 THE BASIC SUBROUTINE

Section Overview

Subroutines are program segments that can be used repeatedly by the main program. They may be stored in an adjacent or in a separate area of memory and *called up* as needed by the main program. They make it unnecessary to rewrite the same sequence of instructions over and over in the main program, thus conserving memory space.

Subroutines are called up in 6800/6802 language by the instruction JSR NAME, where NAME is a variable name representing the address of the subroutine. Each subroutine must end with an RTS (return from subroutine) instruction.

The JSR causes the current address of the main program to be saved in an area of RAM called the *stack*. The RTS causes the program to return to this address.

Microcomputer programs tend to use the same routines repeatedly. If we need to read a keyboard to get an ASCII representation of an alphanumeric character, we will probably need to do it many more times than once. If we need to convert a pure binary number to BCD form once, we will probably need to do it at several points in the program. Program segments that are apt to be used repeatedly throughout a program are generally written as *subroutines*.

A subroutine needs to be written only once, and it is stored in memory at only one place; but it can be called upon and executed any number of times by the main program. A large computer program generally has a whole *library* of subroutines stored in one area of memory and the main program stored in another. The main program *calls* the subroutines as they are needed, and each subroutine passes control back to the main program when it is finished. It is common for a computer to spend most of its time executing subroutines, with the main program doing little more than determining which subroutine comes next.

Figure 9.1 illustrates how a subroutine at address $FE00 is called by the main program with a JSR (Jump to Subroutine, extended mode) instruction at addresses

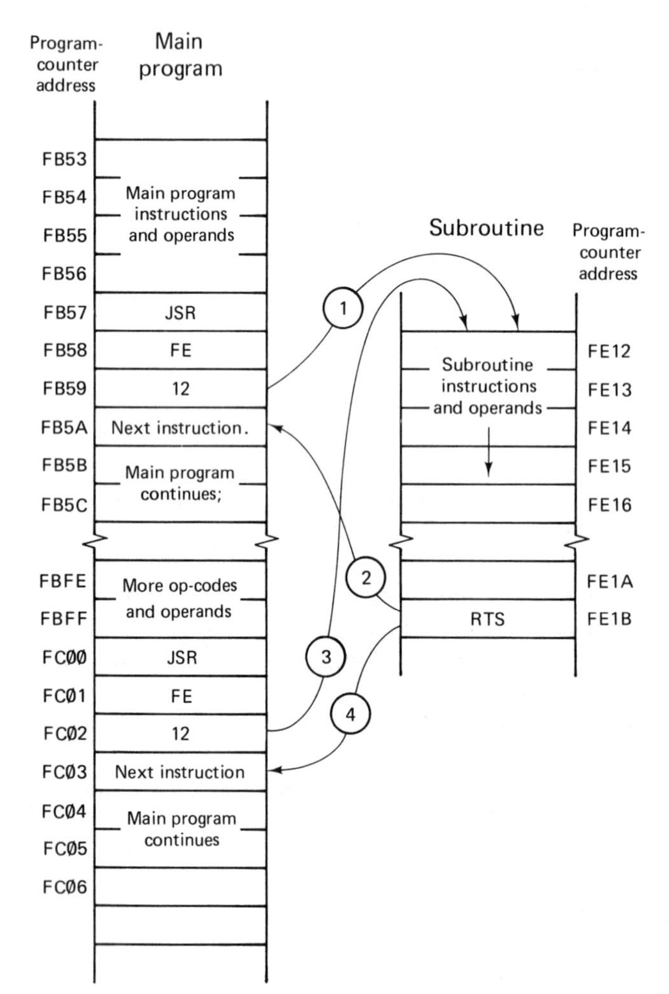

Figure 9.1 A subroutine can be called upon and effectively inserted at several places within the main program.

$FB57, FB58, and FB59. (See Step 1 in the figure.) The subroutine ends with an RTS (Return from Subroutine, inherent mode) instruction at $FE1B, and the program counter automatically resumes the main-program sequence with the next instruction at $FB5A (Step 2 in the figure).

At Step 3 in Figure 9.1 the same subroutine is called again by a second JSR at addresses $FC00, FC01, and FC02. This time the subroutine's RTS instruction sends the program execution back to the next instruction at $FC03 (Step 4 in the figure) upon completion of the subroutine.

Finding the way back: In the example of Figure 9.1, how was the same RTS instruction at the end of the subroutine able to direct the program counter back to address $FB5A the first time and then back to address $FC03 the second time? To answer this question we will need to understand how the JSR and RTS instructions make use of the microprocessor's *stack pointer* register.

The stack pointer is a 16-bit register inside the 6802 microprocessor. It is an *address-pointer* register, which is to say that its contents are generally interpreted as a 16-bit address where the operand data is to be found. We encountered this register in Section 7.3, where we used it to point to the destination addresses in a routine to move a block of data. Now we will use it to point to a special area called the *stack*.

The stack is an area of RAM used to store data, of which each byte will be retrieved in the reverse order from which it was stored. The name comes from the analogy to a stack of cards put down one by one on a table; for example:

Bottom 9 5 10 3 6 8 top

If they are picked up one by one, they will be retrieved in the reverse order, that is

Top 8 6 3 10 5 9 bottom

This is called a LIFO (last-in, first-out) stack.

When a subroutine is called, the stack is used to store the low-order and high-order bytes of the address of the op-code following the JSR instruction in the main program. The JSR instruction causes these 2 bytes to be stored on the stack, in addition to jumping the program counter to the subroutine address.

The RTS instruction at the end of the subroutine retrieves these 2 address bytes and causes them to be loaded into the program counter. This is how the microprocessor is able to return to the main program and pick up right where it left off when the subroutine was called.

Loading the stack pointer: If a program is going to incorporate subroutines, an area of RAM must be set aside to be used as the stack, and the stack pointer register must be loaded to point to the highest stack address. The 6802 processor has internal RAM from addresses $0000 to $007F, so we may wish to initialize the stack pointer at the start of the program by

```
LDS     #$007F
```

A JSR will then cause the low-order byte of the return address to be stored at address $007F, and the high-order byte to be stored at $007E. The stack pointer will be automatically decremented again to point to the next free stack location, $007D.

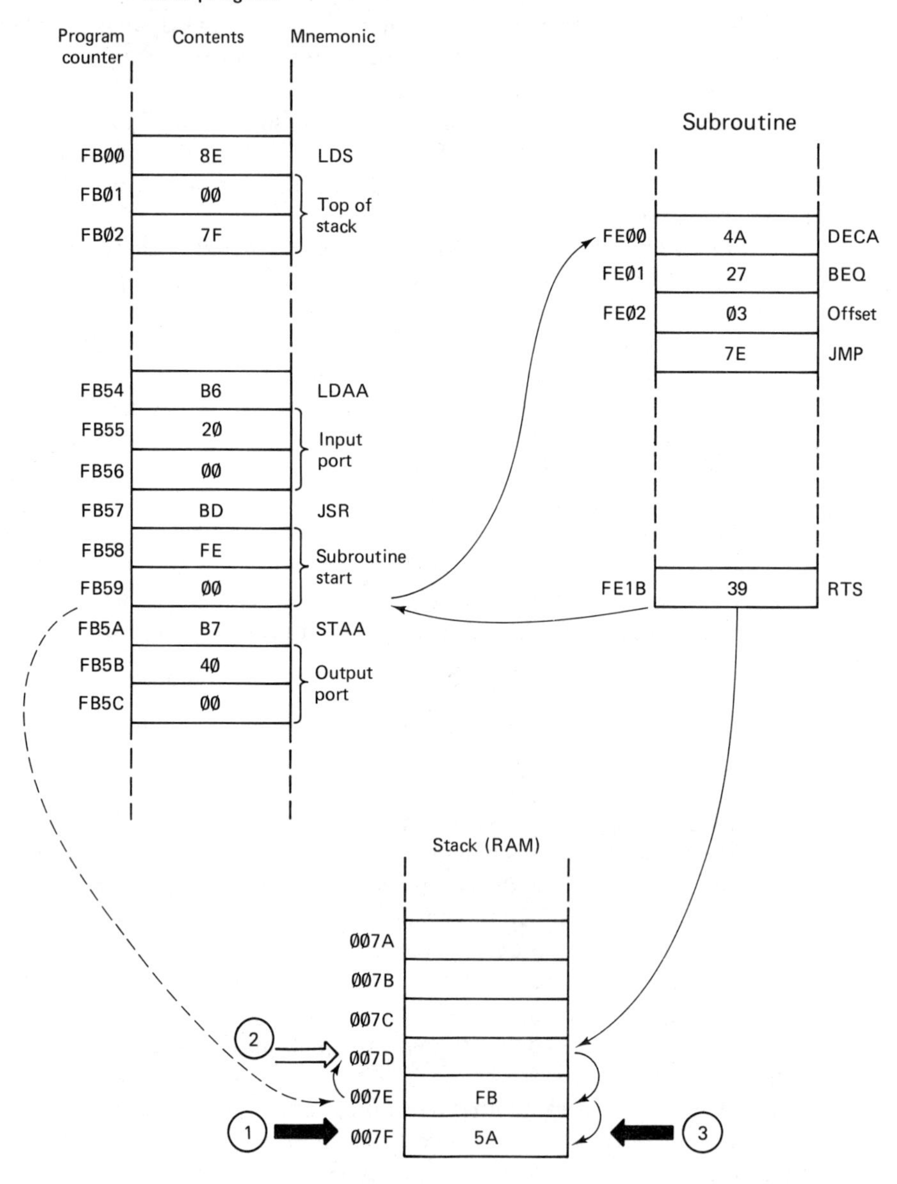

Figure 9.2 The stack is a RAM area which holds the address to return to when the subroutine is over. The stack pointer is a 16-bit register that points to 1 before the JSR, 2 after the JSR, and 3 after the RTS.

Figure 9.2 illustrates the function of the stack pointer and stack RAM in a subroutine call and return.

9.2 STACKING DATA

Section Overview

A subroutine will generally change the data in registers A, B, X, and CCR, so data stored in any of these registers by the main program will be lost upon returning from the subroutine. To make the subroutine less destructive of main-program data, we can "push" A and B on the stack RAM at the beginning of the subroutine and "pull" them back at the end of the subroutine. The X and CCR registers cannot be pushed directly on the stack, but they can be saved there by first transferring them to register A and then pushing A on the stack.

When a microprocessor begins to execute a subroutine, the machine registers (except the program counter and stack pointer) all contain the data they had when the processor left the main program. However, as soon as the subroutine begins to execute, some of these registers, at least, are sure to be changed. Thus, when the processor returns to its main program its internal registers (CCR, A, B, and X in the 6800/6802) are likely to have been altered, and the processor cannot truly pick up where it left off in the main program. To deal with this problem we may do one of the following:

1. Write the main program in such a way that it does not rely on the internal registers to retain any data between a Jump to Subroutine and a Return from Subroutine.
2. Save the internal registers in RAM just before the JSR and retrieve them as the first steps of the main program after RTS.
3. Save the internal registers being used by the subroutine in the stack RAM area, and restore them just before sending the processor back to the main program with an RTS.

Option 1 is unnecessarily limiting. Option 2 is workable, but it requires that *save* and *retrieve* routines be written each time a subroutine is used—and recall that subroutines are meant to be used repeatedly.

Option 3 is thus to be preferred because it minimizes the length of the total program and because it places the burden of register preservation on the subroutines, which are usually obtained from a library already written and debugged. The mainline programmer, relieved of this detail, is thus more free to concentrate on the problem at hand.

Preservation of register data by the subroutine should be the general rule in writing subroutines. If a subroutine will alter the A and CCR registers these should be saved as the first operation of the subroutine, and they should be restored just before the RTS instruction. If a subroutine's speed requirement or other circumstance makes it necessary to suspend this rule, a comment at the beginning of the source code should clearly state what registers may be altered; that is,

```
SUBROUTINE TO CHANGE ASCII 0 - F
TO 7-SEGMENT CODE. A AND CCR
DESTROYED. B AND X PRESERVED.
```

Preserving the registers on the stack is generally easier than saving them in other RAM locations. Consider that a moderately extensive program could easily employ 50 subroutines. If a separate 5 bytes of RAM were reserved to save CCR, A, B, and X for each of them, 250 bytes of RAM would be required. This is not a large amount of memory by today's standards, but it is a large bookkeeping problem to ensure that each subroutine accesses its own, and only its own, RAM bytes.

If each subroutine saves the necessary machine registers on the stack as it is entered and restores them just before the RTS instruction, the same stack area of RAM will be used for all subroutines, and no names or labels will have to be invented to distinguish one subroutine's RAM area from another's. The 6800/6802 processors have only the PSHA and PSHB instructions to *push* (store) data on the stack, and the PULA and PULB instruction to *pull* it back from the stack and load it into the accumulators.

Here is a routine to save all 5 register bytes using only these instructions. The comments given with the routine assume, just for the sake of example, that the stack pointer starts at address $007F, although other uses of the stack (to be discussed later) may just as well have altered it to point to a lower address. Also assumed are 2 bytes of RAM at $0020 and $0021, to facilitate the transfer of data from CCR and X via the A register. These same RAM bytes are to be reused for every subroutine, and they may be located at any otherwise unused area of RAM.

```
        * PRESERVE A, B, CCR, AND X ON STACK
PSHA                    SAVE A AT $007D.
TPA                     GET CCR IN A.
PSHB                    SAVE B AT $007C.
PSHA                    SAVE CCR AT $007B.
STX       RAM           PUT X (MSB) AT $0020, AND
LDAA      RAM+1           X (LSB) AT $0021.
PSHA                    SAVE X (LSB) AT $007A.
LDAA      RAM           GET X (MSB) FROM $0020
PSHA                    AND SAVE AT $0079.
```

This routine should be placed at the beginning of each subroutine. If the subroutine does not alter the contents of X, only the first four instructions are needed. The following routine should be placed immediately before the RTS at the end of the subroutine to restore the machine registers from the stack in LIFO order.

```
          * RESTORE X, CCR, B, AND A FROM STACK
PULA                  GET X (MSB) FROM STACK $0079
STAA      RAM         AND PUT AT $0020.
PULA                  GET X (LSB) FROM $007A, AND
STAA      RAM+1       PUT AT $0021.
LDX       RAM         RESTORE X FROM STACK VIA RAM
PULA                  GET CCR FROM $007B IN A.
PULB                  RESTORE B FROM STACK $007C.
TAP                   RESTORE CCR FROM STACK VIA A.
PULA                  RESTORE A FROM STACK $007D.
```

If only A, B, and CCR are to be preserved, then only the last four instructions of the RESTORE routine should be used. Figure 9.3 illustrates how the same areas of RAM are used to preserve the machine registers for two subroutines. The same example addresses used in the listings are used in the figure. Notice only two moves of the stack pointer are caused by the JSR and RTS instructions. The other five moves are caused by the PRESERVE and RESTORE routines just listed. A great deal of care must be exercised to ensure that bytes are retrieved from the stack in the reverse order from which they were saved, and that every decrement of the stack pointer after entering the subroutine is matched by an increment before leaving it. One of the most common causes of program *crashes* is the subroutine with one more PUSH than PULL, which eventually decrements the stack pointer out of the stack RAM area and causes improper writes in the user-data RAM area.

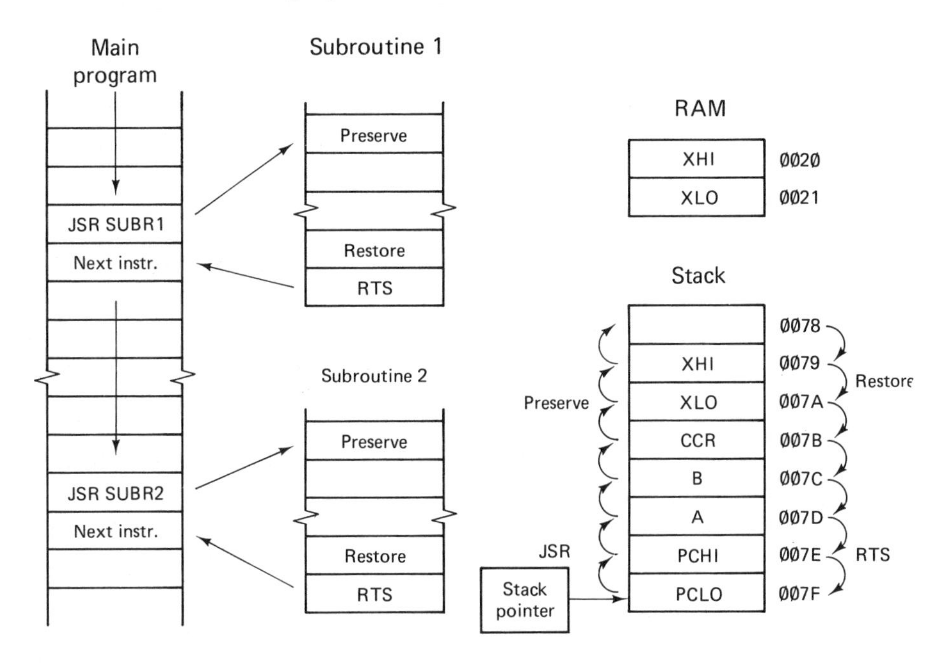

Figure 9.3 Both subroutines use the same area of RAM to preserve the machine registers and restore them before returning to the main program.

REVIEW OF SECTIONS 9.1 AND 9.2

Answers appear at the end of Chapter 9.

1. Subroutines are especially efficient for program segments that are used ______ by the main program.
2. Every subroutine should end with the instruction ________.
3. Distinguish between the *stack* and the *stack pointer*.
4. The stack pointer is loaded with #$Ø1FF. What does it contain after a JSR is executed?
5. Continuing Question 4, what does the stack pointer contain after an RTS instruction is executed?
6. Why is it desirable to save the machine registers in RAM before beginning a subroutine?
7. Why is it better to save these registers on the stack rather than in specially designated RAM areas?
8. How much added time (at 1 μs/cycle) is required to stack and restore A, B, and CCR in a subroutine, using the instructions given in Section 9.2?
9. How is the condition-code register of the 6802 saved on the stack, since there is no PSHP instruction?
10. What would happen to the stack pointer if a subroutine returned to the main program with a JMP or Branch rather than with an RTS?

9.3 NESTING SUBROUTINES

Section Overview

If one subroutine calls up a second subroutine, we say that the second subroutine is *nested* in the first. This is two-level nesting, but nesting to four or more levels is not uncommon. The only requirements are that every subroutine be entered by a JSR and be exited via an RTS and that there be enough RAM stack area to hold the data for the number of levels encountered.

Subroutines should be thought of as the building blocks of computer programs, and they should be used freely from existing libraries and written freely as the elements of new programs. Indeed, it is very common that, in writing a new subroutine, it is found to be beneficial to call upon a second pre-existing subroutine to implement some function required by the new subroutine. Thus the second subroutine will be *nested* within the first. Furthermore, it may be that the second subroutine calls up a third subroutine, producing a three-level nesting of subroutines.

To make a specific example, let us say that a program is being written to control a desktop microcomputer. The program will frequently call upon subroutine 1, SCRNUP, which updates the screen RAM with any new data to be displayed on

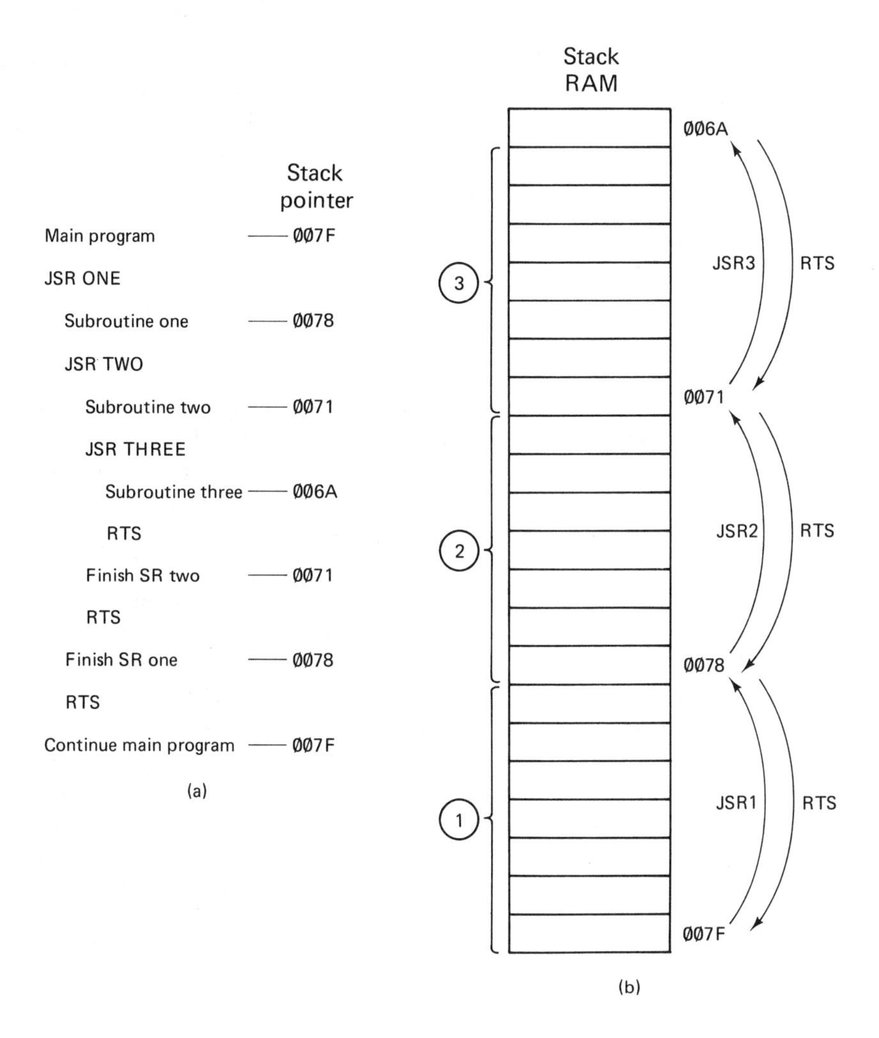

Figure 9.4 Nested subroutines: (a) The main program calls subroutine 1, which calls subroutine 2, which in turn calls subroutine 3. (b) The machine registers are saved on the stack as each subroutine is called, and restored as each RTS is encountered.

the CRT. SCRNUP will, among other things, call upon subroutine 2, READKY, which scans the keyboard and places the ASCII code for any pressed key to accumulator A. READKY will, in the course of its execution, call up *subroutine 3*, CONASC, which converts the row and column number of the key depressed to the ASCII code for that key.

The stack pointer is decremented toward lower RAM addresses as each subroutine is entered and advanced towards its original address as each RTS is encountered. If the advice of the previous section is followed and all the machine registers are stacked at the beginning of each subroutine, a nest of three subroutines will decrement the stack pointer by 3×7, or 21, bytes. This is illustrated in Figure 9.4.

There is no theoretical limit to the depth of subroutine nesting. However, it is uncommon to see nesting more than six levels deep. It is essential that the stack area of RAM extend backward for enough to hold the data stacked by all the successive levels of nested subroutines and that this RAM data not be altered by any other part of the program.

9.4 TIPS ON SUBROUTINE AND STACK USAGE

Section Overview

Data pushed on the stack is stored at successively lower addresses. The stack can be used for temporary storage and to free up machine registers, but it is an ironclad rule that data must be retrieved in the reverse order in which it was stacked, and that every operation which decrements the stack pointer must have a corresponding operation which increments it to avoid losing control of the program.

Parameter passing to subroutines: Often a subroutine is expected to process data generated or obtained for it by the main program. If the data involved is limited to no more than 4 bytes, it can be passed intact from the main program to the subroutine via registers A, B, and X.

If a longer data table is required to be processed, it must be stored in an area of RAM. An example of such a requirement might be a subroutine that converts a table of 10 decimal numbers (in their ASCII representations) into their binary equivalent (5 bytes concatenated to form one 40-bit word). In this case it would probably be easiest to store the file in a reserved area of RAM before entering the subroutine and have the subroutine access this RAM area for its data. The *X* register would be

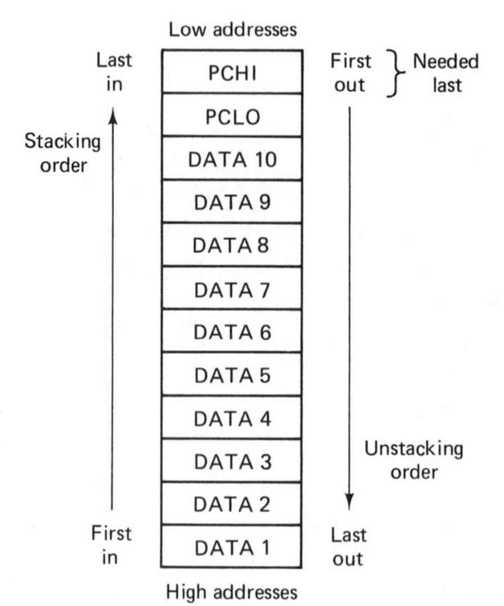

Figure 9.5 A LIFO stack is not suited for passing data to and from a subroutine.

used to index through the RAM area. This has implications for *reentrant coding*, a topic that will be taken up in Section 10.2.

Attempts to use the stack to pass data between the main program and the subroutine would encounter difficulty because the last data saved before leaving the main program is the return address (PCLO and PCHI, stacked by the JSR instruction). This is the last data needed by the subroutine, but it is the first data that would be pulled off the stack within the subroutine, as Figure 9.5 illustrates. Advanced computers sometimes contain a first-in, first-out (FIFO) stack to overcome this problem. Another solution is to have two stack pointers in the microprocessor. One is used by the *system* leaving the second free for handling *user* data.

Accumulators unlimited: It is a very common problem in programming to find that you need another accumulator. Perhaps you need to transfer a byte of data from an input buffer to a file in RAM, but both accumulators already contain important data that must be retained. The solution is to push the contents of one of the accumulators on the stack, use that accumulator to transfer the data, and then pull from the stack to restore the original data to the accumulator.

```
PSHA              Save A.
LDAA    INBUF     Put data from input
STAA    00,X        into data table.
PULA              Restore A.
```

Of course, this technique is useful only if the need for the other accumulator is relatively brief.

Preserving the order of the stack is essential, and a great deal of care must be exercised to ensure that data is always retrieved from the stack in exactly the reverse order as it was laid down. Figure 9.6 shows some valid and invalid stacking techniques. The lines linking pushes and corresponding pulls from the stack will not cross if proper stacking order is observed. The sources and destinations of any *jumps*

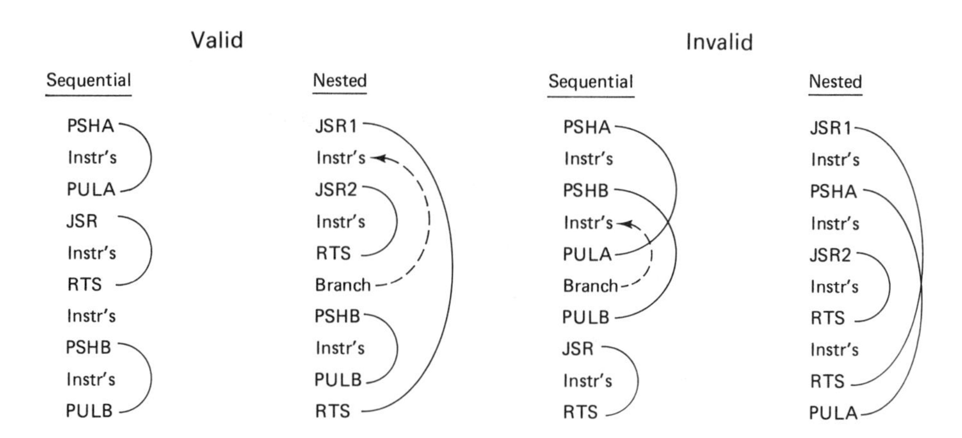

Figure 9.6 Crossed lines linking pushes and corresponding pulls from the stack indicate invalid data obtained from the stack. Branch and Jump instructions (dotted lines) must never cross the solid lines.

or *branches* must also be positioned so that the dotted lines connecting them will not cross any of the stack push/pull lines. Crossings are more difficult to spot in an actual program list because the subroutines are usually listed separately from the main program.

Swapping accumulators A and B, so that the contents of A appear in B and the contents of B appear in A, is not one of the 6802 instructions, but it can be accomplished quickly in the stack. The "lines" of Figure 9.6 are deliberately crossed here because we do not want to restore the same data.

```
PSHA    TO ADR $007F
PSHB    TO ADR $007E
PULA    FROM ADR $007E
PULB    FROM ADR $007F
```

The stack pointer can be "borrowed" for other uses, such as moving a data block, but it must first be saved in RAM and then be replaced before any *pushes* or *pulls* are done on the stack. For example, in the following routine, no instructions such as PSHB or JSR could be placed between the STS RAMSTK and LDS RAMSTK instructions.[1]

```
       SEI              DISALLOW INTERRUPTS WHILE S IS ALTERED.
       STS    RAMSTK    SAVE STACK POINTER IN RAM.
       LDS    #$FCFF    SOURCE BLOCK STARTS AT FD00.
       LDX    #$0100    DESTINATION AT 0100.
NEXT   PULA             GET BYTE (AUTO PREINCREMENT).
       STAA   00,X      SAVE TO PAGE 01.
       INX              NEXT BYTE.
       CPX    #$0200    FINISH PAGE?
       BNE    NEXT      NO: KEEP GOING.
       LDS    RAMSTK    YES: RESTORE STACK POINTER.
       CLI              RE-ENABLE INTERRUPTS.
       JSR    KEYIN     NOW SAFE TO USE STACK AGAIN.
```

Sensing stack condition: PSHA decrements the stack pointer automatically, and PULA increments it, as do DES and INS. However, none of these instructions affects any of the flags in the condition-code register. Also, there is no compare S instruction, so sensing the contents of the stack pointer can be a problem. The STS instruction affects the Z and N flags, so a dummy STS $XXXX instruction can be used to obtain some information about S. The operand addresses $XXXX and XXXX + 1 are any otherwise unusued RAM locations. Z will be set if S = 0 and this can be sensed by the BEQ or BNE instructions. The N flag (sensed by BMI or BPL) will be set if bit 15 of the S register is a logic 1, that is, if S ⩾ $8000.

[1]Since interrupts use the stack, none can be allowed during the use of S as an index register. If an NMI is possible the RAM area at addresses lower than below the data block must be free for use as an NMI stack. Sections 9.5 and 9.6 will discuss interrupts fully.

REVIEW OF SECTIONS 9.3 AND 9.4

Answers appear at end of Chapter 9.

11. A 6802 program contains three levels of nested subroutines. The first and third stack all the machine registers using the PRESERVE routine of Section 9.2, but the second stacks only A, B, and CCR. If the stack pointer starts at $007F, what is the lowest address to which data is written?

12. In Question 11, at what address will A from the second subroutine be saved upon entering the third subroutine?

13. Which 6802 processor registers remain unaffected by the execution of a JSR instruction: A, B, CCR, PC, S, or X?

14. Three bytes of data are placed on a stack in the order 5A, 39, C7. How will they be retrieved if the stack is LIFO? FIFO?

15. What does the JSR instruction do that the JMP instruction does not do?

16. Section 9.4 recommends saving the contents of A on the stack so that A can be used temporarily for other purposes. Why is this preferred to saving it in a specifically designated RAM location?

17. What problem is likely to result from this program sequence?

```
LOOP    LDAB    INBUF
        PSHB
        INCB
        BEQ     LOOP
        PULB
        STAB    00,X
```

18. The S register contains $007F. What does the X register contain after execution of a TSX instruction? Check the programmer's reference before you answer.

19. What address gets the data from accumulator A in this routine? What data is contained in S?

```
LDS     #$0060
PSHA
```

20. Write a 6802 routine that will save A on the stack and branch to an address labeled FULL if the address stored to was $0001.

9.5 BASIC INTERRUPT PROCESSING

Section Overview

A program interrupt can be initiated by a *low* input to the 6802's input pin $\overline{\text{IRQ}}$. An interrupt causes all the machine registers to be stacked and a separate interrupt program to be entered. The processor retrieves its register contents from the stack and resumes its main program when an RTI instruction is encountered at the end of the interrupt routine.

The address of the first instruction of the interrupt routine is stored at address $FFF8 (MSB) and $FFF9 (LSB). The $\overline{\text{IRQ}}$ input will be ignored if the I-mask bit of the CCR is set. The SEI instruction, a processor *reset*, or an interrupt acknowledge sets I.

The word *interrupt* suffers from a bad reputation. In human affairs no one wants to be interrupted, and we are taught that it is rude to interrupt others. In the microprocessor world none of this stigma applies. Interrupts are expected and welcomed because they extend the power of the processor to handle a number of control or computational tasks.

A keyboard-read example: To show the advantage of interrupts, let us consider the example of a desktop microcomputer with the typical keyboard input. The keyboard is here assumed to have a decoder chip included, which automatically converts each keystroke to its 1-byte ASCII equivalent. This key byte is fed to the microcomputer data bus through an input-buffer chip.

Continual-scan input: One way to be sure of catching all the keystrokes is to design the main program so that it continually scans the key buffer. Ten *reads* per second would probably be enough to preclude the possibility of missing anything. This leaves 100 ms, or 100 000 machine cycles, between reads, with each read taking about 20 machine cycles. Obviously, continual scan of an input—or even of several inputs—leaves plenty of time for the processor to complete its primary computational tasks. The problem is that in writing the primary programs, we are inconvenienced by the necessity of getting back to that key-read routine every 100 ms. Often this is not actually such a great inconvenience, and scanned inputs can certainly be used in such cases. However, as the complexity of the primary task increases, it becomes more likely that input scanning will necessitate breaking that task into segments and inserting the scans in between. At this point we begin to think about alternatives.

The interrupt-request input to the microprocessor ($\overline{\text{IRQ}}$ in the case of the 6800/6802) can be used to signal the microprocessor that an input device has data to be read. This relieves the programmer of the burden of having to check continually for inputs. Upon sensing the interrupt request, the processor completes the current instruction in the main program and then jumps to a separate *interrupt routine* to read the input and process the data as required. It then returns to the next instruction in the main program and resumes executing. Figure 9.7 illustrates the interrupt-handling process.

Interrupts are somewhat like subroutines, in that they cause the program counter to leave the main program and jump to a separate routine. They are also similar in that the *return address* for getting back to the main program is automatically saved in the stack area of RAM by the JSR, or in this case the IRQ. Furthermore, subroutines (and interrupt routines) end with an RTS (or RTI—return from interrupt) instruction, which retrieves the return address from the stack.

However, there are several differences between subroutines and interrupts:

- Subroutines are software-initiated by the JSR or BSR instructions; interrupts are hardware-initiated by a *low* level on the $\overline{\text{IRQ}}$ input pin in the case we are discussing. (The initiation of the NMI and SWI is discussed in the next section.)

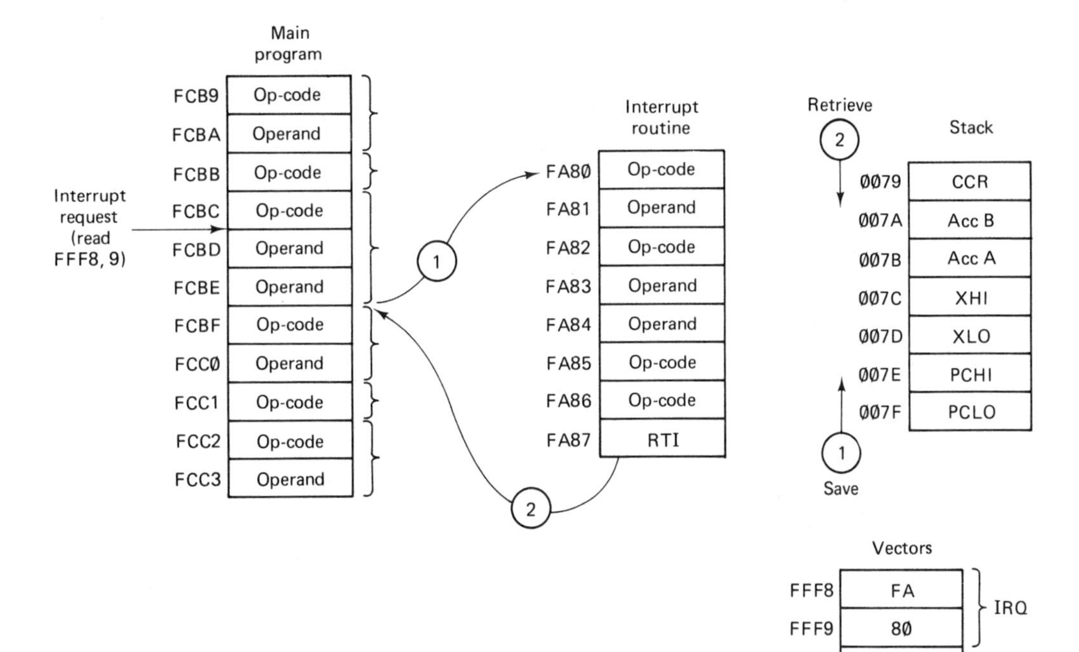

Figure 9.7 The interrupt process. The addresses are given as examples for a typical 6802 system.

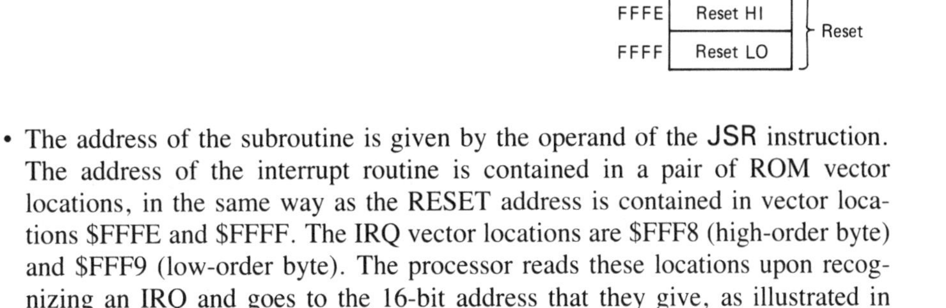

- The address of the subroutine is given by the operand of the JSR instruction. The address of the interrupt routine is contained in a pair of ROM vector locations, in the same way as the RESET address is contained in vector locations $FFFE and $FFFF. The IRQ vector locations are $FFF8 (high-order byte) and $FFF9 (low-order byte). The processor reads these locations upon recognizing an IRQ and goes to the 16-bit address that they give, as illustrated in Figure 9.7.
- A jump to subroutine causes only the 2 bytes of the program counter to be saved on the stack. If the other machine registers need to be saved, it is up to the programmer to do this with PSH instructions. A 6802 interrupt, on the other hand, always causes all seven machine registers to be pushed on the stack. This is generally desirable because the programmer has no way of predicting where the processor may be in the main program when an interrupt request is generated. Therefore, it must be assumed that the CCR, B, A, and X registers contain data that will be needed upon the return to main program.[2]

[2]The PRESERVE A, B, CCR, AND X routine given in Section 9.2 could have been used to do this in 33 extra machine cycles. However, since most interrupt routines are likely to alter all of these registers, it was made an integral part of the IRQ function, where it takes 6 extra machine cycles.

- Interrupts (IRQs but not the NMI; see the next section) can be *masked off* or ignored at the discretion of the programmer. We will discuss this difference in detail.

The interrupt-mask is bit 4 of the 6802's condition-code register. It is called I, but remember that it means Interrupt-*Mask*, not interrupt. When the I bit is set to 1, the processor will not respond to a *low* level at its $\overline{IRQ}$ input; it will keep processing the main program. The I-mask is cleared normally only by the *clear interrupt-mask* (CLI) instruction, or by a restoration of the CCR with the RTI instruction.

The I-mask is set by the following:

1. A processor *reset* (*low* level on the RESET input). This means that an IRQ will not be recognized unless the programmer purposefully clears the I-mask.
2. The *set interrupt-mask* (SEI) instruction. This allows the programmer to forestall interrupts during critical segments of the main program—for example, during timing loops which would be rendered inaccurate by the extra time consumed by the interrupt routine.
3. Recognition of any interrupt. This assures that multiple interrupts will not result from a single *low-level* pulse on the $\overline{IRQ}$ input. The CLI instruction should not subsequently be executed until the $\overline{IRQ}$ line returns to a *high* level. This could be on the order of a second for a human-operated key, so a 1-s delay loop may be required at the end of the interrupt routine but before the RTI. If this is undesirable, a one-shot may be used to assure that the $\overline{IRQ}$ pulse does not outlast the interrupt routine. Figure 9.8 shows an $\overline{IRQ}$-driver circuit.

9.6 INTERRUPT TYPES AND TECHNIQUES

Section Overview

The $\overline{NMI}$ (nonmaskable interrupt) input pin of the 6802 is similar to the $\overline{IRQ}$ except that the I flag will not cause it to be ignored, and it is sensitive to a *high-to-low* voltage transition, not a *low* voltage level.

The software interrupt is initiated by the SWI instruction in a program.

If more than two external devices must be able to interrupt the processor, the processor will have to respond to the interrupt by reading a code on its data bus that tells it which of several interrupt routines to go to. This added flexibility slows down interrupt-response time.

The nonmaskable interrupt (NMI) is initiated by a *high*-to-*low* transition of the 6802's $\overline{NMI}$ input pin. This interrupt is similar to the IRQ, with the following differences:

- The NMI cannot be deferred or masked by the I bit. An NMI sets the I flag, but proceeds even if I was previously set. The main program is left and the NMI routine is entered immediately upon completion of the instruction, during which $\overline{\text{NMI}}$ is brought *low*.
- The $\overline{\text{NMI}}$ input responds to a negative transition (edge triggering), not a negative level. This transition must have a fall time of 100 ns or less. Manual pushbutton switches may be used for the RESET and (with a 1-s software delay) the $\overline{\text{IRQ}}$ inputs, but they invariably bounce and chatter several times on closing and would cause multiple interrupts on the $\overline{\text{NMI}}$ input. Figure 9.9 shows a debounced $\overline{\text{NMI}}$ driver.
- The NMI-routine starting address is fetched from vector location $FFFC (high-order byte) and $FFFD (low-order byte). The same instruction (RTI) is used to return to the main program.

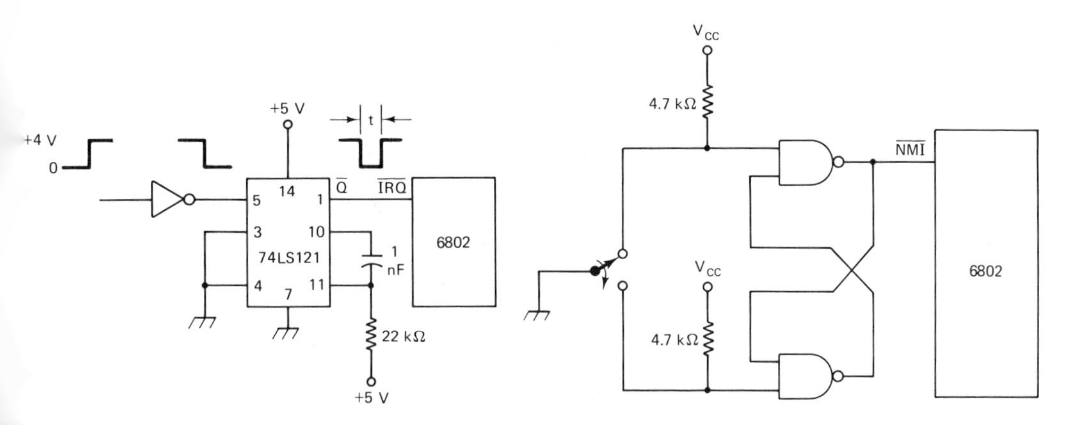

Figure 9.8 A 20-μs interrupt-driver circuit.

Figure 9.9 A switch debounce circuit should be used to drive the $\overline{\text{NMI}}$ input.

The software interrupt instruction (SWI) will cause all 7 bytes of machine register to be pushed on the stack. The program counter will then be loaded from vector locations $FFFA and $FFFB, which are assumed to contain the address of the SWI routine. Notice that the SWI is initiated by the program, not an external hardware signal. SWI also sets the I-mask bit when it is recognized.

Here are a few ways that the SWI can be used:

- An SWI instruction can be inserted in a program at selected points to test how the program would respond if an NMI occurred at that point. The NMI and SWI vectors should contain the same addresses in this case.
- The SWI can be used as a JSR that automatically and very quickly stacks all the machine registers. The RTI (not the RTS) must be used for returns.

- The SWI can be used in program troubleshooting to get a look at the contents of the internal machine registers. The SWI instruction should be inserted just after the critical routine, and the SWI routine should *pull* the contents of the machine registers off the stack and display them.

Multiple hardware interrupt sources can be handled with a little external circuitry and added software. A number of interrupting devices may be connected to the processor's $\overline{\text{IRQ}}$ input using open-collector wired-OR circuitry or a NOR gate. Each interrupting device is then also connected to one of the inputs of a buffer chip (typically an 8-bit 74LS244) which feeds the processor data bus. When the processor senses an $\overline{\text{IRQ}}$ the interrupt routine first reads the buffer chip to determine which of its eight bits has been asserted. The software then directs the processor to the appropriate portion of the routine, based on the source of the IRQ.

The process of *polling* each of the bits from the buffer to determine which interrupt source requested service is time consuming, and delays the speed of response to any of the IRQs. The sources are *prioritized* by the software, so that the bits representing the most critical sources are checked first.

REVIEW OF SECTIONS 9.5 AND 9.6

21. How many bytes are pushed on the stack when a 6802 acknowledges an interrupt?
22. Every interrupt routine must end with the instruction ________.
23. Where does the processor get the address of the start of the IRQ routine?
24. What is the function of the I flag of the CCR?
25. Name two times when a CLI instruction must be used to enable an IRQ.
26. Why is a hardware debounce circuit required to drive the $\overline{\text{NMI}}$ input from a switch?
27. In troubleshooting a program segment that shifts the internal register A, how can you get a look at the contents of A and CCR after the shift?
28. A 6800 programmer needs a subroutine that is very fast but still saves all the machine-register contents from the main program. What can be done?
29. In Question 28, where is the subroutine address stored?
30. What is the disadvantage of polling to handle multiple interrupt sources?

9.7 STREAMLINING THE MEM GAME

The last section of Chapter 7 described a memory-challenge game called MEM. In this section we will improve the game by using subroutines for the time delays and key-input program segments. We will also add an interrupt input, which will stop the loading of data after any desired number of entries.

The only hardware change to the system of Figure 7.6, or the EPROM-based system of Figure 8.5, is the addition of an $\overline{\text{IRQ}}$ input switch, shown in Figure 9.10(a). The program flowchart appears in Figure 9.10(b), and the program listing follows.

MEM Game, Version Two: Program Listing

```
                 1000 * -------------------------------------------------------
                 1010 ********     MEM GAME   --   VERSION TWO      **********
                 1020 *
                 1030 * PROGRAM DEMONSTRATES SUBROUTINES AND INTERRUPT FUNCTION.
                 1040 *
                 1050 * RULES:
                 1060 * 1. PUSH RESET TO START.
                 1070 * 2. PUSH ANY OF THE 4 BUTTONS IN A SEQUENCE TO BE
                 1080 *      REMEMBERED. (OVER .8 SEC GIVES A DOUBLE STROKE)
                 1090 * 3. AFTER 16 ENTRIES ALL LAMPS WILL FLASH TWICE, AND IT
                 1100 *      IS NOW YOUR TURN TO REPEAT THE SEQUENCE.
                 1110 * 4. YOU CAN INTERRUPT THE LOADING SEQUENCE AT LESS THAN
                 1120 *      16 ENTRIES BY PUSHING THE IRQ BUTTON.
                 1130 * 5. THE MACHINE WILL FOLLOW YOUR ENTRIES UNTIL YOU MAKE A
                 1140 *      MISTAKE, THEN DISPLAY YOUR SCORE IN BINARY.
                 1150 * 6. AFTER THE SCORE IS DISPLAYED YOU MAY TRY AGAIN
                 1160 *      TO REPEAT THE SAME SEQUENCE.
                 1170 * 7. IF ANYONE REACHES 16 RIGHT ALL 4 LAMPS FLASH TWICE.
                 1180 *
                 1190         .OR  $FF80   PROGRAM ORIGIN
                 1200         .TA  $4780   ASSEMBLER TARGET ADDRESS
0030-            1210 COUNT  .EQ  $0030   DELAY COUNTER
0031-            1220 TIMES  .EQ  $0031   NUMBER OF MOVES TO REPEAT
2000-            1230 INPUT  .EQ  $2000   INPUT BUFFER ADR
4000-            1240 OUTPUT .EQ  $4000   OUTPUT LATCH ADR
                 1250 *
FF80- CE 00 10   1260        LDX  #$0010  START WITH A 16-MOVE GAME.
FF83- DF 31      1270        STX  TIMES
FF85- CE 00 00   1280        LDX  #$0000  START KEY FILE AT BEGIN OF RAM.
FF88- 8E 00 7F   1290        LDS  #$007F  STACK (FOR IRQ & JSR) AT END OF RAM.
FF8B- BD FF D6   1300 LOAD   JSR  KEYIN   GET KEY VALUE IN ACC A.
FF8E- 81 06      1310        CMPA #06     CODE 06 MEANS IRQ WAS RECEIVED;
FF90- 27 0A      1320        BEQ  FLASH    END INPUTS & FLASH 4 LEDS.
FF92- A7 00      1330        STAA 00,X    PUT KEY VALUE IN FILE.
FF94- BD FF E5   1340        JSR  DELAY   DISPLAY KEY LED 0.8 SEC.
FF97- 08         1350        INX          NEXT FILE SLOT:
FF98- 9C 31      1360        CPX  TIMES    LAST SLOT?
FF9A- 26 EF      1370        BNE  LOAD       NO: LOAD NEXT KEY.
FF9C- 86 0F      1380 FLASH  LDAA #$0F       YES: LIGHT 4 LEDS
FF9E- B7 40 00   1390        STAA OUTPUT          VIA OUTPUT LATCH
FFA1- BD FF E5   1400        JSR  DELAY           FOR 0.8 SEC.
FFA4- 7F 40 00   1410        CLR  OUTPUT   THEN DIM ALL LEDS
FFA7- BD FF E5   1420        JSR  DELAY      FOR 0.8 SEC.
FFAA- B7 40 00   1430        STAA OUTPUT   THEN LIGHT ALL 4 AGAIN
FFAD- BD FF E5   1440        JSR  DELAY      FOR 0.8 SEC.
                 1450 *
FFB0- CE 00 00   1460 PLAY   LDX  #0000   BACK TO START OF KEY FILE
FFB3- BD FF D6   1470 NEXT   JSR  KEYIN   GET KEY VALUE IN ACC A.
FFB6- BD FF E5   1480        JSR  DELAY   DISPLAY KEY LED FOR 0.8 SEC.
FFB9- A1 00      1490        CMPA 00,X    SAME VALUE AS STORED IN FILE?
FFBB- 26 08      1500        BNE  DISPLY   NO: DISPLAY SCORE
FFBD- 08         1510        INX           YES: GO TO NEXT FILE SLOT.
FFBE- 9C 31      1520        CPX  TIMES   LAST FILE SLOT?
FFC0- 26 F1      1530        BNE  NEXT     NO: GET NEXT KEY.
FFC2- 7E FF 9C   1540        JMP  FLASH    YES: FLASH ALL - YOU WON!
FFC5- FF 40 00   1550 DISPLY STX  OUTPUT   SCORE IS IN X, LO BYTE.
FFC8- BD FF E5   1560        JSR  DELAY     DISPLAY IT 0.8 SEC MIN.
                 1570 *
FFCB- B6 20 00   1580 WAIT   LDAA INPUT   WAIT FOR KEY INPUT;
FFCE- 43         1590        COMA
FFCF- 84 0F      1600        ANDA #$0F
FFD1- 27 F8      1610        BEQ  WAIT    PLAY AGAIN IF KEY
FFD3- 7E FF B0   1620        JMP  PLAY      INPUT IS FOUND.
                 1630 *
```

```
                1640 *SUBROUTINE 1 ** GET KEY VALUE IN ACCUM A **
                1650 *
FFD6- B6 20 00  1660 KEYIN   LDAA INPUT    GET KEY;
FFD9- 43        1670         COMA          INVERT SO "PRESSED" = 1.
FFDA- 0E        1680         CLI           LOOK FOR INTERRUPT
FFDB- 01        1690         NOP            FOR A FEW MICROSECS.
FFDC- 0F        1700         SEI           MASK (DISABLE) INTERRUPTS.
FFDD- B7 40 00  1710         STAA OUTPUT   LIGHT LED BY KEY.
FFE0- 84 0F     1720         ANDA #$0F     MASK OFF HI 4 BITS.
FFE2- 27 F2     1730         BEQ  KEYIN    WAIT FOR A VALID KEY.
FFE4- 39        1740         RTS
                1750 *
                1760 *SUBROUTINE 2 ** DELAY FOR 0.8 SECONDS **
                1770 *
FFE5- 7F 00 30  1780 DELAY   CLR  COUNT     DELAY 0.8 SEC VIA NESTED LOOPS,
FFE8- 5C        1790 INNER   INCB           COUNTING ACC B AND
FFE9- 26 FD     1800         BNE  INNER     RAM "COUNT" FROM 00
FFEB- 7C 00 30  1810 OUTER   INC  COUNT     THRU $FF, AND BACK
FFEE- 26 F8     1820         BNE  INNER     TO 00.
FFF0- 39        1830         RTS
                1840 *
                1850 **** INTERRUPT (IRQ) ROUTINE **** STOP LOADING
                1860 *
FFF1- DF 31     1870 ENUFF   STX  TIMES   PUT CURRENT NUMBER OF INPUTS IN "TIMES",
FFF3- 86 06     1880         LDAA #06      REPLACING VALUE 16. SLIP FLAG $06 INTO
FFF5- 97 79     1890         STAA $79       STACK TO BE PICKED UP BY ACC A UPON
FFF7- 3B        1900         RTI           RTI.  MAIN PROG INTERPRETS $06 AS
                1910 *                     "STOP INPUTTING".
                1920 *
                1930         .OR  $FFF8   ROM ADR $FFF8, FFF9 CONTAINS IRQ VECTOR
                1940         .TA  $47F8         (THIS ADR FOR BENEFIT OF ASSEMBLER)
FFF8- FF F1     1950         .DA  ENUFF    WHICH IS ADR OF IRQ ROUTINE "ENUFF"
                1960         .OR  $FFFE   RESET VECTOR
                1970         .TA  $47FE     (ASSEMBLER TARGET ADR)
FFFE- FF 80     1980         .HS  FF80    START ADR OF PROGRAM.
                1990 *
                2000 *                     6802 ADR FF80 = TA 4780 = EPROM 0780
                2010 *                     6802 ADR FFF8 = TA 47F8 = EPROM 07F8
                2020 *
                2030         .EN           END
```

Answers to Chapter 9 Review Questions

1. Repeatedly **2.** RTS **3.** The stack is an area of RAM. The stack pointer is a 16-bit internal machine register.

4. $01FD **5.** $01FF **6.** So they can be restored upon returning to the main program, as though they had never been altered by the subroutine.

7. The same stack area can be reused for many subroutines, and there is no need to keep track of numerous addresses or address names. **8.** 28 μs

9. TPA, then PSHA **10.** It would be driven back two addresses with each JSR, eventually overrunning the RAM area alloted to the stack.

11. $006D **12.** $0071 **13.** A, B, CCR, and X **14.** LIFO = C7, 39, 5A. FIFO = 5A, 39, C7.

15. It saves 2 bytes on the stack, which are the address of the next instruction in the main program.

16. The stack RAM can be reused for other purposes later and does not need to be assigned a name or address, thus minimizing the bookkeeping problem.

17. Each time the BEQ is taken, there will be a PSH stack without a corresponding

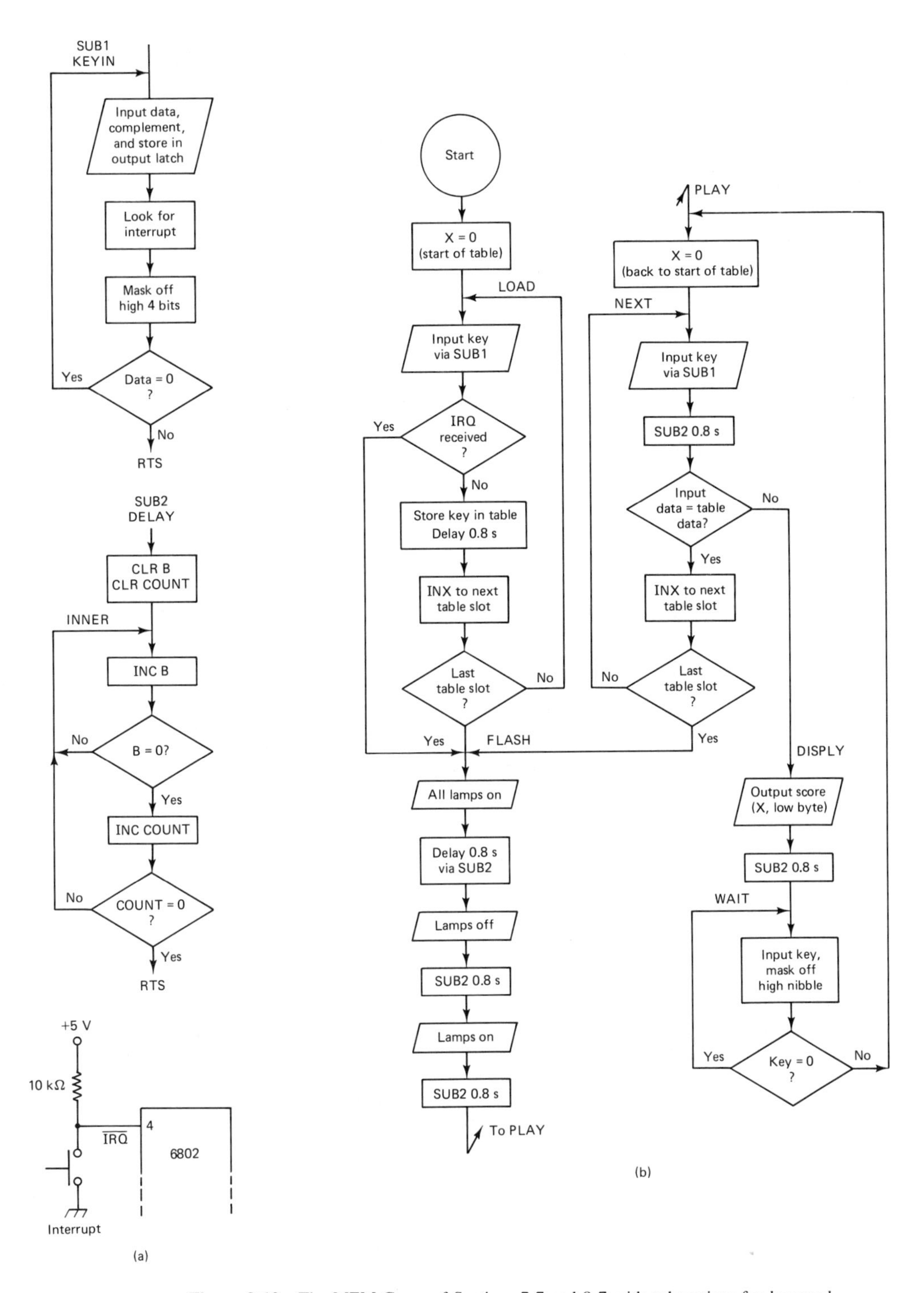

Figure 9.10 The MEM Game of Sections 7.7 and 8.7 with subroutines for *key read* and *time delay*, and an interrupt routine to stop loading at fewer than 15 key inputs. The hardware is that of Figure 8.5.

PUL from stack. This will get the stack pointer out of order, eventually driving it past its alloted RAM area.

18. $0080 **19.** $0060; $005F

20.
```
PSHA
STS     RAMBUF
BEQ     FULL
```

21. 7 **22.** RTI **23.** From addresses $FFF8 and $FFF9.

24. If I = 1 the processor will ignore an IRQ.

25. After a processor reset and after an interrupt acknowledge. **26.** To prevent multiple interrupts from switch bounce. **27.** Temporarily insert an SWI instruction after the shift. **28.** Uses SWI instead of JSR **29.** $FFFA and $FFFB

30. Added hardware and slower interrupt response.

CHAPTER 9 QUESTIONS AND PROBLEMS

Digits after the decimal point refer to section numbers in Chapter 9.

Basic Level

1.1 Write the source code instruction that you would use to jump to a subroutine at address $D500.

2.1 Where does the 6802 get the address to return to when it finishes with the subroutine and goes back to the main program?

3.2 Which, if any, of the instructions in the PRESERVE routine (page 206) alter the bits of the CCR? Is the CCR altered by the PRESERVE routine before it is stacked?

4.2 If a PRESERVE routine saves the 6802 registers in the order

PCLO, PCHI, A, B, CCR, XLO, XHI

In what order must they be retrieved?

5.3 A RAM area from $017F to $0100 is reserved for the 6802 system stack. How many levels of subroutine nesting can be accommodated if each subroutine stacks all seven machine registers?

6.3 What would happen if the body of a subroutine contained a PSHA instruction without a PULA or PULB instruction before the return from subroutine?

7.4 A 6802 subroutine is to be used to convert a 16-bit BCD number to pure binary. How would you pass the BCD bytes to the subroutine?

8.4 What does the following 6802 routine accomplish?

```
STX     $0020
LDAA    $0021
PSHA
LDAA    $0020
PSHA
```

9.5 Write a three-instruction routine that will toggle (change the state of) the I-mask bit of the CCR.

10.6 What advantage does the SWI instruction have over the JSR instruction?

11.6 Draw the schematic digram for the multiple-hardware-interrupt-source system described in Section 9.6.

Advanced Level

12.1 Let us assume that we are writing a 6802 subroutine to move a block of data from one specified page to another. What problem would arise if we used the approach shown in the MOVE DATA routine of List 7.3?

13.1 Look up the 6802's RTS instruction in Appendix B and describe in narrative form what the processor does on each machine cycle.

14.2 Why is it generally better to save the machine registers at the beginning of the subroutine rather than in the main program before the JSR? There are several reasons; give them all.

15.2 When executing a subroutine, why is it desirable to save the machine registers by pushing them on the stack rather than by simply storing them to a selected RAM location? Give two reasons.

16.3 Write the 6802 source code for a routine that will test the stack pointer and jump to another routine called STKERR if the stack pointer is not in the range $04FF to $0480. *Hint:* Use the CPX instruction.

17.4 Section 9.4 recommends sensing the stack pointer condition by storing S to an unused pair of RAM addresses. Why was a ROM address not chosen?

18.4 Write a routine that will continuously read an input buffer at address $2000 and store any nonzero bytes in a file starting at address $9000 and working *upward* to higher addresses. The X register is committed to another function, so S must be used as the pointer.

19.5 Name two ways to prevent multiple IRQ calls from a single actutation of a mechanical switch on the $\overline{\text{IRQ}}$ input.

20.5 What happens if the 6802's $\overline{\text{IRQ}}$ input is pulled *low* momentarily while the I-mask is set and then a CLI is executed? Obtain your answer by writing a test routine for a 6802-based system. Present the assembled code for the test routine.

21.5 A subroutine to read a keyboard and print the corresponding character has already been written starting at address $7B50. A need now arises to have this same routine executed in response to an $\overline{\text{IRQ}}$ input. One programmer proposes that vector addresses $FFF8 and $FFF9 simply be loaded with data $7B and $50, respectively. What is your response to this proposal?

22.6 The $\overline{\text{NMI}}$ input in a 6802 system is not being used, but noise pickup on this line is feared. Write a routine at address $FFF0 that will simply unstack the machine registers again and return the processor to the main program if a spurious NMI is received.

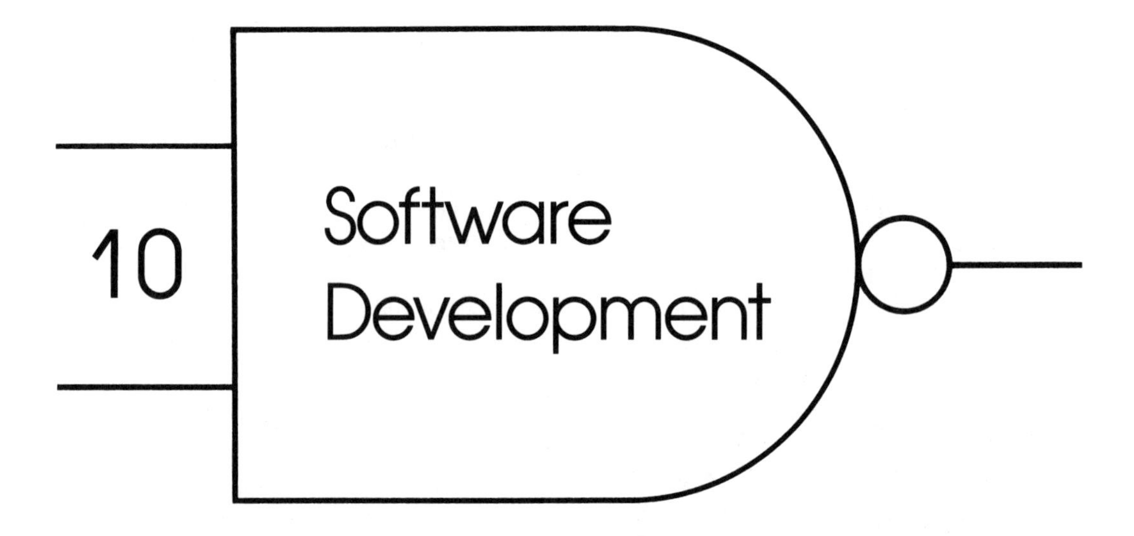

10.1 EDITORS AND SOURCE-CODE GENERATION

Section Overview

Editors are essentially word processors designed for writing, modifying, saving, and retrieving assembly-language source code, and for feeding this code to the assembler. Those who develop and debug microcomputer-based systems will spend more time working with the editor program than with *any* other program or instrument in the lab. Emulators are computers or terminals that allow object code to be run and debugged from the keyboard and CRT.

It often comes as a shock to persons who consider themselves to be electronics technicians or engineers when they realize that they are spending most of their time at a keyboard and CRT generating source code for a computer. This task can be more or less frustrating, depending in large measure on the *editor* program used to generate, save, retrieve, and modify the source code. Indeed, in projects requiring extensive programming and frequent program modification, the choice of an editor program may have more effect on the ultimate success of the project than the choice of the microprocessor itself.

An efficient editor will make program generation and modification a much less difficult task. Here is a list of desirable editor features:

- *Compatibility with the assembler*. This may be achieved by coresidency, which means that the editor which is used to produce the source code and the assembler which translates it into object code are loaded into the host computer's memory simultaneously. Separate editor and assembler programs should not require cumbersome SAVE FILE and RETRIEVE FILE operations each time a source-program change or trial assembly is needed. If they have not been designed to work together, the editor and assembler may use different file formats, so they may not be usable together without an even more cumbersome patch routine.
- *Automatic source-code formatting*. The editor should generate the next line number automatically upon RETURNing from the current line. Label, mnemonic, and operand columns should be reached by a single press of the TAB key. Comments typed to the right of the operand column should not require any special characters or format to set them off from the source code proper.
- *Ease of modification* of the source code is essential, since generally more time is spent troubleshooting and modifying programs than is spent writing them. Here are some questions to ask about an editor program:
  1. Is it easy to insert new lines between existing lines?
  2. Can a line be modified, or is it necessary to delete the old line and type in a whole new one?
  3. Is the editor page-oriented, so that changes can be made and automatically entered anywhere on the screen, or is it line-oriented, so that changes must be made and entered manually, line by line?
- *Linking of two or more source-code listings* may be necessary if the program is too large to be stored at one time in the host machine's RAM. Considering the storage required for source code, comments, object code, and the assembler program itself, it may be impossible to fit a 2K program into a 64K host computer. Linking loaders bring the source code off the disk in sections and store the object code to the disk in sections, allowing the host-computer RAM to be reused. Labels, addresses, and variable values are saved so they can be referenced across sections.

Generating source code with an editor is a task requiring discipline and method. Here are some hints to spare you the pain of learning from direct experience:

- First write down, point by point, the specific functions that the program is to perform, in terms that the user could understand. Then prepare a flowchart. Then write the source code (with comments) out on paper. *Then* begin typing the source code into the terminal.
- *Head each program* with a comment line containing
  1. The file name
  2. The writer's name
  3. The date originally written

4. The date of the last modification, and
5. The current program status, that is, *untested, runs through line 1840*, runs with error in module 3, or similar comments.

- *Type in a complete paragraph of comments* explaining what the program does, on what hardware it is to run, and how it is to be used.
- *Now type in the source-code* mnemonics and comments. Resist the temptation to leave the comments until later. The assembly language alone always seems clear enough while you are writing it, but that is exactly why you should comment it as you write it. It will be pure gibberish to anyone else—it will be gibberish to you in two weeks—unless you supply intelligible comments.
- *Save the source code* to the host-system disk every 20 lines or so, even though the program is not done. One mistake can erase all your work, and there will be no way to retrieve it unless you have been regularly saving the updated version on the disk.

Emulators are host computers that not only assemble source code but allow you to run it without actually having any target-system hardware available. Input signals are derived from the keyboard and from host-system programs that simulate timed events, analog-to-digital converters, and other phenomenon. Outputs are displayed on the host-system CRT. Troubleshooting features are generally included, such as single-step, internal-register display, breakpoint insert, and display of run time between two points in the program. Emulators allow the software team to begin developing and troubleshooting the system program without waiting for the hardware team to complete a prototype system.

In-circuit emulators (ICE) include a patch cord to allow the emulator to replace the target-system processor. Programs developed in the emulator can then be run using selected portions of the system hardware. In this way a program can be developed and debugged step by step in the emulator, and then the target-system hardware can be phased in and debugged, also step by step. Emulators are considered by many to be essential to the development of large microcomputer-based systems.

10.2 MODERN PROGRAMMING PHILOSOPHY

Section Overview

The top-down approach to program development breaks the program into modules, each with a single entry and single exit point. Linear-structured programs follow a single path, as opposed to the difficult-to-follow multiple paths of "rat's nest" programs. Pseudocode is an easier-to-produce alternative to flowcharting, which is especially suited to linear-structured programs.

Position-independent code can be relocated anywhere in a computer's memory. Reentrant subroutines can be called a second time by an interrupting routine before they have been exited the first time.

Top-down programming is a management approach to the development of large programs and is widely recommended. It is based on the old *divide-and-conquer* maxim. The first step in the process is to draw a "flowchart" consisting of a single box, as shown in Figure 10.1(a). In the box write a description of what the program is supposed to do for the user. This is usually stated in terms of the input signals and output responses.

The second step is to split the program into two or three *modules*, each representing one part of the overall program. It is essential that the modules each have only one entry point and one exit point and that they be joined in a simple row or loop, as shown in Figure 10.1(b). Again, the function of each module is specified as clearly as possible. At this stage variable names may be assigned and specifications may be written detailing the inputs and outputs for each module. Now, each of these modules is turned over to a senior programmer, who may again split a module into a series of submodules, as illustrated in Figure 10.1(c). Again, each module may have only one entry and one exit point, and all modules must be connected in a simple row. Many advantages derive from this programming approach.

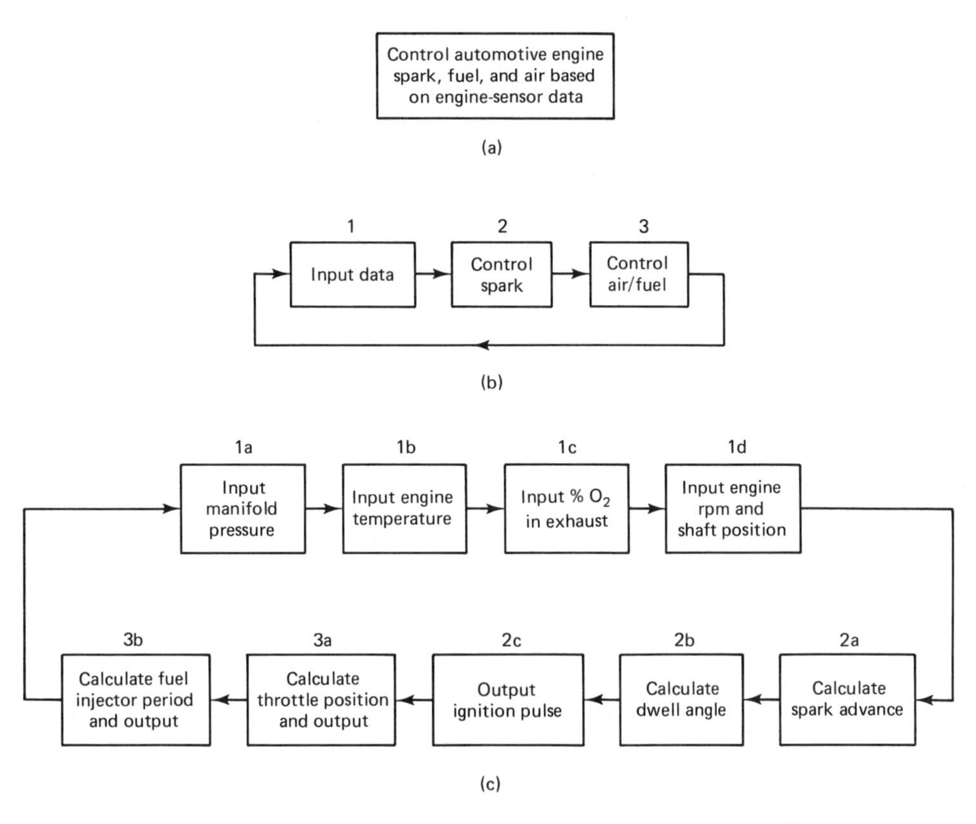

Figure 10.1 Top-down program development splits the total program into a linear or circular series of modules.

- An entire team of programmers can begin working at once. There is no need to wait for Module 1 to be completed to begin work on Modules 2 and 3.
- A library routine can be used where a standard or previously written module is required. This would not, in general, be possible if multiple entry and exit points were allowed.
- Troubleshooting can be done on a module-by-module basis without involving the whole program. This is a matter of supplying the specified inputs with the emulator and getting the program to produce the specified outputs.
- Each programmer needs to understand only one part of the total system. Large systems often involve such an array of concepts—from electronics, optics, mechanics, chemistry, and aerodynamics to thermodynamics and mathematics—that it is unrealistic to expect each person on the team to possess expertise in each area involved.
- Program "maintenance"—by which we mean modification and upgrading—is greatly facilitated, since it often involves only one module, and in any case it can be controlled by the module's input and output specifications. This is very important because large-system programs are constantly being revised and improved, and it is intolerable to have a situation in which a change of a single line of code might cause a "bug" to appear anywhere in a 50 000-word program.

Linear-structured programs are the natural consequence of the top-down programming methodology. Linear programs require that all program elements consist of one of the structures shown in Figure 10.2. Notice that each of these structures has only one entry point and one exit point. Thus the program never splits into multiple paths, and your brain has to follow only one line of thought to understand it. For the benefit of skeptics, it has been proven theoretically and in practice that any solvable programming problem can be solved using only the three basic structures in linear arrangements.

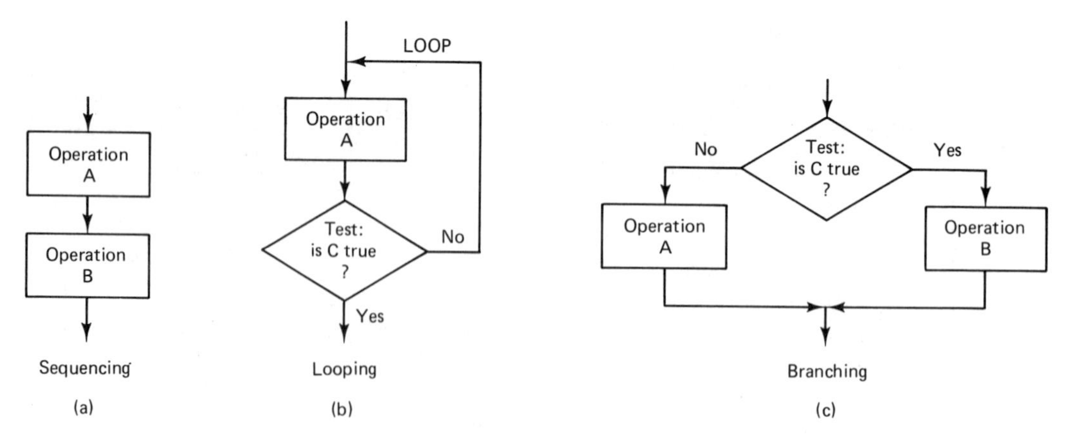

Figure 10.2 Three basic flow structures are sufficient to handle any programming problem. Each has one entry and one exit point.

Ad hoc programming is the polite name given to the process of simply writing programs as they occur to you, without any restrictions on structure or modularity. Usually this produces programs that are called, somewhat less politely, *rat's nests* or *spaghetti code*. The appropriateness of these epithets depends on the environment. There is no doubt that brilliant programmers, using clever tricks obscure to all but themselves, can sometimes write spaghetti code that runs faster and takes less memory than structured code. However, if a bug is found or a modification is needed after this genius has left the company and moved to Brattleboro, pity the poor programmer who has to try to figure this program out.

In general, ad hoc programming makes sense for relatively short programs (a few thousand words) that are going into mass-produced consumer items (such as automobiles or microwave ovens), and in systems where maximum speed is essential. Top-down development of linear-structured code should be the rule for large programs developed by teams of programmers and for programs which are likely to require modification during the life of the product. Figure 10.3 illustrates the difference between rat's nest and modularized-linear programs.

Pseudocode: We have been using flowcharts as a first step to program development and then developing source code from the flowcharts. However easy to read they may be, flowcharts have two serious drawbacks

1. They are difficult and time-consuming to draw and cannot be generated on most computer word processors.
2. They practically invite the programmer to write spaghetti code, since they can portray back branches and patches between program segments so easily.

Pseudocode is being widely used to replace flowcharting because it is easy to generate on a word processor and because it all but forces the writing of linear programs.

Pseudocode is nothing more than the generalized statement of each program step typewritten line by line, rather than enclosed in boxes as it is in flowcharting. In fact, pseudocode could be recopied and placed after the source code to serve as the *comments* section of the program. An example of pseudocode appears at the end of this chapter.

Other alternatives to flowcharting have been developed. The important thing is not which method you use, but rather that you use *some* method of formally planning the sequencing and flow of your program before you start writing the source code itself. Any two-pass approach will greatly enhance the chances that you will be able to write a successful program and that someone else will be able to understand it. As program complexity increases you may find that a four-pass approach produces best results:

1. Define the program functions.
2. Write a generalized flowchart.
3. Write detailed flowcharts.
4. Write the source code.

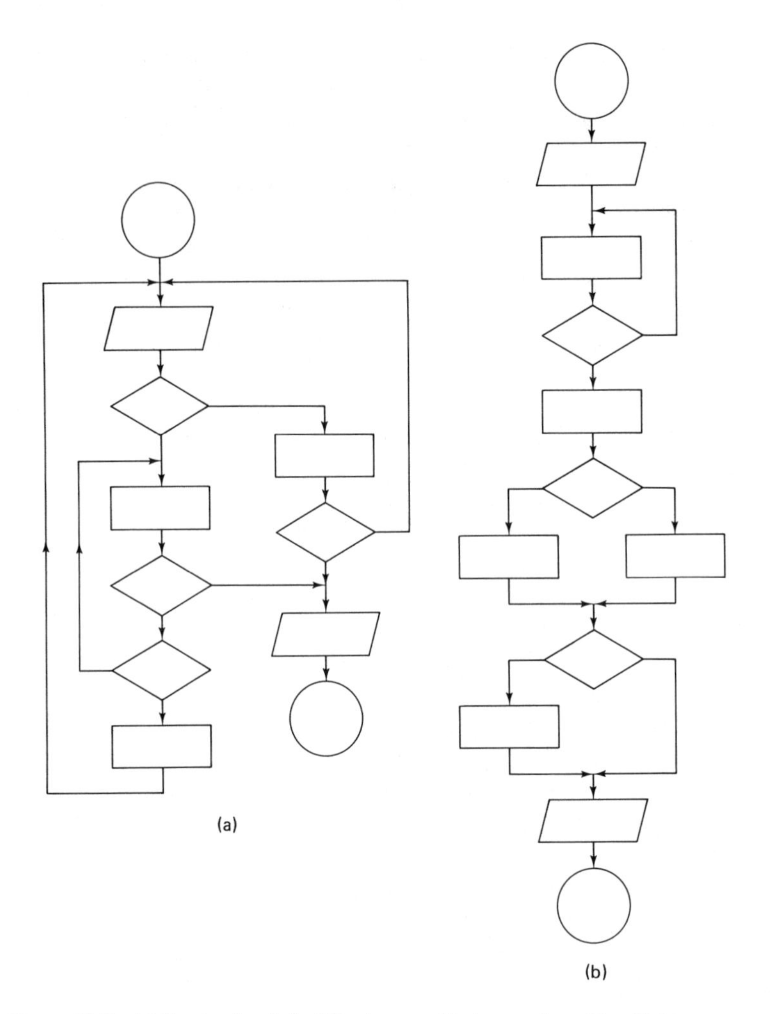

Figure 10.3 (a) Spaghetti code is difficult to troubleshoot and modify. (b) Linear code permits the program to be written and debugged as a series of smaller and more manageable modules.

Position-independent code (PIC) can be moved to any area of the computer's memory and run without modification. This is especially useful when several programs are loaded into RAM from a disk. The same area of RAM may not be free each time, and the program may have to take residence wherever it can find room. The program can be made position-independent or *relocatable* by simply using relative addressing rather than extended addressing for all branches and jumps. In 6800 language this means using BRA in place of JMP. The branch range of -128 or $+127$ bytes is a problem with the 6802. The 6809 processor eliminates this problem with "long branch" instructions.

Reentrant code refers to subroutines that can be called a second time before the first RTS is encountered. This is considered essential in modern programming. For example, let us say that a main program needs to multiply two 16-bit numbers and that it calls a subroutine MULT to do this. Halfway through the subroutine an interrupt request is received. There being no hurry about the multiplication, the interrupt is serviced, and the MULT subroutine is left half-finished for a while. Then, as the processor gets into the interrupt routine, what does it encounter but another call of that same MULT subroutine. Is this reentry of an unfinished routine going to cause a problem?

If the intermediate results of MULT are stored in fixed RAM locations, the second multiply will overwrite the partial results of the first (unfinished) multiply, and there certainly will be a problem. Reentrant subroutines cannot use fixed-address RAM storage. If the subroutine saves all of its partial results on the stack, the interrupt will simply store its intermediate results farther back in the stack and retrieve them before turning the program back to the original multiply problem. The original multiply partial results are not disturbed. Reentrant code then uses the stack, or another indexed RAM area, rather than defined RAM buffers for data storage.

Recursive routines are subroutines that can call themselves. For example, if a line in the MULT subroutine read

```
JSR MULT
```

and if this could be executed without tying the processor in a knot, the routine MULT would be called recursive. Recursive routines are impressive but not generally useful enough for us to go into detail about them here.

REVIEW OF SECTIONS 10.1 AND 10.2

Answers appear at the end of Chapter 10.

1. Source code is entered into a host computer using a program called a(n) ____.
2. Our source programs are so long that we can't fit them into our host computer's RAM all at one time. We need a ________ ________.
3. Name the two worst sins of omission you can commit when writing source code with an editor.
4. What can you use to run and debug object code once you have it assembled?
5. Which programming approach, *top-down* or *ad hoc*, is likely to produce the faster and more compact code?
6. Which programming approach is better suited to large programs requiring a team of programmers?
7. Name the three essential program structures.
8. Give two advantages of pseudocode over flowcharting.
9. Position-independent code is achieved simply by avoiding the use of what?
10. What restriction must be observed when writing reentrant subroutines?

10.3 STACK AND INTERNAL DATA-MOVE OPERATIONS

Section Overview

The next three sections continue the catalog of 6800/6802 instructions begun in Chapter 8. To locate a particular instruction quickly, use the index at the back of the book. Also, do not fail to compare these detailed descriptions with the programmer's reference sheet (last flyleaf in this book.) You will find that most of the detail information is condensed into the programmer's reference.

LDS Load Stack pointer register

Operation. 16-bit (2-byte) operand data is copied into stack pointer.

Addressing modes, machine cycles (~), example forms, and comments

Immediate	3~	8E 00 7F	LDS	#$7F	Data $7F placed in 16-bit stack register.
Direct	4~	9E 00	LDS	0	Stack pointer loaded from address $0000 (MS byte) and $0001 (LS byte).
Extended	5~	BE F8 D0	LDS	$F8D0	($F8D0) → SP_{HI}, ($F8D1) → SP_{LO}.
Indexed	6~	AE 00	LDS	0,X	SP_{HI} loaded from address pointed to by *X*. SP_{LO} loaded from address *X* + 1. *X* not changed.

Flags. N = S_{15} (MS bit of S), Z = 1 if S = 0, V cleared.

STS Store Stack pointer register

Operation. 16 bits from SP are copied to two successive operand addresses.

Addressing modes, machine cycles (~), example forms, and comments

Direct	5~	9F 10	STS	STACK	Previously STACK EQU $10. SP stored to $0010, $0011.
Extended	6~	BF 01 00	STS	256	SP_{HI} → ($0100), SP_{LO} → ($0101).
Indexed	7~	AF 00	STS	0,X	SP_{HI} → (X), SP_{LO} → (X + 1)[1]

[1](X) means "8-bit data at address pointed to by X register." (X + 1) means "8-bit data at the next successive address.

Flags. N is set to the state of the highest-order bit of SP (SP_{15}). Z is set if SP is zero, cleared otherwise. V is always cleared (= 0).

Pushes and Pulls on the Stack

PSHA Push accumulator A on stack

PSHB Push accumulator B on stack

Operation. The data from the accumulator is copied to the address pointed to by the stack pointer register. The value in the stack pointer is then decremented (decreased by 1).

Addressing mode, machine cycles (~), example forms, and comments

Inherent	4~	36	PSHA	*A* → (SP); SP − 1 → SP
Inherent	4~	37	PSHB	*B* saved on stack; Stack pointer decremented.

Flags. No flags are affected by push-on-stack operations.

PULA Pull accumulator A from stack

PULB Pull accumulator B from stack

Operation. The contents of the 16-bit stack-pointer register are first increment by 1. Then the data byte at the address pointed to by the stack pointer is loaded into the accumulator.

Addressing mode, machine cycles (~), example forms, and comments

Inherent	4~	32	PULA	(SP + 1) → SP; (SP) → A.
Inherent	4~	33	PULB	Stack pointer incremented; B loaded from stack.

Flags. No flags are affected by pull-from-stack operations.

Internal Data Moves

TAB Transfer A to B

TBA Transfer B to A

Operation. The data from one accumulator is copied to the other, destroying the data in the second accumulator and leaving both accumulators with the same data.

Addressing mode, machine cycles (~), example forms, and comments

Inherent	2~	16	TAB	A → B; A unchanged, B overwritten.
Inherent	2~	17	TBA	B → A; B unchanged, A overwritten.

Flags. N equals bit 7 of result, Z = 1 if result = 0, V cleared. *Note*: To exchange, or "swap," the contents of A and B, use

```
PSHA
TBA
PULB
```

The stack pointer must contain a valid RAM address to use the PSH and PUL instructions.

TPA Transfer Processor status register to accumulator A.

Operation. The processor status register (P), also called the condition-code register (CCR), is transferred to accumulator A.

Addressing mode, machine cycles (~), example form, and comment

Inherent	2~	07	TPA	CCR → A; CCR unchanged.

Flags. The CCR is comprised of the flag bits, and none of these is changed by the TPA instruction.
The CCR bits are:

7	6	5	4	3	2	1	0
1	1	H	I	N	Z	V	C

Bits 6 and 7 always read as 1s and cannot be altered.

The TPA instruction is generally used to gain access to the status register so that its contents can be displayed or modified. For example, if a latch at an address named LATCH drove six discrete LEDs on lines D0 through D5, the flag bits could be displayed (while single-stepping through instructions) with the code

```
TPA
STAA LATCH
```

TAP Transfer Accumulator A to Processor status register

Operation. The contents of accumulator A (bits 0 through 5 only) are copied to the 6 flag bits of the condition-code register. Bits 6 and 7 are fixed as logic 1s. The current CCR bits are lost.

Addressing mode, machine cycles (~), example form, and comment

Inherent	2~	06	TAP	$A_{0\text{-}5}$ → CCR; A unchanged.

Flags. All six flags may be altered by this instruction. It should be used only after careful consideration because it can easily "crash" a program by causing a wrong branch to be taken or an undesired interrupt to be serviced.

TSX Transfer Stack pointer to X register

TXS Transfer X register to Stack pointer

Operation. The X register and the stack pointer are set to reference the same data. This means that S will contain *one less* than X, since data pulled from the stack resides at an address *one higher* than the initial contents of S.

Addressing mode, machine cycles (~), example forms, and comments

Inherent	4~	30	TSX	$S+1 \rightarrow X$; S unchanged.
Inherent	4~	25	TXS	$X-1 \rightarrow S$; X unchanged.

Flags. No flags are affected by either of these instructions. *Note:* To exchange, or swap, X and S, use the routine:

```
STX   RAM    Save X at RAM and RAM + 1
DES          Subtract 1 from S to cancel
TSX          the +1 from TSX
LDS   RAM    Put original S into X.
             Put old X into S.
```

Two free bytes of RAM are required.

10.4 ARITHMETIC INSTRUCTIONS

Section Overview

All *add* and *subtract* instructions set the C flag after the operation if a carry or borrow has been generated. The ADC and SBC instructions also consider the C flag status *before* the operation, adding in 1 for a prior carry or subtracting 1 for a prior borrow. This is useful in multibyte arithmetic.

***Compare* instructions set the CCR flags as a subtract would but do not alter the register contents. *Test* instructions are compare-to-zero instructions.**

***Increments* and *decrements* are similar to add 1 and subtract 1, respectively, but they do not affect the carry flag as adds and subtracts would.**

ADDA Add operand to accumulator A

ADDB Add operand to accumulator B

Operation. The 8-bit operand is added to the accumulator. The sum is stored in the accumulator. The operand data is not altered. The status of the C flag prior to the instruction is of no consequence. If the sum exceeds 8 binary bits, the ninth bit is stored in the C flag.

Addressing modes, machine cycles (~), and object codes

	Immediate	Direct	Extended	Indexed
ADDA	8B (2~)	9B (3~)	BB (4~)	AB (5~)
ADDB	CB (2~)	DB (3~)	FB (4~)	EB (5~)
	(+ operand byte)	(+ address byte)	(+ 2 address bytes)	(+ 1 offset byte)

Flags. N = bit 7 of result; Z = 1 if accumulator bits are all Ø; V set if the sum exceeds decimal 127, which would upset the sign bit in signed arithmetic; C set if sum exceeds decimal 255; H set if a carry occurs from bit 3 to bit 4. H is sensed by the DAA instruction and is only used in BCD additions.

ADCA Add operand plus Carry to accumulator A

ADCB Add operand plus Carry to accumulator B

Operation. The binary operand is added to the accumulator data, and if C is *set* prior to the instruction, an additional 1 is added. The sum, which equals ACCUM + OPERAND + CARRY, is stored in the accumulator. After the operation C is *set* if the sum exceeds decimal 255 and cleared otherwise.

Addressing modes, machine cycles (~), and object codes

	Immediate	Direct	Extended	Indexed
ADCA	89 (2~)	99 (3~)	B9 (4~)	A9 (5~)
ADCB	C9 (2~)	D9 (3~)	F9 (4~)	E9 (5~)

Flags. N = b_7, Z = 1 if result = Ø, C = 1 if result > $FF, V = 1 if carry from bit 6 to bit 7 overwrites sign bit 7, H = 1 if bit 3 carries to bit 4 (for use in decimal addition).

Multiple-precision addition is facilitated by the add-with-carry instructions. Figure 10.4(a) illustrates a simple 16-bit addition capable of sums up to decimal 65 535. Figure 10.4(b) shows a more general routine that can be extended beyond 3 to any number of bytes for any desired capacity and level of precision.

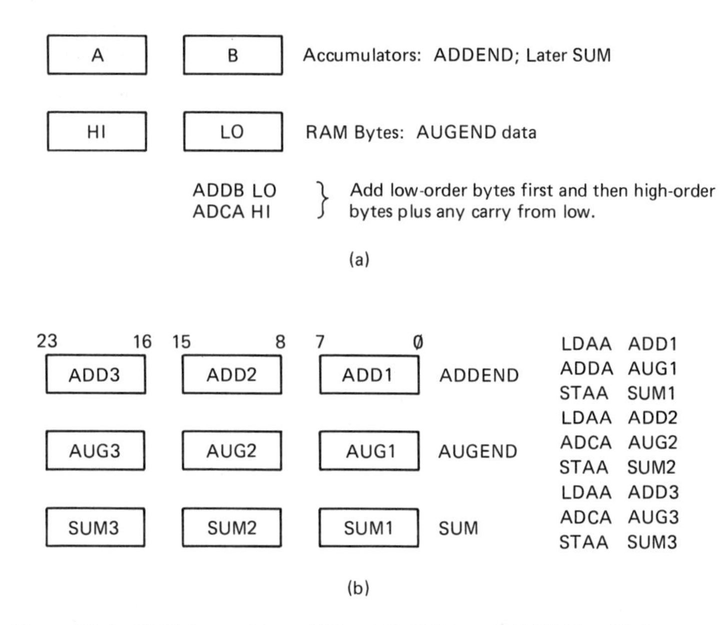

Figure 10.4 Multiple-precision addition: (a) 16-bit, and (b) 24-bit, with the same method easily extended to any number of bits.

ABA Add accumulator B to accumulator A

Operation. Data from B is added to A. The sum is stored in A. The data in B is not altered. Carry is not included in the addition, but C is set if the sum exceeds $FF.

Inherent mode only. Op-code: $1B; machine cycles: **2.**

Flags. Same as ADDA and ADDB.

DAA Decimal Adjust accumulator A

Operation. This instruction is used only after an addition of 2 bytes of BCD data, the sum appearing in accumulator A. The addition, of course, follows the rules of straight binary. The DAA corrects the sum to provide a valid BCD result in accumulator A. DAA follows these rules:

- Add $Ø6 to A if the low-order nibble exceeds 9 or if H = 1. *Note*: If H = 1 and the low-order nibble is greater than 3, the input data could not have been valid BCD; the addition of 6 is not done in this case.
- Also add $6Ø to A and set the C flag if (1) the high-order nibble exceeds 9, or (2) the high-order nibble equals 9 *and* the low-order nibble exceeds 9, or (3) the C flag has been set by the addition. *Note*: DAA does not work with the subtract instructions or with accumulator B.

Immediate mode only. Op-code: $19; machine cycles: **2.**

Flags. N = bit 7; Z = 1 if A = 0; V = 1 if carry from bit 6 to bit 7; C set if result greater than decimal 99 but not cleared if set by preceding addition.

SUBA Subtract operand from accumulator A

SUBB Subtract operand from accumulator B

Operation. Binary subtract the operand (subtrahend) from the accumulator (minuend), storing the result (difference) in the accumulator. Set the C (borrow) flag if the operand is larger than the accumulator.

Addressing modes, machine cycles (~), and object codes

	Immediate	Direct	Extended	Indexed
SUBA	80 (2~)	90 (3~)	B0 (4~)	A0 (5~)
SUBB	C0 (2~)	D0 (3~)	F0 (4~)	E0 (5~)

Flags. N = bit 7, Z = 1 if accumulator = 0, V = 1 if bit 6 borrows from bit 7, C = 1 if bit 7 requires a borrow.

SBCA Subtract operand and previous borrow from accumulator A

SBCB Subtract operand and previous borrow from accumulator B

Operation. Same as SUBA and SUBB, except that an additional 1 is subtracted from the accumulator if there is a borrow condition (C = 1) prior to the subtraction. This facilitates multiple-precision subtraction. Here is a 16-bit example:

```
SUBB  LO    Subtract low-order bytes.
SBCA  HI    Subtract high-order bytes and
            previous borrow, if generated.
```

Addressing modes, machine cycles (~), and object codes

	Immediate	Direct	Extended	Indexed
SBCA	82 (2~)	92 (3~)	B2 (4~)	A2 (5~)
SBCB	C2 (2~)	D2 (3~)	F2 (4~)	E2 (5~)

Flags are the same as for SUBA and SUBB.

SBA Subtract accumulator B from accumulator A

Operation. Binary subtract; input borrow not considered.

Inherent mode only. Op-code $10; machine cycles: **2.**

Flags are the same as for SUBA and SUBB.

CMPA, CMPB Compare operand to accumulator

Operation. These instructions are "dummy subtracts." They set the CCR flags exactly as SUBA and SUBB would, but the accumulator and operand contents are left unchanged.

Addressing modes, machine cycles (~), and object codes

	Immediate	Direct	Extended	Indexed
CMPA	81 (2~)	91 (3~)	B1 (4~)	A1 (5~)
CMPB	C1 (2~)	D1 (3~)	F1 (4~)	E1 (5~)

Flags are the same as for SUBA and SUBB, as if the "dummy" subtraction actually took place.

CBA Compare accumulators

Operation. The CCR flags are set as if B had been subtracted from A, but the accumulators are left unchanged.

Inherent mode only. Op-code: $11; machine cycles: **2.**

Flags. N = bit 7 of hypothetical result, Z = 1 if result = 0, V = 1 if bit 6 borrows from bit 7, C = 1 if bit 7 borrows from next higher bit.

CPX Compare operand to X register

Operation. The CCR flags are set as if the 2-byte operand (high-byte, low-byte order) were subtracted from the 16-bit X register. The X-register contents are not changed.

Addressing modes, machine cycles (~), and object codes

	Immediate	Direct	Extended	Indexed
CPX	8C (3~)	9C (4~)	BC (5~)	AC (6~)

Flags. N = 1 if bit 15 of result would equal 1, Z = 1 if result would equal zero, V = 1 if subtraction of low-order bytes would cause a borrow from the high-order byte. C is not affected.

TSTA, TSTB Test accumulator

Operation. These instructions are equivalent to CMPA #0 and CMPB #0, respectively. Their only advantage is the saving of a byte of program memory, since the operand $00 is not required.

Inherent mode only. TSTA = $4D, TSTB = $5D; machine cycles: **2.**

Flags. N follows state of accumulator bit 7, Z = 1 if accumulator = 0, V is cleared always, and C is cleared always.

TST Test memory contents

Operation. This is a valuable instruction which permits any memory location to be examined for zero contents (Z = 1) and for the status of bit 7 (N flag). The convenience is that this can be done in a single instruction without disturbing either accumulator.

Extended mode. $7D (6~)

Indexed mode. $6D (7~)

Flags. N = operand bit 7, Z = 1 if operand = 0, V = 0 always, C = 0 always.

INCA, INCB, INC Increment accumulator or memory contents

Operation. *One* is added to the 8-bit data in the accumulator or memory location. Note that C is *not set*, for example, by INCA when A = $FF, although C is set by ADDA #1 in that case. Rollaround occurs, so that value $FF is followed by value $00.

Addressing modes, machine cycles (~), and object codes

	Inherent		Extended	Indexed
INCA	4C (2~)	INC	7C (6~)	6C (7~)
INCB	5C (2~)			

Flags. N = bit 7 of result, Z = 1 if result = 0, V = 1 if result = $80 (carry from bit 6 to bit 7 occurred).

DECA, DECB, DEC Decrement accumulator or memory contents

Operation. *One* is subtracted from the straight binary data in the accumulator or memory. Note that the C flag is not affected, even if 1 is subtracted from $00. Value $00 rolls around to $FF on decrementing.

Addressing modes, machine cycles (~), and object codes

	Inherent		Extended	Indexed
DECA	4A (2~)	DEC	7A (6~)	6A (7~)
DECB	5A (2~)			

Flags. N = bit 7 of result, Z = 1 if result = 0, V = 1 if result = $7F (borrow from bit 7 occurred).

INX, DEX Increment X, Decrement X

Operation. *One* is added to (INX) or subtracted (DEX) from the 16-bit binary value in the X register. The only flag affected is Z, which is *set* if the result equals zero. Rollaround occurs between $FFFF and $0000.

Inherent mode only. INX = $08, DEX = $09; machine cycles: **4.**

Flags. Z = 1 if 16 bits all equal 0. Other flags are unaffected.

INS, DES Increment, Decrement Stack pointer

Operation. Same as INX and DEX, except that *no flags are affected.*

Inherent mode only. INS = $31, DES = $34; machine cycles: **4**

Flags. No flags are affected. Use dummy STS instruction to permit testing of N and Z flags.

REVIEW OF SECTIONS 10.3 AND 10.4

Answers appear at the end of Chapter 10.

11. The stack pointer contains $0067. Accumulator A contains $CB. What data is stored where by PSHA?
12. The stack pointer contains $005F. What data is transferred to where by PULB?
13. How can the current status-register bits be saved on the stack?
14. If the C flag is *set* and A contains $01, what is the result of executing ADCA #$FE? Give A and the flag bits set.

15. If the C flag is set and A contains $81, what is the result of executing SBCA #$01? Give A and the flag bits set.
16. Using the technique of Figure 10.4(b), how many 6800 machine cycles would be required for additions with the capacity of 10 decimal digits? Use direct addressing.
17. What is the result, in the H flag and in A, when BCD 49 is in A and BCD 27 is added to it with ADDA #$27?
18. Continuing Question 17, what is done if DAA is then executed?
19. After executing CBA, the C flag is set. Can anything be told about the relative unsigned values of A and B?
20. After executing CPX #$3A59, the N flag equals 1. What is the smallest possible hex value in X?

10.5 MISCELLANEOUS 6800 INSTRUCTIONS

Section Overview

This section completes our catalog of 6800/6802 instructions. In addition to the simple NOP and JMP, it covers software flag control, subroutine instructions, and interrupt instructions.

NOP No Operation

Operation. Although this instruction performs no operation on internal or external data or on the CCR flags, it does cause the program counter to increment by 1 (to the next instruction) and its execution does take up two cycles of machine time. Blocks of NOPs can be used as placeholders for routines removed or to be inserted later. They can also be placed inside a loop to slow execution time.

Inherent mode only. Op-code = $01; machine cycles: **2**.

Flags. No flags are affected.

JMP Jump program execution to operand address

Operation. The program counter is reloaded with the 16-bit value from the 2 bytes following the op-code (extended mode) or from the X register plus offset (indexed mode). Program execution continues from the new program-counter address.

Extended mode. Op-code: $7E; machine cycles: **3**.

Indexed mode. Op-code: $6E; machine cycles: **4**.

Flags. No flags are affected by this instruction.

SEC Set Carry flag (CCR0 = 1)

CLC Clear Carry flag (CCR0 = 0)

SEV Set Overflow flag (CCR1 = 1)

CLV Clear Overflow flag (CCR1 = 0)

Operation. The indicated status-register bit is set or cleared. All operations are inherent mode and take two machine cycles.

SEC = $0D   CLC = $0C   SEV = $0B   CLV = $0A

JSR Jump to Subroutine

Operation. The address of the next main-program instruction is saved on the stack, leaving the stack pointer decremented by 2. Program execution then jumps to a subroutine at an address either given by the two operand bytes (extended mode) or given by the X register plus offset byte.

Extended mode. Op-code: $BD; machine cycles: **9**.

Indexed mode. Op-code: $AD; machine cycles: **8**.

RTS Return from Subroutine

Operation. The program counter is reloaded from the stack, leaving the stack pointer incremented by 2. Program execution then continues from the new program-counter address. Any decrements of the stack pointer occurring between a JSR and an RTS must be undone by an equal number of increments to assure a proper return to the main program: a PUL for each PSH and an RTI for each interrupt acknowledge.

Inherent mode only. Op-code $39; machine cycles: **5**.

Flags. No flags are affected by this instruction.

SWI Software Interrupt request

Operation. This instruction is executed regardless of the state of the I-mask bit. First it causes all seven machine-register bytes to be stored on the stack. Assuming an initial stack pointer of $007F, the order of storage is as follows:

Stack Pointer	Data
$007F	PC, low-order byte
$007E	PC, high-order byte
$007D	X, low-order byte
$007C	X, high-order byte
$007B	A Accumulator
$007A	B Accumulator
$0079	CCR (status register, P)
$0078	final stack pointer contents

It then *sets* the I-Mask bit, disabling acknowledgment of future IRQs until the bit is cleared by either the RTI or CLI instruction. Finally, the instruction loads the program counter from vector addresses $FFFA (PC-LO) and $FFFB (PC-HI), and execution of the software-interrupt routine proceeds from the new program-counter address.

Inherent mode only. Op-code $3F; machine cycles: **12.**

Flags. The *I*-Mask flag is *set*. No other flags are affected.

RTI Return from Interrupt

Operation. The 7 machine-register bytes are restored from the stack in reverse order as they were saved by an IRQ, NMI, or SWI. In the example given earlier under SWI, the stack pointer starts with $0078 and finishes with $007F. Main-program execution then resumes from the address reloaded into the program counter. It is essential that this instruction be used to terminate a routine entered by instruction SWI or hardware inputs IRQ or NMI. It should not be used for any other purpose. It is also essential that any decrements to the stack pointer occurring after the interrupt command be offset by corresponding increments before the RTI. Thus every *push* must have a *pull*, and every JSR must meet with an RTS.

Inherent mode only. Op-code $3B; machine cycles: **10.**

Flags. All six status flags are affected, since the CCR is reloaded from the stack.

SEI, CLI Set, Clear Interrupt-mask bit

Operation. The I-mask flag is bit 4 of the CCR. If it is *set* (equal to 1), any inputs to the $\overline{\text{IRQ}}$ pin will be ignored. The I-mask is *set* by

- A processor *reset*,
- An acknowledgment of SWI, IRQ, or NMI, or
- The SEI instruction.

The IRQ can be reenabled only by the CLI instruction.

Inherent mode only. SEI = $0F; CLI = $0E; machine cycles: **2.**

Flags. Only the I flag is affected.

IRQ Interrupt Request

Operation. This is not a 6800 instruction but rather a response to a hardware input—in particular a *low* level on pin 4. The response is the same as that described under SWI, except that

- IRQ is ignored if the I-mask bit in the CCR is *set*.
- The new program-counter address is obtained from vector locations $FFF8 (high-order) and $FFF9 (low-order).

It takes 12 machine cycles and does not begin until the current instruction (during which pin 4 goes *low*) is completed.

NMI Nonmaskable Interrupt

Operation. This is not an instruction but a response to a *high-to-low* level change on the 6800 pin 6. It is serviced each time a negative transition occurs, making hardware debounce of the pin-6 input mandatory. The new PC address is obtained from vector locations $FFFC and FFFD.

WAI Wait for Interrupt

Operation. This instruction causes the 7 machine-register bytes to be saved on the stack. The processor's VMA output then goes *low* and the BA (Bus Available) output goes *high*, whereas all tristate outputs go to their floating states. The processor ceases executing instructions indefinitely, until a hardware NMI or, if the *I*-mask is clear, an IRQ is received. The interrupt then proceeds in 3 cycles instead of 12, since the stacking is already done. This instruction is used to speed up the interrupt response or to turn the processor's memory and peripherals over to another device when it has no tasks of its own to process.

Inherent mode only. Op-code $3E; machine cycles: **9**.

Flags. The I-mask is set when an interrupt occurs.

10.6 ARITHMETIC ROUTINES

Section Overview

This section presents four elementary programs: converting from binary to decimal, converting from decimal to binary, multiplying two 8-bit binary numbers, and dividing a 16-bit by an 8-bit binary number. These are all unsigned, minimum-word-size operations. Once the basic concepts are understood, writing routines for longer words should be a manageable task.

Binary-to-BCD conversion can be accomplished by adding decimally the weight of each bit of a binary word (1, 2, 4, 8, . . .) as it is shifted *right* into the carry flag. The addition is simply skipped on 0 bits. Here is a pseudocode program list:

```
PROGRAM TO CONVERT SEVEN BINARY BITS TO
THREE BCD DIGITS; UNSIGNED.
  BCDHI, BCDLO BYTES = 0 TO START.
  WEIGHT = 1 FOR LOWEST-ORDER BINARY BIT.
* SHIFT BINARY BITS RIGHT.
    IF RIGHT BIT = 1, ADD WEIGHT TO BCDLO, DECIMALLY.
    IF BCD ADDITION CAUSES CARRY, INCREMENT BCDHI.
    IF BINARY WORD = 0, ALL 1s ADDED, SO QUIT.
  DOUBLE VALUE OF WEIGHT, DECIMALLY
  GO BACK TO SHIFT.
* QUIT
```

List 10.1 gives the program in 6800/6802 code.

```
                 0900 * LIST 10-1. RUNS ON DEVELOPMENT SYST, 6 JUL '87. METZGER
                 1000 * BINARY TO BCD CONVERSION; DECIMAL 127 MAXIMUM.
                 1010 *
0000-            1020 WEIGHT .EQ  $0000    RAM BUFFER HOLDS PLACE VALUE.
0001-            1030 BIN    .EQ  $0001    BINARY INPUT BUFFER.
0002-            1040 BCDHI  .EQ  $0002    BUFFER FOR RESULT (IF > 99).
0003-            1050 BCDLO  .EQ  $0003    HOLDS RESULT LO 2 DIGITS.
                 1060 *
                 1070        .OR  $4000
                 1080 *
4000- 7F 00 03   1085        CLR  BCDLO    RAM HOLDS 1'S AND 10'S.
4003- 7F 00 02   1090        CLR  BCDHI    RAM HOLDS 100S.
4006- 4F         1100        CLRA          A ACCUMULATES RESULT.
4007- C6 01      1110        LDAB #1       START BINARY PLACE WEIGHT
4009- D7 00      1120        STAB WEIGHT    AT 1 FOR BIT 0.
400B- D6 01      1130        LDAB BIN      B HOLDS BINARY INPUT.
400D- 54         1140 SHIFT  LSRB          NEXT MOST-SIGNIFICANT BIT
400E- 24 0C      1150        BCC  SKIP      IN CARRY = 1?
4010- 96 03      1160        LDAA BCDLO
4012- 9B 00      1170        ADDA WEIGHT    YES: ADD BINARY WEIGHT.
4014- 19         1180        DAA             DECIMALLY,
4015- 97 03      1190        STAA BCDLO      AND SAVE.
4017- 24 03      1200        BCC  SKIP
4019- 7C 00 02   1210        INC  BCDHI    ADD 1 TO 3RD DIGIT IF EXCEEDS 99.
401C- 5D         1220 SKIP   TSTB          QUIT IF NO MORE BITS IN BINARY BYTE.
401D- 27 09      1230        BEQ  QUIT
401F- 96 00      1240        LDAA WEIGHT   DOUBLE VALUE OF
4021- 9B 00      1250        ADDA WEIGHT    BIT WEIGHT
4023- 19         1260        DAA            DECIMALLY, TO MAX OF 64.
4024- 97 00      1270        STAA WEIGHT   SAVE VALUE OF NEXT MS BIT.
4026- 20 E5      1280        BRA  SHIFT    CHECK FOR ANOTHER ADD.
4028- 20 FE      1290 QUIT   BRA  QUIT
                 1300        .EN
```

There are other algorithms for binary-to-BCD conversion, but this one gives us a chance to try the 6802's decimal addition mode.

BCD-to-Binary conversion is handled quite differently from the reverse conversion. We will make use of the following facts.

1. The value of the low-order nibble of BCD is the same in the binary system: Ø through 9 are valid in either system.
2. The value of the high-order nibble of BCD is 10 times its digit value; that is, a 4 in the left BCD digit is worth 4 × 10, or 40. In hex, this is $28.
3. A reduction of the left BCD digit to its digit value can be achieved by four shifts right. A multiplication by 8 can be achieved by three shifts left (total, one shift right). A multiplication by 2 can be achieved by one shift left (total three shifts right).
4. The total value of the 2-digit BCD number can be obtained by adding (in binary) the binary value of the low-order digit plus the high-order digit times 8 (one shift right) plus the high-order digit times 2 (three shifts right).

List 10-2 gives the program in 6800 code.

```
                  0900 * LIST 10-2. RUNS ON DEVELOPMENT SYST, 6 JUL '87. METZGER
                  1000 * CONVERT 2 BCD DIGITS TO BINARY.
                  1010 *
0000-             1020 BCD     .EQ  $0000    BCD INPUT DATA.
0001-             1030 BIN     .EQ  $0001    BINARY OUTPUT BUFFER.
                  1040 *
                  1050         .OR  $4000
                  1060 *
4000- D6 00       1070         LDAB BCD      GET 2 BCD DIGITS IN B.
4002- 17          1080         TBA           "A" WILL CONTAIN BINARY TOTAL.
4003- 84 0F       1090         ANDA #$0F     HERE'S THE LOW DIGIT ONLY IN A.
4005- C4 F0       1100         ANDB #$F0     THE HIGH DIGIT ONLY
4007- 54          1110         LSRB           TIMES 8
4008- 1B          1120         ABA            ADDED TO TOTAL
4009- 54          1130         LSRB          NOW THE HIGH-ORDER
400A- 54          1140         LSRB           DIGIT TIMES 2
400B- 1B          1150         ABA            IS ADDED TO "A".
400C- 97 01       1160         STAA BIN      RESULT.
400E- 20 FE       1165 QUIT    BRA  QUIT
                  1170         .EN
```

The multiply routine uses a variation of the system that we were all taught in elementary school, which is to sum a number of partial products, one from each multiplier digit. The partial products are shifted left according to the place value of each multiplier digit. Our routine will differ from the familiar decimal routine in three ways.

1. We start with the most-significant binary digit and work toward the least-significant.
2. We add each partial product to the accumulating final product as it is obtained and then shift the final product left before working on the next partial product. The first (most-significant) multiplication experiences seven shifts; the last, none.
3. The partial ''products'' simply equal the multiplicand if the multiplier digit is a 1 or equal Ø if the multiplier digit is a Ø. There are no other possibilities.

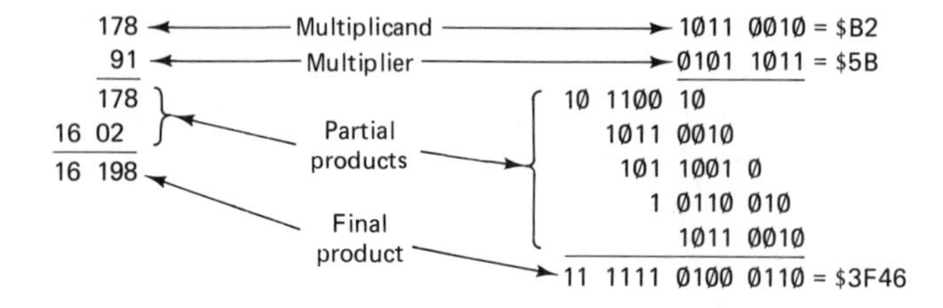

Figure 10.5 An illustration of the technique used to multiply two binary bytes. The partial products are shifted copies of the multiplicand for multiplier 1 bits and zero for multiplier Ø bits.

Figure 10.5 illustrates traditional decimal and our binary multiplication algorithms. Here is a pseudocode list for an 8-bit by 8-bit unsigned multiply with 16-bit result:

```
SET COUNTER TO 8
* SHIFT MULTIPLIER BYTE LEFT 1 BIT.
    IF C SHOWS MS BIT = 1, ADD MULTIPLICAND
      TO LOW-ORDER BYTE OF RESULT.
      IF ADDITION CAUSES A CARRY, INC HIGH BYTE.
    SHIFT 16-BIT ACCUMULATING RESULT LEFT.
    DECREMENT COUNTER.
    IF COUNTER NOT ZERO, SHIFT MULTIPLIER AGAIN.
```

List 10.3 gives the 6800 program.

```
                0900 * LIST 10-3 RUNS ON DEVELOPMENT SYSTEM 6 JUL '87. METZGER
                1000 * MULTIPLY TWO BINARY BYTES.
                1010 *
0000-           1020 PLIER  .EQ  $0000   MULTIPLIER (UNSIGNED BINARY <256)
0001-           1030 CAND   .EQ  $0001   MULTIPLICAND BUFFER; SAME RANGE.
0002-           1040 PRODHI .EQ  $0002   HIGH-ORDER BYTE OF PRODUCT.
0003-           1050 PRODLO .EQ  $0003   LOW-ORDER ANSWER BUFFER.
0004-           1060 COUNT  .EQ  $0004   RAM COUNTS OFF 8 BITS SHIFT & ADD.
                1070 *
                1080        .OR  $0010
                1090        .TA  $4010
                1100 *
0010- 86 08     1110        LDAA ##$8    PREPARE TO SHIFT THROUGH 8 BITS.
0012- 97 04     1120        STAA COUNT
0014- 4F        1130        CLRA         "A" IS MS BYTE OF ACCUMULATING
0015- 5F        1140        CLRB          RESULT; B IS LS BYTE.
0016- 78 00 00  1150 SHIFT  ASL  PLIER   SHIFT MULTIPLIER MS BIT INTO CARRY.
0019- 24 05     1160        BCC  SKIP    NO ADDITION FOR ZERO BITS.
001B- DB 01     1170        ADDB CAND    IF BIT = 1 ADD MULTIPLICAND TO RESULT.
001D- 24 01     1180        BCC  SKIP    IF ADD CAUSES A CARRY
001F- 4C        1190        INCA          ADD 1 TO MS BYTE OF RESULT.
0020- 7A 00 04  1194 SKIP   DEC  COUNT   DONE 8 SHIFTS YET?
0023- 27 04     1196        BEQ  DONE     NO: KEEP AT IT.
0025- 58        1200        ASLB         BIT 7 OF B GOES THROUGH C FLAG
0026- 49        1210        ROLA          INTO A0 (BIT 8 OF RESULT).
0027- 20 ED     1215        BRA  SHIFT
0029- 97 02     1240 DONE   STAA PRODHI  SAVE RESULT.
002B- D7 03     1250        STAB PRODLO
002D- 20 FE     1255 QUIT   BRA  QUIT
                1260        .EN
```

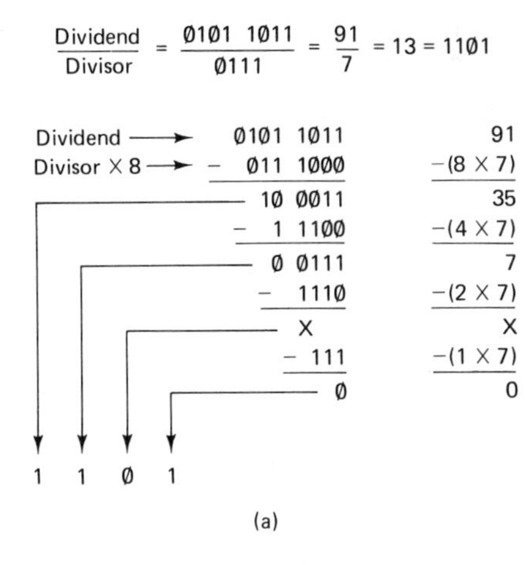

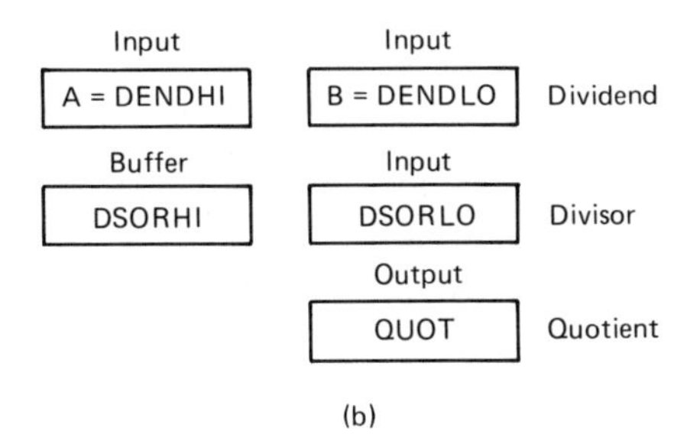

Figure 10.6 (a) An illustration of the shift-and-subtract technique used in binary division. An 8-bit dividend and a 4-bit divisor are shown for simplicity. (b) Register assignments for the 16-bit by 8-bit division routine.

The division routine handles 16-bit dividends and 8-bit divisors and quotients. It uses successive subtractions to determine the 1 or Ø status of each bit of the quotient, starting with the most-significant bit. The first step is to multiply the divisor by 128 and determine if this is larger than the dividend. If it is, bit 7 of the quotient is Ø. Then a check is made with 64 times the divisor, and so on. Figure 10.6 shows the technique for an 8-bit dividend and 4-bit divisor. Here is a pseudocode list for the program. The program listing follows (List 10.4).

```
DIVIDE 16-BIT DIVIDEND FROM DENDHI, DENDLO BY
8-BIT DIVISOR FROM DIVSOR. PLACE 8-BIT
QUOTIENT IN QUOT.
    GET DIVIDEND IN A (HI) AND B (LO).
    SHIFT DIVISOR 8 BITS LEFT IN 2-BYTE RAM DSOR.
  * SHIFT DIVISOR RIGHT 1 BIT (WEIGHT, STARTING 128 TIMES).
    SAVE DIVIDEND IN CASE SHIFTED DIVISOR IS LARGER.
    SUBTRACT WEIGHTED DIVISOR FROM DIVIDEND (16 BITS EACH).
      IF NO BORROW, SET A 1 BIT AND ALLOW SUBTRACTION.
      IF BORROW, SET A Ø BIT AND UNDO SUBTRACTION.
    SHIFT 1 OR Ø LEFT INTO QUOTIENT, AND PREVIOUS BITS LEFT.
    LOOP BACK TO SHIFT RIGHT UNTIL 8 LOOPS DONE.
```

```
                    0900 * LIST 10-4 RUNS ON DEVELOPMENT SYSTEM 6 JUL '87. METZGER
                    1000 * BINARY DIVIDE: 16-BIT DIVIDEND BY 8-BIT DIVISOR.
                    1010 *
0000-               1020 DENDHI .EQ   $0000    DIVIDEND INPUT DATA.
0001-               1030 DENDLO .EQ   $0001
0003-               1040 DIVSOR .EQ   $0003    DIVISOR INPUT DATA.
0004-               1050 QUOT   .EQ   $0004    QUOTIENT RESULT BUFFER.
0005-               1060 DSORHI .EQ   $0005    TWO BYTES NEEDED TO SHIFT DIVISOR
0006-               1070 DSORLO .EQ   $0006     LEFT, INCR WEIGHT AT EACH SUBTR.
0007-               1080 COUNT  .EQ   $0007    RAM COUNTS OFF 8 SHIFTS.
                    1090        .OR   $0010
                    1100        .TA   $4010
                    1110 *
0010- 86 08         1120        LDAA #8        PREPARE TO COUNT OFF
0012- 97 07         1130        STAA COUNT      8 TRIAL SUBTRACTS AND SHIFTS.
0014- 7F 00 06      1140        CLR  DSORLO
0017- 96 03         1150        LDAA DIVSOR    GET INPUT DATA (DIVISOR).
0019- 97 05         1160        STAA DSORHI    SHIFT DIVISOR 8 BITS LEFT.
001B- 96 00         1170        LDAA DENDHI    GET DIVIDEND IN A AND B.
001D- D6 01         1180        LDAB DENDLO
001F- 74 00 05      1190 SHIFT  LSR  DSORHI    SHIFT 16-BIT DIVISOR RIGHT
0022- 76 00 06      1200        ROR  DSORLO     THROUGH CARRY.
0025- 36            1210        PSHA           SAVE DIVIDEND.
0026- 37            1220        PSHB
0027- D0 06         1230        SUBB DSORLO    SUBTRACT DIVISOR WEIGHTED
0029- 92 05         1240        SBCA DSORHI     128 TO 1 TIMES.
002B- 25 05         1250        BCS  NOBIT     IF BORROW RESULTS, UNDO SUBTR.
002D- 31            1260        INS             IF NO BORROW, RESTORE STACK
002E- 31            1270        INS             POINTER, LEAVE DIVISOR A:B
002F- 0D            1280        SEC             REDUCED, AND PREPARE TO
0030- 20 03         1290        BRA  RECORD     RECORD A "1" BIT.
                    1300 *
0032- 33            1310 NOBIT  PULB           RESTORE DIVISORS.
0033- 32            1320        PULA
0034- 0C            1330        CLC            PREPARE TO RECORD A "0" BIT.
                    1340 *
0035- 79 00 04      1350 RECORD ROL  QUOT      ROTATE CARRY INTO QUOTIENT,
0038- 7A 00 07      1360        DEC  COUNT      MORE SIGNIF BITS LEFTWARD.
003B- 26 E2         1370        BNE  SHIFT     SHIFT & SUBTR 8 TIMES.
003D- 20 FE         1380 QUIT   BRA  QUIT       AND YOU'RE DONE.
                    1390        .EN
```

REVIEW OF SECTIONS 10.5 AND 10.6

21. Why are there two different instructions, RTS and RTI? They seem to do the same thing.
22. When JMP, indexed mode, is executed, does the program resume at the address *contained* in X, or is the new address contained in the bytes pointed to by X and X + 1?
23. There is no instruction to set the H flag (CCR5). Devise a routine to do it without additions.
24. A processor encounters a WAI instruction. What will make it come out of the WAIT state?
25. In the BINARY-TO-BCD program, what is the binary number stored in WEIGHT on the last pass through the loop?
26. In the BCD-TO-BINARY program, how is the left BCD digit multiplied by ten?

27. In the BINARY MULTIPLY program, why is a shift left used for accumulator *B*, whereas a *rotate* left is used for *A*?
28. On a 1-MHz clock, the binary multiply will take about 45, 100, 250, or 700 μs?
29. In the BINARY DIVIDE program, the *pushes* on stack are not matched by *pulls* from stack unless BCS NOBIT is taken. What happens to the stack pointer if BCS NOBIT is not taken?
30. What quotient will be obtained if $FECD is divided by $12 using the BINARY DIVIDE routine?

10.7 A COMPARISON OF PROGRAMMING STYLES

Section Overview

This section presents the hardware for a fast-paced game called Air Raid. Then two different programs for the game are presented. The first was developed in an ad-hoc manner and makes liberal use of back branches and irregular structures. The flowchart gives visual expression to the term rat's nest.

The second program was developd using a top-down modularizing approach. The flow was kept fairly linear, and the first draft of the program was a pseudocode list, not a flowchart. Both of these programs represent honest efforts; neither has been poorly done deliberately to make the other look good. A reasonably fair appraisal of older and newer programming styles should thus be possible.

Air Raid simulates a duel between four ground-based antiaircraft batteries (represented by four pushbuttons) and planes flying overhead (represented by a left-to-right shifting light in a row of eight LEDs above the buttons). The hardware is very similar to that for the MEM game. The only difference is the 74LS373 output latch, which drives eight LEDs, whereas the 74LS75 drove four LEDs. Figure 10.7 shows the system diagram.

The rules for Air Raid become obvious to the player after a few minutes of experimentation, but we will define them precisely here as a first step in developing the program.

1. The "plane" LED shifts from left (position 7) to right (position 0) through the carry bit (no light) and then back to bit 7. It dwells in each position for about 0.8 s at first and speeds up after each shift, reaching 0.4 s after 14 passes.
2. A "battery" button is located under LEDs 6, 4, 2, and 0. Touching a button during the third quarter of the plane's dwell over that button will hit the plane, lighting the whole sky (all eight LEDs) red as the plane explodes and raising your score by 2.
3. The plane will attempt to shoot the battery at a point three-fourths of the way through its dwell. It will be able to make a hit only one-fourth of the time. If

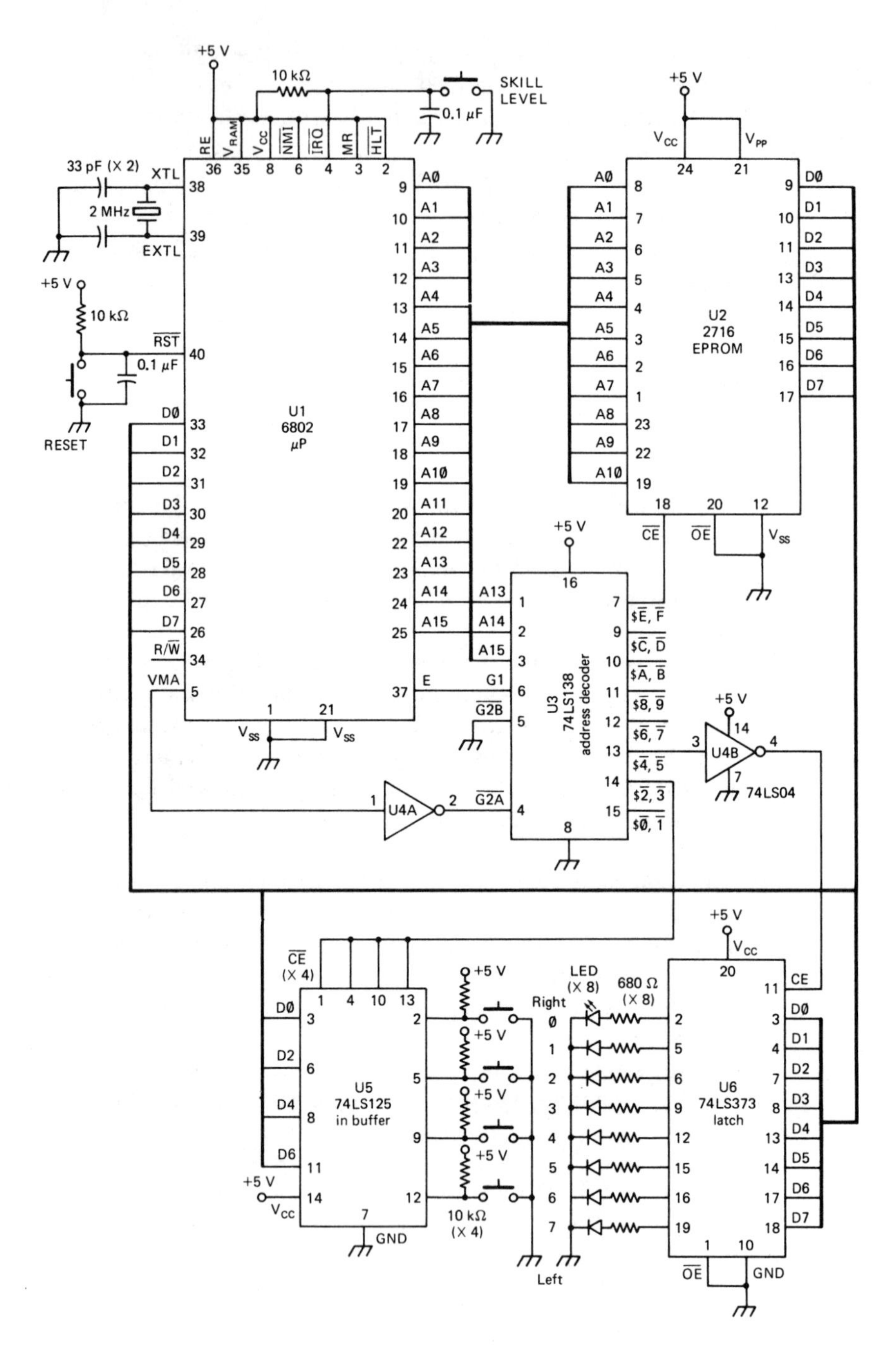

Figure 10.7 Hardware for the Air Raid Game.

the battery is hit, the LED above it will flicker for about 1 second. That LED will then remain dark for all subsequent flyovers, indicating a dead battery.

4. The battery is protected from being shot if the button is pushed during the second or third quarter of the plane's dwell. (The plane is too busy evading your shot.)
5. You cannot make another shot until one complete dwell time has elapsed from the previous shot.
6. You can make the game harder by pressing the IRQ button (but not during a "hit" flash). This will make each hit count 3 points instead of 2, but the plane will be vulnerable to your shot for only four-fifths as long a period. You can press IRQ up to four times for hits that count up to 6 points each and a live time (near the 3/4 dwell point) only 1/20 of a dwell.
7. The game ends when all your batteries have been shot out. Your score is then displayed in binary.

The ad hoc, unstructured version of Air Raid is flowcharted in Figure 10.8. The program listing for this version follows (List 10.5).

Air Raid "Rat's Nest": Program Listing

```
1000 * ------------------------------------------------------------
1010 *                         -- AIR RAID GAME --
1020 *           ILLUSTRATING "RAT'S NEST" PROGRAM STYLE
1030 *  COPYRIGHT 1984                            BY D. L. METZGER
1040 *-------------------------------------------------------------
1050 *
1060          .OR  $FD00   PROGRAM ORIGIN
1070          .TA  $4000   ASSEMBLER TARGET ADR
0000-            1080 MASK   .EQ  0        DARKENS LEDS OVER SHOT-OUT BATTERIES
0001-            1090 TIMBUF .EQ  1        DECR EACH FLIGHT TO SPEED GAME
0002-            1100 CHARGE .EQ  2        HOLDS TIME OF VALID SHOT. DECR BY IRQ
0003-            1110 SCORE  .EQ  3        INCREASED BY BATT HITTING PLANE
0004-            1120 SHTCTR .EQ  4        PICKED UP FROM CHARGE; COUNTED DOWN
0005-            1130 POS    .EQ  5        HOLDS POSITION OF PLANE; BIT 7 = LEFT
0006-            1140 PBUF   .EQ  6        HOLDS CARRY BIT DURING ROTATE INSTR
0007-            1150 OUTBUF .EQ  7        HOLDS CURRENT PLANE POS; 0 IF BATT OUT
0008-            1160 TARGET .EQ  8        PICKED UP FROM OUTBUF; IS BATT HIT?
0009-            1170 SHTBUF .EQ  9        HOLDS POS OF SHOT; COMPARE TO TARGET
000A-            1180 LUCK   .EQ 10        PLANE HITS BATT EVERY 4TH CHANCE
000B-            1190 FLASHR .EQ 11        COUNTS OUT 10 FLASHES WHEN BATT HIT
                 1200 *                    FLIGHT COUNTER IS ACCUM B
2000-            1210 INPUT  .EQ $2000     QUAD SWITCH BUFFER
4000-            1220 OUTPUT .EQ $4000     OCTAL LATCH
                 1230 *
FD00- 86 FF      1240        LDAA #$FF     INITIALIZE TIME OF
FD02- 97 02      1250        STAA CHARGE    VALID SHOT.
FD04- 97 00      1260        STAA MASK     MASK NO POSITIONS TO START.
FD06- 86 80      1270        LDAA #$80     INIT TIME DELAY 0.7 SEC
FD08- 97 01      1280        STAA TIMBUF    PER SHIFT.
FD0A- 7F 00 03   1290        CLR  SCORE    ZERO SCORE TO START.
                 1300 *
FD0D- C6 FF      1310 FLY    LDAB #$FF     START OUTER LOOP;
FD0F- 7F 00 04   1320        CLR  SHTCTR    FLIGHT COUNTER = 255,
FD12- 7F 00 05   1330        CLR  POS       NO SHOT TAKEN,
FD15- 86 01      1340        LDAA #01       PLANE IS IN
FD17- 97 06      1350        STAA PBUF      CARRY POSITION.
FD19- 7C 00 0A   1360 SHIFT  INC  LUCK     MIDDLE LOOP: PLANE'S LUCK
FD1C- 7D 00 01   1370        TST  TIMBUF    CHANGES.
```

```
FD1F- 27 Ø3      138Ø        BEQ  FAST     SPEED UP SHIFTS UNLESS
FD21- 7A ØØ Ø1 139Ø        DEC  TIMBUF    ALREADY AT FASTEST.
FD24- 96 Ø6      14ØØ FAST   LDAA PBUF     ROTATE THRU CARRY,
FD26- Ø6         141Ø        TAP            USING CARRY FLAG
FD27- 76 ØØ Ø5 142Ø        ROR  POS        SAVED IN RAM PBUF.
FD2A- Ø7         143Ø        TPA
FD2B- 97 Ø6      144Ø        STAA PBUF
                 145Ø *
FD2D- 96 Ø5      146Ø LOOP   LDAA POS      MASK OUT DISPLAY OF
FD2F- 94 ØØ      147Ø        ANDA MASK      PLANE LED OVER
FD31- 97 Ø7      148Ø        STAA OUTBUF    DEAD BATTERY.
FD33- 5A         149Ø        DECB          DECR FLIGHT COUNTER;
FD34- 26 Ø3      15ØØ        BNE  CONT1     IF = Ø, SHIFT PLANE
FD36- 7E FD 19 151Ø        JMP  SHIFT      RIGHT.
FD39- C1 8Ø      152Ø CONT1  CMPB #$8Ø     IF FLIGHT COUNTER > $8Ø
FD3B- 26 Ø7      153Ø        BNE  COLD      PLANE IS NOT VULNERABLE.
FD3D- 96 Ø7      154Ø HOT    LDAA OUTBUF   IF FLT CTR = $8Ø, PLANE IS
FD3F- 97 Ø8      155Ø        STAA TARGET    VULNERABLE; PUT PLANE POS
FD41- 7E FD 47 156Ø        JMP  CONT2      IN TARGET, & BATTERY FIRES!
FD44- 7F ØØ Ø8 157Ø COLD   CLR  TARGET     (NO TARGET IN THIS CASE)
FD47- 7D ØØ Ø4 158Ø CONT2  TST  SHTCTR    IF SHOT TIME HAS RUN OUT,
FD4A- 27 Ø6      159Ø        BEQ  KEYIN     LOOK FOR NEW SHOT.
FD4C- 7A ØØ Ø4 16ØØ        DEC  SHTCTR    OTHERWISE RUN SHOT COUNTER
FD4F- 7E FD 63 161Ø        JMP  LIVE       DOWN & TEST IF STILL LIVE.
                 162Ø *
FD52- B6 2Ø ØØ 163Ø KEYIN  LDAA INPUT     PICK UP KEY AND INVERT
FD55- 43         164Ø        COMA           SO "PRESSED" = LOGIC 1.
FD56- 84 55      165Ø        ANDA #$55     MASK OUT BITS NOT OVER KEY.
FD58- 97 Ø9      166Ø        STAA SHTBUF   STORE BINARY 1 WHERE BATT SHOOTS.
FD5A- 26 Ø3      167Ø        BNE  SHOOT    IF NOT ZERO, BATT IS SHOOTING.
FD5C- 7E FD 86 168Ø        JMP  CHK2      IF ZERO PLANE TRIES TO BOMB YOU.
FD5F- 96 Ø2      169Ø SHOOT  LDAA CHARGE   START SHOT COUNTER.
FD61- 97 Ø4      17ØØ        STAA SHTCTR    WITH VALUE $FF.
FD63- 96 Ø4      171Ø LIVE   LDAA SHTCTR   SHOT IS LIVE AS LONG AS SHT CTR
FD65- 81 CØ      172Ø        CMPA #$CØ      IS > OR = $CØ.
FD67- 24 Ø3      173Ø        BCC  HIT      IF LIVE CHECK BATT HIT PLANE?
FD69- 7E FD 86 174Ø        JMP  CHK2      IF NOT, CHECK PLANE HIT BATT?
FD6C- 96 Ø8      175Ø HIT    LDAA TARGET
FD6E- 91 Ø9      176Ø        CMPA SHTBUF    IF PLANE IS OVER
FD7Ø- 27 Ø3      177Ø        BEQ  HITYES     SHOOTING BATTERY,
FD72- 7E FD 86 178Ø        JMP  CHK2         (IF NOT, ARE YOU HIT?)
FD75- 7C ØØ Ø3 179Ø HITYES INC  SCORE     YOU HIT HIM; SCORE!
FD78- 86 FF      18ØØ        LDAA #$FF     FLASH ALL 8 LEDS FOR
FD7A- B7 4Ø ØØ 181Ø        STAA OUTPUT    Ø.8 SEC, DELAY VIA X.
FD7D- CE 6Ø ØØ 182Ø        LDX  #$6ØØØ    (GOSH, THE WHOLE
FD8Ø- Ø9         183Ø BACK1  DEX             SKY WENT RED WHEN
FD81- 26 FD      184Ø        BNE  BACK1      I HIT HIM)
FD83- 7E FD ØD 185Ø        JMP  FLY       FLY A NEW PLANE.
                 186Ø *
FD86- 96 Ø8      187Ø CHK2   LDAA TARGET   IF PLANE IS OUTSIDE
FD88- 84 55      188Ø        ANDA #$55      TARGET ZONE, YOU'RE SAFE.
FD8A- 26 Ø3      189Ø        BNE  BOMBED
FD8C- 7E FD EØ 19ØØ        JMP  SAFE
FD8F- 7D ØØ Ø4 191Ø BOMBED TST  SHTCTR    PLANE BOMBS BATTERY.
FD92- 2A Ø9      192Ø        BPL  OHNO     IS SHTCTR > OR = $8Ø?
FD94- 96 Ø8      193Ø WHO    LDAA TARGET    YES: SOMETHING IS PROTECTED.
FD96- 91 Ø9      194Ø        CMPA SHTBUF    IF PLANE NOT OVER SHOT, THEN
FD98- 26 Ø3      195Ø        BNE  OHNO      BATT IS VULNERABLE.
FD9A- 7E FD EØ 196Ø        JMP  SAFE     SAFE IF TARGET IS OVER SHOT.
```

```
FD9D- 7D ØØ Ø8 197Ø OHNO   TST  TARGET  IF TARGET & SHOT BOTH = Ø, WE'RE
FDAØ- 26 Ø3    198Ø        BNE  HELP     SAFE, OF COURSE (PLANE IN CARRY)
FDA2- 7E FD EØ 199Ø        JMP  SAFE
FDA5- 96 ØA    2ØØØ HELP   LDAA LUCK    CHECK LAST 2 BITS OF LUCK COUNTER.
FDA7- 84 Ø3    2Ø1Ø        ANDA #Ø3      IF THEY = ØØ, YOU'RE BOMBED.
FDA9- 27 Ø3    2Ø2Ø        BEQ  OUCH         (SORRY)
FDAB- 7E FD EØ 2Ø3Ø        JMP  SAFE
FDAE- 97 ØB    2Ø4Ø OUCH   STAA FLASHR  SET FLASHER TO COUNT 1Ø FLICKERS.
FDBØ- 96 Ø8    2Ø5Ø CYCLE  LDAA TARGET  PUT PLANE POSITION IN ACC A.
FDB2- 7F 4Ø ØØ 2Ø6Ø        CLR  OUTPUT  TURN OFF LED.
FDB5- CE ØD ØØ 2Ø7Ø        LDX  #$ØDØØ
FDB8- Ø9       2Ø8Ø BACK2  DEX          DELAY 5Ø MILLISEC.
FDB9- 26 FD    2Ø9Ø        BNE  BACK2
FDBB- B7 4Ø ØØ 21ØØ        STAA OUTPUT  LIGHT LED OVER BOMBED BATTERY.
FDBE- CE ØD ØØ 211Ø        LDX  #$ØDØØ
FDC1- Ø9       212Ø BACK3  DEX          DELAY 5Ø MS.
FDC2- 26 FD    213Ø        BNE  BACK3
FDC4- 7A ØØ ØB 214Ø        DEC  FLASHR  AFTER 1Ø FLASHES, YOU'RE
FDC7- 27 Ø3    215Ø        BEQ  BLASTD   BLASTED AWAY.
FDC9- 7E FD BØ 216Ø        JMP  CYCLE
FDCC- 96 ØØ    217Ø BLASTD LDAA MASK    REMOVE THE BLASTED BATTERY
FDCE- 9Ø Ø8    218Ø        SUBA TARGET   FROM THE MASK.
FDDØ- 97 ØØ    219Ø        STAA MASK
FDD2- 84 55    22ØØ        ANDA #$55    NOT COUNTING BITS WHERE NO BATTS,
FDD4- 27 Ø4    221Ø        BEQ  DISPLY   IF ALL BATTERIES OUT, END GAME.
FDD6- 5F       222Ø        CLRB         IF NOT, START FLIGHT COUNTER AT ØØ
FDD7- 7E FD 19 223Ø        JMP  SHIFT    AND SHIFT PLANE TO NEXT POSN.
               224Ø *
FDDA- 96 Ø3    225Ø DISPLY LDAA SCORE   SHOW SCORE IN BINARY ON
FDDC- B7 4Ø ØØ 226Ø        STAA OUTPUT   8 LEDS,
FDDF- 3E       227Ø        WAI           AND HALT.
               228Ø *
FDEØ- 96 Ø1    229Ø SAFE   LDAA TIMBUF  PICK UP DELAY TIME
FDE2- 27 Ø4    23ØØ WAIT   BEQ  FINISH   DELAY 18 US PER LOOP.
FDE4- 4A       231Ø        DECA          TIMES "A" LOOPS.
FDE5- 7E FD E2 232Ø        JMP  WAIT
FDE8- ØE       233Ø FINISH CLI          LOOK FOR INTERRUPT
FDE9- Ø1       234Ø        NOP
FDEA- ØF       235Ø        SEI          NO MORE INTERRUPTS.
FDEB- 7E FD 2D 236Ø        JMP  LOOP    CHECK FOR HITS AGAIN.
               237Ø *
               238Ø *                   INTERRUPT ROUTINE (IRQ):
FDEE- ØF       239Ø BOOST  SEI          ACCEPT NO MORE INTERRUPTS.
FDEF- 96 Ø2    24ØØ        LDAA CHARGE
FDF1- 8Ø ØA    241Ø        SUBA #1Ø     REDUCE CHARGE (LIVE-SHOT)
FDF3- 97 Ø2    242Ø        STAA CHARGE   TIME BY TEN (DECIMAL).
FDF5- 96 Ø3    243Ø        LDAA SCORE
FDF7- 8B Ø5    244Ø        ADDA #5      ADD FIVE TO SCORE.
FDF9- 97 Ø3    245Ø        STAA SCORE
FDFB- ØE       246Ø        CLI          READY FOR ANOTHER INTERRUPT.
FDFC- 3B       247Ø        RTI
               248Ø        .OR  $FFF8   IRQ VECTOR
               249Ø        .TA  $42F8    POINTS TO
FFF8- FD EE    25ØØ        .DA  BOOST     BOOST ROUTINE.
               251Ø        .OR  $FFFE   RESET VECTOR
               252Ø        .TA  $42FE    POINTS TO
FFFE- FD ØØ    253Ø        .HS  FDØØ      START OF PROGRAM.
               254Ø        .EN          END
```

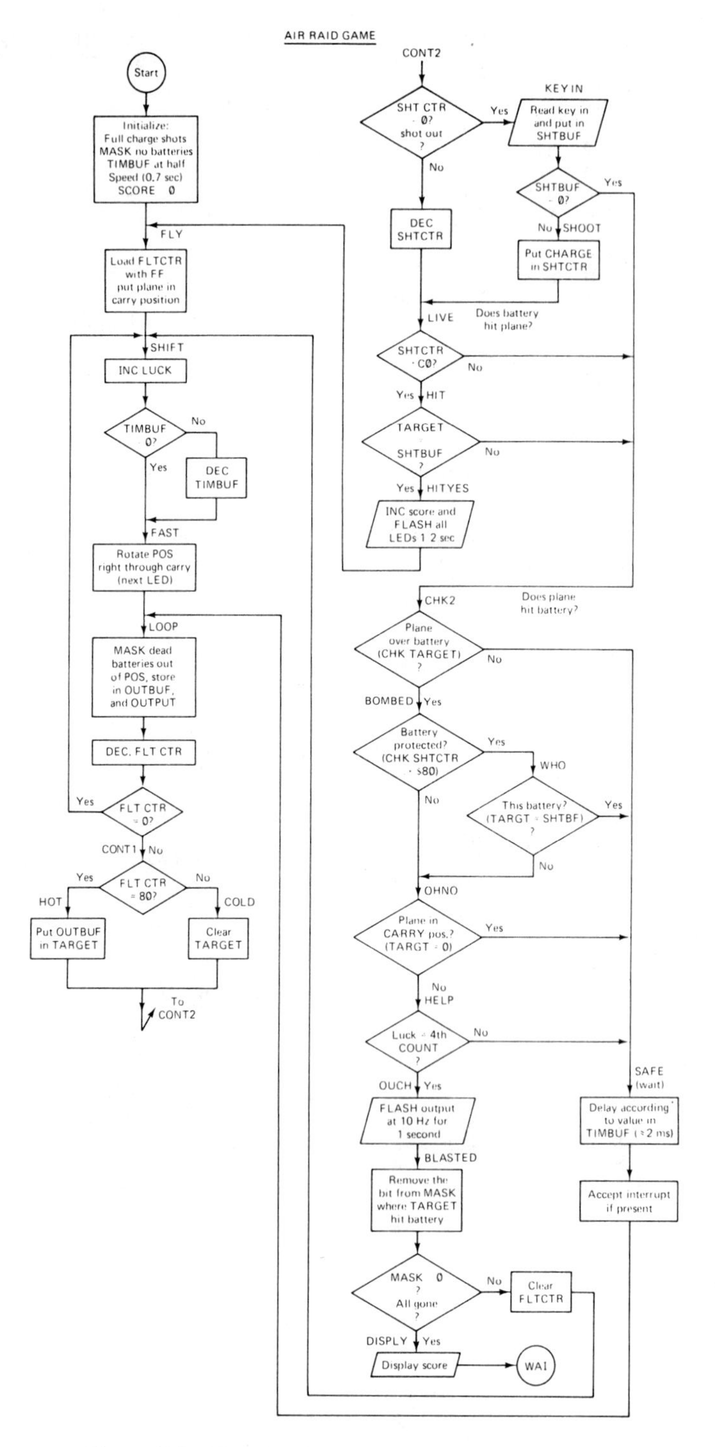

Figure 10.8 Flowchart for the rat's nest version of Air Raid.

The top-down approach to writing the Air Raid program begins in Figure 10.9. Here is a pseudocode listing of the sequence of functions required, followed by the 6800 program listing for the top-down version (List 10.6).

```
         AIR RAID GAME ILLUSTRATING TOP-DOWN MODULAR PROGRAMMING
MODULE 0 - INITIALIZE
  START PLANE IN CARRY POSITION, SCORE = 0, 2 POINTS PER HIT.
  MASK NO BATTERIES ($FF), FULL CHARGE IN SHOT, FULL DELAY.
MODULE 1 - CHECK PLANE POSITION AND TIME.
  DECREMENT DWELL COUNTER. SHIFT PLANE AFTER 256 COUNTS.
    DO FOLLOWING AFTER 256 COUNTS:
      ROTATE PLANE POSITION RIGHT THROUGH CARRY.
      MASK PLANE OUT IF OVER DEAD BATTERY.
      LIGHT PLANE LED IF NOT MASKED.
      CHANGE "LUCK" WHICH LETS BATTERY ESCAPE HIT.
      SPEED GAME BY DECREMENTING DELAY BUFFER.
  DELAY (3 MS MAX) BY DECREMENTING FROM DELAY VALUE.
MODULE 2 - CHECK STATUS OF SHOT FROM BATTERY.
  DECREMENT SHOT COUNTER. BYPASS FOLLOWING
    STEPS IF NOT ZERO:
      READ KEY INPUT. IF KEY PRESSED, PUT CHARGE IN
        SHOT COUNTER.
MODULE 3 - DOES BATTERY HIT PLANE? FLASH 8 LEDS IF SO.
  BYPASS UNLESS DWELL DECREMENTED TO EQUAL $40.
  BYPASS IF BATTERY IS ALREADY DEAD.
  BYPASS IF SHOT COUNTER DECREMENTED BELOW $C0.
    FLASH 8 LEDS AND INCREMENT SCORE. YOU GOT HIM.
MODULE 4 - DOES PLANE HIT BATTERY?
  BYPASS UNLESS DWELL COUNT = $41.
  BYPASS UNLESS PLANE IS OVER LIVE BATTERY.
  BYPASS UNLESS PLANE POSITION MATCHES SHOT POSITION.
  BYPASS UNLESS LUCK IS MULTIPLE OF 4.
  BYPASS UNLESS SHOT COUNTER STILL $80 OR MORE.
    FLASH PLANE 10 TIMES AND KILL BATTERY VIA MASK.
MODULE 5 - DISPLAY SCORE IN BINARY.
  BYPASS UNLESS ALL BATTERIES ARE DEAD.
    DISPLAY SCORE AND HALT.
  LOOK FOR IRQ.
  LOOP BACK TO MODULE 1.
IRQ ROUTINE
REDUCE CHARGE TO MAKE HIT TIME NARROWER.
INCREASE POINTS-PER-HIT BY 1.
RETURN TO MODULE 5.
```

Little will be gained by simply trying to read through these two representative program listings. Building the system will be fun and instructive in other areas, but it will not help you to appreciate the difference between the two programming styles represented. However, if you have two groups in the class set up two working systems

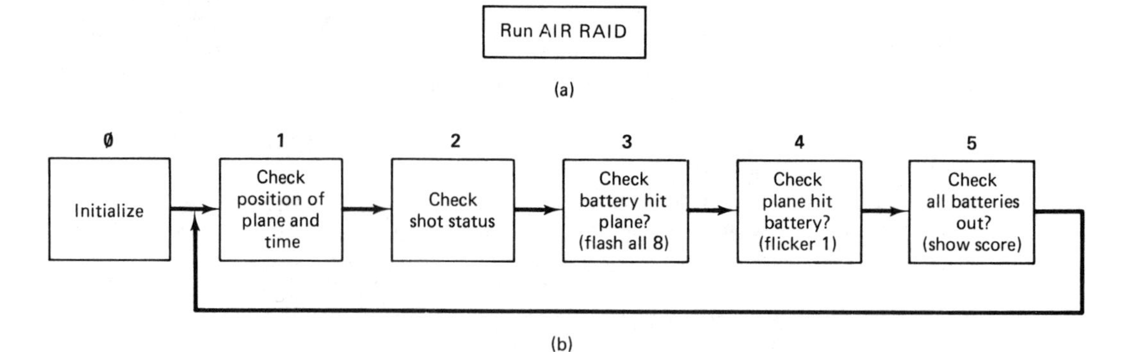

Ø Score = Ø, Luck = Ø, Mask = FF, Delay = FF, Position = Carry, Shot position = Ø, Shot Count = Ø, Dwell Count = Ø, Charge = FF, Points = Ø2.

1 (a) If dwell count is over, shift right and restart dwell count.
  (b) Light output LED unless battery under that LED is shot out.
  (c) Delay and speed up for next dwell.

2 (a) If shot count is over, read shot input. Decrement shot count unless it's already zero.
  (b) Store shot position and store charge in shot count.

3 If plane is over battery, and battery is not shot out, and shot count ≥ CØ, and dwell count = 4Ø, then plane is hit.
  (a) Flash all LEDs twice.
  (b) Increase score.

4 If plane is over battery, and battery is not shot out, and shot count > 7F, and luck count is a multiple of 4, then battery is hit.
  (a) Flash one LED 10 times.
  (b) Mask out display of this LED.

5 If Mask = 1Ø1Ø 1Ø1Ø, then display score and halt. Otherwise, accept interrupts.

(c)

Figure 10.9 A top-down approach to writing a new linear-structured version of Air Raid.

running the two opposing programs, and then vie to implement system modifications, you may begin to understand. Here are some suggestions.

1. Change the program so that a shot-out battery is "repaired" after ten overflights of the planes.
2. Add sound effects: a stacatto of 100-Hz bursts when the plane shoots a battery and a whine from 400 Hz down to 200 Hz when a plane crashes.

Air Raid Linear Structure: Program Listing

```
1ØØØ *---------------------------------------------------------
1Ø1Ø *    AIR RAID GAME    COPYRIGHT 1984 BY D. L. METZGER
1Ø2Ø *
1Ø3Ø * ILLUSTRATING TOP-DOWN LINEAR PROGRAM STRUCTURE.
1Ø4Ø * FOUR ANTIAIRCRAFT BATTERIES SHOOT DOWN  WAVES OF
1Ø5Ø *  ATTACKING PLANES. HOW MANY HITS CAN YOU SCORE BEFORE
1Ø6Ø *  THEY WIPE OUT ALL OF YOUR BATTERIES?
1Ø7Ø *---------------------------------------------------------
```

```
               1080 *                -- RULES  OF  THE  GAME  --
               1090 * PLANE FLYING ACROSS INDICATED BY ROW OF 8 LIGHTS.
               1100 * PLANE DWELLS ON EACH LIGHT ABOUT 0.8 SEC TO START,
               1110 *   SPEEDING UP ON  EACH PASS.
               1120 * PLANE IS VULNERABLE TO BE SHOT DOWN DURING THE THIRD
               1130 *  QUARTER OF EACH DWELL PERIOD. TWO POINTS PER HIT.
               1140 * PLANE SHOOTS AT BATTERY AT 3/4 THROUGH
               1150 *  DWELL PERIOD.  BATTERY (BUTTON) PROTECTED
               1160 *  FROM BEING SHOT BY PLANE FOR 1/2 A
               1170 *   DWELL PERIOD AFTER SHOT. IRQ BUTTON REDUCES PLANE-
               1180 *   VULNERABLE AND BATTERY-PROTECTED TIMES BUT ADDS
               1190 *    1 POINT PER HIT TO SCORING.
               1200 * USE 4 TIMES MAX OR SHOTS BECOME DUDS. LED DARKENS
               1210 *  OVER BATTERY AFTER IT IS SHOT OUT. PLANE IS LUCKY
               1220 *  ENOUGH TO HIT BATTERY ONLY EVERY 4TH CHANCE.
               1230 *  SCORE DISPLAYED IN BINARY WHEN ALL BATTERIES  SHOT
               1240 *    OUT.  ******   MODULE 0 -- INITIALIZATION
               1250 *------------------------------------------------
2000-          1260 INPUT  .EQ  $2000
4000-          1270 OUTPUT .EQ  $4000
0000-          1280 SCORE  .EQ  00
0001-          1290 LUCK   .EQ  01
0002-          1300 DELAY  .EQ  02
0003-          1310 MASK   .EQ  03
0004-          1320 DWLCT  .EQ  04
0005-          1330 CHARGE .EQ  05
0006-          1340 SHPOS  .EQ  06
0007-          1350 SHCT   .EQ  07
0008-          1360 POINTS .EQ  08
               1370        .OR  $FD00
               1380        .TA  $4500
               1390 *----------------------------------------------------
               1400 * ACCUM B = PLANE POSITION. MASK DARKENS  LEDS OVER
               1410 * SHOT-OUT BATTERIES. DELAY OF  $FF STARTS WITH ABOUT
               1420 *  0.8 SEC PER SHIFT.  CARRY SAVED ON STACK
               1430 *  FOR ROTATING PLANE POSITION THROUGH CARRY BIT.
               1440 *----------------------------------------------------
FD00- 5F       1450        CLRB
FD01- 7F 00 00 1460        CLR  SCORE
FD04- 7F 00 01 1470        CLR  LUCK
FD07- 7F 00 04 1480        CLR  DWLCT
FD0A- 7F 00 06 1490        CLR  SHPOS
FD0D- 7F 00 07 1500        CLR  SHCT
FD10- 86 FF    1510        LDAA #$FF
FD12- 97 03    1520        STAA MASK
FD14- 97 05    1530        STAA CHARGE
FD16- 97 02    1540        STAA DELAY
FD18- 86 02    1550        LDAA #$02
FD1A- 97 08    1560        STAA POINTS
FD1C- 8E 00 7F 1570        LDS  #$007F
FD1F- 0D       1580        SEC
FD20- 07       1590        TPA
FD21- 36       1600        PSHA
               1610 *----------------------------------------------------
               1620 * MODULE 1 -- CHECK PLANE STATUS, OUTPUT  PLANE POSITION,
               1630 *  AND DELAY.
               1640 * IF DWELL COUNT = 0, ROTATE PLANE RIGHT  THRU CARRY,
               1650 *  MASK IF OVER DEAD BATTERY.  DISPLAY POSITION, CHANGE
               1660 *  LUCK COUNTER FOR MODULE 4,
               1670 *----------------------------------------------------
FD22- 7A 00 04 1680 PLAY   DEC  DWLCT
FD25- 26 12    1690        BNE  WAIT
FD27- 32       1700        PULA
FD28- 06       1710        TAP
FD29- 56       1720        RORB
FD2A- 07       1730        TPA
FD2B- 36       1740        PSHA
FD2C- 37       1750        PSHB
```

```
FD2D- D4 03     1760        ANDB MASK
FD2F- F7 40 00  1770        STAB OUTPUT
FD32- 33        1780        PULB
FD33- 7C 00 01  1790        INC  LUCK
FD36- 7A 00 02  1800        DEC  DELAY
FD39- 96 02     1810 WAIT   LDAA DELAY
FD3B- 4A        1820 LOOP   DECA
FD3C- 26 FD     1830        BNE  LOOP
                1840 *------------------------------------------------------------
                1850 * MODULE 2 -- CHECK SHOT STATUS AND INPUT SHOT IF TIMED OUT
                1860 *------------------------------------------------------------
                1870 * IF SHOT COUNTER NOT = 0, DECREMENT IT.  IF = 0, PLACE
                1880 *  KEY INPUT IN SHOT POSITION MASKING OUT NO-BATTERY
                1890 *  AND DEAD-BATTERY BITS. IF THERE IS A VALID INPUT,
                1900 *   RECHARGE SHOT COUNTER.
                1910 *-----------------------------------------------------------
FD3E- 7D 00 07  1920        TST  SHCT
FD41- 27 03     1930        BEQ  KEY
FD43- 7A 00 07  1940        DEC  SHCT
FD46- 26 10     1950 KEY    BNE  EXIT
FD48- B6 20 00  1960        LDAA INPUT
FD4B- 43        1970        COMA
FD4C- 84 55     1980        ANDA #$55
FD4E- 94 03     1990        ANDA MASK
FD50- 97 06     2000        STAA SHPOS
FD52- 27 04     2010        BEQ  EXIT
FD54- 96 05     2020        LDAA CHARGE
FD56- 97 07     2030        STAA SHCT
FD58- 01        2040 EXIT   NOP
                2050 *------------------------------------------------------------
                2060 * MODULE 3 -- FLASH ALL LEDS IF BATTERY   HITS PLANE.
                2070 *------------------------------------------------------------
                2080 * IF PLANE IS 3/4 WAY THROUGH ITS DWELL, AND SHOT POSITION
                2090 *   = PLANE POSITION, AND MASK SHOWS BATTERY NOT
                2100 *  SHOT OUT, AND SHOT COUNT >= $C0,
                2110 *  THEN FLASH ALL 8 LEDS TWICE (1-SEC DELAYS)
                2120 *   AND ADD TO SCORE.
                2130 *------------------------------------------------------------
                2140 *
FD59- 96 04     2150        LDAA DWLCT
FD5B- 81 40     2160        CMPA #$40
FD5D- 26 2D     2170        BNE  NOHIT
FD5F- D1 06     2180        CMPB SHPOS
FD61- 26 29     2190        BNE  NOHIT
FD63- D5 03     2200        BITB MASK
FD65- 27 25     2210        BEQ  NOHIT
FD67- 96 07     2220        LDAA SHCT
FD69- 81 C0     2230        CMPA #$C0
FD6B- 25 1F     2240        BCS  NOHIT
FD6D- 86 FF     2250        LDAA #$FF
FD6F- B7 40 00  2260        STAA OUTPUT
FD72- CE FF FF  2270        LDX  #$FFFF
FD75- 09        2280 ONE    DEX
FD76- 26 FD     2290        BNE  ONE
FD78- 7F 40 00  2300        CLR  OUTPUT
FD7B- 09        2310 TWO    DEX
FD7C- 26 FD     2320        BNE  TWO
FD7E- 86 FF     2330        LDAA #$FF
FD80- B7 40 00  2340        STAA OUTPUT
FD83- 09        2350 THREE  DEX
FD84- 26 FD     2360        BNE  THREE
FD86- 96 00     2370        LDAA SCORE
FD88- 9B 08     2380        ADDA POINTS
FD8A- 97 00     2390        STAA SCORE
FD8C- 01        2400 NOHIT  NOP
                2410 *------------------------------------------------------------
                2420 * MODULE 4 -- IF PLANE HITS BATTERY FLASH ONE LED TEN TIMES.
                2430 *------------------------------------------------------------
```

```
                    2440 * IF PLANE IS 3/4 THROUGH DWELL, AND PLANE IS OVER ANY
                    2450 *  BATTERY, AND MASK SHOWS BATTERY IS LIVE, AND IF
                    2460 *  THERE IS NO SHOT AT THE PLANE OR SHOT COUNTER
                    2470 * <80, AND LUCK COUNTER LOW TWO BITS ARE ZERO,
                    2480 *  THEN FLICKER PLANE LED 10 TIMES AT 8 HZ AND
                    2490 *  REMOVE BATTERY BIT FORM MASK.
                    2500 *-----------------------------------------------------------
                    2510 *
FD8D- 96 04         2520        LDAA DWLCT
FD8F- 81 41         2530        CMPA #$41
FD91- 26 33         2540        BNE  SAFE
FD93- C5 55         2550        BITB #$55
FD95- 27 2F         2560        BEQ  SAFE
FD97- D5 03         2570        BITB MASK
FD99- 27 2B         2580        BEQ  SAFE
FD9B- D1 06         2590        CMPB SHPOS
FD9D- 26 05         2600        BNE  LUCKY
FD9F- 7D 00 07      2610        TST  SHCT
FDA2- 2B 22         2620        BMI  SAFE
FDA4- 96 01         2630 LUCKY  LDAA LUCK
FDA6- 84 03         2640        ANDA #$03
FDA8- 26 1C         2650        BNE  SAFE
                    2660 *
                    2670 * ACCUM A IS USED TO COUNT OFF TEN FLICKERS OF PLANE LED
                    2680 *
FDAA- 86 0A         2690        LDAA #$0A
FDAC- F7 40 00      2700 CYCLE  STAB OUTPUT
FDAF- CE 0D 00      2710        LDX  #$0D00
FDB2- 09            2720 BACK1  DEX
FDB3- 26 FD         2730        BNE  BACK1
FDB5- 7F 40 00      2740        CLR  OUTPUT
FDB8- CE 0D 00      2750        LDX  #$0D00
FDBB- 09            2760 BACK2  DEX
FDBC- 26 FD         2770        BNE  BACK2
FDBE- 4A            2780        DECA
FDBF- 26 EB         2790        BNE  CYCLE
FDC1- 96 03         2800        LDAA MASK
FDC3- 10            2810        SBA
FDC4- 97 03         2820        STAA MASK
FDC6- 01            2830 SAFE   NOP
                    2840 *
                    2850 *------------------------------ ---------------------
                    2860 * MODULE 5 --   ACCEPT INTERRUPT AND DISPLAY SCORE.
                    2870 *--------------------------------------------------------
                    2880 *
FDC7- 0E            2890        CLI
FDC8- 96 03         2900        LDAA MASK
FDCA- 81 AA         2910        CMPA #$AA
FDCC- 26 08         2920        BNE  GO
FDCE- 96 00         2930        LDAA SCORE
FDD0- B7 40 00      2940        STAA OUTPUT
FDD3- 7E FD D3      2950 HALT   JMP  HALT
FDD6- 0F            2960 GO     SEI
FDD7- 7E FD 22      2970        JMP  PLAY
                    2980 *
                    2990 *--------------------------------------------------------
                    3000 * INTERRUPT ROUTINE (IRQ) - REDUCES CHARGE MAKING TIMING
                    3010 *  MORE CRITICAL AND PLANE HARDER TO HIT. INCREMENTS
                    3020 * POINTS PER HIT. IMPOSSIBLE TO HIT PLANE IF USED MORE
                    3030 *  THAN 4 TIMES.
                    3040 *-----------------------------------------------------------
                    3050 *
FDDA- 0F            3060 BOOST  SEI
FDDB- 96 05         3070        LDAA CHARGE
FDDD- 80 0D         3080        SUBA #$0D
FDDF- 97 05         3090        STAA CHARGE
FDE1- 7C 00 08      3100        INC  POINTS
FDE4- CE C0 00      3110        LDX  #$C000
```

```
FDE7- 09        3120 HOLD   DEX
FDE8- 26 FD     3130        BNE   HOLD
                3140
FDEA- 0E        3150        CLI
FDEB- 3B        3160        RTI
                3170 *
                3180 *-----------------------------------------------------------
                3190 *  INTERRUPT REQUEST AND RESET VECTORS.
                3200 *-----------------------------------------------------------
                3210 *
                3220        .OR   $FFF8
                3230        .TA   $47F8
FFF8- FD DA     3240        .DA   BOOST
                3250        .OR   $FFFE
                3260        .TA   $47FE
FFFE- FD 00     3270        .HS   FD00
                3280 *
                3290 * 6802 SYSTEM ADR         FD00     FFF8
                3300 * APPLE TARGET ADR        4500     47F8
                3310 * WORKSPACE & 2716 ADR    0500     07F8
                3320 *
                3330        .EN
```

Answers to Chapter 10 Review Questions

1. Editor **2.** Linking loader **3.** Failing to comment the source code and failing to save to disk frequently **4.** An emulator **5.** Ad hoc
6. Top-down **7.** Sequencing, looping, and branching
8. Pseudocode is easier to produce, and it forces a linear structure.
9. Extended-mode memory references **10.** Data must be stacked rather than saved in designated RAM locations. **11.** $CB at address $0067
12. The contents of address $0060 are loaded into *B*.
13. Execute TPA, then PSHA. **14.** A = $00, C = 1, Z = 1, V = 1
15. A = $7F, V = 1 **16.** 5-byte addition, requiring 50 cycles
17. H = 1, A = $70
18. Since H = 1, 70 + 06 = 76, the correct decimal answer.
19. B is larger than A. **20.** $BA59 **21.** RTS unstacks 2 bytes; RTI unstacks 7.
22. The former **23.** TPA, ORAA #$20, TAP **24.** An NMI, a valid IRQ, or a processor RESET **25.** 0110 0100
26. The digit times 8 is added to the digit times 2.
27. The rotate picks up the carry output by the shift, making a 16-bit shift in *A-B* concatenated. **28.** 250 μs
29. It is restored by the INS instructions. **30.** $FF

CHAPTER 10 QUESTIONS AND PROBLEMS

Digits after the decimal point refer to section numbers in Chapter 10.

Basic Level

1.1 What happens if there is a momentary power interruption while you are entering source code into the host-system RAM prior to assembly? How can you prevent this disaster?

2.1 When should the comments to the source-code program be entered via the editor?

3.2 What is the commonly used term for software modification and upgrading?

4.2 What is the name given to subroutines that can be called by the main program, interrupted by an IRQ, and then called again by the IRQ routine?

5.3 If *A* contains $9C, what does it contain after executing TAP and then TPA?

6.3 If *S* contains $ABCD, what address is read by the instruction PULB?

7.4 If accumulator A contains $90, what is the status of the C flag after executing SBCA #$1E?

8.4 If A contains $53 and the instruction ADDA #$49 is executed, what will be the result in A? If a DAA is subsequently executed, what will be its effect on each digit of A and on the CCR?

9.4 What differences in effect, other than timing, are there between INCA and ADDA #1?

10.5 A 3-byte JMP instruction at address $3A5A is to be replaced with a 2-byte BRA instruction. How can this be done without reshuffling the addresses of all the program instructions after the JMP?

11.5 The I-mask is cleared and an SWI is encountered. (a) Is the SWI executed? (b) Will a subsequent IRQ be executed?

12.6 Analyze the multiply program of List 10.3 and indicate the maximum time that it might take, assuming a 1-MHz clock.

13.6 Attempts to single-step through the DIVIDE routine of List 10.4 on a computer trainer usually fail because the step and display routines alter the S register. What instructions could be substituted for PSHA, PSHB, PULA and PULB, respectively to solve this problem? What about the two INS instructions?

Advanced Level

14.1 Define: coresident editor, linking loader, and emulator.

15.1 List at least four things you should do in generating software *before* you type in the first instruction mnemonic.

16.2 List five advantages of modular programming.

17.2 List two advantages of pseudocode over flowcharting and one advantage of flowcharting over pseudocode.

18.3 After the instruction LDS #$8000 is executed, what are the contents of the N, V, Z, and C bits of the CCR?

19.4 Explain the difference between ADCA #21 and ADDA #21.

20.4 Write a 6802 routine to subtract a 24-bit binary number at addresses $001A,

$001B, and $001C from a 24-bit number at $0017, 0018, and 0019. Store the result in $0014, $0015, and $0016.

21.5 Assume that the 6802 *S* register contains $007F and an IRQ routine is entered. Write a routine to reload the machine registers from the stack and return to the main program without using the RTI instruction. You may use a RAM buffer area.

22.5 Give two circumstances in which the 6802's WAI instruction may be useful.

23.6 Present a display of the steps by which the BINARY-TO-BCD routine of List 10.1 would convert $B9 to BCD.

24.6 Show how $72 is multiplied by $A6 using the algorithm of Figure 10.5.

25.7 Take a stand for or against pseudocode and linear modularized programming for video games, and defend it.

11 Microcomputer Troubleshooting

11.1 TROUBLESHOOTING PRINCIPLES

Section Overview

Success in troubleshooting microcomputer systems is probably more dependent on your discipline in applying the knowledge you have than on the extent of your knowledge. It is necessary to focus on *one* problem or function of the system, form an expectation of what *should* happen, and then measure what *does* happen.

Malfunctions in microcomputer systems are most frequently caused by operator error and by mechanical failures such as broken cables or dirty contacts. Failures in electronic devices are relatively infrequent. In systems under development the software generally requires much more troubleshooting effort than the electronic hardware.

Effective troubleshooting is a matter of checking out what a device or system is actually doing and comparing that with your expectations of what it should be doing. This makes it plain that you cannot do effective troubleshooting unless you have a pretty clear and detailed idea of what the system you are working on is supposed to be doing. For many of us this is the primary reason for studying electronic devices and microcomputer systems: to understand them well enough to be able to form expectations about what the voltmeter should read or what the oscilloscope trace should look like.

Without these expectations we will find ourselves taking a long series of very impressive measurements on a system, but the measurements will be meaningless because we won't know which ones indicate proper system functioning and which ones indicate aberrations that could lead us closer to the problem. This sequence of *expectation* and *verification* requires such discipline that it is a good idea to jot your expectation down or speak it aloud[1] before verifying it with a measurement. If you first state, "The voltage at V_{CC} pin 8 must be at least 4.90 V dc," you will not be tempted to say that it is "probably OK" if it checks out at 4.85 V. You will stop and beef up the supply wiring until it comes up to your expectations. This may not be the problem, but at least you will have eliminated one potential source of problems. You can thus put that one out of mind and concentrate fully on the next one.

Forming expectations about the signals in a microcomputer system is usually more difficult than it is in other electronic systems, because the signals are usually complex, virtually nonrepetitive, and variable, depending upon data and conditions in other parts of the system. Thus it would probably be of little value to start checking the voltage waveforms on the microprocessor data lines because those waveforms are so complex and variable that we wouldn't be able to tell a good one from a bad one.

In response to this complexity, we have three options:

1. Concentrate on the aspects of the system that are simple, repetitive, and known. This will be the approach of Sections 11.2 and 11.3.
2. Limit the complexity of the system temporarily by supplying a diagnostic program that exercises only one part of the system hardware on a rapidly repeating basis. Sections 11.4 and 11.5 explore this approach.
3. Use specialized test equipment that is capable of handling complex nonrepetitive signals. This is the topic of Chapter 12.

Failure probabilities: Certainly it makes good sense to concentrate your first troubleshooting efforts on the things that are most likely to go wrong. Here is a listing of the most frequent causes of failure in microcomputer systems.

1. *Operator error:* The most common cause of service calls is a lack of understanding of how to use the system. One needs to keep in mind the old adage,

   If it ain't broke, don't fix it.

   Many a technician has torn down a computer looking for a bad chip, only to have someone point out later that the operator was using the wrong access code or had a switch setting wrong.

[1]Vicious rumors to the contrary notwithstanding, this is the reason why computer technicians are often heard talking to themselves.

2. *Connectors and cables* are the next most frequent cause of microcomputer system failure. These include printed-circuit card connectors and IC sockets within the computer as well as external connectors and wires to keyboards, printers, disk drives, and other devices.
3. *Software media,* such as physically or magnetically damaged disks and tapes, rank next in the failure-probability list. Of course, this ranking depends a great deal on the care with which the software media are treated.
4. *IC chips* fail relatively infrequently, but microcomputer systems use so many of them that it is not uncommon for a system failure to be caused by a bad IC.
5. *Power-line noise* or undervoltage may cause sporadic system errors, especially in industrial environments where welders and heavy machines are in use.

Hardware or software? An inoperative microcomputer system usually gives no immediate clue as to whether the problem is caused by hardware or software. Before any specific troubleshooting can begin, this question must be resolved.

- A new system just wired up is likely to contain hardware errors. These must be identified and corrected before software troubleshooting can begin.
- A system that has been functioning but suddenly malfunctions is likely to be experiencing hardware problems.
- If a system works with one program but fails when another program is installed, it is likely that the second program contains software errors.
- Newcomers inexperienced in programming often tend to concentrate on looking for hardware errors, when in fact software troubleshooting takes several times the effort of hardware troubleshooting on most microcomputer projects.

11.2 THE QUICK FIX

Section Overview

Microcomputer systems are complex enough to make a technical approach to the problem a headache worth avoiding if possible. Often enough it will be possible to locate the trouble without getting technically involved with the system. This section provides a collection of hints that may lead you to a lucky "quick fix."

Troubleshooting a microcomputer system can become a very involved process, but with ingenuity, some advance planning, and a little bit of luck you should be able to fix about half of the problems you encounter without getting involved in the complexities of the system at all.

Be prepared: When responsibility for servicing a particular microcomputer is given to you, don't wait for a failure; do a little advance work so you will be prepared for the failure when it comes.

- Mark the system manuals with the date of receipt, the serial numbers of the system components, and the phone number of the selling dealer. If the manuals are of the *user* rather than *service* variety, call or write for the service manuals. File all data pertaining to the system in one place.
- Run the system through several of its elementary operations. Make a record in the manual of the software, procedures, and keystrokes required and the responses observed. This record will relieve you (or your successor) of much frustration when the machine refuses to function and no one is certain if the problem is operator error or a hardware breakdown.
- Leave a sketch in the manual showing the positions of any internal switches and connectors and the slots occupied by any removable circuit cards.
- Feel the ICs in the system; make a note of those that feel stone cold and those that are too hot to touch comfortably with the back of your finger.
- Make sure your service kit contains spares of all the ICs used in the system. Order any new ones required immediately, and if you have a programmer, make a backup copy of any EPROMs contained in the system.
- If the system has diagnostic programs on disk or on EPROMS that were used during system development, lock them and their documentation away against the day of failure. These programs, which seem so useless when the system is running, will be more valuable than gold when it fails.

Divide and conquer: The first rule in troubleshooting is, Form an expectation, and then check it out. The second rule is *Divide and conquer*. The first foe to eliminate is operator error. Don't run the machine yourself; ask the regular operator to run it while you watch. Then try to run it yourself, using the notes you made in an earlier good run.

Next, eliminate the possibility of bad software. Use a backup program that was verified to be good when the system was first purchased and has been under lock and key ever since. With operator error and software eliminated, the problem is isolated to system hardware.

Begin your attack on the hardware by stripping the system down to its minimum configuration. Remove printers, plotters, disk drives, and special-function circuit cards, and see if the processor will operate standing alone. If it does, you may start reinstalling the peripherals, checking as each one is added to see if it was the one causing the failure. If the processor does not work in a minimum-configuration mode, there are still a few possibilities for obtaining a quick fix.

Physical problems are more common than electrical problems.

- Clean the top and bottom of the circuit board with a brush or pressurized air stream. Loose nuts and wire ends have a way of falling in between two conductors.
- Remove printed circuit cards, clean their contact fingers with contact cleaner, and reinsert them firmly.
- Lift the socketed ICs part way out of their sockets and then reseat them to reestablish connections broken by oxidation.

- Flex cables while attempting to run the system. Check especially the points where the wires enter their connector housings.
- Use a plastic probe to flex the circuit boards at several places while operating the system. This may reveal bad solder joints, hairline cracks, or shorts in printed circuit wiring.

Check the supply voltage V_{CC} with a digital voltmeter *at the microprocessor chip* itself, and view it on an oscilloscope. Microprocessors generally draw more current but have more stringent supply-voltage requirements than logic or analog devices. Light-gauge wirewrap wire or thin circuit-board tracks can be responsible for voltage drops of several tenths of a volt along the V_{CC} and ground lines, leaving insufficient voltage at the processor. An at-the-chip voltage which drops below

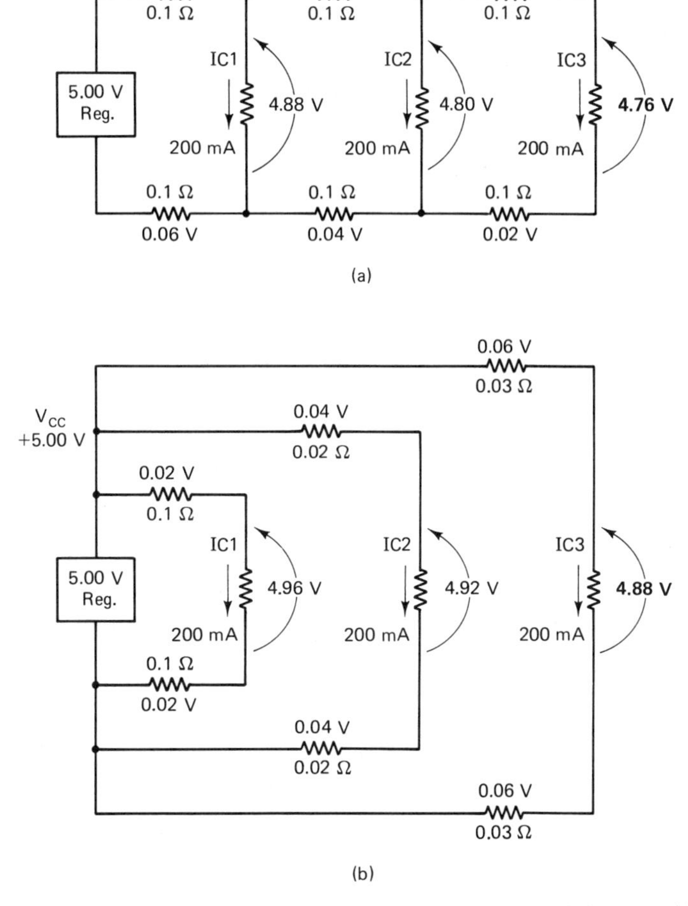

Figure 11.1 The wire length to IC3 is the same in the daisy-chain system (a) as it is in the "tree" system (b), but the voltage loss is twice as much in this typical example.

4.80 V is highly likely to be a source of trouble and should be corrected. Figure 11.1 shows how V_{CC} and ground lines to high-current chips should be wired in a ''tree'' form, rather than daisy chained, to minimize voltage loss.

REVIEW QUESTIONS FOR SECTIONS 11.1 AND 11.2

Answers appear at the end of Chapter 11.

1. Before taking a measurement in troubleshooting, you should have a (n) ______.
2. The most common cause of service calls on microcomputer systems is ______.
3. The microcomputer hardware components that fail most frequently are ______.
4. If a system has both hardware and software errors, which must be fixed first?
5. List three things that should be recorded in the system manual when the system is functioning properly.
6. How can poor contact of IC sockets to IC pins be improved?
7. What is the minimum V_{CC} that should be tolerated for a microprocessor, and where should it be measured?
8. What type of supply wiring is recommended for microprocessor components: daisy-chain or direct-supply?
9. Which peripheral items and/or plug-in cards should be removed when beginning to troubleshoot microcomputer hardware?
10. Who should run the machine when you first arrive to troubleshoot it?

11.3 OSCILLOSCOPE TRACES

Section Overview

An oscilloscope can be used to make preliminary checks on a microcomputer system.

- **The supply should have no more than 0.2-V p-p noise on it, and the clock should show clean 1 and Ø levels.**
- **Control lines such as chip enables may not show a visible pulse, but they should cause the scope to trigger on normal (driven) mode if they are being accessed.**
- **Address and data lines should show voltages that are not frozen at perfect V_{CC} or ground levels. Adjacent lines should show at least subtle waveform differences if they are not shorted. Invalid logic levels on the data bus are normal.**

Most microcomputer lines do not carry the repetitive analog kinds of signals that oscilloscopes were designed to display, but oscilloscopes are commonly available, and they can be pressed into service for certain measurements in microcomputer

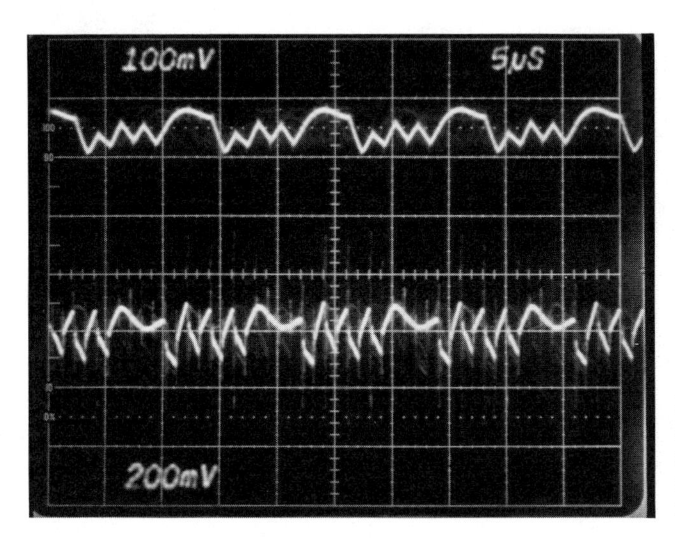

Figure 11.2 An acceptable +5-V supply line (top trace) and a noisy supply line in a microcomputer system; ac coupling of the oscilloscope was used. Note that the scale factor is double on the bottom trace.

troubleshooting. The scope should be dual-trace, triggered-sweep, dc-coupled, with a bandwidth at least three times the microprocessor clock frequency.

Clock and supply: The V_{CC} supply should be observed at several points in the system, including the processor V_{CC} and V_{SS} pins. Use a low-capacitance times-10 probe; do not use a direct probe. The scope-probe ground clip should be no more than a few inches long and should be connected to a system ground point near the measured voltage. If two signals are being displayed, both ground clips must be connected.

It is too much to hope that the power supply will be free of noise, but if the scope shows the noise on the +5.0-V line to be greater than 0.2 V p-p, additional 0.1-μF bypass capacitors should be added to reduce the noise voltage. By adjusting the *sweep speed* and *trigger* controls, it is usually possible to read the period of the major component of the noise. Usually this will be the clock period—in the megahertz range. If it is 60 or 120 Hz, the trouble is in the power supply itself. Figure 11.2 shows a noisy and an acceptable V_{CC} trace on a microprocessor system.

The clock is usually the only repetitive signal that can be counted on to be present at all times in a microprocessor system. It should generally be a relatively clean pulse train, pulling *low* to about 0.1 V and *high* to about +4.0 V. Some processors require clock waveforms that are asymmetrical (unequal *high* and *low* times) or that pull higher than +4.0 V. Most processors also require that the clock have specified maximum rise and fall times (typically from 100 to 10 ns). To measure the rise time, your scope must have a rise time at least three times as fast as the time being measured. Oscilloscope risetime can be calculated from bandwidth:

$$t_r = \frac{0.35}{f_{BW}}$$

For example, a 20-MHz scope has a risetime of 17.5 ns. It would be useful in measuring risetimes down to about 52.5 ns. Figure 11.3 shows how to measure the risetime of a clock waveform.

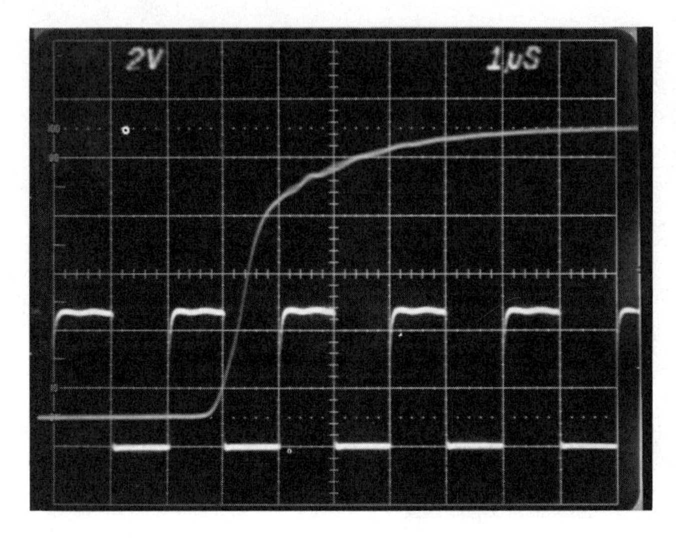

Figure 11.3 A 6802 clock signal (lower trace) at 1 μs/cm, with an expanded view showing a 10%-to-90% risetime of 50 ns (overlay trace, 20 ns/c, vertical uncalibrated).

Control lines such as CE and R/$\overline{\text{W}}$ are good candidates for oscilloscope evaluation. Let us say that a 7-segment display in a microcomputer system is stuck at a fixed (and incorrect) display. We would observe the CE line of the latch that drives the LEDs. We would not expect to see a clean repetitive waveform, but any indication of valid logic-1 levels would show that the latch was being accessed.

If the CE line of the latch were being brought *high* for 1 μs once every 10 ms, the *high* time would be 1/10 000 of the *low* time and would be very difficult to observe on the oscilloscope. Still, the pulse may be detected. There are two ways to do it.

1. Set the scope to *driven* or *normal* sweep mode (not auto). Increase the sweep speed to 10 μs/div and increase the trace brightness. Set to + trigger slope (for the positive CE pulse) and adjust the trigger level control. A good scope will show a 0.1-division-wide pulse. You may even be able to measure its width by sweeping at 1 μs/div and darkening the room to bring out the dim trace.
2. If your scope does not have a vertical delay line, the preceding procedure will probably not work. If the CE pulse comes only once every second, it won't work, regardless. But the scope sweep will be triggered by the CE pulse, even if the pulse is too brief and infrequent to be seen. Set the sweep speed to 10 ms/div, and adjust the trigger controls as outlined previously. If the trace keeps flashing across the screen, it indicates that a triggering pulse is present, even if it cannot be seen. To verify that the triggering is not due to noise, ground the scope's input or connect it to V_{CC}, and observe that the triggering ceases.

Figure 11.4 illustrates how to set the scope to detect infrequent brief pulses.

Address lines should generally show levels below +0.4 V (logic 0) or above +2.4 V (logic 1), although the traces will usually overlap because of the noncyclical nature of the signals. It is not uncommon for some of the higher-numbered address

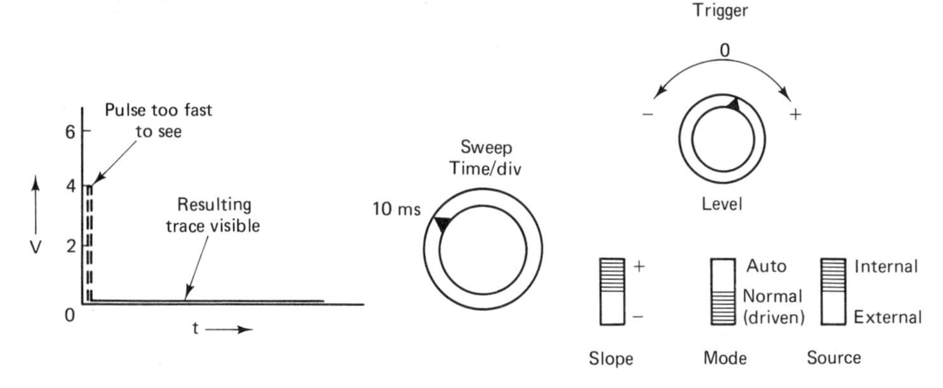

Figure 11.4 Even if a pulse is too brief to be seen, it can be detected with the oscilloscope. Each pulse will cause a 0-V trace line to flash across the screen.

lines to remain at one logic level and not switch. This just indicates that the program being executed is limited to one area of memory. Figure 11.5 shows typical address-line waveforms.

Invalid logic levels (between +0.4 and +2.4 V) are not to be expected on the address lines, unless the processor has tristate address outputs that have been put in a floating state to give some external device control of the address bus.

Two things that are sure indications of trouble on the address lines are voltage levels below +20 mV and voltage levels above +4.8 V. These indicate a line shorted to ground or V_{CC}, respectively.

The most useful scope check of the address lines is to use the two probes of a dual-trace scope to observe adjacent lines, for example A0 and A1 or A8 and A9. Adjacent lines should not show identical traces. For simple tight-loop programs, they may occasionally have identical logic levels throughout the program, but the minor voltage variations during the logic-1 levels should show subtle differences. If they

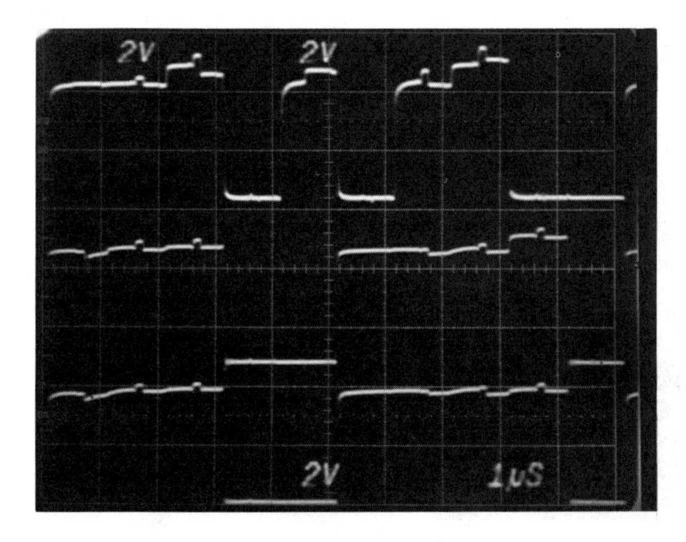

Figure 11.5 Three address-line waveforms on a 6802 system. The center and bottom traces happen to be switching identically, but subtle differences in the *high* levels show that the lines are not shorted.

do not, the adjacent lines are probably shorted. This can be checked with an ohmmeter when the power is turned off. Where pc tracks weave between the pins of the DIP, shorts to the pins on the opposite side are also likely, and should be checked in the same way. For example, in a 24-pin memory the line from pin 11 may weave between pins 13 and 14, and may be shorted to one of them.

Another scope check that is nearly as useful is to observe the same line at different places in the system. Line A5 at the processor should show the same waveform as line A5 at the RAM, unless an address buffer chip separates them. If different waveforms are observed, the line is probably open. Again, an ohmmeter will confirm this.

A word of caution: Many analog VOM ohmmeters impose test currents in the vicinity of 100 mA on the $R \times 1$ range. This can damage microprocessor chips.

Data lines can be checked with the oscilloscope in much the same way as address lines:

1. Do any of the lines fail to switch while a program is being run? It would be a very short and very unusual program that allowed any of the data lines to remain in either the *high* or *low* state. Suspect a short to V_{CC} or ground.
2. Do any adjacent pairs of lines have exactly the same waveform? Suspect a short between these lines.
3. Does the waveform on each line at the processor match the waveform of the corresponding line at the other chips, such as the ROM, RAM, and I/O devices? Radical differences may indicate an open line.

Data-line waveforms often look different from address-line waveforms because invalid logic levels (between +0.4 and +2.4 V) are common on the data lines. Such levels often appear while the processor is performing internal operations, and neither the processor nor any other devices are asserting data on the bus. These are $\overline{\text{VMA}}$ (valid-memory-address-*not*) cycles, and the data bus is free to float to invalid levels during these times. Figure 11.6 shows two typical data-line waveforms.

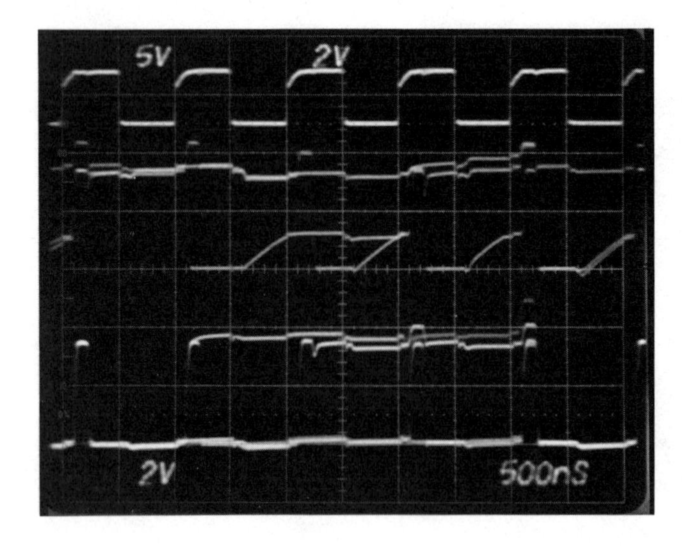

Figure 11.6 A 6802 clock signal (top trace) with a data line driving TTL inputs (center) and a data line driving microprocessor inputs. Invalid levels are common during *low* clock and $\overline{\text{VMA}}$ periods. Overlapping traces indicate that the program does not repeat at the sweep rate of the scope.

floating bus but by a fight for the bus, this can be checked easily. Just jump a 1-kΩ resistor from V_{CC} to the suspected line. If the invalid level is pulled *high*, the line was simply floating. If it remains at an invalid level, there is a problem—probably two devices trying to assert data on the bus at once.

11.4 A BASIC DIAGNOSTIC PROGRAM

Section Overview

Technical troubleshooting begins by removing the main program ROM containing the reset vector and substituting a ROM containing a very simple program loop that will produce predictable and repetitive voltages in the system.

The 6802 produces a static and known set of voltages when the RESET line is brought *low*, and these can be checked with a meter.

If you have reached this point in your troubleshooting efforts without achieving success, you will have to admit that it is not one of your lucky days. It will be necessary to begin a very organized and technical approach to the problem. This is the time when service technicians of the lesser breed will lose their heads and abandon the task, or begin dithering about in random efforts that they vainly hope will allow them to stumble across the problem. Those who are able to keep their nerve will simply forge ahead and do what has to be done.

Making waves: Imagine trying to troubleshoot an audio amplifier while somebody is talking into the microphone. The oscilloscope can't show you what's happening because the waveform keeps changing. We don't do it that way—we inject a steady sine wave, so the scope can display a stable repetitive wave. Microcomputer programs are normally as unpredictable as voice waveforms. To see what's happening we need to generate a stable repetitive waveform. We do that with a test program.

A processor and a ROM comprise a very simple microcomputer system. In fact, they comprise the system that was described at the end of Chapter 3. Our first task will be to reduce the system under service to this simple configuration. The program ROM should be removed and a new ROM containing a test program should be inserted in its place. The test program should be very short and should be designed to cause all, or nearly all, of the address and data lines to switch in an endless loop. One such program is given in Section 3.6. Here is another:

DIAGNOSTIC PROGRAM 2—UP & EPROM

Object code				Cycles	Source code			Comments
FFF0	97	0F		4	LOOP	STAA	$000F	Address lines complement on first and fourth cycles.
FFF2	43			2		COMA		
FFF3	7E	FF	F0	3		JMP	LOOP	Data bus complements on alternate four cycles.
FFFE		FF	F0			(Reset vector)		

This program is a loop only 9 machine cycles long, so it produces voltage patterns that repeat frequently enough to be observed on an oscilloscope at 1 μs/div for a 1-MHz clock. With the help of the 6802 cycle-by-cycle operation summary (Appendix B), we can predict the voltage levels that should be seen at selected cycles on certain lines.

- The processor's VMA output goes *low* on cycle 3, the third cycle of the STAA (direct mode) instruction. We can trigger on this line (negative slope), since it is a unique event in the program; neither of the other instructions causes VMA to go *low*.
- Address lines A15 through A4 will be *high* and A3 through AØ will be *low* during cycle 1, since the processor will be reading address $FFFØ.
- Each data line will be *high* during one cycle 4 and *low* during the next cycle 4, since the data being written is complemented after each write. Invalid logic levels may be seen during the first half of the write cycle because the processor does not assert its data until the second half of the cycle. It may be helpful to display two loops (18 cycles) at 2 μs/div to avoid overlapping traces.
- The $R/\overline{W}$ line should go *low* during write cycle 4.
- The *chip-select* line of the EPROM should be asserted on all cycles except 3 and 4, during which the RAM address $ØØØF is on the bus.

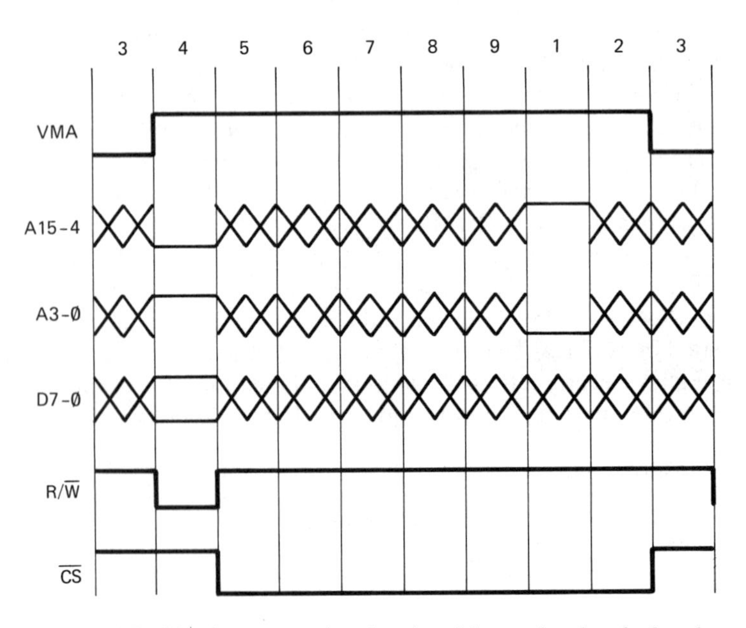

Figure 11.7 Waveform expectations for selected lines and cycles of a 9-cycle processor / EPROM diagnostic program. It was not deemed necessary to form expectations for the periods marked *X*.

Using the above information, we can sketch the expected waveform for the test EPROM in sufficient detail to assure ourselves that the processor, EPROM chip-select decoder, and EPROM are all working properly. Figure 11.7 shows the waveform expectations. Note that we have not taken the trouble to predict every logic level on every cycle. That is not necessary, and it would just be a waste of time. We have identified at least one logic *high* and one *low* state for each address, data, and control line, however.

Curses! Foiled again: Let us say that we have checked the power supply and the clock, observed the address and data lines on the normal program, and found nothing out of order. Then we ''burned'' an EPROM with a diagnostic-loop program, but the scope traces don't look anything like our expectations; mostly they are just an unsynchronized jumble of *high* and *low* levels. Now what are we supposed to do?

One thing that you can do is remove, one by one, all the chips that are connected on the address or data bus and are not essential to the operation of the diagnostic program. These may be RAMs, other ROMs, latches, buffers, or more exotic things like video display generators or DMA controllers. One of these chips may be placing a short on a bus line, causing the processor to lose track of your program. (We didn't want to pull these chips unless we had to, so we saved it until last.)

One thing that you should *not* do is page further back in the chapter hoping for better luck with some other technique. If it won't work with the system stripped down to 3 chips and a three-instruction program loop, it certainly won't work with the whole system trying to run a several-thousand-word program. And the limited system will be a whole lot easier to fix.

More expectations: If the test-loop program still doesn't work, you then make a list of the things that you expect to happen, and you check them out, one by one:

1. The processor will output the reset-vector address $FFFE when the $\overline{\text{RESET}}$ line is pulled *low*. Although VMA will be *low* the EPROM can be made to respond by jumping its $\overline{\text{CS}}$ input to ground. The EPROM will then place the most-significant byte of the program start address ($FF) on the data bus. Hold the *reset* button down and check that all the address lines are *high* except A0, which is *low*, and that the data lines are all *high*.
2. When *reset* is released the processor should call address $FFFF and read the least-significant half of the starting address $F0). It should then call addresses $FFF0 through $FFF5 repeatedly, reading the data given in the program object-code list. We can check some of this by forcing some of the *high* address lines low with the RESET line held *low*. (We would be reluctant to force a *low* line *high*, because this is more likely to cause excessive current and damage to an IC.) As an example, we may jumper A3 and A2 to ground with the RESET line *low*. The address called would be

1111 1111 1111 0010 or $FFF2

According to the program listing, the EPROM should respond by placing $43 on the data bus. We would check data lines D7 through D0 for the pattern

0100 0011

at the EPROM and at the processor pins.

3. The processor should stay in the address range $FFF0 through $FFF5. The 6802 microprocessor is particularly easy to troubleshoot in this regard because its address lines remain active with the address of the next instruction when the processor is halted. (Most other processors let the address lines float in tristate in the *halt* mode.) We can clip the HALT line to ground and read the state of the address lines with the scope. Two or three random *halts* will tell us whether the processor is (a) staying in the intended loop, (b) getting stuck in some other loop, or (c) counting through its entire address range (as it usually does when it gets lost and can't find any executable instructions.)
4. The jumper from the EPROM's $\overline{CS}$ input to ground should now be removed to verify the operation of the chip-select logic.

REVIEW QUESTIONS FOR SECTIONS 11.3 AND 11.4

Answers appear at the end of Chapter 11.

11. An oscilloscope check shows 0.1 V p-p noise pulses at a 1-μs repetition rate on the 5.0-V supply line of a microprocessor. It this normal, or is it a problem?
12. What is the minimum oscilloscope bandwidth required to measure a 10-ns risetime accurately?
13. A $\overline{CE}$ line goes *low* for 1 μs every 5 s. How should the scope trigger mode, slope, level, and sweep speed be set to detect the pulse?
14. Which of these lines is most likely to contain invalid levels (between +0.4 and +2.4 V): A0, D0, E, R/$\overline{W}$, VMA?
15. How can an oscilloscope be used to check for shorts between adjacent address lines?
16. What are the desirable features of a first diagnostic program, in terms of length and function?
17. When should peripheral chips (latches, buffers, and so on) be removed from their sockets, and why?
18. Is it safe to temporarily jumper address and control lines *high* or *low* to force access of certain addresses?
19. What voltages should the 6802 address lines present when the RESET line is held *low*?
20. A malfunctioning 6802 system is *halted* several times and the address lines measured. A9 through A0 vary but A15 through A10 always read

    1101 11

    What range of addresses is the processor carrying?

11.5 EXTENDED DIAGNOSTIC PROGRAMS

Section Overview

Once the basic processor-ROM system has been made to work, simple diagnostic program loops should be written to exercise additional peripheral chips one at a time. Two goals should be to have the loop short enough to be observable on an oscilloscope (25 machine cycles or less) and to have the peripheral produce a response that is visible to the troubleshooter if it is functioning.

Eventually you will find a shorted, open, or misconnected line or a bad IC, and the basic diagnostic loop will work. Now it is time to expand the system to include one of the peripheral chips. Let's say it is a latch that drives a 7-segment display. We can modify our program loop to activate that latch. Let us say that the latch appears at address $C000.

DIAGNOSTIC PROGRAM 3—UP, EPROM, AND LATCH

Object code				Cycles	Source code			Comments
FFF0	B7	C0	00	5	LOOP	STAA	LATCH	Store alternately 1s and
FFF3	43			2		COMA		0s to latch bits.
FFF4	7E	FF	F0	3		JMP	LOOP	Ten-cycle loop.
FFFE		FF	F0			(Reset vector)		

This program will access the latch every 10 μs, assuming a 1-MHz clock, and will alternately place *high* and *low* levels at the latch output bits. We can use the scope to ascertain that the latch's chip enable is asserted every 10 μs and that data at each output switches with a 20-μs cycle time.

Bring in more chips: If the latch test is successful we can bring in another chip, say an input buffer at $A000. We will use a diagnostic program that picks up data from the input side of the buffer and delivers it to the output (LED) side of the latch. Thus the display on the LED should change as the inputs to the buffer are jumpered or switched *high* and *low*.

DIAGNOSTIC PROGRAM 4—UP, EPROM, LATCH, AND BUFFER

Object code				Source code			Comments
FFF0	B6	A0	00	LOOP	LDAA	BUFR	Pick up 8 bits and Send to
FFF3	B6	C0	00		STAA	LATCH	LED latch.
FFF6	7E	FF	F0		JMP	LOOP	
FFFE		FF	F0		(Reset vector)		

Now let's add an external RAM chip and modify the program to test at least one of its memory cells. The RAM is assumed to occupy addresses $2000 through $3FFF.

DIAGNOSTIC PROGRAM 5—UP, EPROM, LATCH, BUFFER, AND RAM

	Object code			Cycles	Source code			Comments
FFE0	B6	A0	00	4	LOOP	LDAA	BUFR	Pick up 8 bits in A.
FFE3	B7	33	55	5		STAA	RAM	Pass to B via random RAM
FFE6	F6	33	55	4		LDAB	RAM	location.
FFE9	F7	C0	00	5		STAB	LATCH	Output to LED.
FFEC	7E	FF	E0	3		JMP	LOOP	
FFFE		FF	E0			(Reset vector)		

This program also lights the LED with a pattern depending on the data at the buffer input, but it requires the RAM to function, or the LED pattern will not change with changing inputs. The loop is 21 cycles long. The RAM's *chip enable* should be asserted for three cycles (two for STAA RAM and one for LDAB RAM) during each loop. The buffer's chip-enable line should be asserted only once each loop (on cycle 4), so this can be used for a stable scope trigger. The RAM should be asserting data on the data bus during the last cycle of the LDAB RAM instruction, which is cycle 13, the ninth cycle after the triggering cycle. This data should be observed to change as the corresponding bit at the input of the buffer is changed.

11.6 MICROCOMPUTER SERVICING TIPS

Section Overview

This section offers a variety of practical tips for troubleshooting and repairing microcomputer equipment. The two major points are as follows:

1. **Don't attack the hardware with a soldering pencil until you are absolutely sure that the problem is not in the software or in one of the connectors or plug-ins.**
2. **If you must unsolder and replace components in a microcomputer system, take the time to learn the new techniques required to avoid damage to the delicate pc boards. Old techniques from the days before high-density circuit boards will do horrible damage.**

Perhaps the most useful troubleshooting tip comes from a few years of observation by the author: many times I have seen technicians waste hours looking for hardware problems when all the time the trouble was with the software. Very seldom have I seen technicians looking for trouble in the software when in fact the problem was in the hardware. Therefore, before you start digging into the hardware be sure that you really do have a hardware problem:

Avoiding further damage should always be your first priority. Microcomputer circuit boards have very delicate copper tracks, and many of the chips may be electrically delicate as well.

- Electrostatic Damage (ESD) to semiconductors may occur at voltages imperceptible to human senses. ESD can cause deterioration of voltage levels or switching speeds, resulting in problems that show up only later, in the field. Board-level service work should be done only on an antistatic work surface, with a grounded soldering pencil, antistatic solder sucker, and operator-grounding wrist strap.
- Before disassembling anything, make a sketch of how it goes back together. Have several small boxes available for safekeeping of screws and other small parts.
- Make it a firm habit to check that the power is off before removing or inserting any connectors, circuit boards, or ICs.
- Also make it a habit to check the orientation of an IC twice when you reinsert it. DIP chips will go in just as easily reversed as the proper way.
- Use a probe that has the metal point guarded by an insulative plastic ridge on either side to protect against shorting adjacent IC pins. A slip of a clip can blow a chip.

Soldered chips are often preferred over socketed chips by circuit-board manufacturers, not only because the sockets are likely to cost more than the chips they hold but also because the socket connections are likely to fail more often than the ICs themselves. Prototype equipment, one-of-a-kind systems, and systems that are likely to require frequent hardware service or modification should, of course, use sockets. Low-cost mass-produced items, however, are likely to have the ICs soldered directly to the board.

A temperature-regulated soldering pencil is essential for removing soldered-in ICs. Ordinary soldering pencils are likely to soften the adhesive that bonds the copper tracks to the board, resulting in serious damage. Use solder wick or a good-quality solder sucker to remove the solder from the pin joints, one at a time. If a pin is particularly troublesome, don't overheat it; give it time to cool and come back to it last.

If you are certain that the IC is bad or if it only costs 75¢, it will be better to clip its pins at the IC body and remove them one at a time rather than trying to wrestle the whole thing out at once. If the pins do not come out easily they may have been clinched before being soldered. Use a knife to straighten the leads while the solder is molten.

Once the IC is out, the solder sucker should be used again to clear any holes that are still partially blocked. A toothpick can be used (gently!) in conjunction with the soldering pencil to clear blocked holes.

In reinserting the IC you may wish to consider putting it in a socket this time. In any case, bend the pins so they match the rows of holes perfectly. Never force the pins into the holes, and especially never force a pin or a wire into a hole that is being heated; with the adhesive weakened this is sure to cause a copper track to come loose from the board. In tough cases it may help to use a fine cutters to clip the pins to points at the ends so they will guide themselves into the holes without catching the copper pads and pulling them up.

Board swapping and chip swapping (if they are in sockets) are the easy ways to narrow down the problem and fix the system without getting into really technical troubleshooting. Don't be ashamed to take the easy way out if multiple systems are available to make it possible. Many microprocessor boards are so complex and delicate that component-level servicing in the field poses an unacceptable risk of further damage to an expensive assembly. It is becoming more common to service at the *board level* in the field and to confine *chip-level* servicing to the OEM (original equipment manufacturer) facility.

Digital multimeters with resistance resolutions down to 10 mΩ are now common items. With a pair of needle-sharp probe points these can be used to trace shorts down to their lowest-resistance point of origin.

REVIEW QUESTIONS FOR SECTIONS 11.5 AND 11.6

21. In Diagnostic Program 3, page 279, with what data does accumulator A start?
22. In Diagnostic Program 3, why complement A and then store it to the latch? Why not just complement the data at the latch directly: COM LATCH.
23. How many cycles will the loop of Diagnostic Program 4, page 279, take?
24. In Diagnostic Program 4, on what cycles (if at all) will the processors R/$\overline{W}$ line go *low*? Take the start of the listing as cycle 1.
25. In Diagnostic Program 5, what changes to the object code would be required to test specifically the last byte of the RAM?
26. Diagnostic Program 5 refuses to run. As a test, the *reset* line is held *low* and the ROM's $\overline{CS}$ line is shorted to ground. The data bus lines are then read with a voltmeter. What should be the logic levels, D7 through D0?
27. Give two reasons why manufacturers often solder ICs in directly instead of using sockets.
28. How would you remove a 74LS00 that is suspected of having failed but is not in a socket?
29. How can shorts between adjacent bus lines be pinpointed?
30. How can you avoid lifting solder pads when replacing soldered-in ICs?

11.7 TROUBLESHOOTING EXERCISES ON THE MEM GAME

This section suggests a number of exercises to gain familiarity with the troubleshooting methods presented in this chapter. The system chosen is the MEM game of Figure 8.5.

1. **Power-supply problems:** Set up the MEM game of Figure 7.6 or 8.5 and run the program. Observe the clock, an address line, and a data line on a multitrace

scope. Now slowly decrease the supply voltage until the system stops functioning. Attempt to *reset* it. Note any symptoms observable on the scope. Remove the 0.1-μF supply-bypass capacitors and insert a 100-μH inductor in series with the +5-V line. Observe the effect on supply-line noise and on system operation.

2. **Pulse catching:** Restore the supply lines to normal and the system to normal operation. Connect the scope to the chip-enable line of the latch and attempt to display the enable pulse. Then use the scheme illustrated in Figure 11.4 to detect the pulse.
3. **Bugs:** Have a friend put a "bug" in the system by
   a. Shorting two adjacent address, data, or buffer-input lines with a fine and nearly invisible wire, or
   b. Opening an address or data line at the processor, memory, address decoder, or latch. Wires broken inside the insulation and clear nail polish on IC pins or wirewrap posts are among the diabolical tricks that can be used to effect an invisible break. Modify Diagnostic Program 4 (page 279) to run on the MEM system. Burn an EPROM with this program and use it, along with the techniques described in Section 11.5, to locate the bug.

Answers to Chapter 11 Review Questions

1. Expectation **2.** Operator error **3.** Cables and connectors. **4.** Hardware **5.** Start-up procedure, switch settings, hot and cold ICs. **6.** Partially remove and then reinsert the IC. **7.** 4.80 V minimum at the microprocessor pins V_{CC} and V_{SS}. **8.** Direct supply **9.** All that are not essential to minimal system operation. **10.** The regular operator **11.** Normal **12.** 105 MHz **13.** Driven mode, negative slope, slightly negative level, 10 ms/div. **14.** D0 **15.** The lines will show exactly identical waveshapes on a dual-trace scope. **16.** It should be a continuous loop, as short as possible, and should cause all address, data, and control lines to switch *high* and *low* at predictable times. **17.** If the diagnostic program fails to run, because one of them may be shorting an address or data line. **18.** Lines can generally be forced *low*, but forcing them *high* is likely to cause damage to the IC. **19.** A0 should be *low*; A1–A15 should be *high*. **20.** $DC00 through $DFFF. **21.** It isn't known and needn't be known. The COMA changes all 1s to 0s and all 0s to 1s on each loop, and that's all we want. **22.** Complement, extended, can only be done to a readable/writable device, and you can't read a latch. **23.** 12 cycles **24.** Only on the ninth cycle. **25.** Change address FFE4 to 3F and FFE5 to FF. **26.** 1011 0110 **27.** Expense and reliability. **28.** Cut the legs; then heat and remove the pins one at a time. **29.** Use a digital ohmmeter on its lowest range and move the probes along the tracks until the lowest resistance reading is found. **30.** Use a temperature-regulated soldering pencil, a solder sucker, and great care.

CHAPTER 11 QUESTIONS AND PROBLEMS

Digits after the decimal point refer to section numbers in Chapter 11.

Basic Level

1.1 Which is the more likely cause of a service call on a commercial microcomputer system, a RAM memory chip or a printed-circuit-card edge connector?

2.1 How can you temporarily simplify the complex waveforms in a microcomputer system so they can be observed on an oscilloscope?

3.2 Quote the two cardinal rules of troubleshooting.

4.2 When a new system is first installed and becomes operational, why is it a good idea to feel all the ICs and make a note of any that are unusually hot or cold?

5.3 A microprocessor system has an 8-MHz clock. Is a 15-MHz oscilloscope adequate for servicing this system?

6.3 In a microcomputer system all of the address lines except A15 are seen to switch from near zero to +4 V when the normal program is run. Line A15 stays near 0 V. Is this a sure indication of a problem with line A15?

7.4 For Diagnostic Program 2 of page 275, predict the states of data lines D0 and D1 during cycle 6.

8.4 What address is on the bus when the 6802's RESET input is held *low*? What may be necessary to force the memory to output the data from this address?

9.4 What does the 6802 address bus do if the data bus continuously reads invalid instructions, such as $00?

10.5 In Diagnostic Program 3, on what cycle will the R/$\overline{W}$ line go *low*? On what cycle will the chip-enable of the latch be asserted?

11.5 Diagnostic Program 5 just fails to fit on the oscilloscope screen at 2 μs/cm sweep (assuming a 1-MHz clock). Suggest a change to reduce the test loop to exactly 20 cycles.

12.6 Once an IC has been removed, how should the socket holes be cleared of solder to accommodate the new IC?

13.6 Name two things to check before plugging an IC back into its socket.

Advanced Level

14.1 Summarize the basic logic behind technical troubleshooting.

15.1 In troubleshooting a microcomputer system you observe V_{CC} at the processor and see 0.5-V p-p noise pulses. You are not sure if this is too high or not. Should you try to eliminate the noise or go on to measure the clock and address signals? Or do you have some other suggestion?

16.2 List four things that you can do to spot and/or fix trouble in cables, connectors, and sockets.

17.2 Wire of No. 30 gauge has a resistance of 0.35 Ω/m. Circuit board tracks 0.020″

wide by 0.001″ thick present 1.3 Ω/m. What length of each of these will cause a voltage drop of 0.1 V when supplying a current of 0.3 A?

18.3 The $\overline{CE}$ input of an input buffer is supposed to be pulled *low* for 1 μs once every 500 ms by an operating program. How can this pulse be detected by an oscilloscope, since it is present for only 0.0002% of the time?

19.3 How can a 1-kΩ resistor be used to determine whether an observed invalid logic level (say, +1.2 V) is caused by a floating bus or a fight for the bus.

20.4 Why is it easier to troubleshoot a microcomputer when it is running a diagnostic program than when it is running its regular program?

21.4 In running Diagnostic Program 2 (page 275), the oscilloscope displays data line D4 over nine cycles. On the first half of cycle 4 this line is at an invalid level of +1.5 V. On the second half an overlapping trace is seen—near 0 V and near +4 V. Explain the reasons for this, and tell whether the trace is normal or abnormal.

22.5 Modify Diagnostic Program 3 so that the latch outputs switch at a rate near 1 Hz so they can be checked by a VOM or LED instead of requiring a scope.

23.5 Modify Diagnostic Program 5 so that it tests all the RAM addresses from $2000 through $3FFF and outputs the data from the input buffer only if all 8K-words are stored and read correctly. How long will it take between the read of the buffer and the store to the latch in this program, assuming a 1-MHz clock?

24.6 Why is it bad practice to push IC or socket pins into a circuit board while heating the pads to clear them of solder?

25.6 Two adjacent circuit-board lines are apparently shorted, since they show 0.3 Ω between them. Yet no short is observable on the board. How can the short be located, since it is not visible?

12 Specialized Microcomputer Test Equipment

12.1 LOGIC PROBES, CLIPS, AND PULSERS

Section Overview

Digital electronic systems and signals are markedly different from their analog predecessors, so it is not surprising that new types of test equipment are needed for digital troubleshooting.

Logic probes can be used to spot ''dead'' lines. They are less versatile than a scope but also are less expensive and more portable. Logic pulsers can be used to force a line *high* even if a saturated output transistor is pulling it *low*.

Before the digital revolution electronic systems were characterized by dc bias levels, steady ac (usually sine-wave) signals, and one, or maybe two, signal paths. Traditional test equipment was designed to meet this need; that is, to measure dc and cyclical ac waves.

Digital systems, on the other hand, have no ''bias'' levels other than V_{CC}, and the signals are practically nonrepetitive, with data paths at least 28 lines wide (16 address, 8 data, and 4 control lines). Small wonder, then, that a variety of new test instruments have been developed to meet the specific needs of digital and microprocessor systems. However, just as buying a diet book won't make you thin, buying a specialized piece of test equipment won't make you an expert troubleshooter. You will need to have the right kind of equipment for the type of servicing you will be doing, and you will need to know how to use it.

Logic probes are handheld devices that give a visual (colored LEDs) or aural (high or low tones) indication of whether an individual line has a logic *low* level, a logic *high* level, or a pulse train switching between *high* and *low* levels. Some logic probes will also detect invalid logic levels and give a separate indication for this condition. Although their small size and low cost make them useful for quick checks out on the job, they find less application in bench servicing because the oscilloscope does everything they can do and a great deal more. An exception might be aural-level indication, which allows you to keep your eyes on the probe point and IC pins and thus move more quickly through a number of line checks.

Logic clips are clipped onto a DIP (dual in-line package) IC and show the logic state of every pin on a number of LEDs built into the clip. Used at random on an unfamiliar system, the resulting pattern of lights will probably be meaningless. However, if a test program has been written for a specific IC in a specific system and if this program causes the LEDs to blink in a sequential pattern, then the logic clip can assist in spotting one bad line among many very quickly.

Logic pulsers are handheld devices that look much like logic probes, but they are designed to output pulses rather than detect them. Figure 12.1 shows the most common situation where a logic pulser is useful. The ICs are assumed to be soldered into the board, so it is not practical to remove them on the merest hunch. The failed device must be identified without removing it.

The line with the scope probe on it shows about 20 mV dc. We are not sure if this is because IC1 has a shorted output, IC2 has a shorted input, or the line itself is shorted to ground. The logic pulser is used to inject a 1-μs current pulse of about

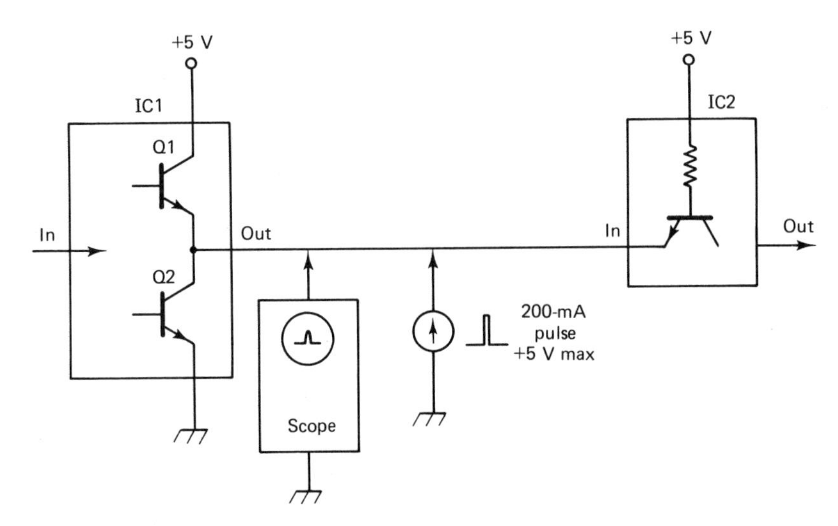

Figure 12.1 A logic pulser brings IC1 out of saturation and delivers an input pulse to IC2.

200 mA to the line. The pulse repetition rate is 1 kHz, so the average current is only $\frac{200}{1000}$, or 0.2, mA—not enough to damage the ICs. The 200 mA is enough, however, to bring pulldown transistor Q2 (within IC1) out of saturation, thus placing a logic *high* on the IC2 input for 1 μs. If the pulser produces valid inputs at IC2 and a valid output is also seen, IC2 is good. IC1 is either bad or has its inputs frozen in order to hold its output *low*. We may be able to use the logic pulser at the IC1 input to see if unfreezing the input will allow the output to go *high*. If we are unable to get the IC1 output to switch *high*, that IC has probably failed.

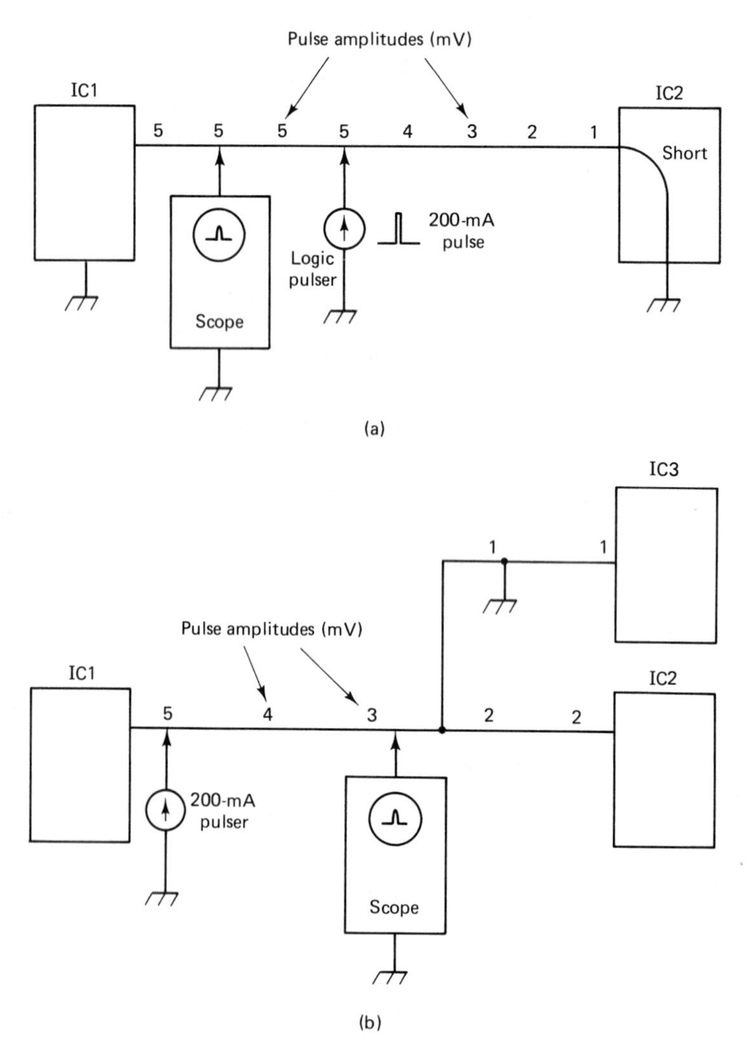

Figure 12.2 Millivolt pulse amplitudes can be traced with a scope to locate shorts in an IC or wire track.

On the other hand, if the logic pulser is unable to produce a valid logic 1 on the line, there is a short, either in the IC1 output, the IC2 input, or along the line itself. A 1-inch copper track on a printed circuit board will typically have a resistance of 25 mΩ. The 200-mA pulses will produce 5 mV across such a resistance—enough to be measured by a good oscilloscope. Figure 12.2 shows how millivolt drops along the copper tracks can be measured and interpreted to locate the short.

12.2 SIGNATURE ANALYZERS

Section Overview

Signature analysis has been vigorously promoted as a painless, any-body-can-do-it method of microcomputer troubleshooting. In truth it is a powerful and easy-to-use technique—once the necessary setup work as been done to take full advantage of it. The catch is that signature analysis is almost useless if the preliminary setup work has not been done, and there is no way to do the setup on a malfunctioning system.

A signature analyzer is a digital test instrument that checks for an expected string of pulses at preselected points in the system under test. It generally has three probes in addition to the ground clip. These are to be connected to the system clock, a "start-strobe" line, and the selected digital test point in the system. A string of serial data bits, one per clock cycle, is stored and displayed as 4 hex digits on the analyzer. (Often the hex b, d, and E are replaced with other characters such as P, H, and U, since the former are liable to be misread at the odd viewing angles encountered in troubleshooting.) This four-digit *signature* is compared with the expected signature, which was previously recorded and entered on the troubleshooting chart. If the signatures match, that section of the system is presumed to be good. If they do not, there is presumed to be a failure in that area.

A number of cautions are necessary concerning this method of troubleshooting.

- The system must be designed with a *start-strobe* line.
- Correct signatures for a number of test points must be taken and recorded using a functioning system. Signature analysis is useless for bringing up new or modified systems.
- The system under test must be functioning at least to the point where the *clock* and *strobe* signals are present. If the strobe is initiated by software, this means that the system must be following its program before the signature analyzer can be used.
- The system under test must be placed in a completely unambiguous state. All inputs must be absolutely standard. This is often not possible. For example, a 12-bit A/D converter, even with the analog inputs shorted, can probably not be guaranteed to output a perfect count of zero.

- Microcomputer system components are usually so interrelated that a failure in one section will cause changes in the data streams in all the other sections. Thus a single failure may cause nearly all of the signatures in the system to change.
- Signatures for a microcomputer system will, in general, change when the program changes. Perhaps the biggest selling point for microcomputer systems is that they can be readily modified to meet changing needs. This same feature means that the recorded signatures will be obsoleted by software changes. In some systems it is possible to avoid this problem by having the system run a special test program while it is being serviced.

If a system is mass produced to the extent of tens of thousands of units, it may become practical to introduce deliberately a large number of potential failures, one by one, into a prototype system and record the signatures produced at each test point by each failure. This would allow the troubleshooter to reference a ''fail'' signature back to a specific source of trouble. Of course, a complete troubleshooting guide of this sort would require many hours of work by an individual whose talents would be in great demand for more immediately profitable development work, so a really efficient signature-analysis troubleshooting system is infrequently implemented.

REVIEW QUESTIONS FOR SECTIONS 12.1 AND 12.2

Answers appear at the end of Chapter 12.

1. Give two reasons why the oscilloscope is not the ''universal test instrument'' in the digital world as it is in the analog world.
2. Which of these cannot be identified by a logic probe?
   (a) Line stuck *high* **(b)** Line stuck *low* **(c)** Adjacent lines shorted **(d)** Line switching from *high* to *low*
3. Logic clips are of real use when employed in conjunction with a ______________ __________.
4. The output of a TTL counter drives the inputs of four TTL logic gates. The output-to-input line seems to be frozen at near-ground level. A logic pulser is applied to the line. What will be seen if the counter is good but is just never being driven out-*high*?
5. Continuing Question 4, what will be seen if the input of one of the logic gates is shorted?
6. Continuing Question 5, how can the particular shorted gate be identified?
7. If a signature analyzer displays four hex digits, does that mean that it is monitoring 16 data lines?
8. How can a signature analyzer be used on a malfunctioning microcomputer system for which no table of expected signatures has been prepared?
9. A signature table shows an expected reading to 3H0C at test point *X*, but the

actual reading is 27U9. Other than the fact that something is wrong, what information do the digits 27U9 convey?

10. How can system firmware be modified and updated without obsoleting the table of signatures already prepared for an instrument?

12.3 A MICROPROCESSOR STATIC EMULATOR

Section Overview

A simple bank of switches, substituted for the microprocessor, can provide steady output signals to call up many of the processor's peripheral devices and read and write data to them. The expected responses can be checked with a voltmeter, since they do not flit by in a microsecond, as they do when driven by the processor.

This technique works well for ROMs, static RAMs, latches, and buffers, but it is not well suited for testing the more advanced peripheral chips, which are usually dynamic parts that will not operate under static conditions.

A microprocessor static emulator is a simple and inexpensive test device that you can build yourself. A static emulator is plugged in in place of the microprocessor and uses a number of switches to output steady logic-0 or logic-1 levels to the control, address, and (sometimes) data buses. Figure 12.3 shows a static emulator for the 6802 processor.

As an example of how the static emulator can be used, let us say that we want to check out the operation of a RAM at addresses $2000 through $3FFF. We might set the address switches to

0010 0011 0100 0101

which is $2345, an address chosen at random. Then we set VMA = 1, $R/\overline{W}$ = 0, and E = 1. The RAM should now be reading the data bus. Let us set the data switches to $67, or binary

0110 0111

Now we can scope the RAM's chip-select, write-enable, address, and data lines to see if they are set up as expected. Then we float the emulator's data lines and switch $R/\overline{W}$ to logic 1. The RAM should now be asserting data $67 on the bus. Switching the emulator to a different address should cause different data to be asserted.

The static emulator is most useful for troubleshooting completely dead systems and for bringing up new systems in the early stages of development. It can be used to exercise ROMs, latches, buffers, and static RAMs. It cannot generally be used to exercise fully dynamic RAMs or advanced interface chips with dynamic internal circuitry. Dynamic parts, it will be recalled, use charge storage to retain data. These charges must be refreshed, typically every few milliseconds.

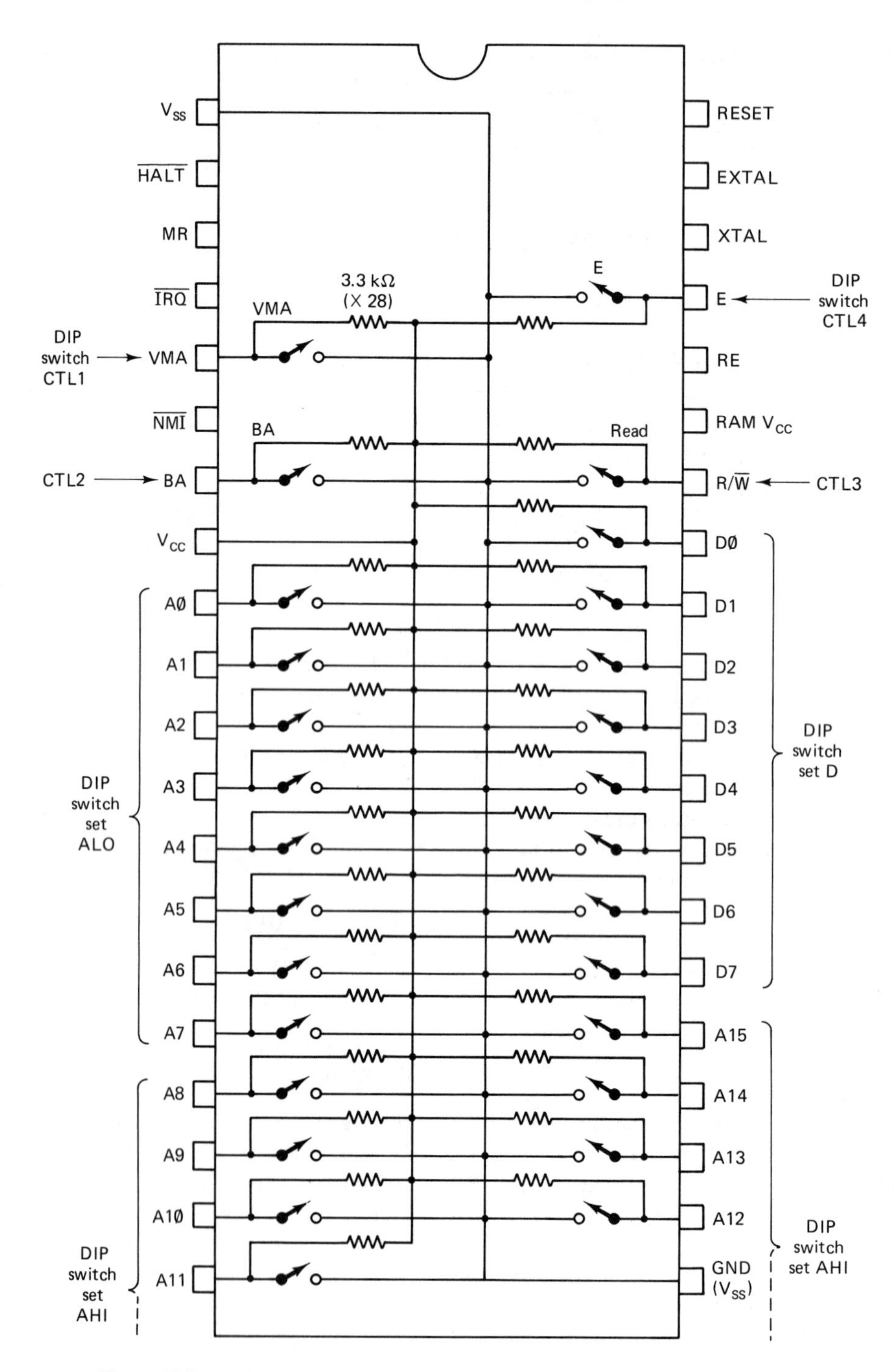

Figure 12.3 A static emulator for the 6082 replaces the processor chip and allows steady address, data, and control signals to be applied.

12.4 SINGLE-STEPPING THE 6802

Section Overview

The great frustration of troubleshooting microcomputers is that they won't hold still so you can see what they are doing. An error comes and is gone in a microsecond, and we never even saw it. Test-program loops are one way to tame this flighty situation. The static emulator is another. Single-step operation of the processor is a third.

Single-stepping the 6802 is accomplished by bringing the HALT line *high* only long enough for it to begin one instruction and then bringing it *low* again. The 6802 will finish the instruction it has started before it obeys the new HALT command.

A single-step circuit for the 6800/6802 is shown in Figure 12.4. It uses only two TTL ICs and can be added to any existing 6800 or 6802 system. A single instruction, of however many machine cycles, will be executed each time S3 is pressed. Here is how the circuit works.

When S3 is pressed, U1A and U1B debounce the switch and pin 10 of U1C is held *high*. The *E*-clock, which continues to run while $\overline{\text{HALT}}$ is *low*, is then passed, inverted, to the trigger input of U2A. Since *J* and *K* are both *high*, U2A toggles (changes states) at each negative edge on pin 1 (positive edge of *E*). Recall that a *JK* flip-flop toggles on trigger if *J* and *K* are both 1, holds state if *J* and *K* are 0, sets $Q = 1$ if *J* is 1 and *K* is 0, and resets or clears ($Q = 0$) if $J = 0$ and $K = 1$.

The first rise of *E* thus *sets* the QA output, driving the $\overline{\text{HALT}}$ line of the processor *high*. The second rise of *E* switches this line *low* again, one machine cycle later. This fall of the U2A output (pin 12) triggers U2B to *set*, and its $\overline{Q}$ output (pin 8) goes *low*. The *J* input of U2A is now *low* while *K* is *high*, so subsequent *E* pulses (after the first two) only hold U2A in the *reset* state, in which it already is. With $\overline{\text{HALT}}$ *low* the processor completes its current instruction and halts.

Tracing your steps: Single-stepping is of little value if you can't see where you are stepping. Here are some ways of finding out where you are:

- The 6802 keeps the address of the next instruction asserted on its address bus when it is halted. Monitoring the least-significant 3 or 4 bits with voltmeters or scope probes should be enough to let you know if the processor is following the program listing. If you can get a 40-pin logic clip or bring the address lines out to a dummy chip of dimensions to match your logic clip, you can monitor all 16 address lines. Note that the 6802 data lines and all the bus lines of the 6800, 6809, 6502, Z80, and most other processors *float* during *halt* and cannot be read.
- You can often deduce, by following the program listing, that a certain line (chip-enable, R/$\overline{\text{W}}$, latch output) is supposed to switch *high* or *low* on the fifth or eighth or some other instruction. Even if the pulse is transitory, the scope-triggering scheme of Figure 11.4 can be used to spot the pulse.

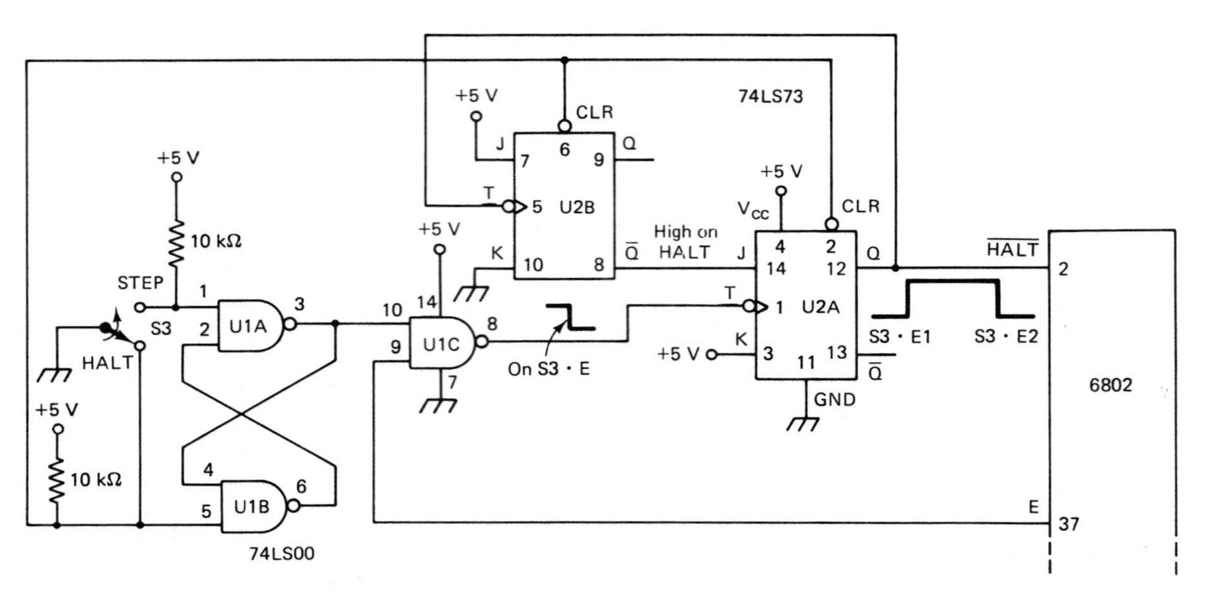

Figure 12.4 A circuit to cause the 6802 to execute one instruction each time the *step* button is pushed. The addresses will remain on the bus, since the 6802 address lines are not tristate.

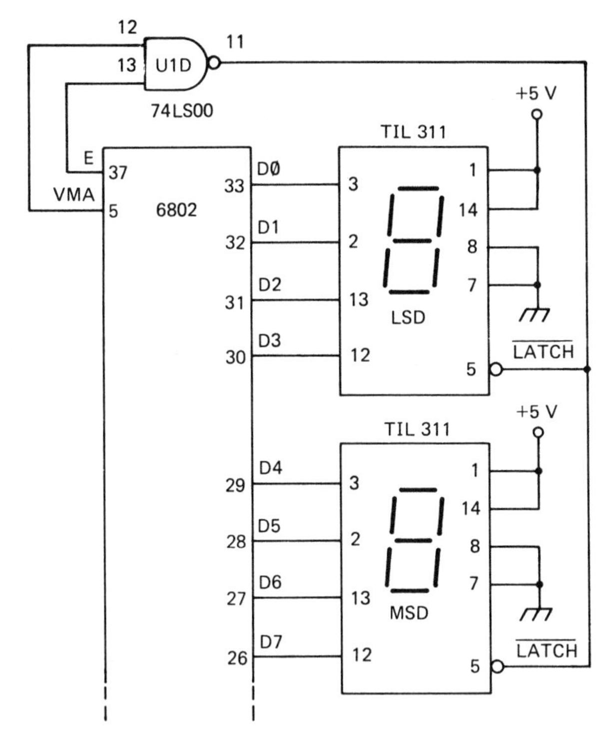

Figure 12.5 Two TIL-311s latch and display the last word on the data bus while the processor is halted. This is usually the operand data or the op code of the next instruction (see Appendix B).

- An IC containing a hexadecimal LED display, decoding logic, and a latch is available and can be used to capture and display the last bits on 4 lines. It is ironic that these things cost more than a microprocessor chip, but then they cost quite a bit less than a logic analyzer. Figure 12.5 shows the TIL-311 LEDs connected to display the data bus of a 6802. The connection to display address lines is similar.

REVIEW OF SECTIONS 12.3 AND 12.4

Answers appear at the end of Chapter 12.

11. How should the static emulator's *data* switches be set to allow a selected ROM address to assert data on a bus?
12. If the static emulator pulls a data line *high*, how can another device be allowed to pull that line *low*?
13. In the system of Figure 3.7 a *write* is to be made to the 6810 using the static emulator. What positions of switches A8, R/$\overline{W}$, and E will cause a write?
14. Why can the static emulator not be relied upon to store data in a dynamic RAM?
15. In the single-step circuit of Figure 12.4, what is the logic level of U2A, pin 12, when S3 is in the HALT position, and what makes it so?

16. How many times does U2A recognize its *T* input, per step?
17. How many times does U2B recognize its *T* input, per step?
18. Why can't the 6802 data output be checked with a voltmeter during the *halt* condition?
19. How can a TIL-311 IC display the 6802 data bus when a voltmeter can't?
20. When you troubleshoot by single-stepping, what are you looking for? In other words, when will you say, ''Here's the problem''?

12.5 LOGIC ANALYZERS: DISPLAY MODES

Section Overview

A logic analyzer is basically a RAM memory; one thousand 32-bit words is a typical size. The memory is continually storing data from the microprocessor buses. The memory is a FIFO—as each new word is stored, an old word (1000 words old) is discarded. A sophisticated triggering system is used to select just which 1000-sample section of the microprocessor's program to store.

The stored program segment is most conveniently displayed in disassembled form—instruction mnemonics are written on the analyzer's CRT in place of the 8-bit op-codes, and the operands and addresses are displayed on the same line in hex. Straight hex, binary, and oscilloscope-style graphical displays are also available for more detailed examination of system performance.

Logic analyzers are undoubtedly the most versatile among microcomputer test instruments. They are also the most expensive—$3000 to $10,000 covering the price range of what would be considered ''basic'' units.

Essentially, a logic analyzer is a fast RAM memory with a lot of controls on how it is loaded and on how its contents are displayed. A typical analyzer might have a capacity of 1000 words, each 32 bits wide. In operation, 16 bits would be loaded from the address lines, 8 from the data lines, 4 from control lines (VMA, $R/\overline{W}$, $\overline{IRQ}$, and $\overline{NMI}$, for example), and 4 from logic probes attached to points in the system selected by the user (latch outputs, for example). Usually each word of the analyzer's memory stores bus data from one VMA-true clock cycle. The data stored by the logic analyzer may be displayed in a variety of formats.

- *Waveform display* makes the analyzer look like a 10- or 20-channel oscilloscope. This option is most useful with fast internal clocking, described later.
- *Binary display* shows each line as a column of 1s and Øs on the CRT. The display is 32 columns wide—1 column for each bit. The columns are 1000 words high (1 word for each stored word), although the CRT can display only about 20 of these at a time. The binary display mode is useful for spotting lines

```
STATE  STS                      ADR       DAT
TRIG  0011 1111111000011111 00100000
00001 0011 1111111000100000 00100000
00002 0111 1111001010011100 00100000
00003 0111 1111001010011101 00011100
00004 0011 1111001010011110 00011100
00005 0011 1111001010111010 00011100
00006 0111 1111001010111010 10000110
00007 0111 1111001010111011 00000001
00008 0111 1111001010111100 10110111
00009 0111 1111001010111101 11100100
00010 0111 1111001010111110 00100011
00011 0011 1110010000100011 00100011
00012 0110 1110010000100011 00000001
00013 0111 1111001010111111 10100110
Position Cursor: f1        Cursor (f2): fast
```

Figure 12.6 A binary display on a logic analyzer, revealing a possible short between adjacent lines A3 and A4.

that are frozen *high* or *low* and shorts between adjacent lines. Figure 12.6 shows a binary display on a logic analyzer.

- *Hex display* consolidates the address and data bits of the binary display into 1 hex digit per 4 binary digits. This format is useful for following object-code assembly lists of the program. Figure 12.7 shows a hex display for the last part of the *mulitply* program of List 10.3.
- *Octal*, *decimal*, and *ASCII character* equivalents of selected groups of binary digits may also be displayed on some analyzers.
- *Disassembly mode* is the easiest to use, by far. It requires that the analyzer be equipped with a special pod and software package to convert the stored object code into the assembly-language mnemonics for your processor. Usually it is necessary to purchase a separate pod and software package for each micropro-

```
STATE  BIN  STS  ADR DAT
-0004 0000 0111 E023  27
-0003 0000 0111 E024  04
-0002 0000 0011 E025  04
-0001 0000 0011 E029  04
TRIG  0000 0111 E029  97
00001 0000 0111 E02A  02
00002 0000 0011 0002  02
00003 0000 0110 0002  45
00004 0000 0111 E02B  D7
00005 0000 0111 E02C  03
00006 0000 0011 0003  03
00007 0000 0110 0003  67
00008 0000 0111 E02D  20
00009 0000 0111 E02E  FE
Position Cursor: f1        Cursor (f2): slow
```

Figure 12.7 (a) A hex readout from the *multiply* routine of page 248, displayed on a logic analyzer. The program resides on page $E0 rather than 00. The display starts with BEQ DONE, and the branch is taken. The second status (STS) bit is VMA, and the fourth is R/$\overline{W}$.

```
STATE  PC   DISASSEMBLY   BUS ACTIVITY
-0041 E023 BEQ    E029
-0037 E025 ASLB
-0035 E026 ROLA
-0033 E027 BRA    E016
-0029 E016 ASL    0000    0000->80 0000<-00
-0023 E019 BCC    E020
-0019 E01B ADDB   01      0001->A3
-0016 E01D BCC    E020
-0012 E01F INCA
-0010 E020 DEC    0004    0004->01 0004<-00
-0004 E023 BEQ    E029
TRIG  E029 STAA   02      0002<-45
00004 E02B STAB   03      0003<-67
00008 E02D BRA    E02D
Position Cursor: f1       Cursor (f2): slow
```

Figure 12.8 A disassembly display on the logic analyzer. The program segment is the same as that shown in Figure 12.7, except that BEQ DONE is *not* taken the first time it is displayed.

cessor type that you want your analyzer to "support", as the manufacturers term it. In selecting a logic analyzer the availability of disassembly support for the processors you use or are likely to use in the future should be a prime consideration.

Of course, if you think about it for a moment, you will see that the analyzer software has no way of knowing the labels your program has assigned to operands and addresses, so it must display these simply in hex. This and the fact that there are no comments means that you will still need the assembly listing to make much sense out of the analyzer's display; but there is still no question that a disassembly list is easier to read than a hex list. Figure 12.8 shows an example.

Internal versus external clock: Using the system clock, the logic analyzer stores one word per processor machine cycle (excepting VMA-not cycles, where no valid data appears on the bus). Using a fast internal clock, the memory can store up to 50 samples per clock cycle from each line. This allows the analyzer to display

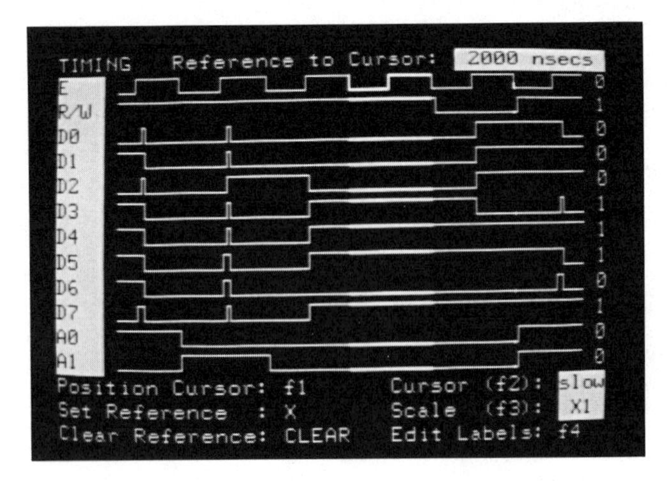

Figure 12.9 A timing display on a logic analyzer.

timing differences between lines and pinpoint timing problems. Figure 12.9 shows a timing display.

12.6 LOGIC ANALYZERS: TRIGGERING MODES

Section Overview

Logic-analyzer triggering is all-important to the successful use of this instrument. A specific combination of address, data, timers, counters, and peripheral data bits can be selected to tell the logic analyzer which segment of the target-system program to store and display.

Data qualification allows storage of selected program portions only, making it possible to cover longer program segments.

Triggering parameters must be programmed into the logic analyzer by the user to ensure that the desired portion of the target program is retained in its memory. If the analyzer stores 1000 machine cycles, this is only 1 ms, assuming a 1-MHz clock. In deciding how to trigger, it is helpful to understand that data is being stored in the analyzer memory continuously as long as it is running. As each new machine-cycle word is stored, an old word (stored 1000 cycles earlier) is lost. The trigger selection determines when this storage process is to stop.

Triggering is programmed by keying in specific address, data, and/or control states that will cause storage to stop. *Trigger delay* is commonly set to half the memory length—500 words for a 1000-word memory. This causes storage to stop 500 cycles after the triggering event and allows examination of events leading to the trigger and events following it.

- *Random trigger* is accomplished by setting all the trigger conditions to *X* (don't care) and pressing the *start* button. This allows you to see if the processor is acquiring data and following instructions or if it is stuck in an endless loop.
- *Reset trigger* is relatively easy to set up. For the 6802 the address condition is set to $FFFF, R/$\overline{W}$ is set to 1 (read), and VMA is set to 1. The other conditions are set to *X*, and the *start* button is pressed. The analyzer will trigger when the processor is *reset*, and the first 500 machine cycles of the program can be followed.
- *Specific addresses and data* can be entered as trigger conditions to permit tracing of later sections of the program. Remember that many microprocessors use instruction prefetch. They actually read the op-code of the next instruction during the last machine cycle of the current instruction. In the following program segment, for example, it would be necessary to trigger on address $FF4B, not address $FF4A, to catch the processor as it comes out of the loop. Address $FF4A is accessed by prefetch on every pass through the loop.

```
FF45  7A  00 00  LOOP  DEC   COUNT
FF48  26  FB           BNE   LOOP
FF4A  B6  C0 00        LDAA  KEYBD
```

- *Sequences of events* or time delays can be programmed into the trigger conditions of some logic analyzers. Examples might be, Trigger on the first valid read of address $F9B1 after a write is made to the latch at address $4000 or Trigger 600 μs after the eighth write of data $00 to any of the addresses $2FXX; that is, $2F00 to $2FFF. A few evenings with the logic analyzer user's manual and an hour or so on the bench with it will make the usefulness of these advanced triggering techniques more clear.

Data qualification allows more efficient use of the analyzer's limited memory and display by storing only data that qualifies under certain rules programmed in by the user. For example, you may wish to qualify (store) only the first and 256th pass through a 256-pass loop, since storing all passes would fill the entire 1000-word memory with trivial data and prevent your seeing the entry and exit on the same run through the program. As another example, you may wish to store only the writes to an output latch, so that a large number of output bytes can be examined without wasting storage memory on the program that produces the outputs.

Setup storage is a useful feature found on more-expensive logic analyzers. An EEPROM or a floppy disk is used to save complex *sequence-of-events* or *data-qualification* setups so they can be retrieved effortlessly. This is more important than you might at first realize, because if you are using a logic analyzer, you will need to concentrate on how the target system is running its program. The last thing you need then is the distraction of figuring out a new triggering program to key into the logic analyzer.

Data comparison is a logic-analyzer feature most useful in hardware troubleshooting of mass-produced systems. It makes use of two memories. One is filled with good data taken from a system that is known to be functioning properly. The other is loaded from the system under service. The logic analyzer automatically spots and displays any differences.

It bears repeating that software bugs and not hardware bugs are the great bane of microcomputer system developers. The main use of a logic analyzer is in software troubleshooting. You have a program printout and you know (or you had better learn) how the processor should proceed from one address to another in executing that program. You then display on the logic analyzer what the processor is *actually* doing. As in all troubleshooting, when you find a reality that does not match your expectation, you have all but found the problem.

REVIEW OF SECTIONS 12.5 AND 12.6

21. What is the central component of a logic analyzer?
22. Which logic analyzer display mode is most useful for spotting adjacent lines shorted together?
23. Which mode is best for scanning for software errors?
24. Which display mode and which triggering mode would be used to spot timing errors?

25. A logic analyzer has a 2000-word memory and is clocked internally every 40 ns. How many machine cycles will it display in the waveform mode if the target system has a 1-MHz clock?
26. If a logic analyzer advertisement says that it "supports" the 8086 microprocessor, what does it mean?
27. How would you set up the trigger controls of a logic analyzer to trigger on the first store of any data to page Ø?
28. A delay loop occupies target-system addresses $Ø13B through $Ø147. What address would you trigger on to see how the processor exits from the loop?
29. What logic-analyzer feature would you use to store only the first and last passes through a loop without storing all the passes in between?
30. If a logic analyzer triggers on address $FF5Ø, can it examine previous activity, for example, at address $FF4Ø?

12.7 TROUBLESHOOTING WITH SINGLE STEP

This section suggests a number of troubleshooting exercises utilizing the techniques presented in this chapter. Before beginning be sure to read Appendix A, *Microcomputer Wiring Hints*.

1. Set up the Air Raid system of Figure 10.7 or the MEM game of Figure 8.5, and add the single-step hardware of Figure 12.4.
2. Connect voltmeters and/or scope probes to lines AØ, A1, and A2, and predict from the program listing what each logic level should be after RESET is pushed and then after each press of STEP.
3. Connect a TIL-311 display first to address lines AØ through A3 and then to data lines DØ through D3, and predict what it should read at each step until the program hits its first delay loop.
4. Write and enter a diagnostic-loop program that accesses the input buffers and the output latches. Single-step through the program and use the scope-trigger technique of Figure 11.4 to detect the access of the input buffer on the predicted step.

Answers to Chapter 12 Review Questions

1. It displays 2 or 4 data lines, whereas microprocessor data paths are at least 28 bits wide, and it displays repetitive signals, whereas microcomputer signals are generally nonrepetitive.
2. (c) Adjacent lines shorted. **3.** Test program
4. The pulser will force the line *high*.
5. The line voltage will pulse to only a few millivolts positive.
6. The lowest millivoltage will appear at the input of the shorted chip.
7. No. It monitors 16 serial bits as they go by on one data line. **8.** It can't.

9. If 27U9 appears in the table with an interpretation, it could tell you exactly what the trouble is. If not, no other information can be inferred.
10. Provide a separate *test-program* ROM for use while doing signature analysis.
11. All switches in the *open* position. **12.** The pull *high* is through 3.3 kΩ. Any device pulling *low* would have to sink only 1.5 mA.
13. Open A8, close R/$\overline{W}$, open *E*.
14. DRAM data must be rewritten every few milliseconds to avoid data loss.
15. The *Q* output is *low*, cleared by S3 asserting the U2A $\overline{CLR}$ input *low*.
16. Twice **17.** Once
18. The 6802 data lines float (high-*Z*) during halt.
19. The TIL-311 contains an internal 4-bit latch, which holds the data asserted during the last VMA-true cycle.
20. When a line switches to a level different from what you have predicted by interpreting the program listing. **21.** A fast RAM memory.
22. Binary display **23.** Disassembly mode
24. Waveform display, internal clock. **25.** 80
26. The analyzer is capable of interpreting the op-code bits and displaying the corresponding instruction mnemonics.
27. Address = \$ØØXX, data = XX, VMA = 1, R/$\overline{W}$ = Ø
28. \$Ø149 **29.** Data qualification **30.** Yes, because its memory is a FIFO and triggering simply *stops* the continuous store process.

CHAPTER 12 QUESTIONS AND PROBLEMS

Digits after the decimal point refer to section numbers in Chapter 12.

Basic Level

1.1 What advantages does a logic probe have over an oscilloscope?

2.1 A logic pulser is applied to a line that appears to be shorted to ground. What voltage will be seen on the line if the problem is actually due to a continuously active TTL output pulling *low*?

3.2 Why do signature analyzers often make use of the characters *H* and *U* in their displays?

4.2 Is signature analysis more applicable to newly prototyped systems, custom one-of-a-kind systems, or mass-produced systems?

5.3 How should the static emulator's data switches be set when the address and control lines are calling up an EPROM location?

6.3 When using the static emulator to load and read out a static RAM, what precautions should be observed in changing addresses to avoid false writes?

7.4 Does the 6802 halt immediately at the end of the current machine cycle or at the end of the current instruction when $\overline{HALT}$ is brought *low*?

8.4 In the circuit of Figure 12.4, what are the states of U2A's *J* and *K* inputs with the switch at HALT, and what does this prepare U2A to do?

9.5 A logic analyzer has a 1000-word memory and samples its input lines every 200 ns via an internal clock. It is used on a 6802 system having a 0.8-MHz *E*-clock. How many 6802 machine cycles can be stored?

10.5 Which logic-analyzer display mode is most useful for spotting software errors? Which for timing problems? Which for hardware faults such as shorted lines?

11.6 If a logic analyzer's trigger section is set for binary address 0010 0110 XXXX XXXX, data $00, and R/$\overline{W}$ = 0, describe in words what will trigger it.

12.6 How should a logic analyzer's triggering be set up to wait for an interrupt request and display the events leading to it and following it?

Advanced Level

13.1 Why can't an ordinary pulse generator be used to perform the function of a logic pulser?

14.2 How can an improper signature at a specific point in a microcomputer system be analyzed to lead the troubleshooter to the source of the problem?

15.3 Describe how the static emulator of Figure 12.3 could be used to output data $AB to a latch at address $6000.

16.4 Return to Diagnostic Program 5 on page 280. Assume that the single-step circuit of Figure 12.4 and the data-bus display of Figure 12.5 have been added to the system, and list the addresses and data seen at each press of the STEP button for one pass through the loop. Consult Appendix B.

17.5 Analyze the program segment shown in Figure 12.8 to determine its function. Write comments to the source code.

18.5 Can a logic analyzer display waveform characteristics such as risetime, noise, and invalid logic levels? Explain how or why not.

19.6 Does the logic-analyzer trigger setup determine when data storage is to start or when it is to stop? Explain how data both preceding and following the trigger event can be captured.

20.6 Define the terms *data qualification* and *instruction prefetch*, giving the implications for troubleshooting with a logic analyzer.

PART III MICROCOMPUTER INTERFACING

13 Analog/Digital Conversions

13.1 D-TO-A CONVERSION METHODS

Section Overview

Digital-to-analog (D-to-A) converters produce an analog output voltage that is proportional to the numerical (usually straight-binary) value of a digital input signal. There are two basic techniques for achieving D-to-A conversion: current summing and ladder converters. However, D-to-A converters in integrated-circuit form are so universally employed that system users and designers need have little concern for the internal circuitry.

A few real-world quantities are inherently discrete or digital, but most are continuously variable or analog. Table 13.1 lists a selection of digital and analog quantities that might be subjects for measurement or control by microcomputer-based systems. Obviously, for a computer to deal with the analog quantities, they will have to be converted into digital form. As a first step toward analog-to-digital conversion and as a useful interfacing technique in its own right, we initially discuss digital-to-analog conversion.

Current summing offers the simplest approach to D-to-A conversion. The basic circuit for a 4-bit converter is given in Figure 13.1. Each binary bit, when in

TABLE 1.1 SELECTED DIGITAL AND ANALOG PARAMETERS

Digital
1. Number of spectators in a stadium.
2. Floor-stop command for an elevator.
3. Total amount of a shopping bill.
4. Number of items produced by an assembly line.
Analog
1. Length, distance, area, and volume.
2. Temperature
3. Air pressure, water pressure, oil pressure.
4. Weight
5. Time
6. Fuel consumption in miles per gallon or liters/100 km.
7. Light intensity.
8. Electrical voltage, current, and resistance.
9. Magnetic field strength.
10. Relative humidity of air.

a logic-1 state, contributes a current proportional to its binary place value in the summing resistor.

The scheme shown in Figure 13.1(a) is simple, but it requires that the output *low* voltages from the latch be exactly 0 V and that the output *high* voltages at each bit be stable and substantially equal. These conditions are not likely to prevail. The summing resistor must also be so low in value that the maximum voltage developed across it will be negligible compared to the latch output voltages. This condition can

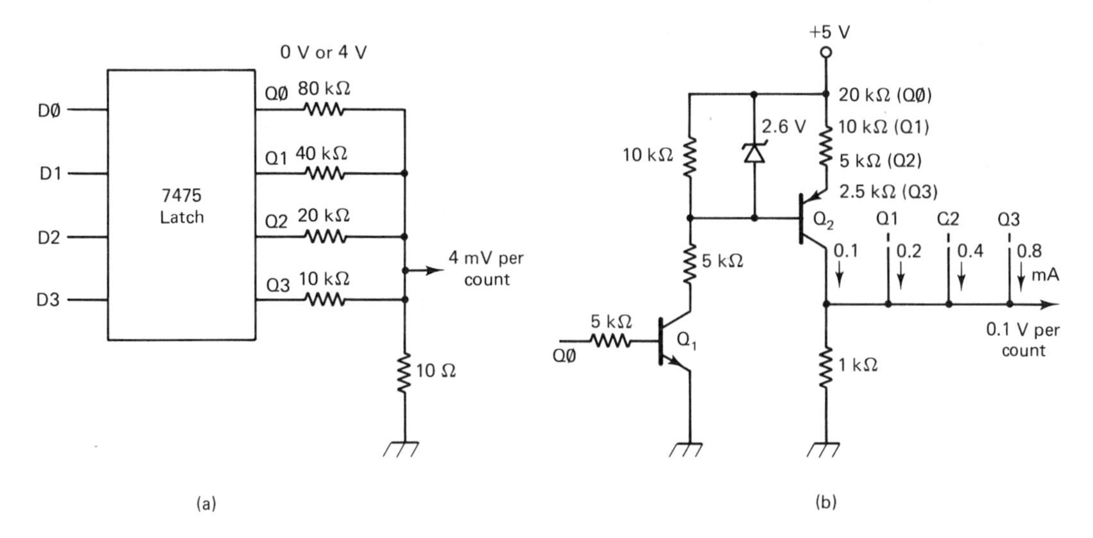

Figure 13.1 Current-summing D-to-A converters using (a) voltage sources and (b) current sources.

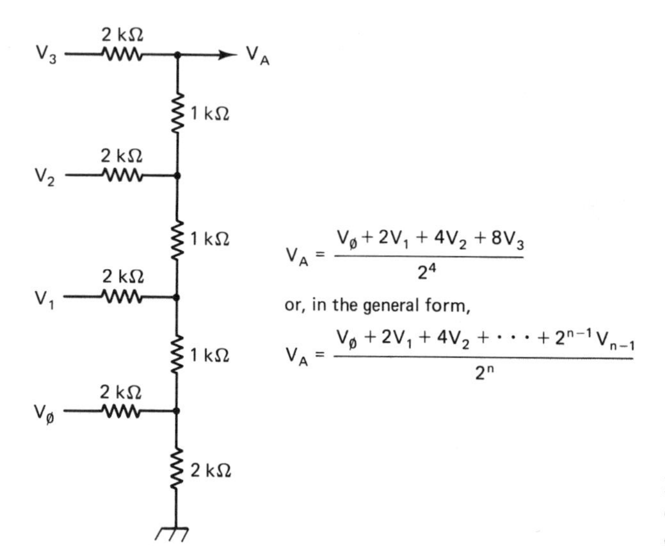

Figure 13.2 The ladder type D-to-A converter avoids the need for widely differing resistance values.

be met, but it will probably necessitate an op-amp to boost the analog output voltage to a usable level.

The circuit of Figure 13.1(b), shown for only one of the four digital outputs, permits larger output voltages and removes the requirement for stable latch output voltage by providing a current source Q_2, which is turned *on* by switch Q_1 when a latch output goes *high*.

The ladder circuit of Figure 13.2 provides D-to-A conversion without the widely differing resistor values of the current-summing circuit. It also requires that the latch output voltages ($V_\emptyset$ through V_3) be essentially equal and that the *low*-level digital outputs be essentially 0 V. Circuit-analysis buffs may find it an interesting exercise to prove that the output voltages are as predicted by the formula for each of the 15 nontrivial digital inputs. The ladder can, of course, be extended to 5, 8, or any number of digital inputs. The resistors can also be any values, provided that they are kept in the 2-to-1 ratios shown.

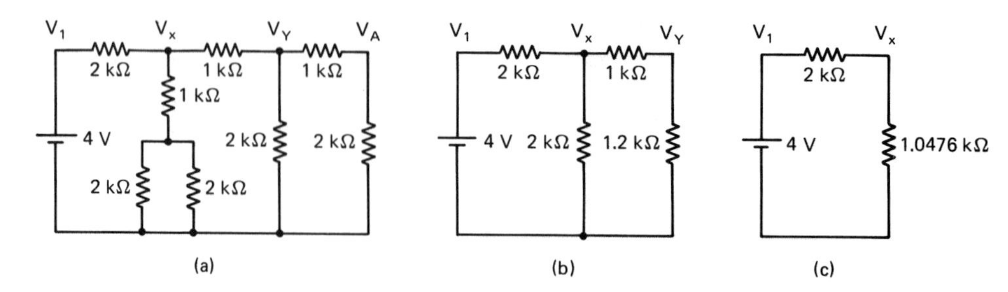

Figure 13.3 Solution for Example 8.1. Circuits (a), (b), and (c) are successive simplifications of the circuit of Figure 13.2.

Example 8.1

Calculate the output voltage of the ladder D-to-A converter with $V_1 = 4.0$ V and all other inputs equal to 0. Use the formula and verify the result by circuit analysis.

Solution

By the formula given with Figure 13.2,

$$V_A = \frac{2V_1}{2^4} = \frac{2(4)}{16} = 0.5 \text{ V}$$

By circuit analysis, referring to the redrawings of Figure 13.3:

$$\text{(c):} \quad V_X = 4\,\frac{1.0476}{3.0476} = 1.375 \text{ V}$$

$$\text{(b):} \quad V_Y = 1.375\,\frac{1.2}{2.2} = 0.750 \text{ V}$$

$$\text{(a):} \quad V_A = 0.750\,\frac{2}{3} = 0.50 \text{ V}$$

13.2 D-TO-A CONVERTER SPECIFICATIONS

Section Overview

Digital-to-analog converters are specified by their resolution (in bits), their accuracy, their linearity (deviation from ideal straight-line output), and their response speed.

Resolution: The D-to-A converters of Figure 13.1 and Figure 13.2 use 4 binary input digits and can produce 2^4, or 16, different analog output states. Such a converter has an inherent error possibility of about $\pm\frac{1}{32}$ of full scale, or about $\pm 3\%$. Figure 13.4 illustrates why this error is $\pm 1/32$ and not $\pm 1/16$, as you might at first expect.

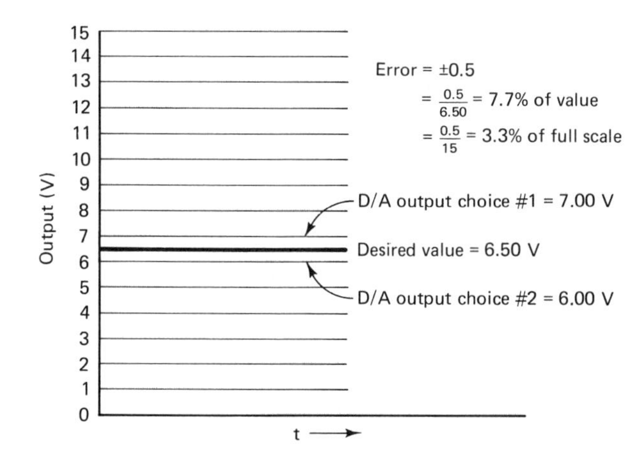

Figure 13.4 The D-to-A conversion dilemma. Accuracy is never better than ± ½ count.

Here is a table of D-to-A converter resolutions by bits, their maximum number of output states, and their inherent error liability, assuming perfectly matched resistors and perfect switches.

Bits	Maximum Count	Error Liability
4	15	±3.3%
5	31	±1.6%
6	63	±0.8%
7	127	±0.4%
8	255	±0.2%
9	511	±0.1%
10	1023	±0.05%
11	2047	±0.02%
12	4095	±0.01%

Linearity expresses the closeness of the *input count* versus *output voltage* graph to the ideal striaght-line graph. It is usually specified as *nonlinearity*, which is the greatest deviation of output voltage from the ideal, expressed as a percent of full-scale output.

Caution should be exercised in comparing linearity specifications from different manufacturers because there are three variants of the linearity specification. They are, from the most stringent to the most forgiving, *endpoint* linearity, *zero-based* linearity, and what is variously called *best-fit*, *independent*, or *floating* linearity. Figure 13.5 illustrates these three forms and the varying results that can be obtained from them. Endpoint linearity specifications in the range from 1% to 0.01% are typical.

Accuracy: Linearity depends upon resistor matching within the D-to-A converter circuitry and is not correctable or adjustable for IC devices. Accuracy is usually specified as *full-scale error*, and is, for a moment at least, adjustable by external resistances to zero. For mass-produced instruments with no provision for adjustment,

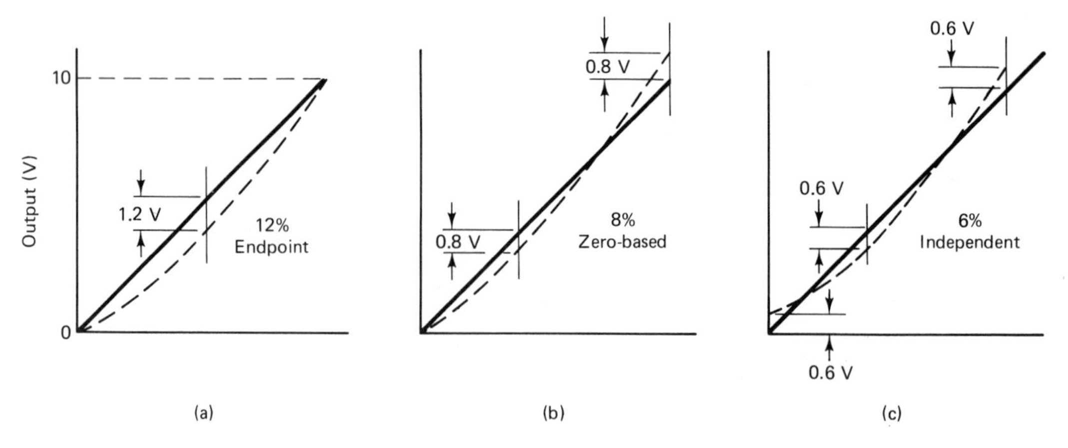

Figure 13.5 (a) Endpoint linearity, (b) zero-based linearity, and (c) independent linearity.

the *full-scale*-error specification is most important, but where a calibration adjustment is provided, the temperature coefficient of full-scale error is more relevant.

Settling time is the time required for the output voltage to stabilize to within a range represented by one count, after the application of the digital input. Times of 1 μs and less are typical.

Monotonicity is a guarantee that increasing input count will in no case produce a decreasing output voltage level. A small error in the resistors fed by the more-significant bits could overwhelm the effects of inputs from the less-significant bits, resulting in a drop in voltage between the counts.

0111 1111 and 1000 000

It is possible to have a D-to-A converter with 10-bit resolution but 9-bit monotonicity.

REVIEW OF SECTIONS 13.1 AND 13.2

Answers appear at the end of Chapter 13.

1. Is compass bearing (direction) an inherently digital or inherently analog quantity?
2. Calculate the output for the circuit of Figure 13.1(a) if $Q_0 = +4.0$ V, $Q_1 = 0$, $Q_2 = +4.0$ V, and $Q_3 = +4.0$ V.
3. Use circuit analysis to calculate V_A for the circuit of Figure 13.2 if $V_3 = 4.0$ V and $V_2 = V_1 = V_0 = 0$. (No need to cheat by using the formula. This one is easy.)
4. What justifies the complexity of the current-source circuitry of Figure 13.1(b)?
5. A D-to-A conversion with a maximum error of $\pm 0.5\%$ is required. What is the fewest number of bits that the converter may have?
6. A D-to-A converter must produce 500 distinct output levels. How many bits resolution must it have?
7. A manufacturer of D-to-A converters has been specifying a best-fit linearity of 0.25%. Would the endpoint linearity figure be *less than*, essentially *equal to*, or *greater than* 0.25%?
8. A D-to-A converter has full-scale error essentially equal to zero by adjustment and endpoint linearity of 1%, with the greatest deviation at midscale. What is the worst-case error as a percentage of reading?
9. In Figure 13.1, what percent decrease in the 10-kΩ resistor would produce an output voltage change equal to that caused by switching Q_0 from 0 to +4 V?
10. In an 8-bit circuit similar to Figure 13.1, what percent deviation of the resistor for the most-significant bit would destroy the monotonicity of the converter?

13.3 A-TO-D CONVERSION METHODS

Section Overview

Analog-to-digital (A-to-D) conversion can be accomplished by using a microcomputer to feed a D-to-A converter. The program for the microcomputer can be developed in several ways.

1. **Countup of a D-to-A converter until its output matches the analog input. This technique is relatively slow, taking perhaps a few milliseconds for an 8-bit conversion.**
2. **Successive approximation, establishing the proper digital input to a D-to-A converter bit by bit. This is similar to the first method, but it is faster, requiring typically 100 μs for an 8-bit conversion.**
3. **Continuous conversion has the digital value count up or down to follow a changing analog input. It is quick to follow a slowly changing analog input but slow at finding the right digital value when switched to a new waveform.**

A count-up A-to-D converter driven by a microprocessor is shown in Figure 13.6. It consists of an 8-bit D-to-A converter whose output is counted up on a 256-step staircase. This output is compared to an analog input signal. After each increasing count is output to the D-to-A converter, the comparator output is checked. The count which makes the D-to-A output voltage exceed the analog input voltage is stored as the digital representation of the analog signal.

Figure 13.7 gives a flowchart of the A/D count-up-conversion routine. The program listing follows (List 13.1).

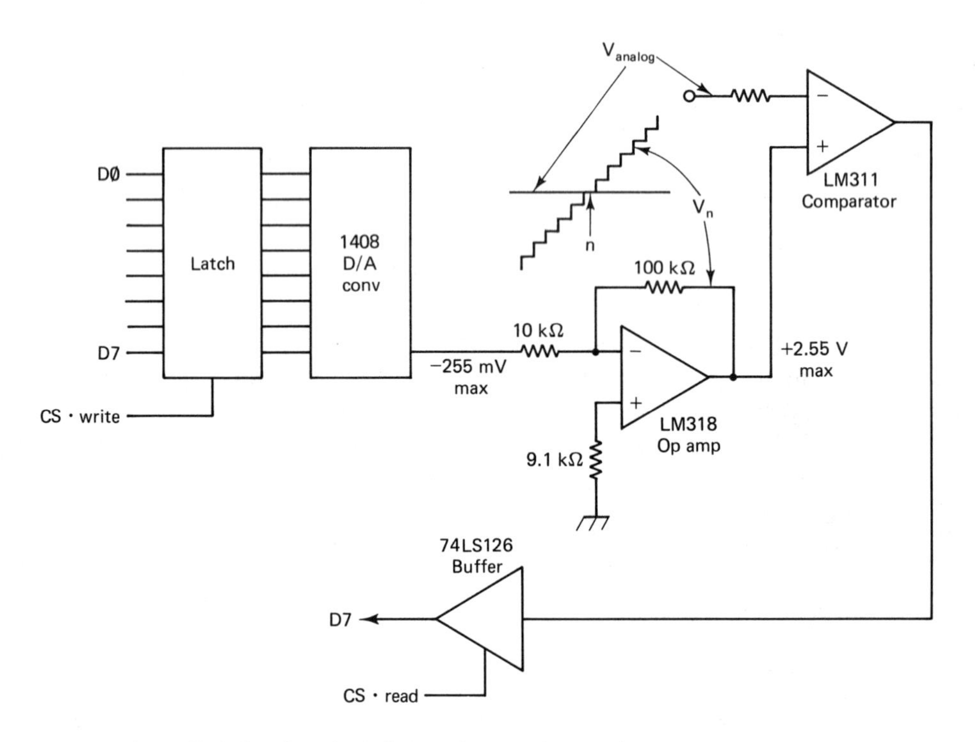

Figure 13.6 Interface circuit for A-to-D conversion. A microprocessor drives a D-to-A converter until its output matches the analog input.

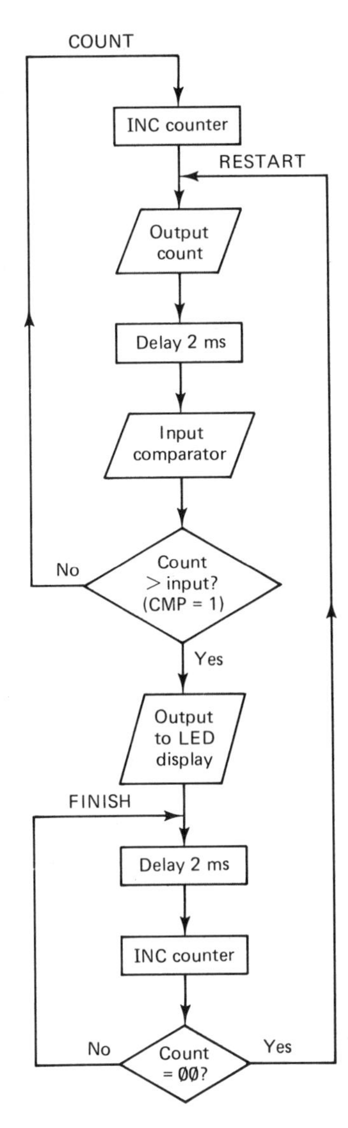

Figure 13.7 Flowchart for a *countup* A-to-D conversion routine using the hardware of Figure 13.6.

```
                1000 * COUNTUP A-TO-D CONVERSION.  LIST 13-1
                1005 * SAMPLE PROGRAM; NOT RUN OR TESTED. JULY, '87.
                1010 * SEE HARDWARE OF FIG. 13-6.     D. METZGER.
                1020 *
                1030          .OR  $FF81
                1040          .TA  $4081
9000-           1050 LATCH    .EQ  $9000
A000-           1060 COMPAR   .EQ  $A000
8000-           1070 DISPLY   .EQ  $8000
                1080 *
FF81- 86 FF     1090 CONVRT  LDAA #$FF     START AT BOTTOM OF
FF83- 4C        1100 COUNT   INCA           RAMP (A = 0).
FF84- B7 90 00  1110         STAA LATCH    OUTPUT TO D/A CONVERTER.
```

```
FF87- C6 80     1120           LDAB #$80    DELAY APPROX 2 MS
FF89- 5A        1130 WAIT      DECB          TO GIVE ANALOG OUTPUT
FF8A- 26 FD     1140           BNE  WAIT     AND COMPARATOR TIME TO ACT.
FF8C- F6 A0 00  1150           LDAB COMPAR  COMPARATOR GIVES "1"AT BIT 7
FF8F- 2A F2     1160           BPL  COUNT    WHEN COUNT = ANALOG INPUT.
FF91- B7 80 00  1170           STAA DISPLY  DISPLAY DIGITAL COUNT.
FF94- C6 80     1180 LOOP      LDAB #$80    KEEP COUNT AND DELAY
FF96- 5A        1190 WAIT2     DECB          GOING AFTER COMPARATOR = 1
FF97- 26 FD     1200           BNE  WAIT2    TO KEEP ALL RAMPS
FF99- 4C        1210           INCA          EQUAL.
FF9A- 26 F8     1220           BNE  LOOP    DO ANOTHER LOOP
FF9C- 27 E5     1230           BEQ  COUNT    UNTIL RAMP TOPS OUT
                1240           .EN
```

The successive-approximation conversion scheme is illustrated and flowcharted in Figure 13.8. It uses the identical hardware of the count-up scheme (Figure 13.6), but the software makes it much faster. The technique is to set all bits to 0 except the most significant, which is set to 1. The comparator is then checked to see if the D-to-A output is too high. If it is, the bit is changed back to 0. Then the next-most-significant bit is set or cleared in the same way.

The number of program loops required to complete a conversion is equal to the number of bits of resolution, as compared with the count-up scheme, in which the number of program loops required equals the number of counts of resolution. List 13.2 is a 6802 program listing for successive-approximation A-to-D conversion.

Continuous A-to-D conversion achieves speed by starting its count from the last successful conversion and counting up or down as required to match the new analog value. This speed advantage is lost when the converter meets a new analog voltage or one that has changed abruptly and radically. Figure 13.8(c) shows the continuous conversion scheme. Again, the hardware is the same, as given in Figure 13.6.

```
                1000 * LIST 13-2   SUCCESSIVE APPROXIMATION A/D CONVERTER
                1010 *                FIGURE 13-6 HARDWARE.
                1015 * EXAMPLE PROGRAM - UNTESTED.  10 JULY '87. METZGER.
                1020 *
                1030            .OR   $FF81
                1040            .TA   $4081
0010-           1050 MASK       .EQ   $0010    RAM HOLDS CURRENT BIT TO BE DECIDED
0011-           1060 OLDCNT     .EQ   $0011    TO RETRIEVE OLD IF NEW VALUE TOO HI.
4000-           1070 CMPIN      .EQ   $4000    INPUT 1 BIT FROM COMPARATOR.
6000-           1080 DOUT       .EQ   $6000    LATCH 8 BITS OUT.
                1090 *
FF81- 4F        1100 START      CLRA           ACCUM "A" IS COUNT TO OUTPUT.
FF82- C6 80     1110            LDAB #$80      MASK HOLDS BIT IN QUESTION.
FF84- D7 10     1120            STAB MASK       START WITH M.S. BIT.
FF86- 97 11     1130 NEXBIT     STAA OLDCNT    SAVE OLD COUNT.
FF88- 9A 10     1140            ORAA MASK      SET A BIT = 1 FOR NEW COUNT.
FF8A- B7 60 00  1150            STAA DOUT      TRIAL OUTPUT TO D/A CONV.
FF8D- F6 40 00  1160            LDAB CMPIN     COMPARATOR STILL SHOW COUNT
FF90- 2A 02     1170            BPL  LOW        TOO LOW?
FF92- 96 11     1180            LDAA OLDCNT    IF TOO HIGH, RETRIEVE OLDCNT.
FF94- 74 00 10  1190 LOW        LSR  MASK      MOVE "1" TO NEXT L.S. BIT.
FF97- 7D 00 10  1200            TST  MASK      IF 8 BITS HAVE BEEN TESTED
FF9A- 26 EA     1210            BNE  NEXBIT     MASK WILL = 0.
FF9C- 20 E3     1220            BRA  START      AND YOU'RE DONE.
                1230            .EN
```

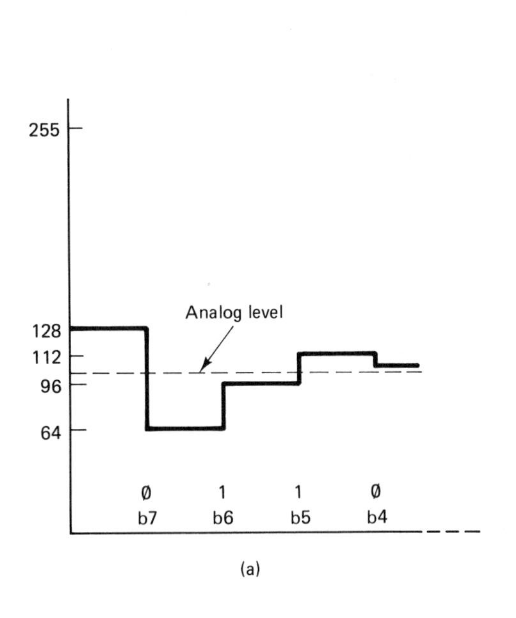

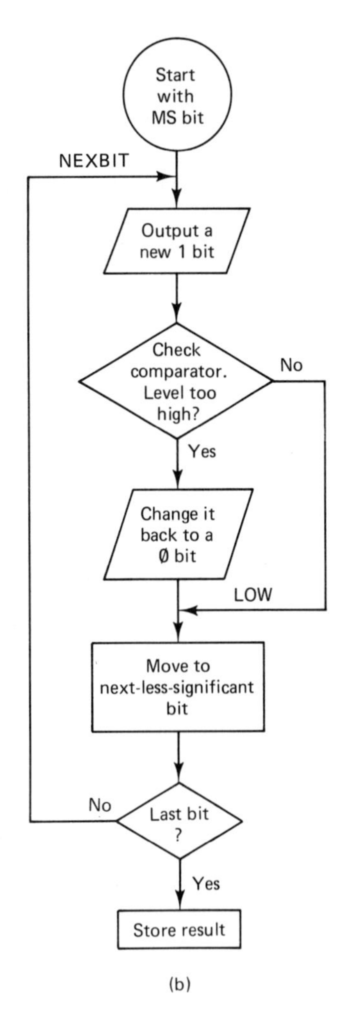

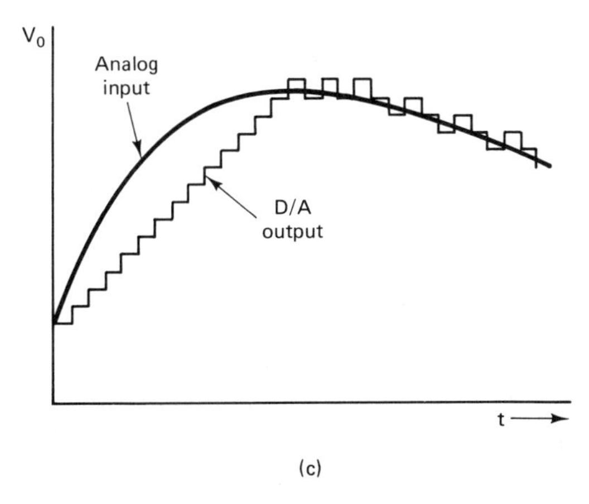

Figure 13.8 (a) Successive approximation completes an 8-bit A/D conversion in 8 passes through the program loop. (b) A program flowchart for successive-approximation conversion. (c) Continuous A-to-D conversion.

13.4 STAND-ALONE A-TO-D CONVERTERS

Section Overview

Processor-driven A-to-D converters are relatively slow, and they tie up the processor with peripheral tasks. There are several other options, which require little attention from the processor.

1. **Processor-independent A-to-D converters use a counter or successive-approximation scheme, but the counter is integrated on the A-to-D chip.**
2. **Flash converters use comparators and logic gates instead of counter schemes and are extremely fast.**
3. **Dual-slope converters use a clock counter to compare two time intervals, one proportional to the input voltage and the other proportional to a reference voltage.**

Processor-independent A-to-D converters are available in IC form for only a few dollars per unit. They contain the counting and comparator circuitry as well as the D-to-A converter, thus relieving the processor and the programmer of the count-and-check task. They usually employ the successive-approximation conversion scheme and have a clock input and an output that interrupts the processor when a conversion is completed. They are commonly available with resolutions from 8 to 12 bits and conversion times on the order of 100 μs. A typical low-cost stand-alone A-to-D converter is the ADC 0804, shown in Figure 13.9. This IC converts a 0- to 5-V analog input to an 8-bit binary output with a conversion time of 100 μs and an error of less than 0.5%. It operates from a single +5-V supply, and requires a 2.50-V reference source, which may be supplied by the LM336 IC regulator. The ADC 0804 generates an $\overline{IRQ}$ signal for the processor as each conversion is done. The eight digital outputs remain in a high-impedance state until a read signal ($\overline{RD}$) is applied to pin 2. Figure 13.9 shows a stand-alone test circuit for the ADC 0804. It converts an analog voltage level to an 8-bit binary number, and displays it on 8 discrete LEDs.

Flash converters (also called asynchronous converters and parallel converters) use a large number of comparators, combining their outputs through logic gates to produce binary outputs corresponding to digital inputs. The number of comparators required is one less than the number of digital output counts, so an 8-bit flash converter requires 255 comparators. This is not beyond the capability of present-day IC technology, but such circuit complexity does place flash converters in a more expensive bracket than the previously described clocked converters. They are currently available with 6-bit to 8-bit resolution and conversion response times of 100 ns, so they can be used to digitize analog signals up to several megahertz. Because of their speed they are much more likely to respond to noise spikes by reporting incorrect digital values than other slower types of A-to-D converters. They require no clock input.

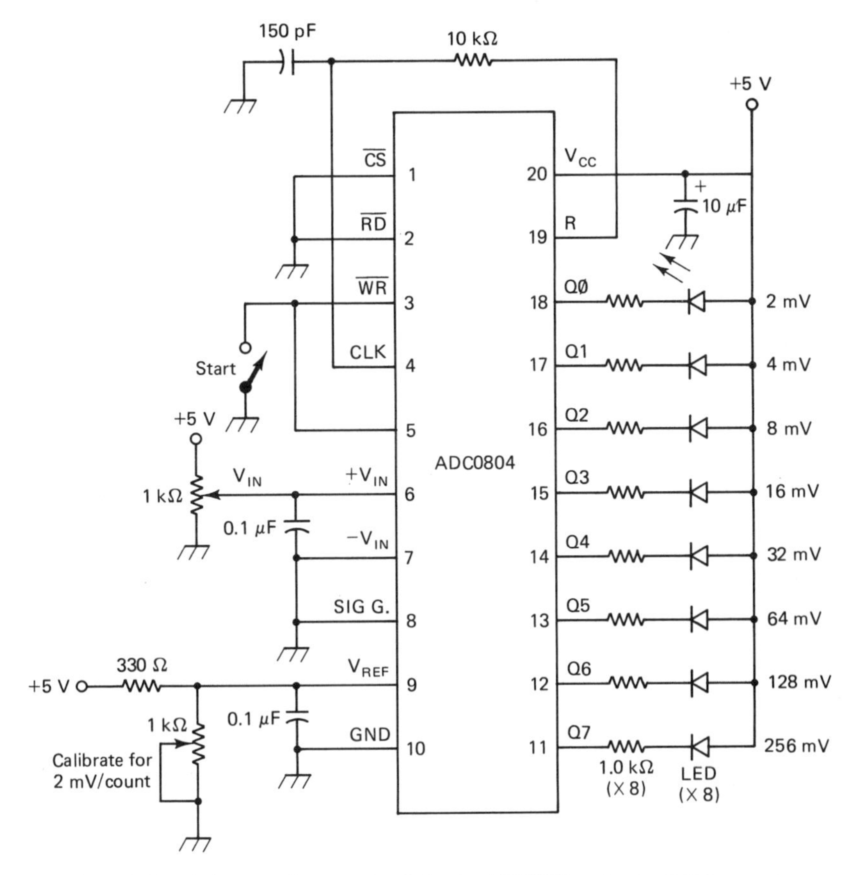

Figure 13.9 A free-standing test circuit for the ADC0804 analog-to-digital converter.

Figure 13.10 shows a simplified A-to-D flash converter with its truth table and logic expressions.

The dual-slope A-to-D converter of Figure 13.11 is the basis of most commercial digital voltmeters. It can be made highly accurate, since its basic function is to measure the ratio between the input voltage and an internal reference voltage. The tolerance and stability of all the resistors, capacitors, and semiconductors in the circuit do not affect the final accuracy of the conversion—only the accuracy of the reference supply is important.

At the start of the conversion cycle the flip-flop is reset, so Q_2 is turned *on* by the *low* level at its gate, whereas Q_1 is turned *off* by +5 V at its gate. The integrator thus generates a negative ramp at a rate of 5 V/s for a +0.5-V input. As soon as this ramp crosses zero, the comparator output goes positive, enabling the AND gate and starting the counter. Let us assume that the counter is a three-stage BCD type counting to 999. Also, let the clock rate be 5 kHz. The integrator will thus reach −1 V output in the 0.2 s it takes for the counter to reach maximum count.

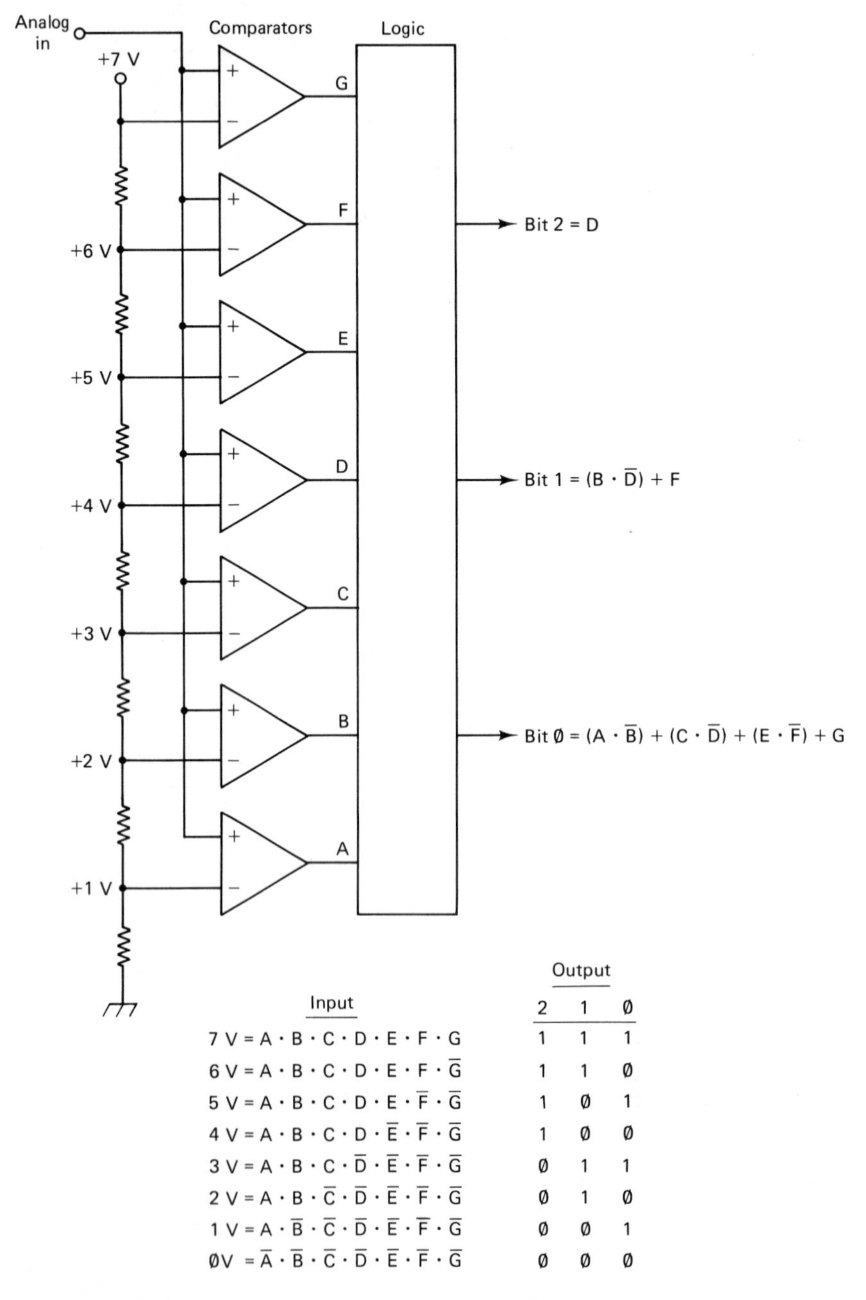

Input	Output 2	Output 1	Output Ø
7 V = A · B · C · D · E · F · G	1	1	1
6 V = A · B · C · D · E · F · $\overline{G}$	1	1	Ø
5 V = A · B · C · D · E · $\overline{F}$ · $\overline{G}$	1	Ø	1
4 V = A · B · C · D · $\overline{E}$ · $\overline{F}$ · $\overline{G}$	1	Ø	Ø
3 V = A · B · C · $\overline{D}$ · $\overline{E}$ · $\overline{F}$ · $\overline{G}$	Ø	1	1
2 V = A · B · $\overline{C}$ · $\overline{D}$ · $\overline{E}$ · $\overline{F}$ · $\overline{G}$	Ø	1	Ø
1 V = A · $\overline{B}$ · $\overline{C}$ · $\overline{D}$ · $\overline{E}$ · $\overline{F}$ · $\overline{G}$	Ø	Ø	1
ØV = $\overline{A}$ · $\overline{B}$ · $\overline{C}$ · $\overline{D}$ · $\overline{E}$ · $\overline{F}$ · $\overline{G}$	Ø	Ø	Ø

Figure 13.10 A flash converter of n bits recognizes 2^n input levels and requires $2^n - 1$ comparators plus decoding logic.

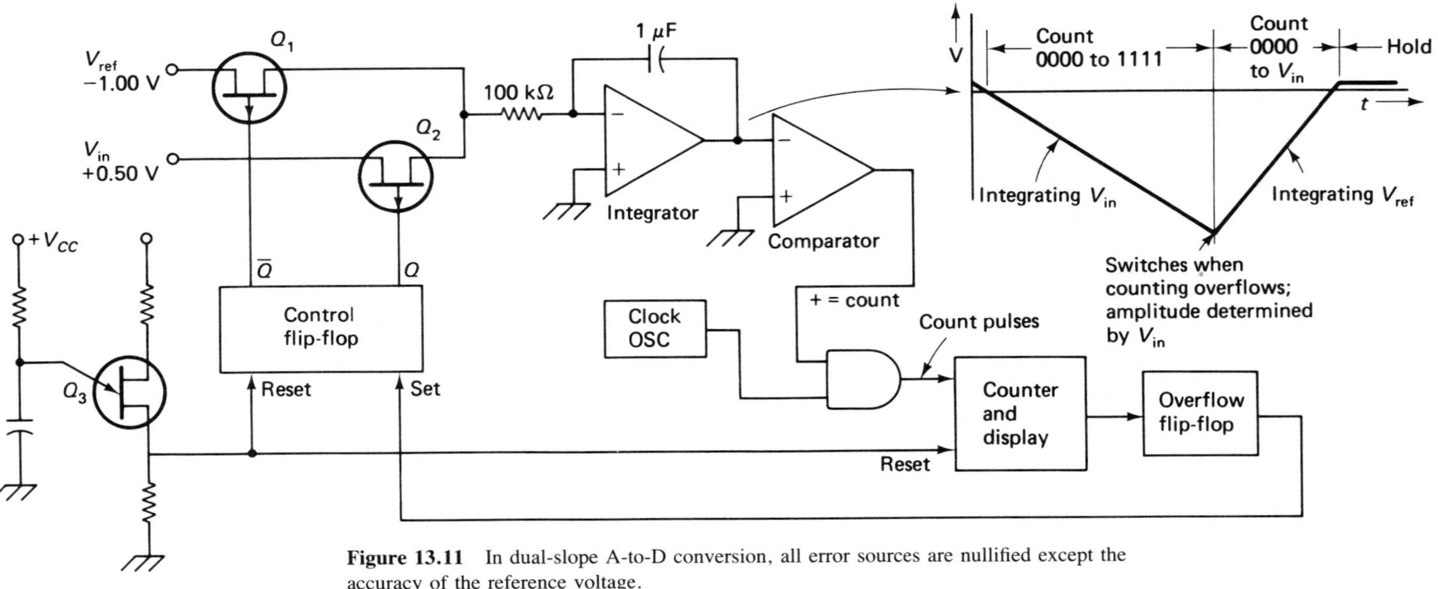

Figure 13.11 In dual-slope A-to-D conversion, all error sources are nullified except the accuracy of the reference voltage.

The highest-order bit of the counter feeds an overflow flip-flop, which is *set* as the counter advances from 999 to 000. This sets the control flip-flop, disconnecting V_{in} and connecting V_{ref} to the integrator via Q_1.

Note here that the time required for the count from 000 through 999 and back to 000 depends upon the clock frequency and is relatively constant. However, the slope of the integrator output, and hence the voltage reached as the counter rolls back to 000, depends directly upon V_{in}.

Now the counter begins counting again up from zero, but this time the integrator output is rising at a rate of 10 V/s, as determined by V_{ref}. The run back to zero from −1 V will take 0.1 s, and at 5 kHz the counter will read 500 at the zero crossing. The counter will stop at this point because a positive integrator output will set the comparator output negative, disabling the AND gate to the counter. This value will be displayed until the UJT Q_3 fires (after a second or so), resetting the control flip-flop and starting a new cycle.

If the integrator's resistor or capacitor should be a little low in value, the negative ramp would be a little steeper, and the peak voltage would be a little higher, but this would be of no consequence because the positive ramp would be a little steeper by the same percentage. Likewise, if the oscillator should run a little slow, the time to reach the negative peak would be a little longer, but the time to return to zero would be longer by the same percentage, and the final count would be unaffected.

REVIEW OF SECTIONS 13.3 AND 13.4

Answers appear at the end of Chapter 13.

11. If a microcomputer program for 6-bit A-to-D conversion takes 1 ms per conversion using the count-up technique, how long would 8-bit conversions take by the same technique?

12. If 6-bit conversion by successive approximation takes 1 ms, how long will 8-bit conversion take by the same technique?

13. A microcomputer controlled a single A-to-D converter and was able to obtain a stable digital value in about 1 ms, leaving the processor ample time for its other tasks. The system was changed to have the converter switch between two analog inputs, and now each conversion seems to be taking about 100 ms, monopolizing the processor's time. The conversion scheme used apparently was ________.

14. Which type of A-to-D conversion is *not* based on finding the correct count to input to a D-to-A converter?

15. Assuming a 1-MHz clock, approximately how long will the successive-approximation A-to-D conversion routine listed on page 312 take per conversion: 100 μs, 300 μs, 1 ms, or 3 ms?

16. Dual-slope A-to-D conversion removes all the variables from conversion accuracy except ________.

17. How many comparators would be required in a 10-bit flash converter?

18. Which type of A-to-D converter is most susceptible to noise?

19. In Figure 13.6, why does the input buffer feed data line D7, rather than one of the other data lines?
20. Which type of A-to-D converter is normally used in digital voltmeters?

13.5 DIGITAL RECORDING OF ANALOG DATA

Section Overview

Digital recording of audio material offers superb durability, but a single popular song can require several megabytes of storage space. Companding is a technique to widen the range of analog voltage levels representable by a given number of digital states and thus improve the dynamic range of the recording.

Analog data, representing voice, music, video images, and telemetered data on temperature, pressure, position, and other machine functions, is increasingly being converted to digital form for transmission and recording. The advantage is that digital data does not deteriorate. Plastic records may become scratchy, and cassette tapes may exhibit nonlinearity or noise, but the playback from a compact disk will always be just as good as the first time because the data is recorded on it digitally and then reconverted into sound by a D-to-A converter.

The amount of memory required to store analog data digitally can be large enough to make it a problem. Let us consider the example of a 300-s-long musical recording with the moderately high frequency response of 10 kHz.

Sampling rate: If an analog signal of 10 kHz is sampled at a 10-kHz rate, no output at 10 kHz results. This is illustrated in Figure 13.12(a). A sampling rate of 20 kHz will result, on the average, in a 10-kHz output that is -3dB down from lower-frequency outputs. A signal frequency one-half the sampling rate is called the Nyquist frequency. A comparison of Figure 13-12(b) and (c) shows how the 10-kHz information could be lost even with a sample rate of 20 kHz, depending on phasing, so in practice the sampling rate must be higher than twice the highest signal frequency.

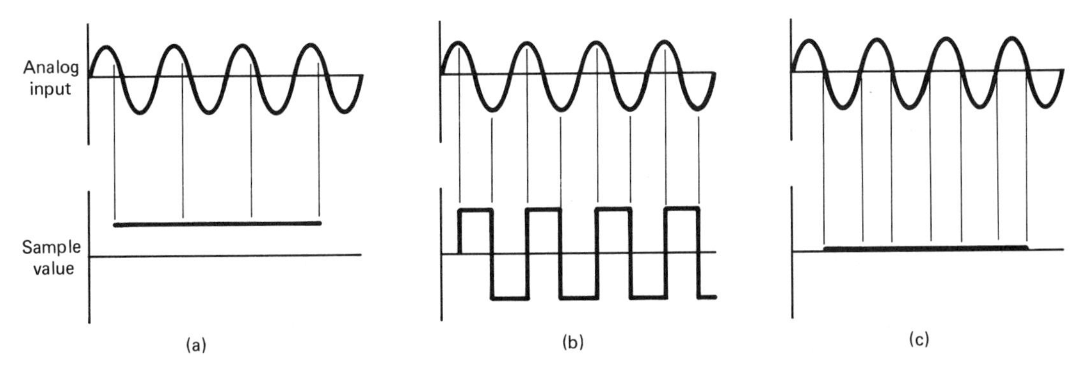

Figure 13.12 Sampling rates of f or even $2f$ are inadequate for reliable digital recording of analog data.

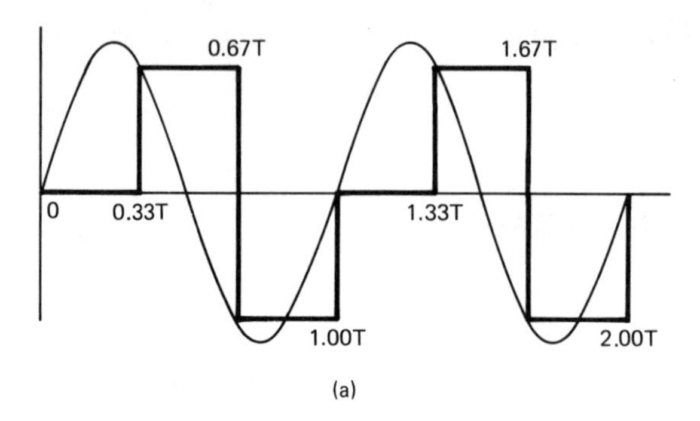

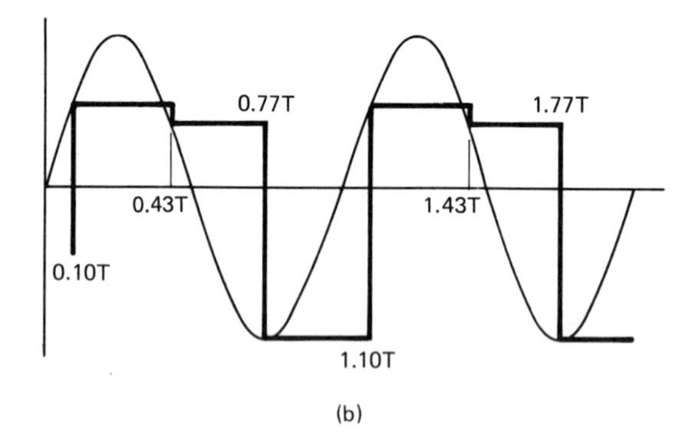

Figure 13.13 Two phasings of sampling a sine wave of frequency f at a sample rate of $3f$. After filtering, the output at f is relatively unattenuated at any phasing.

Figure 13.13 shows that a sampling rate of 30 kHz guarantees a substantial output for a 10-kHz input, regardless of the phasing of the sampling and input. For a 300-s recording, 30 000 samples per second requires

$$(300 \text{ s})\ (30\ 000/\text{s}) = 9 \times 10^6$$

or 9 million words of data storage.

Aliasing: If signals higher in frequency than half the sampling rate are allowed into the ADC circuit a spurious output may result at a low frequency which is the difference between the signal frequency and the sampling rate. This undesired signal is called an alias, and is represented by the heavy line in Figure 13.14. Aliasing can be prevented by passing the analog signal to be digitized through a low-pass filter with a cutoff frequency below one-half the sampling frequency before sending it to the converter.

Dynamic range: An audio level of 20 dB makes you strain to hear it. An audio level of 80 dB is approaching the uncomfortably loud. Although most popular music keeps a pretty consistent level of loudness, classical music often utilizes the 60-dB range between *pianissimo* and *double forte* to achieve its effects. Sixty decibels is a voltage ratio of 1000 to 1. A recording medium must therefore be capable of producing, for example, levels of 10 mV without getting lost in the noise and then levels of 10 V that do not saturate the capabilities of the system.

If we reproduce the 10-mV audio levels with 16 discrete steps, as shown in Figure 13.15(a), the 10-V levels will require 16 000 steps. This places a requirement

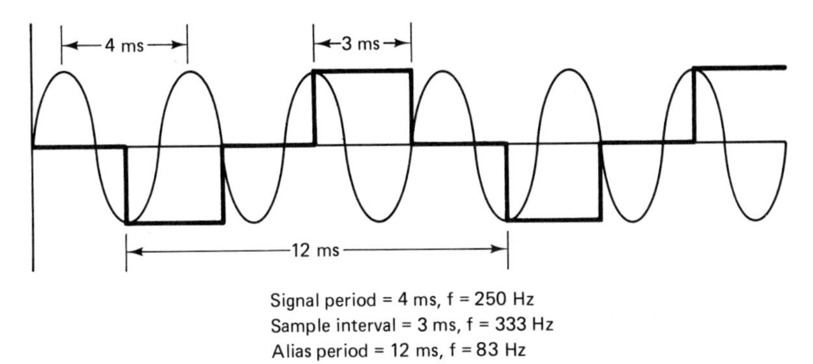

Figure 13.14 Signals higher than half the sampling frequency must be filtered out before being applied to an ADC to avoid the production of a false signal called an *alias*.

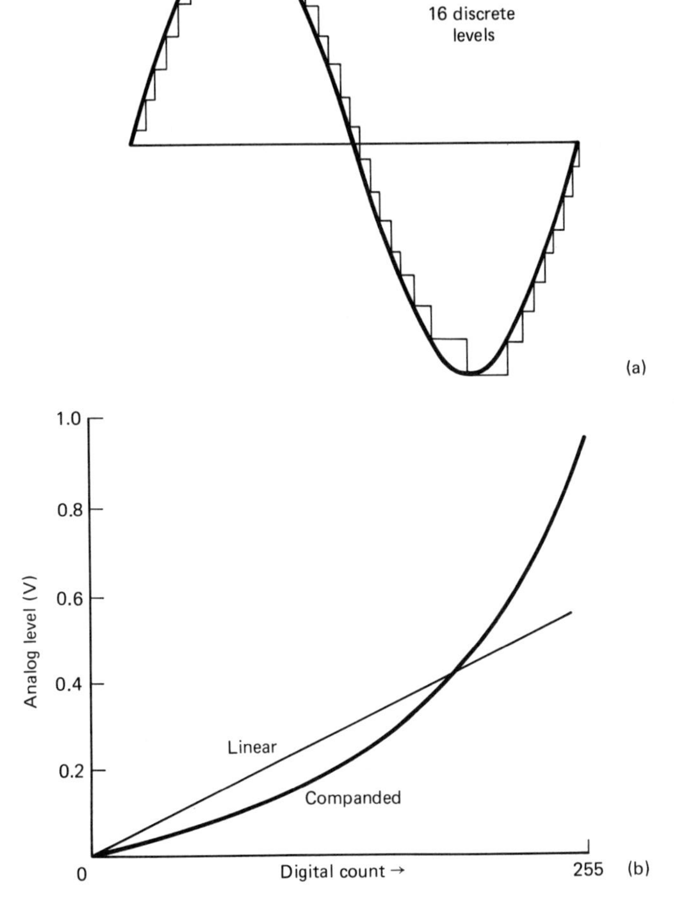

Figure 13.15 (a) A rough approximation of a sine wave can be made from 16 discrete levels, requiring 4 bits per sample. (b) Companding increases the weight of each digital count at higher amplitudes to permit a larger analog range to be stored in a given digital word size.

of 14-bit word length on the digital recording system, since $2^{14} = 16\,384$. The total number of bits required to store a 300-s musical passage with a 10-kHz bandwidth and a 60-dB dynamic range is then

$$(9 \times 10^6 \text{ words})(14 \text{ bits/word}) = 126 \text{ Mbits}$$

This is about 16 megabytes.

Companding is a technique for providing large dynamic ranges with fewer bits per word. The word companding, of course, is a hybrid of *compressing* and *expanding*. In a companded system a digital count at the high-amplitude range produces relatively more output change than an equal count at the low end. Figure 13.15 illustrates linear versus companding systems. With 30 dB of companding, the 60-dB dynamic range of our previous example could be achieved with about half of the memory previously required, because each word would need only about half as many bits to cover the same analog range.

13.6 LINEARIZATION BY LOOKUP TABLE

Section Overview

Analog transducers often produce outputs that do not vary linearly with the measured parameter. However, as long as each analog output does correspond to a unique value of the measured parameter, the analog signal can be converted to a digital value, and that digital value can be part of an address, directing the microcomputer to data that displays the corresponding parameter value. This table lookup of the displayed value completely bypasses the problem of transducer nonlinearity.

A thermistor is a convenient and inexpensive device for measuring temperature, but its resistance change with temperature is markedly nonlinear, as shown in Figure 13.16(a). When used as part of an analog meter, such nonlinearity usually results in a meter scale that is compressed at one end.

A microprocessor-based instrument can use a nonlinear transducer and effectively linearize it by a software device called a lookup table. Figure 13.16(b) shows that each analog voltage across the thermistor corresponds to a specific binary count from an A-to-D converter. These counts can be used to direct a microprocessor to a specific word of data in a table stored in ROM. Each word in the table contains the code bits necessary to display the temperature corresponding to the original analog voltage received.

The hardware for a 6802-based thermometer using an independent A-to-D converter and a thermistor input is given in Figure 13.17. Figure 13.18 shows the program flowchart, and the program listing follows (List 13.3). Only a few elements of the lookup table are given because the table is extensive and the contents depend on the thermistor characteristics.

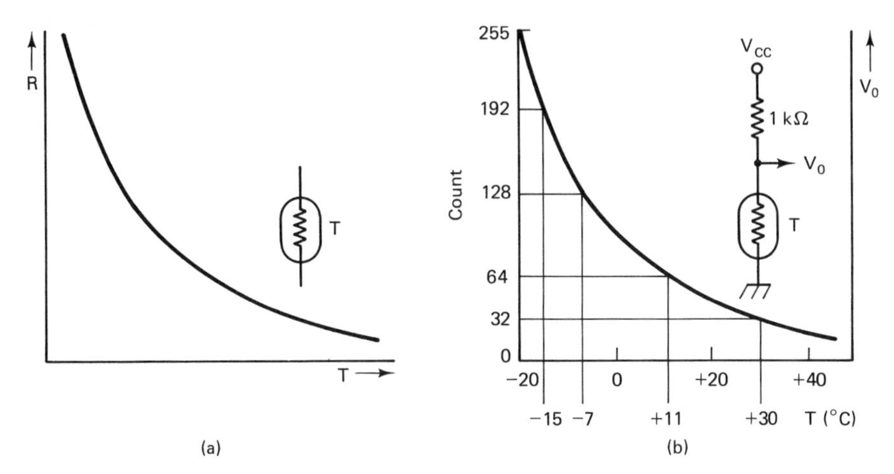

Figure 13.16 (a) A thermistor's resistance varies nonlinearly with temperature, but (b) each output voltage from the voltage divider does correspond to a unique temperature.

```
                    1000 * LIST 13-3 * THERMISTOR LINEARIZER * FOR FIG. 13-17 HDW.
                    1010 * TESTED AND RUNNING, 10 JULY '87, BY D. METZGER.
                    1020 *
                    1030        .OR   $FF00
                    1040        .TA   $4100
0000-               1050 XBUF   .EQ   $0000    INDIRECT ADDRESS FOR X IS $0000, $0001.
8000-               1060 ADC    .EQ   $8000    ADR OF ADC 0804 A-TO-D CONVERTER.
9000-               1070 DISPLY .EQ   $9000    ADR OF LATCH FEEDING 8 LEDS; BCD OUTPUT
                    1080 *
FF00- 86 FE         1090        LDAA  #$FE     ESTABLISH START OF TABLE AT $FE00.
FF02- 97 00         1100        STAA  XBUF      TABLE CONTAINS BCD BYTES TO DISPLAY.
FF04- B6 80 00      1110 CONVRT LDAA  ADC      GET COUNT OF ANALOG INPUT.
FF07- 97 01         1120        STAA  XBUF+1   PUT IN LOW BYTE OF X BUFFER.
FF09- DE 00         1130        LDX   XBUF     POINT TO BYTE IN TABLE.
FF0B- A6 00         1140        LDAA  0,X      PICK UP BYTE AND
FF0D- B7 90 00      1150        STAA  DISPLY   SEND TO LATCH.
FF10- 09            1160 WAIT   DEX            WAIT A SECOND BEFORE
FF11- 26 FD         1170        BNE   WAIT      NEXT CONVERSION.
FF13- 20 EF         1180        BRA   CONVRT   DO IT AGAIN.
                    1190 *
                    1200        .OR   $FE00    START OF TABLE.
                    1210        .TA   $4000    HOST RAM STORAGE AREA.
FE00- 99 98 97
FE03- 96 95 94
FE06- 93 92 91
FE09- 90 89 88
FE0C- 87 86 85
FE0F- 84 83 82
FE12- 81 80         1220        .HS   99989796959493929190898887868584838281 80
FE14- 79 78 77
FE17- 76 75 74
FE1A- 73 72 71
FE1D- 70 69 68
FE20- 67 66 65
FE23- 64 63 62
FE26- 61 60         1230        .HS   79787776757473727170696867666564636261 60
                    1240 * LINEAR TABLE FROM 99 TO 60 STORED FOR TEST.
                    1250        .OR   $FFFE
                    1260        .TA   $41FE
FFFE- FF 00         1270        .HS   FF00     RESET VECTOR
                    1280        .EN
```

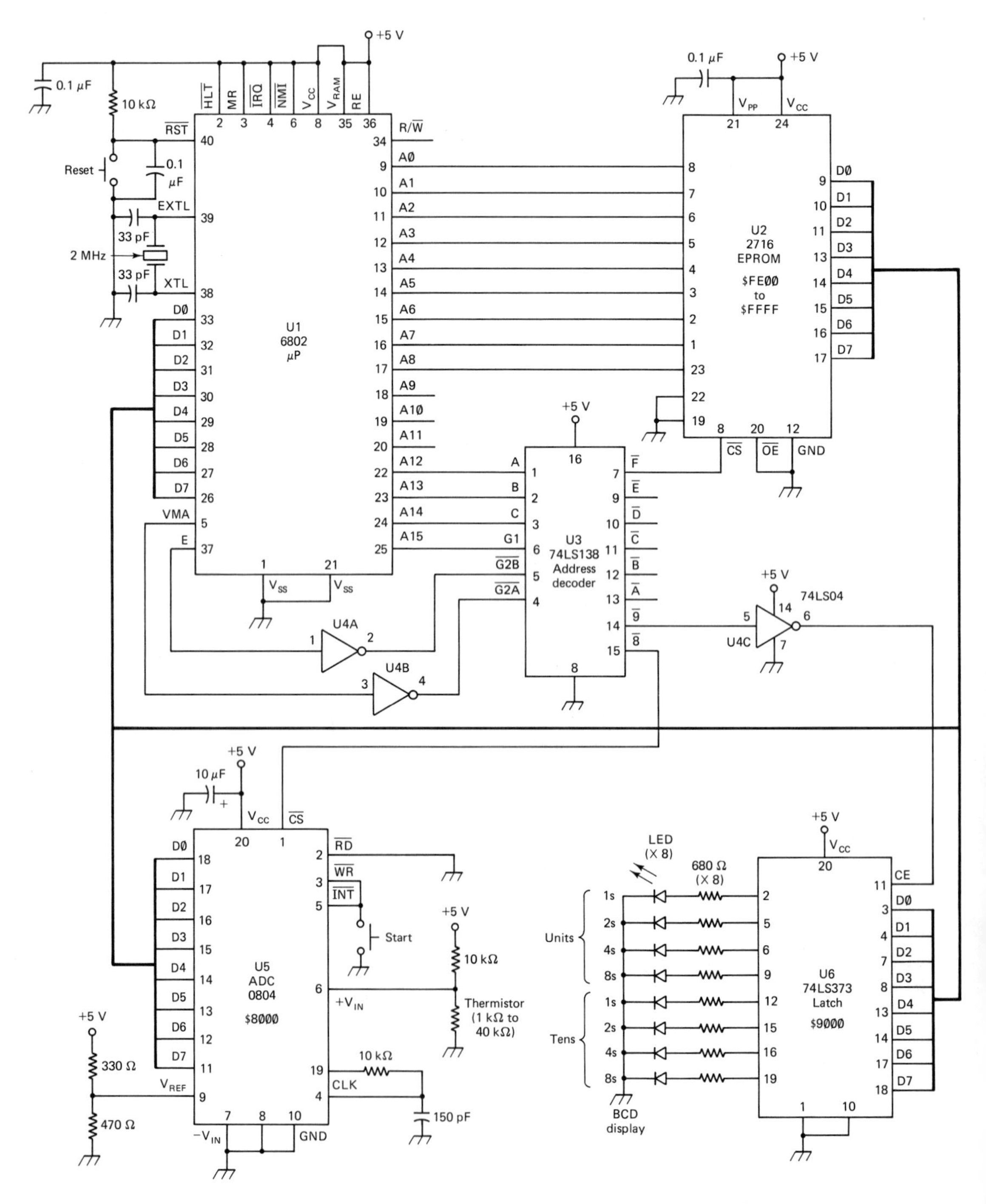

Figure 13.17 A 6802-based system for linearizing a thermistor to display temperature from an LED–segment data table.

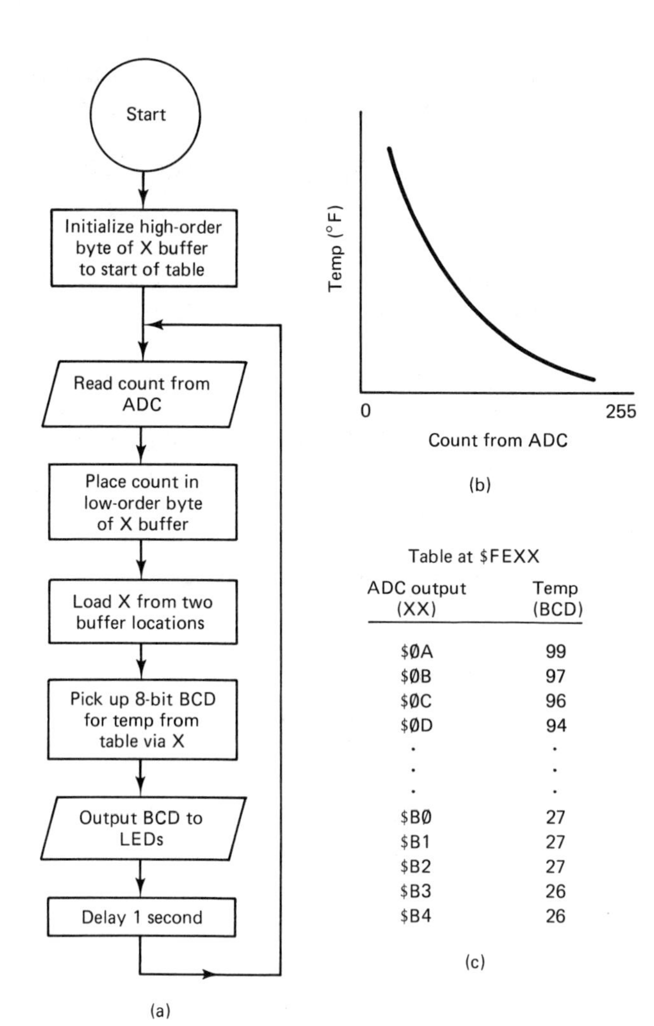

ADC output (XX)	Temp (BCD)
$ØA	99
$ØB	97
$ØC	96
$ØD	94
.	.
.	.
.	.
$BØ	27
$B1	27
$B2	27
$B3	26
$B4	26

Figure 13.8 (a) Successive approximation completes an 8-bit A/D conversion in 8 passes through the program loop. (b) A program flowchart for successive-approximation conversion. (c) Continuous A-to-D conversion.

REVIEW OF SECTIONS 13.5 AND 13.6

21. How many bytes of memory would be required to store digitally a single black-and-white picture composed of picture elements in a 400 × 300 array, with 16 levels of gray (from black to white) possible for each element?

22. How many bytes of memory would be required to digitally store a ½-h video program consisting of 30 complete pictures (as in Question 21) every second?

23. An A-to-D converter has a conversion time of 200 μs. What is the fastest input signal it can sample with less than 3 dB attenuation, 200 Hz, 600 Hz, 1500 Hz, or 5 kHz?
24. An audio signal is stored digitally without companding. Sixteen levels are allowed for the softest signal recorded, and the loudest signal is 30 dB stronger than the softest. How many bits are required per word of storage?
25. An 8-bit digital recording uses companding. If the count from 200 to 201 produces a 10-mV output change, the count from 20 to 21 will produce (less, the same, or more) output change.
26. A thermometer uses a thermistor input and a microprocessor lookup table to generate the display. The range covered is −100°F to +250°F. The A-to-D converter is 8-bit. In use it is noted that the display of an increasing temperature rises as follows:
    123 124 125 127 128 129 131
    Explain the skips at 126 and 130.
27. Would a 9-bit A-to-D converter in the instrument described in Question 26 completely eliminate the problem of skipped digits?
28. In the Thermistor Linearizer program, what hex byte should be entered in the table at address $FE4Ø, using the thermistor graphed in Figure 13.16(b)?
29. If you were in charge of production of the thermometer of Figure 13.17, would you consider changing the ADC-0804 for a processor-driven 1408-based converter? Why or why not?
30. In the vein of Question 29, would you use 7447 decoders or software decoding to drive a pair of 7-segment LEDs for decimal display?

13.7 A PROGRAMMABLE WAVEFORM GENERATOR

The linearized A-to-D converter system of the previous section is real enough and can be built, but obtaining and entering the 200 or more bytes of the lookup table make it considerably more time-consuming than the usual lab project. This section presents the hardware for a programmable waveform generator, along with two programs to illustrate how the output waveshape can be controlled by software. These programs require the entry of fewer than 50 bytes, and are more suitable for laboratory projects.

The hardware uses a 6802 processor and 6116 RAM for program storage. A 74LS373 latch drives a 1408 eight-bit D-to-A converter, which outputs a wave shape determined by the program. Figure 13.19 shows the system hardware.

A ramp-generator program is flowcharted in Figure 13.20 and listed in List 13.4. It is easy to see how a second *decrementing* loop could be added to change the ramps to triangles.

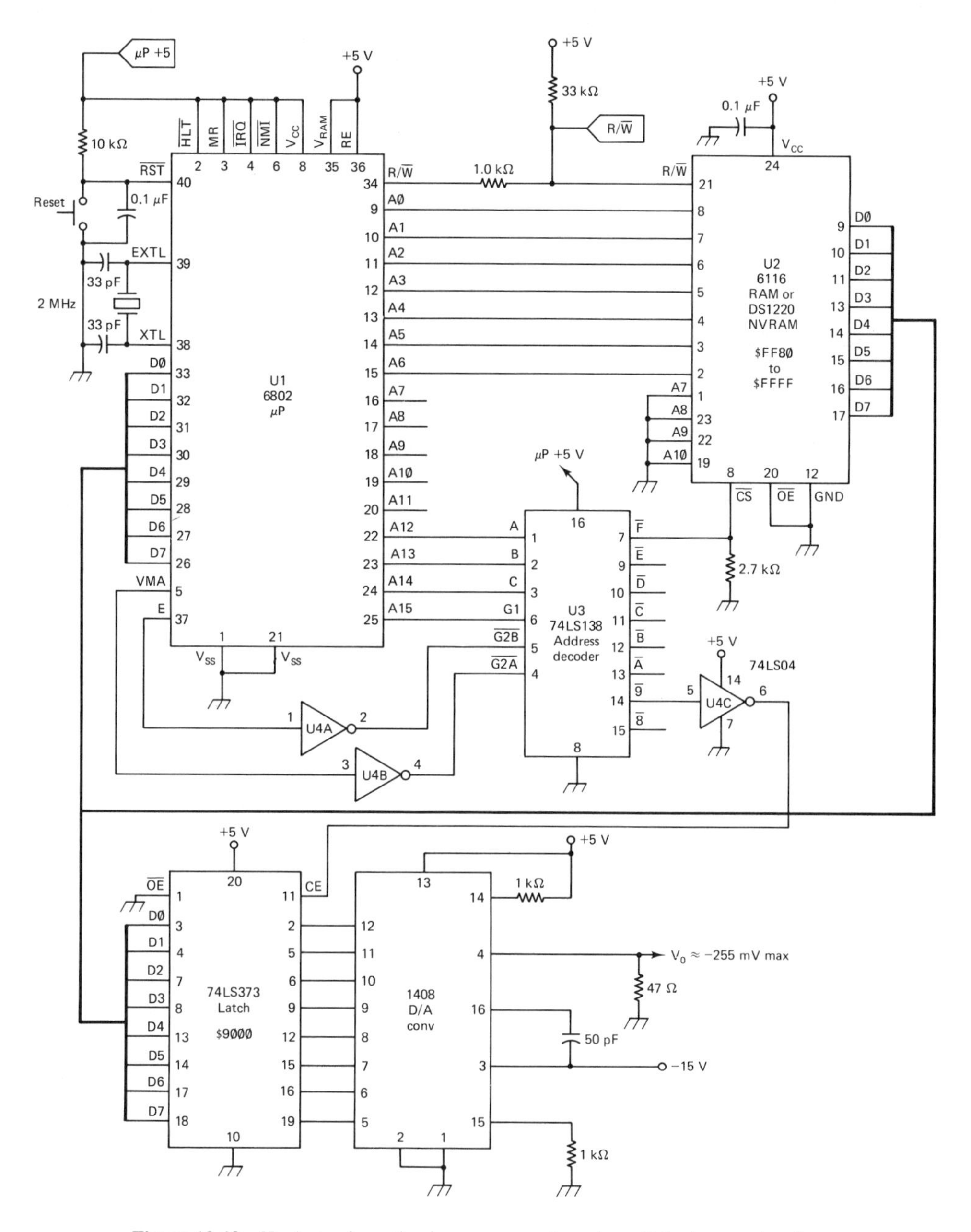

Figure 13.19 Hardware for a simple ramp generator using a D-to-A converter. See Figure 6.14 for an EPROM configuration.

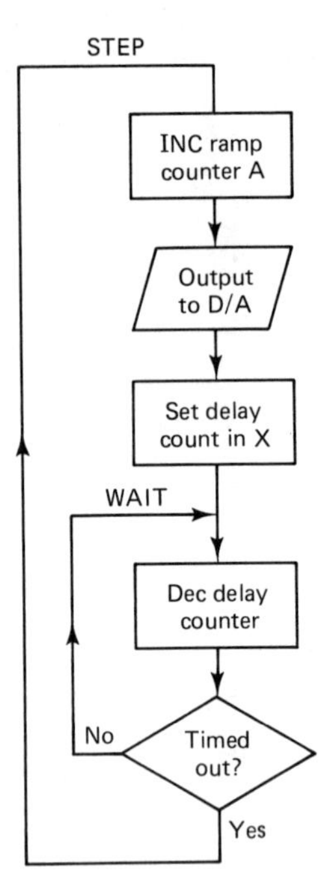

Figure 13.20 Flowchart for the the ramp-generator project.

```
                  1000 * LIST 13-4  RAMP GENERATOR.  FIG 13-19 HARDWARE
                  1010 * TESTED AND RUNNING, 10 JUL '87.  D. METZGER
                  1020 *
                  1030        .OR  $FF80
                  1040        .TA  $4080
9000-             1050 LATCH  .EQ  $9000
0001-             1060 COUNT  .EQ  1         SET UP FOR FAST RAMPS.
                  1070 *
FF80- 4C          1080 STEP   INCA           "A" HOLDS COUNT 0 - 255
FF81- B7 90 00    1090        STAA LATCH      TO D/A CONV VIA LATCH.
FF84- CE 00 01    1100        LDX  #COUNT    DELAY 16 US TO 1 SEC PER
FF87- 09          1110 WAIT   DEX             STEP BY COUNTINIG DOWN X.
FF88- 26 FD       1120        BNE  WAIT
FF8A- 20 F4       1130        BRA  STEP      NEXT STEP.
                  1140 *
                  1150        .OR  $FFFE     RESET VECTOR
                  1160        .TA  $40FE
FFFE- FF 80       1170        .HS  FF80
                  1180        .EN
```

A sine-like waveform can be generated by the same hardware of Figure 13.19 by switching to a program that varies its step sizes, as shown in Figure 13.21. The program list follows (List 13.5).

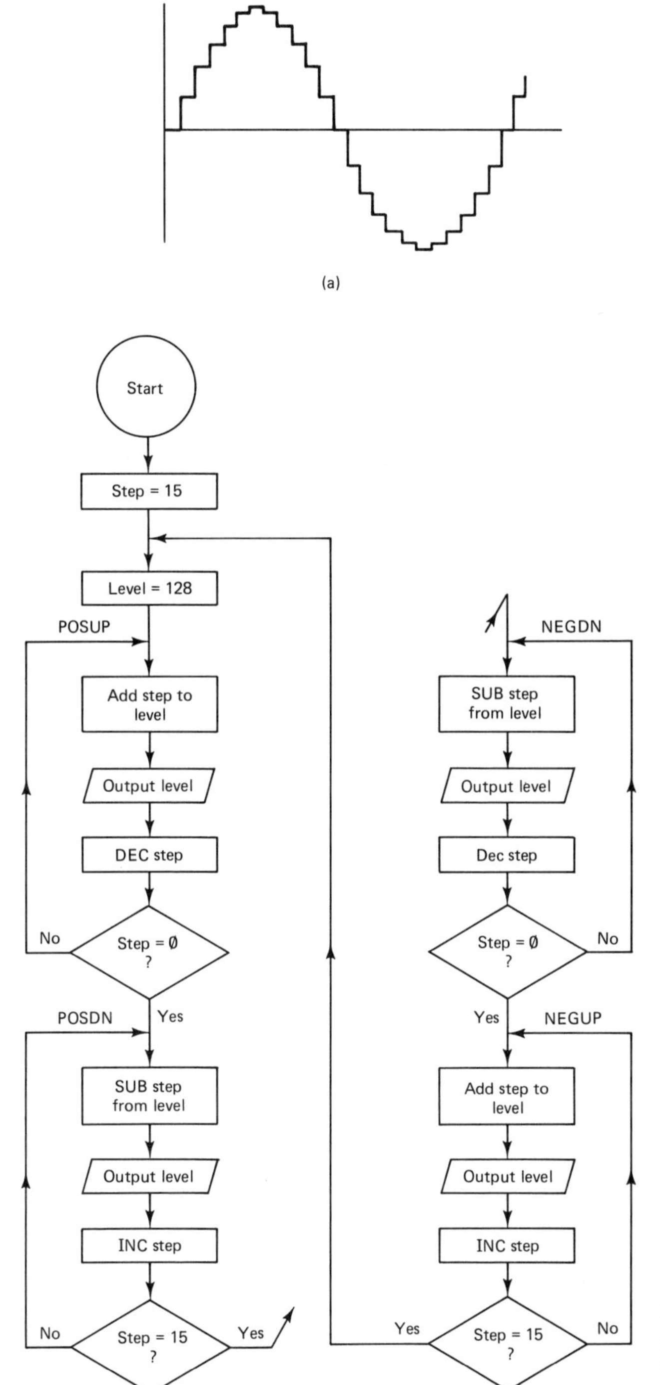

Figure 13.21 A simple program generates a sinelike waveform. The actual number of levels is twice that shown in (a). The program flowchart is shown in (b).

```
                 1000 * LIST 13-5 ** SINE-LIKE WAVEFORM FROM 8-BIT
                 1010 *   D/A CONVERTER. FIG. 13-19 HARDWARE.
                 1015 *   TESTED; PRODUCES 540-HZ SINE. 10 JUL '87. METZGER.
                 1020 *
                 1030          .OR  $FF90
                 1040          .TA  $4090
9000-            1050 LATCH    .EQ  $9000
                 1060 *
FF90- C6 0F      1070 START    LDAB #15      STEP SIZE STARTS AT 15.
FF92- 86 80      1080          LDAA #128     LEVEL STARTS MIDWAY: 0 - 255.
FF94- 1B         1090 POSUP    ABA           ADD STEP TO LEVEL AND OUTPUT
FF95- B7 90 00   1100          STAA LATCH     ON POS HALF CYCLE, UP SLOPE.
FF98- 5A         1110          DECB          CUT DOWN STEP SIZE.
FF99- C1 00      1120          CMPB #0       (INSTR ONLY MATCHES TIME OF POSDN)
FF9B- 26 F7      1130          BNE  POSUP    GO UP UNTIL STEP SIZE = 0.
FF9D- 10         1140 POSDN    SBA           GO DOWN ON POSITIVE HALF CYCLE,
FF9E- B7 90 00   1150          STAA LATCH     WITH INCREASING STEP
FFA1- 5C         1160          INCB           SIZES; LAST STEP SIZE IS
FFA2- C1 10      1170          CMPB #16       15; THEN START REDUCING STEP
FFA4- 26 F7      1180          BNE  POSDN     SIZES ON NEG HALF CYCLE.
FFA6- 5A         1190          DECB
FFA7- 10         1200 NEGDN    SBA           SAME SEQUENCES ON NEGATIVE
FFA8- B7 90 00   1210          STAA LATCH     HALF CYCLE,
FFAB- 5A         1220          DECB           DOWN AND UP SLOPES,
FFAC- C1 00      1230          CMPB #0        UNTIL STEP SIZE GOES FROM
FFAE- 26 F7      1240          BNE  NEGDN     15 TO 0, AND
FFB0- 1B         1250 NEGUP    ABA            BACK TO 15 AGAIN.
FFB1- B7 90 00   1260          STAA LATCH
FFB4- 5C         1270          INCB
FFB5- C1 10      1280          CMPB #16
FFB7- 26 F7      1290          BNE  NEGUP    THEN START OVER.
FFB9- 20 D5      1300          BRA  START
                 1310          .OR  $FFFE
                 1320          .TA  $40FE
FFFE- FF 90      1330          .HS  FF90     RESET VECTOR
                 1340          .EN
```

Answers to Chapter 13 Review Questions

1. Analog **2.** +52 mV **3.** 2.0 V
4. The digital driving signals can be any valid logic 1 or ∅ levels.
5. 7 bits **6.** 9 bits **7.** Greater than 0.25% **8.** 2%
9. 11% **10.** 0.78% **11.** 4 ms **12.** 1.33 ms
13. Continuous conversion. **14.** Flash or parallel converter.
15. 300 μs **16.** The reference supply voltage **17.** 1023
18. Flash converters. **19.** D7 is easy to check with the BMI and BPL instructions.
20. Dual-slope conversion. **21.** 60 kilobytes **22.** 3.24 gigabytes
23. 1500 Hz **24.** 9 bits **25.** less **26.** There are only 256 possible elements in the lookup table and 350 possible temperature numbers. The resolution of the instrument is slightly less than 1°F.
27. No. The nonlinearity of the thermistor would mean that two binary counts would have the same display at low temperatures, and one binary count would skip across numbers at high temperatures. **28.** $11
29. Perhaps. There is no need for extra speed or processor time, and the 1408 converter may be less expensive. **30.** Software decoding. Replacing hardware with software usually results in a cost savings if production volume is high.

CHAPTER 13 QUESTIONS AND PROBLEMS

Digits after the decimal point refer to section numbers in Chapter 13.

Basic Level

1.1 If an 8-bit D-to-A converter of the type shown in Figure 13.1 delivers 1.00 mA for bit 7, what current does it deliver for bit Ø?

2.1 How many resistors are required for a 10-bit ladder-type D-to-A converter?

3.2 A D-to-A converter with a maximum error of ±1% of full scale is required. What is the minimum number of bits it may have?

4.2 One D-to-A converter specifies 1% endpoint linearity. A competing type specifies 1% independent linearity. Which is the more linear converter?

5.3 Which is the faster A-to-D conversion technique, count-up or successive approximation?

6.3 The type 741 op-amp has an output slew rate on the order of 0.5 V/s and a frequency response on the order of 10 kHz in a gain-of-10 circuit. What would happen in the circuit of Figure 13.6 if 741s were used for the comparator and amplifier?

7.4 What is the fastest type of A-to-D converter?

8.4 What is the only parameter affecting the accuracy of a dual-slope A-to-D converter?

9.5 Is a sampling rate of 5 kHz adequate for digital storage of audio waveforms up to 5 kHz frequency?

10.5 In a companded digital recording system, an instantaneous voltage change from 100 mV to 150 mV translates to a change of 10 binary counts. Will a voltage change from 300 mV to 350 mV translate to a digital change of more than 10 or less than 10 counts?

11.6 Read the graph of Figure 13.16 and estimate the number of Celsius degrees per binary count, **(a)** at −15° C and **(b)** at +30° C.

12.6 What technique of A-to-D conversion is used by the Thermistor Linearizer program of List 13.3?

Advanced Level

13.1 Compare the maximum outputs of the D-to-A converter circuits of Figures 13.1(a) and 13.2 by assuming all inputs to be 4.0 V. What component value change could raise the output level of the inferior circuit? What problem would be worsened by this step?

14.1 Use the equation given with Figure 13.2 to calculate the output of an 8-bit ladder D-to-A converter for input counts of 100 and 200. Assume a +4.00-V input. Does this type of converter exhibit any inherent nonlinearity?

15.2 Define in your own words: resolution, accuracy, and monotonicity.

16.2 Explain in your own words the difference between endpoint, zero-based, and independent linearity.

17.3 The *count-up* A-to-D conversion program charted in Figure 13.7 completes the conversion after the first diamond decision block. What is the purpose, then, of the last three blocks comprising the FINISH loop?

18.3 Is continuous A/D conversion faster or slower than successive-approximation A/D conversion? (This is not a simple yes or no question. Explain.)

19.4 Why do parallel converters tend to have less resolution than other types?

20.4 Referring to Figure 13.11, explain the effect on the integrator slopes if: **(a)** the 1-μF capacitors change to 1.01 μF, **(b)** the clock oscillator changes from 5 kHz to 5.05 kHz, and **(c)** V_{ref} changes from -1.00 V to -1.01 V.

21.5 Reference the calculations appearing in Section 13.5 to determine the time in seconds that can be recorded in 32K-bytes of memory if a 30-dB volume range is to be accommodated and 8 discrete levels are allowed for the lowest-volume signals. The maximum frequency is to be 3500 Hz.

22.6 Suggest how the circuit of Figure 13.17 could be modified so that the upper and lower temperature limits of the system could be adjusted, still spreading the 256 digital counts over the chosen range of temperatures.

23.7 Analyze the program of List 13.5 and make a table showing the succession of binary counts (in hex) that is output.

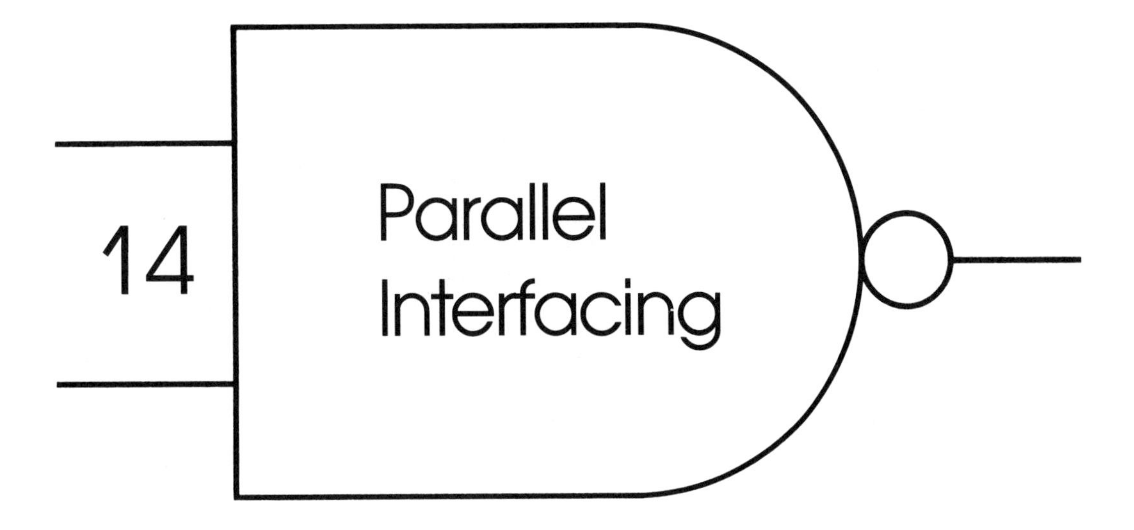

14.1 THE 6522 PARALLEL INTERFACE CHIP

Section Overview

The 6522 is a 40-pin IC that contains two 8-bit I/O ports. Each bit of each port can be individually set (by programming) to function as an output latch or as an input buffer. The 6522 I/O ports are accessed by the microprocessor at four consecutive addresses:

- **The lowest address accesses the Port B latches and buffers.**
- **The second address accesses the Port A latches and buffers.**
- **The third address accesses Data-Direction Register B, which defines which bits of Port B are inputs (0 bit) and which are outputs (1 bit).**
- **The fourth address accesses Data-Direction Register A, which similarly configures Port A for inputs or outputs.**

Ports versus latches and buffers: A latch can output binary data from a microprocessor, and a buffer can input data to the processor, but the processor can command a *port* to function either as an output latch or as an input buffer from one

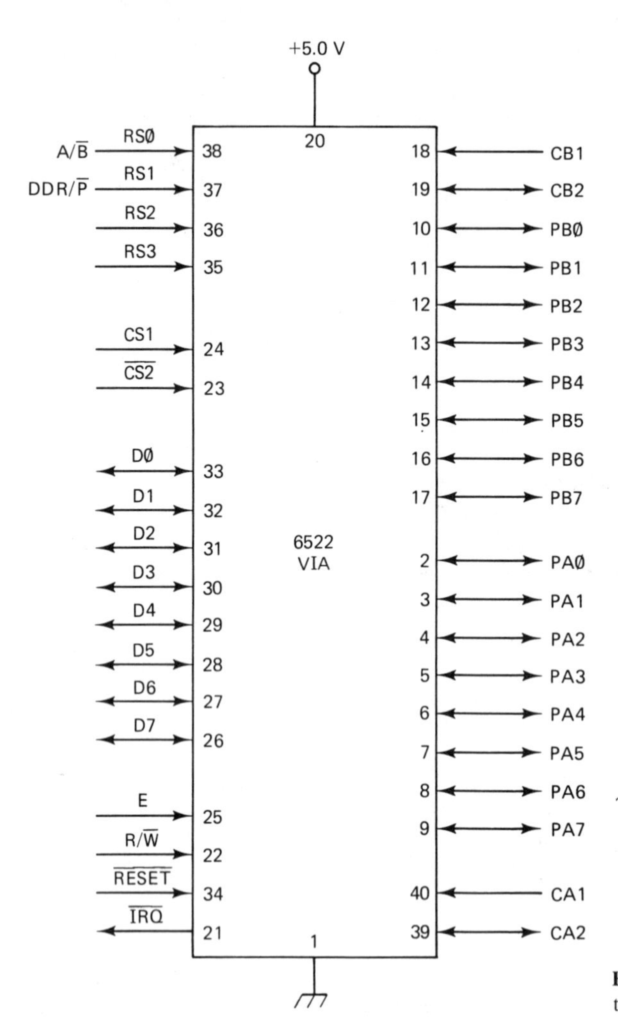

Figure 14.1 Basic parallel I/O functions for the 6522 Versatile Interface Adapter.

instruction to the next. The 6522, called a *versatile interface adapter* (VIA) by its manufacturers, contains two ports in a 40-pin package. Figure 14.1 shows the VIA pin functions relevant to the basic operation of the I/O ports examined in this section. The VIA contains handshake, interrupt, and timer functions also, and we will use these features and the pins relating to them in later sections of this and the next chapter.

The ports are bitwise programmable, which means that each bit can be set up to be either an output latch or an input buffer. This is done by writing an 8-bit control word into a data-direction register (DDR) in the 6522, one of which is associated with each port. A binary 1 in a DDR bit position sets up the corresponding port pin as an output. A Ø bit in the DDR configures the corresponding line of the port as an input. Here are some examples.

- LDAA #$FF Port *A* is configured for 8 output bits.
  STAA DDRA

- LDAA #$F0 Port *B* is configured for outputs on bits 7–4,
  STAA DDRB inputs on bits 3–0.

- CLR DDRA Port *A* is reconfigured for all inputs.

Addressing the ports. The VIA ports A and B and their corresponding data-direction registers are accessed at four consecutive addresses by the microprocessor. This is accomplished by two *register-select* inputs to the VIA:

- RS0 is normally connected to address bus line A0. If RS0 = 0 the B-side (PB or DDRB) is selected. If A0 = 1 the A-side (PA or DDRA) is selected.
- RS1 is normally connected to address bus line A1. If RS1 = 0, one of the ports (PB or PA) is selected. If RS1 = 1, one of the data-direction registers is selected.

There are two more *register-select* inputs, RS2 and RS3. We are assuming for the moment that these are held *low*.

Two chip-select inputs are provided (CS1, *high* active, and $\overline{\text{CS2}}$, *low* active) to be used in determining the base address at which the VIA appears. These are generally fed by high-order address lines from the processor, either directly or through address-decoding logic. If the base address is $8000, the I/O registers would be at the following addresses:

$8000	Port B
$8001	Port A
$8002	Data direction register B
$8003	Data direction register A

The bits programmed as outputs continue to assert whatever data was last written to them, even though the VIA is not selected. This is the basic output-latch function. The bits programmed as inputs only connect the port pins to the data bus pins while the chip is selected, which is the basic function of a buffer.

Notice from Figure 14.1 that the VIA is a clocked device and must be connected to the 6802's *E*-clock. (This clock would be called Φ_2 in 6800 or 6502 microprocessor systems.) The *E* signal is made a condition of chip-select for most writable devices in 6802 systems, but it must not be made a condition for chip-select of the 6522. This is because the VIA requires a 180-ns minimum address-setup time (with *E low*) in which to activate the one of its 15 registers that is being accessed. When *E* makes its positive transition the register read or write is enabled to proceed.

Input and output operations cannot be done simultaneously, so the R/$\overline{\text{W}}$ line must be connected to specify input (read) or output (write).

The processor's reset line must also be connected to the VIA reset input to initialize it for proper operation. Asserting a *low* on reset clears all VIA registers to 0s, so the VIA "comes up" with both ports configured for all inputs.

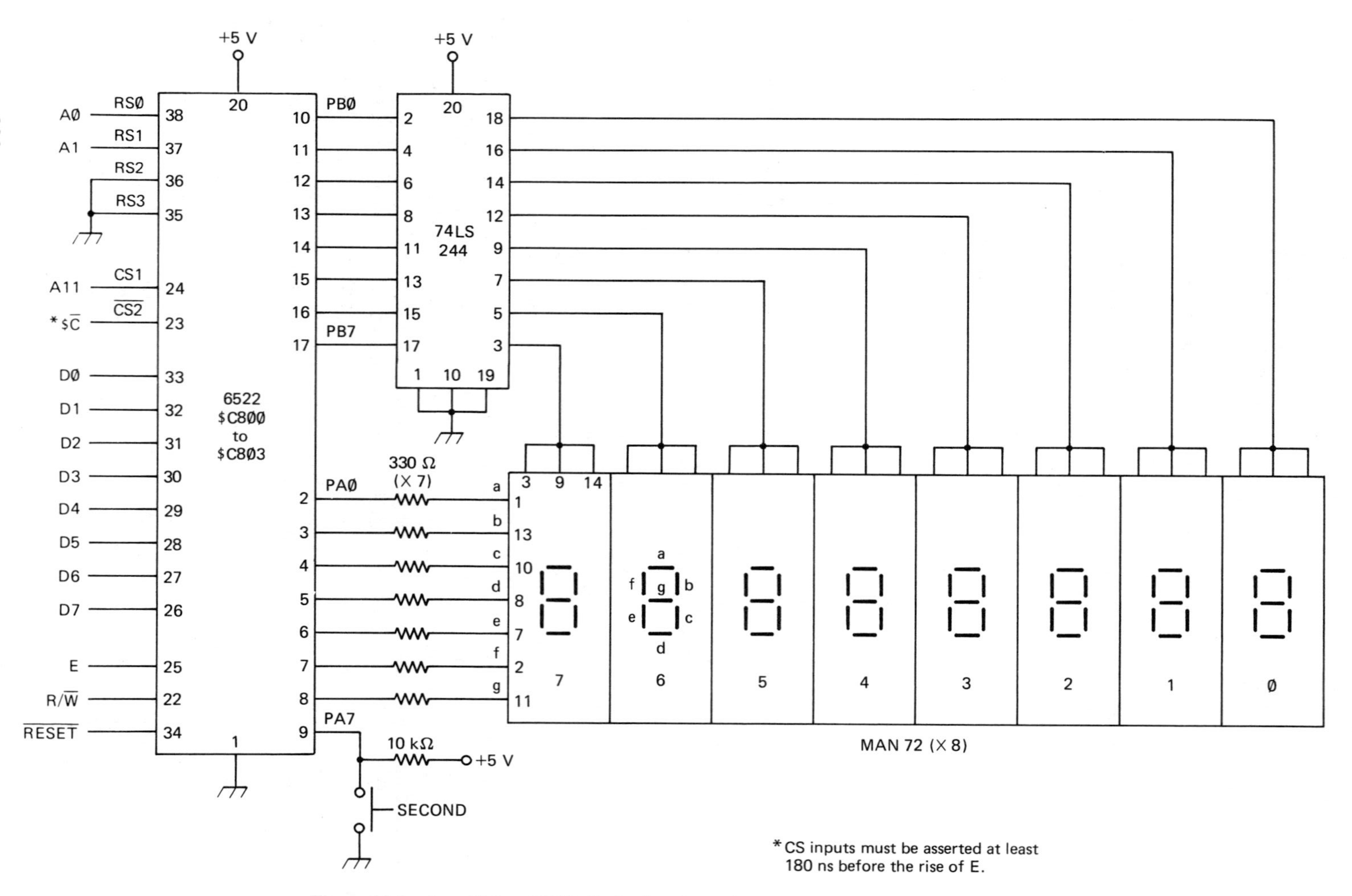

Figure 14.2 A multiplexed LED display interface using the VIA. An input is provided via Port *A*, bit 7. The processor hardware of Figure 6.14 or Figure 7.6 may be used to drive this interface.

14.2 A MULTIPLEXED DISPLAY WITH THE 6522

Section Overview

This section shows how a 6522 can be interfaced to a 6802 microprocessor to provide 8 characters of 7-segment display. The characters are software decoded, so no additional hardware is needed, and the display is not limited to the decimal or hexadecimal digits.

Multiplexing a display: The human eye sees flashes that occur faster than about 25 Hz as continuous light. The time per flash at this rate is 1/25 s, or 40 ms. A microcomputer can execute about 10 000 instructions in this time—literally "quicker than a flash." These facts are often used to produce a multicharacter display by strobing, or flashing each character in sequence for a few milliseconds each. The flashing display appears continuous to a human, and the processor has ample time to switch the display in addition to its other duties.

Figure 14.2 shows the hardware for an 8-LED multiplexed display driven by a 6522. Both ports are configured as outputs. A 74LS244 buffer is needed to assure adequate pull-up current, since Port B is specified to source only 1 mA. Port B scans across the 8 LEDs, pulling the anodes of one at a time *high*. Port A outputs the segment codes to pull the cathodes *low* for the segments to be lit on each LED. Bit 7 of port A is not needed for an LED cathode, so it is configured as an input and used to shift the display to 8 new characters in a table.

The hardware can be interfaced to the processor and program memory of Figure 6.14 or 7.6. In each system, internal RAM resides at $0000 through $007F. Program memory is decoded for a high-end range of 128 bytes from $FF80 through $FFFF. The range $C800 through $CFFF is used for the 6522. The logic for base address $C800 is

$$VIA = A11 \cdot \$C$$

The program flowchart is given in Figure 14.3. The display program with two sets of 8 characters in the table follows (List 14.1). We'll let you try to figure out what the display shows.

```
LIST 14-1      1000 * MULTIPLEX DISPLAY ** 8-LED MESSAGE ** D. METZGER
               1010 *  USE HARDWARE OF FIG 14-2.  RUNS; 15 JUL '87.
               1020 *
               1030          .OR  $FF80     SYSTEM STARTING ADR.
               1040          .TA  $4080     ASSEMBLER RAM STORAGE.
C800-          1050 PB       .EQ  $C800
C801-          1060 PA       .EQ  $C801     PORTS A & B.
C802-          1070 DDRB     .EQ  $C802
C803-          1080 DDRA     .EQ  $C803     DATA-DIRECTION REGISTERS.
0010-          1090 XBUF     .EQ  $10       TWO BYTES TO SAVE X.
0012-          1100 DLYRAM   .EQ  $12       DELAY COUNTER.
0013-          1110 COUNT    .EQ  $13       SCAN COUNTER.
               1120 *
```

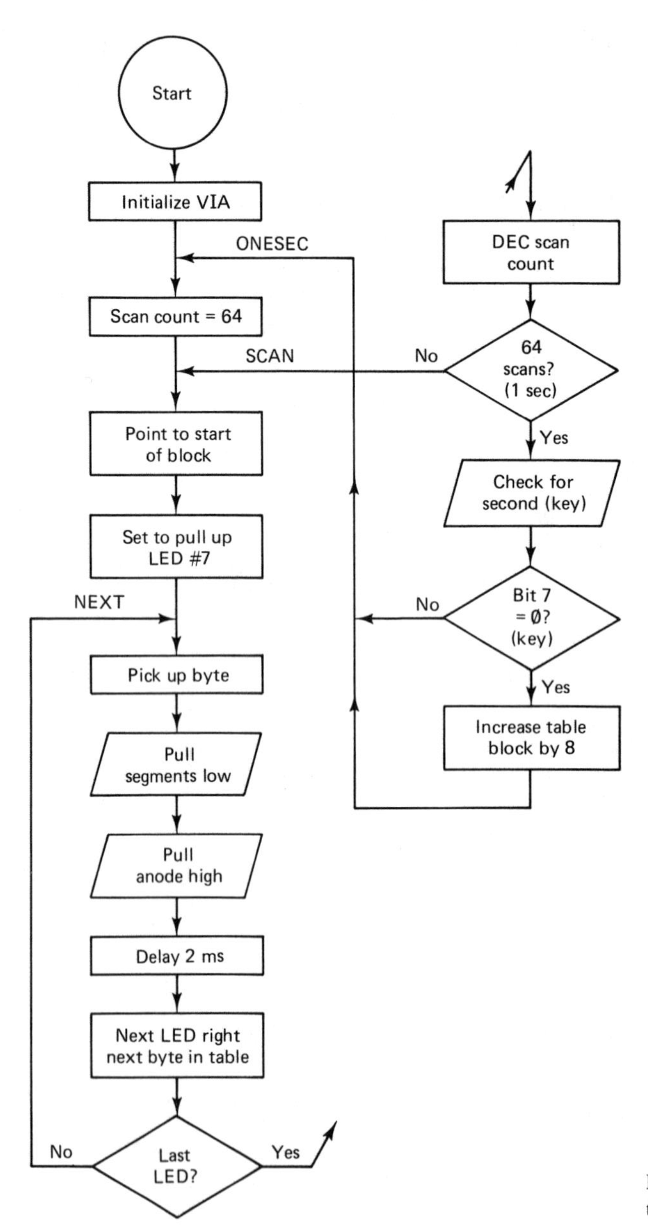

Figure 14.3 The program flowchart for the LED display interface of Figure 14.2.

```
FF80- 86 FF      1130         LDAA #$FF    PORT B IS ALL OUTPUTS TO
FF82- B7 C8 02   1140         STAA DDRB     PULL 8 ANODES HIGH.
FF85- 86 7F      1150         LDAA #$7F    PORT A PULLS 7 CATHODES LOW;
FF87- B7 C8 03   1160         STAA DDRA     BIT 7 IS INPUT: NEXT DISPLY.
FF8A- CE FF D0   1170         LDX  #$FFD0  START ADR OF FIRST MESSAGE.
FF8D- DF 10      1180         STX  XBUF    RAM BYTES POINT TO STARTING ADR.
FF8F- 86 40      1190 ONESEC  LDAA #64     GOING TO SCAN DISPLAY 64
FF91- 97 13      1200         STAA COUNT    TIMES, THEN CHECK KEY.
FF93- DE 10      1210 SCAN    LDX  XBUF    POINT TO CURRENT MESSAGE.
FF95- C6 80      1220         LDAB #$80    FIRST PULL LEFT LED ANODES HI.
```

```
FF97- A6 00     1230 NEXT   LDAA 0,X      GET CHARACTER FROM FILE.
FF99- B7 C8 01  1240        STAA PA       PULL CATHODES LOW.
FF9C- F7 C8 00  1250        STAB PB       PULL ANODE HIGH.
FF9F- 7F 00 12  1260        CLR  DLYRAM   DELAY FOR 2.5 MS.
FFA2- 7C 00 12  1270 WAIT   INC  DLYRAM
FFA5- 2A FB     1280        BPL  WAIT
FFA7- 08        1290        INX           POINT TO NEXT CHARACTER.
FFA8- 54        1300        LSRB          READY ANODE NEXT LED RIGHT.
FFA9- 26 EC     1310        BNE  NEXT     FALL THRU IF ALL 8 LEDS DONE.
FFAB- 7A 00 13  1320        DEC  COUNT    64 SCANS TAKE ABOUT 1 SEC.
FFAE- 26 E3     1330        BNE  SCAN     DISPLAY ALL 8 LEDS 64 TIMES.
FFB0- B6 C8 01  1340        LDAA PA       CHECK INPUT EVERY SECOND.
FFB3- 2B DA     1350        BMI  ONESEC   NO INPUT: KEEP SCANNING.
FFB5- 86 08     1360        LDAA #8        INPUT: ADD 8 TO LO BYTE OF
FFB7- 9B 11     1370        ADDA XBUF+1    XBUF TO POINT TO NEXT BLOCK
FFB9- 97 11     1380        STAA XBUF+1    OF FILE AND
FFBB- 20 D2     1390        BRA  ONESEC    BEGIN NEW DISPLAY.
                1400 *
                1410        .OR  $FFD0    START OF DATA FILE.
                1420        .TA  $40D0    HOST COMPUTER CORRESPONDING AREA.
                1430 * DATA LISTS FOR TWO MESSAGES FOLLOW.
FFD0- 7F 09 08
FFD3- 63 06 7F
FFD6- 08 7F     1440        .HS  7F090863067F087F
FFD8- 2B 79 46
FFDB- 06 7F 21
FFDE- 08 11     1450        .HS  2B7946067F210811
                1460 *
                1470        .OR  $FFFE
                1480        .TA  $40FE
FFFE- FF 80     1490        .HS  FF80
                1500        .EN
```

REVIEW OF SECTIONS 14.1 AND 14.2

Answers appear at the end of Chapter 14.

1. In a 6522 VIA, data \$3B is written to DDRA. Which pin numbers on the package are inputs by this operation?
2. List in order the binary states of VIA pins 22, 23, 24, 37, and 38 when data is being input via Port B.
3. If VIA pin 24 is tied to 6802 address line A14, VIA pin 23 is connected to A15, and RS0 and RS1 are tied, respectively, to A0 and A1, what is the lowest address that will access Data-Direction Register A?
4. What is the purpose of the 74LS244 in Figure 14.2?
5. What character appears on what LED in Figure 14.2 if \$20 is output to port B and \$47 is output to port A?
6. In the Multiplex Display program, which instruction determines whether all 8 LEDs have been scanned?
7. Which instruction decides when to display the second set of 8 characters?
8. What character does the second byte of the list (\$09) cause to be displayed? (Consult Figure 14.2.)
9. Why was bit 7 of Port A chosen as the key input?
10. What is the advantage of multiplexing a display over driving it with fixed latches?

14.3 HANDSHAKING WITH THE VIA

Section Overview

Handshaking is a scheme for positively assuring the transfer of data between a sender and a receiver. Full handshaking requires two lines in addition to the data lines. The sender uses one line to signal the receiver that it has new data to send. The receiver uses the other line to signal back to the sender when it has read the current data and is ready to accept new data. The external device, be it sender or receiver, often generates an interrupt signal for the processor when it is ready for another data transfer.

The data-transfer problem: When *reset*, the 6522 comes up with both ports configured as inputs. Let us say that we have a keyboard connected to Port A, and we want the processor to read the 8-bit code for each key as it comes in and store it in a RAM area. We soon realize that we have two problems.

1. How can we free the processor from spending all its time checking for data at Port A? It should only need to read the port once when a key has actually been depressed.
2. How will the processor recognize when the data at Port A is actually new data, so that it doesn't read 2 or 20 keystrokes when only 1 was intended?

The solution to both of these problems is provided by the 6522's *handshake* mode of input. Figure 14.4 shows the lines involved in handshaking with the Port A input. The keyboard (or other input device) is assumed to have a low-active *data-ready* output, which is asserted each time a key is pressed. The VIA responds to this

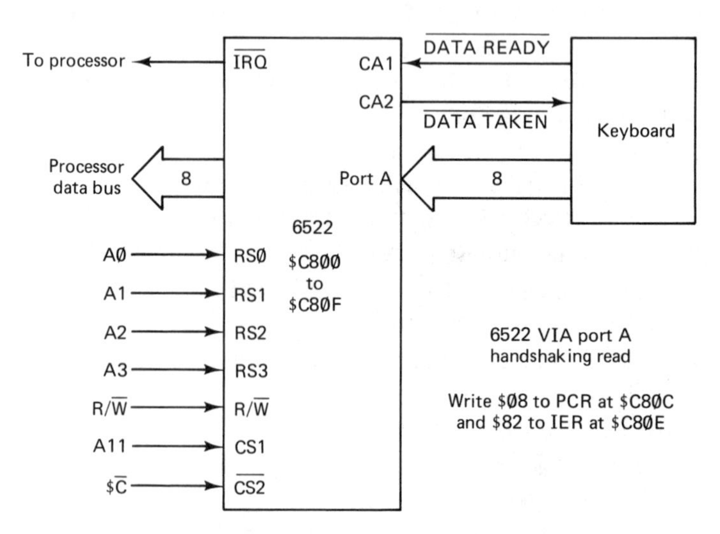

Figure 14.4 The Peripheral Control Register and the Interrupt-Enable Register must be programmed to configure the VIA for handshake operation.

data-ready input at its CA1 line by asserting its low-active $\overline{IRQ}$ output. The 6522 $\overline{IRQ}$ output is connected directly to the processor's $\overline{IRQ}$ input, and the interrupt routine causes the processor to read the data from Port A via the VIA's connection to the data bus.

The processor's *read* of the VIA causes low-active output CA2 to be asserted, and the keyboard is assumed to have logic that interprets this signal to mean "data taken." The keyboard should respond by removing the *data-ready* signal it had sent to the VIA and clearing its data output to make way for the next key press.

Programming the VIA for handshaking: The VIA does not "come up" ready for handshaking operation. Two registers within the VIA must be programmed with the proper bits to activate the handshake lines CA1 and CA2 and the interrupt output $\overline{IRQ}$. Selecting these registers requires that register-select lines RS3, RS2, RS1, and RS0 be connected to the corresponding lines of the system address bus.

Of course, with 4 register-select lines to the 6522, there are fully 2^4, or 16, registers within it. (There was good reason for naming it a *versatile* interface adapter.) We have seen the two *port* registers and the two *data-direction* registers, and we will examine a few of the functions of the two which control the *handshake* functions now. Some of the timer-register functions will be seen in the next chapter, but an exhaustive exposition of all of the features of the VIA would hardly be practical. This is what manufacturers' data books and application notes are for. You may find it helpful to look ahead to Figure 15.7 on page 367, which summarizes the VIA features from time to time in the following discussions.

The Peripheral Control Register (PCR is register 12 and is accessed at address $C80C in the system of Figure 14.4. To program CA1 as a negative-edge-sensitive *data-ready* input and CA2 as a negative-true *data-taken* output, the nibble $8 must be written into the least-significant 4 bits of this register. The most-significant nibble controls the CB1 and CB2 lines, as we will see in a moment.

Each of the 16 possible low-order nibbles in the PCR (register 12) programs a different function for CA1 and CA2. Here are some other useful nibbles:

- $0 Both CA1 and CA2 are negative-edge interrupt-triggering inputs.
- $D CA1 is an interrupt-triggering *data-ready* input. CA2 asserts a fixed *low*-level output.
- $F CA1 is the same as for $D. CA2 outputs a *high*.

The Interrupt Enable Register (IER) is register 14 and is accessed at address $C80E in Figure 14.4. To enable the interrupt from *data-ready* input CA1, a 1 must be written into bits 7 and 1 of this register.

```
LDAA   #$82    Set a 1 in bits 1 and 7.
STAA   $C80E   Bit 1 of IER goes high.
```

Interrupts from CA2 (when it is configured as an input) are enabled by setting a 1 in bits 0 and 7 of the IER with byte $81. The details of the IER involve VIA functions not yet covered, so a complete discussion is deferred to Section 15.4.

14.4 ADDITIONAL FEATURES OF THE VIA PORTS

Section Overview

Port A of the 6522 can be an input or an output, but it can handshake only in the input mode. Port B can handshake in the input or output modes and has the added capability of sourcing a drive current of 1 mA as an output, whereas Port A can source only 0.1 mA.

A *transparent* buffer is one that applies to the data bus the input data present at the time of access. A *latched* buffer seizes the input data when a strobe input is asserted and holds that data for presentation when the chip is accessed.

Port A can be set up for handshaking only in the read mode. Port B can be configured for handshaking in the read mode or the write mode.

The setup for Port B write handshaking is similar to the setup for Port A read handshaking. The VIA lines involved are shown in Figure 14.5. Here are the steps required.

```
LDAA    #$FF     Store all 1s in DDRB to
STAA    $C802      configure it for outputs.
LDAA    #$80     Force bit 7 of Peripheral Control
ORAA    $C80C      Register high for handshake with
STAA    $C80C      CB2 output, CB1 input.
LDAA    #$10     Force bit 4 of Interrupt-Enable
ORAA    $C80E      Register high to enable interrupt
STAA    $C80E      from CB1, data taken.
```

Write handshaking via Port B proceeds as follows:

1. The processor writes a byte to the 6522 Port B address ($C800 in Figure 14.5). This causes the data to be asserted on the Port B pins. The CB2 handshake output also is asserted *low*, telling the peripheral device to read the data.
2. The peripheral now processes the data sent to it by the processor via the VIA. Let's say that the peripheral is a 20-character-per-second printer, so it takes 50 ms to get ready for the next byte. It then asserts a *low* on CB1 to tell the VIA that it is finished and is ready for new data. The *low* to CB1 causes an $\overline{\text{IRQ}}$ to be output from the VIA to the processor. The processor's interrupt-handling routine will then cause a new byte of data to be written to the VIA.

Differences in the VIA buses: The VIA inputs that are driven by the processor present a high resistive impedance and a maximum capacitive loading of 7 pF each. These include D0–D7, RS0–RS3, the CS lines, R/$\overline{\text{W}}$ and *E*.

The Port A and Port B pins, when programmed as inputs, present one standard TTL load. This is to say that they tend to pull *high* and deliver about +1.6 mA to any device that pulls them *low*.

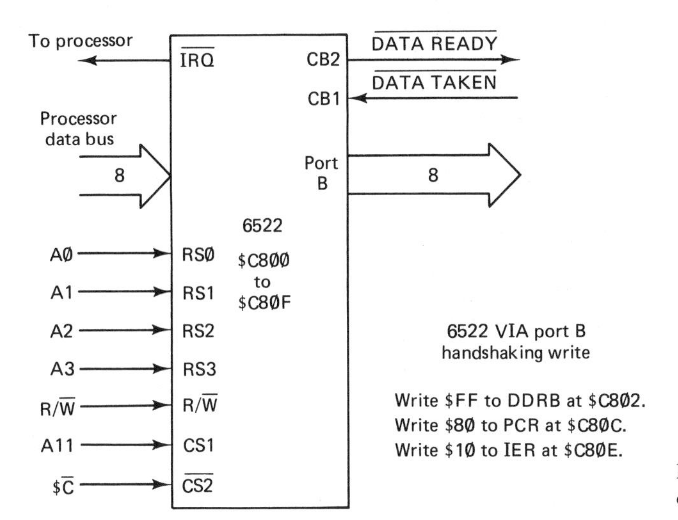

Figure 14.5 Write handshaking on Port B of the VIA.

When programmed as outputs the Port A and Port B lines can drive one standard TTL load, which is to say that they are guaranteed to be able to sink at least 1.6 mA when outputting a logic 0. When outputting a logic 1, however, the Port A and Port B lines exhibit a difference. Port A lines are guaranteed to source only 0.1 mA to a resistive load while pulling *high*. Port B lines are guaranteed to source at least 1.0 mA while pulling *high*, making them more suited for feeding transistor relay or lamp drivers. It is common to see measured output capabilities far beyond the specified limits, but as we have emphasized before, it is very poor engineering practice to make use of performance that is not guaranteed by the chip manufacturer.

If Port B is programmed for outputs and the processor executes a read of the port, the data that was written to Port B will simply be read back, regardless of the actual state of the Port B pins. However, if Port A is configured as an output, a read of Port A will pick up the logic levels actually at the pins. A load that pulls up or down strongly may overpower the outputs, resulting in an actual level different from the level written.

Figure 14.6 illustrates the differences between Ports A and B.

Transparent versus latched inputs: The 6522 input ports are normally *transparent*. This means that a read of the data bus ''sees'' straight through to the data at the port input pins when the read operation occurs. However, the inputs can be programmed to become nontransparent, or latched. This means that they catch and hold the input data present when control input CA1 is asserted. It is this latched data that is read when the port is read.

The latched-input function is enabled by the VIA's register 11, the Auxiliary Control Register (ACR). It is accessed at address $C80B in our example system. Bit 0 of the ACR is set to 1 to enable latching of Port A inputs, and bit 1 is set to 1 to latch Port B inputs. A VIA *reset* clears all register bits to 0, disabling the latching feature.

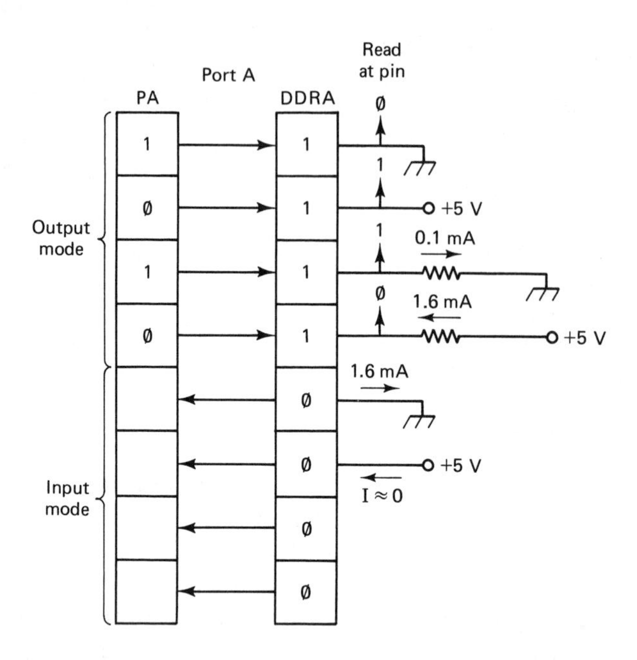

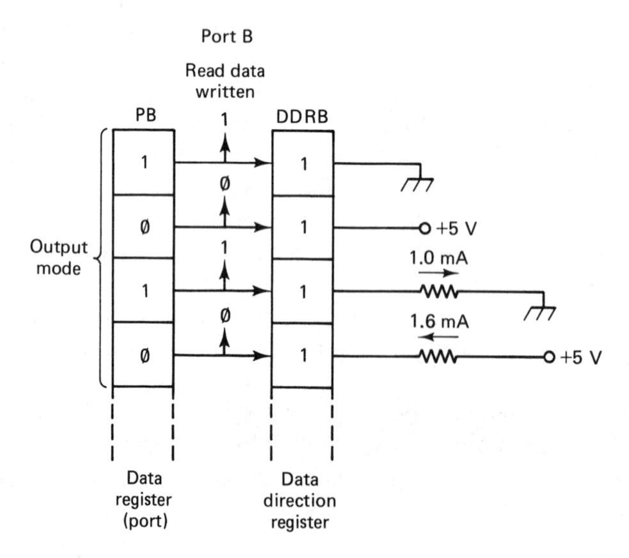

Figure 14.6 Ports A and B of the 6522 exhibit somewhat different properties in the output mode.

REVIEW OF SECTIONS 14.3 AND 14.4

Answers appear at the end of Chapter 14.

11. On a processor-read handshake with the 6522, which handshake signal causes the $\overline{\text{IRQ}}$ line to be pulled *low*?

12. When does the peripheral device send a second byte of data to the VIA?

13. List the binary levels on RSØ through RS3 that will select the 6522's Peripheral Control Register.
14. In storing bits to the Interrupt Enable Register, why do we first OR the accumulator contents with the IER contents before storing?
15. Immediately after a *reset*, what are the operating modes of the 6522 handshake lines?
16. In write handshaking on Port B, which handshake line is asserted just prior to the next write?
17. A resistor from a Port A input to ground is going to be used to pull the line to 0.4 V in the absence of any other pull *high*, thus defaulting to a logic-Ø input. What is the highest safe resistor value?
18. Which port should be used to output 0.5 mA to the base of a transistor that controls a relay?
19. Which port should be used as an output if it is desired to have the program detect and report any output loads that are shorted?
20. Is the familiar 74LS244 buffer transparent?

14.5 STEPPER MOTORS

Section Overview

A common application of parallel output interfacing is the stepper motor. Stepper motors permit precise positioning without the complication and expense of position-sensing and feedback-control systems that are required with regular dc or ac servo motors. They do not just simply run when a voltage is applied to them. They have at least four (and usually more) input wires, and the binary inputs to these must be changed intelligently to a new pattern each time the motor shaft is to rotate to a new position. They are thus ideally suited for control by a microprocessor. Their major disadvantage is that they provide no feedback to the controller to verify their current position. If they are once overloaded or stalled the controller can continue with an erroneous assumption of the system's actual position.

A conventional motor rotates when voltage is applied to it. For most motors the number of turns made depends on the amount of the voltage applied and on the frictional and inertial resistance encountered. Determining the exact position of a device driven by a conventional motor is therefore impossible without the addition of some sort of position-indicating device. Stepper motors, in contrast, are commanded to rotate a fixed fraction of a turn each time a new command is issued to them by a microcomputer. Computing the exact position of a device driven by a stepper motor is a simple matter, assuming that the motor is free to move and has not been stalled by meeting an immovable object.

A simple stepper motor is diagrammed in Figure 14.7. There are four stationary

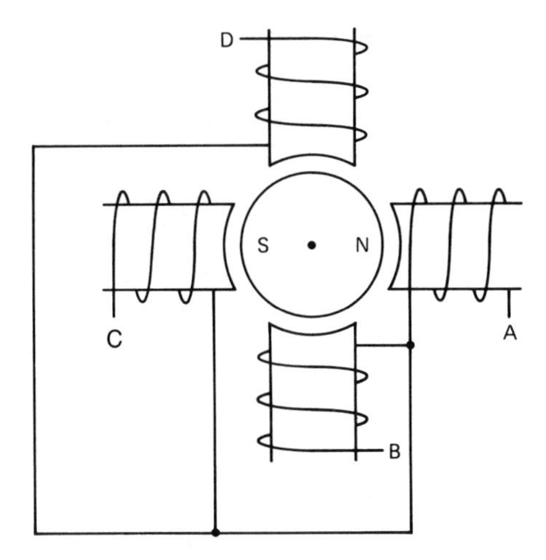

Figure 14.7 Schematic representation of a four-pole stepper motor.

electromagnets positioned around a rotatable armature, which is permanently magnetized. The coils are energized in clockwise or counterclockwise sequence to make the rotor turn clockwise or counterclockwise in 90° increments. There are no brushes or sparking contacts, so stepper motors are ideal for environments containing flammable vapors.

Currents of 100 mA and more are required for the coils of even small stepper motors, so external current-driver circuitry is required between the microprocessor's parallel output interface and the motor itself. Figure 14.8 shows an elementary current driver for the four-coil stepper motor. The diodes across the coils are to prevent inductive kickback voltage from damaging the transistors when the currents in the coils are switched off. More complex driver circuits might switch to a lower source voltage when the motor is not stepping to minimize power waste while still providing a braking force to hold the rest position. To drive the motor clockwise, the sequence of data in the following list must be applied by the latch:

	D3	D2	D1	DØ
0°	Ø	Ø	Ø	1
−90°	Ø	Ø	1	Ø
−180°	Ø	1	Ø	Ø
−270°	1	Ø	Ø	Ø

To rotate the motor counterclockwise the sequence is as follows:

	D3	D2	D1	DØ
0°	Ø	Ø	Ø	1
+90°	1	Ø	Ø	Ø
+180°	Ø	1	Ø	Ø
+270°	Ø	Ø	1	Ø

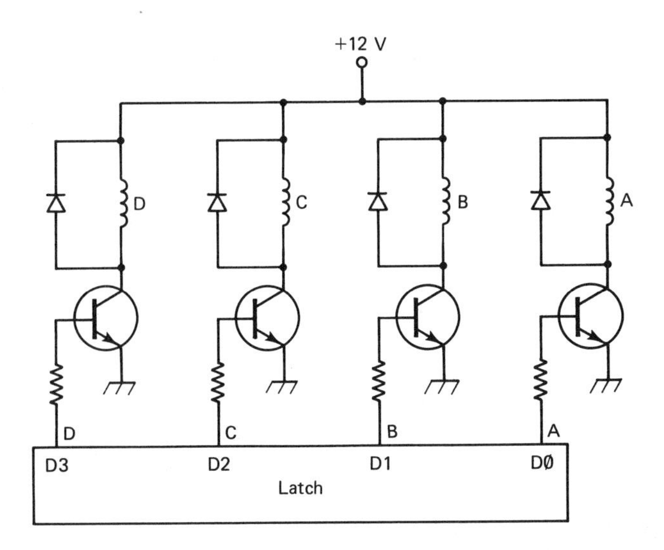

Figure 14.8 A driver circuit for a four-winding stepper motor.

Of course the microcomputer program should provide a delay between the output of one binary pattern and the next. For a rotation speed of 600 rpm, which is 10 revolutions per second, the time between 90° steps is ¼(100 ms), or 25 ms.

Finer control of stepper motor position may be accomplished in three ways:

1. Use mechanical gear reduction to decrease the amount of rotation per step. A 9-to-1 ratio would produce 10° of rotation per step. This will also increase output torque and decrease output speed. Gear reduction is very commonly used.
2. Use more coils around the rotor. Eight coils would interface nicely with 8-bit latches.
3. Energize two adjacent coils simultaneously to provide steps halfway between the regular single-coil steps. The pattern for clockwise rotation of the motor of Figure 14.8 would now be as follows:

	D3	D2	D1	DØ
0°	Ø	Ø	Ø	1
−45°	Ø	Ø	1	1
−90°	Ø	Ø	1	Ø
−135°	Ø	1	1	Ø
−180°	Ø	1	Ø	Ø
−225°	1	1	Ø	Ø
−270°	1	Ø	Ø	Ø
−315°	1	Ø	Ø	1

The most economical stepper motor will minimize the mechanics and copper coil windings and make maximum use of software and integrated electronic devices. Figure 14.9 represents a stepper motor that has only two coils. Four positions of the

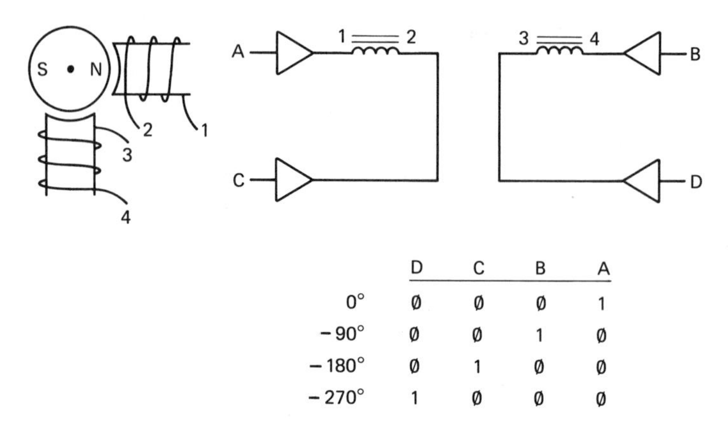

	D	C	B	A
0°	Ø	Ø	Ø	1
−90°	Ø	Ø	1	Ø
−180°	Ø	1	Ø	Ø
−270°	1	Ø	Ø	Ø

Figure 14.9 A two-winding stepper motor can provide four positions per revolution if dual drivers are used to reverse the current in the coils.

magnetic field are obtained by reversing the current through the coils to obtain a 180° shift in the field. The driver circuitry uses bridge amplifiers to reverse the coil current. This motor can also achieve 8 steps per revolution by energizing both coils between 90° steps.

14.6 THE IEEE-488 PARALLEL BUS STANDARD

Section Overview

Another common example of parallel interfacing is the IEEE-488 bus, also called the GPIB, or General Purpose Interface Bus. It allows instruments and computers to interchange 8-bit digital data in a form that is standardized among instrument manufacturers. The devices on the bus at any one time consist of one *talker*, one or more *listeners*, and one *controller*, which may be the microcomputer. The bus consists of 8 data lines, 8 handshake and control lines, and 8 ground or shield lines.

The General-Purpose Interface Bus (GPIB) was originally developed by Hewlett-Packard Corporation and is sometimes referred to as the HP bus. In 1975 it was adopted by the Institute of Electrical and Electronics Engineers as their standard No. 488, hence the term IEEE-488 bus. The standard provides for 8-bit parallel data transfer among as many as 15 instruments over distances as great as 20 m. The data is asynchronous—that is, it is not sent at any particular clock frequency—but data rates as high as 1 MHz are possible.

A block diagram of three instruments interfaced by the GPIB is given in Figure 14.10. The system represented is an automatic frequency-response plotter for an amplifier. The processor controls the frequency of a programmable function generator and feeds the frequency data to the horizontal axis of the plotter. A digital voltmeter

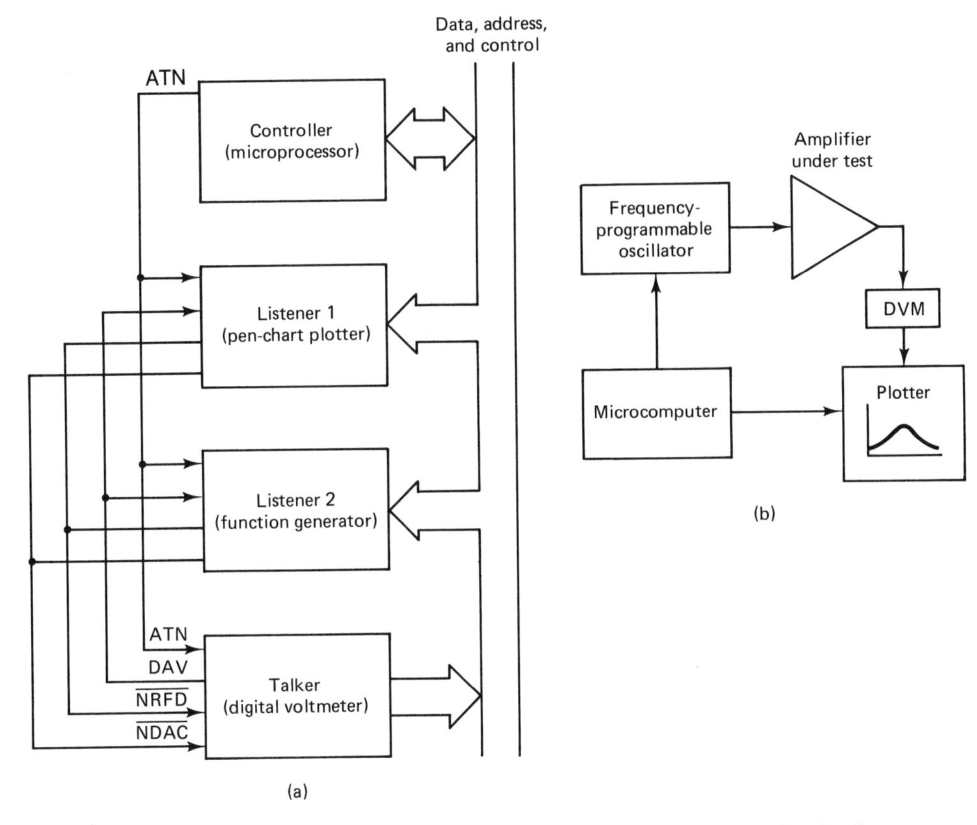

Figure 14.10 A representative system composed of individual instruments interfaced via an IEEE–488 bus. This example shows an automatic frequency-response plotter for an amplifier.

monitors the amplifier output and feeds the vertical input of the plotter under processor control.

The microprocessor is the *controller*. It sends commands and addresses via the data bus to indicate which device is to talk and which devices are to listen during the next exchange of data. The controller asserts the ATN (attention) line to indicate to the other devices that the bits on the bus represent commands and addresses, not data. More than one controller may reside on the bus, but only one of them is permitted to assert address and control signals at a time. An inactive controller asserts SRQ (Service Request) and waits for REN (Remote Enable) before assuming control.

Handshaking on the GPIB uses three lines to accommodate the fact that several devices may be *listening* while one is *talking*. All the listeners must be finished with the last data and ready for new data before the talker can assert new data, and all the listeners must indicate that the new data has been received before the talker can remove the data.

Figure 14.11 shows how several *listeners* are *wired-or* connected on their handshake outputs. Any one device can pull the line *low* if it is not ready, and the transaction will not continue until all devices are ready.

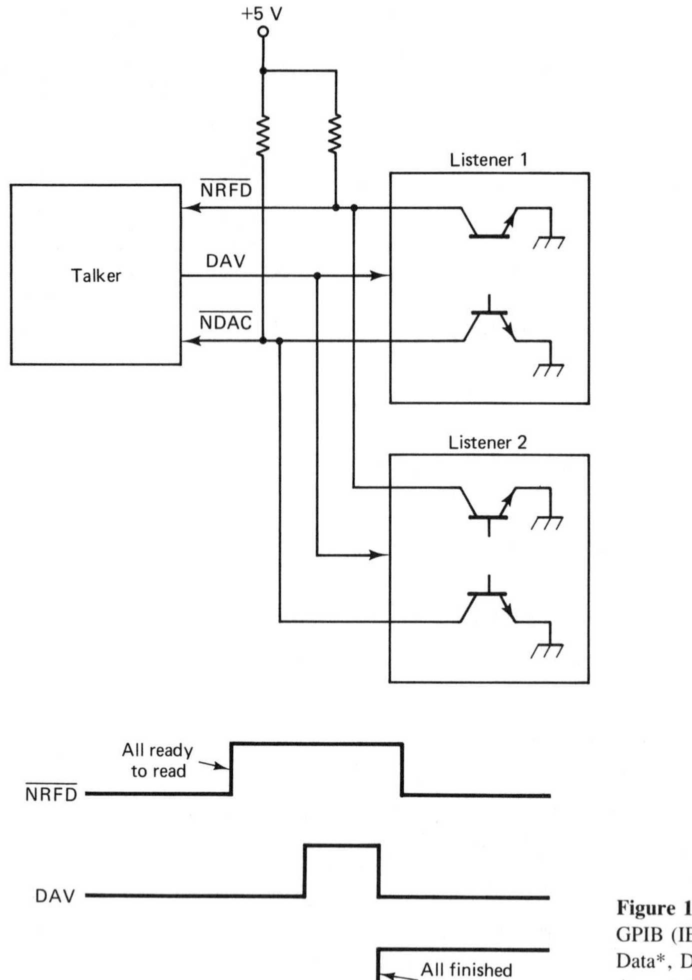

Figure 14.11 Handshake signals for the GPIB (IEEE–488 bus): Not Ready For Data*, Data Available, and No Data Accepted.

The pinout of the GPIB connector is given in Figure 14.12. All the signals are TTL-level compatible. Notice that the Data Input/Output pins are inconveniently labeled DIO1 through DIO8, so that processor line PA5, for example, would have to be connected to GPIB line DIO6.

12 Shield	11 ATN	10 SRQ	9 IFC	8 NDAC	7 NRFD	6 DAV	5 EOI	4 DIO4	3 DIO3	2 DIO2	1 DIO1
24 Signal GND	23 GND	22 GND	21 GND	20 GND	19 GND	18 GND	17 REN	16 DIO8	15 DIO7	14 DIO6	13 DIO5

Figure 14.12 Pinout for the GPIB (IEEE 488) connector.

The five control lines have the following functions:

- ATN Attention; asserted by controller. Commands talkers to get off the bus. Next data on bus is address and command information.
- EOI End Or Identify; asserted by controller or talker to indicate end of a sequence of data or by any device whose address is called during controller polling.
- IFC Interface Clear; asserted by controller to remove all talkers from the bus.
- REN Remote Enable; asserted by controller to permit another controller to take over the bus.
- SRQ Service Request; asserted by any device to interrupt current transactions and cause system to talk or listen to interrupting device.

REVIEW OF SECTIONS 14.5 AND 14.6

21. Why might a stepper motor be preferred over a dc servo motor in positioning the pen on a chart recorder?
22. Why would a stepper motor not be used to drive an instrument-cooling fan?
23. A 6-coil stepper motor similar to the one in Figure 14.8 is to rotate at 1000 rpm. What delay should be placed between each output to the latch?
24. How can four magnetizing directions be achieved with a 2-coil stepper motor?
25. What is the maximum number of instruments that can be listening on a GPIB bus at one time?
26. What else do the GPIB data lines carry besides data?
27. What signal tells the instruments on the GPIB to listen for address and command information?
28. Which handshake signal(s) are controlled by the talking device?
29. Which handshake signal(s) are controlled by the listening device?
30. Which GPIB pin is asserted when a talking device is finished outputting data?

14.7 A TESTER FOR TTL LOGIC ICs

This section presents a simple function tester for TTL logic ICs in 14-pin packages. The tester uses the basic 6802 microcomputer system of Figure 6.14 with a 6522 interface providing 12 programmable I/O lines for chip test and four lines to drive a 7-segment-LED chip identifier.

The program is expandable to cover any number of ICs. The listing given (List 14.2) provides for types 7400 and 7402, and a data-development table is given for the type 7404. We will leave it to you to add the 7404 test to the program and to develop tests for other popular chips such as the 7408, 7401, 7420, and 7486.

In operation the program simply runs through its whole library of tests, regardless of what chip type is in the test socket. If the chip passes all the tests for a

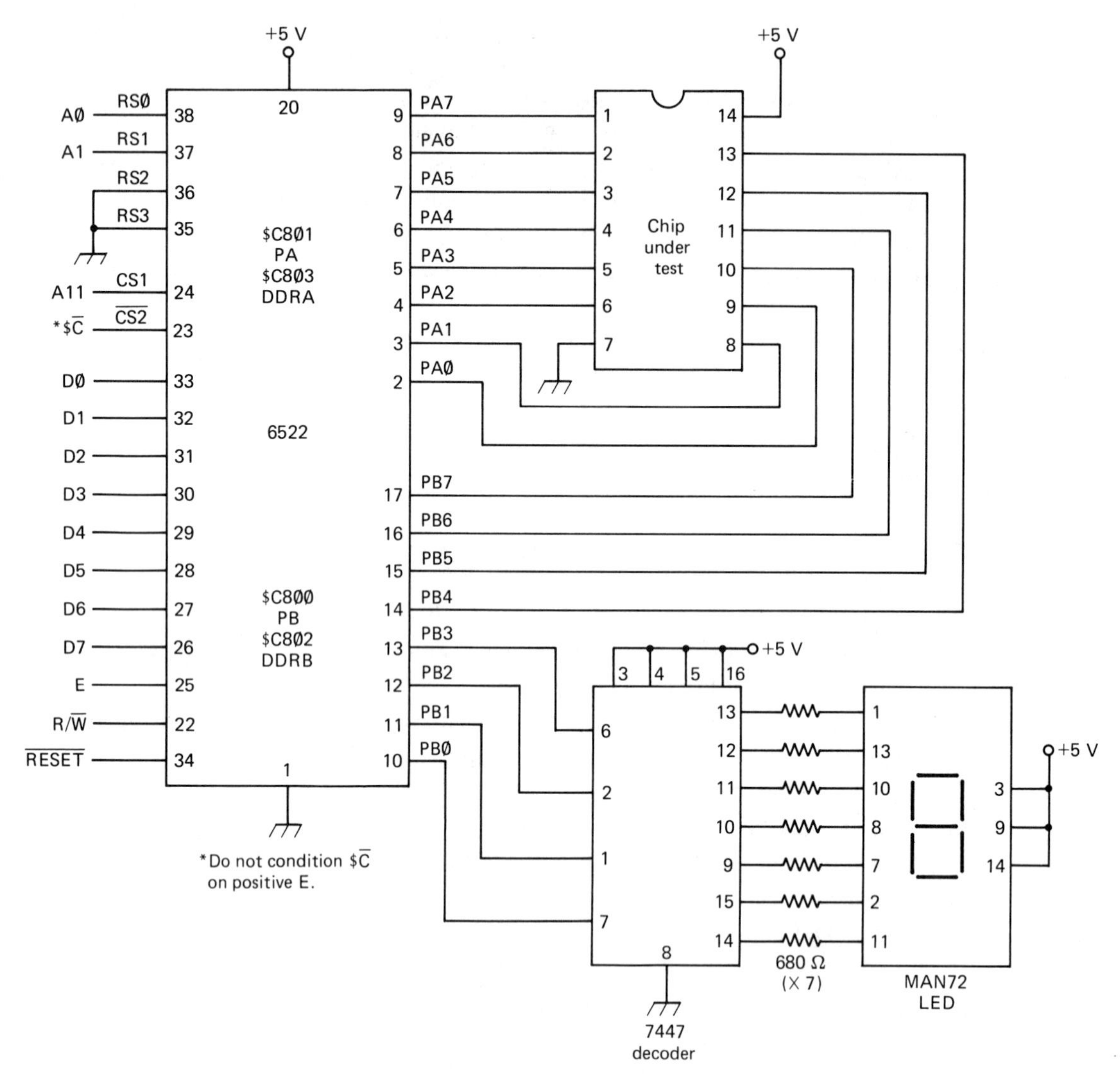

Figure 14.13 Hardware interface for the TTL IC Tester. The 6802 processor of Figure 6.14 or 7.6 can be used to control the 6522 VIA.

particular type, the corresponding number (0 through 9 only, unfortunately) is displayed. If none of the tests is passed, the display remains blank.

Figure 14.13 shows the tester hardware which may be interfaced to the basic processor and RAM of Figure 7.6, or to the EPROM-based system of Figure 6.14. The 6522 CS inputs must be asserted 180 ns *before* E goes *high*, so G2B of the 74LS138 must be grounded, not driven by an inverter from E as these Figures show. Note how absolutely indispensible the bitwise I/O programmability of the 6522 is to the operation of the tester.

```
                1000 * LIST 14-2 ** IC TESTER ** RUNS 15 JUL '87. METZGER
                1010 *  ROUTINES HERE FOR TYPES 7400 AND 7402.
                1020 * USE FIG 6.14 OR FIG 7.6 CPU.  FIG 14.13 INTERFACE.
                1030 * "E" MUST BE REMOVED FROM VIA CHIP-SELECT LOGIC.
                1040 *
                1050          .OR   $FF80
                1060          .TA   $4080
C800-           1070 PB       .EQ   $C800
C801-           1080 PA       .EQ   $C801    VIA PORTS
C802-           1090 DDRB     .EQ   $C802
C803-           1100 DDRA     .EQ   $C803    DATA-DIRECTION REGISTERS
                1110 *
FF80- CE BF D9  1120          LDX   #$BFD9   DATA-DIRECTION BITS FOR 7400
FF83- FF C8 02  1130          STX   DDRB      TO DDRB AND DDRA VIA X BYTES.
FF86- CE BF D9  1140          LDX   #$BFD9   TEST 7400 GATES FOR TRUTH
FF89- FF C8 00  1150          STX   PB        1 AND 1 = 0 ?
FF8C- BC C8 00  1160          CPX   PB       PB & PA OUTPUTS AS EXPECTED?
FF8F- 26 0B     1170          BNE   NEXTA     NO: TRY NEXT TYPE.
FF91- CE 50 76  1180          LDX   #$5076   TEST VARIOUS 7400 GATES FOR
FF94- FF C8 00  1190          STX   PB        0 & 1 = 1 OR 0 & 0 = 1.
FF97- BC C8 00  1200          CPX   PB       IF ALL TESTS PASS, ITS A 7400;
FF9A- 27 FE     1210 OKA      BEQ   OKA       DISPLAY 2ND DIGIT OF "$5076".
FF9C- CE 6F 6F  1220 NEXTA    LDX   #$6F6F   DATA DIRECTION BITS FOR 7402.
FF9F- FF C8 02  1230          STX   DDRB      (6F TO DDRB, 6F TO DDRA)
FFA2- CE 2F 2B  1240          LDX   #$2F2B   TEST 7402 FOR TRUTH OF
FFA5- FF C8 00  1250          STX   PB        1 OR 0 = 0 (PB AND PA)
FFA8- BC C8 00  1260          CPX   PB       ALL OUTPUTS GOOD?
FFAB- 26 0B     1270          BNE   NEXTB    NO: TRY NEXT TYPE.
FFAD- CE 92 90  1280          LDX   #$9290   TEST 7402 FOR TRUTH OF
FFB0- FF C8 00  1290          STX   PB        0 OR 0 = 1.
FFB3- BC C8 00  1300          CPX   PB       IF BOTH TESTS GOOD, DISPLAY "2"
FFB6- 27 FE     1310 OKB      BEQ   OKB       FROM BYTES "$9290".
                1320 *
                1330 * PLACE DATA FOR OTHER CHIP TYPES HERE.
                1340 *
FFB8- 86 FF     1350 NEXTB    LDAA  #$FF     PLACE AT END OF CHIP TESTS
FFBA- B7 C8 00  1360          STAA  PB        TO BLANK DISPLAY LED
FFBD- 20 FE     1370 HALT     BRA   HALT      AND HALT.
                1380          .OR   $FFFE
                1390          .TA   $40FE
FFFE- FF 80     1400          .HS   FF80
                1410          .EN
```

TABLE 14-1 IC TESTER DATA DEVELOPMENT

	7			PA				0	7			PB				0	Hex	
IC pins →	1	2	3	4	5	6	8	9	10	11	12	13	D	I	S	P	*A*	*B*
7400																		
DDR	1	1	0	1	1	0	0	1	1	0	1	1	1	1	1	1	D9	BF
1 · 1 → 0	1	1	0	1	1	0	0	1	1	0	1	1	1	1	1	1	D9	BF
0 · 1 → 1	0	1	1	1	0	1	1	0	0	1	0	1	0	0	0	0	76	50
7402																		
DDR	0	1	1	0	1	1	1	1	0	1	1	0	1	1	1	1	6F	6F
1 + 0 → 0	0	0	1	0	1	0	1	1	0	0	1	0	1	1	1	1	2B	2F
0 + 0 → 1	1	0	0	1	0	0	0	0	1	0	0	1	0	0	1	0	90	92
7404																		
DDR	1	0	1	0	1	0	0	1	0	1	0	1	1	1	1	1	A9	5F
1 → 0	1	0	1	0	1	0	0	1	0	1	0	1	1	1	1	1	A9	5F
0 → 1	0	1	0	1	0	1	1	0	1	0	1	0	0	1	0	0	56	A4

Answers to Chapter 14 Review Questions

1. Pins 4, 8, and 9 **2.** 1, Ø, 1, Ø, Ø **3.** $4ØØ3
4. It provides increased current drive for the LEDs.
5. A capitol L on LED 5 **6.** BNE NEXT **7.** BMI ONESEC
8. H **9.** It can be checked easily with the BMI instruction. **10.** In Figure 14.2, two multiplexed ports do the work of eight fixed ports.
11. The *data ready* on CA1. **12.** When the VIA asserts *data taken* on CA2.
13. RSØ = Ø, RS1 = Ø, RS2 = 1, RS3 = 1
14. To avoid writing Øs where 1s are already stored, and thus disabling other functions in the register. **15.** They are disabled.
16. CA1 **17.** 250 Ω **18.** Port B **19.** Port A
20. Yes **21.** No position-sensing hardware would be required.
22. Accurate position control is not needed, only high-speed rotation. **23.** 10 ms
24. Reverse the currents through the coils to gain two more directions.
25. 14 **26.** Instrument addresses and commands from the controller.
27. Assertion of the ATN (attention) control line. **28.** DAV
29. NRFD and NDAC **30.** EOI, pin 5

CHAPTER 14 QUESTIONS AND PROBLEMS

Digits after the decimal point refer to section numbers in Chapter 14.

Basic Level

1.1 A 6522 VIA is wired so that its Port A is accessed at address $CCØ1. List the addresses of Port B, Data-Direction Register A, and Data-Direction Register B.

2.1 In the preceding VIA interface, how would Port B be programmed for all inputs?

3.2 Referring to Figure 14.2, what data bytes must be written to Ports A and B, respectively, to produce the display L on LED3?

4.2 Interpret the data file at the end of List 14.1 to predict the message that will be displayed by the hardware of Figure 14.2.

5.3 How are VIA pins CA1 and CA2 programmed to become handshake lines for Port A?

6.3 What are the functions assigned to the VIA pins CA1 and CA2 by the Peripheral Control Register as the VIA "comes up" from a reset?

7.3 How many addressable internal registers does the 6522 have? Look ahead to Figure 15.7 and list their names.

8.4 Tell whether input and output handshaking are possible for **(a)** the VIA Port A and **(b)** the VIA port B.

9.4 VIA ports A and B are configured for all outputs. Will a read of Port A always reflect the data that was previously written to Port A? Answer the same question for Port B.

10.5 How can four different rotor positions be obtained in a stepper motor with only 2 field coils?

11.5 What advantage does the 2-coil stepper motor of Figure 14.9 have over the 4-coil design of Figure 14.7?

12.6 In Figure 14.11, when does the $\overline{\text{NRFD}}$ line go *high*?

13.6 In Figure 14.10, how does the controller identify which of the two listeners is to receive the next data on the data bus?

Advanced Level

14.1 Write the 6802 source code to program a VIA's Port A for all inputs, read a byte of data, increment it, change the Port to all outputs, and output the new data. Calculate the time required if *E* is 1 MHz.

15.1 Write the 6802 source code for a routine that outputs all possible binary combinations via Port B and reads Port A after each new output. The routine should count the number of times that the A input data equals the B data just output, store this number at Port B, and halt.

16.2 Explain how the hardware of Figure 14.2 and the program of List 14.1 could be modified using two 74LS154 ICs to drive 32 display LEDs with one VIA. Include a partial sketch.

17.3 Explain the problems that can occur **(a)** when inputting data and **(b)** when outputting data if handshaking is not used.

18.3 Using the addresses assigned in Figure 14.4, write a 6802 source-code routine that will hold CA2 *high* for 1 ms, *low* for 1 ms, and then will wait for CA1 to go *low*, whereupon the 6802 will acknowledge an IRQ input.

19.4 Explain why the processor's $\overline{\text{IRQ}}$ input is used in handshaking with the VIA. Couldn't the data transfer be accomplished more easily by the main program?

20.4 Describe the difference between a transparent input buffer and a latched input buffer. How can the 6522 be configured for one or the other?

21.5 When a system using a stepper motor is first turned on, the controlled device must first be returned to a "home" position. Explain why this is so and why it is not necessary in a system using a dc servo motor.

22.5 Write the 6802 source code for a program to rotate the stepper motor of Figure 14.8 counterclockwise at a speed of 100 rev/min. Assume a 1-MHz clock.

23.6 Explain how the talker on an IEEE-488 bus decides when to assert new data and when to remove that data from the bus.

24.6 In an IEEE-488 system containing more than one controller, how can the second controller take over the bus from the first controller?

25.7 In the format of Table 14.1, list the hex digits to configure the IC tester of Figure 14.13 for a 7410 IC, and give the digits required to test it for the condition $1 \cdot 1 \cdot 1 \rightarrow \emptyset$.

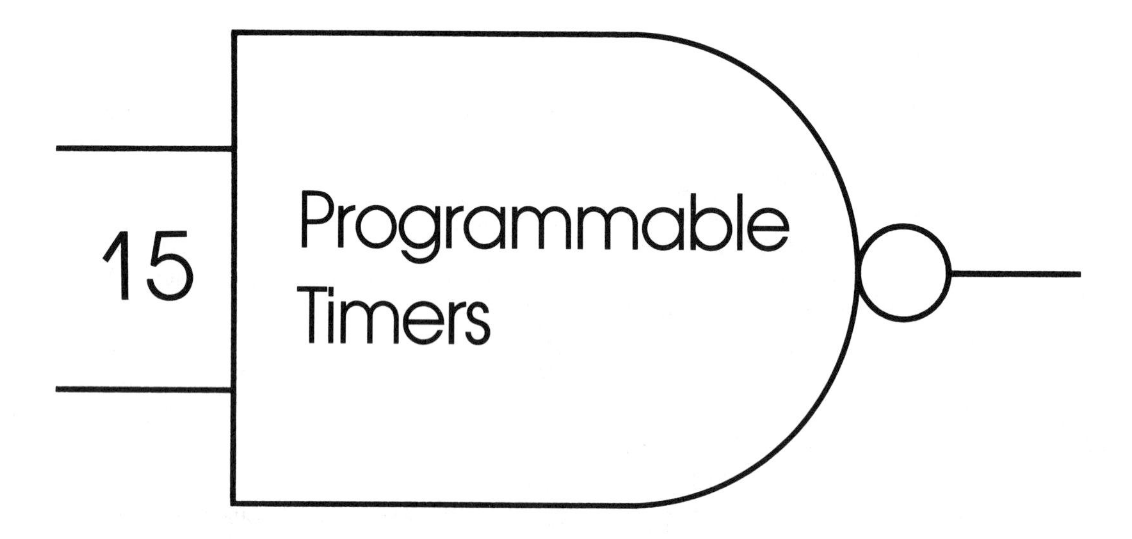

15.1 BASIC 6522 TIMER OPERATION

Section Overview

Hardware timers free the microprocessor from running time-delay loops so it can be devoted to more "intelligent" tasks. Where two independently variable time delays must run simultaneously, it is not even possible to handle the timing with program-delay loops.

Timers are binary counters that count down at the *E*-clock rate and generate an interrupt for the processor when a count of zero is reached. The initial count is loaded to the counter from the processor via the data bus.

The need for hardware timers: In Chapter 4 we wrote a few programs to generate various tones by holding the processor in a delay loop for half an audio cycle and then switching the speaker-driving voltage. If you will glance back at these programs you will see that they completely tie up the processor so it can do nothing else. If an urgent interrupt request comes along, either the tone will have to be abandoned or the interrupt will have to wait.

An even more untenable situation arises when two time delays must be running at once. Consider the case of a microprocessor controller for a 4-cylinder fuel-injected engine. The dwell time for 1 cylinder (which is the charging time for the ignition-coil inductance) may be going on while the fuel-jet opening time is going on for another. The processor alone cannot handle both jobs, yet it is expected also to be monitoring engine temperature, manifold pressure, fuel/air ratio, and other factors.

The solution is to take the simple timing tasks away from the processor and give them to an external binary counter. When the counter *times out*, it will notify the processor with an interrupt signal, and the processor will do what needs to be done. In the meantime the processor will be free to handle the more complex computational and control tasks.

Two 16-bit timers are contained in the 6522 Versatile Interface Adapter. They can each be configured for different functions, but we will first examine how timer T1 operates in its normal mode—the mode it comes up in upon resetting the VIA. This is called the *one-shot mode* because the timer needs to be reloaded by the processor after each interrupt output is generated.

T1 is 16 bits wide, but it must be loaded via the data bus, so it is broken into two byte-sized registers. These are VIA register 4 (T1C-L, low-order byte) and register 5 (T1C-H, high-order byte). If the VIA chip-select lines decode to place its base address at $C8ØØ, the timer T1 can be loaded and read at addresses $C8Ø4 (low-order byte) and $C8Ø5 (high-order byte). Notice that this low/high byte order is reversed from the usual practice in 6800-based systems. This is because the 6522 is a member of the 6502 microprocessor family, and low-byte/high-byte is the usual order in that family. Figure 15.1 illustrates the 6522 timer T1 registers and the 6522 interface to access them. Figure 15.7 shows how these registers relate to the other 6522 registers.

Each of the 6522 functions must be specifically enabled to generate an interrupt if an $\overline{\text{IRQ}}$ output is desired. This is done by loading a 1 into the designated bits of the Interrupt-Enable Register (IER, register 14). To enable the timer-1 interrupt the designated bits are 7 and 6, so $CØ must be written to address $C8ØE. Section 15.4 explains the IER in detail.

To use the timer you must first calculate the initial count required. Let us say that we want a 5-ms delay in a system with a 0.5-MHz clock. The count required is

$$N = \frac{\text{delay}}{\text{clock period}} = \frac{5 \text{ ms}}{2 \ \mu\text{s}} = 2500 = \$9C4$$

The value $C4 must be stored to the low-order byte of T1 (T1C-L, at address $C8Ø4 in our system). Then the value $Ø9 must be stored to T1C-H at address $C8Ø5. The write to T1C-H automatically starts the countdown. When both registers reach zero the $\overline{\text{IRQ}}$ output of the 6522 is asserted *low*. It is assumed that the processor will respond to this interrupt by servicing the timed device.

If another time delay is to be done, the processor will write the delay count into T1C-L and T1C-H. The write to T1C-H will restore the $\overline{\text{IRQ}}$ line to a *high* (inactive) level. If another interval is not to be timed, a dummy read of T1C-L should be done, as this will also reset the $\overline{\text{IRQ}}$ output to a *high* level. Also be aware that if

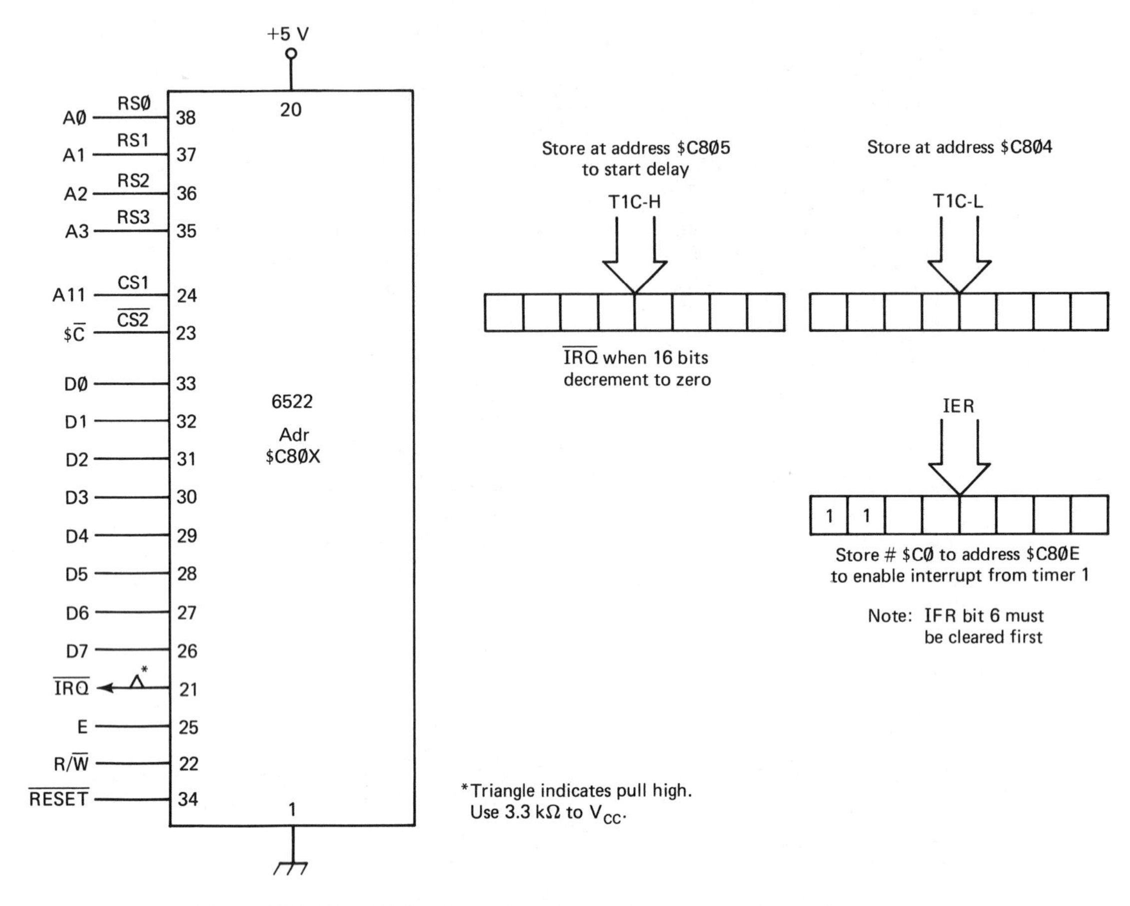

Figure 15.1 Timer T1 has high-order and low-order counter registers. The Interrupt-Enable Register permits T1 to generate an IRQ when it counts down to zero.

the desired delay is less than 256 counts, $ØØ must still be written to T1C-H, because it is this write which starts the timer.

As a point of interest, the longest time delay that can be obtained by T1 without processor intervention is (2^{16}) (1 μs), or 66 ms on a 1-MHz clock. Of course, longer delays can be made by simply having the processor count out a number of 66-ms time intervals, reloading T1C-L and T1C-H with $FFFF after each one, and letting it do its calculations in the 66-ms intervals.

It may also be of some interest to know that T1C-H and T1C-L can be read by the processor as they are counting down, so it is possible to find out how much time is left. After the counters *time out* (i.e., reach zero) they *roll around* to $FFFF and continue counting ($FFFE, . . .), so the 2's complement value in the timer can be read to determine how long the $\overline{IRQ}$ has been asserted.

15.2 ADVANCED TIMER CONFIGURATIONS

Section Overview

This section explains how the 6522 timer T1 can be configured to produce continuous and precisely timed $\overline{\text{IRQ}}$ outputs without processor intervention. It also shows how an additional output timed by T1 can be obtained from a port B line, PB7. The various modes of T1 are selected by setting or clearing specific bits in the Auxiliary Control Register (ACR, register 11).

T1 free-run mode: In the one-shot mode the timer generates a precisely determined delay between the write to T1C-H and the assertion of $\overline{\text{IRQ}}$. However, the time it takes for the processor to respond to the interrupt and to reload the timer may depend on how busy the processor was when it received the interrupt. Perhaps it was in the middle of a critical task and masked out the interrupt for a short period with the SEI instruction. A series of time delays therefore might not produce perfectly regular intervals. Perfectly regular intervals can be produced by taking the timer-reload function away from the processor and giving it to a new pair of registers, the T1 *latches*, low- and high-order bytes (T1L-L and T1L-H). Figure 15.2 illustrates the relationship of the T1 counters and latches.

The T1 latches are registers 6 (low-order byte) and 7 (high-order byte) of the VIA. These latches are automatically loaded with the bits written by the processor into the corresponding T1 counters, so they are normally loaded by writing to (in our example system) addresses $C8Ø4 (low-order byte) and $C8Ø5 (high-order byte).

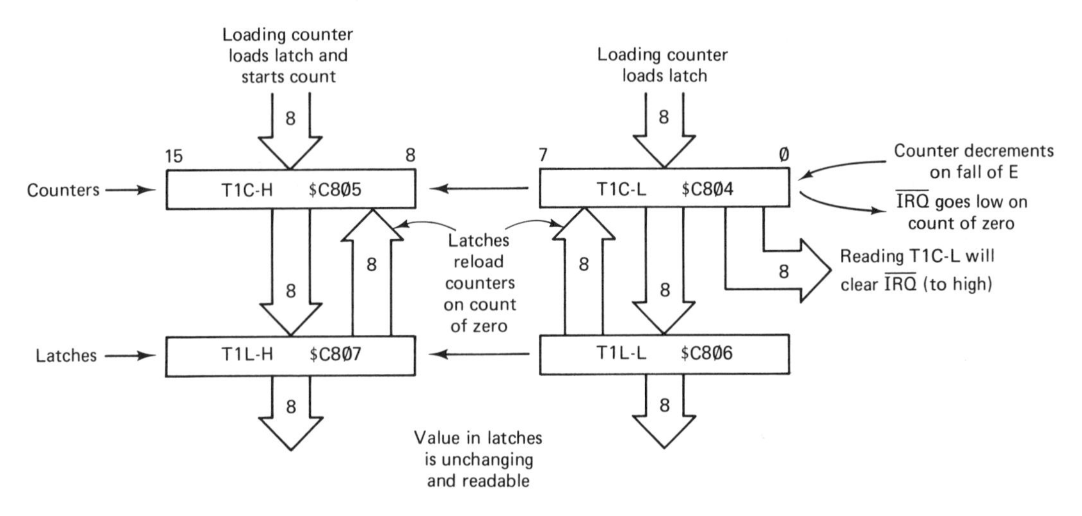

Figure 15.2 The free-run mode of timer T1. The counters are reloaded automatically from the latches as each interval times out.

The data in the latches does not change as the counters decrement, so addresses \$C8Ø6 (low) and \$C8Ø7 (high) can be read at any time to find the current timer interval.

In the free-run mode, the counters are automatically reloaded from the latches when the decrementing count reaches zero, and a new time interval begins. The $\overline{IRQ}$ is asserted when the count reaches zero but the processor can take its time in responding to the interrupt without delaying the regular count intervals. The only requirement is that the processor get around to servicing the interrupt before the counters time out a second time. The $\overline{IRQ}$ can be reset to a *high* level without disturbing the timing by reading the low-order counter.

Timer T1 is configured for the free-run mode by writing a 1 into bit 6 of the Auxiliary Control Register, which is at address \$C8ØB in our system. This can be done without disturbing the other bits by a logic OR with a byte having a 1 in the bit-6 position and Øs in all other bit positions. List 15.1 shows a program segment to initialize T1 for 100-ms free-running interrupts and an interrupt routine to shift an LED pattern driven by Port B at 10 shifts per second.

```
                  1000 * LIST 15-1   RUNS, 15 JULY '87.   D. METZGER
                  1010 * SET 6522 VIA TIMER T1 TO SHIFT 1 OF 8
                  1020 *  LIT LEDS RIGHT AT 100-MS RATE, VIA PORT B.
                  1030 *
                  1040 * NOTE: RS2, RS3, AND IRQ* MUST BE CONNECTED TO
                  1050 *        THE 6802 AS IN FIG 15.11. LEDS ON PORT B.
                  1060          .OR  $FF80
                  1070          .TA  $4080
C80B-             1080 ACR      .EQ  $C80B     AUX CONTROL REG.
C805-             1090 T1CH     .EQ  $C805     TIMER 1 COUNTER, HIGH.
C804-             1100 T1CL     .EQ  $C804     TIMER 1 COUNTER, LOW.
C800-             1110 PB       .EQ  $C800     PORT B PULLS LEDS LOW.
C802-             1120 DDRB     .EQ  $C802     DATA DIRECTION B.
C80E-             1130 IER      .EQ  $C80E     INTERRUPT-ENABLE REG.
0010-             1140 BUFR     .EQ  $0010     RAM PASSES LED PATTERN TO IRQRTN.
                  1150 *
FF80- C6 FF       1160          LDAB #$FF      SET PORT B FOR ALL OUTPUTS.
FF82- F7 C8 02    1170          STAB DDRB
FF85- C6 FE       1180          LDAB #$FE      READY TO PULL 1 LED LO ON IRQ.
FF87- D7 10       1190          STAB BUFR       SUBRTN WILL SHIFT LO RAM BIT.
FF89- 8E 00 7F    1200          LDS  #$007F    INITIALIZE STACK AREA FOR IRQ.
FF8C- 86 C0       1210          LDAA #$C0      WRITE 1'S TO BITS 7 AND 6 TO
FF8E- B7 C8 0E    1220          STAA IER        ENABLE IRQ FROM T1.
FF91- 86 40       1230          LDAA #$40      BIT 6 = 1 TO ESTABLISH FREE-RUN
FF93- BA C8 0B    1240          ORAA ACR        MODE OF TIMER 1 VIA AUX CTRL
FF96- B7 C8 0B    1250          STAA ACR        REG. ORAA SAVES OTHER BITS.
FF99- 86 C3       1260          LDAA #$C3      ABOUT 50,000 COUNTS AT 2 US EACH
FF9B- B7 C8 05    1270          STAA T1CH       (IGNORE LOW-ORDER BYTE).
FF9E- 0E          1280 WAIT     CLI            ACCEPT INTERRUPT.
FF9F- 20 FD       1290          BRA  WAIT      PROCESSOR IS FREE; VIA COUNTS TIME.
                  1300 *
                  1310 * IRQ ROUTINE
                  1320          .OR  $FFC0
                  1330          .TA  $40C0
FFC0- D6 10       1340          LDAB BUFR      GET CURRENT LED PATTERN.
FFC2- 0D          1350          SEC            SHIFT BINARY 1 TO BIT 0.
FFC3- 59          1360          ROLB           ONE BIT IN B IS A 0.
FFC4- 25 01       1370          BCS  SKIP      IF THE 0 SHIFTS INTO C
FFC6- 59          1380          ROLB            SEND IT AROUND TO BIT 0.
```

```
FFC7- F7 C8 00 1390 SKIP    STAB PB       IF NOT, ONE SHIFT IS ENOUGH.
FFCA- B6 C8 04 1400         LDAA T1CL     DUMMY READ TO RETURN IRQ LINE HI.
FFCD- D7 10    1410         STAB BUFR     SAVE NEW PATTERN. (B WILL BE UNSTACKED)
FFCF- 3B       1420         RTI           BACK TO MAIN PROG; CLI/WAIT LOOP.
               1430         .OR  $FFF8
               1440         .TA  $40F8    IRQ VECTOR AT FFF8, THEN SWI, NMI, RST
FFF8- FF C0 FF
FFFB- 80 FF 80
FFFE- FF 80    1450         .HS  FFC0FF80FF80FF80
               1460         .EN
```

An extra hardware output for the T1 timer can be configured to appear at Port B, bit 7. This configuration requires that PB7 be set as an output by writing a 1 to DDRB7 (at address $C8Ø2) and that a 1 be ORed to ACR bit 7 (at address $C8ØB). There are two forms that the output at PB7 can take.

- With T1 in the one-shot mode (ACR6 = Ø), the PB7 output remains *low* during the countdown from the loaded value to zero and remains *high* from the count-of-zero to the next processor load of the counter high-order byte, which starts the timer.
- With T1 in the free-run mode (ACR6 = 1) the PB7 output remains *low* during the first timing period and all subsequent odd-numbered periods. It stays *high* during all even-numbered timing periods, thus producing a timed square-wave output if the T1 latches are not reloaded.

The PB7 output configuration for the 6522 is illustrated in Figure 15.3.

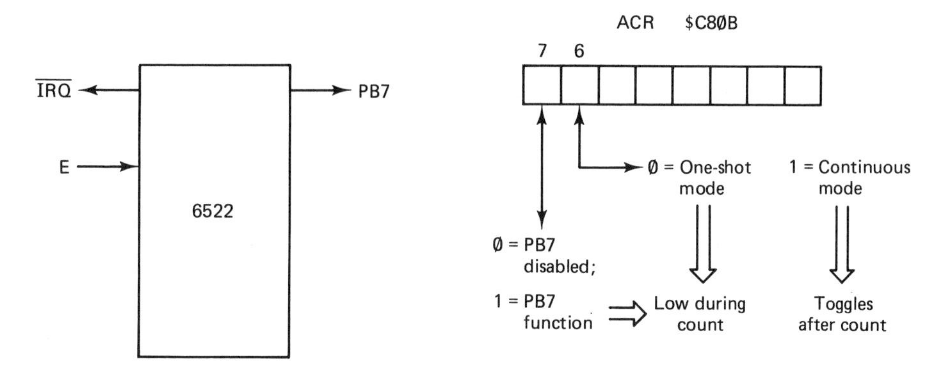

Figure 15.3 The Auxiliary Control Register can program timer T1 for single or continuous timing intervals and for outputs to the 6522 Port B, pin 7.

REVIEW OF SECTIONS 15.1 AND 15.2

Answers appear at the end of Chapter 15.

1. Why is it undesirable to use processor software loops to generate time delays in more complex systems?

2. A microwave oven is being designed to include a clock, a cook timer, and an alarm timer, all of which must operate simultaneously. How can a program be written to allow the processor to handle all three delays?
3. When does the T1 timer start running?
4. What step is necessary to permit timer T1 to assert the 6522's $\overline{IRQ}$ pin when it times out?
5. Upon receiving an IRQ, the processor finishes its current calculation. Then it reads T1C-L as $37 and T1C-H as $D9. What does this mean? (The clock is 1 MHz.)
6. How can the T1 free-run mode produce regular intervals when the one-shot mode cannot?
7. The processor writes $12 to T1C-L and $34 to T1C-H. After 7000 *E*-clocks, the processor reads T1L-H. What value does it read?
8. The Auxiliary Control Register has $8Ø written to it. What is the T1 mode and the status of PB7?
9. How can bit Ø of the ACR be set to 1 without changing any of the other seven ACR bits?
10. How can T1 be configured to produce a 1-kHz square wave at PB7? (Assume a 1-MHz clock.)

15.3 THE VIA TIMER 2

Section Overview

The 6522 contains a second 16-bit timer T2, which has two modes of operation. The first is a one-shot mode in which *E*-clocks count down from a preloaded value and the $\overline{IRQ}$ is asserted when 0 is reached. The second mode is similar, except that the timer decrements in response to negative inputs to the PB6 pin instead of the *E*-clock. This permits counting of irregularly timed events, and it makes possible the generation of very long delays by feeding the input of T2 from the output of T1.

Timer 2 one-shot mode: Timer 2 consists of a low-order byte T2C-L, which is register 8 of the VIA, and high-order byte T2C-H, which is register 9. The timer comes up in the one-shot mode in which the count down is started by writing to the high-order byte, the same as timer T1. The program should load the low-order byte first and then load the high-order byte to start the decrements, which occur at the rate of the *E*-clock.

Upon decrementing to 0, T2 causes the $\overline{IRQ}$ output to go *low* if the enabling bits for T2 have been written into the Interrupt-Enable Register. The enabling bits are 7 and 5; write $AØ to address $C8ØE. Details of the IER appear in the next section.

The $\overline{IRQ}$ can be returned to a *high* level by a read of T2C-L or by the next write to T2C-H, which also restarts the timing interval. T2 must be reloaded by the processor; there is no free-run mode in which it is reloaded automatically. Only two addresses are used to access T2; there are no readable latches to retrieve the original count. Reading T2C-H and T2C-L will obtain the current count, which is always decrementing. The decrementing continues past $ØØØØ to $FFFF, $FFFE, and so on. The 6522's single $\overline{IRQ}$ line is the only hardware output from T2; there is no output possible on the Port B bus as there is for T1.

The T2 pulse-counting mode is entered by changing ACR bit 5 from Ø to 1. This can be done without affecting other ACR bits by ORing $2Ø with the ACR and writing the result back to the ACR. In the pulse-counting mode T2 does not decrement on the negative edge of *E* but on the negative edge of an externally applied pulse that is applied to T2 through Port B pin PB6. Of course, PB6 should be configured as an input by writing a Ø to DDRB6.

Each negative pulse at PB6 will decrement the T2 counter by one, and the pulses need not occur regularly nor at any minimum frequency. The maximum frequency, however, is ¼ of the *E*-clock rate. When the counter decrements to 0, the $\overline{IRQ}$ line will be asserted, signaling that the preloaded number of pulses has been counted. If the counter is preloaded with $FFFF the 1's complement of the value in the counter will give the number of PB6 pulses received. (One's complement is simply the result of changing all binary 1s to Øs and all Øs to 1s; the 6800 COM instruction.)

Long time delays, up to 143 minutes on a 1-MHz clock, can be obtained by cascading the two counters. The T1 output at PB7 is fed to the PB6 input of T2, as shown in Figure 15.4. The $\overline{IRQ}$ output must be enabled for T2 but not for T1, so $AØ is written to the IER at address $C8ØE after a *reset*.

T1 should be set for free-running square-wave outputs on PB7 (ACR7 = 1, ACR6 = 1), and T2 should be set for pulse counting on PB6 (ACR5 = 1). Thus data $EØ should be written to address $C8ØB.

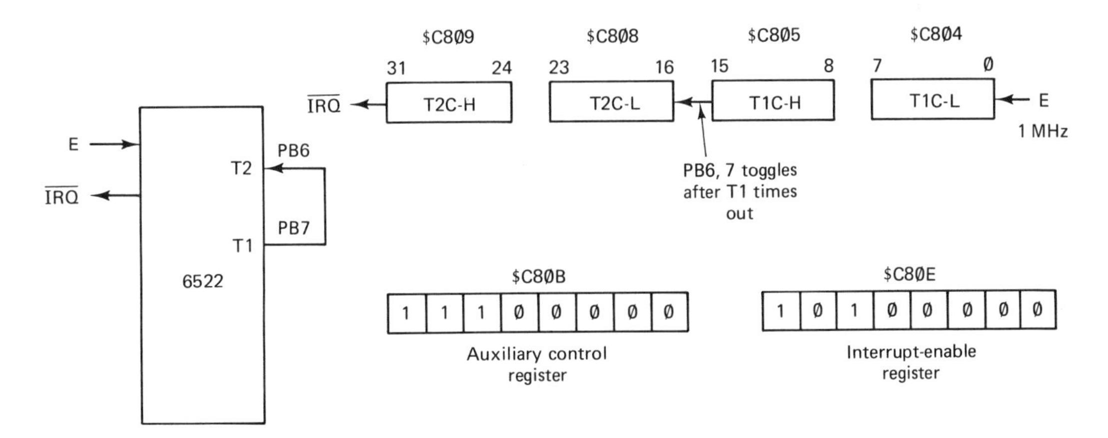

Figure 15.4 Delays as long as 143 minutes (on a 1-MHz clock) can be obtained by feeding the T2 counter input from the T1 counter output.

15.4 COORDINATING THE VIA FUNCTIONS

Section Overview

The 6522 VIA contains two ports, two timers, and a shift register, each of which is capable of generating an interrupt on a single $\overline{IRQ}$ output line. This section explains how the processor can specify which functions can cause an interrupt via the Interrupt-Enable Register and how it can determine where an interrupt came from by reading the Interrupt Flag Register.

A brief review of the Auxiliary Control Register and VIA Peripheral Control Register is also included.

The Interrupt-Enable Register. There are seven separate events that can activate the VIA's interrupt output to the processor's $\overline{IRQ}$ input:

- Port *A* handshake input CA1 and CA2 if input-configured.
- Port *B* handshake input CB1 and CB2 if input-configured.
- Timer T1 timed out.
- Timer T2 timed out.
- Shift register, eight shifts done (summary in Figure 15.7).

Upon receiving an interrupt the processor's interrupt-service routine will first have to determine the source of the interrupt. In industrial environments spurious interrupts caused by noise are unavoidable, and the software should be written to double-check that the interrupt is valid. The interrupt-identification problem may then be simplified by disabling all the interrupt sources that are not to be serviced during a particular program segment. If the number of enabled interrupt sources can be reduced to one at a time, the problem of interrupt origin is solved.

The Interrupt-Enable Register is register 14 of the 6522 and is accessed at address $C80E in our system. Figure 15.5 shows the functions of each bit in the IER. Since individual bits of the IER must be set (to enable) or cleared (to disable) frequently without disturbing the status of the other bits, a unique method of writing to the IER has been implemented by the 6522 hardware. You will recall that an individual bit can be set using the OR instruction and cleared using the AND instruction without disturbing the other bits. Here are two examples, the first setting ACR0 (on the left) and the second clearing it (on the right).

```
LDAA   #$01          LDAA   #$FE
ORAA   ACR           ANDA   ACR
STAA   ACR           STAA   ACR
```

These functions can be implemented more directly in the IER. Here are the rules:

1. The bits written to the IER ($C80E) in bit positions 0 through 6 do not appear

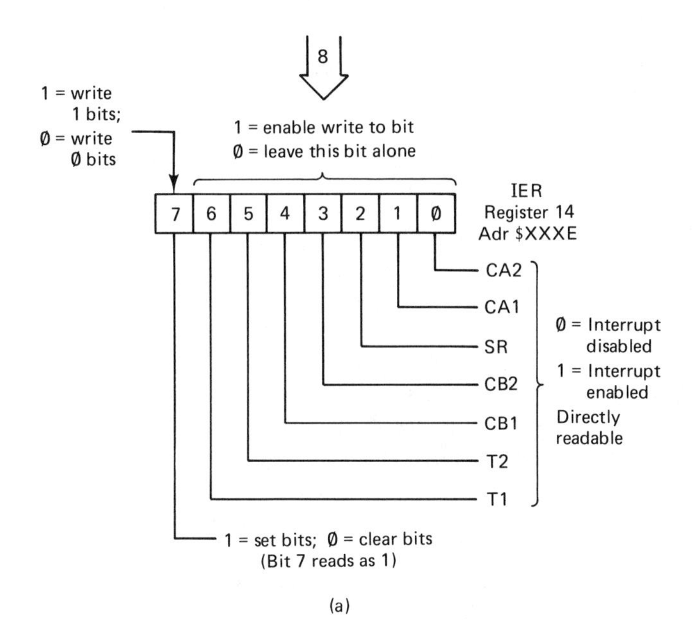

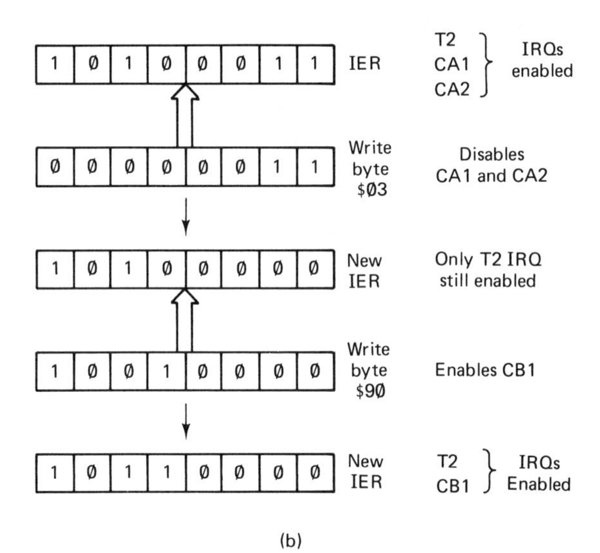

Figure 15.5 (a) Writing to the IER enables or disables any of seven interrupt sources in the VIA. (b) An example disable and enable procedure, to be read from top to bottom.

directly in the IER. Instead, these are write-enabling bits; a 1 specifies *write to this bit* and a Ø specifies *leave this bit alone*.

2. The bit written to the IER address, bit-7 position, specifies whether 1s (enable interrupt) or Øs (disable interrupt) are to be written to the write-enabled bits.

Thus IERØ is set by writing binary 1ØØØ ØØØ1 (or $81) to the IER address, and IERØ is cleared by writing binary ØØØØ ØØØ1 (or $Ø1) to the IER.

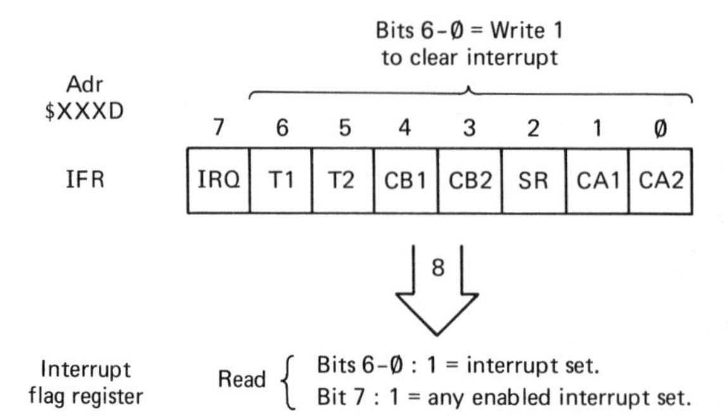

Figure 15.6 The Interrupt Flag Register of the VIA can be read to determine which, if any, of the interrupt sources are requesting an interrupt.

The Interrupt Flag Register (IFR) is register 13 of the VIA and is accessed at address $C80D in our system. Each bit in the IFR corresponds to one of the interrupt sources enabled or disabled by the IER. The interrupt source sets its IFR flag bit whenever the interrupt conditions are fulfilled. If the corresponding IER enable bit is *set*, the VIA's output $\overline{\text{IRQ}}$ is also asserted.

Bit 7 of the IFR is *set* if any interrupt flag is *set* while its corresponding interrupt-enable bit is also set. In other words, IFR7 = 1 if the $\overline{\text{IRQ}}$ output has been asserted. This bit gives the processor a fast way to poll for interrupts at its convenience, rather than actually responding to a hardware interrupt input. The IFR is read and the BMI instruction is used to take the program to an interrupt-sorting routine if IFR7 is *set*. The IFR functions are illustrated in Figure 15.6.

Interrupt sorting is done by examining the flag bits one at a time and sending the processor to the appropirate interrupt-service address when a *set* flag is found. Implicit in this process is the idea of *interrupt priorities*. If the processor gets behind and two or more interrupt flags are up at one time, the processor will obviously have to service one of them first. It would be best if multiple interrupts could be serviced in some rational order of priorities—the most crucial first, the most time-consuming last, or whatever is decided by the software designer.

The interrupt priority hierarchy is the order in which the IFR flag bits are examined. If a LSR is used to shift bits into the processor's carry flag, the natural priority assigned by the 6522 is

CA2 CA1 SR CB2 CB1 T2 T1

If the ASL is used to shift bits into C for testing, the order is

T1 T2 CB1 CB2 SR CA1 CA2

Any other order would require a rather lengthy series of bit-mask and test-for-nonzero routines.

In any case, the first order of business in checking for the interrupt source must be clearing bit 7 (the *any-enabled-interrupt* flag) and the flag bits for any interrupts disabled by the IER. To clear bits directly in the IFR, we must make use of the peculiar fact that IFR bits are *cleared* by writing a logic 1 to them and left unaffected if a Ø is written to them. The 1s are thought of as a *bit-select* code for the function *clear* by writing to IFR. Bit 7 cannot be cleared in this way. Of course there is no

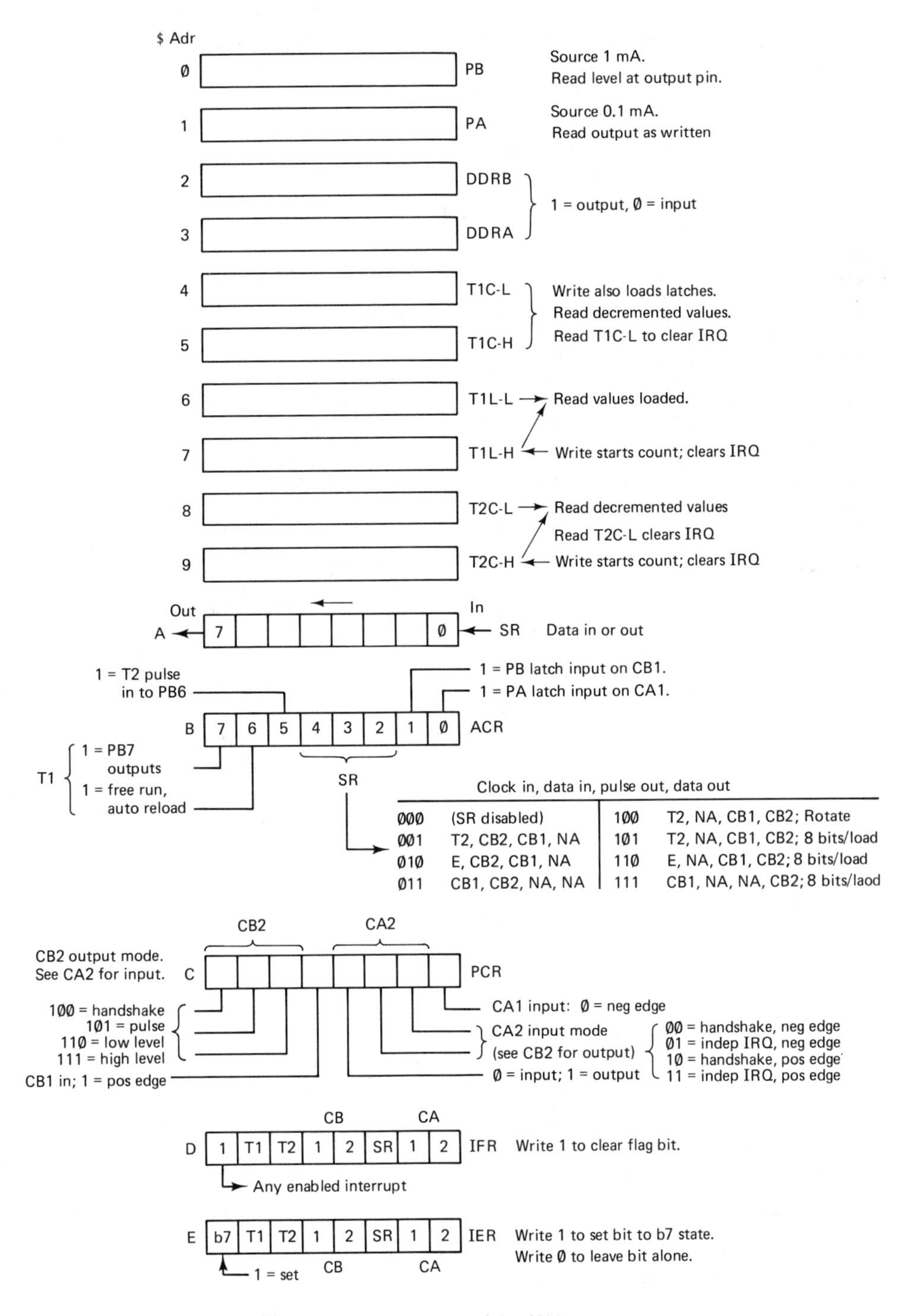

Figure 15.7 A summary of the 6522 registers.

need to *set* bits in the IFR by writing to the register directly, so no means is provided to do this. Here is a routine to sort interrupts for the VIA.

```
                    1000 * LIST 15-2 ** INTERRUPT SORT FOR VIA. SEE FIG 15.6.
                    1010 * PROGRAM SEGMENT; NOT TO RUN.   16 JULY '87. METZGER
                    1020 *
                    1030            .OR  $FF80
                    1040            .TA  $4080
C80D-               1050 IFR        .EQ  $C80D     INTERRUPT FLAG REGISTER.
C80E-               1060 IER        .EQ  $C80E     INTERRUPT ENABLE REGISTER.
FFD0-               1070 TABLE      .EQ  $FFD0     TABLE OF IRQ-ROUTINE ADDRESSES.
                    1080 *
FF80- F6 C8 0D 1090            LDAB IFR       SKIP SORT ROUTINE IF
FF83- 2A 18    1100            BPL  NOIRQ     FLAG 7 = 0 (NO ACTIVE IRQ)
FF85- CE FF D0 1110            LDX  #TABLE    X POINTS TO TABLE CONTAINING
               1120 *                          ADDRESS OF ROUTINE 0.
FF88- B6 C8 0E 1130            LDAA IER       ENABLED IRQS = 1; BIT 7 =1.
FF8B- B4 C8 0D 1140            ANDA IFR       1S FOR VALID IRQS AND B7 ONLY.
FF8E- 84 7F    1150            ANDA #$7F      CLEAR BIT 7 SO NO BITS WILL
               1160 *                          MEAN NO INTERRUPTS.
FF90- 44       1170 NEXT       LSRA           SHIFT INT FLAG INTO CARRY.
FF91- 25 06    1180            BCS  DOIT       CARRY SET = SERVICE INTERRUPT.
FF93- 27 08    1190            BEQ  NOIRQ     NO "1" BITS = NO IRQ CALL.
FF95- 08       1200            INX            POINT X TO NEXT ADR IN TABLE OF
FF96- 08       1210            INX             SERVICE ADDRESSES
FF97- 20 F7    1220            BRA  NEXT       AND CHECK NEXT IRQ FLAG.
FF99- EE 00    1230 DOIT       LDX  0,X       GET ADDRESS OF SERVICE ROUTINE FROM
FF9B- 6E 00    1240            JMP  0,X        TABLE INTO X; THEN GO TO THAT ADR.
FF9D- 01       1250 NOIRQ      NOP            EXIT SORT; RETURN TO MAIN PROG.
               1260            .EN
```

A summary of the 6522 registers and operations appears in Figure 15.7. The shift-register permits serial data to be received or transmitted via CB1 and CB2 at a selected clock rate. Chapter 16 deals with a peripheral chip dedicated entirely to serial data communication, so we will not cover the 6522 shift register in detail.

REVIEW OF SECTIONS 15.3 AND 15.4

Answers appear at the end of Chapter 15.

11. What value should be loaded to what address to cause a 20-ms delay via the T2, assuming a 1-MHz clock?

12. How can T2 be set up for a second delay period equal to the first?

13. What data must be stored to what registers to set up the 6522 for an $\overline{IRQ}$ output after 48 negative pulses into PB6?

14. How can the 6522 be configured to produce time delays on the order of 10 min?

15. How many interrupt-generating functions are there in the 6522, and how many interrupt output lines are there?

16. What hex byte should be written to what register to configure T1, CA1, and CB1 to produce interrupt outputs?

17. Continuing Question 16, what hex byte should be written to disable the interrupts from CA1 and CB1, leaving T1 enabled?

18. What does it mean if the IFR is nonzero (BEQ not taken) but bit 7 of the IFR is zero (BPL taken)?
19. What is achieved by writing $FF to the IFR?
20. What is achieved by ANDing the IFR bits with the IER bits?

15.5 WRITING TO NONVOLATILE MEMORY

Section Overview

This section provides an illustration of how the parallel ports and the timer of the 6522 may be used in conjunction with one another. Imagine an automotive monitoring system in which engine rpm data is to be recorded every half second during a test-track run. This will require pulse counting and 0.5-s timing functions from the VIA. The data are to be stored in an EPROM to permit a compact no-moving-part system with easy transfer of data to a laboratory for analysis. This will require data latching and 50-ms timing for the programming pulses as each byte is written to the EPROM.

Three other IC solutions to the problem of retaining data after power-down are discussed in this section.

An automotive data logger using the VIA and a 2716 EPROM is shown in Figure 15.8. Timer T2 counts downward once per engine revolution via negative pulses at the PB6 input. T2 is initially loaded with $FFFF, so the value read from T2C-L after 0.5 s must be complemented and multiplied by 2 to obtain revolutions per second.

The 0.5-s intervals are obtained by configuring T1 for an IRQ after a 50-ms one-shot period. The IRQ routine counts the number of 50-ms periods and restarts the timer after periods one through eight. After the ninth period the IRQ routine configures the VIA for a *low* output at PB7 during the tenth 50-ms period, and sets the period counter back to zero.

Port A and the low-order three bits of Port B are used as latches to hold the address on the EPROM during programming. This address is held in a two-byte RAM buffer and transferred to PA and PB via an accumulator. The high-order five bits are masked off before being stored to Port B. The address is incremented (with carry to high-order byte) after each 0.5-s interval. Data latch U3 holds the data to be stored during the programming interval.

A generalized flowchart of the timing and data-storage routine is given in Figure 15.9. The routine is interrupt-driven at 50-ms intervals, leaving the processor mostly free for other tasks.

The Electrically Erasable Programmable Read-Only Memory (EEPROM or E^2PROM) provides another means of nonvolatile data storage. Early versions of the EEPROM, represented by the 2817 (a 2K $\times$ 8-bit chip) required that external latches hold fixed address and data inputs to the memory while a specially shaped 21-V, 10-ms programming pulse was applied. These extra hardware requirements

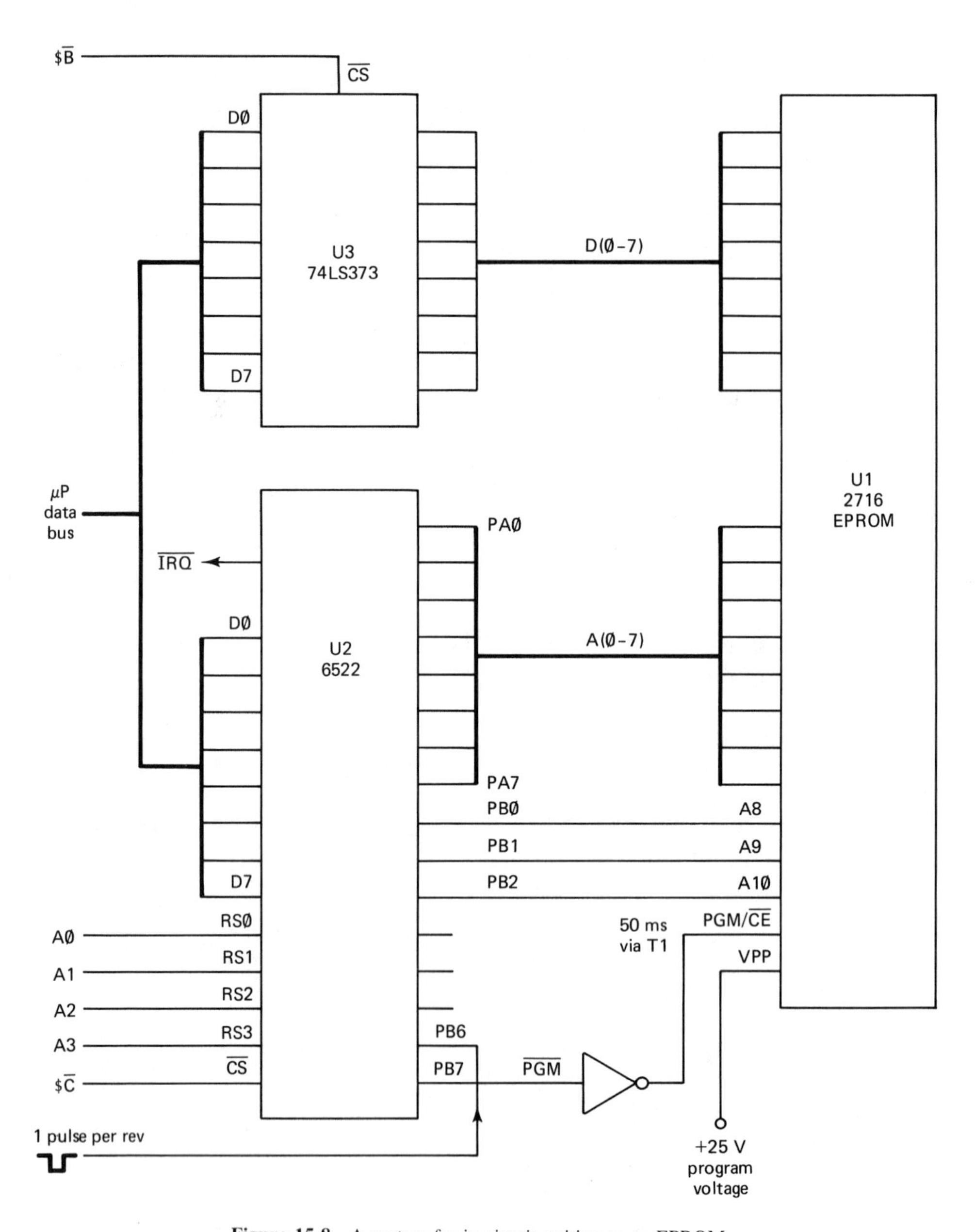

Figure 15.8 A system for in-circuit writing to an EPROM.

limited the application of the early EEPROMS. Newer versions, represented by the 2817A, have internal latches and require only the standard +5-V supply. The write-time is still 10 ms, during which time the 2817A bus lines float and it cannot be read or written to. A BUSY output is asserted during this write time. Limited life is characteristic of both versions, with a minimum of 10 000 write cycles guaranteed.

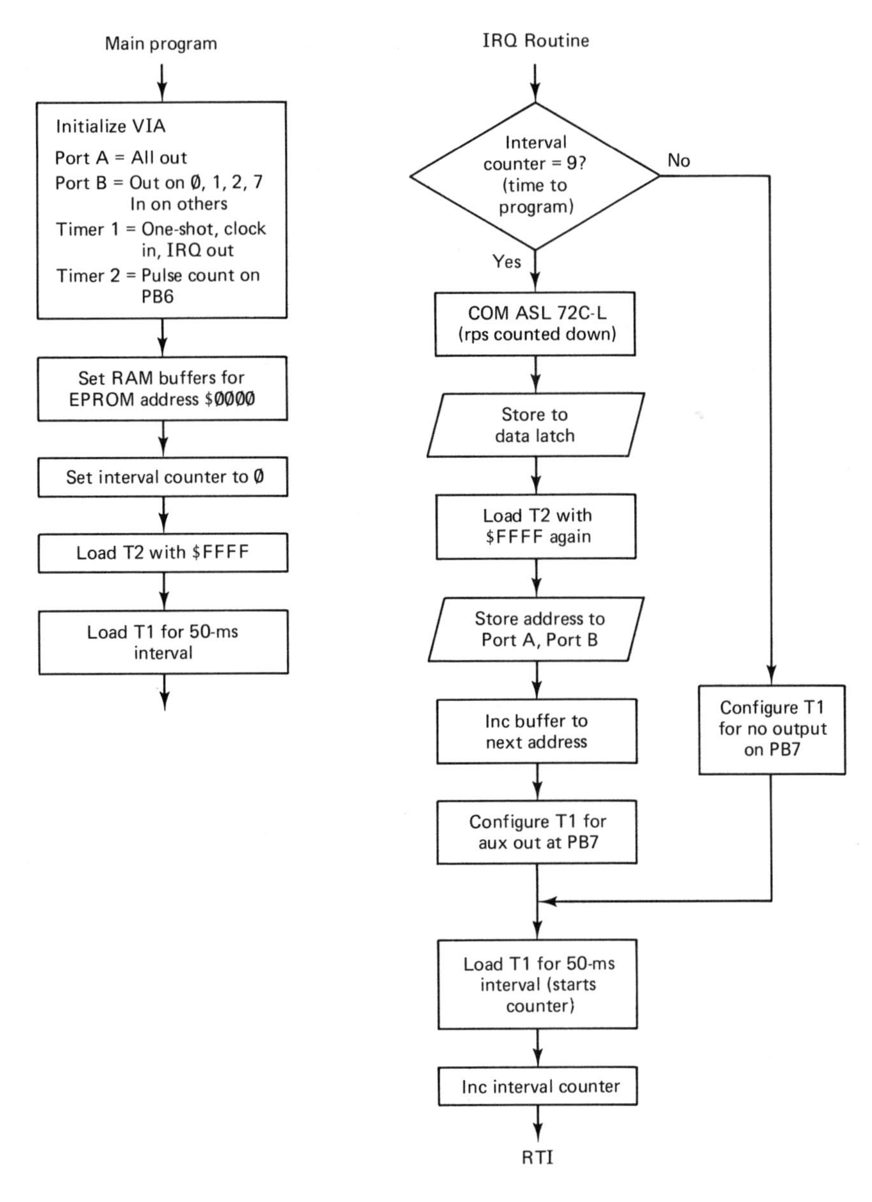

Figure 15.9 Flowchart for the automotive rpm logger of Figure 15.8.

The 28-pin configuration of the 2817A is not pin-compatible with the 2716 EPROM and 6116 RAM pinouts. Xicor produces a type 2816A, and National Semiconductor markets a type 9816A E^2PROM, which are 2716 pin compatible but lack the BUSY output. An attempt to read a previously written key data byte must be made to determine whether these chips are busy or ready. Eight-kilobyte memories such as the 2764 EPROM and the 6264 RAM have replaced the 2K chips in most applications, and the new 2864A EEPROMs are essentially pin-compatible with these 28-pin packages.

Shadow RAMs are used to eliminate the 10-ms *busy* time during which EEPROMs cannot be accessed. A shadow RAM consists of a separate static RAM and a matching EEPROM in one package. The computer system accesses the RAM, which has read and write times of a fraction of a microsecond. Data from the RAM is transferred to the EEPROM upon the assertion of a *nonvolatile store* input to the chip. The entire RAM contents are stored in less than 10 ms. Write-cycle life to the EEPROM is limited, as it is in other EEPROMs.

Battery-backed RAMs, or NOVRAMS, present perhaps the easiest solution to the nonvolatile-storage problem. They consist of a low-power static RAM with a lithium battery and control circuitry integrated in the package. When V_{CC} is lost by power-down or removal of the IC from its socket the chip goes into a standby mode in which data is retained by the internal battery for up to 10 years if necessary. Reading and writing to a NOVRAM is as fast and simple as for an ordinary static RAM. The pinouts for the Dallas Semiconductor DS1220 (2K × 8) and DS1225 (8K × 8) NOVRAMs are completely compatible with the 6116 and 6264 standard RAMs.

15.6 A SPEECH SYNTHESIZER INTERFACE

Section Overview

This section shows how the VIA can be used to make a SP0256 speech-generator IC count from 1 to 10. The versatility of the VIA is demonstrated by

- **configuring the bitwise-programmable Port B for one input (to select an English or Spanish counting sequence) and 7 outputs (six for the speech codes and one for a *beep* tone);**
- **using one of the VIA timers to count two-second intervals between spoken words, and the other to produce a beep tone to replace every third count;**
- **using one of the VIA control lines to signal the processor when a new speech component (called an *allophone*) is needed, another to start the selected allophone, and a third to allow the user to start the counting sequence.**

This reliance on the VIA leaves the microprocessor itself essentially free for other tasks, in this case the control of 8 LEDs connected to Port A of the VIA.

The SP0256 speech generator produces a discrete speech component called an allophone, lasting from 40 to 420 ms, in response to a *low* pulse on its *address load* ($\overline{ALD}$) input. The speech sound produced is determined by the 6-bit *address* placed on its inputs A1 through A6. Note that this "address" is supplied from a data table via the data bus, and is not a memory address in the sense that we have been using the word.

When the allophone is completed the SP0256 asserts its standby (SBY) output to call for another address so it can produce its next allophone. A single word such as *six* may require a succession of seven allophones. The system can be used to teach the Spanish counting sequence because every third word is replaced by a *beep* tone, prompting the learner to say the word. The missing words change with each run through the program.

The microprocessor spends less than 1% of its time controlling the speech generator. To demonstrate this, eight LEDs are connected to Port A and lit in a shifting pattern under direct processor control. The change in pace of the lights when the speech program interrupts the processor is not even noticeable.

Here is a pseudocode list of the English/Spanish counting program. The actual program appears in List 15.3, and the system hardware is given in Figure 15.10.

```
        * ONE-TO-TEN COUNT AT 2-SEC INTERVALS; ENGLISH OR SPANISH
MAIN PROGRAM
  INITIALIZE VIA.
    Configure Port A for all outputs to LEDs (DDRA).
    Configure Port B for input on bit 6, outputs all others (DDRB).
    Configure CA2 as independent output for ALD (PCR).
    Configure CB1 as IRQ input for SBY allophone request (PCR & IER).
    Configure CB2 as IRQ input for start switch (PCR & IER).
  SHIFT LIGHT PATTERN TO DEMONSTRATE PROCESSOR IS FREE.
    Enable IRQ.
    Shift lit LED and delay for 0.2 sec.
INTERRUPT ROUTINE
  TEST SOURCE OF INTERRUPT.
    IF source is CB2 then go to START; ELSE go to WATNOW.
  START (begin English or Spanish word strings at 2-s intervals.)
    Read PB6; Set X buffer to start of English or Spanish data.
    Set interval counter to 20.
    Start T2 for IRQ in 100 ms.
    Enable T2 interrupts via IER.
    Return to MAIN.
  WATNOW (Wait or make next sound?)
    Stop T1 from producing beep on PB7 (if it is).
    IF source of IRQ is T2 then go to TIME; ELSE
      Go to SPEAK subroutine (IRQ was SBY via CB1).
      Return to MAIN.
  TIME (100 ms done; 2 sec done or not?)
    Decrement interval counter.
IF interval counter is Ø, continue; ELSE return to MAIN.
    Reset counter to 20.
    Decrement SKIPPER (skips every third word).
    IF skipper = Ø, go to BEEP; ELSE
      Go to SPEAK subroutine.
      Return to MAIN.
```

BEEP tone starts on PB7 via T1.
Reset skipper to 3.
Point to next in table.
Check for end flags.
Return to MAIN.

SPEAK – A SUBROUTINE TO SUPPLY DATA TO SPEECH GENERATOR.

Fetch allophone from table and point to next in table.
IF end-of-table flag (ten counts done) then
Stop continuous generation of 100-ms IRQs.
Return from subroutine.
IF end-of-word flag (wait 2-s interval) then
Return from subroutine.
ELSE (speaking in progress—keep it going.)
Output allophone to speech processor.
Output pulse via CA2 to $\overline{ALD}$ to start speech.
Return from subroutine.

```
                  1000 * LIST 15-3 ** VOICE COUNTER ** FOR FIG 15.10 HARDWARE **
                  1010 *  TESTED AND WORKING, 25 AUG,'87, BY D. METZGER.
                  1020 *
                  1030         .OR  $FE00
                  1040         .TA  $4000
6800-             1050 PB      .EQ  $6800
6801-             1060 PA      .EQ  $6801
6802-             1070 DDRB    .EQ  $6802
6803-             1080 DDRA    .EQ  $6803
6805-             1090 T1CH    .EQ  $6805
6809-             1100 T2CH    .EQ  $6809
680B-             1110 ACR     .EQ  $680B
680C-             1120 PCR     .EQ  $680C
680D-             1130 IFR     .EQ  $680D
680E-             1140 IER     .EQ  $680E
0010-             1150 XBUF    .EQ  $0010
0012-             1160 TENTH   .EQ  $0012
0013-             1170 SKIPR   .EQ  $0013
                  1180 *
FE00- 86 FF       1190 INIT    LDAA ##$FF     INITIALIZE:
FE02- B7 68 03    1200         STAA DDRA       PORT A ALL OUTPUTS.
FE05- C6 BF       1210         LDAB ##$BF
FE07- F7 68 02    1220         STAB DDRB       PORT B OUT, BUT PB6 IN.
FE0A- 86 3E       1230         LDAA ##$3E      0011 1110 = $3E = HI OUT CA2. ($3C = L0
)
FE0C- B7 68 0C    1240         STAA PCR        SET CB1 = POS, CB2 = NEG, IRQ INPUTS.
FE0F- 86 88       1250         LDAA ##$88      ENABLE CB2 (START) INTERRUPT ONLY.
FE11- B7 68 0E    1260         STAA IER
FE14- 8E 00 7F    1270         LDS  ##$007F
FE17- 86 03       1280         LDAA #3        SKIP EVERY 3RD DIGIT; BEEP INSTEAD.
FE19- 97 13       1290         STAA SKIPR
                  1300 *
FE1B- 86 7F       1310 LITES   LDAA ##$7F     PULL 1 LED LO (ON)
FE1D- 0E          1320 WAIT    CLI            SHIFT LED AROUND AT ABOUT 4 SHIFT/SEC
FE1E- B7 68 01    1330         STAA PA         TO SHOW THAT PROCESSOR IS NOT
FE21- CE 40 00    1340         LDX  ##$4000     BUSY WITH SPEECH GENERATION.
FE24- 09          1350 DELAY   DEX
FE25- 26 FD       1360         BNE  DELAY
FE27- 0D          1370         SEC
FE28- 46          1380         RORA            (NEXT LED RIGHT)
FE29- 25 F2       1390         BCS  WAIT       SKIP OVER CARRY,
FE2B- 20 EE       1400         BRA  LITES      (BIT 0 TO BIT 7)
                  1410 *
                  1420 * INTERRUPT ROUTINE: 3 POSSIBLE SOURCES.
FE2D- B6 68 0D    1430 INTRPT  LDAA IFR       CHECK IRQ SOURCE
```

```
FE30- 84 08      1440        ANDA #$08      CB2 (START SWITCH)?
FE32- 27 1F      1450        BEQ  WATNOW     NO: THEN FIND WHAT.
FE34- CE FF 00 1460          LDX  #$FF00      (SET START OF ENGL FILE)
FE37- B6 68 00 1470 START    LDAA PB         YES: READ PORT B, BIT 6;
FE3A- 84 40      1480        ANDA #$40        (1 = ENGL, 0 = SPAN)
FE3C- 26 03      1490        BNE  ENGL        BIT 6 = 1; KEEP ENGL.
FE3E- CE FF 80 1500          LDX  #$FF80      BIT 6 = 0; SET SPAN.
FE41- DF 10      1510 ENGL   STX  XBUF
FE43- 86 14      1520        LDAA #20       RAM BUFFER COUNTS OFF
FE45- 97 12      1530        STAA TENTH      TENTHS OF A SEC.
FE47- 86 C3      1540        LDAA #$C3      START T2 COUNTING FOR IRQ
FE49- B7 68 09 1550          STAA T2CH       AFTER 100 MS (0.5 MHZ CLK)
FE4C- 86 A0      1560        LDAA #$A0       1010 0000 ENABLES
FE4E- B7 68 0E 1570          STAA IER         T2 INTERRUPTS.
FE51- 20 43      1580        BRA  RETURN
FE53- 7F 68 0B 1600 WATNOW   CLR  ACR       KILL TONE OUT PB7 FROM T1.
FE56- B6 68 0D 1610          LDAA IFR       CHECK IRQ SOURCE.
FE59- 84 20      1620        ANDA #$20       T2 (100 MS UP)?
FE5B- 26 05      1630        BNE  TIME        YES: SEE IF 2 SEC UP.
FE5D- BD FE 9C 1640          JSR  SPEAK       NO: IRQ CAME FROM SBY HI,
FE60- 20 34      1650        BRA  RETURN       (SPEECH-LOAD REQUEST).
                 1660 *
FE62- 86 C3      1670 TIME   LDAA #$C3      START ANOTHER 100-MS TIME TO IRQ.
FE64- B7 68 09 1680          STAA T2CH
FE67- 7A 00 12 1690          DEC  TENTH     COUNT OFF ANOTHER .1 SEC DONE.
FE6A- 27 02      1700        BEQ  CONTIN    2 SEC DONE?
FE6C- 20 28      1710        BRA  RETURN     NO: RETURN TO MAIN PROG.
FE6E- 86 14      1720 CONTIN LDAA #20        YES: SET UP ANOTHER 2 SEC.
FE70- 97 12      1730        STAA TENTH
FE72- 7A 00 13 1740          DEC  SKIPR     IS THIS THE 3RD WORD?
FE75- 27 05      1750        BEQ  BEEP       YES: MAKE TONE.
FE77- BD FE 9C 1760          JSR  SPEAK      NO: START WORD.
FE7A- 20 1A      1770        BRA  RETURN    BACK TO MAIN.
FE7C- 86 C0      1780 BEEP   LDAA #$C0      SET T1 FOR CONTIN SQR WAVE
FE7E- B7 68 0B 1790          STAA ACR        OUTPUTS ON PB7.
FE81- 86 02      1800        LDAA #$02      512 COUNTS/HALF CYCLE =
FE83- B7 68 05 1810          STAA T1CH       488 HZ ON 0.5 MHZ CLK.
FE86- 86 03      1820        LDAA #3        SET TO SKIP THIRD
FE88- 97 13      1830        STAA SKIPR      WORD AGAIN.
FE8A- DE 10      1840        LDX  XBUF      GET READY FOR NEXT WORD.
FE8C- 08         1850 LOOK   INX            LOOK FOR END CODE $7E OR $7F.
FE8D- 6D 00      1860        TST  0,X        IS BYTE > $7F?
FE8F- 2A FB      1870        BPL  LOOK        NO: CHECK NEXT BYTE.
FE91- DF 10      1880        STX  XBUF        YES: POINT TO END CODE; WAIT
FE93- BD FE 9C 1890          JSR  SPEAK         IF END-WORD; HALT IF END-FILE.
FE96- 86 FF      1900 RETURN LDAA #$FF      WRITE 1'S TO ALL IFR FLAGS
FE98- B7 68 0D 1910          STAA IFR        TO CLEAR ALL OLD INTERRUPTS.
FE9B- 3B         1920        RTI            BACK TO MAIN PROG.
                 1930 * SUBROUTINE TO SUPPLY ALLOPHONES TO SPEECH GEN.
FE9C- DE 10      1940 SPEAK  LDX  XBUF
FE9E- A6 00      1950        LDAA 0,X       GET ALLOPHONE BYTE.
FEA0- 08         1960        INX
FEA1- DF 10      1970        STX  XBUF
FEA3- 81 FF      1980        CMPA #$FF       END OF FILE CODE?
FEA5- 26 0A      1990        BNE  WORD         NO: CHECK IF END OF WORD.
FEA7- C6 30      2000        LDAB #$30         YES: DISABLE 100-MS IRQS,
FEA9- F7 68 0E 2010          STAB IER            AND CB1 (SBY) INTERRUPTS.
FEAC- 7F 68 0B 2020          CLR  ACR        KILL BEEP TONE.
FEAF- 20 17      2030        BRA  QUIET
FEB1- 81 FE      2040 WORD   CMPA #$FE      END-OF-WORD CODE?
FEB3- 27 13      2050        BEQ  QUIET      YES: SHUT OFF SPEECH.
FEB5- B7 68 00 2060          STAA PB         NO: OUTPUT ALLOPHONE.
FEB8- 86 3C      2070        LDAA #$3C
FEBA- B7 68 0C 2080          STAA PCR       LOW PULSE OUT CA2
FEBD- 86 3E      2090        LDAA #$3E       TO START ALLOPHONE;
FEBF- B7 68 0C 2100          STAA PCR        THEN RETURN CA2 HIGH.
FEC2- 86 90      2110        LDAA #$90
FEC4- B7 68 0E 2120          STAA IER       ENABLE IRQ FROM CB1 (SBY).
```

```
FEC7- 39          2130          RTS
                  2140 *
FEC8- 7F 68 00    2150 QUIET    CLR  PB       OUTPUT PAUSE ALLOPHONE (00)
FECB- 86 3C       2160          LDAA ##$3C     TO SHUT THE THING UP.
FECD- B7 68 0C    2170          STAA PCR      PULSE CA2 (ALD*) LOW,
FED0- 86 3E       2180          LDAA ##$3E     THEN HIGH TO LOAD ALLOPHONE.
FED2- B7 68 0C    2190          STAA PCR
FED5- 86 10       2200          LDAA ##$10    DISABLE IRQ FROM CB1 (SBY).
FED7- B7 68 0E    2210          STAA IER       (HAVE TO WAIT UNTIL 2 SEC UP)
FEDA- 39          2220          RTS
                  2230 *
                  2240          .OR $FF00          * ENGLISH FILE
                  2250          .TA $4100
FF00- 2E 0F 0B
FF03- FE          2260          .HS 2E0F0BFE           * ONE
FF04- 0D 1F 1F
FF07- FE          2270          .HS 0D1F1FFE           * TWO
FF08- 1D 0E 13
FF0B- FE          2280          .HS 1D0E13FE           * THREE
FF0C- 28 28 3A
FF0F- FE          2290          .HS 28283AFE           * FOUR
FF10- 28 28 06
FF13- 23 FE       2300          .HS 28280623FE         * FIVE
FF15- 37 37 0C
FF18- 0C 02 29
FF1B- 37 FE       2310          .HS 37370C0C022937FE   * SIX
FF1D- 37 37 07
FF20- 07 23 0C
FF23- 0B FE       2320          .HS 3737070723 0C0BFE  * SEVEN
FF25- 14 02 0D
FF28- FE          2330          .HS 14020DFE           * EIGHT
FF29- 0B 18 06
FF2C- 0B FE       2340          .HS 0B18060BFE         * NINE
FF2E- 0D 07 07
FF31- 0B FF       2350          .HS 0D07070BFF         * TEN
                  2360 *
                  2370          .OR $FF80          * SPANISH FILE
                  2380          .TA $4180
FF80- 16 16 38
FF83- 35 FE       2390          .HS 16163835FE
FF85- 21 35 37
FF88- 37 FE       2400          .HS 21353737FE          * DOS
FF8A- 0D 33 14
FF8D- 37 37 FE    2410          .HS 0D33143737FE        * TRES
FF90- 08 16 18
FF93- 0D 0E 35
FF96- FE          2420          .HS 0816180D0E35FE      * CUATRO
FF97- 37 37 13
FF9A- 2C 08 35
FF9D- FE          2430          .HS 3737132C0835FE      * CINCO
FF9E- 37 37 14
FFA1- 37 37 FE    2440          .HS 3737143737FE        * SEIS
FFA4- 37 37 13
FFA7- 14 0D 14
FFAA- FE          2450          .HS 373713140D14FE      * SIETE
FFAB- 35 32 35
FFAE- FE          2460          .HS 353235FE            * OCHO
FFAF- 38 1F 14
FFB2- 23 14 FE    2470          .HS 381F142314FE        * NUEVE
FFB5- 21 13 14
FFB8- 37 37 FF    2480          .HS 2113143737FF        * DIEZ
                  2490 *
                  2500          .OR $FFF8
                  2510          .TA $41F8
FFF8- FE 2D       2520          .DA INTRPT    * INTERRUPT ROUTINE
FFFA- FE 00       2530          .DA INIT      *
FFFC- FE 00       2540          .DA INIT
FFFE- FE 00       2550          .DA INIT      * RESET
                  2560          .EN
```

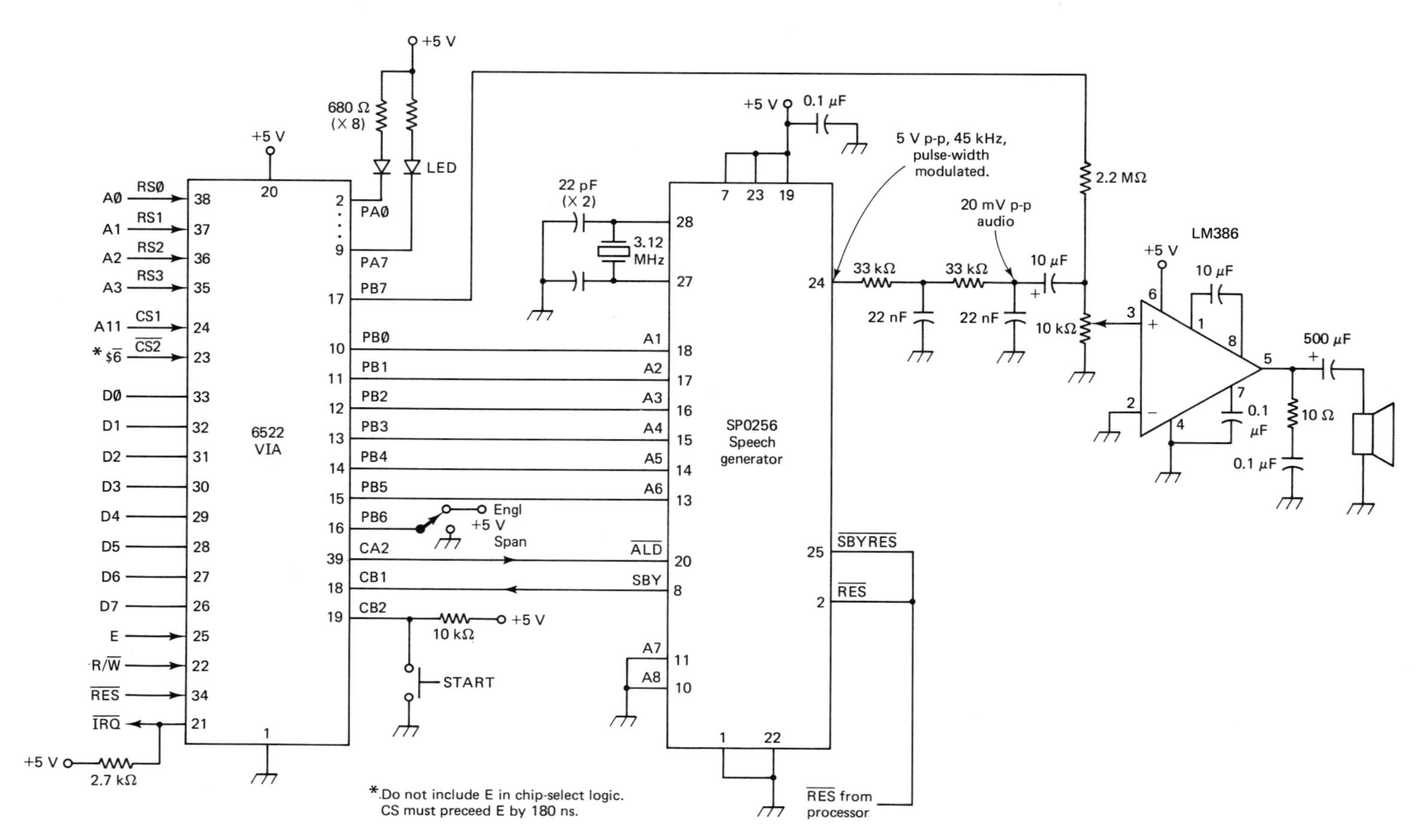

Figure 15.10 The VIA handles interrupts, pulse generation, interval timing, and I/O interfacing for a speech generator.

REVIEW OF SECTIONS 15.5 AND 15.6

21. How long does it take to change the contents of one address in a 2817 E^{2}PROM?
22. What data is contained in a 2716 byte that has been erased?
23. Besides application of the write-data bits, what function must be executed to program a 2716?
24. What major advantage does the NOVRAM have over the shadow RAM?
25. Why can the 2817 address and data lines not be tied directly to the processor buses?
26. Referring to Figure 15.8, what system address is used to select the EPROM address to be written?
27. Again referring to Figure 15.8, when does U2 assert an $\overline{IRQ}$ to the processor?
28. What is the solution to the problem of reading an E^{2}PROM during a write cycle?
29. What is the purpose of the LEDs in Figure 15.10?
30. When is the processor interrupted in the system of Figure 15.10?

15.7 A TWO-VOICED PROGRAMMABLE TONE GENERATOR

Section Overview

A single tone is somewhat limited in the effects it can produce. Harmony in music and realism in synthesized human voices is obtained by generating and mixing several tones of different frequencies. This section describes a relatively simple project in which the VIA timers each produce time delays for the half-cycles of two independent tones, leaving the processor free to control the outputs, select the tones, and monitor their durations.

The hardware for a two-voiced tone generator is diagrammed in Figure 15.11. It can be interfaced to any 6800 or 6802 microcomputer or, with different programming, to a 6502- or 6809-based system. One tone is output from PB7, which is switched under direct control of timer T1 in the free-run mode. The other tone is output from PBØ under processor control when T2 signals via an $\overline{IRQ}$ that its half-cycle time is up. Since only one timer must be serviced by the processor, there is only one interrupt source enabled, and we are spared the considerable trouble of having to determine the source of the interrupt.

The software is quite simple. The timer counts for producing the desired tones are stored in sets of four bytes in the following order:

T1C-L   T1C-H   T2C-L   T2C-H

These bytes determine the time per audio half-cycle for *voice* 1 and *voice* 2. The voice-2 output at PBØ must be toggled by the processor, and the delay count must be reloaded upon timeout of T2 and consequent $\overline{IRQ}$ signal.

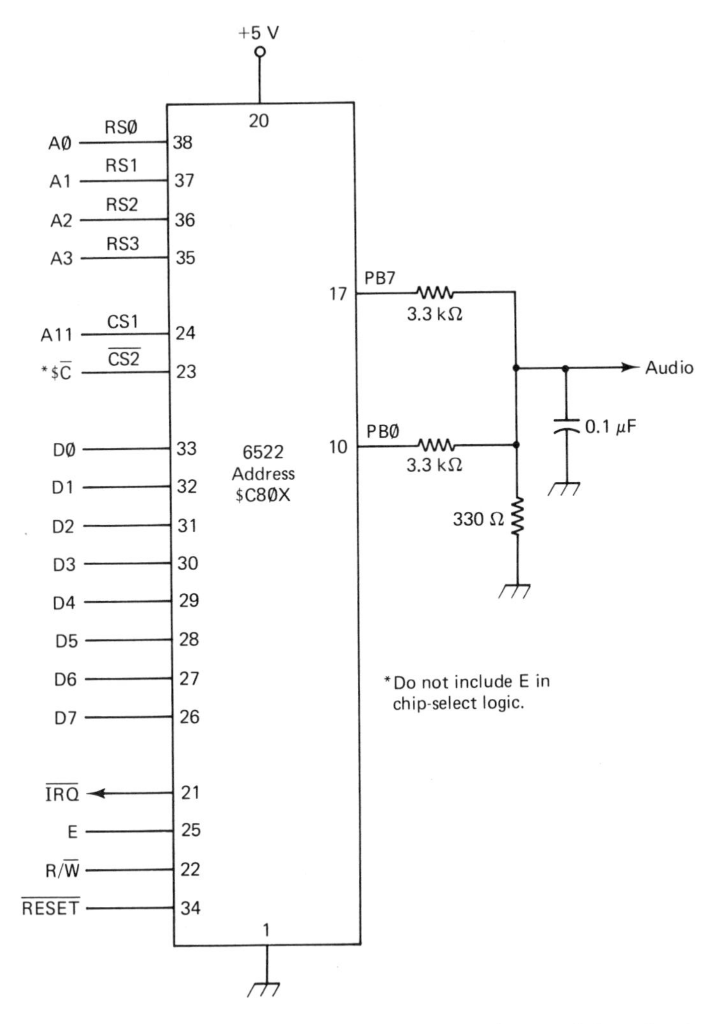

Figure 15.11 The timers in the 6522 can be used to produce two independent tones —the basis of harmony.

Each pair of notes is held for about 0.5 s by a nested delay loop counted by the processor. The times out for interrupt servicing affect the tone lengths, but not noticeably.

As an example, *A* above middle *C* is 440 Hz and requires a delay of 1136 μs per half cycle. On a processor with a 2-μs *E*-clock period, T1 would require to be loaded with 568 (which is $238) to produce this delay. Thus T1C-L must be loaded with $38 and T1C-H must be loaded with $Ø2.

As the following program shows (List 15.4), the IRQ services add 23 clock cycles, or 46 μs, to the countdown time for T2, making the delay count slightly more difficult. As a second example, this time with the T2 output, middle *C* is 261.6 Hz and requires a delay of 1911 μs per half-cycle.

$$1911\ \mu s \text{ per half-cycle} - 46\ \mu s \text{ per IRQ}$$
$$= 1865\ \mu s \text{ or } 932 \text{ counts per half-cycle}$$

This is $3A4, so T2C-L must be loaded with $A4 and T2C-H with $Ø3 to produce middle C.

```
                 1000 * LIST 15-4   **   TWO-VOICED TONES USING
                 1010 *   THE 6522 TIMERS.   FIG. 15-11 HARDWARE.
                 1015 * RUNS, 17 JULY, '87. *   D. METZGER
                 1020 *
                 1030          .OR  $FF80
                 1040          .TA  $4080
C802-            1050 DDRB     .EQ  $C802
C80B-            1060 ACR      .EQ  $C80B
C80E-            1070 IER      .EQ  $C80E
C804-            1080 T1CL     .EQ  $C804
C805-            1090 T1CH     .EQ  $C805
C808-            1100 T2CL     .EQ  $C808
C809-            1110 T2CH     .EQ  $C809
C800-            1120 PB       .EQ  $C800
                 1130 *
FF80- 8E 00 7F   1140          LDS  #$007F  PREPARE FOR INTERRUPTS,
FF83- 86 FF      1150          LDAA #$FF    PORT B ALL OUTPUTS.
FF85- B7 C8 02   1160          STAA DDRB
FF88- 86 C0      1170          LDAA #$C0    ACR7 AND 6 HIGH FOR PB7
FF8A- B7 C8 0B   1180          STAA ACR      FREE-RUN OUTPUTS.
FF8D- 86 A0      1190          LDAA #$A0    IER 7 AND 5 HIGH FOR
FF8F- B7 C8 0E   1200          STAA IER      INTERRUPTS FROM T2.
FF92- CE FF E0   1210          LDX  #$FFE0  START OF NOTE FILE.
FF95- 0E         1220          CLI          ENABLE INTERRUPTS.
FF96- A6 00      1230 SING     LDAA 0,X     LOAD TIMER 1
FF98- B7 C8 04   1240          STAA T1CL    LOW BYTE
FF9B- A6 01      1250          LDAA 1,X
FF9D- B7 C8 05   1260          STAA T1CH    HIGH BYTE
FFA0- 43         1270          COMA         FLAG BYTE FF MEANS
FFA1- 27 FE      1280 HALT     BEQ  HALT     YOU'RE DONE.
FFA3- A6 02      1290          LDAA 2,X     LOAD TIMER 2
FFA5- B7 C8 08   1300          STAA T2CL    LOW BYTE
FFA8- 97 02      1310          STAA $02     (SAVE FOR IRQ)
FFAA- E6 03      1320          LDAB 3,X
FFAC- F7 C8 09   1330          STAB T2CH    HIGH BYTE
FFAF- D7 03      1340          STAB $03     (SAVE FOR IRQ)
FFB1- DF 30      1350          STX  $0030   SAVE FILE POINTER IN RAM.
FFB3- CE 7A 12   1360          LDX  #$7A12  500,000-US DELAY / 8 CYCLES
FFB6- 09         1370 WAIT     DEX           X 2 US/CYCLE = 31,250 OR
FFB7- 26 FD      1380          BNE  WAIT     $7A12 DELAY COUNTS.
FFB9- DE 30      1390          LDX  $0030   RESTORE FILE POINTER.
FFBB- 08         1400          INX
FFBC- 08         1410          INX
FFBD- 08         1420          INX          POINT TO NEXT TWO TONES.
FFBE- 08         1430          INX
FFBF- 20 D5      1440          BRA  SING    NEW TONES EVERY 1/2 SEC.
                 1450 *
                 1460 * INTERRUPT-REQUEST SERVICE (12 CYCLES FOR IRQ)
FFC1- 96 02      1470 INTR     LDAA $02     3 CY  RETRIEVE TIMER COUNT.
FFC3- B7 C8 08   1480          STAA T2CL    5 CY    23 CY X 2 US = 46 US.
FFC6- D6 03      1490          LDAB $03     3 CY
FFC8- F7 C8 09   1500          STAB T2CH    5 CY  START COUNTING T2
FFCB- 86 01      1510          LDAA #$01    TOGGLE PB BIT 0
FFCD- B8 C8 00   1520          EORA PB
FFD0- B7 C8 00   1530          STAA PB
```

```
FFD3- 0E         1540          CLI           READY FOR NEXT HALF CYCLE.
FFD4- 3B         1550          RTI
                 1560 * NOTE FILE FOLLOWS.
                 1570          .OR   $FFE0
                 1580          .TA   $40E0
FFE0- 38 02 A4
FFE3- 03 00 03
FFE6- 80 04      1590          .HS   3802A40300038004
                 1600          .OR   $FFF8
                 1610          .TA   $40F8
FFF8- FF C1      1620          .DA   INTR          * DATA DIRECTIVE SETS IRQ VECTOR
FFFA- FF 80 FF
FFFD- 80 FF 80   1630          .HS   FF80FF80FF80  * NMI, SWI, RESET VECTORS
                 1640          .EN
```

Answers to Chapter 15 Review Questions

1. Because they use valuable processor time for noncomputational tasks.
2. The processor alone can't do it. At least two timers are needed.
3. When the high-order byte of the counter ($C805) is written to.
4. The interrupt-enable register must have 1s written to bits 7 and 6 (store $C0 to address $C80E) **5.** The $\overline{IRQ}$ was asserted 9.9 ms ago.
6. In the free-run mode the timers are reloaded automatically when the timer reaches zero. In the one-shot mode the timer must wait for an undetermined length of time for the processor to do this. **7.** $34
8. One-shot mode; PB7 goes *low* during count.
9. LDAA #$01, ORAA ACR, STAA ACR
10. Write $C0 to the ACR; write $F4 to T1C-L; write $01 to T1C-H.
11. $20 to address $C808; $4E to address $C809. **12.** The same values must be written to T2C-L and T2C-H by the processor.
13. $20 to ACR, $A0 to IER, $30 to T2C-L, $00 to T2C-H
14. Set T1 for free-running outputs at PB7 and T2 for counting pulses into PB6. Tie PB7 to PB6, cascading the counters. **15.** Seven, for one output.
16. $D2 to the IER at $C80E. **17.** $12 (an argument could be made for $3F).
18. One of the interrupt flags is set, but it has been disabled from generating an $\overline{IRQ}$ by the IER. **19.** All interrupt flags are cleared.
20. Only enabled interrupts still produce 1 bits in the result. **21.** 10 ms **22.** $FF
23. $\overline{CE}$ must be brought *high* for 50 ms and V_{pp} must have +25V applied.
24. Unlimited write-cycle life.
25. During the 10-ms erase/write cycle, these lines are latched at fixed levels; meanwhile the processor bus lines must be free to switch.
26. $C001 (low byte) and $C000 (high byte).
27. When T1 has counted a 50-ms delay.
28. Shadow RAMs **29.** To demonstrate that the processor spends a neglible fraction of its time on the speech program.
30. $\overline{IRQ}$ is asserted when the START button is pushed, when a 100-ms interval is over, and when a speech allophone is finished.

CHAPTER 15 QUESTIONS AND PROBLEMS

Digits after the decimal point refer to the section number in Chapter 15.

Basic Level

1.1 In the one-shot mode, how does the VIA timer T1 notify the processor that the timing period is ended?

2.1 Refer to Figure 15.1 and assume a 6802 system with a 1-MHz clock. What data should be written to what address to cause a time delay of approximately 12 ms?

3.2 How are the VIA latches for T1 loaded?

4.2 In the free-run mode of timer T1, when does a new time interval begin?

5.3 How is the 6522's timer T2 changed from a timer (counting *E*-clocks) to an event counter (counting negative transitions on PB6)?

6.3 Timer T2 has timed out and pulled the $\overline{\text{IRQ}}$ line *low*. How can $\overline{\text{IRQ}}$ be returned *high* without restarting the timer?

7.4 What bytes should be written to the 6522's IER register to disable an interrupt from T1 and enable an interrupt from T2?

8.4 What bytes should be written to the 6522's PCR and IER to set up CA1, CA2, CB1, and CB2 all as separate low-active interrupt inputs?

9.5 How long does it take to write a byte of data to a 2817A EEPROM?

10.5 Under what conditions will an EEPROM assert its BUSY output?

11.5 List the advantages and disadvantages of an EEPROM compared to a NOVRAM.

Advanced Level

12.1 Write an interrupt-service routine in 6802 source code that will reset the 6522 T1 timer for another 10-ms delay each time an IRQ is received. The routine must count the number of 10-ms intervals and output data $55 to a latch at address $C800 after 100 such intervals.

13.1 In a 1-MHz 6802/6522 system, T1C-L is loaded with data $4A and T1C-H with $91. When the timer generates an $\overline{\text{IRQ}}$ signal, the processor is in the midst of a critical routine and the I-mask bit is set. By the time the processor clears I and reads the VIA registers, T1C-L contains $C3 and T1C-H contains $FD. What is the total time since T1 began counting?

14.2 Write a 6802 routine to set a VIA at base address $A100 to produce a continuous 10-Hz square wave from PB7. Assume a 1-MHz clock.

15.2 Explain why the timer one-shot mode does not produce perfectly regular intervals.

16.3 Write the 6802 source code for a routine to configure T2 to count negative transitions of PB6 and to signal the processor with an $\overline{\text{IRQ}}$ when this count reaches 99. The VIA is at base address $A100.

17.3 Write the 6802 source code for a routine to set up the 6522 for a delay of 30 min. Refer to Figure 15.4 and assume a 1-MHz clock.

18.4 The 6522's IER has only T1 and T2 enabled as interrupt sources. Write the source code for a 6802 interrupt-service routine that will determine whether the interrupt was caused by T2 (first priority) or T1, and send the processor to a routine called SERVT1 or SERVT2, as required.

19.4 A read of the 6522's IFR shows data $60. What does this mean, in terms of interrupt sources?

20.5 Write the 6802 source code for a routine that will sense a BUSY signal from an EEPROM via the VIA's CA1 input, hold the byte to be written and the address in a 3-byte buffer area, set timer T2 for an interrupt after a 12-ms delay, and return control to the main program. Assume CA1 and T2 are already configured.

21.5 Complete the above project in Problem 20 by writing a routine to sense the timeout of T2 and transfer the byte to the address as saved in RAM.

22.5 What is a *Shadow RAM*, and why is it used?

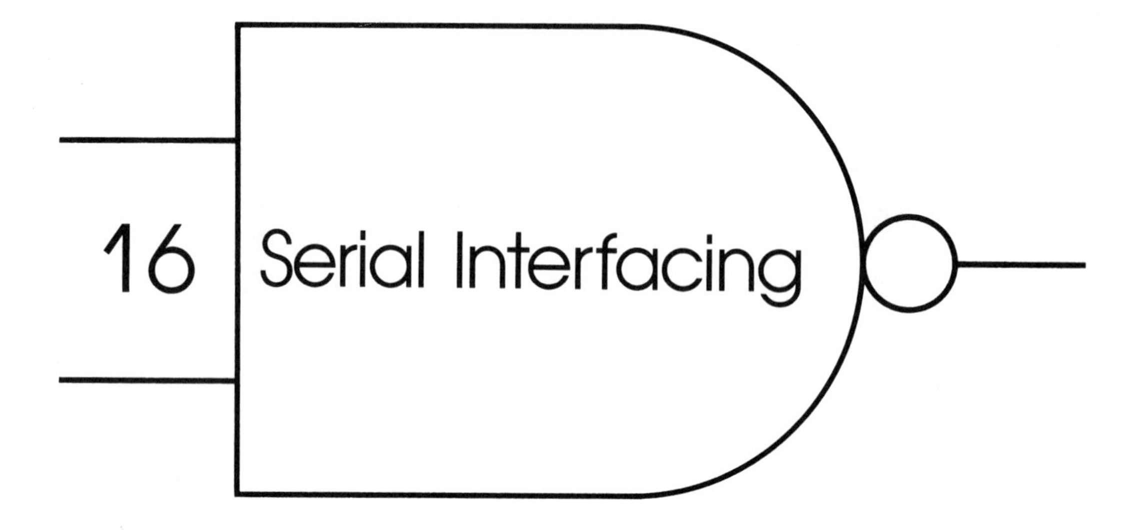

16.1 SERIAL DATA AND THE RS-232

Section Overview

Computerized control and data-management systems require extensive digital communication between the processor and external devices such as keyboards, pen plotters, CRT displays, voltmeters, and A-to-D converters. Often the required data rate is only a few bytes per second—a sharp contrast to the computer's million-byte-per-second data bus. In such cases it is more economical to send the 8 bits of a data byte sequentially on a single transmission line, repacking the serialized bits into a parallel byte at the receiving end. A standard form for serial-data-transmission hardware has been established by specification RS-232C, and this is widely followed.

Baud rate: Two-wire transmission lines have been employed for years to handle audio communications. Such lines are inexpensive, and they are easy to install and repair. The highest audio frequency preserved in telephone voice-communication circuits is typically 3300 Hz, so we might expect such circuits to handle square digital pulses at about one-third that rate, or 1000 bits per second. This is a data rate (theoretically) of 125 bytes per second, more than fast enough for most keyboard or instrument inputs and for most printer, plotter, or motor- and valve-control outputs.

A byte of data is sent serially by switching the signal line *high* or *low* to correspond with the data bits as they are shifted rightward out of a register. The shift rate is the maximum theoretical bit rate, that is, if each data byte followed immediately on the heels of the previous one. This maximum bit rate is called the *baud* rate in honor of Baudot, an early worker in coded data transmission. In our example, each bit caused the line to remain *high* or *low* for 1 ms, and the baud rate was 1 kHz or, more properly, 1000 bits per second. Standard baud rates used in serial data transmission, and their bit periods are:

75	(13.3 ms)	1 200	(833 μs)
110	(9.1 ms)	2 400	(417 μs)
150	(6.7 ms)	4 800	(208 μs)
300	(3.3 ms)	9 600	(104 μs)
600	(1.67 ms)	19 200	(52 μs)

The bit stream: Serial data communication is often asynchronous. This means that there is no common clock to keep the bits from the transmitter and reads at the receiver in step. Since there is only one signal line, synchronization of transmitter and receiver must be accomplished by a signal on that line. Here is one way that it is done.

The *rest* state of the transmission line when no data is being transmitted is a logic-1 level. The beginning of a byte of data is signaled by a transition to a logic-0 level. This level lasts for one bit-time interval and is called the *start bit*. It tells the receiver to begin timing and reading the coming data bits. The eight time intervals following the start bit convey the logic-1 or logic-0 levels of bits 0 through 7 of the transmitted data byte. Figure 16.1 illustrates the serial bit stream.

If the transmitted byte were $55, the bit stream would be

S	**0**	**1**	**2**	**3**	**4**	**5**	**6**	**7**	**Bit function**
0	1	0	1	0	1	0	1	0	Bit value

The first 0 is the start bit. The remaining bits are printed in reverse of their usual place-value order. Notice that this particular byte, at a rate of 1000 baud, constitutes a short burst of a 500-Hz square wave. However, the bit stream for data byte $66 is

0 0 1 1 0 0 1 1 0

which is 2 cycles of a 250-Hz square wave, and the bit stream for data byte $0 is a single negative pulse 9 ms long:

0 0 0 0 0 0 0 0 0

Parity: A tenth bit, called the *parity bit*, is often transmitted after the data bits. This bit serves as a check to catch errors in the received data. In *odd parity* the transmitter counts the number of 1 bits in the transmitted byte as it sends them out. If the number of 1s is 0, 2, 4, or 8, the parity bit is set to logic 1. If the number of 1s is odd, the tenth bit is set to 0. Thus the total number of 1 bits is always odd.

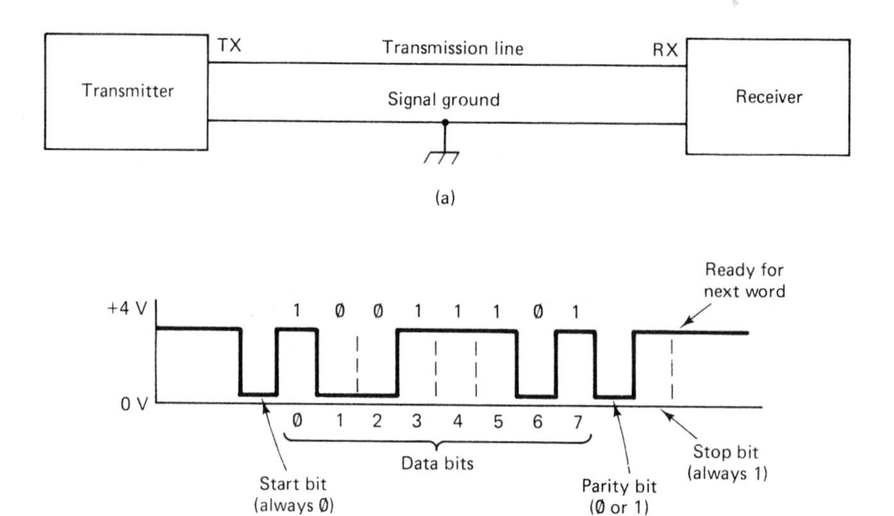

Figure 16.1 Serial data transmission requires only a single wire and ground (a). Eight data bits plus start, stop, and parity check bits are sent as a series of voltage levels on the single line. (b).

In *even parity* the tenth bit is a 1 if the previous bit stream contains an odd number of 1 bits, so the total number of 1 bits (excluding *start* and *stop* bits) is always even. Of course, before serial data communication can begin, the transmitter and receiver must agree on a baud rate and on whether even parity, odd parity, or no parity bit is being used. This agreement can be arrived at manually (by switch settings) or, in more advanced systems, by a software routine executed before data is sent.

The end of a data word is indicated by one or two *stop* bits which cause the line to remain at the logic-1 state for at least one bit period. The line will then remain at the logic-1 level until the next byte is to be transmitted. The logic-0 *start bit* signals this event.

A clock at the receiver starts running halfway through the received start bit and causes the state of the line to be read at the ends of the next 10 bit times. Thus each bit is read at the middle of its bit period if the transmitter and receiver baud rates are exactly the same. There are 10 bit intervals from the middle of the *start* bit to the middle of the *stop* bit and an allowance of ½ interval on each read, so the maximum timing difference between transmitter and receiver is

$$\frac{\pm\ 1/2 \text{ interval}}{10 \text{ intervals}} = \pm 5\%$$

This assumes no distortion of the transmitted square wave by stray capacitance on the line, an assumption that is often not justified. Baud-rate generators are usually crystal controlled to ensure that the receiver and transmitter remain in step for the required 10 bit periods.

Specification RS-232C was established by the Electronic Industries Association and has become the accepted hardware standard for serial data communication. Briefly, the specification requires the following:

- Logic 1 (called *mark*) between −5 and −15 V transmitted.
- Logic 0 (called *space*) between +5 and +15 V transmitted.
- Transmitter output not greater than +25 V open circuit. Source impedance high enough so that a short between any two lines will cause a fault current not greater than 0.5 A. Signal switching-transition rate not greater than 30 V/μs.
- Receiver load between 3000 and 7000 Ω, noninductive. Shunt capacitance up to 2.5 nF permitted. Voltages from −3 to −25 V must be recognized as a logic 1. Voltages from +3 to +25 V must be recognized as a logic 0.
- Connector pinout as shown in Figure 16.2.

Many manufacturers have taken the liberty of altering the functions of some of the RS-232C connector pins for their own specialized applications. In addition, many of the specified pins are optional lines intended for handshaking and modem (phone-line) applications. It is common to see an RS-232C hookup with pins 2 and 3 tied together for the signal, pin 7 tied to a signal-ground line, pin 1 connected to the system chassis, and all other pins unused. It is also common (and frustrating) to find that a supposedly standard RS-232 interface connection is not at all standard, and works only with equipment containing specialized lines and signal functions defined by a maverick manufacturer.

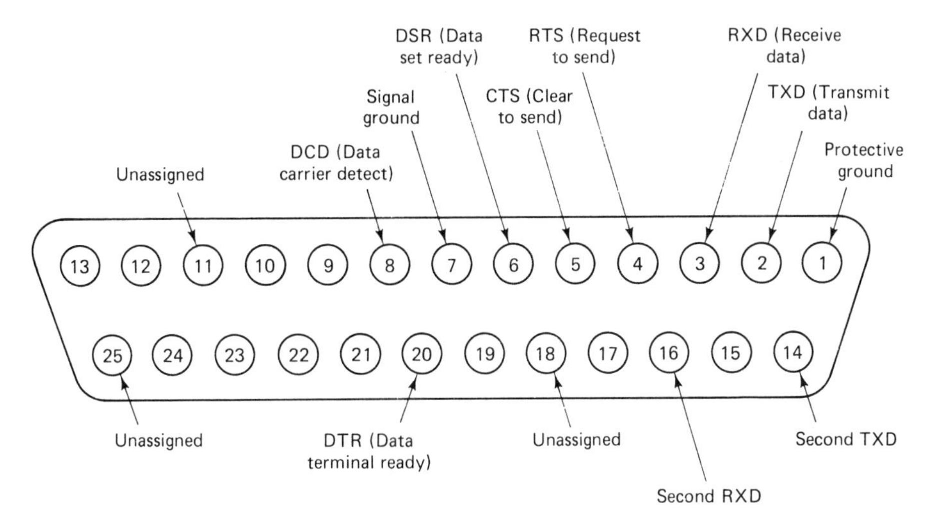

Figure 16.2 The DB-25 connector for the RS-232C serial data transmission system.

16.2 SERIAL INTERFACE HARDWARE—THE ACIA

Section Overview

The type 6850 IC can be programmed to transmit and to receive serial data in a variety of format and parity options, as specified by writing to the control register. The processor simply stores data to an 8-bit transmit-data register, and the 6850 converts it to a serial bit string. Likewise, the received serial data is automatically packed into a receive-data register, which need only be read by the processor. The 6850 status register informs the processor when the next data byte can be read or written and reports on errors in received data. The 6850 operates on TTL-level signals so level-converter circuitry is required to adapt it to RS-232C voltage levels.

The 6850 serial-interface IC takes most of the work out of interfacing micro-computers with serial data lines. Motorola calls the 6850 an ACIA, for Asynchronous Communications Interface Adapter. Other manufacturers refer to similar chips as UARTs, for Universal Asynchronous Receiver/Transmitter. Figure 16.3 shows the pinout of the 6850. The chip is activated by asserting the three chip-select lines, CSØ, CS1, and $\overline{\text{CS2}}$. These are normally driven by selected processor address lines, either directly or through decoding logic, to place the ACIA at the desired addresses in the memory map.

The receive and transmit clock inputs are driven by a baud-rate generator, which supplies a square wave at the baud rate or a selected multiple thereof. The *E* and

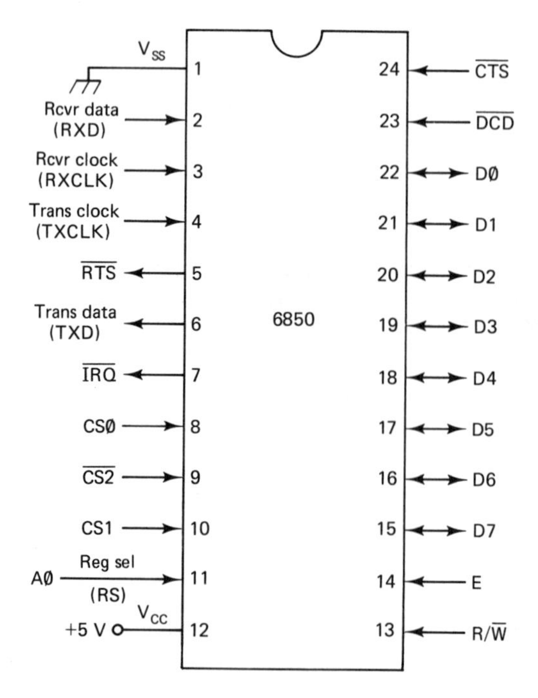

Figure 16.3 Pin functions for the 6850 Asynchronous Communications Interface Adapter.

R/$\overline{W}$ inputs and the 8 data lines are connected directly to the corresponding lines of the processor.

The ACIA control register (CR) is used to program the function, format, and interrupt enables of the 6850. This is a *write-only* register. Its contents cannot be read back. It is selected by bringing R/$\overline{W}$ *low* and the RS (register-select) input *low*. The RS input is usually connected to AØ, so the control register is selected by writing to the ACIA base address. The Transmit and Receive-Data Registers (TDR and RDR) are accessed at the ACIA base address *plus one* (RS1 *high*, AØ = 1).

Bits Ø and 1 of the control register must both be written with 1s during system initialization to *reset* the ACIA. After that they select the baud rate as the TX and RX clock rate (ØØ), the clock rate divided by 16 (Ø1) or divided by 64 (1Ø).

Control Bit		Function
1	**Ø**	
Ø	Ø	Baud rate = CLK
Ø	1	Baud = CLK ÷ 16
1	Ø	Baud = CLK ÷ 64
1	1	Master reset

For example, if we programmed for divide by 64 and required a baud rate of 1200, the TX and RX clocks required would be

$$1200\text{ Hz} \times 64 = 76\ 800\text{ Hz}$$

Crystals for such low frequencies are available, but they are expensive. It would probably be more economical to use a 4.9152-MHz oscillator and an external divide-by-64 circuit to produce the 76.8-kHz clock.

The format of the bit string is selected by control-register bits 2, 3, and 4, according to the following table.

Control Bit			Word Format
4	**3**	**2**	
Ø	Ø	Ø	7 data, even parity, 2 stop
Ø	Ø	1	7 data, odd parity, 2 stop
Ø	1	Ø	7 data, even parity, 1 stop
Ø	1	1	7 data, odd parity, 1 stop
1	Ø	Ø	8 data, no parity, 2 stop
1	Ø	1	8 data, no parity, 1 stop
1	1	Ø	8 data, even parity, 1 stop
1	1	1	8 data, odd parity, 1 stop

All formats have one *low*-level start bit preceding the bits listed in the table. The example bit string shown in Figure 16.2 uses format 1 1 1 (8 data, odd parity). In the 7-data-bit modes, bit 7 from the processor is stripped off before serial data transmission. The 7-bit modes are useful because many user codes, such as ASCII, do not contain more than 2^7 (128) symbols and have no need of the eighth bit.

Interrupt and Handshake Lines: The ACIA transmitter can accept new data from the processor into its Transmit-Data Register (TDR) while the current data is being shifted out. Recall that the serial transmission of a byte (at 1200 baud) takes about 1000 times as long as the parallel transfer from the processor to the ACIA. The processor is generally busy with other tasks while the serial transmission is going on. When the Transmit Data Register dumps its contents to the output shift register an IRQ can be generated by the ACIA. The processor then has several thousand machine cycles to respond to the interrupt and load the next byte into the TDR.

To avoid a situation in which the transmitter sends data when no receiver is accepting it, a form of handshaking can be used. The transmitter can assert the ACIA's Request-To-Send output ($\overline{\text{RTS}}$) and look for a Clear-To-Send reply from the receiver (on the ACIA's $\overline{\text{CTS}}$ input) before initiating the transmit sequence.

A *break* signal (continuous logic Ø output on the transmit-data line) can also be programmed to inform the receiver of an end of message. All three of the preceding functions are configured by bits 5 and 6 of the ACIA Control Register according to the following table.

Control Bits		**$\overline{\text{RTS}}$**	**$\overline{\text{IRQ}}$ when**	**Break =**
CR6	**CR5**	**Output**	**TDR empty**	**(Ø on TD line)**
Ø	Ø	Low	Disabled	—
Ø	1	Low	Enabled	—
1	Ø	High	Disabled	—
1	1	Low	Disabled	Asserted

Figure 16.4 illustrates ACIA handshake and interrupt capabilities.

The receiver section of the ACIA can also be programmed to generate an interrupt for its processor. A logic 1 written to CR bit 7 enables $\overline{\text{IRQ}}$ outputs from any of three receiver functions:

1. Receive Data Register full. This simply means that a full data byte has been shifted in and is available to be read by the receiving processor.
2. Receiver overrun. This is an error condition in which a new serial data byte is shifted in before the old one is read by the processor, resulting in the loss of the old data.
3. Data Carrier Detect ($\overline{\text{DCD}}$) input transition from *low* to *high* level. This signal is asserted *low* while a modem (telephone interface) is receiving a transmitted signal. A *low* to *high* transition would indicate loss of the transmitted signal.

Later sections of this chapter will provide more details about interrupt prioritizing and modem operation.

Level converters are necessary to translate the TTL voltages of the ACIA to the ±12-V levels of the RS-232-C transmission line and back. The MC1488 IC contains four TTL to ±12-V level converters, and the MC1489 contains four ±12-V to TTL converters. Both ICs come in 14-pin DIP packages.

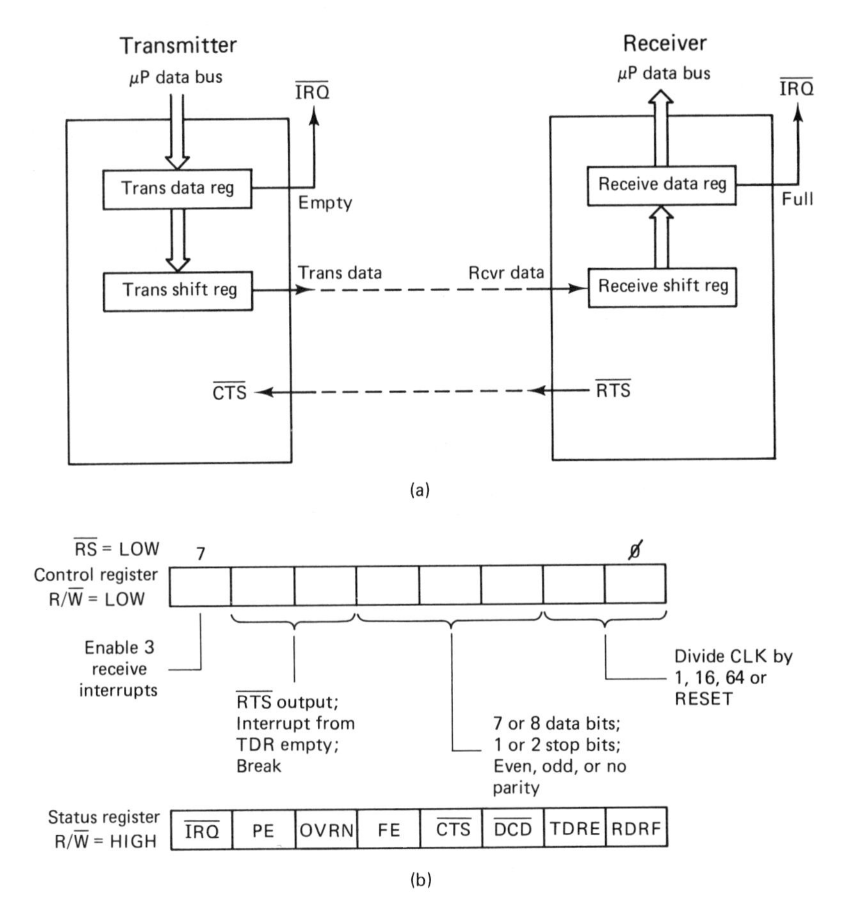

Figure 16.4 The ACIA holds data in the data register while other data is being shifted in or out of the shift register.

REVIEW OF SECTIONS 16.1 AND 16.2

Answers appear at the end of Chapter 16.

1. How long will it take to transmit 2 kilobytes of data at 600 baud if each word consists of a start bit, 8 data bits, a parity bit, and 1 stop bit?
2. Odd parity is being used. Assign the parity bits to these data bytes: \$97, \$3A, \$D4.
3. We have crystals available for 2.0, 3.58, 4.0, and 5.0 MHz. Can any of these be used with a binary divider to produce a clock rate within 2% of 1200 baud? If so, how many divide-by-2 stages will be needed?
4. An RS-232 transmitter source impedance is 500 Ω, and receiver shunt capacitance is the maximum specified by RS-232C. What is the highest standard baud rate

that can be used if signal-time constant must be not more than 10% of bit-interval time?

5. An ACIA has $\overline{CS2}$ connected to the output of a 74LS138, which goes *low* for addresses $6000 through $7FFF. CS0 is tied to A12, CS1 is tied to A11, and RS is tied to A0. What is the lowest address that will access the ACIA's Transmit-Data Register?
6. What hex data word should be written to the control register initially to *reset* the ACIA? Use 0 bits wherever functions are not required.
7. What hex data word should be written to use the internal $\frac{1}{64}$ clock divider for baud rate; configure for 7 data bits, odd parity, 1 stop bit; deactivate the transmitter IRQ, RTS, and *break* outputs, and enable the receiver interrupts?
8. What hex data should be written to configure the ACIA for no division of the baud-rate clock; 8 data bits, even parity; interrupt generated when transmitter ready to accept another byte; receiver interrupt disabled?
9. Why does the processor store data to the ACIA's Transmit-Data Register rather than directly to the output shift register?
10. Which pin of the ACIA determines whether the processor accesses the Transmit-Data Register or the Receive-Data Register?

16.3 SOFTWARE FOR THE 6850 SERIAL INTERFACE

Section Overview

In addition to the write-only Control Register, the ACIA has a read-only Status Register. The 8 status bits can be read and interpreted by the processor to determine when the ACIA transmitter or receiver is ready for another data transfer or if an error has been detected in received data. Bit masking or shifting must be used to read the status bits individually.

Receiving data: The serial bit stream is sent out the transmitting 6850's TXD pin at TTL levels and is picked up at TTL levels on the receiver's RXD input pin. Data shifted in at this pin is read, bit by bit, at the baud rate selected by the RXCLK input and clock divider bits of the control register. As soon as all bits are shifted in to the Receive Shift Register, the data is transferred to the ACIA's Receive-Data Register (RDR). From there the data is simply read by the processor at the ACIA base address plus one (RS input, tied to A0, equal to 1). The RDR is at the same address as the TDR; R/$\overline{W}$ is *high* for the former and *low* for the latter. In the 7-bit-data modes, the processor reads bit 7 as logic 0.

The ACIA Status Register (SR) is read at the base address (A0 = 0). This is the same address as the CR. The R/$\overline{W}$ line is *low* for CR (write only) and *high* for SR (read only). Here is a list of SR bits and their functions.

Status Bit	Function indicated by logic 1
Ø	*Receive-Data Register Full*; new data available to read; cleared by read of RDR. $\overline{\text{IRQ}}$ if enabled.
1	*Transmit-Data Register Empty*; ready to accept another byte; cleared by write to TDR. $\overline{\text{IRQ}}$ if enabled.
2	*Data Carrier lost*; a modem interface signal; follows state of data carrier detect ($\overline{\text{DCD}}$) input; IRQ asserted if *high* and receive interrupts enabled; IRQ cleared by read of SR, then read of RDR.
3	*Not Clear To Send*; a handshake signal; follows state of $\overline{\text{CTS}}$ input. If bit 3 is in the 1 state bit 1 is inhibited from going to the 1 state.
4	*Framing Error*; *stop bit not received.* This bit could indicate the transmission of a *break* signal, faulty agreement of receiver and transmitter in format or timing, or noisy or interrupted signal.
5	*Receiver Overrun*; second word transferred into receive data register before first word read out by receiving processor; IRQ asserted if receiver interrupts enabled. IRQ and bit cleared by read of RDR.
6	*Parity Error*; number of 1 bits in RDR not even (or odd) as selected by bits 2, 3, and 4 of control register.
7	*Interrupt request active*; set to 1 if $\overline{\text{IRQ}}$ output is *low*; cleared by read of RDR or write to TDR.

Transmit software for the ACIA is relatively simple. The only thing necessary is to ensure that the Transmit-Data Register is empty before storing each byte. This can be done in either of two ways:

1. Enable only the ACIA transmit interrupt by writing binary Ø1 to control register bits 6 and 5, respectively. Then any assertion of the processor's $\overline{\text{IRQ}}$ input means that the TDR is ready for more data and that the Clear To Send input ($\overline{\text{CTS}}$) is *low*, presumably indicating that the receiver is ready for us to send more data.
2. Read the status register before storing each byte. Mask off all but bit 1 (ANDA #$Ø2) and do not transmit until bit 1 = 1 (BNE TRANS). This alternative may be more attractive if the processor's $\overline{\text{IRQ}}$ is already involved with VIA ports and timers and further complication of the "where did the interrupt come from" problem seems intolerable.

Receive software is generally somewhat more involved. If we can assure that the processor will have time to check the status register at least as frequently as serial words are being transmitted, we might simply mask off all but bit Ø and test for SRØ = 1 (RDR full), ignoring the possibility of receive errors. To keep the processor free for other tasks and to minimize the possibility of faulty data, a receive program such as this should be implemented:

1. Enable receiver interrupts (control bit 7 = 1) in main-program initialization.
2. Output Request To Send *low* (control bit 6 = Ø), telling transmitter "Clear To Send" in main program.

3. Respond to IRQ from ACIA by reading the status register.
4. Mask off all but bits 2, 4, 5, and 6. If any of these is a 1, an error has occurred. Remove $\overline{\text{RTS}}$ output and abort run or ask transmitter for a rerun.
5. If bits 2, 4, 5, and 6 are all 0, check that bit 0 is a logic 1. If so, read Receive-Data Register and file data. If not, interrupt did not come from ACIA. Check other interrupt sources.

16.4 MODEMS

Section Overview

Modems provide a means for transmitting digital data serially over long distances using transmission lines and systems originally designed for voice communication. The system most commonly employed is frequency shifting of a continuous audio tone; one frequency represents a logic 0, the other, a logic 1. In *duplex* operation two frequency-shifting tones appear on a single line; one transmits in one direction while the other transmits simultaneously in the opposite direction.

Long distance: Parallel data transmission is usually limited to distances of a few feet because of the effects of stray capacitance on the fast pulses involved. Direct transmission of serial digital data is confined to dedicated lines with lengths of only a few hundred feet because the dc levels involved will not pass through standard speech amplifiers, which typically limit transmitted frequencies to the range 300 to 3300 Hz.

A *modem* is a modulator-demodulator. To *modulate* means to control the amplitude, frequency, or phase of a continuous signal (called the carrier) in accordance with an information signal to be transmitted. In practical terms a digital modem is a device for exchanging data between remote computers or terminals via telephone lines. Direct-coupled modems are connected by wires to the telephone line. Acoustically coupled modems operate with a standard telephone handset, which is placed in a cradle with the earpiece near a modem microphone and the mouthpiece near a modem speaker. On the computer side, the modem will interface to a serial transceiver (UART or ACIA) either at TTL levels or via an RS-232C line. A typical data rate is 1200 baud.

Frequency-shift keying (FSK) is a form of frequency modulation that is commonly used by low-speed modems. Figure 16.5 illustrates FSK. In this system a logic 1 (called *mark*) is represented by the presence of a sine-wave frequency of 1270 Hz. A logic 0 (called *space*) is indicated by a shift to 1070 Hz. At 300 baud only three or four cycles are sent during each bit interval, but the receiver's demodulator, using a circuit called a *phase-locked loop*, is able to convert these frequency shifts back into level shifts that can be read by the ACIA.

Frequency modulation is used in preference to amplitude modulation because it is less vulnerable to noise. Poor connections and noise on signal lines cause output level changes similar to amplitude modulation, but only genuine changes in frequency

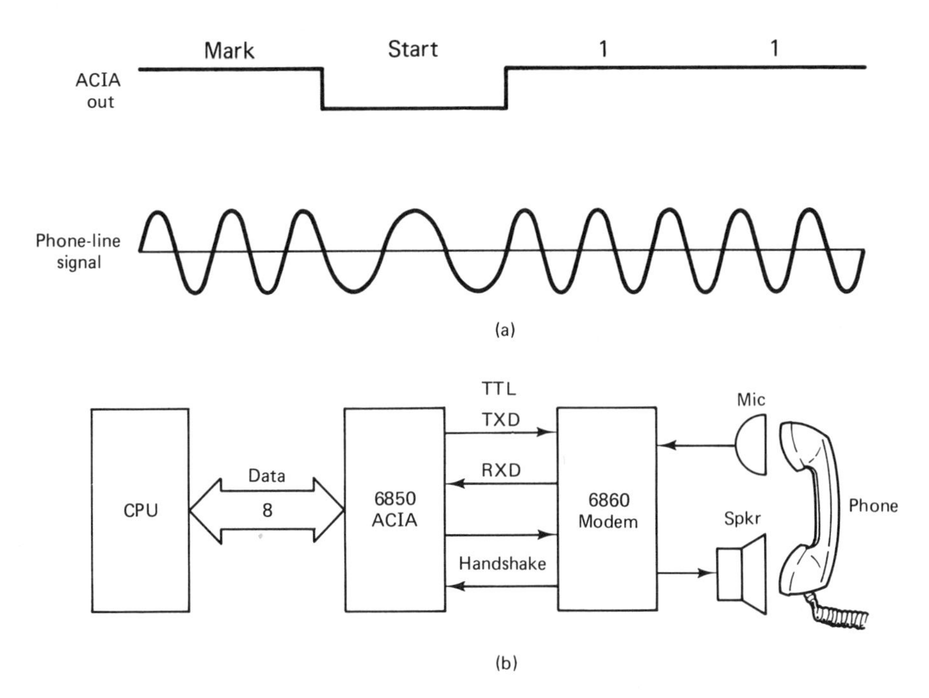

Figure 16.5 (a) A modem changes the dc levels of the RS-232 signals into frequency-shifting audio tones for transmission over the telephone lines. A modem system is illustrated in (b).

are likely to trigger the FM demodulator. *Phase-shift keying* (PSK) is sometimes used to achieve higher transmission speeds. In PSK, the transmitted sine wave is shifted abruptly in phase or not shifted to represent digital logic states, as illustrated in Figure 16.6.

In simplex operation data flows in only one direction at a time on the long-distance line. This presents a problem because if the receiver detects a data error, it has no way of notifying the transmitter of this until the transmitter is finished talking, and the transmission direction is turned around.

In duplex operation two carrier signals are applied to the phone line simultaneously, as shown in Figure 16.7. Generally the calling station transmits on the low-frequency carrier and receives on the high-frequency channel, whereas the answering station transmits on the high frequency and receives on the low. The two frequency bands are kept separate by selective filters at the receivers.

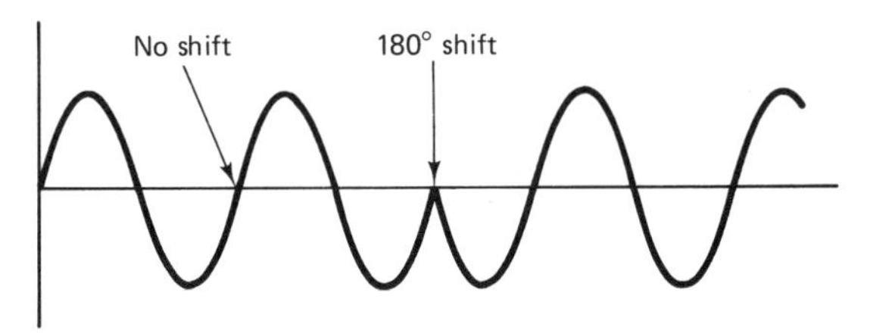

Figure 16.6 Higher baud rates can be achieved by using phase shifting rather than frequency shifting in a modem.

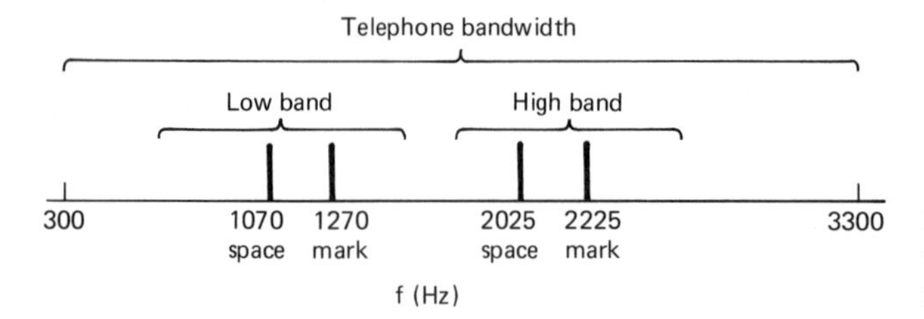

Figure 16.7 Duplex modem operation uses two bands of audio frequencies; one for transmitting and one for receiving.

In full duplex operation data is sent in both directions simultaneously. This may be done simply for *echoing*, which is an elementary technique for data-error detection. In echoing, an exact copy of the received data is sent back to the transmitting station for comparison and correction if necessary.

In the half-duplex mode, data is transmitted in only one direction at a time, although two channels are available. The receiver can use the second channel to signal the transmitter if an error is detected or if a *wait* is needed.

Protocol is a set of conventions for data exchange, which, although arbitrary, must be agreed upon by the transmitter and the receiver before meaningful communication can take place. Some of the protocol items are as follows:

- Baud rate, parity convention, word length, and stop bits.
- Simplex or duplex operation.
- Handshake signals: RTS, CTS, DCD.
- Echo or checksum for error detection.
- Abort and notify, or rerun upon receive error.
- Interrupt, data-turnaround, and data-end signals.

Smart modems may be capable of automatically answering a call, checking the incoming data to determine the protocol standards of the calling device and automatically adjusting to them. Dumb modems use fixed or switch-selected protocol and require manual answering of calls and initiation of data transfer.

REVIEW OF SECTIONS 16.3 AND 16.4

Answers appear at the end of Chapter 16.

11. An ACIA's control register is accessed by storing to address $58FØ. How is its status register accessed?
12. How may the transmitting processor be notified when it is time to dump another byte to the ACIA?
13. Where does the transmitter's CTS input signal come from?
14. Why is SR bit 1 not allowed to go *high* unless bit 3 is *low*?
15. What flag is raised if the transmitting processor writes data to its ACIA faster than the receiving processor reads it from its ACIA?

16. Write the 6800 source code for getting only bits 2, 4, 5, and 6 from the ACIA Status Register into the processor's accumulator A.
17. What does a modem modulate and demodulate?
18. Why are digital signals not sent directly over phone lines?
19. What is duplex operation of a modem and how is it achieved?
20. What is echoing?

16.5 DIGITAL TAPE RECORDING

Section Overview

Magnetic tape is widely used for the economical storage of large quantities of digital data. High-quality commercial units may write bits directly on the tape, but digital data may also be stored on magnetic tape using a frequency-modulation technique, similar to that used in modems. The disadvantage of magnetic-tape storage is serial access. It may take many seconds to find the data segment required.

The Kansas City standard was an early format for recording digital data on cassette tape, and may still be of interest for illustrating basic serial data-handling techniques. In this format, bits are stored sequentially, least-significant bit first, and bytes are stored sequentially, lowest address first. Logic-Ø bits are represented by 4 cycles of a 1200-Hz tone, and logic-1 bits are represented by 8 cycles of a 2400-Hz tone, as illustrated in Figure 16.8. Each byte is preceded by a Ø-level *start* bit and followed by two 1-level stop bits. Each byte of data requires a record time of

$$(4 \text{ cycles/bit}) (1/1200 \text{ s/cycle}) (11 \text{ bits/byte}) = 37 \text{ ms/byte}$$

Two kilobytes of memory can thus be stored in

$$(2048 \text{ bytes}) (37 \text{ ms/byte}) = 75 \text{ s}$$

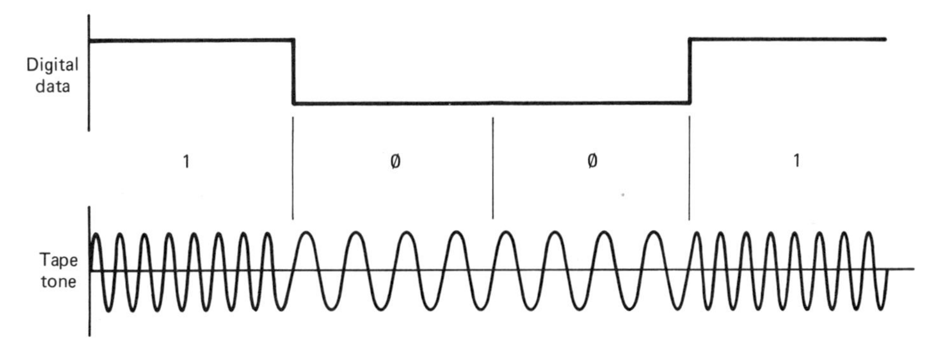

Figure 16.8 The Kansas City format for recording digital data on cassette tape. The frequency-shift scheme is similar to that used in modems.

and a 60-min tape costing only a few dollars can hold 96 kilobytes of data in this format.

A data record starts with a 5-s leader of all logic-1 bits, followed by an identification code byte to distinguish among various records that may be stored on the same tape. Four bytes are then recorded, which give the starting address of the data to be recorded and the ending address. The recorded data follows, and the record concludes with a *checksum* byte. The checksum is the 2's complement summation of the 4 address bytes plus all data bytes. When reading back, the 4 address bytes plus the data bytes plus the checksum byte should total to zero. If errors are present, there is only 1 chance in 256 that the checksum byte will be correct by accident.

The tape read can be accomplished by using a phase-locked loop to convert the audio frequencies to logic levels, but it is possible and more economical to replace hardware with software and read the audio cycles directly, as illustrated in Figure 16.9. With this technique, only a simple transistor buffer is needed between the tape-recorder output and the microprocessor input. The audio signal is monitored by the processor and a time-delay routine of 520 μs is started as soon as it goes positive. Then the audio level is read again. If the audio frequency is 1200 Hz, a logic Ø will be read; if it is 2400 Hz, a logic 1 will be read. After a 208-μs delay, the processor is set to monitoring the level again for the next positive transition.

This read loop is completed 4 times per bit. If a logic 1 is read 3 or 4 times, a 1 bit is shifted into the received-byte register. If a logic 1 is read Ø or 1 time, a Ø bit is shifted in. If a logic 1 is read 2 times, an error is reported. The bit-read routine is repeated 8 times, with a new bit shifting in each time until the byte is complete. Then the byte is stored in RAM and the stop bits and the start bit for the next byte are monitored. Figure 16.10 is a flowchart of the Kansas City tape-read routine. Tape speed and microprocessor delay loops are not critical because the timing is restarted from the positive audio transitions occurring every 833 μs.

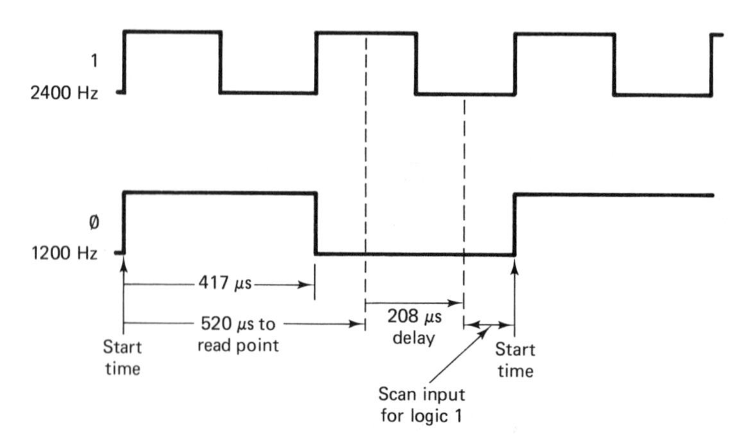

Figure 16.9 Timing for the 6802 program to read tape in the Kansas City format. A logic Ø will read Ø after a 520-μs delay; a logic 1 will read 1.

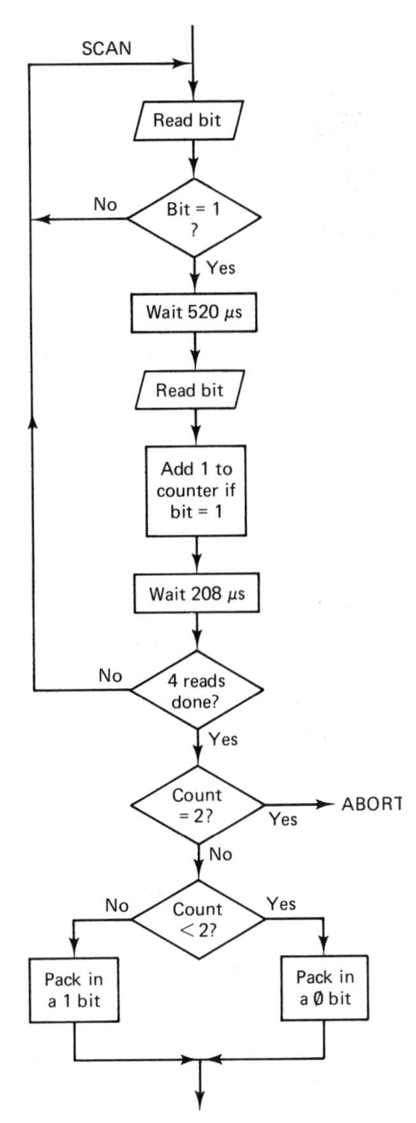

Figure 16.10 Flowchart for the Kansas City tape-read program.

16.6 A BAR-CODE SCANNER

Section Overview

Many supermarkets now use optical scanners to pick coded information directly off the product label. This section examines the bar-code format and the requirements on the microcomputer for inputting and processing its data.

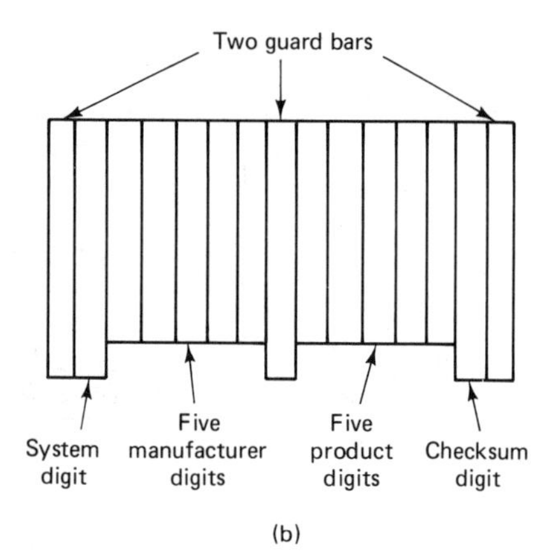

Figure 16.11 (a) A typical UPC (Universal Product Code) symbol. (b) The symbol contains four information areas and three pairs of guard bars.

The universal product code (UPC) is a 10-digit decimal number that (ideally) should be unique for every market product. The first five digits identify the manufacturer. The second five digits specify the type of product. No pricing information is contained in the UPC, because prices are likely to fluctuate between the time of printing the label and the time of sale. Rather, pricing is stored in the computer at the point of sale and is looked up and reported to the cash register. Inventory and reordering can also be handled by the computer without the need for physically checking the stock on the shelves—assuming that all stock that leaves the store leaves over the counters.

The 10-digit UPC comprises the major part of the rectangular *bar code*, an example of which is shown in Figure 16.11. The bar code pattern is scanned from left to right. The 5 ''manufacturer'' digits on the left are separated from the 5 ''product'' digits on the right by two guard bars, which are slightly longer than the rest. The 10 decimal digits are generally printed directly under their corresponding bar codes.

The 10 digits are flanked on the left by a system-code digit, whose bars are also slightly elongated. This code digit is also printed at the left side of the symbol. It is a 0 for grocery items and a 3 for drug and health-related items.

On the right is a decimal checksum digit (also elongated), which usually appears only in bar-code and not in numeral form. This digit is used to detect errors, as is explained shortly. The entire 12 digits of bars are flanked with two narrow guard bars on each side.

Bars and spaces: Each digit is represented by two dark *bars* and two light *spaces*. Each bar and each space may be 1, 2, 3, or 4 *elements* wide. An element is approximately 0.013 inches in width. A dark element is interpreted as a logic 1; a light element is a logic Ø. Each digit is 7 elements wide and is encoded as follows:

Decimal Digit	Left 6 Digits (Odd Parity)	Right 6 Digits (Even Parity)
0	000 1101	111 0010
1	001 1001	110 0110
2	001 0011	110 1100
3	011 1101	100 0010
4	010 0011	101 1100
5	011 0001	100 1110
6	010 1111	101 0000
7	011 1011	100 0100
8	011 0111	100 1000
9	000 1011	111 0100

Notice that the right-digit codes are the complements of the left-digit codes. Dark bars are always 1, 2, 3, or 4 elements wide, representing binary 1, 11, 111, or 1111, respectively. White spaces are likewise 1, 2, 3, or 4 elements wide, representing 0, 00, 000, or 0000. Left-side digits always begin with a 0 (white unit) and end with a 1 (dark unit). Right-side digits always begin with a 1 (dark) and end with a 0 (white). Also note again that each digit is composed of two white spaces (1, 2, 3, or 4 elements wide) and two dark bars (1, 2, 3, or 4 elements wide). Figure 16.12 shows an example bar code illustrating all 10 digits.

The checksum character is not simply the sum of the other eleven digits. It is obtained by the following procedure:

1. Starting with the system digit on the left, sum every other digit, working toward the right; the total of 6 decimal digits.
2. Multiply this total by 3.

Figure 16.12 A UPC sample symbol containing all 10 digits.

3. Starting with the leftmost manufacturer digit, sum every alternate digit, working toward the right; the total of 5 decimal digits.
4. Add the results of Step 2 and Step 3.
5. Subtract 10 from the result of Step 4 repeatedly until a number between 0 and 9 results. This is the checksum digit, encoded in the rightmost position of the bar code.

Word length: If the left- and right-hand guard bars are stripped off, the bar code produces an 89-bit word. With the guard bars, there are 95 bits per word.

Each bar-code symbol consists of 30 dark bars; 2 for each of 12 digits and 2 for each of 3 guard-bar pairs. The number of white spaces between the 30 dark bars is 29. There are thus 30 + 29, or 59, measurable widths in a UPC symbol. Each width, ideally, has one of four values.

The software problem in reading a UPC symbol is not a simple one. It is compounded by two factors.

1. The scanning rate is not constant. In some systems the scan is achieved by pulling the package across a fixed light source, resulting in extremely wide variations in time per element. In newer systems a moving light beam scans an essentially stationary package, but size variations on the order of 2:1 make the scan time per unit nonconstant. The program must be able to accept serial data without knowing the baud rate in advance. Assuming a fastest scan rate of 60 inches per second, the shortest time per element is

$$\frac{0.013 \text{ in./element}}{60 \text{ in./s}} = 216 \ \mu\text{s/element}$$

Assuming a slowest scan rate of 6 inches per second, the longest time per element is 2.17 ms.

2. The printing and reading tolerances on the UPC bars are quite broad. Thus if element time on a particular scan were determined to be 1.00 ms, we would structure the program to accept dark-bar and white space as follows:

0.50–1.49 ms	1 unit wide
1.50–2.49 ms	2 units wide
2.50–3.49 ms	3 units wide
3.50–4.49 ms	4 units wide

Of course, these times would have to be variable to accommodate scan rates from about 0.2 to 2 ms per element.

The program for reading the bar code may proceed by storing the scan times for the 30 bars and 29 spaces in 59 consecutive RAM locations. The maximum binary value of 255 should represent a slightly "fat" 4-element bar at the slowest allowable scan rate, which is about

$$(2.17 \text{ ms/element})\ (4.5 \text{ elements}) = 9.8 \text{ ms/bar} \qquad (\text{max})$$

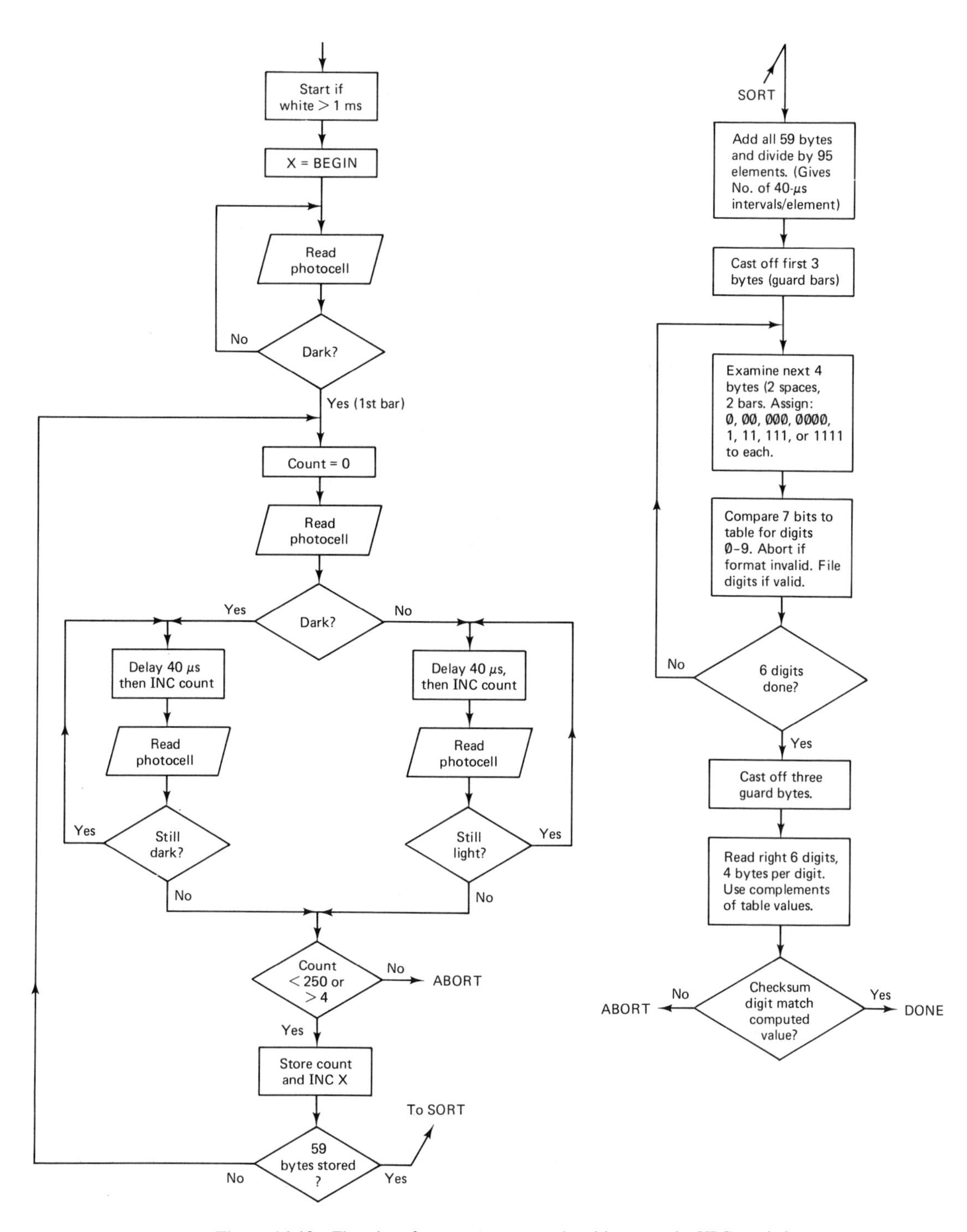

Figure 16.13 Flowchart for a program to read and interpret the UPC symbol.

The time per bar could thus be measured with each binary count representing

$$\frac{9800\ \mu s}{255} = 38.4\ \text{us}, \quad \text{or about} \quad 40\ \mu s$$

The count for a 1-element bar at the fastest scanning rate would be

$$\frac{217\ \mu s/\text{element}}{40\ \mu s} = 5\ \text{counts/element (min)}$$

Figure 16.13 presents an abbreviated flowchart for a UPC bar-code scanner.

REVIEW OF SECTIONS 16.5 AND 16.6

21. Give the major advantage and the major disadvantage of storing digital data on magnetic tape.

22. Does the Kansas City system use amplitude modulation, frequency modulation, phase modulation, or direct recording?

23. What is the baud rate of the Kansas City tape format?

24. A checksum byte is computed by adding the 128 bytes of a test record. Each byte is $55 to produce a test pattern of alternating 1s and Øs. What is the checksum byte?

25. In Figure 16.10 what operation makes sure that the 520-μs time interval does not start on the even-numbered *high* levels when a 1 is being received?

26. The UPC symbol on a package of cookies reads 95210 13574. The symbol on a bottle of catsup reads 95210 37063. What information can you glean?

27. How is product price obtained from the UPC symbol?

28. A UPC is 0 27100 39810 *X*. Calculate *X*.

29. Turn the bar code of Figure 16.12 upside down and read the first 6 digits. Write *X* for any that are not valid codes.

30. Total scan time for 59 bars and spaces of a UPC symbol was 4307 forty-microsecond units. The first white space was recorded as 98 forty-microsecond units in the first byte of the RAM file. What binary representation should be assigned to this space?

16.7 A SERIAL DATA LINK

Connecting two computers with a serial data link is not difficult. Figure 16.14 shows a system using two 6850 ACIAs that can be used for communication between two 6802 computers. The processor, memory, and address decoder of Figure 7.6 or 6.14 may be used. The data line carries TTL levels, not RS-232C levels, and should be limited to a length of about 30 ft. The baud-rate clocks are taken directly from the 0.5 MHz *E*-clock to minimize hardware. Using the internal divide-by-64 function, the data rate is 7812 baud—nonstandard, but serviceable.

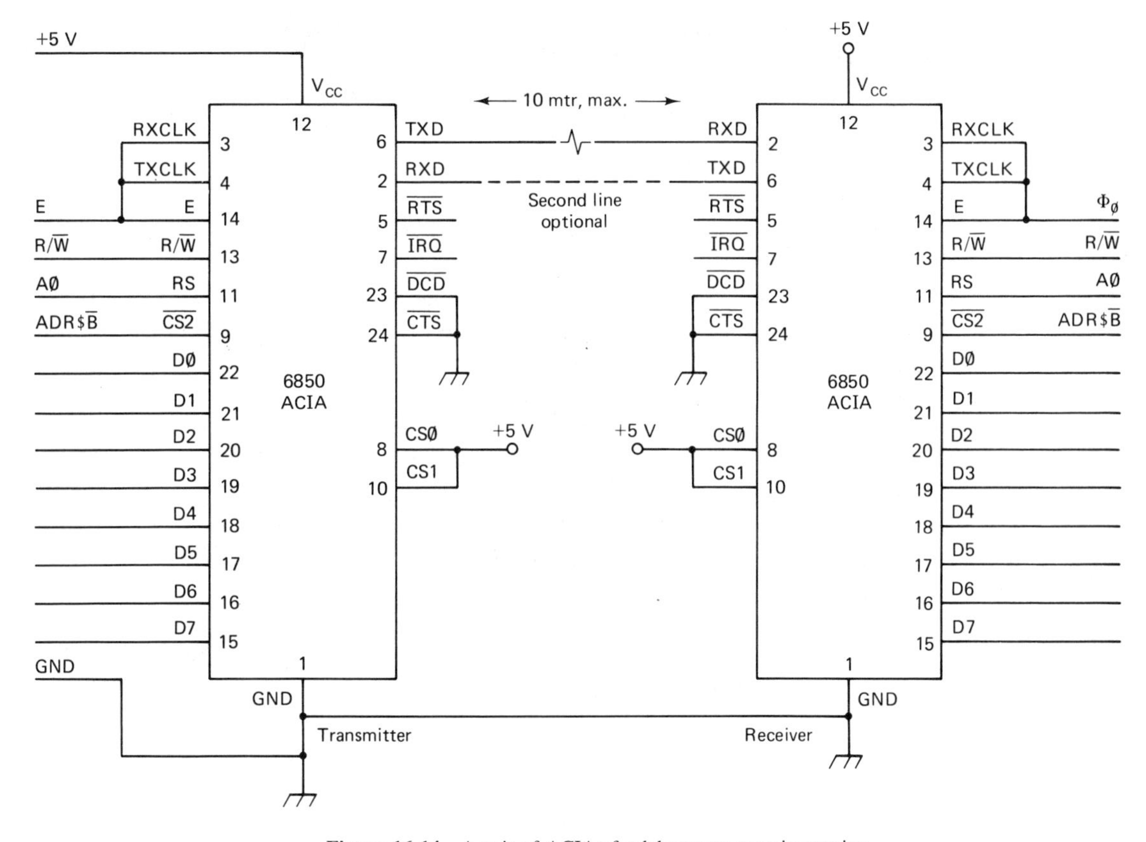

Figure 16.14 A pair of ACIAs for laboratory experimentation.

Handshake and interrupt lines are not used in this elementary application because the two computers will be programmed to do nothing but transmit and receive data.

A transmit test should be conducted first. The transmitting processor is programmed to configure the ACIA and then have it continually transmit the same byte. This produces a repetitive output signal which can be viewed on the oscilloscope and interpreted. List 16.1 shows a sample Transmit-test program.

```
                1000 * LIST 16-1   **  ACIA TRANSMIT TEST.  SENDS BIT
                1010 *  STREAM 0 1001 0111 0 1 * FIG 16-14 HARDWARE.
                1020 * TESTED AND RUNNING, 27 JULY '87 BY D. METZGER
                1030 *
                1040          .OR  $FF80
                1050          .TA  $4080
B000-           1060 SERCR    .EQ  $B000     SERIAL CONTROL/STATUS REG.
B001-           1070 TDR      .EQ  $B001     TRANSMIT DATA REGISTER.
                1080 *
FF80- 86 03     1090          LDAA #$03      MASTER RESET ACIA BY
FF82- B7 B0 00  1100          STAA SERCR      WRITE TO CONTROL REG.
FF85- 86 1E     1110          LDAA #$1E      CONFIGURE FOR 8 BITS,
FF87- B7 B0 00  1120          STAA SERCR      ODD PARITY, 128 US/BIT.
FF8A- B6 B0 00  1130 WAIT     LDAA SERCR     READ STATUS REGISTER; CHECK
FF8D- 84 02     1140          ANDA #$02       BIT #1 FOR TDR EMPTY.
```

```
FF8F- 27 F9      1150       BEQ  WAIT      WAIT UNTIL IT IS EMPTY.
FF91- 86 E9      1160       LDAA #$E9       THEN SEND DATA STREAM.
FF93- B7 B0 01   1170       STAA TDR       (ANY DATA WILL DO - $E9
FF96- 20 F2      1180       BRA  WAIT       IS JUST AN EXAMPLE)
                 1190       .OR  $FFFE
                 1200       .TA  $40FE
FFFE- FF 80      1210       .HS  FF80
                 1220       .EN
```

Transferring a data file from one computer to another is not that much more difficult than conducting the transmit test. We will leave the development of the transmit program to you. Here are some hints:

1. Start with the Transmit-test program. Use LDAA 0,X to pick up the data. Load X with the first address of your data file before the *wait* loop, and increment X after each *store* to the TDR.
2. Use an end-of-file flag (data $FF will do) to stop transmitting when the entire file has been sent. Check for data $FF after the LDAA 0,X and *halt* if $FF is encountered.

Receiving a file requires initializing to the same data format and checking the SR0 bit for new data before reading receive-data register.

```
*ACIA RECEIVE-FILE PROGRAM

        LDAA    #$03       Master Reset.
        STAA    SERCR
        LDAA    #$1E       8 bit, odd parity
        STAA    SERCR
        LDX     #$FFC0     Start of File.
HOLD    LDAA    SERSR      Check Status Bit 0.
        ANDA    #$01
        BEQ     HOLD       Wait until data present.
        LDAA    RDR        Get data.
        CMPA    #$FF       Halt on end-of-file.
HALT    BEQ     HALT
        STAA    0,X        Store in file.
        INX                Next file address.
        BRA     HOLD       Wait for next byte.
```

Try transferring an ASCII message between two computers using the ACIAs.

Answers to Chapter 16 Review Questions

1. 37.5 s **2.** 0, 1, 1 **3.** 5.0 MHz divided in 12 stages gives 1221 Hz, a 1.7% error. **4.** 19 200 baud

5. $7801 **6.** $03 **7.** $C7 **8.** $38

9. This allows a second byte to be stored while the first byte is being transmitted.

10. R/$\overline{W}$, pin 13.

11. By loading from address $58FØ.

12. Status bit 1 goes *high* and an IRQ is generated.

13. The receiver's $\overline{RTS}$ output. **14.** Data should not be transmitted unless the receiver is indicating "Clear To Send." **15.** Receiver overrun, SR5.

16. LDAA STATR; ANDA #$74 **17.** An audio-frequency carrier signal.

18. The dc and low-frequency components of RS-232-type signals will not pass the phone-line bandwidth, which cuts off below 300 Hz. **19.** Transmission in two directions simultaneously by using two frequency-modulated audio carriers.

20. An error-checking scheme in which received data is retransmitted to the sending station.

21. Advantage: high storage volume at low cost; disadvantage: serial access; slow to find data needed.

22. Frequency modulation **23.** 300 baud **24.** $8Ø

25. Wait 208 μs **26.** They came from the same manufacturer.

27. It is looked up in a table by the computer at the point of sale. **28.** $X = 3$

29. None are valid because they are all even parity. **30.** Binary ØØ

CHAPTER 16 QUESTIONS AND PROBLEMS

Digits after the decimal point refer to section numbers in Chapter 16.

Basic Level

1.1 Define the terms *serial* and *baud*.

2.1 Even parity is being used, with bits Ø through 6 conveying ASCII data, and bit 7 being the parity bit. Are any of the following received data bytes in error?

FØ A6 7B ØØ 19 3C 87 ED 24 Ø5

3.2 What hex data word should be written to the ACIA control register to set a data rate of 300 baud with 8 data, no parity, and 2 stop bits? An interrupt should be generated each time the transmitter is ready to send a new byte. The receiver section of the ACIA is not to be used. The transmitter clock signal is 4.8 kHz.

4.2 What is a *break* signal in serial data transmission?

5.3 If a 6850 is chip-selected at base address $9ØØØ, at what addresses are its Transmit Data Register, Receive Data Register, Control Register, and Status Register normally located?

6.3 Which bits of the status register indicate an error condition if they are logic 1?

7.4 Define the terms *modem*, FSK, and PSK.

8.4 Define the terms *protocol* and *smart modem*.

9.5 How are logic 1 and logic Ø represented in the Kansas City tape format?

10.5 In the routine of Figure 16.10, how many reads of the tape-player output level are made for each bit of received data?

11.6 In the UPC bar code, how many different widths of bars are there?

12.6 If a UPC code is 0 75128 39150, what is the checksum digit?

13.6 How many elements wide is a UPC bar symbol?

Advanced Level

14.1 Explain how a receiver of serial data knows when to read the first bit and each succeeding bit in a stream of data.

15.1 Compare the logic-0 and logic-1 voltages in TTL and in RS-232 systems.

16.2 How is it possible for the transmitting microcomputer to write a new byte of data to the ACIA while the previous byte is being sent?

17.2 Describe a *receiver overrun* condition in serial data transmission.

18.3 Write the source code for a 6802 interrupt-service routine that checks the ACIA status register and sends a file byte (pointed to by X) if the Transmit Data Register is empty and reads a byte into accumulator A if the Receive Data Register is full.

19.4 Distinguish among simplex, half-duplex, and full-duplex data communication.

20.4 What is the advantage of FM over AM in data communication? What is the advantage of PSK over FSK?

21.5 In reading a Kansas City tape recording, how does the reading computer determine the RAM address at which to begin storing the data and the length of the data record to be stored?

22.5 How is the checksum byte computed in a Kansas City recording?

23.6 Prepare a detailed flowchart for a routine to examine a file of 6 bytes (which are *left* bar-code digits) and build a 6-byte file containing the corresponding decimal values.

24.6 Write the 6802 source code for the routine flowcharted for Problem 23.

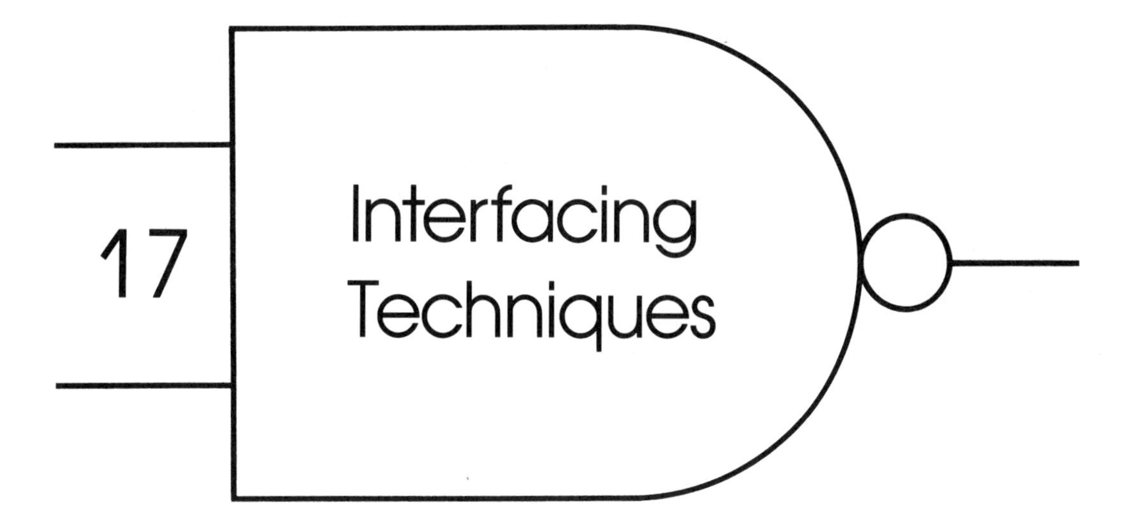

17.1 BUS BUFFERING

Section Overview

There are several considerations that may make it necessary for a hardware buffer IC to be placed between a microprocessor bus and the processor itself or a peripheral chip. This section examines these considerations and the types of buffers available.

Bus loading. Microprocessor components are generally capable of sinking at least 1.6 mA (*low* output state) and sourcing at least 0.1 mA (*high* output state). One standard TTL input will require the sinking of 1.6 mA, thus exhausting the microprocessor's output capability. Four LS-series TTL inputs will typically require the sinking of a total of 1.6 mA (0.4 mA each) and can be driven reliably with a microprocessor output. Standard TTL devices are hardly ever connected to microprocessor buses for this reason.

Buffers of the LS-series are generally capable of sinking 24 mA at *low* output, so they can drive over 50 LS-series devices or 15 standard TTL devices.

Microprocessor inputs and 74-HC- and 74HCT-series gates typically present an input loading of 10 pF each. Their dc current requirements are negligible. Heavier capacitive loading of a microprocessor output simply slows down the output response, so a maximum-load specification is difficult to give. As a rule of thumb, a 1-MHz system should be capable of driving 10 CMOS or microprocessor loads, whereas a 2-MHz system should be limited to three such loads.

Bus buffers should be used when more than 4 LS-TTL or 10 high-impedance devices must be placed on the system bus and when capacitive loading makes desired 2-MHz system operation impossible. The 74LS244 is an 8-line buffer, 2 of which are suitable for 16-line address buffering. The data bus is bidirectional, so a bidirectional buffer is required. The 74LS245 is such a buffer. Its *direction* input is fed from the processor's R/$\overline{W}$ line to allow processor data out during *write* cycles.

The 74HC244 and 74HC245 buffers, while somewhat more expensive, have high-impedance inputs and output drive capabilities as good as LS-TTL. They can often be added to an existing system without further exacerbating the bus-loading problem. The 74HCT series has input threshold voltages nearly the same as its TTL counterparts and is generally a direct "drop-in" replacement for 74LS-TTL with less loading. Figure 17.1 illustrates microprocessor bus buffering.

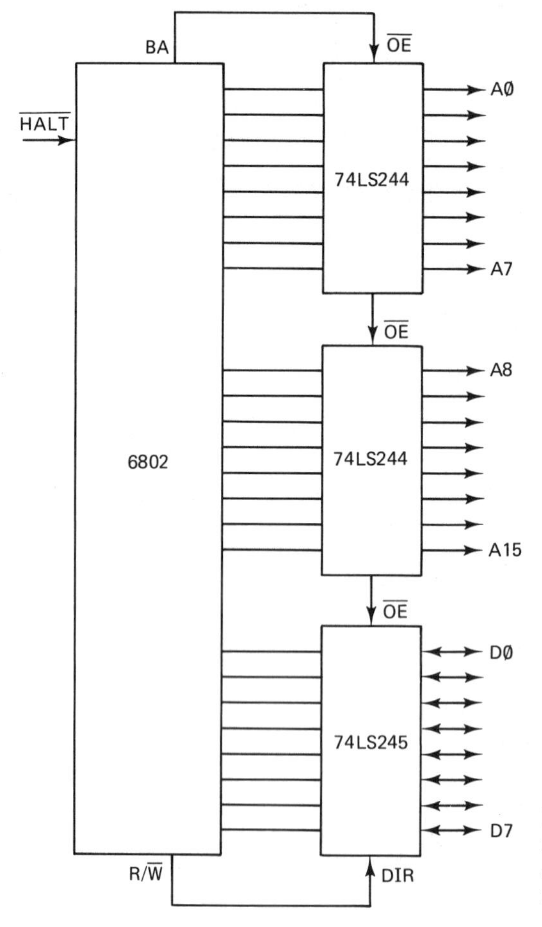

Figure 17.1 Address and data lines may require buffers to drive a large number of devices. The data bus requires a bidirectional buffer.

Isolation is another reason for using bus buffers. The 6802 address bus, for example, cannot be placed in a tristate condition. To allow another device to take control of the bus, 6802 can be placed in the HALT state and the address-bus buffers can be placed in the high-Z output state. This is also illustrated in Figure 17.1.

Damage protection of expensive systems is a final reason for using buffers to feed peripheral devices. Address and data signals are generally of too high a speed to be carried on external lines, but if a computer's peripheral input and output lines must be brought outside the instrument cabinet, they should first be fed through buffers. This will limit capacitive loading by long external lines and limit damage from misapplied external voltages to the buffer chips.

17.2 DIRECT MEMORY ACCESS

Section Overview

If a large amount of data needs to be transferred to or from a microprocessor's memory in the shortest possible time, it is more efficient to take the processor off the address and data buses and let an external *DMA controller* generate the addresses and transfer the data. In the case of the 6802 this can permit 16K of data to be transferred in 16 ms, compared to 330 ms under processor control. Other DMA modes allow data to be "slipped in" to the memory without actually stopping the processor.

A high-resolution graphic image on a CRT can easily require 100 000 bits of RAM storage to maintain it. If animation of the image is a goal, this amount of data may have to be rewritten 10 or 20 times a second. The write rate required might then be 200 000 bytes per second, or 1 write every 5 μs. Data transfer under processor control is not capable of such speed. The write routine would require 18 μs per byte, even assuming that a hardware interrupt or end-of-table flag could be used to eliminate repeated tests for the ending address:

```
        SEI                         No interrupt while stack is altered.
LOOP    PULA            4~          Get byte from table.
        INX             4~          Point to next CRT RAM.
        STAA    0,X     6~          Byte 00 is end of
        BNE     LOOP    4~          file flag.
        CLI
```

Disk drives also require very high data-transfer rates, so that simply reading the data would require all the processor's attention, leaving it no time for controlling the disk drive or for computing checksums and other housekeeping tasks associated with storing the data.

Direct Memory Access (DMA) speeds up the process of getting data into and out of memory and allows it to be accomplished without continual supervision by the processor. There are three rather different forms of DMA, and there are variations within some of the forms and confusions in the names given to them. We will attempt to sort these out now.

Continuous DMA, also called *block-transfer*, or *processor-halted* DMA, is the simplest to implement. It is diagrammed in Figure 17.2. In this system the DMA controller *halts* the processor at the end of its current instruction and places its address and data buses and VMA and R/$\overline{W}$ in a floating state, either directly or by means of tristate buffers.

The DMA controller contains an address register and a data-counter register. These are preloaded by the processor with the starting address of the data block to be transferred and the number of bytes to be transferred, respectively. The DMA transfer is initiated by a service request from an external device (CRT end of vertical scan or disk-drive data stream, for example). The DMA controller then calls up addresses at the *E*-clock rate and reads or writes via the data bus as instructed. The data source or data destination can be fixed addresses (for example, an input buffer or output latch), or both can be areas in memory that the DMA increments or decrements through.

The Motorola 6844, a typical DMA controller, has four address registers and data counters to handle multiple files. It can transfer data at a rate of 1 megabyte per second in the processor-halt mode. A set of control registers configures it for a variety of functions.

Cycle stealing: In many applications there is strong objection to stopping the processor for some noticeable fraction of a second while DMA is going on. Screen flicker on a CRT and delays in handling critical interrupts are typical problems. Still,

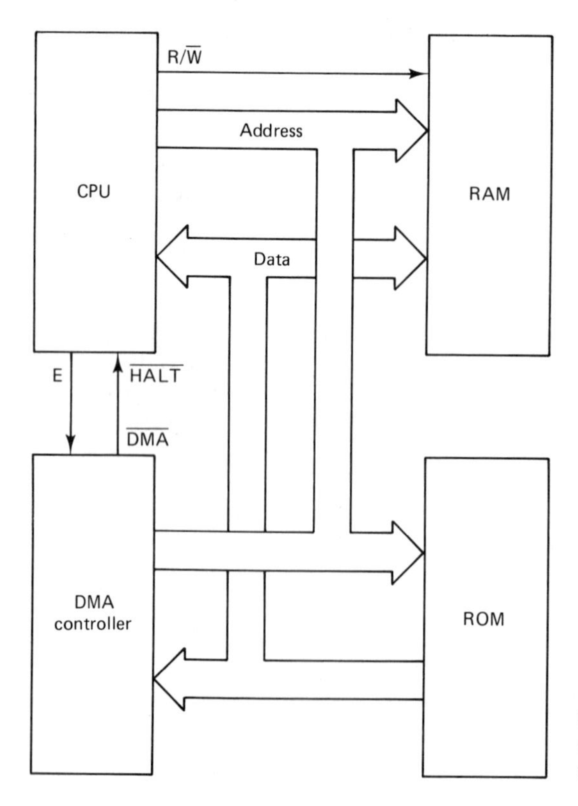

Figure 17.2 The processor can be halted and its buses tristated to allow the DMA device to access memory.

it may be possible to transfer the required data only under DMA control because not enough processor time is available to do it under program control. Several techniques are available to sneak a few bytes at a time into or out of memory while the processor is not using the bus:

- Halt-and-restart is the crudest of these techniques. It is the same as processor-halt, described earlier, except that only a few bytes are transferred under DMA control at a time. The processor is then allowed to return to its program for a fixed amount of time. Any trade-off ratio is possible, of course, but faster DMA rates mean slower processing rates with this technique. The 6800 also has one ''dead'' cycle on each *halt* and each *restart*, further impairing performance with this method.
- Many processors have special cycle-stretching inputs, which suspend the processor's clock for perhaps 5 or 10 μs upon receiving a DMA request signal. The 6802 has no such input, but the input for several other processors is given:

6800	TSC	(Three-state control)
6809	$\overline{\text{DMA}}$	(DMA input)
8085	HOLD	
Z80	$\overline{\text{WAIT}}$	

  One or more bytes may be transferred under DMA control in this way without actually halting the processor or waiting for the completion of the current instruction. The processor is slowed down, but only slightly in comparison with the previous DMA method. Clock suspension is limited to a few microseconds because the processors are dynamic parts and will lose internal data if placed in a static condition. The cycle-stretching DMA technique is illustrated in Figure 17.3.
- Most processors have a number of instructions that contain VMA-not cycles or the equivalent. These are time intervals in which the processor is working on data internally and is not using the address or data buses. These cycles can truly be stolen for DMA activity. Generally, a single $\overline{\text{VMA}}$ cycle does not allow

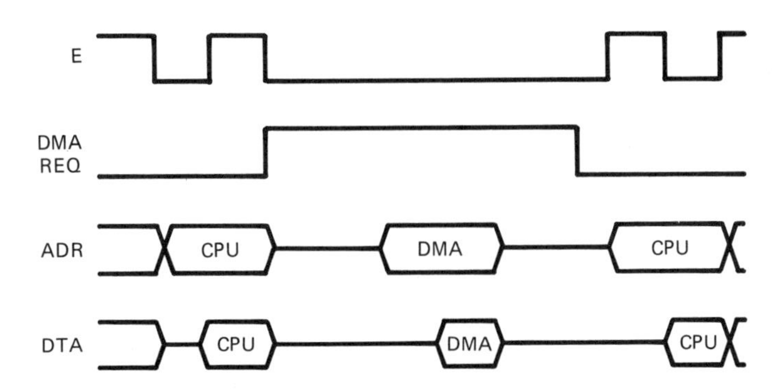

Figure 17.3 A machine cycle can be stretched to allow the DMA device to access memory between processor accesses of memory.

enough time for the DMA controller to recognize its opportunity and slip in a data byte, but the 6800/6802 processor's indexed-mode instructions, JSR, TSX, and TXS, and all the *branch* instructions have two consecutive $\overline{\text{VMA}}$ cycles, which allows adequate time for a DMA cycle. Of course, if one of these instructions doesn't happen to come up for 50 μs and a disk read requires a DMA cycle in the next 10 μs, this form of cycle stealing may not be adequate.

Interleaved, or multiplexed, DMA makes use of the fact that data reads and writes by a 6800-family processor are only done in the last half of the *E*-clock cycle. If memories are used that are, roughly speaking, twice as fast as the processor, then the DMA can access the memory during the first half of the clock cycle and the processor can access it during the second. This technique makes DMA *transparent* to the processor. No aspect of the main program, not even timing, is affected. The only penalty is that slightly more expensive RAM and ROM chips may have to be purchased to obtain the necessary memory speed. Also note that the address and data buses must be buffered so that the processor can be disconnected from the memory during the first (*low*) half of the *E* cycle. Figure 17.4 illustrates interleaved DMA timing.

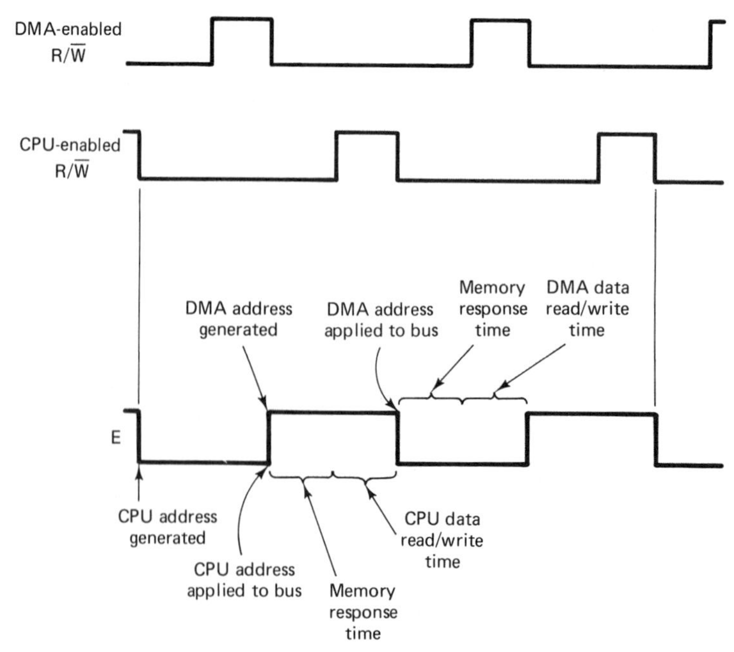

Figure 17.4 If the memory is fast enough, processor and DMA accesses of memory can be interleaved on alternate half cycles of the clock.

REVIEW OF SECTIONS 17.1 AND 17.2

Answers appear at the end of Chapter 17.

1. Someone unwittingly replaced the four LS-TTL devices on a 6802 address bus with four standard TTL devices of the same type number. What is likely to be observed on an oscilloscope connected to these lines?
2. Why are 74LS244s used for address-bus buffering, but the 74LS245 is used for data-bus buffering?
3. A 6802 system uses one 74LS244 for data input and three 74LS373s for data output. An addition is planned that will add two more 8-bit output latches. How can this be handled?
4. Someone has attempted to transmit the E-clock signal from a 6802 microprocessor to another system by connecting it to a 10-ft cable. The 6802 stops working. How will you solve the problem?
5. Why can a DMA controller transfer data faster than a microprocessor?
6. In continuous DMA what is the processor doing while the DMA controller is transferring data?
7. Where does The DMA controller get the address that it calls up when it takes over the address bus?
8. Which two types of DMA do not slow down the processing of the main program?
9. What hardware addition (besides the DMA controller) would be necessary to implement DMA on a 6802 system?
10. In cycle stretching, how is the extra time for DMA cycles obtained?

17.3 INTERFACING TO DYNAMIC RAMS

Section Overview

Dynamic RAMs are generally less expensive, consume less power, and contain more bits per chip than static RAMs, but they require additional circuitry for data accessing and for data refreshing, which must occur every 2 to 4 ms. DRAMs operate with an extremely fast series of timing pulses, making noise pickup and physical wiring placement more critical than in the case of static RAMs. For these reasons, DRAMs are most commonly used in mass-produced systems with storage volumes of 128 kilobytes or more.

A DRAM memory cell consists of an integrated capacitor with a value on the order of 1 pF. By comparison, a static RAM typically requires 8 integrated transistors per bit. The metal-oxide dielectric of this capacitor is such a good insulator that a charge can be held in it reliably for 2 to 4 ms, whereupon the cell must be read and rewritten to restore its full charge.

Rows and columns: DRAM capacitors are arranged in an array of rows and columns, as depicted in Figure 17.5. Although a 3 × 3 array is shown for simplicity, a real DRAM might contain 64 kilobits, arranged as a 256-row by 256-column array. An individual cell is read by turning on one entire row of capacitor-accessing MOS transistors and then selecting one of the column-sense amplifier outputs to feed the data-output line. Since a read depletes the charge on a whole row of capacitors a write-back is performed on each read by switching the amplifier output back to recharge the row of capacitors.

A write to a particular capacitor is accomplished by placing the desired 1 or 0 level on the data line, selecting a row and a column, and closing the *refresh/write* switches. In practice, the row- and column-select lines are driven by decoder circuits, so 8 *row* lines can feed 256 rows.

An entire row is refreshed at one time in an operation that is essentially the same as a *read*. A complete refresh of a 64K DRAM then takes the equivalent of 256 machine cycles, 1 cycle for each of 256 columns. Even if the processor were halted every 2 ms for DRAM refreshing, assuming a 2-MHz system and 0.5 μs per refresh cycle, the percentage of time lost for refresh would be only

$$\frac{(256 \text{ cycles})\ (0.5\ \mu\text{s/cycle})}{(2 \text{ ms})} = 6.4\%$$

Of course, there are more efficient methods, employing cycle stealing and interleaving, that can make DRAM refresh completely transparent to the processor.

A typical dynamic RAM IC, the 4164, containing 65 536 words of 1 bit per word, is illustrated in Figure 17.6. Eight of these chips provide 64K of RAM for a processor with an 8-bit data bus, such as the 8088. Sixteen of them provide 64K words (128K bytes) for a 16-bit processor like the 68000. The practice of storing each bit of a word in a different chip may seem strange if your experience has been limited to EPROMs and the newer static RAMs. However, this practice was common with static RAMs a few years ago (the 2102 RAM was a popular example), and it is still widely followed with DRAMs.

Address multiplexing: You will notice that the 4164 has only 8 address pins, yet accesses 65 536 addresses. This is accomplished by multiplexing the address inputs. First the *row-address strobe* (RAS) is asserted, and the low-order address byte is applied to the DRAM. Then the *column-address* strobe (CAS) is asserted, and the high-order address byte is applied. The entire process of latching the two halves of the address into the DRAM requires on the order of 100 ns, so there is ample time in the usual processor clock cycle to complete the read or write.

DRAM Controllers: In addition to the address multiplexer, a DRAM memory bank needs a number of other control functions.

- A timing system to generate the Row- and Column-Address Strobe signals, the multiplexer high-byte/low-byte switchover timing, and a *data-valid* signal to prevent reads during refresh.
- A refresh timer to initiate refreshing at 2-ms intervals (4-ms in some newer DRAMs).
- A refresh counter to call up each row in succession during refresh.

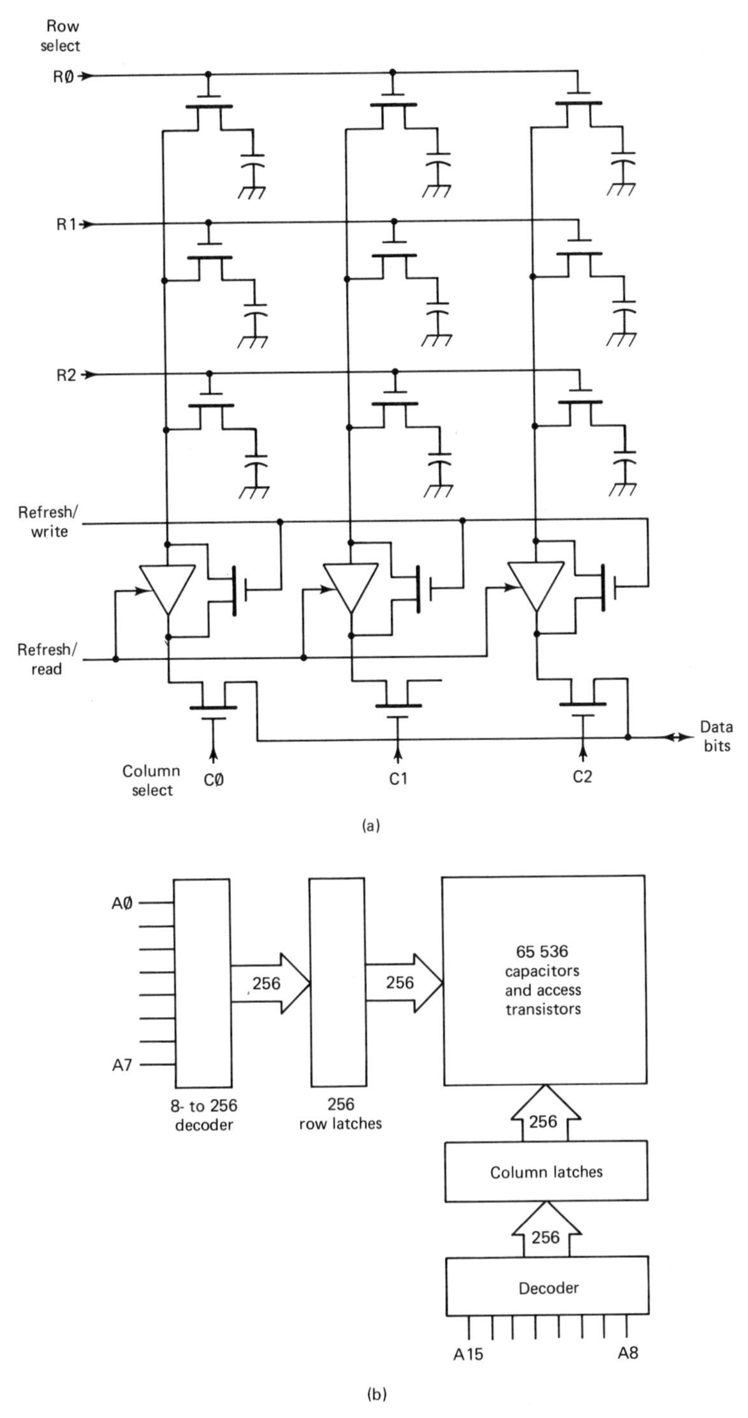

Figure 17.5 (a) Internal dynamic RAM cell structure. (b) Organization of a 64 kilobit DRAM memory.

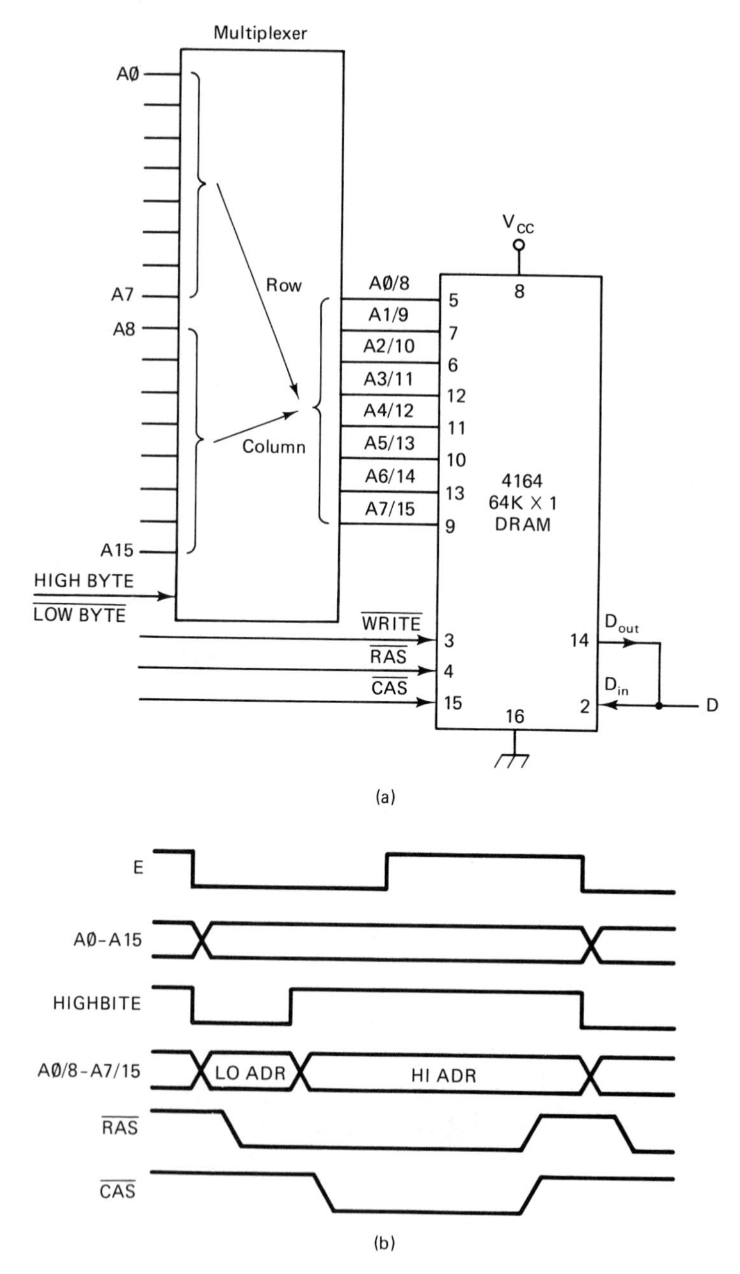

Figure 17.6 (a) The address lines of a dynamic RAM are multiplexed. (b) The low-order address lines are latched first when *Row Address Strobe* is asserted.

• An arbitration unit to resolve conflicts, as, for example, when the refresh timer demands a refresh cycle while the processor is in the middle of a read-modify-write instruction.

A DRAM controller provides all of these functions in one IC, usually a 28- or 40-pin package. These controllers tend to be quite specific as to the types of DRAMs and the types of processors that they support, so we will not go into detail on any of them. It is also a trend worth noting that static RAMs, limited to 4096 bits per chip until a few years ago, are now readily available with 262 144 bits per chip. Also, the cost advantage of DRAMs over SRAMs, once a ratio of 10:1, has dropped to about 3:1.

Figure 17.7 shows the interface circuitry for a processor, DRAM controller, and bank of eight DRAM chips.

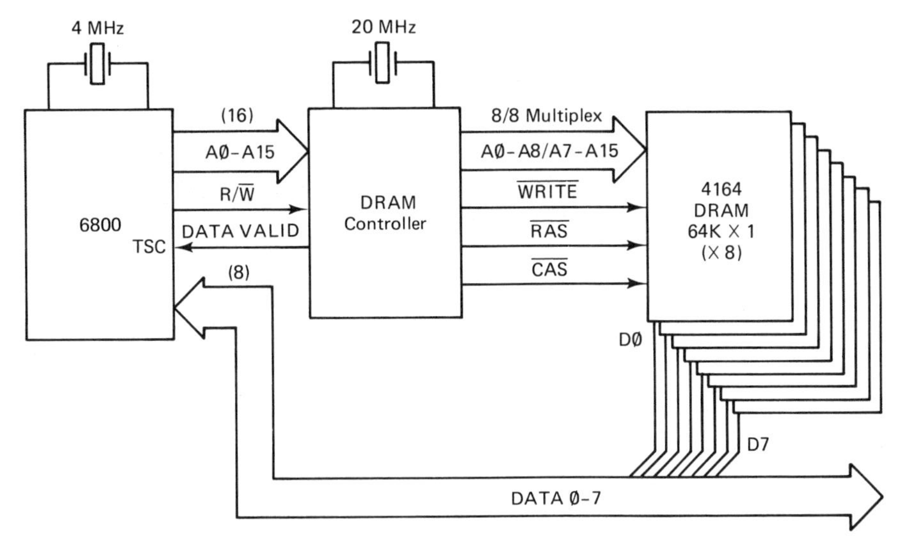

Figure 17.7 A dynamic RAM controller and an array of 8 DRAM chips.

17.4 NOISE FILTERING AND SIGNAL CONDITIONING

Section Overview

Ideally, microcomputer signals are pulses, switching from 0 V to +4 V cleanly in a few nanoseconds. Actually, noise pickup and consequent misreading of data bits is the bane of the microcomputer hardware technician. This section discusses a variety of techniques for minimizing noise and restoring the ideal logic *low* and *high* levels.

Speed Kills: In high-density packages, adjacent printed-circuit tracks or wire-wrap runs are going to be unavoidably coupled by stray capacitance. Such capacitance can be minimized by trying to alternate high-speed signal lines with power-supply, ground, and static-control lines when designing the circuit board and by running component-side wiring at right angles to solder-side wiring. Still, a few picofarads of stray coupling is inevitable.

Figure 17.8(a) shows a prime candidate for noise pickup. An open-collector output drives a 74LS-TTL input. A low-impedance, fast-falling pulse on an adjacent line is coupled by 2 pF of stray capacitance. The resulting negative pulse to the 74LS-TTL input has a time constant of 20 ns. The LS-TTL part has a propagation delay of 10 ns and will surely respond to the noise pickup. Here are some tips for avoiding the noise pickup.

1. Drive the input from an active output instead of an open-collector output, as shown in Figure 17.8(b). The 120-Ω pullup resistor will get the noise spike down to less than 1 ns—too fast for the LS-TTL input to respond to.
2. Put some capacitance from the noisy input to ground. As shown in Figure 17.8(c), 6 pF will cut the noise pulse from −4 V to −1 V. Actually, the line and its transistors probably already present more than 6 pF of stray capacitance to ground, which is why noise pickup isn't a more frequent problem than it is. The stray capacitance between problem lines is probably more like 5 to 10 pF. Up to 100 pF can be tried on a noisy line, but at some point the required response speed of the line will be lost.
3. Part swapping can sometimes provide a solderless solution to noise problems in systems where the design has not been verified. If the noise source IC is a fast part that doesn't need to be so fast, you can try replacing it with a 74L-series part. The rise and fall times of these parts are about 30 ns, making noise

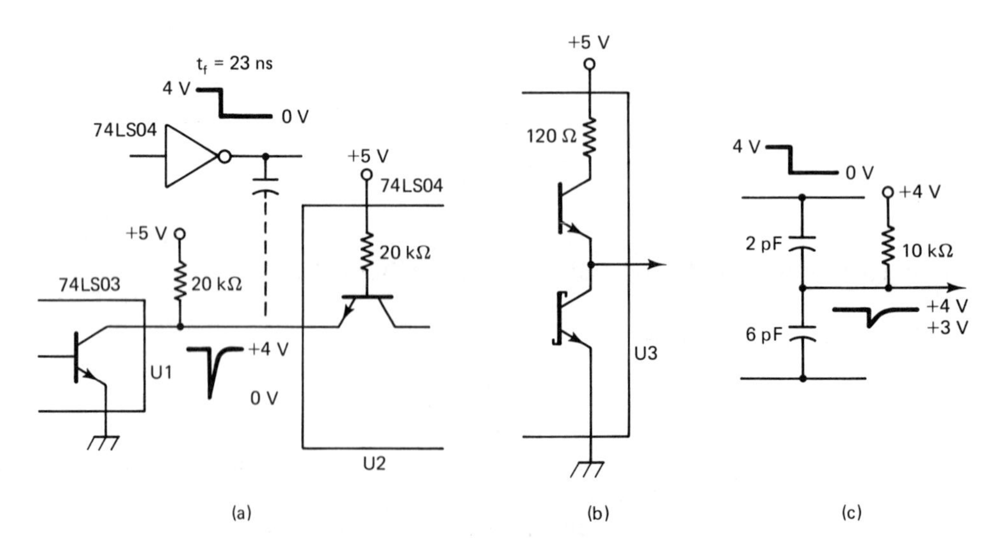

Figure 17.8 Stray capacitive coupling can inject noise pulses into a line driven by an open-collector device (a). Driving the line with an active-output device (b) and adding a small capacitance to ground (c) will reduce noise pickup.

transfer to adjacent lines less severe. U2 can also be replaced. Try a standard 7400 part for stronger pullup (4 kΩ instead of 20 kΩ) or a 74L00 part for more-sluggish response to short noise pulses. The 7400 may exceed dc loading limits and the output of the 74L00 may prove too slow, but it's worth a try. Of course, in servicing production equipment whose design has been proven by many operational units, it would be inviting timing problems to change from one logic series to another.

Power-line noise on the +5-V supplies and on the 60-Hz line to ground is an endless problem, and any good digital technician should keep a pocket full of 0.1-μF capacitors (of the appropriate voltage ratings) to solder across the lines when tracking down noise problems. Heavy-gage wire (#18) should be used for V_{CC} and ground distribution. A system that requires 0.5 A average may require pulses over 2 A from the 5-V supply during switching transients.

Input signal lines should be filtered to ensure that they cannot pass pulses much faster than the intended signals if noise pickup is likely to be a problem. With high-impedance MOS inputs such filters are relatively easy to design. Figure 17.9

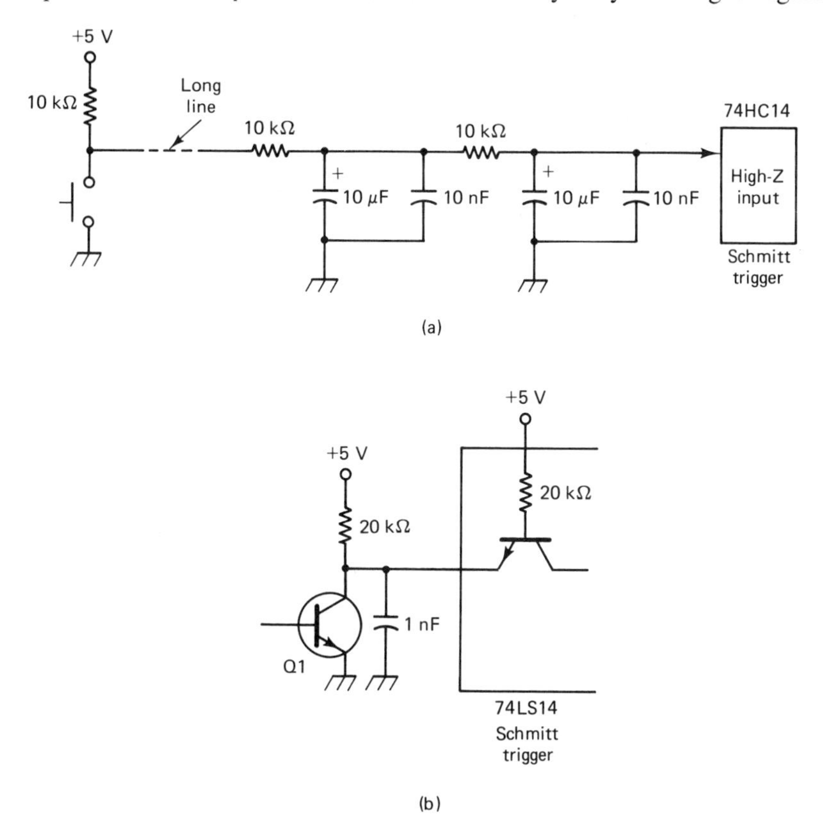

Figure 17.9 (a) Long input lines to high-*Z* devices should be filtered to prevent noise pickup. (b) TTL devices are harder to filter. See text for component value calculations.

shows a two-section filter for a keyswitch input. Activations at a rate of up to five per second are to be accepted but bounces lasting up to 20 ms are to be rejected. A time constant of 100 ms is chosen:

$$C = \tau/R = 100\ \text{m}/10\text{k} = 10\ \mu\text{F}$$

Since large-value capacitors do not effectively bypass submicrosecond noise pulses, each 10-μF electrolytic capacitor is shunted with a 10 nF ceramic or mylar bypass.

TTL inputs are harder to filter because resistances in series will prevent the input from being pulled directly to ground. A simple capacitor across the input is the best expedient. If 1-MHz noise is to be screened out but 20-kilobaud serial data is to be accepted in Figure 17.9(b), a time constant between 1 μs and 50 μs should be selected. Choosing 10 μs

$$C = \tau/R = 10\ \mu\text{s}/10\ \text{k}\Omega = 1\ \text{nF}$$

Schmitt-trigger inputs are recommended to avoid "chattering" from noise pickup as the R-C charging curve passes through the *low*- to *high*-threshold.

Optical coupling is widely used in heavy industrial environments to provide input to microcomputer-based instruments. There are two main reasons for this.

1. The optical sensors used do not respond to submicrosecond noise pulses, so the problem of filtering noise out of the input lines is effectively bypassed.
2. The optical couplers remove all electrical connection between the input device and the microcomputer. The fear is thus laid to rest that an expensive microcomputer system may be totally destroyed by someone hooking the 440-V power line where they should have hooked the position-sensing switch.

Figure 17.10 shows two optical-isolation circuits. The first uses an LED to render a photodiode conductive and thus turn on the output transistor. Photodiodes are quite fast, and this system is capable of switching at rates above 1 MHz. The second system uses a photoresistive cell in place of the photodiode. Photoresistive cells have

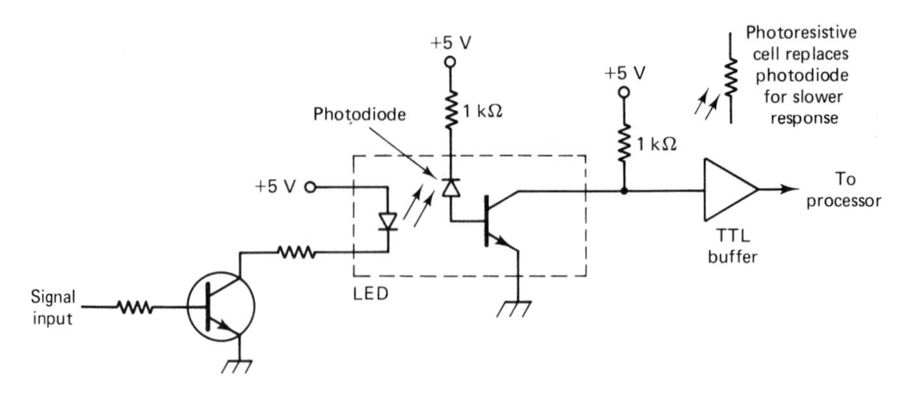

Figure 17.10 Optical coupling minimizes potential for damage to electronic systems from misapplied input voltages. Proper selection of photodectors can screen out fast noise pulses.

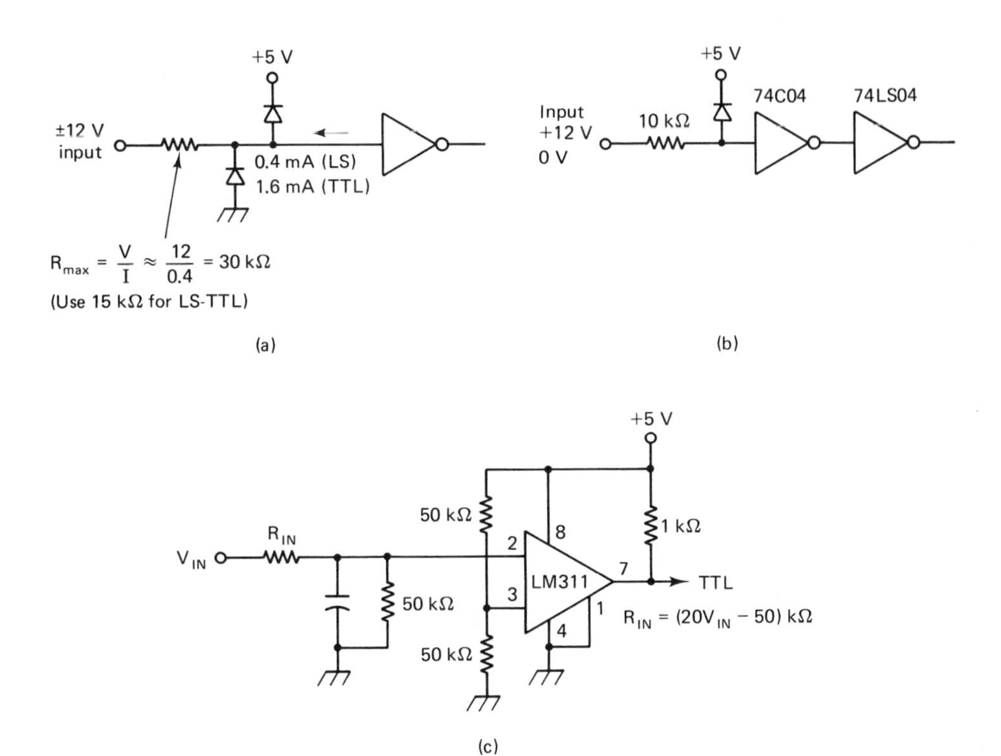

Figure 17.11 Three circuits for converting non-TTL voltage levels to TTL-level signals.

response times on the order of 10 ms to 100 ms, so this circuit is only useful for slow-speed input signals. In fact it is most often used when spurious high-speed pulses would otherwise cause false inputs.

Non-TTL input levels can be used to drive TTL inputs through one of the circuits presented in Figure 17.11. The first circuit simply clamps higher-input voltages at +5.6 V and −0.6 V. The resistor must be capable of pulling the TTL input to 0 V by sinking the 0.4-mA LS-TTL current on the low-voltage input level.

The second circuit uses a CMOS driver to feed the TTL input because the input resistor could not pull the TTL input to 0 V while sinking 0.4 mA from the LS-TTL input. The third circuit uses a differential comparator to allow a small-input signal to produce a TTL input.

REVIEW OF SECTIONS 17.3 AND 17.4

Answers appear at the end of Chapter 17.

11. Calculate an estimation of the resistance of the dielectric between the DRAM memory cells. Is it closer to 1 M Ω, 100 GΩ, or 1 TΩ?
12. The type 41256 DRAM is a 256-kilobit memory, arranged along the same lines as the 4164. How many cycles will it take to refresh it?

13. Referring to Figure 17.5, what is the condition of the sense amplifier outputs when data is being written to the DRAM?
14. How can the 4164 DRAM allow access of 65 536 addresses when its package has only 16 pins?
15. How many 4164 ICs would be required to provide 256 kilobytes of storage?
16. The 4416 is a 16K × 4 bit DRAM. If a refresh cycle takes 250 ns, how long would the processor have to be halted to do a complete refresh?
17. In Figure 17.8(a), why is noise pickup only a problem when U1 outputs a logic *high* level?
18. The circuit of Figure 17.9(b) uses a 5-nF capacitor in place of the one shown. What is the risetime of the input pulse when Q_1 turns off?
19. An auto-chassis welder is microcomputer controlled. The part-in-position sensor switch is being fed through an optical coupler because of a severe noise problem from the welders. What type of photodetectors should be used?
20. How can a signal which switches between +1.0 V and +0.2 V be used to drive a TTL input?

17.5 DISK-DRIVE INTERFACES

Section Overview

Floppy disks answer the need for high-volume storage of digital data. Their low cost and fast access time to a number of different files on the same disk has made them the data-transfer medium of choice for desktop computers and programmable instruments.

Floppy disks consist of a square plastic envelope, either 8 in. or $5\frac{1}{4}$ in. on a side, which contains a disk coated with a magnetically sensitive material. The newer $3\frac{1}{2}$-in. floppy disks are encased in a rigid plastic case and hold about twice as much data as a $5\frac{1}{4}$-in. disk, in spite of their smaller size. The disk is placed in a disk drive, which spins it at 5 or 6 revolutions per second when data is to be read or written. The disk drive contains a magnetic read/record head, which rides on the surface of the spinning disk and can be moved from the edge toward the center under control of a stepper motor. Data bits are recorded as changes in magnetization on concentric rings, called **tracks**. There are typically 35 or 40 tracks on a $5\frac{1}{4}$-in. disk and 77 tracks on a 3½ or 8-in. disk. They are numbered consecutively from the outside, starting with 0. Spacing between the tracks is generally 0.02 in. Figure 7.12 illustrates the 5¼-in. disk features.

Track format: Each track on a 8-in. disk is divided into 26 sectors, with an unrecorded gap between sectors. Each sector contains 128 bytes (1024 bits) of data. A little arithmetic will show that the total storage capacity of an 8-in. floppy disk in the format described is 256,256 bytes.

The $5\frac{1}{4}$-in. disks use a variety of formats. There may be 8 to 16 sectors per track, and each sector may contain 512 or 256 bytes. Total storage capacity is 143

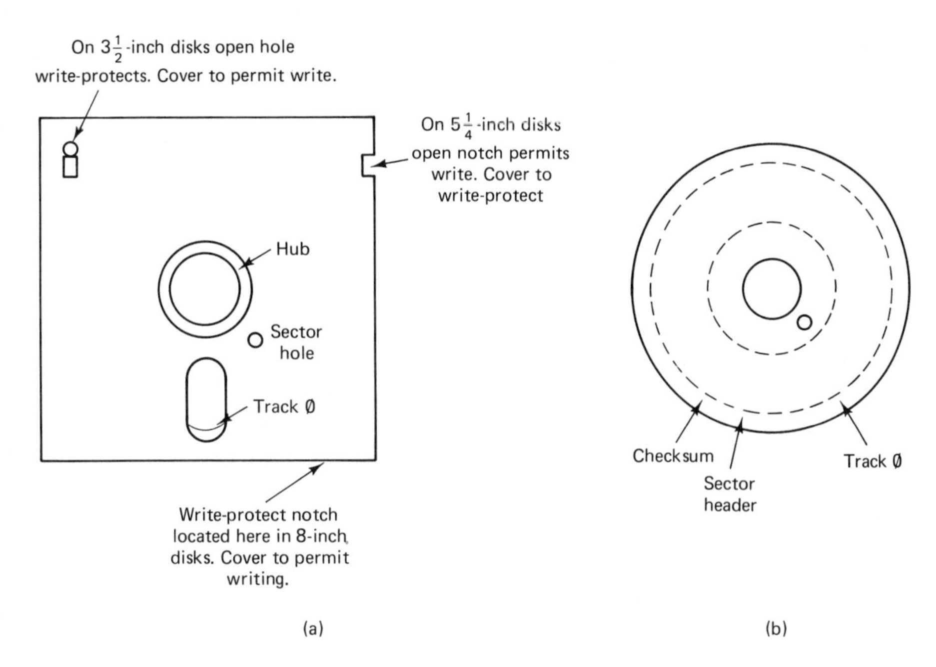

Figure 17.12 Floppy disk format.

to 163 kilobytes, depending on format. The disk contains a single small hole located just inside the inner track to mark the position of sector 0 of each track. The other sectors may be identified by data read with reference to the first sector. This is called *soft sectoring*. An earlier disk format used separate holes to mark each sector and was called *hard sectoring*.

Each sector starts with one or more *sector header* bytes giving the track, sector number, and, in some cases, the file length or RAM addresses where the file is to be stored. Each sector ends with a checksum byte, which is used to catch read errors.[1] Data is always accessed in blocks of bytes called *records*. A record usually corresponds to one sector. Even if a file contains only 10 bytes, an entire record must be devoted to storing it; the remaining bytes in that record are then unusable. This overhead and the storage requirements of DOS (described next) limit the usable storage capacity of a disk to perhaps 60% to 80% of its unformatted capacity.

DOS: Several of the tracks on a disk are usually devoted to storing a program called a Disk-Operating System, or DOS (pronounced to rhyme with *boss*). The DOS program provides instructions for saving, reading, deleting, naming, and obtaining a catalog list of files stored or to be stored. It also handles the details of finding a chain of free sectors to store your program and keeping an ordered list of the sectors used by each program to be used in reading the program back. DOS is sometimes referred to as the *system*, and a computer may be provided with a *system disk*, or a disk with *system tracks*. A disk without system tracks will operate only in conjunction with another disk that contains DOS.

[1]A Cyclical Redundancy Check (CRC) is a more sophisticated method which not only checks for errors but corrects them, provided that there is a maximum of one error per record.

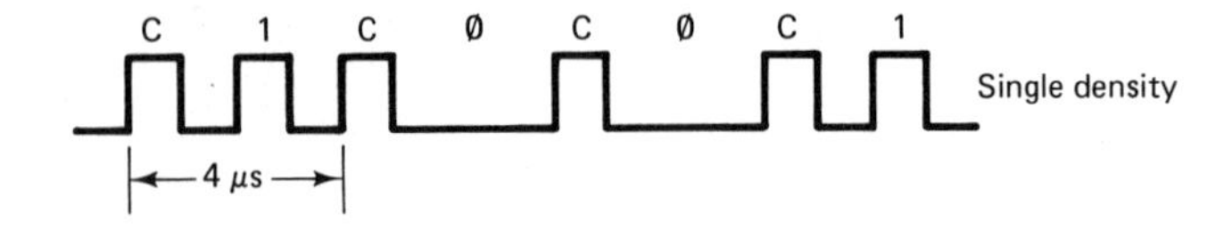

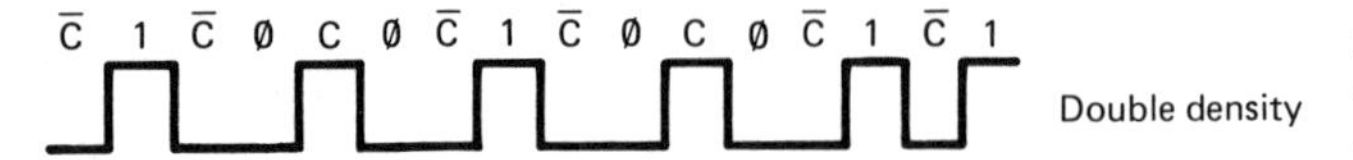

Figure 17.13 Single-density disk recording places data bits between clock pulses. Double-density recording omits the clock pulses if they would fall next to a data pulse.

It is perhaps unfortunate that most DOS programs are the property of a single computer manufacturer, and programs recorded under one DOS will not read under another DOS, even though the two machines use the same processor. There are three major exceptions to this incompatibility problem. The first is a DOS called CP/M, which stands for Control Program/Microprocessor. CP/M is usable with 8080 and Z80 microprocessors and is not limited to a single manufacturer. The second is MS DOS, written by Microsoft and used in the IBM PC and compatible machines. A newer DOS for general use on larger systems (such as a 68000-based computer) is UNIX, licensed by Bell Laboratories.

Bits are recorded serially on the disk, interleaved with clock bits, which appear regularly every 4 μs. The clock bits are necessary to maintain synchronization, since an unrecorded gap of hundreds of 1 or Ø bits could otherwise allow the disk and the processor to get out of step on a read. This system is called frequency modulation (FM) because a string of 1 bits produce a 500-kHz signal, while a string of Øs produces 250 kHz.

Some disk formats use double-density recording, in which the clock bits are omitted except where two Ø bits appear together. This allows the data pulses to be placed closer together. Figure 17.13 illustrates single- and double-density recording.

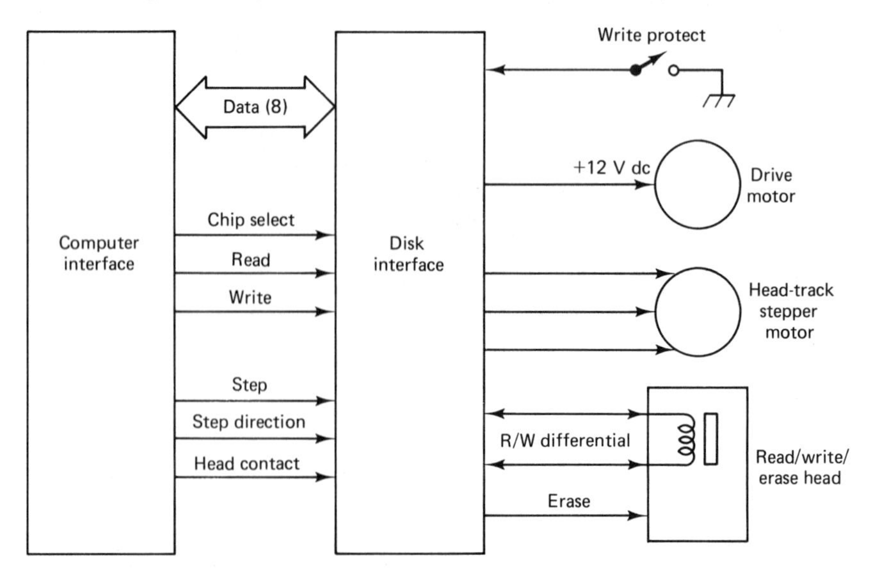

Figure 17.14 Typical disk-drive control signals.

The disk-drive electronics amplifies the millivolt-level signal from the read head to digital levels, strips off the clock pulses, and packs the serial bits into bytes for reading by the processor. Floppy disk controller ICs containing this electronics are readily available. Figure 17.14 shows the major interface lines required between the disk-drive hardware and its electronics and between the computer and the disk system.

17.6 KEYBOARD INTERFACING

Section Overview

Keyboards are available in a number of forms. Some are sold as a unit and have electronics on board to output a 7-bit ASCII code and an $\overline{\text{IRQ}}$ signal in response to a key press. This section presents a general method for reading an unencoded keypad consisting of an array of normally open switches in rows and columns. The basic technique is to have the processor scan the rows consecutively, looking for one that has a depressed key in it. Then that row is examined to determine which specific key was pressed. The keys are numbered consecutively, and a counter is incremented up to the number of the depressed key.

A matrix keyboard consists, conceptually, of a number of parallel horizontal conductors passing over, but not touching, a number of parallel vertical conductors. A key is located at each intersection of a horizontal and a vertical conductor. Depressing a key causes the two conductors to touch. Figure 17.15 illustrates the form of one 8 × 8 matrix keyboard used in a popular low-cost computer. As you can see, not all the possible 64 intersections are used, and the physical layout (a standard QWERTY typewriter keyboard) is quite different from the conceptual layout.

Reading the keyboard requires two 8-bit ports—one output latch and one input buffer. A set of 8 resistors to V_{CC} pull all the vertical conductors (columns) *high* when the keyboard is not being scanned. Port *A* reads all 1s in this case, indicating no key is pressed.

To scan the keyboard, one of the horizontal *row* conductors is pulled *low* by a Ø bit from output Port *B*, while the other 7 rows are held *high* by outputting 1 bits. Let us say that the *low* row is the bottom, or PBØ row. The PB output will be $FE. If none of the bottom row of keys is pressed, a read of Port *A* will still show all 1 bits ($FF), indicating that if any key is pressed, its number is $Ø8 or higher. We therefore add $Ø8 to our *keycount* buffer and proceed to pull the PB1 row *low* (output $FD).

Let us say that key *D* in the second row is pressed. A read of Port *A* will show 1111 1Ø11, or $FB. This is the first non-FF value and indicates that a key is depressed; in this case, key $ØA. We identify this number by shifting the port-*A* bits right into the C flag. If C = 1, we add 1 to the *keycount*. This happens twice, as bits Ø and 1 are shifted right and *keycount* reaches value $ØA. The third shift results in C = Ø, which is our cue to report the value $ØA as the number of the depressed key. Of

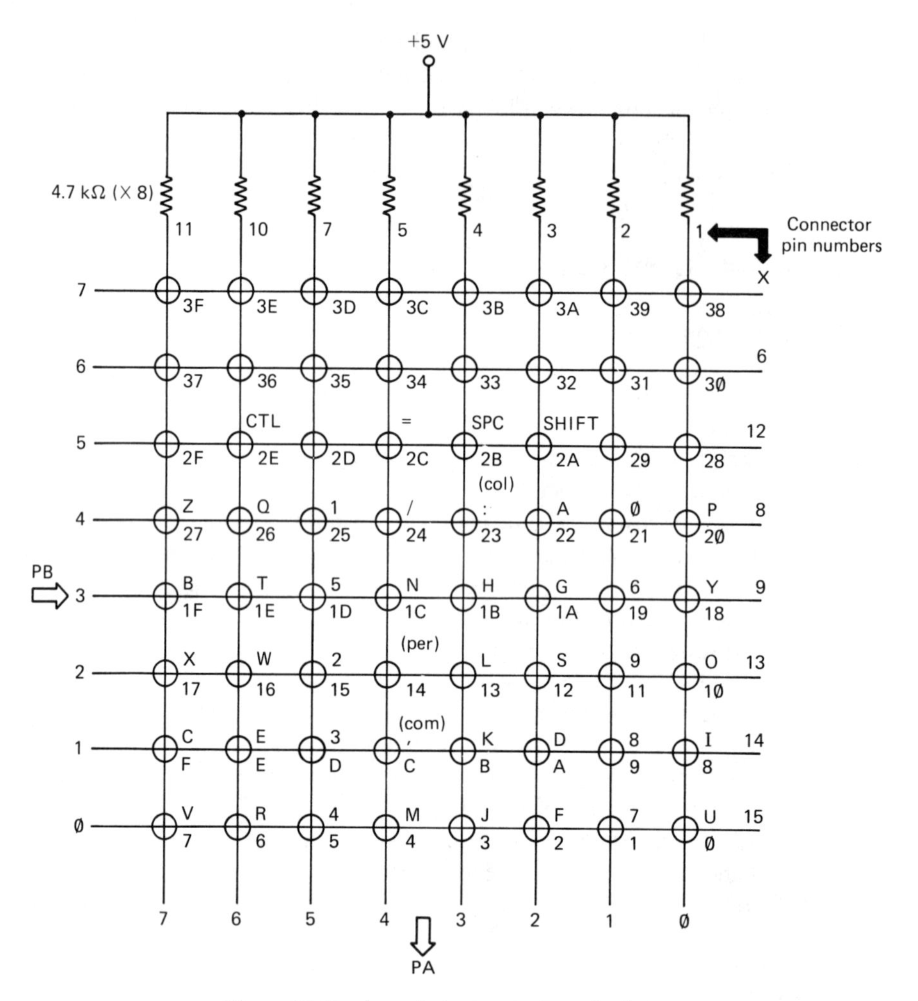

Figure 17.15 A matrix keyboard schematic diagram.

course, if we go up all the rows and *keycount* reaches 64 (or \$40) with never a 0 in the C flag, we know that no key has been pressed. Figure 17.16 shows a flowchart for the KEYSCAN routine.

In truth, the rows, once pulled *low*, are left *low* for the remainder of the scan, just because it is a little easier to do and causes no harm. Also, the column bits, read all or mostly *high* by Port *A*, are complemented to all or mostly *low* so that a test for zero can be used to identify a row with no key pressed.

A complete keyboard routine would probably use the KEYSCAN as a subroutine. The overall routine would provide software debounce, a provision to guard against multiple reads of a single keypress, and a lookup table to find the ASCII code corresponding to

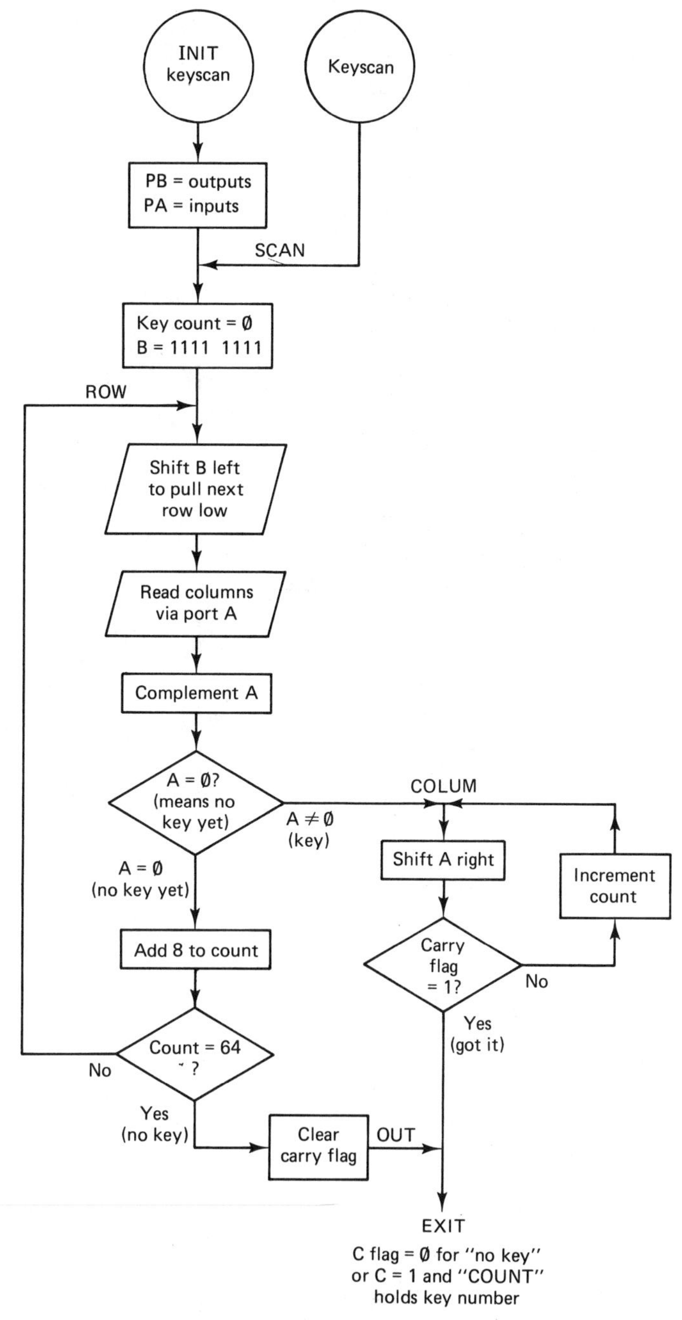

Figure 17.16 A subroutine for reading the keyboard of Figure 17.15 and reporting the keynumber, \$ØØ through \$3F.

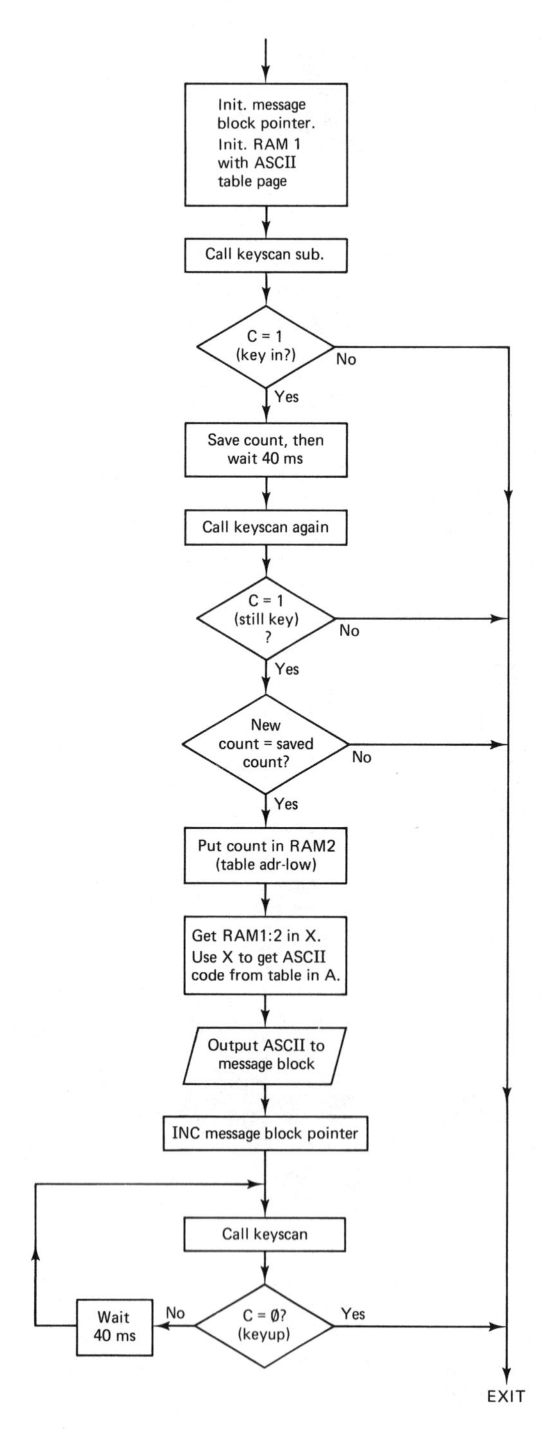

Figure 17.17 Flowchart for a routine to read and debounce the keyboard and create a file of ASCII code in RAM from keys typed.

each key number. A flowchart for the complete routine appears in Figure 17.17. Here are the details:

- Software debounce can be achieved by reading a key with KEYSCAN, waiting about 40 ms, and then reading again. If the two reported keys are identical, it was a valid key. If not, it was probably a noise pulse.
- A KEYUP routine calls the KEYSCAN subroutine and refuses to accept a new key until after KEYSCAN reports no key pressed.
- The ASCII lookup is done with indirect addressing. A RAM address is loaded with the page number of the ASCII table. The next consecutive address is loaded with the key number from the KEYSCAN routine. The X register, loaded high-byte and low-byte from these RAM locations, now points to a location in the table containing the proper ASCII code.

The routine is charted to build a new Message Block of ASCII characters input via the keyboard. In the next chapter we will see how this file can be displayed on a video screen. The 40-ms delays can be processor-counted subroutines, but if they use an external timer the processor can be freed for other duties during most of the key-read time.

REVIEW OF SECTIONS 17.5 AND 17.6

21. Is the linear distance per bit on a $5\frac{1}{4}$-in. floppy disk (outer track) closer to 0.01 in., 0.002 in., or 0.0005 in.?

22. If a sector holds 512 bytes, how many sectors are needed to hold a 26-kilobyte program?

23. Where does a computer get the detailed instructions it needs to operate the disk's record head, drive motor, and stepper motor?

24. What system is used to catch read errors on floppy disks?

25. What would the read-head output waveform look like if a string of $FF bytes were recorded at double density and no clock pulses were removed?

26. On the keyboard of Figure 17.15, a letter *T* is pressed. What are the port *A* and port *B* data bytes that sense this key?

27. Calculate an estimate of how long the KEYSCAN routine of Figure 17.16 will take if no key is depressed. Assuming a 6800 processor on a 1-MHz clock, is the time closer to 75 μs, 250 μs, 950 μs, or 5 ms?

28. Why does KEYSCAN stop at keycount = $30? The keyboard has no keys at positions $30 and above.

29. In Figure 17.17, what is the purpose of the second call of the KEYSCAN subroutine?

30. In Figure 17.15, the "column" conductors all have pullup resistors, but the "row" conductors do not. Why is this?

17.7 A KEYBOARD-READ PROGRAM

Keyboards such as the one depicted in Figure 17.15 are readily available and can be read with a 6522 VIA providing the input and output ports. This section gives the hardware details and the program to read the keyboard and display the key number in binary on a set of discrete LEDs. The debounce and lookup-table routines are not included at this point. They will be added in the next Chapter when a complete TV Typewriter is described.

The project hardware is diagrammed in Figure 17.18. The keyboard is shown in more detail in Figure 17.15. Once the keyscan is completed, Port *B* is used to light six LEDs displaying the key number in binary. These two functions, *pull low during key scan* and *light LEDs*, do not conflict, except that they cannot be done simultaneously.

The program (List 17.1) follows the flowchart of Figure 17.16, except that a procedure is added at the end to light the six LEDs with the binary key number for 12 ms and then automatically return to the KEYSCAN routine. KEYSCAN will light the LEDs but only for a small percentage of the time, which should be almost unnoticeable.

```
                    1000 * LIST 17-1 ** READ 8 X 8 MATRIX KEYBOARD.
                    1010 *  DISPLAY KEY NUMBER ON 8 LEDS IN BINARY.
                    1020 *  SEE FIG. 17-18 FOR HARDWARE.
                    1030 *  TESTED AND RUNNING: 27 JULY '87 BY D. METZGER
                    1040          .OR  $FF80
                    1050          .TA  $4080
C800-               1060 PB       .EQ  $C800
C801-               1070 PA       .EQ  $C801
C802-               1080 DDRB     .EQ  $C802
C803-               1090 DDRA     .EQ  $C803
0010-               1100 COUNT    .EQ  $0010     COUNTS UP TO KEY NO.
0011-               1110 KEYNO    .EQ  $0011     BUFFER TO SAVE KEY NUMBER.
0012-               1120 DELAY    .EQ  $0012     COUNTER FOR 12-MS DELAY.
                    1130 *
FF80- 7F C8 03      1140          CLR  DDRA      PORT A ALL INPUTS.
FF83- C6 FF         1150          LDAB #$FF
FF85- F7 C8 02      1160          STAB DDRB      PORT B ALL OUTPUTS.
FF88- 7F 00 10      1170 SCAN     CLR  COUNT     KEYCOUNT STARTS AT 0.
FF8B- C6 FF         1180          LDAB #$FF
FF8D- 58            1190 ROW      ASLB           PULL NEXT ROW FROM BOTTOM LOW.
FF8E- F7 C8 00      1200          STAB PB
FF91- B6 C8 01      1210          LDAA PA        READ COLUMNS.
FF94- 43            1220          COMA           KEY PRESS NOW MAKES A 1;
FF95- 26 0D         1230          BNE  COLUM     ANY "1" MEANS A KEY PRESSED.
FF97- 96 10         1240          LDAA COUNT     IF NO KEY WE DON'T NEED "A"
FF99- 8B 08         1250          ADDA #8         SO USE IT TO ADD A ROW'S
FF9B- 97 10         1260          STAA COUNT      WORTH OF KEY NUMBERS.
FF9D- 81 40         1270          CMPA #64       LAST ROW?
FF9F- 26 EC         1280          BNE  ROW       NO: CHECK NEXT ROW.
FFA1- 0C            1290          CLC            YES: THEN NO KEY.
FFA2- 20 0C         1300          BRA  OUT
                    1310 *
FFA4- 44            1320 COLUM    LSRA           KEY IN THIS ROW. FIND COLUMN.
FFA5- 25 05         1330          BCS  GOTIT     "1" SHIFTED IN - YOU GOT IT.
FFA7- 7C 00 10      1340          INC  COUNT     NO "1" BIT;
FFAA- 20 F8         1350          BRA  COLUM      THEN KEEP SHIFTING.
FFAC- 96 10         1360 GOTIT    LDAA COUNT     IF YOU HAVE A KEY,
FFAE- 97 11         1370          STAA KEYNO      SAVE ITS NUMBER.
```

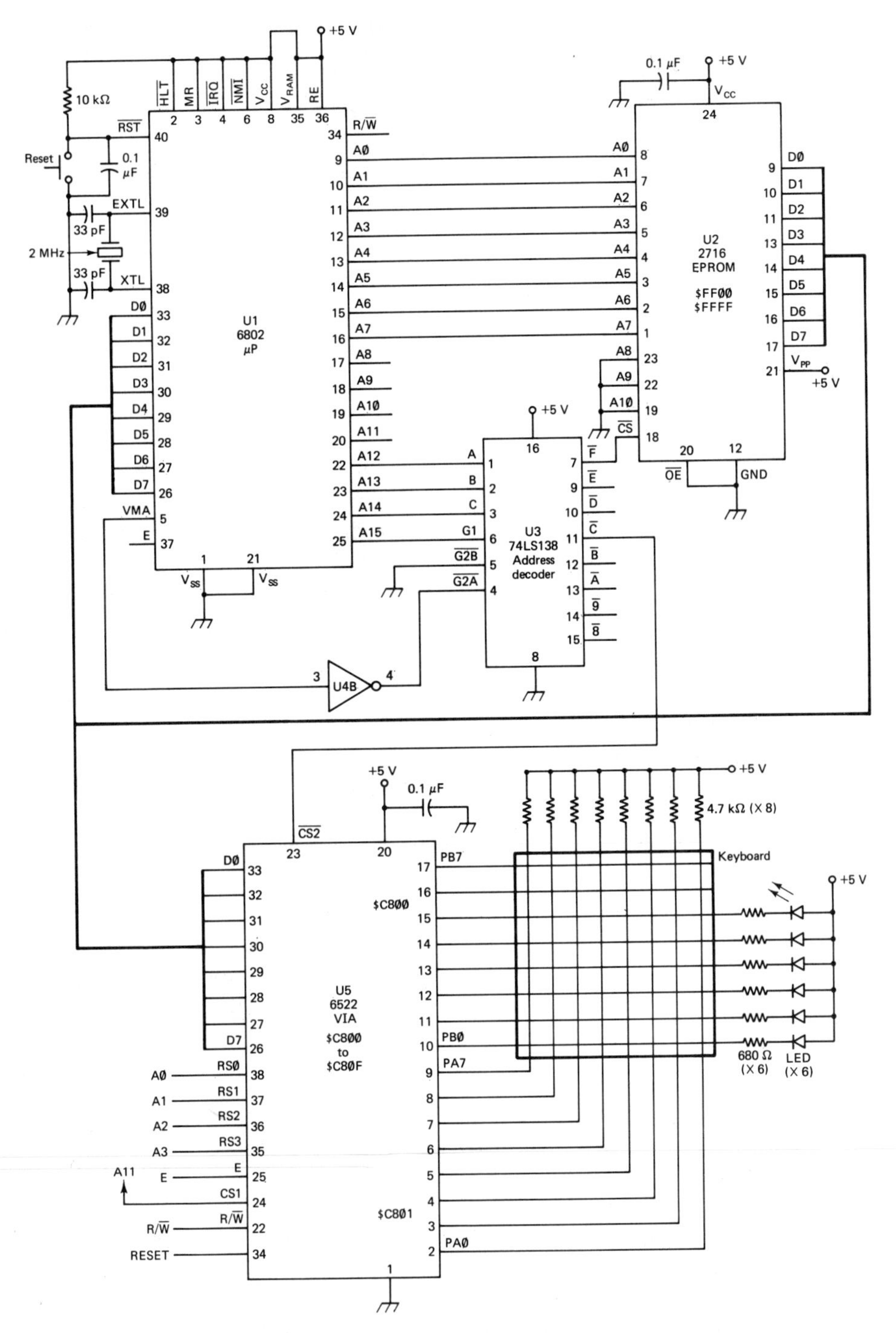

Figure 17.18 Hardware to read the keyboard (using the routine of Figure 17.16) and display the key number in binary.

```
FFB0- 96 11     1380 OUT   LDAA KEYNO   OUTPUT KEY NUMBER TO LEDS;
FFB2- 43        1390       COMA          PULL LED LOW ON ACTIVE BITS.
FFB3- B7 C8 00  1400       STAA PB      (OLD NO. IF NO NEW KEY)
FFB6- 7C 00 12  1410 WAIT  INC  DELAY   HOLD THEM LIT FOR 12 MS
FFB9- 7A 00 12  1420       DEC  DELAY    (ON A 1/2-MHZ CLOCK)
FFBC- 7C 00 12  1430       INC  DELAY
FFBF- 26 F5     1440       BNE  WAIT
FFC1- 20 C5     1450       BRA  SCAN    AND LOOK FOR KEY AGAIN.
                1460       .OR  $FFFE
                1470       .TA  $40FE
FFFE- FF 80     1480       .HS  FF80
                1490       .EN
```

Answers to Chapter 17 Review Questions

1. The *low* data levels may not go below +0.4 V, as they must to be interpreted properly. **2.** The data bus both reads and writes, so bidirectional buffers are needed. The address bus writes only. **3.** One way is to buffer the data bus with a 74LS245. An easier way is to replace the 74LS373s with 74HCT373s.

4. Feed E into a noninverting 74LS-series buffer, and connect the output of the buffer to the cable.

5. The DMA controller does not have to read and interpret program steps. Its "program" is fixed. **6.** Nothing. The processor is halted.

7. The DMA controller has internal address registers, which are preloaded by the processor and incremented or decremented with every DMA cycle. **8.** Cycle stealing from VMA-not cycles and interleaved or multiplexed DMA. **9.** Tristate address buffer. **10.** Clock pulses are lengthened. **11.** 10 GΩ

12. 512 cycles **13.** High-impedance state. **14.** The 16-bit address is applied in 2 bytes to the same set of pins; low-order byte with $\overline{RAS}$ asserted and high-order byte with both $\overline{RAS}$ and $\overline{CAS}$ asserted. **15.** 32 **16.** 32 μs **17.** With U1 at logic-0 output, the saturated transistor to ground gives the line a low impedance. **18.** 50 μs **19.** Photoresistive cells

20. Use the comparator circuit of Figure 17.11(c).

21. 0.0005 in. **22.** 52 Sectors **23.** From a program called DOS, usually stored on several tracks of a "system" disk.

24. A checksum byte at the end of each sector. **25.** A dc logic-1 level, which cannot be maintained across the magnetic pickup coil. **26.** Port B outputs $F7 (or $F0 in the program of Figure 17.17). Port A inputs $BF. **27.** 250 μs

28. That's probably a good idea. It would take 60 μs or so off the keyscan routine. **29.** It is a software debounce and noise filter. Two identical key reads 40 ms apart are required to accept a key as valid. **30.** The "column" lines are inputs and need pullup to prevent noise pickup. The "row" lines are active outputs.

CHAPTER 17 QUESTIONS AND PROBLEMS

Digits after the decimal point refer to section numbers in Chapter 17.

Basic Level

1.1 How many LS-TTL devices and how many CMOS devices can be driven by a microprocessor output?

2.1 How is a 74LS244 buffer placed in the *high-Z output* state?

3.2 How long will it take a 1-MHz 6844 DMA controller to transfer 10 000 bytes from one RAM area to another in the continuous DMA mode?

4.2 Will cycle-stretching DMA affect a microprocessor's execution of a timing loop? Will it affect a 6522's timer function?

5.3 What is the basic storage element in a DRAM?

6.3 How many address pins would you expect to find on a 256K DRAM chip?

7.3 How many refresh cycles are required to refresh a 256K DRAM?

8.4 Give two reasons for the popularity of optical couplers in heavy industrial environments.

9.4 In Figure 17.11(a) calculate the value of the input resistor needed to drive a standard TLL input from input levels of ± 6 V.

10.5 Give the maximum number of tracks, sectors, and records on a $5\frac{1}{4}$-in. floppy disk.

11.5 What do the letters DOS stand for?

12.6 The keyboard of Figure 17.15 produces data \$F7 at Port *A* when data \$FB is output from Port *B*. What key has been pressed?

13.6 In Figure 17.15, the Q key is pressed. Referring to Figure 17.16, how many passes through the "Add 8" box and how many passes through the "Shift *A*" box will be made in reading this key?

Advanced Level

14.1 Why is a single 74LS244 not suitable as a data bus buffer for a 6802 microprocessor?

15.1 Give at least three reasons for using bus buffers in microcomputer systems.

16.2 Explain the differences between cycle stretching, cycle stealing, and interleaving types of DMA.

17.2 What is transparent DMA? Does stealing of $\overline{\text{VMA}}$ cycles produce transparent DMA?

18.3 List the signal inputs and outputs for a typical DRAM controller chip.

19.4 List three circuit-board layout techniques that will minimize the pickup of noise by stray capacitive coupling.

20.4 In Figure 17.11(c) calculate the value of R_{IN} if trigger voltage V_{IN} is to be 3.0 V. Draw a new diagram with component values labeled for $V_{IN} = 1.0$ V.

21.5 Explain how single-density disk recording prevents a long string of 1s or 0s from demanding a dc output voltage from the disk read head.

22.5 Explain how double-density recording doubles the data rate without doubling the baud rate of the data stream from the disk.

23.6 How can a matrix keyboard avoid picking up noise spikes and key contact bounces as valid key inputs?

24.6 Explain how the number of the key pressed can be used to obtain the ASCII code for the character corresponding to that key. For example, key $16 is W, and the ASCII byte $57 must be obtained.

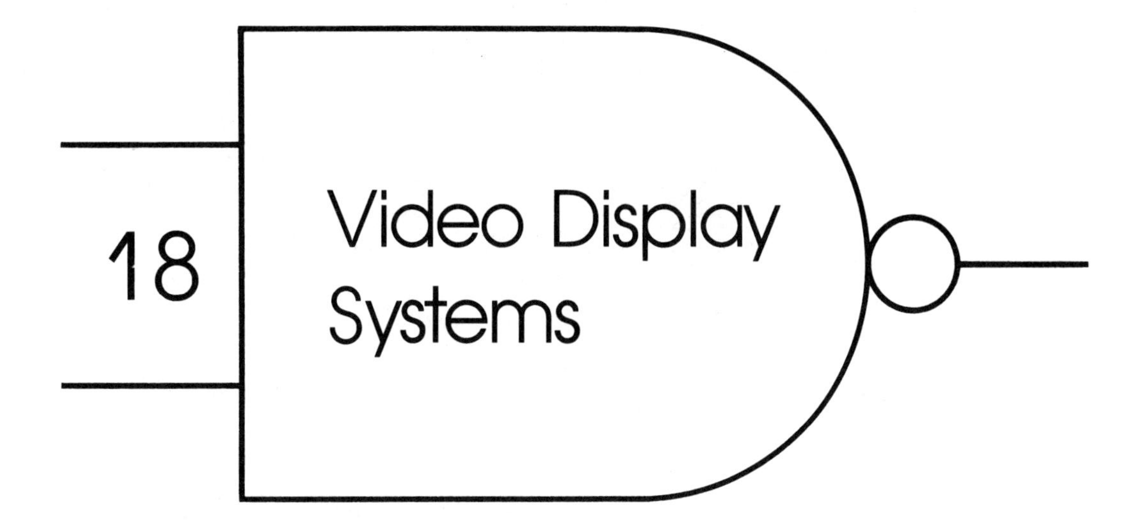

18.1 TELEVISION IMAGE RECONSTRUCTION

Section Overview

Most computer video displays use systems that are very similar to commercial television receivers. This involves rapid horizontal scanning of an electron beam across the inside face of a CRT, coupled with slower vertical scanning. The resulting point of light covers the entire screen area 60 times per second, so the image appears continuous. Images are produced by modulating the intensity of the electron beam: stronger for white (bright-glowing) areas, weaker for gray, and shut off for black.

CRT structure: A monochrome cathode-ray tube (CRT), which is the heart of most video display systems, is diagrammed in Figure 18.1. The ''cathode rays'' are really just a stream of electrons in a vacuum. The name was given 100 years ago, before electrons had been discovered. A *cathode* is any negatively charged conductor, and the source of the electron beam in a CRT is still called a cathode.

The cathode is heated by a filament, which is no more sophisticated than a light bulb and makes the orange glow visible inside the neck of a CRT. The heat agitates the molecules of the cathode surface so violently that millions of electrons are thrown

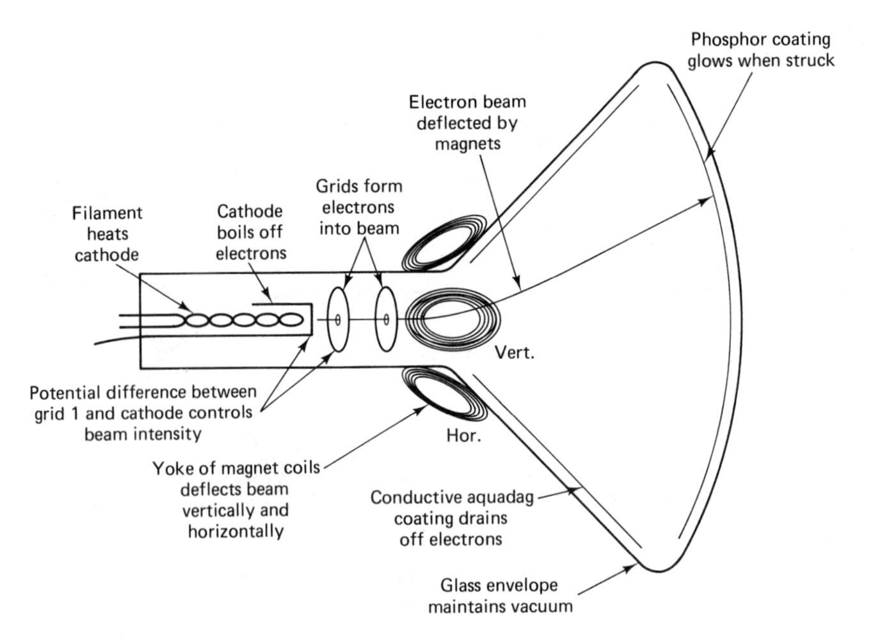

Figure 18.1 A single beam of electrons is deflected by electromagnets to produce the image on a monochrome CRT (cathode-ray tube).

off into the evacuated space inside the CRT. The negative charge on the cathode repels the electrons away.

The *grids* were named for the configuration of their counterparts in radio vacuum tubes, circa 1910. In a CRT they are not grids but disks with a pinhole in the center to allow only a fine beam of electrons to pass. The intensity of the electron beam is varied by changing the relative voltage between the cathode and the nearest grid. Grid 1 may be about 10 V more negative for a bright spot on the CRT and about 20 V more negative for a dark spot.

The electrons are accelerated toward the CRT face by a high positive voltage applied to a conductive coating on the inside of the CRT body. This voltage is very high—on the order of +10 kV—and is usually carried by a thick red wire attached to the back side of the CRT body. The inside face of the CRT is coated with a phosphor material, which glows (white, green, or orange in most computer applications) when struck by the electron beam.

Scanning the CRT: The electron beam is deflected rapidly across the CRT face from left to right, making a horizontal line. The horizontal sweep rate is 15 750 lines per second, allowing 63.5 μs per line. Approximately 80% of the horizontal time is on-screen left-to-right sweep time. About 20% is reserved for right-to-left retrace, or *flyback*, during which time the electron beam is extinguished.

Simultaneously, the beam is being deflected vertically, from top to bottom, but at a much slower rate. This causes the single-point electron beam to cover the entire *field* of the CRT, making a pattern called a *raster*. The vertical sweep frequency is 60 Hz. The number of lines per field is

$$\frac{15\ 750\ \text{lines/s}}{60\ \text{fields/s}} = 262.5\ \text{lines/field}$$

Actually the number of lines appearing on the screen is usually about 90% of this number, because about 10% of the vertical scan time is reserved for vertical retrace, during which the screen is also darkened. Figure 18.2 illustrates the raster-scan process.

Interlacing is used in commercial television and in some computer displays to improve the definition of the image without increasing the scanning rate. The trick is to have the lines of the odd- and even-numbered fields fall in between one another and to have different video information on adjacent lines. A complete picture (called a *frame*) is then composed of two 262.5-line fields, and it contains 525 lines. The frame rate is 30 per second. A flicker rate of 30 Hz is noticeable and annoying, but the field rate—and hence the flicker rate—is 60 Hz, which appears to a human as not flickering at all.

Magnetic deflection is used in essentially all television and computer video displays. Electrostatic deflection is used in oscilloscopes where display geometry is critical, but magnetic deflection, being more forceful, allows the electron-path length, and consequently the depth of the CRT cabinet, to be minimized. The deflection electromagnets fit over the neck of the CRT and are called, collectively, the *yoke*.

The buildup of current in the horizontal deflection coil stores considerable energy. The rapid elimination of this current during the brief flyback period causes

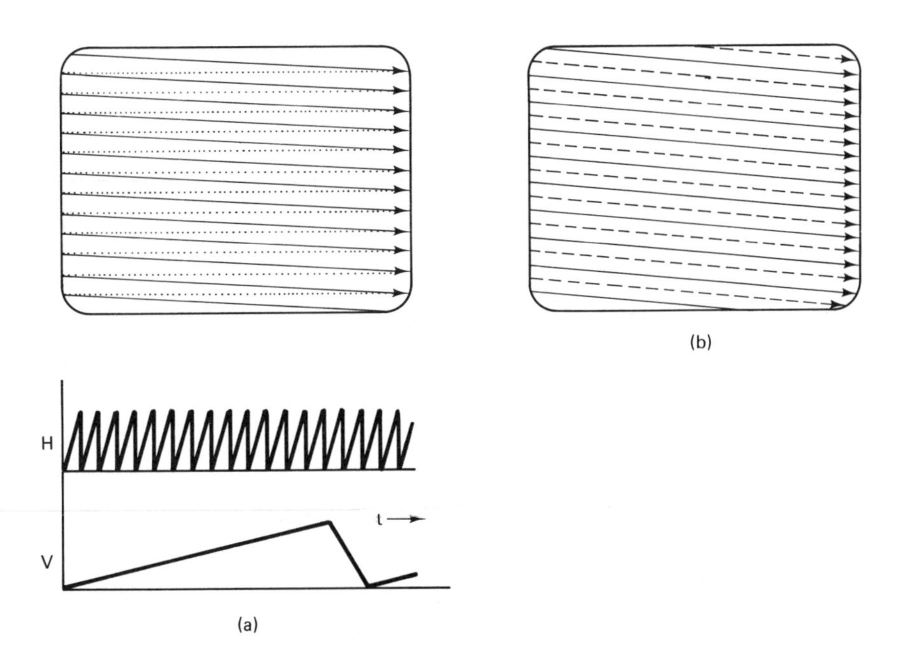

Figure 18.2 (a) A *raster* is produced by scanning the electron beam rapidly from left to right while moving it more slowly from top to bottom. (Dotted lines represent retrace from right to left, with beam off.) (b) Interlaced scanning fits alternating scans between one another to double the line density. (Dashed lines represent second scan.)

a very high inductive kickback voltage to be generated. This voltage, suitably processed by a flyback transformer, a damper diode, and a high-voltage rectifier, is used as the positive accelerating voltage for the CRT. Thus a failure of the horizontal deflection system will also cause loss of high voltage and complete darkening of the CRT.

18.2 THE VIDEO SIGNAL

Section Overview

A composite video signal contains, on one conductor, synchronizing signals to keep the vertical and horizontal sweep oscillators in step with the received video information, the video information itself, and (in the case of color) the color information. RGB monitors use three separate conductors to convey the three constituent colors of the image.

The detail or resolution of a video image depends ultimately on the data rate or bandwidth of the signal-generation, storage, and processing equipment feeding it.

A composite video signal is diagrammed in Figure 18.3 and photographed from an oscilloscope in Figure 18.4. The more-negative levels produce dark areas on the CRT and the more-positive areas produce light areas. At the end of each horizontal sweep the signal goes "blacker than black" for about 5 μs. This most-negative signal is clipped off by a *sync separator* circuit and used to initiate the

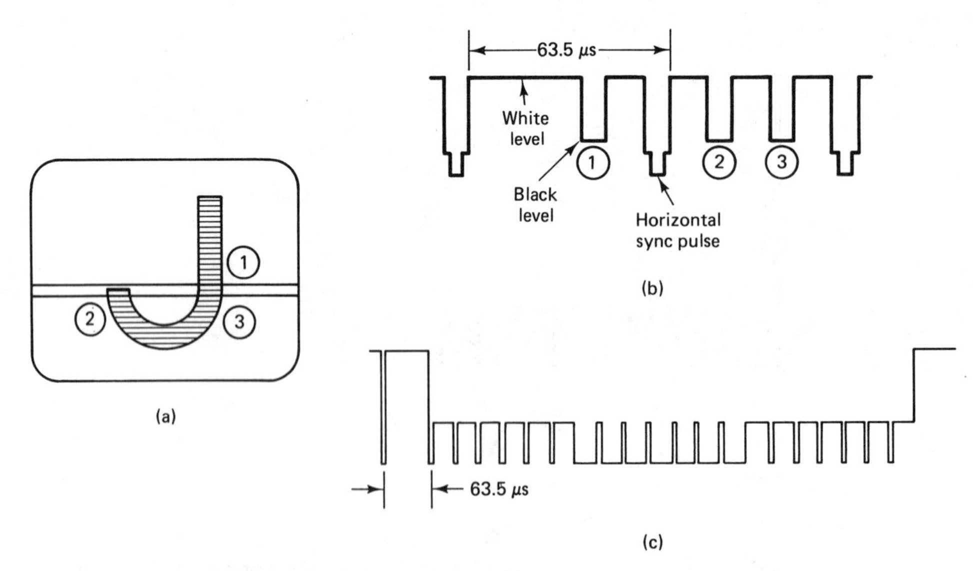

Figure 18.3 The composite video signal for the two scanning lines at (a) are depicted at (b). The last horizontal scanning line and the vertical-retrace pulse are shown at (c).

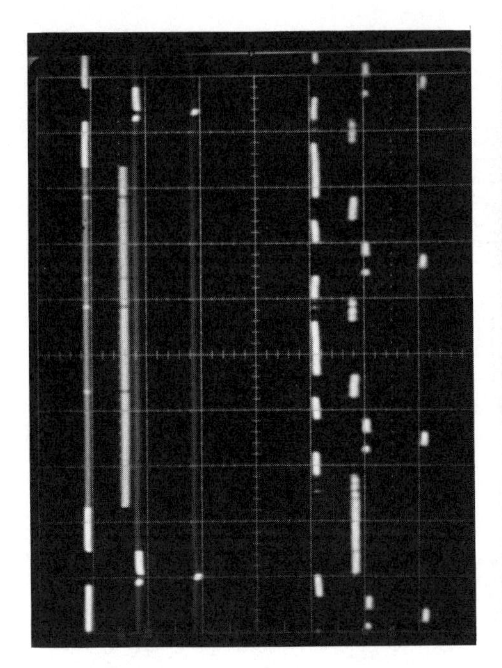

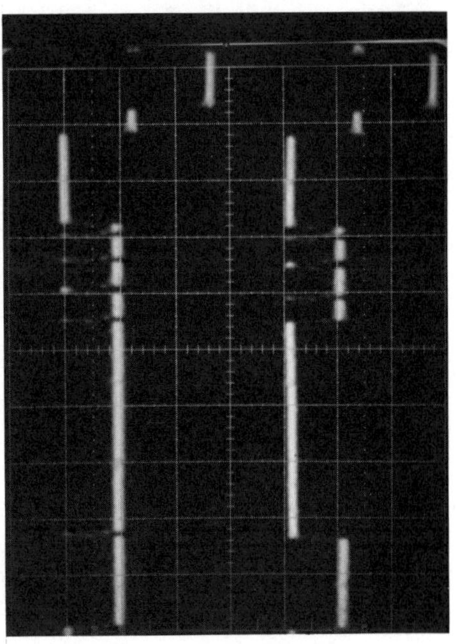

Figure 18.4 Four composite video displays from the "house" drawing of Figure 20.15 (page 519). For each trace a white screen is at the top, a black screen in the middle, and sync pulses are at the bottom. (a) Top trace: an entire field of 262½ horizontal lines at 2 ms/cm. The vertical sync pulse is at 1.0 cm; the white border stops at 2.3 cm; the roof ridgepole white is at 4.4 cm, eave at 5.8 cm, window top at 6.4 cm, window bottom at 7.4 cm, house bottom at 7.8 cm. The white border resumes at 8.4 cm and the next vertical sync pulse is at 9.4 cm.) Bottom trace, at 20 μs/cm, shows three horizontal lines near the eave of the house, including four horizontal sync pulses. (b) Top trace: one horizontal line just above the eave, at 5 μs/cm, showing the left gable at 1.6 cm, right gable at 5.4 cm, window at 5.9 and 6.5 cm, back of house at 6.9 cm, and white border from 7.1 to 8.7 cm. A horizontal sync pulse appears from 9.2 to 10.1 cm. Bottom trace: the next horizontal line, also at 5 μs/cm, showing a long white interval for the horizontal line of the eave.

horizontal flyback. In television textbooks you may see similar diagrams presented upside down from the way they appear here, reflecting the signal levels transmitted by television stations. That is called *positive sync*. Our diagrams show the levels sent from a computer to a video monitor, which are typically 1 V p-p, negative sync, to a 75-Ω load.

The blacker-than-black sync level predominates for about 20 horizontal sweep periods (over a millisecond) at the bottom of each vertical sweep. An integrator accumulates enough output from the sync separator in this time to trigger a retrace of the vertical sweep generator. Figure 18.5 shows a block diagram of the main circuits of a monochrome video monitor.

Color signals are encoded as amplitude and phase changes on a special color subcarrier. The frequency of this subcarrier is 3.58 MHz, and the amplitude and

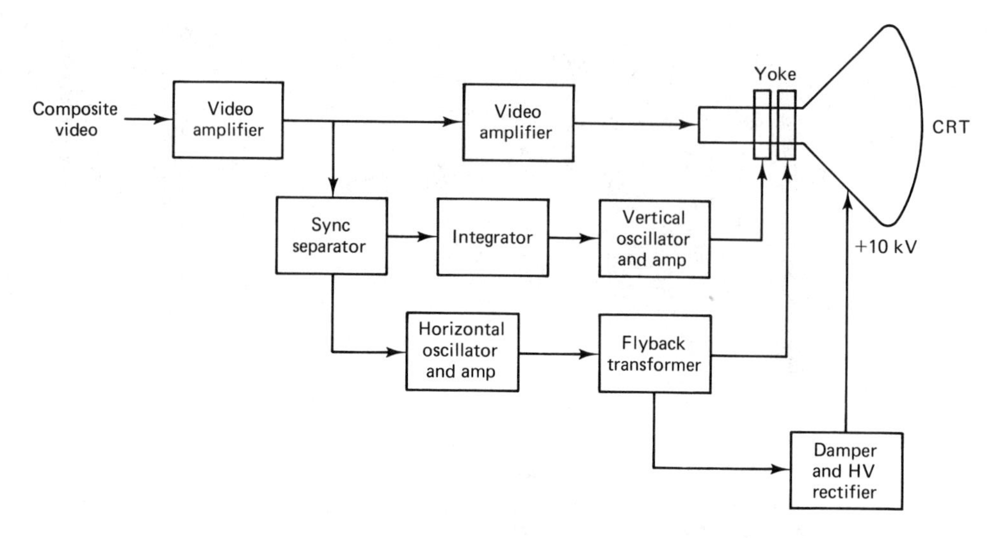

Figure 18.5 A block diagram for a monochrome video display monitor circuit. The high accelerating voltage for the CRT is generated by inductive kickback during horizontal retraces.

phase changes to it are also conveyed on the single composite-video line if a color display is being generated.

A color CRT uses three electron beams, each positioned at such an angle that its electrons strike phosphor of only one of the three types on the screen face—red-glowing, green-glowing, or blue-glowing. Any color can be made up of combinations of these primary lights.[1] Figure 18.6 depicts a typical color CRT. The intensity level for each electron beam is determined by fairly complex analog circuitry, which decodes the two color signals and the monochrome black/white signal from the composite video signal.

Some computers are now using RGB monitors, in which three separate signals, for the red, green, and blue electron guns, are sent to the video monitor. This removes the requirement for decoding all this information from the composite video signal and makes possible higher-definition color displays.

Character generation: When alphanumeric characters are displayed on a video monitor, the constraints of the horizontal and vertical scanning system must be observed as well as the constraints of the digital data-storage system. The CRT beam cannot be moved around at will like a pencil on paper. The beam must be turned *on* and *off* as it sweeps each line through a row of characters. Figure 18.7 shows the characters My produced in a 5 × 8 matrix with 2 spaces between each character. If the CRT is to display 80 columns of such characters (a typical number) and if one digital bit is dedicated to turning the beam *on* (1) or *off* (0) at each point, the number of bits per line is

[1]Red, blue, and yellow are the primary pigments, or color-subtraction agents. Red, green, and blue are the primary lights, or color-addition agents.

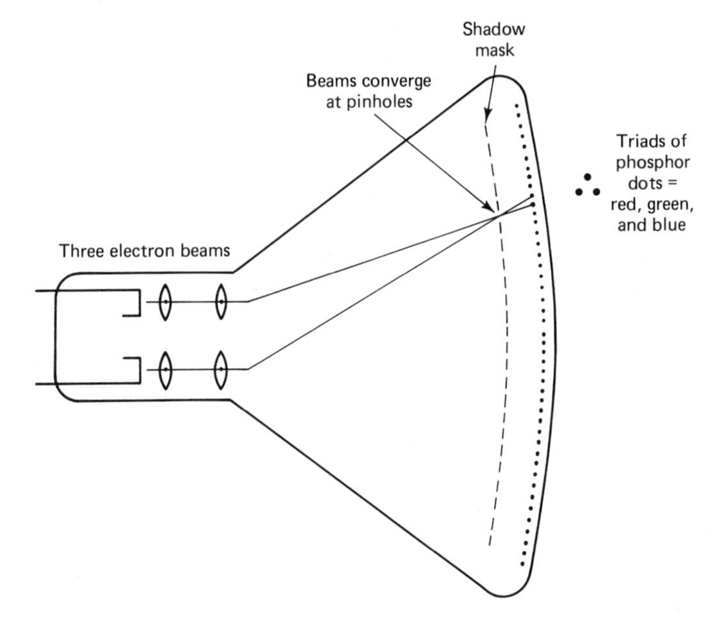

Figure 18.6 A color CRT uses three electron beams angled in such a way that each beam strikes only those phosphor areas that glow red, green, or blue, respectively.

$$(7 \text{ bits/character})\ (80 \text{ characters/line}) = 560 \text{ bits/line}$$

If we assume an on-screen time that is 70% of the 63.5-μs horizontal sweep time, there will be 44 μs in which to output 560 bits. The bit rate required is

$$\frac{560 \text{ bits/line}}{44\ \mu\text{s/line}} = 12.7 \text{ MHz}$$

This data rate is beyond the capability of most microprocessors or even of most DMA controllers.

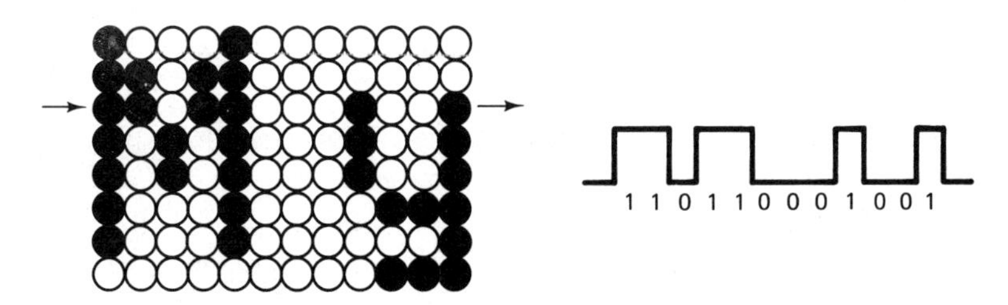

Figure 18.7 A segment of a video scanning line through the letters *My*.

The total number of bits required to store a screen containing 20 rows of characters at 8 lines per character is

$$(560 \text{ bits/line})(8 \text{ lines/row})(20 \text{ rows/screen}) = 89\,600 \text{ bits/screen}$$

This is 11 200 bytes. Yet the amount of information represented on the screen is far less than 11 kilobytes, as we will now see.

There are fewer than 128 alphanumeric characters, counting capital and small letters, numerals, and special symbols; so a 7-bit code is sufficient to identify each one. There are 80×20, or 1600, characters on the screen as described, so they can be specified by an amount of data:

$$(7 \text{ bits/character})(1600 \text{ characters/screen}) = 11\,200 \text{ bits}$$

This is 1400 bytes; exactly one-eighth the amount required to store the screen bit by bit.

Direct storage of each picture element (pixel) is required for screen graphics, and it is used, for example, in Apple's Macintosh computer to provide a wide array of type styles and the ability to mix graphics and text on one screen display. However, most small computers use character generators to minimize the memory and speed requirements of displaying a screen full of alphanumeric characters. We shall devote the remainder of this chapter to one such chip, the Motorola 6847 Video Display Generator.

REVIEW OF SECTIONS 18.1 AND 18.2

Answers appear at the end of Chapter 18.

1. What is a cathode ray?
2. What is a raster?
3. What becomes of the electrons inside the CRT once they hit the phosphor screen?
4. What becomes of the energy stored in the horizontal deflection yoke and flyback transformer during horizontal retrace?
5. A proposed picture-phone system uses a 120-line interlaced scan with 24 frames per second. What are the vertical and horizontal scanning rates?
6. What permits the sync signals to be separated from the video signals on a composite video conductor?
7. What is the significance of the frequency 3.58 MHz?
8. What is an RGB monitor?
9. A video graphics screen displays 192×128 pixels. How many bytes are required to store one screen?
10. In Question 9, what is the data rate required to produce the 192-pixel horizontal resolution using a standard television scan and assuming 70% on-screen time for the horizontal sweep?

18.3 THE 6847 VIDEO DISPLAY GENERATOR

Section Overview

The 6847 is a 40-pin IC that generates composite color video signals. It reads data from a screen memory and generates alphanumeric or graphical figures on the CRT screen. The 6847 has an on-board character generator for one alphanumeric mode and provides, in addition, six color-graphics modes and four monochrome-graphics modes of varying resolutions.

Video interfacing has become immensely less difficult with the introduction of low-cost LSI chips dedicated to this purpose. Motorola's 6847 VDG (Video display generator) is available for about $10 and is remarkably easy to use. We will first describe the VDG pins and functions and then see how it can be connected with a few support chips to produce a fixed video display independent of the computer. Finally, we will interface the video system to the computer, allowing the computer to control the display.

The VDG operates independently of the microprocessor most of the time. It functions more like a microprocessor itself than like a peripheral device. For example, it has a 13-line address bus, which outputs addresses to its own display memory. This memory contains data that the VDG reads, via its data bus, to determine what characters or graphics patterns to display.

A scan of the CRT is made 60 times per second, and the VDG reads the entire display RAM at that rate to keep the video display visible. By contrast, the processor accesses the screen RAM only occasionally, when a new byte needs to be written as the result of a key press, for example.

The VDG has 12 display modes, which are specified in detail by Figure 18.8. We will have occasion to use only *Alphanumeric Internal*, *Semigraphics Four*, and *Resolution Graphics Two* in this chapter, so you may wish to concentrate your attention on these three. The display mode is selected by 8 input pins to the 6847, as listed in the rightmost column of Figure 18.8. Each box in the figure represents one display ''element.'' Each element contains from 24 pixels (Semigraphics Four) to 1 pixel (Resolution Graphics Six). An alphanumeric element is one character. You should be able to confirm the following:

- For the monochrome (resolution graphics) modes, the number of elements on-screen equals the number of memory bits required.
- For the color graphics modes, the number of memory bits is twice the number of elements, since each element needs 2 bits to specify one of four colors.
- For the alphanumeric mode the number of memory bytes required equals the number of characters displayed.
- In the semigraphic modes each byte represents an 8-by-12-pixel area, the same space required for one alphanumeric character. Each area is limited to a single color. *Semigraphics Four* allows eight colors plus black.

VDG Mode	Elements on Screen	Display Element Configuration (per byte). Each box is one element.	Data Bit Functions 7 6 5 4 3 2 1 Ø	Colors (B = Black)	Memory Bytes Required	Control Word 35 G/$\overline{A}$	34 S/$\overline{A}$	31 EX/$\overline{IN}$	27 GM2	29 GM1	30 GMØ	39 CSS	32 INV
Alphanumeric Internal	32 X 16	5 X 7-pixel character	X X A A A A A A (6-bit ASCII)	1 + B	512	Ø	Ø	Ø	X	X	X	A	A
Alphanumeric External	32 X 16	8 X 12-pixel	L L L L L L L L (one of 12 rows)	1 + B	512	Ø	Ø	1	X	X	X	A	A
Semigraphics Four	64 X 32	4; L_3 L_2 6; L_1 $L_Ø$	$XC_2C_1C_0$ $L_3L_2L_1L_0$ (4 elements, one color)	8 + B	512	Ø	1	Ø	X	X	X	X	X
Semigraphics Six	64 X 48	4; L_5 L_4 4; L_3 L_2; L_1 $L_Ø$	$C_1C_0L_5L_4$ $L_3L_2L_1L_0$ (6 elements, one color)	4 + B	512	Ø	1	1	X	X	X	A	X
Color Graphics One	64 X 64	4; E_3 E_2 E_1 $E_Ø$ 3 (16 X 3 pixels)	C_1C_0 C_1C_0 C_1C_0 C_1C_0 E_3 E_2 E_1 E_0	4	1024	1	X	X	Ø	Ø	Ø	A	X
Resolution Graphics One	128 X 64	3; 7 6 5 4 3 2 1 Ø 3; L L	$L_7L_6L_5L_4$ $L_3L_2L_1L_0$	1 + B	1024	1	X	X	Ø	Ø	1	A	X
Color Graphics Two	128 X 64	3; E_3 E_2 E_1 $E_Ø$ 3	C_1C_0 C_1C_0 C_1C_0 C_1C_0 E_3 E_2 E_1 E_0	4	2048	1	X	X	Ø	1	Ø	A	X
Resolution Graphics Two	128 X 96	2; 7 6 5 4 3 2 1 Ø 2; L L	$L_7L_6L_5L_4$ $L_3L_2L_1L_0$	1 + B	1536	1	X	X	Ø	1	1	A	X
Color Graphics Three	128 X 96	2; 2; E_3 $E_Ø$	C_1C_0 C_1C_0 C_1C_0 C_1C_0 E_3 E_2 E_1 E_0	4	3072	1	X	X	1	Ø	Ø	A	X
Resolution Graphics Three	128 X 192	2; 1; L_7 $L_Ø$	$L_7L_6L_5L_4$ $L_3L_2L_1L_0$	1 + B	3072	1	X	X	1	Ø	1	A	X
Color Graphics Six	128 X 192	2; 1; E_3 $E_Ø$	C_1C_0 C_1C_0 C_1C_0 C_1C_0 E_3 E_2 E_1 E_0	4	6144	1	X	X	1	1	Ø	A	X
Resolution Graphics Six	256 X 192	1; 1; L_7 $L_Ø$	$L_7L_6L_5L_4$ $L_3L_2L_1L_0$	1 + B	6144	1	X	X	1	1	1	A	X

Notes:

1. X indicates an inactive input or bit. No change for 1 or Ø.
2. A indicates an active bit.
   - Data bits form alphanumeric characters by ASCII code.
   - Control bit INV gives bright character on black background for INV = Ø, dark character on bright background for INV = 1.
   - Control bit CSS selects one of two color sets when active.

	CSS = Ø	CSS = 1
Alphanumerics:	Green/Black	Orange/Black
Graphics (except Semi 4):	Green, Yellow, Blue, Red	Buff, Cyan, Magenta, Orange

3. L indicates illumination; Ø = dark, 1 = bright.
4. C indicates color-determining bit. E = element color.

C_2	C_1	C_0	Color (E)
Ø	Ø	Ø	Green
Ø	Ø	1	Yellow
Ø	1	Ø	Blue
Ø	1	1	Red
1	Ø	Ø	Buff (brownish yellow)
1	Ø	1	Cyan (greenish blue)
1	1	Ø	Magenta (purplish red)
1	1	1	Orange

Figure 18.8 Display modes for the 6847 Video Display Generator. We will concentrate on the Alphanumeric Internal, Semigraphics Four, and Resolution Graphics Two modes.

Note that the semigraphic modes allow *black*, which the color-graphic modes do not allow.

The format of the CRT display is shown in Figure 18.9 for the alphanumeric and semigraphic modes. The other modes are similar, except that the screen-corner memory addresses go to higher values.

The pinout of the 6847 package is given with Figure 18.10. Of the 40 pins, 31 are readily accounted for:

- 13 pins: address outputs (We will use only the least-significant 9 at first, since we will need only 512 memory bytes.)
- 8 pins: data inputs from display memory.
- 8 pins: mode select, as listed in Figure 18.8.
- 2 pins: Power supply (+5.0 V) and ground.

We will now examine the functions of the nine remaining pins.

$\overline{\text{HS}}$ (horizontal sync) and $\overline{\text{RP}}$ (row preset) are outputs used primarily to implement external character generation, such as lowercase letters, non-English alphabets, and special symbols. We will not be using them, so they can be left unconnected.

CLK is an input that must be fed with TTL-level clock signal of 3.579545 MHz, which is the exact frequency of the television color subcarrier. The duty cycle (*high* time as a percent of cycle time) must be near 50% to make adjacent dots on the screen the same size, since a pixel is output on each half cycle.

$\overline{\text{FS}}$ (field sync) is an output that goes *low* for about 2.0 ms during the vertical blanking period at the end of each scan of the screen. This happens 60 times per

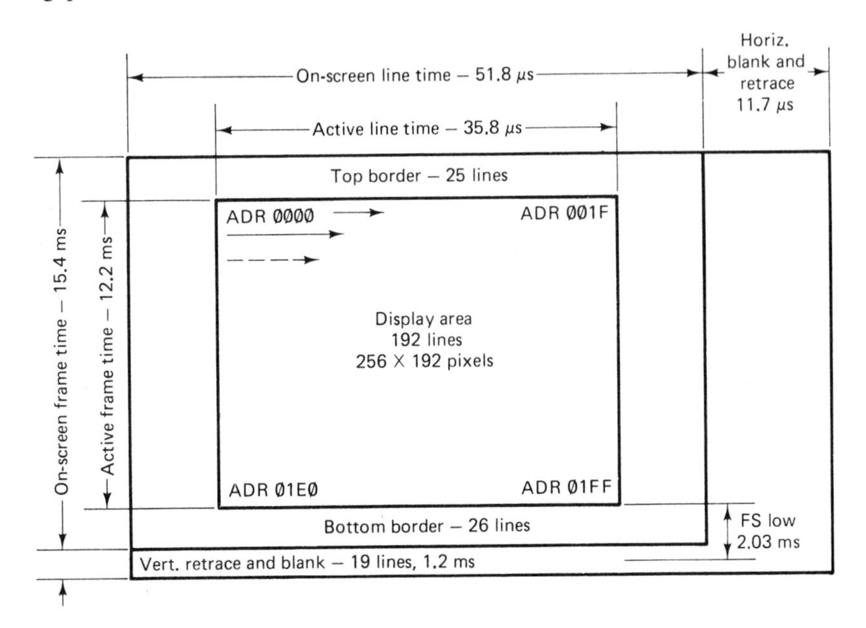

Figure 18.9 Screen-display specifications for the 6847 VDG. The corner addresses are given for the alphanumeric and semigraphics modes, with a base address of $0000.

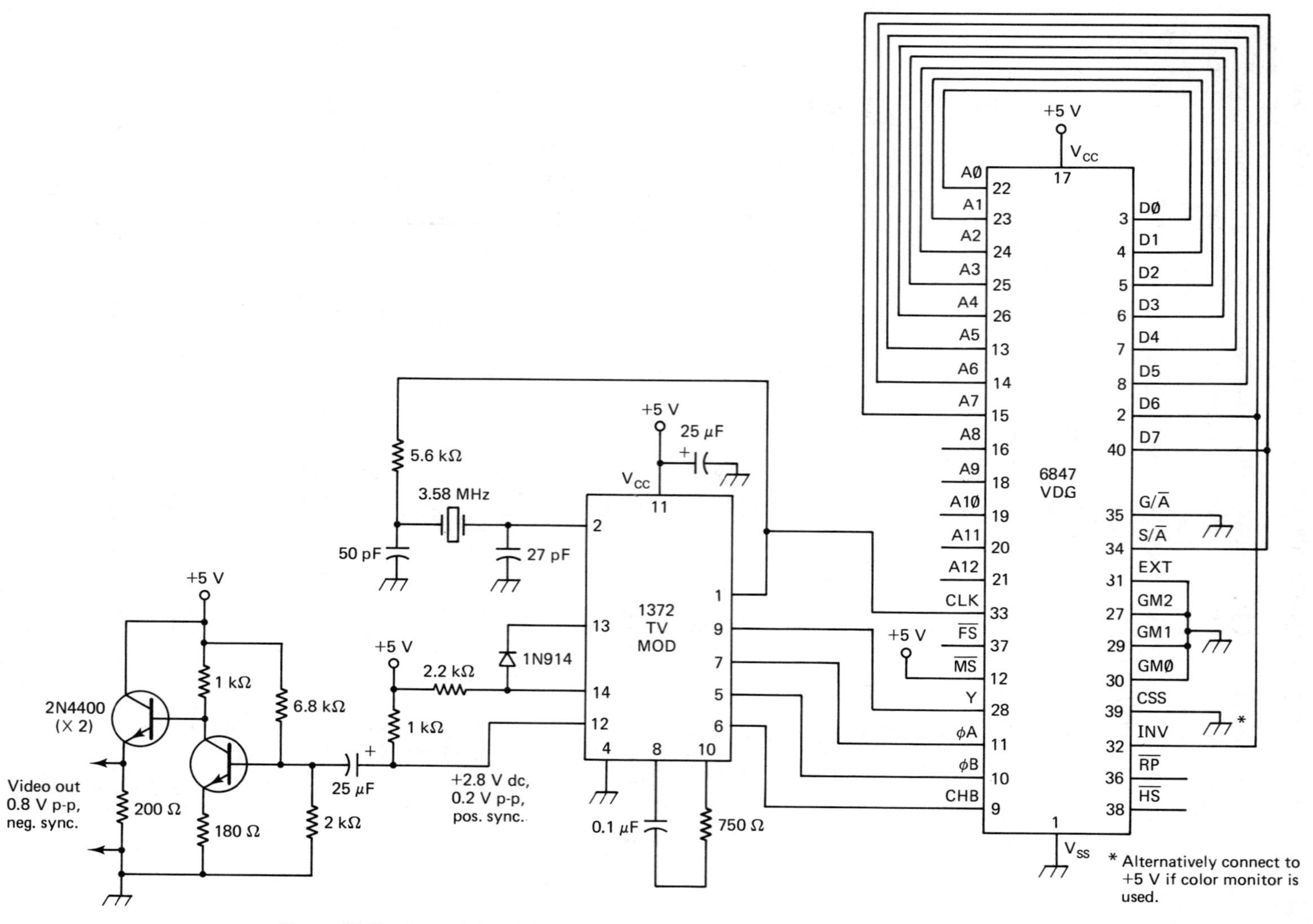

Figure 18.10 A stand-alone VDG test circuit that will display the 6847's font of Alphanumeric Internal characters and its Semigraphics Four color combinations.

second. The VDG is not accessing its display memory during these times, so you can use $\overline{\text{FS}}$ to signal the processor (via its $\overline{\text{IRQ}}$ input) that a memory write can be performed without interference.

$\overline{\text{MS}}$ (memory select) is an input that tells the VDG to stop accessing memory and let its address outputs float in the high-Z state. It can be used to interrupt the display while the processor writes to memory. There is an 80-ns delay from the time $\overline{\text{MS}}$ goes low to the time that the 6847 floats its address lines. The processor must, therefore, assert $\overline{\text{MS}}$ at least 80 ns *before* it asserts write signals to the display memory. Failure to observe this precaution will result in fights for the bus and spurious characters on the screen. The problem can be sidestepped by writing to the display memory only when $\overline{\text{FS}}$ is *low*.

The remaining four pins of the 6847 are video outputs. Y is the standard black-and-white composite video-and-sync signal. The sync peaks are at about +1 V and the white level is at about 0.5 V. Color information is provided via outputs ΦA and ΦB, which are normally connected directly to corresponding pins of a type-1372 color TV modulator IC. CHB (chroma bias) also interfaces directly to the 1372.

18.4 STAND-ALONE VDG TEST CIRCUITS

Section Overview

This section presents two test circuits for the 6847 Video Display Generator. The first uses the VDG's own address outputs to feed its data inputs, thereby obtaining a sample of every possible character or graphics pattern. The second circuit accesses a programmed EPROM that contains a message to be displayed.

Character and graphics displays can be made to appear on the video screen by simply connecting the low-order eight address lines of the VDG to the data lines. The mode-select pins can be hard-wired to select alphanumeric or various graphics modes as they are changed from V_{CC} to ground. Since the 6847's ASCII character set (Figure 18.11) contains only 2^6, or 64, characters, two data lines are free to control the display mode.

Figure 18.10 shows a 6847 wired for a fixed test display of normal and inverse alphanumeric characters on the top half of the screen and semigraphic-four patterns on the bottom half. Address line A8 and the various mode-select inputs can be rewired to +5 V or ground to obtain samples of the other graphics display modes.

The 1372 and two-stage transistor amplifier shown with Figure 18.10 provide the 1-V p-p, negative-sync composite video signal required by most color and monochrome video monitors. If a monitor is not available, an ordinary TV receiver can be used by replacing the video amplifier with the rf carrier-generator circuit of Figure 18.12. It must be understood that the quality of the display (especially in color) will be very much poorer if the rf circuit is used.

A fixed message display or graphic image can be placed on the screen by connecting a 2716 EPROM to the VDG, as shown in Figure 18.12. You can program

Hex	ASCII	Hex	ASCII	Hex	ASCII	Hex	ASCII
00	@	10	P	20	blank	30	0
01	A	11	Q	21	!	31	1
02	B	12	R	22	" (quote)	32	2
03	C	13	S	23	#	33	3
04	D	14	T	24	$	34	4
05	E	15	U	25	%	35	5
06	F	16	V	26	&	36	6
07	G	17	W	27	' (apostrophe)	37	7
08	H	18	X	28	(	38	8
09	I	19	Y	29	)	39	9
0A	J	1A	Z	2A	*	3A	: (colon)
0B	K	1B	[	2B	+	3B	; (semicolon)
0C	L	1C	\	2C	, (comma)	3C	<
0D	M	1D	]	2D	- (hyphen)	3D	=
0E	N	1E	↑	2E	. (period)	3E	>
0F	O	1F	←	2F	/ (solidus)	3F	?

Figure 18.11 The ASCII character subset implemented by the 6847 in the Alphanumeric Internal mode.

the first 512 bytes of the EPROM with your message using Figure 18.11 to select the hex code for the desired ASCII characters. You can mix normal and inverse characters (dark letter on light background) by connecting the INV pin to the D6 output and adding $40 to the ASCII code for any characters to be inverted. You can switch to the alternate-color characters by feeding the CSS pin from D7 and adding $80 to any characters that are to appear in orange, rather than green.

You can mix graphics with alphanumerics by instead connecting $\overline{A}$/S input pin 34 to D7 and using bytes between $80 and $FF to produce semigraphic-four patterns. Use Figure 18.8 to verify that the following bytes will produce the blocks of color listed, each block being one thirty-second of a screen wide and one-sixteenth of a screen high.

Data	Color
8F	Green
9F	Yellow
AF	Blue
BF	Red

Data	Color
CF	Yellow-brown
DF	Blue-green
EF	Red-violet
FF	Orange

Data	Display
X0	Black
X3	Bottom half-block
XC	Top half-block
X5	Right half-block

REVIEW OF SECTIONS 18.3 AND 18.4

Answers appear at the end of Chapter 18.

11. How does the 6847 determine what characters to display on the video screen?

12. How many bytes of RAM are required for the color-graphics mode in which each element is 2 × 2 pixels and may assume any of 4 colors? (The full screen is 256 × 192 pixels.) Calculate it; don't cheat and look it up.

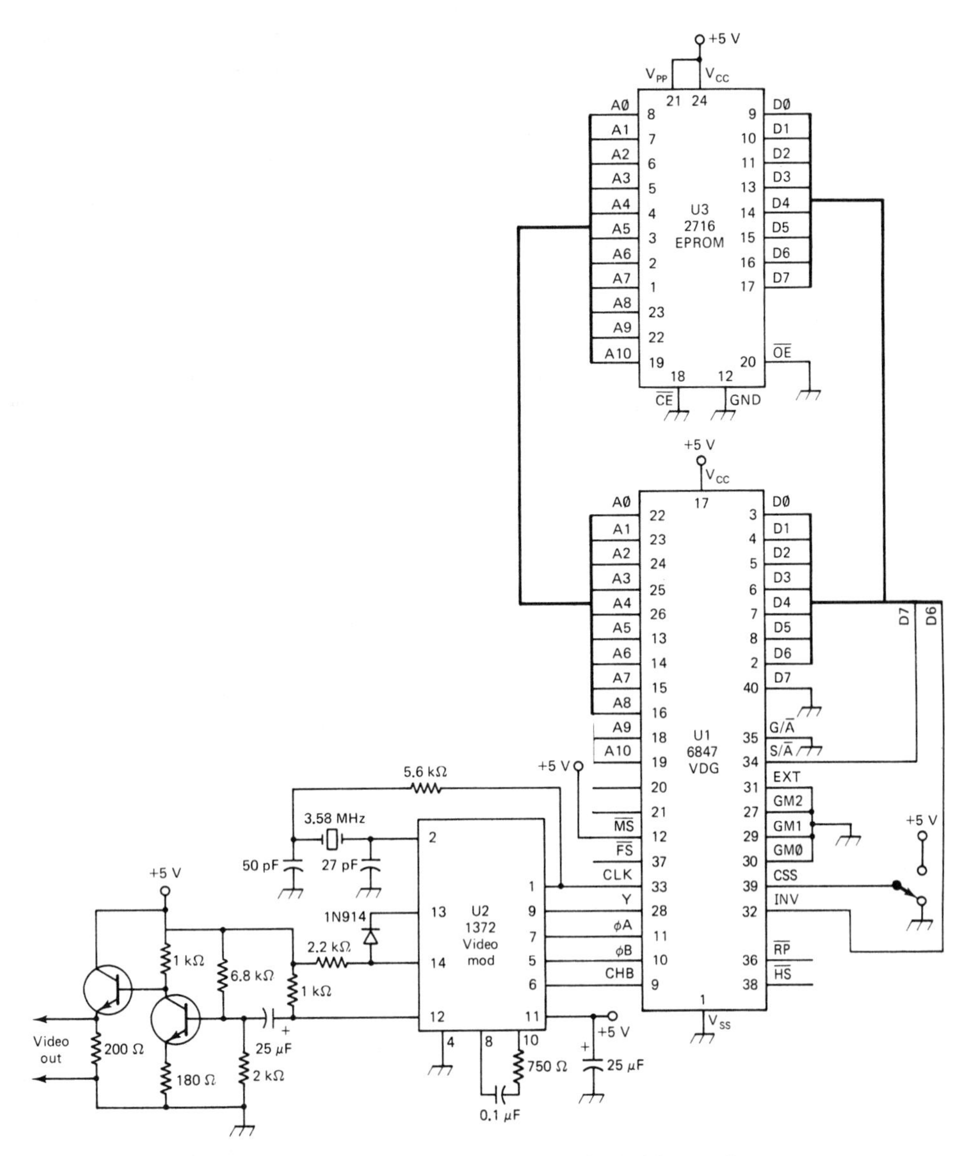

Figure 18.12 A stand-alone circuit that will display a fixed alphanumeric message or Semigraphics Four color pattern on the CRT.

13. In Resolution Graphics Two of the 6847, what bit of what byte address represents the upper right corner of the screen?
14. In Color-Graphics Three mode, what bits in what byte will place a blue dot at the bottom left corner of the screen?
15. What 6847 output line can be used to inform the processor that the VDG will not be using the display memory for a while?

16. What control word should be applied to the 6847 to select Color-Graphics Six with the buff/orange color set? See Figure 18.8, use Ø for any X bits, and answer in hex.
17. Using the circuit of Figure 18.12, what byte at what EPROM address will place an inverse letter X in the third space on the fifth line of the CRT?
18. In the Semigraphics Four mode, using Figure 18.12, what byte should be loaded to what addresses to produce a 4-pixel-wide red line down the left edge of the screen?
19. Use Figure 18.9 to calculate the number of bits displayable in monochrome, and the number of bits that could be transmitted at the same data rate during the border and blanking times.
20. How would you modify the circuit of Figure 18.12 to provide four separate screen displays at will, from one 2716 EPROM?

18.5 INTERFACING THE VDG TO THE COMPUTER

Section Overview

This section presents hardware diagrams and programs necessary to allow the computer to write data to the VDG's display RAM. The hardware consists primarily of a pair of tristate buffers, which connect the processor's address and data buses to the display RAM only when new data is being written. The programs vary with their objectives, but an essential part of all is an interrupt routine that writes data to the display RAM only after the VDG has signaled that it is floating its address bus.

The video interface is shown in Figure 18.13. It can be used with the basic microcomputer system of Figure 7.6 or with any microcomputer system employing the 6800, 6802, 6809, or 6502 processors and permitting access to the address and data buses and the $R/\overline{W}$ and E (or $\Phi 2$) lines. This includes a large number of personal computers and educational "single-board" computers or "evaluation kits" from the microprocessor manufacturers. The computer should be programmed to load the display RAM by responding to an $\overline{IRQ}$ generated by the video display generator's $\overline{FS}$ output. The I-mask bit should be *set* unless the processor has data to load to the display RAM.

The basic system shown uses two octal buffers and feeds only 7 data bits and 9 address bits to the display RAM. This is adequate for the Alphanumeric-Internal and Semigraphics Four modes, since each uses only 2^9, or 512, bytes of memory and neither uses bit 7 of the byte. Data line 6 is tied to the INV input, so it can be used to produce inverse characters in the alphanumeric mode.

When interfacing the video display a common problem is interference from the microcomputer's clock. This can be minimized by feeding the V_{CC} lines through low-

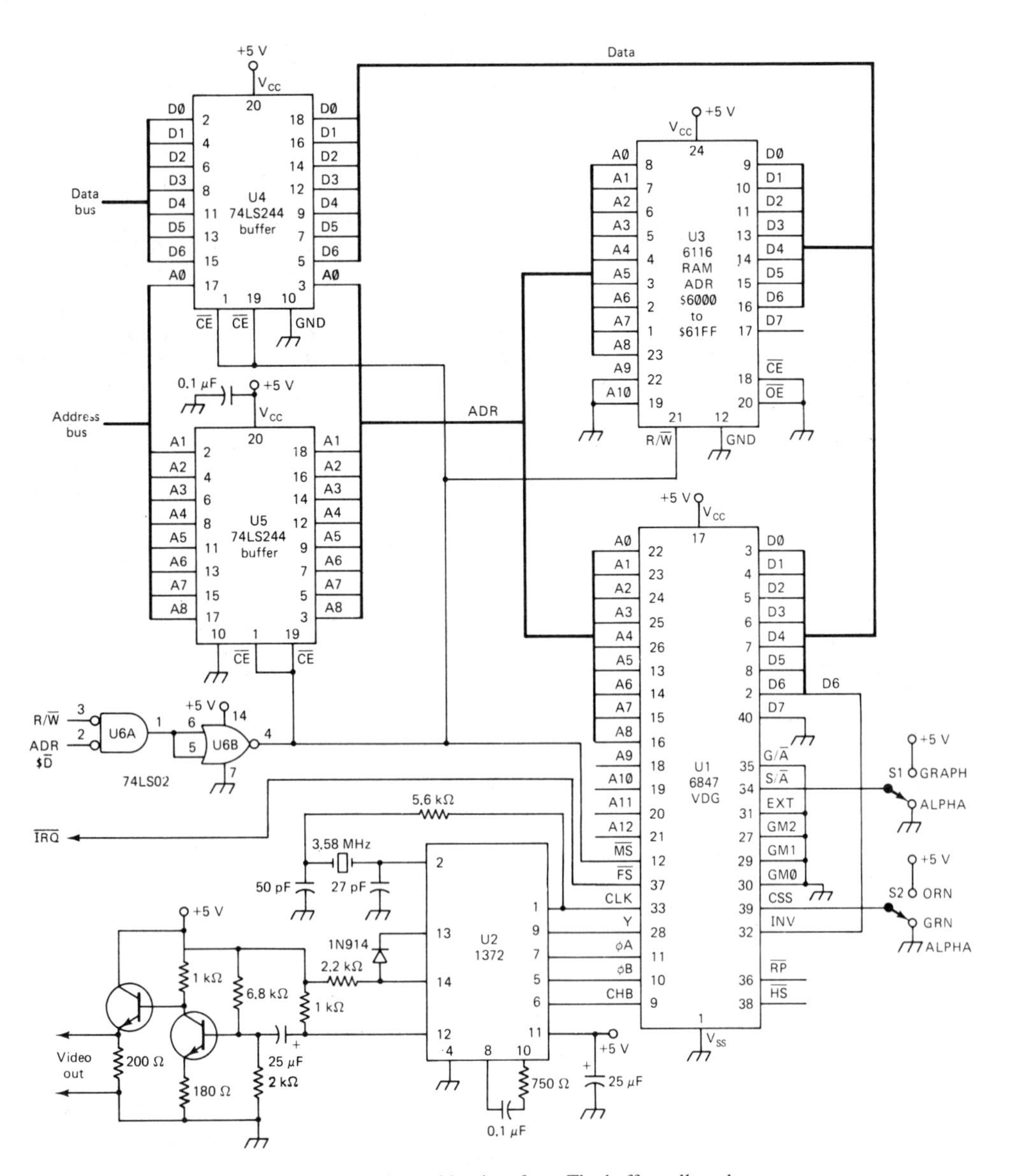

Figure 18.13 A computer-driven video interface. The buffers allow the computer to write to the display RAM during vertical retraces. The computer circuit of Figure 3.7 may be interfaced to this circuit.

resistance (less than 0.1 Ω) chokes of about 10 μH, keeping all leads short, and distributing several 0.1-μF capacitors along the V_{CC} lines to ground.

Here are two test programs for the video display. The first displays the 6847's font of characters and semigraphic-four color patterns in a changing pattern, and the second produces a changing pattern of diagonal color lines on the CRT.

```
                1000 * LIST 18-1 * VIDEO GENERATOR CHARACTER DISPLAY.
                1010 * FOR HARDWARE FIG. 18-13 & FIG 6-14 CPU.
                1020 * TESTED & RUNNING, 29 JUL, '87 BY D. METZGER.
                1030          .OR  $FF80
                1040          .TA  $4080
0000-           1050 RAM1     .EQ  $0000     INNER LOOP DELAY COUNTER
0001-           1060 RAM2     .EQ  $0001     SET FOR NUMBER OF INNER LOOPS.
0002-           1070 CHAR     .EQ  $0002     HOLDS ASCII CHARACTER, PASSES TO SUBRTN.
0003-           1080 XBUF     .EQ  $0003     SAVES X FOR USE IN IRQ ROUTINE.
D000-           1090 VBEG     .EQ  $D000     START OF SCREEN DISPLAY RAM.
D200-           1100 VEND     .EQ  $D200     ONE PAST END OF SCREEN RAM.
                1110 *
FF80- CE D0 00  1120 START    LDX  #VBEG     MAIN ROUTINE POINTS TO SCREEN
FF83- DF 03     1130          STX  XBUF       (SAVE IN RAM AT 0003:0004)
FF85- 8E 00 7F  1140          LDS  #$007F     START; INIT STACK FOR IRQ & SUBR.
FF88- 86 20     1150          LDAA #$20      SET FOR 32 PASSES THRU DELAY LOOP,
FF8A- 97 01     1160          STAA RAM2       * 256 * 10 CY * 2 US = 0.16 SEC.
FF8C- 0E        1170 IDLE     CLI            PROCESSOR WAITS FOR VERT RETRACE
FF8D- 20 FD     1180          BRA  IDLE       IRQ BEFORE WRITING TO RAM.
                1190 *
FF8F- 96 02     1200 INTER    LDAA CHAR      INTERRUPT ROUTINE STORES NEXT ASCII
FF91- DE 03     1210          LDX  XBUF       CHARACTER TO NEXT SCREEN ADDRESS.
FF93- A7 00     1220          STAA 0,X        (X CAME TO SUBR WITH VALUE FROM MAIN)
FF95- 7C 00 02  1230          INC  CHAR      ADVANCE TO NEXT CHARACTER AND
FF98- 08        1240          INX             NEXT SCREEN LOCATION.
FF99- DF 03     1250          STX  XBUF      SAVE X IN RAM SO STACK DOESN'T EAT IT.
FF9B- 8C D2 00  1260          CPX  #VEND     PASSED END OF VIDEO RAM?
FF9E- 26 05     1270          BNE  NEXT       NO: GO TO NEXT LOCATION.
FFA0- CE D0 00  1280          LDX  #VBEG      YES: POINT X TO BEGINNING AGAIN.
FFA3- DF 03     1290          STX  XBUF
FFA5- BD FF A9  1300 NEXT     JSR  DELAY     SLOW WRITING DOWN SO EYE CAN FOLLOW.
FFA8- 3B        1310          RTI            AFTER DELAY WAIT FOR RETRACE TO WRITE.
                1320 *
FFA9- D6 01     1330 DELAY    LDAB RAM2      DELAY SUBROUTINE. INNER LOOP COUNTS
FFAB- 7A 00 00  1340 WAIT     DEC  RAM1       IN RAM1; 5.12 MS/RUN ON 0.5-MHZ
FFAE- 26 FB     1350          BNE  WAIT        CLOCK.
FFB0- 5A        1360          DECB           OUTER LOOP COUNTS TIMES FROM RAM2
FFB1- 26 F8     1370          BNE  WAIT       IN ACCUM B. MAX 1.3 SEC.
FFB3- 39        1380          RTS
                1390 *
                1400          .OR  $FFF8
                1410          .TA  $40F8
FFF8- FF 8F     1420          .DA  INTER     INTERRUPT VECTOR
FFFA- FF 80 FF
FFFD- 80 FF 80  1430          .HS  FF80FF80FF80  * SWI, NMI, RES VECTORS
                1440          .EN
```

```
                1000 * LIST 18-2.  VIDEO DAZZLER: COLOR STRIPES ON CRT.
                1010 *  USE FIG 18-13 HARDWARE; S1 IN GRAPHICS MODE.
                1020 * TESTED AND RUNNING 29 JULY '87 BY D. METZGER.
                1030 *
                1040          .OR  $FF80
                1050          .TA  $4080
0000-           1060 HIX      .EQ  $0000     RAM BUFFERS TO MANIPULATE
0001-           1070 LOX      .EQ  $0001      VIDEO ADDRESSES.
0003-           1080 CELL     .EQ  $0003     RAM HOLDS GRAPHICS COLOR & FORM.
0004-           1090 COUNT    .EQ  $0004     FOR 5-MS TIME DELAY.
                1100 *
FF80- 8E 00 4F  1110          LDS  #$004F    INITIALIZE STACK.
FF83- CE D0 00  1120 START    LDX  #$D000    CLEAR SCREEN TO ALL DARK;
FF86- 6F 00     1130 CLEAR    CLR  0,X        SEMIMGRAPHICS FOUR MODE
FF88- 08        1140          INX             OF 6847 VDG.
FF89- 8C D2 00  1150          CPX  #$D200     (LOWER RIGHT CORNER OF CRT)
FF8C- 26 F8     1160          BNE  CLEAR
```

```
FF8E- C6 1F      1170        LDAB #$1F      SET GAP BETWEEN CHANGED CELLS;
FF90- CE D0 00 1180 RUN      LDX  #$D000     (VARY AT WILL) : TOP OF CRT.
FF93- DF 00      1190        STX  HIX
FF95- DE 00      1200 LOOP   LDX  HIX
FF97- 96 03      1210        LDAA CELL      WRITE TO CRT AT ADDRESS X WITH
FF99- A7 00      1220        STAA 0,X        COLOR PATTERN FROM CELL.
FF9B- 7C 00 04 1230 WAIT     INC  COUNT     5 MS/WRITE * 32*16 CHAR * 1 WRITE
FF9E- 26 FB      1240        BNE  WAIT       PER 15 CHAR AVG * 30 PASSES
FFA0- 37         1250        PSHB             = 5 SEC PER SCREEN CYCLE.
FFA1- DB 01      1260        ADDB LOX       ADVANCE POSITION ON CRT BY "B".
FFA3- D7 01      1270        STAB LOX
FFA5- 33         1280        PULB           RETRIEVE B VALUE.
FFA6- 24 03      1290        BCC  NOCARY    ADVANCE X HIBYTE IF CARRY FROM
FFA8- 7C 00 00 1300          INC  HIX        ADDING TO LOBYTE.
FFAB- 96 00      1310 NOCARY LDAA HIX       HAS X PASSED END POSITION
FFAD- 81 D2      1320        CMPA #$D2       OF VIDEO RAM?
FFAF- 27 02      1330        BEQ  PAGE       YES: START ANOTHER LINE.
FFB1- 20 E2      1340        BRA  LOOP       NO: NEXT CELL.
FFB3- 7C 00 03 1350 PAGE     INC  CELL      CHANGE CELL CONTENTS
FFB6- 5A         1360        DECB           DECREASE GAP BETWEEN CELLS
FFB7- C1 01      1370        CMPB #1         WRITTEN TO MIN OF 1.
FFB9- 26 D5      1380        BNE  RUN       MAKE ANOTHER LINE.
FFBB- 20 C6      1390        BRA  START     DARKEN SCREEN AND GO AGAIN.
                 1400        .OR  $FFFE
                 1410        .TA  $40FE
FFFE- FF 80      1420        .HS  FF80
                 1430        .EN
```

18.6 A VIDEO GRAPHICS PROGRAM

Section Overview

This section presents a design project with the objective of drawing a straight line between any two points on the CRT, using the 6847's resolution graphics two mode. This will involve modifying the hardware to handle the extra data and address lines required and developing the software to identify the bytes and bits that need to be written into the display RAM. A line-drawing routine is, of course, the basis for more-complete graphics programs, but more important is the insight this program provides to the nature of more-sophisticated programming of system peripherals.

Resolution Graphics Two of the 6847 produces images 128 elements wide by 96 elements high, for a total of 12 288 elements. The elements may be identified by their X and Y coordinates, as shown in Figure 18.14(a) with hex values. These are controlled by a display memory of 1536 bytes, or 12 288 bits. The bits are accessed by sequential addresses, increasing across the page and then down to the next line, the way a book is read or a TV screen is scanned. This is illustrated in Figure 18.14(b). The basic problem in writing a program to draw a line between two points is to translate points from the (X, Y)-coordinates of the screen to the *address-bit* coordinates of the display memory.

The Y-coordinate is not difficult to translate. Inspection of Figure 18.14 reveals that the Y byte is identical with the second and third nibbles of the display address. A simple 16-bit-wide shift, executed four times, will complete the Y conversion.

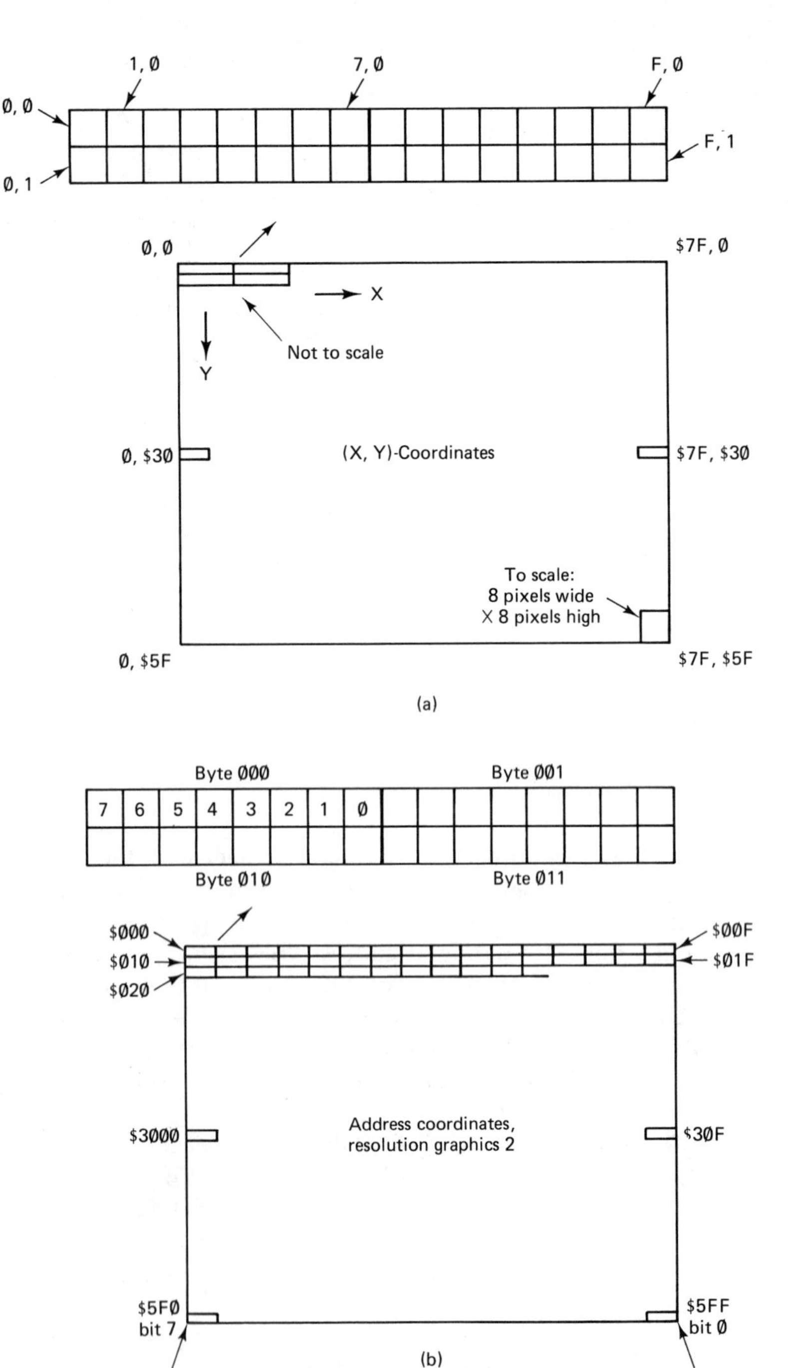

Figure 18.14 A video line-drawing program must translate the given and calculated (X,Y)-coordinates (a) into corresponding address-and-bit coordinates in the display-RAM address area (b).

The X-coordinate byte dictates, by its bits 3, 4, 5, and 6, one of 16 bytes across the screen. These 4 bits of X are, in fact, the first (lowest-order) nibble of the display address. Three shifts right will place these bits in the proper first-nibble position, where they can be combined with the second- and third-nibble bits from Y. The byte address is now completely specified.

The bit in the byte is specified by the low-order 3 bits of the X-coordinate. If these three X bits are 000, the leftmost (7) bit of the display-address byte must be set. If they are 111 the rightmost bit (seven shifts to the right) must be set. The procedure is to mask off all but the lowest 3 bits of X and count down (0 to 7 counts) as a bit is shifted from position 7 (to position 0 at most) in the address-buffer byte.

The values of X and Y that form a straight line between endpoints X_1, Y_1 and X_2, Y_2 can be fround readily from well-worn mathematical techniques. Referring to Figure 18.15, a value of Y can be calculated for each integer value of X between endpoint values X_1 and X_2 using

$$Y = \frac{Y_2 - Y_1}{X_2 - X_1}(X - X_1) + Y_1 \qquad (18.1)$$

The flowchart for the line-drawing program is given in Figure 18.16. As Equation 18.1 indicates, the program will require a multiplication and division. We saw in Chapter 10 how the 6802 can perform these operations, but the routines do tend to clutter the program. Since we are on the verge of introducing the 6809 microprocessor (which has a multiply instruction in its instruction set), we will defer the actual writing of the program until Chapter 20, when we will implement it in 6809 code.

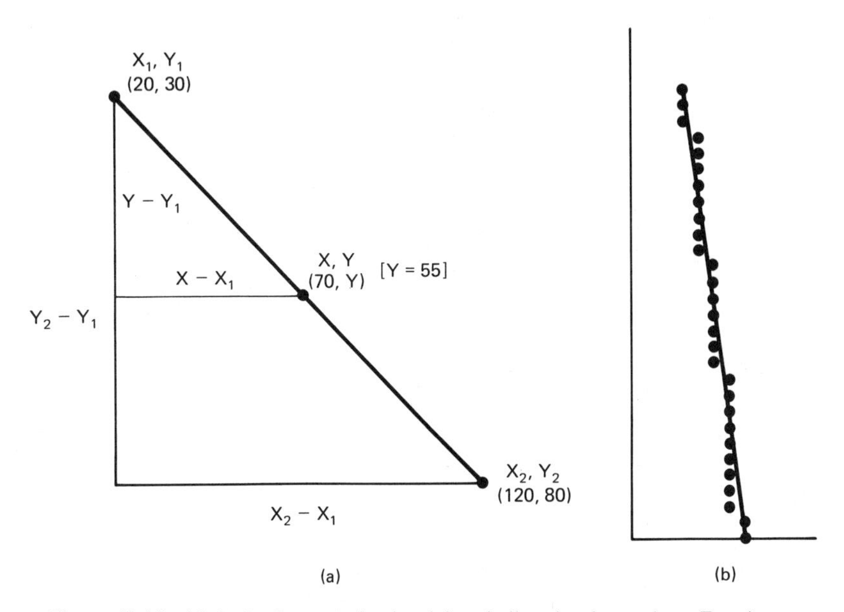

Figure 18.15 (a) A simple proportion involving similar triangles produces Equation 18.1 for calculating Y-values for each X-value between X_1 and X_2. (b) For steep slopes, several Y-steps must be inserted for one X-value.

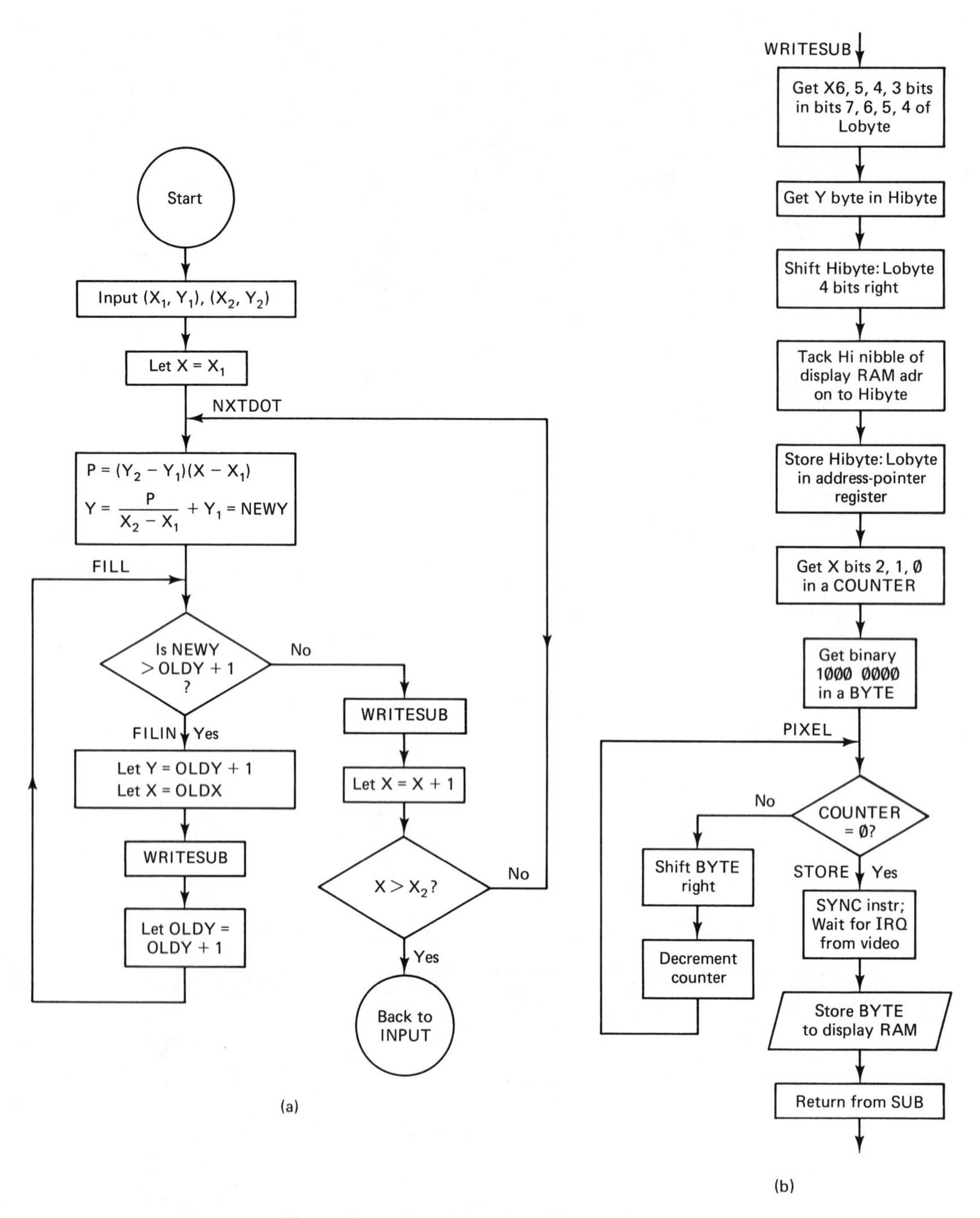

Figure 18.16 Flowchart for the video line-drawing program.

REVIEW OF SECTIONS 18.5 and 18.6

21. How many buffer lines need to be added to the system of Figure 18.13 to utilize the Resolution Graphics Two mode?
22. If a write is done to the display RAM without waiting for an $\overline{IRQ}$ signal, what may result?
23. In the circuit of Figure 18.13, what is the address for the character at the bottom left corner of the CRT?
24. The character-font display program of page 454 is modified with

FONT LDAA #$01; STAA CHAR

before the IDLE label and

LDAA #$1B; CMPA CHAR; BEQ FONT

after the IDLE. What does this do?
25. In Figure 18.14, what bit of what address corresponds to the *X*, *Y* point $3C, $41?
26. What (*X*, *Y*)-coordinates correspond to bit 5, address $43D?
27. Does a 1 bit in Resolution Graphics produce a light or dark spot?
28. What is the purpose of the FILL loop in Figure 18.16(a)?
29. How would the program of Figure 18.16 handle an upward-sloping line (say, Y1 = 2, Y2 = 1)?
30. Why are the X/Y-to-address conversion and store-to-display-RAM [Figure 18.16(b)] placed in a subroutine?

18.7 A TV TYPEWRITER

This section presents the program promised in Chapter 17. The program reads an 8 × 8 keyboard, consults a table to convert the key number to its ASCII code, and stores that code at the next sequential location in the display RAM. The hardware has already been presented in Figure 17.18 (basic processor and keyboard), and Figure 18.13 (video system). The program listing follows (List 18.3).

```
               1000 * LIST 18-3 ** TV TYPEWRITER.  READS 8X8 KEYBOARD;
               1010 *  DISPLAYS CHARACTERS ON VIDEO SCREEN. SEE FIG 17-18
               1020 *   FOR KEYBOARD, FIG 18-13 FOR VIDEO SYSTEM.
               1030 *  TESTED AND WORKING,  7 OCT, '87 BY D. METZGER
               1040 *
               1050          .OR   $FF40     START OF PROGRAM.
               1060          .TA   $4740     HOST SYSTEM RAM STORAGE AREA.
C800-          1070 PB       .EQ   $C800     PORT B OUT TO KEY ROWS.
C801-          1080 PA       .EQ   $C801     PORT A IN FROM KEY COLUMNS.
C802-          1090 DDRB     .EQ   $C802     DATA DIRECTION REGISTERS;
C803-          1100 DDRA     .EQ   $C803      1S = OUT, 0S = IN.
0008-          1110 COUNT    .EQ   $0008     COUNTS UP TO KEY NO.
0011-          1120 KEYNO    .EQ   $0011     BUFFER TO SAVE KEY NUMBER.
0012-          1130 DELAY    .EQ   $0012     COUNTER FOR 5-MS DELAY.
```

```
0014-             1140 VID      .EQ  $0014    & $0015 HOLDS VIDEO ADDRESS.
                  1150 *
FF40- 7F C8 03 1160 INIT     CLR  DDRA     PORT A ALL INPUTS.
FF43- C6 FF    1170          LDAB #$FF
FF45- F7 C8 02 1180          STAB DDRB     PORT B ALL OUTPUTS.
FF48- 8E 00 7F 1190          LDS  #$7F     START STACK AT END OF INTERNAL RAM.
                  1200 *
FF4B- CE D0 00 1210          LDX  #$D000   POINT TO UPPER LEFT OF SCREEN
FF4E- 86 20    1220          LDAA #$20      ASCII FOR BLANK IS $20.
FF50- A7 00    1230 CLEAN    STAA 0,X      CLEAR SCREEN
FF52- 08       1240          INX            ADVANCING FROM UPPER LEFT
FF53- 8C D2 00 1250          CPX  #$D200    TO LOWER RIGHT; 512 BYTES.
FF56- 26 F8    1260          BNE  CLEAN
                  1270 *
FF58- CE D0 00 1280 START    LDX  #$D000   POINT TO UPPER LEFT OF VIDEO SCREEN.
FF5B- DF 14    1290          STX  VID
FF5D- BD FF 92 1300 READ     JSR  KEY      GET NUMBER OF KEY IN ACCUM A.
FF60- 97 11    1310          STAA KEYNO    SAVE KEYNUMBER TO COMPARE AFTER DELAY.
FF62- 7F 00 12 1320          CLR  DELAY    8.6 MS DELAY: 256 LOOPS THRU
FF65- 6D 00    1330 WAIT     TST  0,X        7 US  (NO FUNCTION - JUST TAKES TIME)
FF67- 7A 00 12 1340          DEC  DELAY      6 US     (2 US/CYCLE)
FF6A- 26 F9    1350          BNE  WAIT       4 US  DELAY DEBOUNCES KEY.
FF6C- BD FF 92 1360          JSR  KEY      READ KEY AGAIN.
FF6F- 91 11    1370          CMPA KEYNO     SAME?
FF71- 26 EA    1380          BNE  READ       NO: KEEP LOOKING.
FF73- 7F C8 00 1390 NOTUP    CLR  PB         YES: WAIT FOR KEY UP; PULL
FF76- B6 C8 01 1400          LDAA PA          ALL ROWS LOW AND LOOK FOR
FF79- 81 FF    1410          CMPA #$FF        ALL COLUMNS TO BE HI.
FF7B- 26 F6    1420          BNE  NOTUP       ANY LO INPUT MEANS KEY STILL DOWN.
                  1430 *
FF7D- 86 FF    1440          LDAA #$FF     NOW GET ASCII BYTE FROM TABLE.
FF7F- 97 10    1450          STAA KEYNO-1    1) POINT X TO NUMBER OF
FF81- DE 10    1460          LDX  KEYNO-1       KEY IN TABLE.
FF83- A6 00    1470          LDAA 0,X        2) GET ASCII FROM TABLE INTO A.
FF85- DE 14    1480          LDX  VID        3) POINT X TO CURRENT VIDEO
FF87- 0E       1490          CLI                MEMORY BYTE.
FF88- 3E       1500          WAI             4) WAIT FOR VERT RETRACE INTERRUPT.
                  1510 * PROCESSOR HALTS UNTIL IRQ RECEIVED.
FF89- 0F       1520          SEI           NO INTERRUPTS UNTIL NEXT VALID KEY.
FF8A- 08       1530          INX           POINT X TO NEXT SCREEN ADDRESS.
FF8B- DF 14    1540          STX  VID
FF8D- 20 CE    1550          BRA  READ     LOOK FOR NEXT KEY.
                  1560 *
FF8F- A7 00    1570 IREQ     STAA 0,X      WRITE TO SCREEN. (I-MASK SET WHEN
FF91- 3B       1580          RTI            IRQ RECOGNIZED TO PREVENT MULTIPLES.)
                  1590 *
FF92- 7F 00 08 1600 KEY      CLR  COUNT    KEYREAD SUBROUTINE.
FF95- C6 FF    1610          LDAB #$FF     KEY NUMBER STARTS AT 00.
FF97- 58       1620 ROW      ASLB          PULL NEXT ROW FROM BOTTOM LOW.
FF98- F7 C8 00 1630          STAB PB
FF9B- B6 C8 01 1640          LDAA PA       READ COLUMNS.
FF9E- 43       1650          COMA          KEY PRESS NOW MAKES A 1;
FF9F- 26 0D    1660          BNE  COLUM    ANY "1" MEANS A KEY PRESSED.
FFA1- 96 08    1670          LDAA COUNT    IF NO KEY WE DON'T NEED "A"
FFA3- 8B 08    1680          ADDA #8        SO USE IT TO ADD A ROW'S
FFA5- 97 08    1690          STAA COUNT     WORTH OF KEY NUMBERS.
FFA7- 81 40    1700          CMPA #64      LAST ROW?
FFA9- 26 EC    1710          BNE  ROW      NO: CHECK NEXT ROW.
FFAB- 0C       1720          CLC           YES: THEN NO KEY.
FFAC- 20 E4    1730          BRA  KEY       JUST KEEP LOOKING.
                  1740 *
FFAE- 44       1750 COLUM    LSRA          KEY IN THIS ROW. FIND COLUMN.
FFAF- 25 05    1760          BCS  GOTIT    "1" SHIFTED IN - YOU GOT IT.
FFB1- 7C 00 08 1770          INC  COUNT    NO "1" BIT;
FFB4- 20 F8    1780          BRA  COLUM     THEN KEEP SHIFTING.
```

```
FFB6- 96 08      1790 GOTIT  LDAA COUNT   IF YOU HAVE A KEY,
FFB8- 39         1800        RTS           GO BACK TO MAIN PROGRAM.
                 1810 *
                 1820        .OR  $FFF8   IRQ, SWI, NMI, & RESET
                 1830        .TA  $47F8
FFF8- FF 8F      1840        .DA  IREQ
FFFA- FF 40      1850        .DA  INIT
FFFC- FF 40      1860        .DA  INIT
FFFE- FF 40      1870        .DA  INIT
                 1880 *
                 1890        .OR  $FF00   ASCII TABLE
                 1900        .TA  $4700
FF00- 55 37 46
FF03- 4A 4D 34
FF06- 52 56 49
FF09- 38 44 4B
FF0C- 2C 33 45
FF0F- 43 4F 39
FF12- 53 4C 2E
FF15- 32 57 58 1910          .AS  \U7FJM4RVI8DK,3ECO9SL.2WX\
FF18- 59 36 47
FF1B- 48 4E 35
FF1E- 54 42 50
FF21- 30 41 3A
FF24- 2F 31 51
FF27- 5A 20 20
FF2A- 20 20 3D
FF2D- 20 20 20 1920          .AS  \Y6GHN5TBP0A:/1QZ    =    \
               1930          .EN
```

Answers to Chapter 18 Review Questions

1. A beam of electrons. **2.** The pattern of scanning lines on a CRT. **3.** They are collected by the aquadag coating and returned to the cathode through the HV power supply. **4.** It is used to produce the positive high voltage.

5. Vertical, 48 Hz; horizontal 2880 Hz. **6.** The sync signals are more negative than any part of the video signal. **7.** Color information is encoded by amplitude and phase modulation of a subcarrier at this frequency. **8.** One in which the three colors are sent separately, rather than as a composite video signal.

9. 3072 bytes **10.** 4.3 MHz **11.** It reads a special RAM, the display memory. **12.** 3072 bytes **13.** Bit 0 of byte $000F **14.** Bits 7 and 6 of byte $0BE0 must be 1 and 0, respectively. **15.** $\overline{FS}$, pin 37 **16.** $9A **17.** Data $58 at address $0082.

18. Data $BA to addresses $0000, 0020, 0040, . . .,0100, . . .,01E0. **19.** 49 152 bits displayable; 69 800 bits transmittable during blank time. **20.** Switch A9 and A10 of the EPROM from +5V to ground manually. The four different combinations will access four 512-word blocks of memory. **21.** Three lines; two address, one data.

22. A fight for the bus, and spurious characters on the screen. **23.** $D1E0 **24.** Only the alphabet (*A–Z*) is displayed. **25.** Bit 3, address $417

26. X = $6A, Y = $43 **27.** Light spot **28.** It fills in missing *Y* lines, as shown in Figure 18.15(b). **29.** The program would need to be modified to handle this. **30.** Because they are used at two places in the main program.

CHAPTER 18 QUESTIONS AND PROBLEMS

Digits after the decimal point refer to section numbers in Chapter 18.

Basic Level

1.1 What do the letters CRT stand for?

2.1 How can the beam of a CRT be made less bright? Answer in terms of grid and cathode voltages.

3.1 To what part of a television-type CRT is the positive high voltage applied?

4.2 What three distinct types of information are carried on a single line in a composite color TV signal?

5.2 State one advantage and one disadvantage of using a character generator instead of direct bit-by-bit storage for alphanumeric video displays.

6.3 In the Semigraphics Four mode of the 6847, what byte will produce a buff color in the top two elements, leaving the bottom two elements dark?

7.3 Do the data pins of the 6847 only read data, only write data, or can they do both? Answer the same question for the 6847 address bus.

8.4 For the test circuit of Figure 18.10, list the first four characters that will appear on the third line of text.

9.4 In Figure 18.12, what byte must be stored to what EPROM address to place an 8×12-pixel red square at the right end of the screen, seventh row down (of 16 rows)?

10.5 In which display modes of the 6847 is data bit 7 not used? Which modes of the 6847 are available in Figure 18.13?

11.5 The $\overline{\text{ADR ENABLE}}$ input (pin 2 of U6A) in Figure 18.13 is decoded as

$$\text{VMA}\cdot\text{A15}\cdot\text{A14}\cdot\overline{\text{A13}}$$

In the alphanumeric mode, what is the base (lowest) address of the bottom right character on the screen?

12.6 Use Equation 18.1 to find the value of Y corresponding to X = \$3D on a line between the points (\$27,\$1F) and (\$4B,\$69).

13.6 In Figure 8.16(b) what is the purpose of the loop labeled "PIXEL"?

Advanced Level

14.1 Explain what interlaced scanning is and why it is used.

15.1 Explain the purpose of each of the following items in a TV receiver or CRT monitor: yoke, flyback transformer, damper diode.

16.2 Explain how a color CRT uses a shadow mask to produce color images.

17.2 An oscilloscope is to be driven by a character generator to display 10 lines of text at 20 characters per line. A noninterlaced raster scan as described in Figure 18.7 is to be used, with a frame rate of 48 Hz. Allow 3 dark scan lines between text lines, and allow 20% retrace time for both vertical and horizontal scans. Calculate the maximum data rate required.

18.3 Explain the functions of the Y, $\overline{\text{FS}}$, and $\overline{\text{MS}}$ pins of the 6847 VDG.

19.3 Using the following facts, prove that the 6847's active frame time is 12.2 ms, as shown in Figure 18.9. Show the calculations.

a. Scan rate = 60 Hz.

b. One frame consists of 525 lines, interlaced on alternate scans.

c. Active scans for 6847 consist of 192 lines.

20.4 Describe the wiring changes that would be required in Figure 18.12 to produce a display in the Color Graphics Two mode.

21.5 Explain why the display RAM's address lines cannot be driven directly from the processor's address bus in the circuit of Figure 18.13.

22.5 Explain the purpose of the $\overline{\text{IRQ}}$ output and the R/$\overline{\text{W}}$ and ADR $\overline{\text{\$D}}$ inputs in Figure 18.13.

23.6 Use the bit-shift techniques described in Section 18.6 to translate the (*X*, *Y*)-coordinates $53, $4A to Resolution Graphics Two address and bit coordinates.

24.6 What is the purpose of the FILIN routine in Figure 18.16(a)?

PART IV SOME ADVANCED MICROPROCESSORS

19.1 THE 6809 PROGRAMMER'S MODEL

Section Overview

The 6809 has 14 bytes of internal register, compared to 9 bytes for the 6800 and 6802. Major differences from the 6800 are:

- **There are two identical index registers, X and Y.**
- **There are two stack pointers; S for the computer's monitor program or operating *System* and U for the microcomputer programmer, or *User*.**
- **There is a *direct page* register DP, which holds the high-order byte in direct-mode addressing. This allows a 2-byte access to any address instead of to zero-page addresses only.**
- **The A and B accumulators concatenated can be referenced together as a single 16-bit accumulator, D. This allows several 16-bit arithmetic and data-move instructions.**

The 6800 microprocessor was introduced in 1974, about three years before the first personal computers and six years before microcomputers began to appear in automobiles and appliances. Considering this, it is remarkable that not only is the 6800/6802 still considered a viable processor, but new single-chip microcomputers with essentially the same architecture and instruction set are still being brought out.

What is not at all remarkable is the fact that a few years' experience revealed shortcomings in the 6800 design. Some of these you will have noticed from your work in the first three parts of this text.

- There is only one index register. There should be at least two to handle source and destination files.
- There should be instructions to add data to the index register and to push X on the stack.
- There should be two stack pointers: one for the operating system or monitor and one that is free for user applications.
- Eight-bit arithmetic gives only about $\frac{1}{2}$% accuracy, which is inadequate for many control applications. Sixteen-bit arithmetic should be possible with a single instruction.
- Relative addressing should be able to cover the entire address range to facilitate position-independent code.

All these additions and more were made in the 6809 microprocessor, which was introduced in 1978.

The programmer's model for the 6809 is shown in Figure 19.1. The two accumulators A and B can be referenced separately as in the 6800, but they can also be referenced together as the 16-bit double accumulator, D. Thus

```
ADDA $89AB
```

adds the data from address $89AB to accumulator A, the same as in the 6800. But

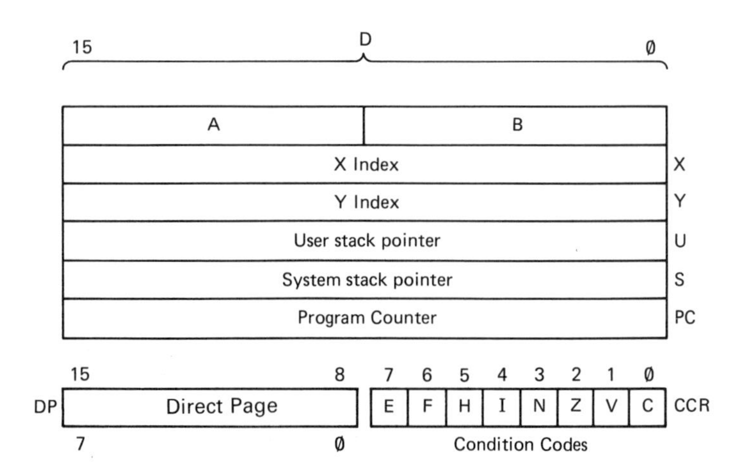

Figure 19.1 The 6809 Programmer's model. *D* is a 16-bit accumulator consisting of high-order byte *A* and low-order byte *B*.

ADDD $89AB

adds the 2 bytes from addresses $89AB (high-order) and $89AC (low-order) to the double accumulator D, consisting of A (high-order) and B (low-order). Remember that D is not a separate register. Changing D changes A and B; changing A changes the high-order byte of D.

Only three types of instructions are available for the double accumulator:

- Loads and stores.
- Adds, subtracts, and compares.
- Transfers to or from and exchanges with one of the other 16-bit registers (X, Y, S, U, or PC). An *exchange* (EXG) is a data swap between two registers.

Logic operations (AND, OR, EOR, COM, NEG) are not available for D. The SHIFTS and ROTATES are not available, nor are INC, DEC, TST, BIT, and the carry-input operations ADC and SBC.

The index registers X and Y are functionally identical and much the same as X in the 6800. There are many added capabilities which are explained in the following sections.

The System stack pointer S is also called the hardware stack pointer because it points to the RAM area used to save the machine registers when a hardware (or software) interrupt is acknowledged. The system stack is also used to save PC when a JSR or BSR is executed. One difference from the 6800/6802 is that the 6809 stack pointers decrement and then store, so they end up pointing to the last stored byte. The 6800 stack pointer stores and then decrements, so it ends up pointing to the next free RAM location. This is illustrated in Figure 19.2. The User stack, U, can be used for PUSH and PULL operations, or as another index register.

The Direct Page register holds the high-order address byte in direct addressing, so the program need only specify the low-order byte. With the 6802 the high-order byte was always $00, so direct addressing was usable only on page zero. The 6809

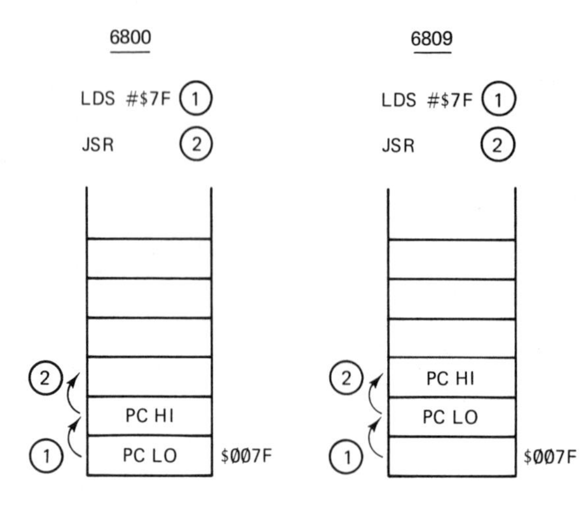

Figure 19.2 The 6800/6802 stack pointer uses postdecrement when saving data and preincrement when retrieving it. The 6809 uses predecrement on *push* and postincrement on *pull*.

comes up from *reset* with DP = $00, so direct addressing also references page 0. However, DP can be loaded subsequently with any value to allow the faster direct-addressing mode to reach any area of memory.

There is no LOAD DP instruction. DP must be loaded by transfer from A or B using

```
LDAA  #$F8
TFR   A,DP
```

19.2 THE 6809 PACKAGE PINOUT

Section Overview

Most of the pins of the 6809 have exact counterparts in the 6802. Exceptions are as follows:

1. **FIRQ: Fast Interrupt Request input.**
2. **BA and BS: Bus Available and Bus Status outputs to indicate one of four processor states, *normal*, *interrupt-acknowledge*, *sync*, or *bus grant*.**
3. **DMA/BREQ: Direct Memory Access or Bus Request input.**
4. ***Q*: Quadrature clock output; leads *E* by 90°.**

Also noteworthy is the fact that the 6809 address lines go to a high-*Z* state when the processor is not using the bus.

The pinout of the 6809 microprocessor is given in Figure 19.3. The pin numbers are different in almost every case, but the following pin functions are the same in the 6809 as in the 6802:

- D0 through D7.
- A0 through A15, although the 6809 address lines go tristate under *halt* or *bus available*, whereas the 6802 address lines remain active.
- $\overline{\text{IRQ}}$ is negative-level active and sends the processor to the vector address at $FFF8, FFF9 in both chips.
- $\overline{\text{NMI}}$ is negative-transition active with vector at $FFFC, FFFD in both chips.
- BA is a normally *low* output which goes *high* upon *halt* or when the processor is a *wait* state. This is the same in both processors. (In the 6809 the address outputs float when BA is *high*, and there is an additional cause for BA to go *high* as listed later under $\overline{\text{DMA/BREQ}}$.)
- MR (6802) and MRDY (6809) are similar inputs that are pulled *low* to stretch the processor clock period, thus allowing more access time for slow memory devices. There are subtle differences between them, but normally they are simply tied *high*.
- The R/$\overline{\text{W}}$ (read/write-not) and *E* (clock) outputs are the same for both processors.

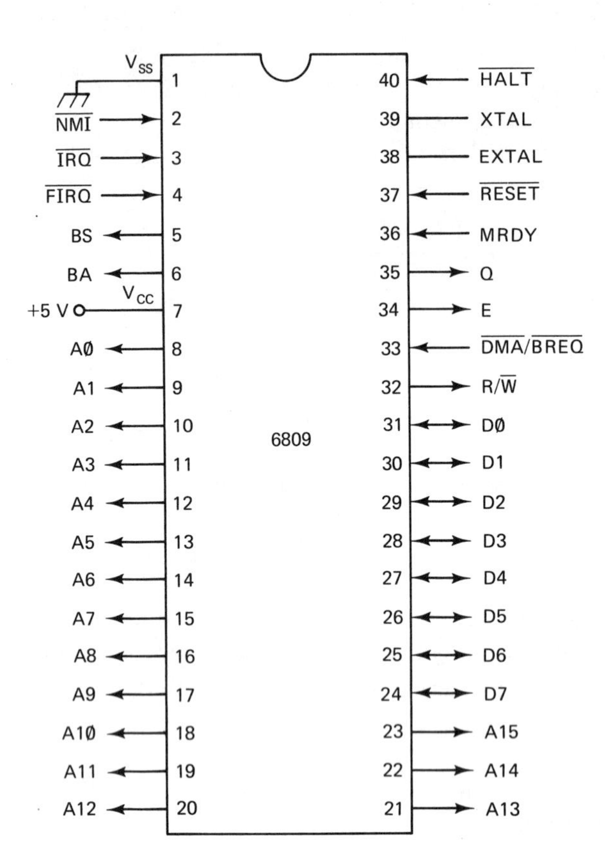

Figure 19.3 Pinout for the 6809 microprocessor.

E is *low* for the first (address) half-cycle and *high* for the second (data) half-cycle.

- XTAL and EXTAL use the same circuit in both processors. The *E* frequency is one-fourth of the crystal frequency. An external TTL signal of frequency $4f_E$ may be applied to EXTAL if XTAL is grounded, as in the 6802. Note that the pin positions are reversed in the two processors.
- The $\overline{\text{HALT}}$ and $\overline{\text{RESET}}$ inputs are the same in both processors. The $\overline{\text{RESET}}$ line activates the *reset* when the voltage falls below 0.8 V but does not return the processor to the *run* mode until it rises to 4.0 V. This wide hysteresis zone allows a simple *R-C* network and switch to be used for reset but will also necessitate an external pullup resistor if $\overline{\text{RESET}}$ is driven by a TTL gate.

Missing pins: Four 6802 pins are not found on the 6809:

- The V_{CC} for the 32 standby-RAM bytes and the RE (internal RAM-Enable) input do not appear on the 6809 because there is no internal RAM.
- The VMA pin is eliminated on the 6809 by a clever trick. Whenever a VMA-not cycle is in effect, the 6809 outputs address $FFFF. This is the *reset* vector

address, which is always ROM, so there is no harm in calling it up when the processor is not using the bus.

- The 6809 has only one V_{SS} (ground) pin. The 6802 has two.

Extra pins: The 6809 has four pins that are not found on the 6802:

- The $\overline{FIRQ}$ is a level-sensitive interrupt input that stacks 3 bytes instead of the usual 12 for 6809 interrupts—a fast interrupt. Section 20.3 provides more details.
- The $\overline{DMA/BREQ}$ input is used to suspend processor operation for direct-memory access, as, for example, in dynamic RAM refreshing, or to allow another processor to request access to the address and data buses.
- The BS (bus status) output is decoded with the BA (bus available) output to signal one of four processor states:

BA	BS	Processor State Acknowledged
0	0	Run
0	1	Interrupt or Reset
1	0	Sync
1	1	Halt or Bus grant

- The Q output is a second clock signal: it is discussed in detail in Section 19.6.

REVIEW OF SECTIONS 19.1 AND 19.2

Answers appear at the end of Chapter 19.

1. Explain any differences between the X and Y index registers.
2. Explain any differences between the S and U stack pointers.
3. What does accumulator D contain after executing

```
LDA   #$12
LDD   #$ABCD
LDB   #$34
```

4. What does this routine do?

```
TFR   X,D
ADD   #256
TFR   D,X
```

5. How can you invert the logic sense of all 16 bits of the D register—that is, change 1s to 0s and 0s to 1s?
6. What does this routine do? Why is the first instruction there?

```
EXG   DP,A
LDA   #$C2
EXG   DP,A
```

7. What is the difference in the address output drivers of the 6809 compared to the 6802?
8. What does the 6809 do during VMA-not cycles, since it has no VMA output pin?
9. How many bytes of RAM are integrated on the 6809 chip?
10. Why is the FIRQ faster than the IRQ?

19.3 6809 VERSIONS OF 6800 INSTRUCTIONS

Section Overview

Several common 6800/6802 instructions appear to be missing from the 6809 instruction set. Generally this is the result of consolidating functions under one instruction to make room for more advanced 6809 instructions. For example, six 6800 instructions to set and clear the C, I, and V flags are replaced by ORing and ANDing, respectively, of the condition-code register with an immediate byte.

The index registers and stack pointers are incremented, decremented, and otherwise manipulated with the LEA (Load Effective Address) instructions.

Several 6809 instructions are functionally the same as certain 6802 instructions but have different asembly-language mnemonics.

The 6809 does not have special instructions to set or clear specific bits of the condition-code register as the 6802 does. Instead the function is accomplished by bit masking and setting with AND and OR instructions. For example, to *set* the carry flag (bit 0) an OR #$Ø1 is used.

```
       E F H I N Z V C
CCR    Ø 1 Ø 1 Ø Ø Ø Ø      ORCC #1 = SEC
#$01   Ø Ø Ø Ø Ø Ø Ø 1
       _______________
$51    Ø 1 Ø 1 Ø Ø Ø 1
```

To clear both the C and I-mask flags in one operation, an AND is used.

```
       E F H I N Z V C
CCR    Ø 1 Ø 1 Ø Ø Ø 1      ANDCC #$EE = CLC, CLI
#$EE   1 1 1 Ø 1 1 1 Ø
       _______________
$4Ø    Ø 1 Ø Ø Ø Ø Ø Ø
```

Most 6809 assemblers will accept statements such as CLC and CLI and produce the machine-language bytes for ANDCC #$EE. Notice that all 8 CCR bits can be set and cleared with the ORCC and ANDCC instructions. The 6800 has set and clear instructions for only 3 bits. Here is a list of 6809 instructions to set and clear the CCR bits. The F and E flags are explained in Section 20.3.

```
SEC   ORCC #$01        CLC   ANDCC #$FE
SEV   ORCC #$02        CLV   ANDCC #$FD
SEZ   ORCC #$04        CLZ   ANDCC #$FB
SEN   ORCC #$08        CLN   ANDCC #$F7
SEI   ORCC #$10        CLI   ANDCC #$EF
SEH   ORCC #$20        CLH   ANDCC #$DF
SEF   ORCC #$40        CLF   ANDCC #$BF
SEE   ORCC #$80        CLE   ANDCC #$7F
```

Load Effective Address (LEA) instructions: The 6809 uses a number of indirect addressing modes, so the simple term *address* could refer to an indirect address (which holds another address) or to an *effective address*, which is the address containing the final operand data.

The four pointer registers X, Y, S, and U are intended to hold operand addresses, so loading one of these registers is called *load effective address* in 6809 terminology. This name may seem roundabout to a hardware technician, but it is perfectly straightforward to a programmer.

The LEA instructions include the capability of adding a signed constant value to the operand, and the operand can be the present contents of the register to be loaded. Thus to add 1 to the X register, the assembly language is

```
LEAX 1,X
```

which means load X with the old value of X, plus 1. Here are the increment and decrement instructions for the four pointer registers.

```
INX   LEAX  1,X        DEX   LEAX  -1,X
INY   LEAY  1,Y        DEY   LEAY  -1,Y
INS   LEAS  1,S        DES   LEAS  -1,S
INU   LEAU  1,U        DEU   LEAU  -1,U
```

As in the 6802, the Z flag is set if X or Y reach zero, but it is not affected if S or U reach zero. A dummy STS or STU can be used to sense a zero condition of S or U.

The signed constant added to the pointer register can be any value from −32 768 to +32 767.

- If the constant is from −16 to +15 (5-bit offset) the LEA takes 2 bytes and 5 cycles.
- If the constant is from −128 to +127 (8-bit offset) the LEA takes 3 bytes and 5 cycles.
- If the constant is from −32 768 to +32 767 (16-bit offset) the LEA takes 4 bytes and 8 cycles.

Variable offsets: The 2's complement value in accumulator A, B, or D can also be added to a pointer register with the LEA instructions.

LEAX A,X	Adds accumulator A to X.
LEAY D,Y	Adds accumulator D to Y.
LEAY B,X	Transfers X + B to Y.

These instructions are always 2 bytes long. They take 5 cycles for an A or B offset and 8 cycles for a D offset. They are useful for obtaining random access to a file. For example, A might pick up the key number from a keyboard, and X might have the base address of a table of ASCII bytes, arranged by key number. The correct byte would be pointed to by LEAX A,X. For initializing the value of a pointer register, the simple immediate-mode *load* instructions should be used—for example, LDX #$69B0.

Renamed instructions: The following instructions have simply been renamed, except for TBA and TAB, which require an extra *test* instruction to set the flags as they are set by the 6800.

FUNCTIONALLY EQUIVALENT INSTRUCTIONS

6800	6809	6800	6809
LDAA	LDA	PSHA	PSHS A
LDAB	LDB	PSHB	PSHS B
STAA	STA	PULA	PULS A
STAB	STB	PULB	PULS B
ORAA	ORA	TAB	TFR A,B TST A
ORAB	ORB	TBA	TFR B,A TST A
CPX	CMPX	TAP	TFR A,CC
TSX	TFR S,X	TPA	TFR CC,A
TXS	TFR X,S	WAI	CWAI #$FF

Missing instructions: Three 6800 instructions involving the A and B registers are missing in 6809 code and must be synthesized using the stack.

6800	6809 Equivalent	
ABA	PSHS B	ADDA ,S+
SBA	PSHS B	SUBA ,S+
CBA	PSHS B	CMPA ,S+

The meaning of the S+ structure is explained in Section 20.2.

19.4 NEW INSTRUCTIONS FOR THE 6809

Section Overview

The 6809 has an ABX instruction, which adds accumulator B to the X register for easy random file access. The 16-bit double-accumulator instruction set is limited, but individual A and B instructions can often syn-

thesize the 16-bit functions. A Sign Extend (SEX) instruction converts 8-bit 2's complement signed numbers to fill 16-bit registers by extending the 1 or Ø of bit 7 to fill bits 8 through 15.

There is an 8-bit by 8-bit unsigned multiply instruction with a 16-bit result. There is no divide instruction, but the double accumulator makes a 16-bit by 8-bit divide routine relatively simple.

The ABX instruction causes the value in accumulator B to be added to the value in pointer register X in unsigned (straight binary) arithmetic, with the result appearing in X. This instruction affects no flags of the CCR. There are no similar instructions for the A accumulator or Y pointer register.

Note that LEAX B,X also adds B to X, but in this case B is treated as a 2's complement signed value, and the Z flag is set or cleared according to the final value of X.

The SEX instruction (Sign Extend) converts an 8-bit 2's complement number in B into a 16-bit 2's complement number in D. This is done by copying bit B7 (the sign bit) into the 8 high-order bits of D.

Double-byte operations: Loads, stores, adds, subtracts, and compares can be done directly in the double accumulator. Here are instruction sequences for synthesizing other functions in accumulator D.

To Achieve	Use these steps		Comments
ANDD M	ANDB M+1	ANDA M	*Z* set if high-order byte = Ø.
ORD M	ORB M+1	ORA M	*Z* set if high-order byte = Ø.
COMD	COMB	COMA	*Z* set if high-order byte = Ø.
NEGD	COMA	NEGB	Flags set on low-order byte only.
ASLD[a]	ASLB	ROLA	Flags set on high-order byte only.
ASRD	ASRA	RORB	Flags set on low-order byte only.
LSRD	LSRA	RORB	Flags set on low-order byte only.
ROLD	ROLB	ROLA	Flags set on high-order byte, except *C* is rotated.
RORD	RORA	RORB	Flags set on low-order byte, except *C* is rotated.
INCD	ADDD #1		*C* flag set by ADD but not by INC.
DECD	SUBD #1		*C* flag set by ADD but not by DEL.

[a]LSL is the same as ASL.

Figure 19.4 illustrates the 16-bit rotate procedures. If it is desired that the N and Z flags be set according to the result in D, a dummy ADDD #Ø instruction can be added.

The multiply instruction (MUL) performs an unsigned binary multiplication of the bytes in A and B, storing the result in D. The process requires one byte and 11 cycles, compared to 23 bytes and (typically) 220 cycles for the 6800 routine of Section 10.6. MUL sets the C flag to the value of B7. This allows the 16-bit product to be rounded up or down to 8 significant bits by using the sequence

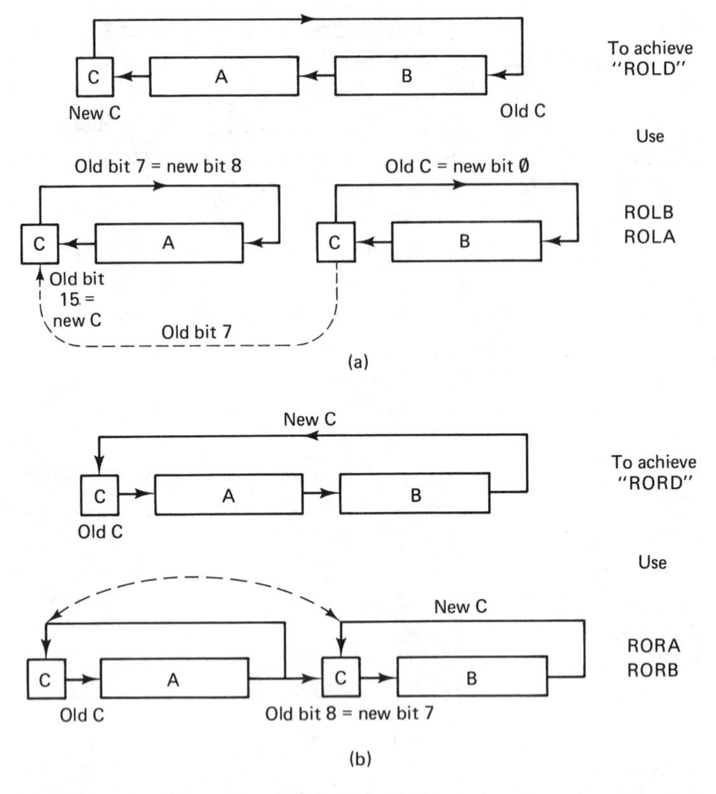

Figure 19.4 Rotate left and rotate right of the 16-bit register *D* can be synthesized by combining two 8-bit rotate instructions.

```
        MUL
        ADCA #Ø
```

Accumulator A will now contain the product, with AØ representing groups of 256, A1 representing groups of 512, and so on.

Note that the multiplier and multiplicand data are lost after MUL. To prevent this an STD (Store D) to a RAM buffer can be used before the MUL.

Divide: The 6809 does not have a *divide* instruction, but a relatively short routine can be written to provide this function. The routine builds the quotient in B as it shifts the dividend left out of D.

DIVIDE 16-BIT DIVDND BY 8-BIT DIVSOR. QUOTIENT APPEARS IN *B*.

```
        LDA     #8
        STA     COUNT     Count off 8 subtract-&-shifts.
        LDD     DIVDND    Get 16-bit dividend.
SHIFT   ASLB              Shift dividend left
        ROLA                in B and A.
        CMPA    DIVSOR    Will subtraction work?
        BCS     SKIP      Borrow says no. Skip it.
        SUBA    DIVSOR    Subtract divisor from shifted dividend.
```

```
        INCB                    Set bit in B (quotient) to be shifted left.
SKIP    DEC       COUNT         Count off another shift.
        BNE       SHIFT         Keep going until 8 shift and subtracts.
```

This division routine does not check for division by zero, and it assumes that the quotient will be less than 256. Testing for the former is simple. The programmer may decide to abort or to "max out" the quotient at $FF if divide by zero is encountered.

Testing for conditions that will produce a quotient greater than 255 is also simple. This will be so if the high-order byte of the dividend (accumulator A) equals or exceeds the divisor (RAM byte DIVSOR). The test is

```
CMPA    DIVSOR     Check A − RAM.
BHS     NO GOOD    Won't work if A ≥ RAM.
```

If the division would be "no good," a solution may be to *scale* the dividend down—that is, shift it right to divide it by 2 and then repeat the test. Several shifts may be necessary to enable a division. The quotient can then be shifted left from B into A (multiplying it by factors of two) to rescale the result.

REVIEW OF SECTIONS 19.3 AND 19.4

Answers appear at the end of Chapter 19.

11. What machine instruction sets the interrupt mask bit in a 6809?

12. How does the 6809 clear the C, E, and F flags?

13. What 6809 instruction will decrement Y four times?

14. What 6809 instruction should be used to initialize S to point to address $Ø1FF.

15. What is the 6809 equivalent for the 6800 instruction PULB?

16. Accumulator A contains $12 and B contains $A7. What does D contain after execution of the SEX instruction?

17. Why is NEGD not achieved by NEGA, NEGB?

18. How would you mask all the odd-numbered bits of D to zero?

19. If A contains $7D and B contains $B1, what do D, Z, and C contain after executing MUL?

20. How can the *Divide 16-bit* routine be enabled to handle $AB38 ÷ $95?

19.5 POSTBYTE POWER FOR EXPANDING INSTRUCTIONS

Section Overview

The 6809 can transfer the contents of any machine register to any other machine register of the same size with a single instruction called TFR. It can also exchange (swap) between any pair of 8-bit or any pair of 16-

bit registers with one basic EXG instruction. This section explains the structure of the *postbyte* that makes it possible.

Any or all of 8 machine registers can be pushed on or pulled off of the stack with a single instruction. This is also done with a postbyte.

Finally, the 6809 *branch* instructions can reach any address in memory, and a third type of extra byte permits this.

TFR and EXG: The contents of any 8-bit register of the 6809 can be transferred to any other 8-bit register; A, B, CC (condition-code), or DP (direct page). The data in the source register is left unchanged by the transfer. An *exchange* of data between any two 8-bit registers is also possible; an operation not provided in the 6800.

Transfers and exchanges between any two 16-bit registers (D, X, Y, S, U, or PC) are also possible, using the same TFR and EXG instruction types. The TFR op code is $1F and the EXG op code is $1E. The source and destination registers in either case are specified by a byte following the op-code, which is called a *postbyte*. This byte may be conceived of as part of the op-code, distinguishing among the different instructions, for example, EXG X,D and EXG B,DP. On the other hand, it may be conceived of as an operand byte specifying different operand registers for a single EXG instruction.

The high-order nibble of the TFR/EXG postbyte specifies the source register, and the low-order nibble specifies the destination register, as shown in Figure 19.5. Using unlisted hex digits and mixing left- and right-column hex digits (16-bit and 8-bit registers) is undefined and may crash the program or corrupt data. The distinction between *source* and *destination* applies only to TFR instructions. EXG A,B is functionally the same as EXG B,A.

Stacking: The 6809's *push* and *pull* instructions can save or retrieve any or all of the internal machine registers with a single instruction. The registers to be stored or retrieved are specified by the 8 bits of a postbyte that follows the op-code. Figure 19.6 shows the postbyte register assignments for each bit. A 1 specifies *push* or *pull* this register; a Ø specifies no action on this register.

The PSH and PUL instructions each require 2 bytes and 5 cycles plus one cycle for each register pushed or pulled. Note that it is necessary to specify which stack pointer to use. PSHA is not a valid instruction; it must be in such a form as PSHS A

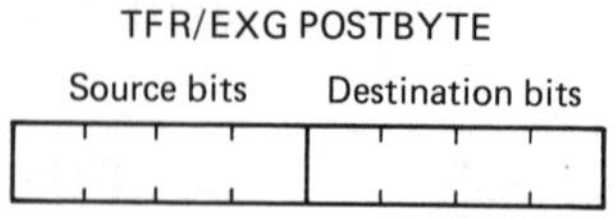

TFR/EXG POSTBYTE

Source bits   Destination bits

16-bit registers	8-bit registers
$Ø = D	$8 = A
$1 = X	$9 = B
$2 = Y	$A = CC
$3 = U	$B = DP
$4 = S	
$5 = PC	

Figure 19.5 A *postbyte* following the op-code byte specifies which registers are to be transferred or exchanged. Mixing 16-bit and 8-bit registers in the same instruction is not allowed.

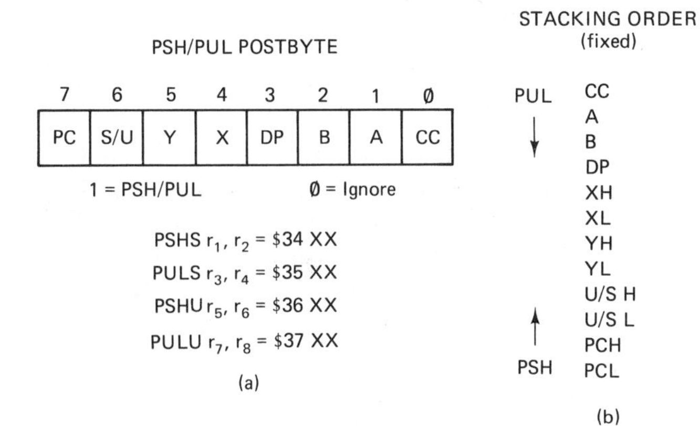

Figure 19.6 (a) A logic 1 in the postbyte following a *push* or *pull* op-code identifies a register to be saved or retrieved. (b) As many as 8 registers, involving 12 bytes, can be listed as operands in one instruction, but they are always pushed or pulled in the same order.

or PULU A,X,CC. The assembler will structure the necessary postbyte. In the second instruction, for example, the complete object code would be \$37, \$13.

The stacking order is always as listed in Figure 19.6, regardless of the order in which the registers are listed in the source code. If the system stack is invoked (PSHS or PULS), then bit 6 of the postbyte refers to the U register. If the op-code references the user stack, then the postbyte bit 6 references the system stack pointer.

The Long Branch: The 6809 has all of the 6800's branch instructions, plus a new one and extra names for two of the old ones:

BRN — Branch never; a 2-byte, 3-cycle NOP; useful in fine-tuning time delays or to substitute for an active branch during program debugging.

BHS — Branch if accumulator higher than or same as operand (unsigned). This instruction is identical with BCC at the op-code level.

BLO — Branch if accumulator lower than operand (unsigned). This instruction is identical with BCS at the op-code level.

Modern programming methodology minimizes the use of branches and keeps the branch-from and branch-to points generally within the range of -127 to $+128$ bytes when they are used. However, to maintain position independence, a longer branch is sometimes required. For such cases the 6809 has a long-branch version of each branch instruction: LBEQ for BEQ, for example. The range of the long branches is $-32\ 768$ to $+32\ 767$, with wraparound at the FFFF—ØØØØ junction. This need not be of concern to the programmer, however, since branch operands are nearly always specified by address label, leaving the assembler program to calculate the branch operand. Long-branch op-codes consist of 2 bytes—the first is called a prebyte and is always \$1Ø, and the second is the normal short-branch op-code. The long-branch operand is always a 2-byte 2's complement number, so long-branch instructions require 4 bytes. Exceptions are LBRA (long-branch always), which has the 1-byte op-code \$16 and the usual 2-byte operand, and LBSR, which is also only 3 bytes long.

19.6 6809 TIMING AND MEMORY INTERFACING

Section Overview

This section describes the clock signals and circuits for the 6809 and 6809E microprocessors. The 6809E uses an external clock-generator circuit and has three differing pin assignments to facilitate use of the processor with multiprocessor systems or systems with critically clocked peripherals.

Address and data bus loading for the 6809 is similar to that for the 6802. A simple memory interface that can be used to run test loops is described.

Two clock signals are provided for the 6809. The first is *E* (enable), which is the same as *E* for the 6802 or Φ2 for the 6800. The second is *Q* (quadrature), which leads *E* by one-quarter of a cycle and has essentially the same 50% duty cycle. Figure 19.7 shows the relationship between *E* and *Q*. Here are some of the uses of the *E* and *Q* signals:

- Peripherals can be selected with *E* OR *Q* logic. This allows the first quarter of the cycle for the new address to stabilize on the bus, leaving three-fourths of a cycle for device callup and data transfer. With the 6802 it was necessary to select writable devices with *E-true* logic to avoid false writes as the address lines switched immediately after the fall of *E*. This left only half a cycle for device access.
- The 6809 asserts data on a write cycle no later than 200 ns after the rise of *Q*. On a 1-MHz clock this leaves a data-valid time of 545 ns. The 6802 asserts write data no later than 225 ns after the rise of *E*, leaving a data-valid time of 295 ns. The 6809 thus provides considerably more access time to writable devices.
- Data is normally latched on the fall of *E*, which marks the end of the cycle, as it is in the 6802. However, it may be desirable to latch it on the fall of *Q*, which is approximately the midpoint of the *data-valid* time. The logic com-

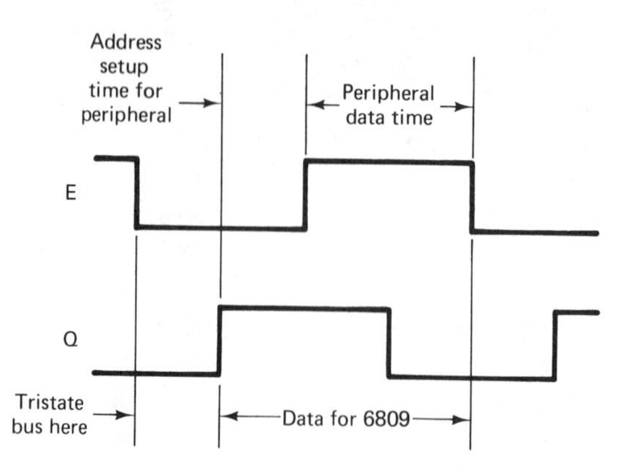

Figure 19.7 The 6809 has an *E* (enable) clock signal and a *Q* (quadrature) clock signal, which leads *E* by 90°.

bination E AND Q marks a time *period* during which data is valid (if the system speed is not too fast and the memory speed is not too slow).

Bus loading. The 6809 address bus can drive one standard Schottky TTL load plus 90 pF of capacitance at 1 MHz. This makes it safe to drive five LS-TTL devices plus eight CMOS or 6800-series devices without buffering. The data bus can drive up to 130 pF, making it safe to put five LS-TTL peripherals plus as many as 12 CMOS or 6800-series peripherals on the 6809 data bus. Figure 19.8 shows the system timing for the 6809.

The 6809E version requires an external clock generator. A simple one is shown in Figure 19.9. The use of an external clock generator allows complete access to all

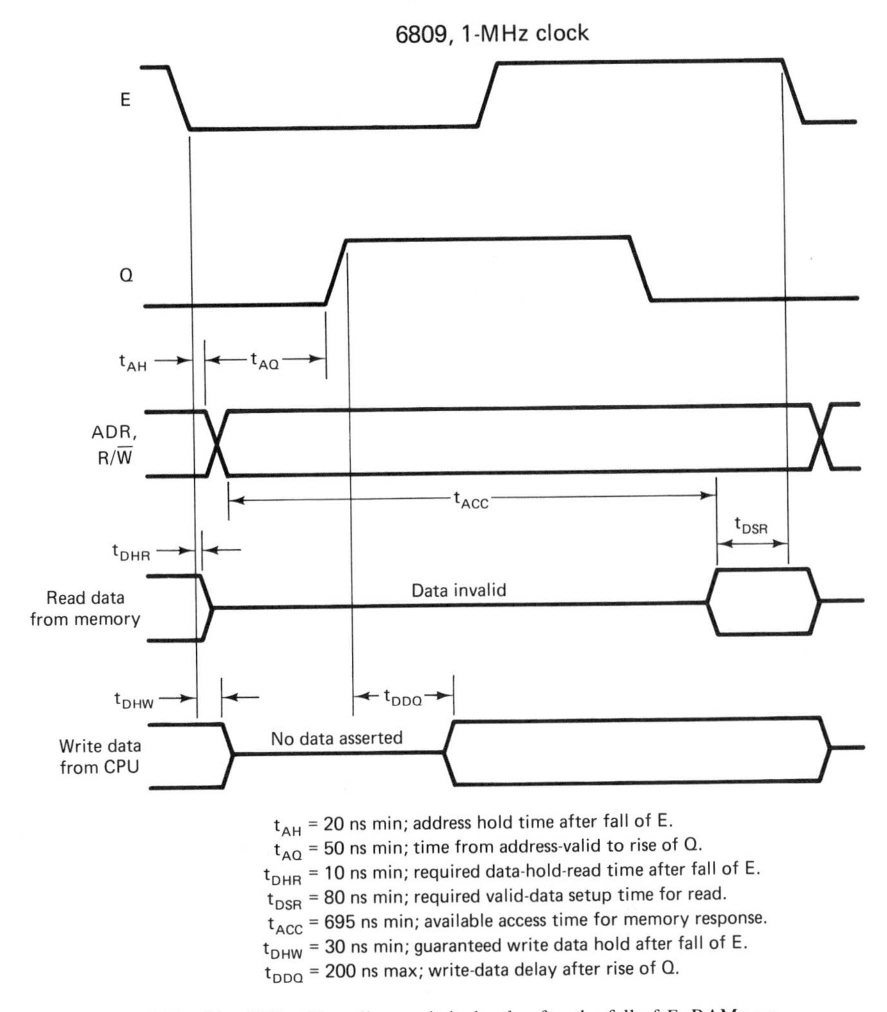

Figure 19.8 The 6809 address lines switch shortly after the fall of E. RAMs are generally enabled on the rise of Q, leaving three-fourths of a cycle for memory access. The 6809 asserts data shortly after the rise of Q on a write cycle.

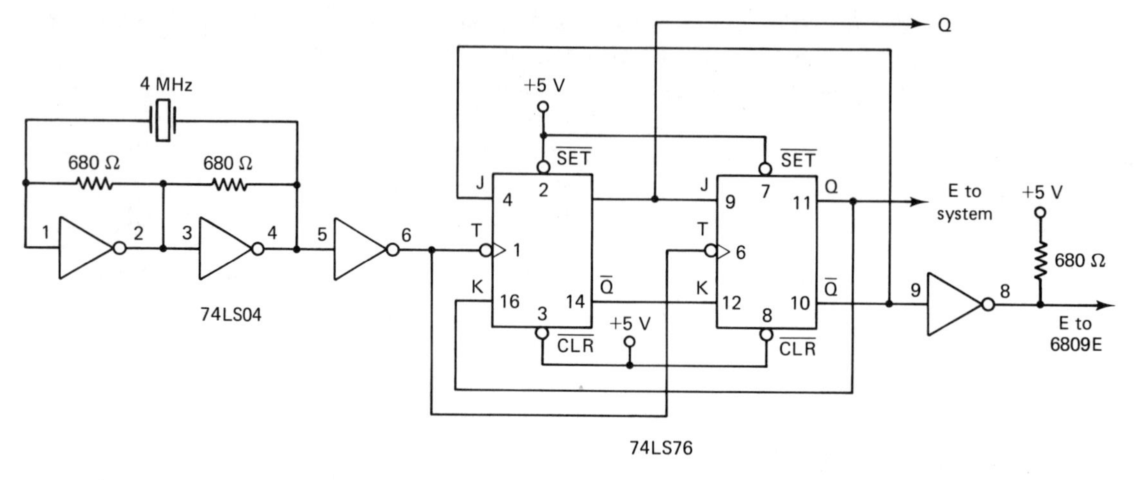

Figure 19.9 A clock circuit for the 6809E, supplying *E* and *Q* signals.

timing points and provides for several processors or peripherals to run from the same clock or clocks related in frequency or phase.

The *E* and *Q* pins of the 6809E are *inputs*, not outputs. The XTAL, EXTAL, and MRDY pins of the 6809 are not required for the 6809E, so three new pins are added:

- AVMA (advanced valid memory address) is an output that goes *low* to indicate impending $\overline{\text{VMA}}$ cycles and *high* to indicate impending cycles on which the processor will use the address and data busses. The AVMA signal is three-fourths of a cycle ahead of the processor's actual VMA state, so it gives advance notice to DMA devices to get on or get off the bus.
- LIC (last instruction cycle) output goes *high* during the last clock cycle of each instruction and goes *low* to mark the beginning of the next instruction.
- TSC is an input that immediately causes the 6809E bus to go tristate when it is brought *high*. It is used to allow a number of processors or DMA devices to share the bus. It must be tied *low* if not used.

The $\overline{\text{DMA/BREQ}}$ input of the 6809 is replaced by a BUSY output on the 6809E. This line goes *high* to indicate that a noninterruptable operation is in process and the TSC input must not be asserted. The other devices on the bus are expected to sense BUSY before asserting TSC. Figure 19.10 shows the pin changes for the 6809E.

A simple 6809 system suitable for running test loops and checking system timing signals is shown in Figure 19.11. The RAM memory programmer of Figure 3.10 or Figure 5.9 can be used to enter a test program. Here are two test programs; one to produce a 9-cycle test loop and a second (List 19.1) to flash an LED at an observable rate.

* 6809 Test Loop, Nine Machine cycles, with write.

Object Code		Source Code		Comments
0110	86 01	LDA	#1	Get $01 in direct page register.
0112	1F 8B	TFR	A,DP	
0114	0C 70	LOOP INC	$170	Read, modify, write. (6 ~)
0116	7E 0E	JMP	LOOP	Endless loop of 9 cycles (3 ~)

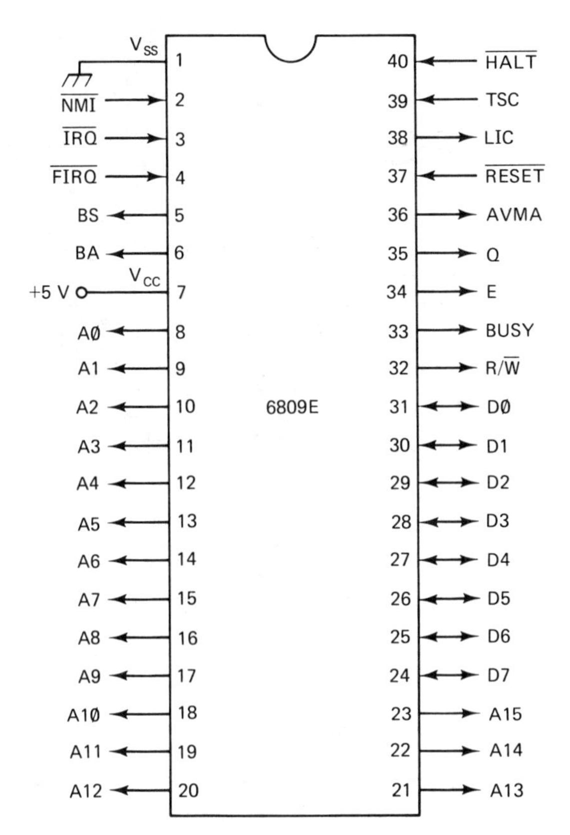

Figure 19.10 Pinout for the 6809E microprocessor.

```
                1000 * LIST 19-1 ** LED FLASHER - 6809. FIG 19.11 HARDWARE.
                1010 * TIE LED THRU 1 K FROM +5 V TO 6809 PIN 11.
                1020 * UNTESTED -    31 JULY '87 -  D. METZGER.
                1024          .OR  $0110
                1026          .TA  $4110
                1030 *
0110- CC FF FF  1040          LDD  #$FFFF  LOOP KEEPS A3 LOW EXCEPT FOR
0113- 83 00 01  1050 LOA3     SUBD #1       TWO VMA* CYCLES. LIGHTS LED.
0116- 26 FB     1060          BNE  LOA3    7 CY * 2 US/CY * 65536 LOOPS
                1070 *                      = 0.92 SEC LIT.
0118- 10 8E FF
011B- FF        1075          LDY  #$FFFF  LOOP KEEPS A3 HI, 8 CY/LOOP.
011C- 31 3F     1080 HIA3     LEAY -1,Y    LED DARK: 8 CY * 2 US * 65536
011E- 26 FC     1090          BNE  HIA3     = 1.05 SEC DARK.
0120- 20 F1     1100          BRA  LOA3    (LOOPS MADE DIFFERENT JUST TO
                1110          .OR  $017E    ILLUSTRATE D ACCUM & LEAY INSTR)
                1120          .TA  $417E
017E- 01 10     1130          .HS  0110    RESET VECTOR.
                1140          .EN
```

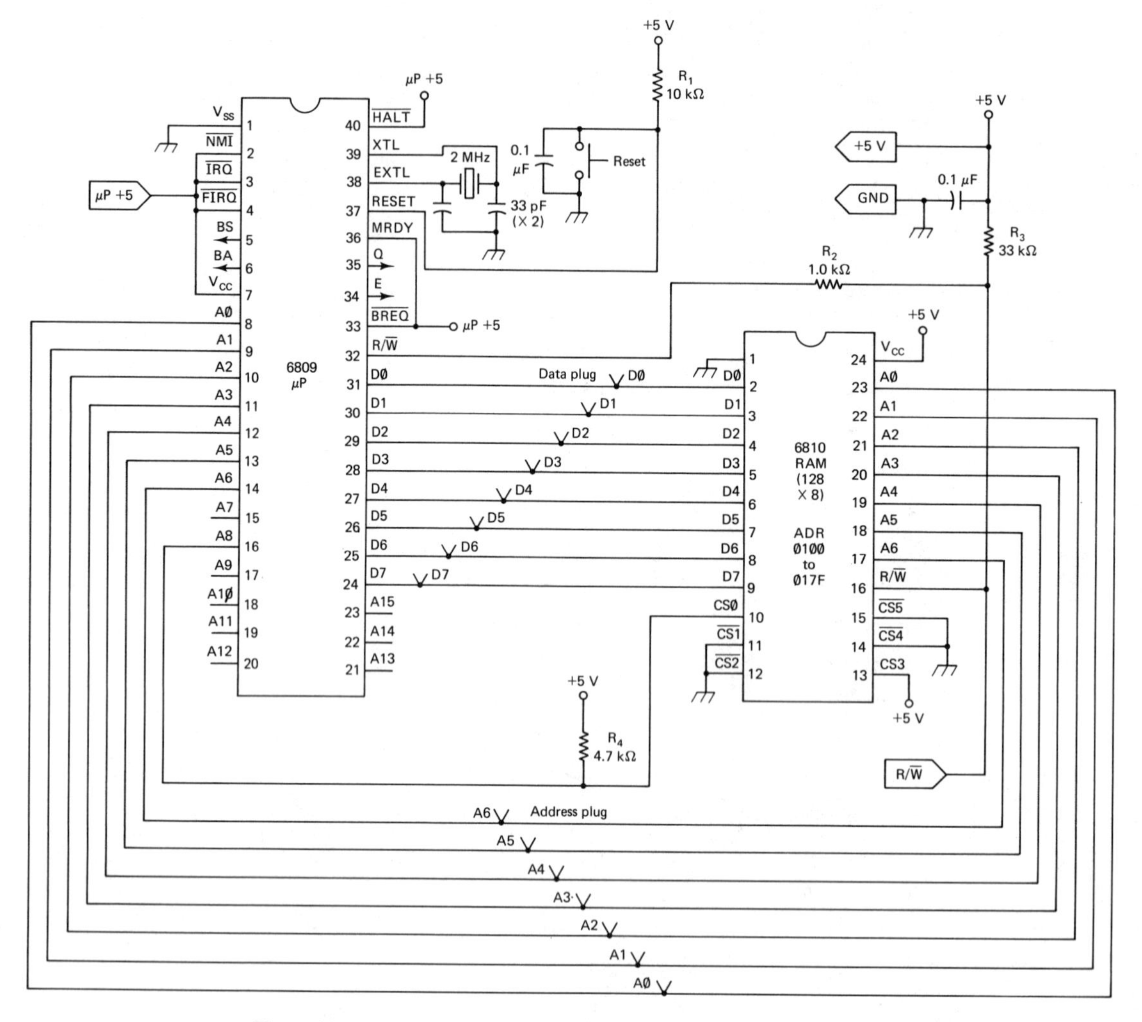

Figure 19.11 A simple system for running short test loops on the 6809. The programming circuit of Figure 3.10 or Figure 5.9 may be used to load programs into the RAM.

REVIEW OF SECTIONS 19.5 AND 19.6

21. What is the machine code to transfer the A accumulator to the X register?

22. What is the machine code to transfer the program counter to the Y register?

23. What is the assembly code to save A, B, X, and CC on the user stack?

24. What is the hex code for the instruction in Question 23?

25. What will be the next instruction address if the branch is taken in the following instruction?

```
5A32    10 27 15 A4    LBEQ    NEWAD
```

26. Is it possible to branch from address \$1000 to address \$F000 with a long branch? How?
27. During 6809 interval $\overline{E} \cdot \overline{Q}$, are address or data bits valid?
28. Which bits are valid during $E \cdot \overline{Q}$?
29. Which 6809E pin advises a DMA device of a coming opportunity to take over the bus for a cycle?
30. How many machine cycles would the nine-cycle test loop have taken if direct addressing had not been used in the loop?

19.7 MORE IMPROVEMENTS ON THE MEM GAME

This section presents a 6809/EPROM version of the MEM game that was first presented in Chapter 7. The program includes sound effects as well as lights. It was developed as a 6809 familiarization exercise by two student technicians, Richard Stone and Darin Gaynier. The hardware diagram is given in Figure 19.12, and the program listing follows (List 19.2).

```
                  1000 ************************************************************
                  1010 *                                                          *
                  1020 *  LIST 19-2. MEM GAME WITH SOUND FOR 6809, FIG 19-12      *
                  1030 *   WRITTEN AND TESTED BY RICH STONE & DARRIN GAYNIER      *
                  1040 *   7 JAN '85. REVIEWED & TESTED 7 AUG '87 BY METZGER.     *
                  1050 *                                                          *
                  1060 ************************************************************
                  1070          .OR $FE00
                  1080          .TA $4000
0016-             1090 LONG     .EQ $0016     2-BYTE BUFFER FOR FILE LENGTH.
5000-             1100 LATCH    .EQ $5000     4 LED OUTPUTS ON BITS 0 - 3.
7000-             1110 SPKR     .EQ $7000     SPEAKER OUTPUT ON BIT 0 OF LATCH.
3000-             1120 INPUT    .EQ $3000     4 SWITCHES ON BITS 0 - 3.
                  1130 *
FE00- 10 CE 07
FE03- FF          1140 START    LDS    #$07FF SET STACK TO TOP OF RAM.
FE04- 86 0F       1150          LDA    #$0F
FE06- B7 50 00    1160          STA    LATCH  LIGHT ALL LEDS
FE09- BD FF 09    1170          JSR    DELAY   FOR A SECOND;
FE0C- 7F 50 00    1180          CLR    LATCH   THEN DARKEN THEM
FE0F- BD FF 09    1190          JSR    DELAY   FOR A SECOND,
FE12- B7 50 00    1200          STA    LATCH   AND LIGHT THEM AGAIN
FE15- BD FF 09    1210          JSR    DELAY   FOR A SECOND;
FE18- 7F 50 00    1220          CLR    LATCH   THEN LEAVE THEM DARK.
FE1B- 86 50       1230          LDA    #$50   A SERIES OF TONES STARTS THE GAME.
FE1D- BD FE CB    1240          JSR    TGEN
FE20- 86 70       1250          LDA    #$70
FE22- BD FE CB    1260          JSR    TGEN
FE25- 86 60       1270          LDA    #$60
FE27- BD FE CB    1280          JSR    TGEN
FE2A- 86 40       1290          LDA    #$40
FE2C- BD FE CB    1300          JSR    TGEN
FE2F- 86 3F       1310          LDA    #$3F
FE31- B7 50 00    1320          STA    LATCH  LIGHT ALL LEDS
FE34- BD FE CB    1330          JSR    TGEN   (END OF BELLS AND WHISTLES)
FE37- CC 00 0F    1340          LDD    #15    SET FILE LENGTH TO 15 KEY PRESSES
FE3A- DD 16       1350          STD    LONG    (DEFAULT IF NO FIRQ RECEIVED)
                  1360 *
                  1370 * FIRST LOAD A FILE WITH UP TO 15 PUSHES OF THE KEYS.
```

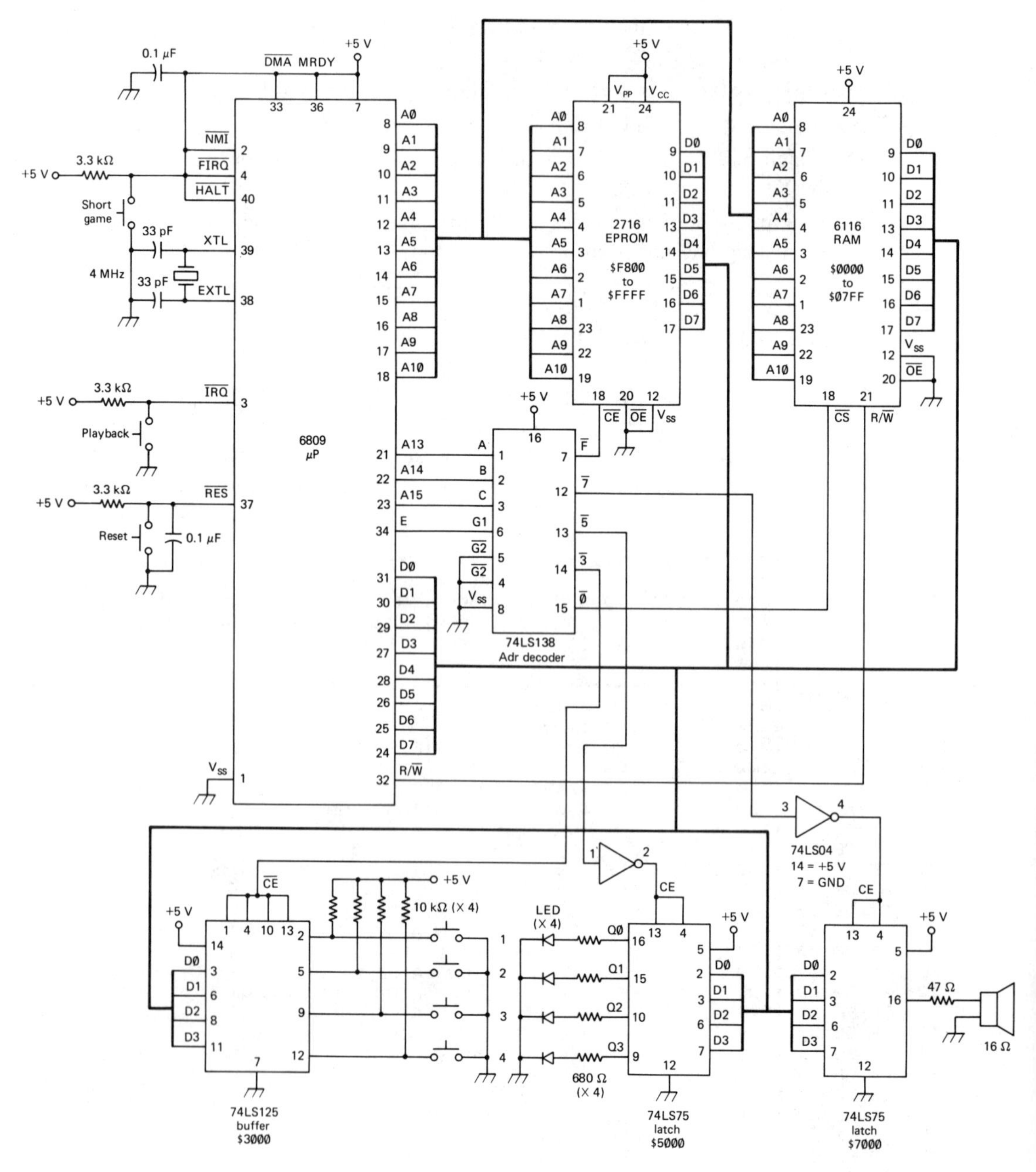

Figure 19.12 Hardware for 6809 version of the MEM game, first presented in Chapter 7 for the 6802.

```
FE3C- 10 8E 00
FE3F- 00       1380        LDY   #$0000 START FILE AT BOTTOM OF RAM.
FE40- 1C BF    1390 ENTER  ANDCC #$BF   ENABLE FIRQ INTERRUPT TO SHORTEN GAME.
FE42- 1A 40    1400        ORCC  #$40     DISABLE FIRQ.
FE44- 81 33    1410        CMPA  #$33   JUST BACK FROM FIRQ? ($33 IS CODE)
FE46- 27 19    1420        BEQ   FINISH  YES:GO TO PLAY TO REPEAT SERIES.
FE48- B6 30 00 1430        LDA   INPUT  LOOK FOR KEY.
FE4B- 43       1440        COMA           (PRESSED = 1 BIT)
FE4C- 84 0F    1450        ANDA  #$0F     (MASK OFF HIGH 4 BITS)
FE4E- 4D       1460        TSTA         IS ACCUM = 0? (NO KEY)
FE4F- 27 EF    1470        BEQ   ENTER   YES: KEEP LOOKING.
FE51- A7 A0    1480        STA   ,Y+     NO: SAVE KEY & POINT TO NEXT IN FILE.
FE53- B7 50 00 1490        STA   LATCH  LIGHT LED BY PRESSED KEY.
FE56- BD FF 09 1500        JSR   DELAY  WAIT A SEC BEFORE MAKING TONE.
FE59- BD FE CB 1510        JSR   TGEN   SUBRTN SENSES KEY NO. TO SET PITCH.
FE5C- 10 9C 16 1520        CMPY  LONG   ALL KEYS DONE?
FE5F- 26 DF    1530        BNE   ENTER   NO: GET ANOTHER KEY.
               1540 *
FEE5- 86 58    2140 TONE3  LDA   #$58
FEE7- 20 02    2150        BRA   SOUND
FEE9- 86 48    2160 TONE4  LDA   #$48
FEEB- 8E 01 10 2170 SOUND  LDX   #$0110 SET COUNTER FOR NUMBER OF CYCLES.
FEEE- 1F 89    2180 BEEP   TFR   A,B    MAKE A COPY OF REQUIRED DELAY.
FEF0- F7 70 00 2190        STB   SPKR   SET SPEAKER STATE FOR 1ST HALF CYCLE.
FEF3- 5A       2200 FIRST  DECB
FEF4- 12       2210        NOP          DELAY LOOP HOLDS .5 TO 2 MS OR SO.
FEF5- 26 FC    2220        BNE   FIRST
FEF7- 1F 89    2230        TFR   A,B    GET DELAY COUNT BACK IN ACCUM B.
FEF9- 5C       2240        INCB         CHANGE STATE OF BIT 0.
FEFA- F7 70 00 2250        STB   SPKR   OUTPUT NEW STATE.
FEFD- 5A       2260 SECND  DECB
FEFE- 12       2270        NOP          AGAIN FOR .5 - 2 MS OR SO.
FEFF- 26 FC    2280        BNE   SECND   FOR SECOND HALF CYCLE.
FF01- 30 1F    2290        LEAX  -1,X   COUNT DOWN NUMBER OF CYCLES.
FF03- 26 E9    2300        BNE   BEEP    MAKE MORE CYCLES UNTIL X = 0.
FF05- 7F 50 00 2310        CLR   LATCH  DARKEN ALL LEDS.
FF08- 39       2320        RTS          BACK TO MAIN PROGRAM.
               2330 *****************************************************
               2340 * DELAY SUBROUTINE - ABOUT 1 SECOND.
FF09- 8E 00 00 2350 DELAY  LDX   #0     FULL COUNT: 65536 LOOPS.
FF0C- 30 1F    2360 CDOWN  LEAX  -1,X
FF0E- 26 FC    2370        BNE   CDOWN
FF10- 39       2380        RTS
               2390 *****************************************************
               2400 * MACHINE PLAYBACK INTERRUPT.  LETS YOU SEE SEQUENCE.
FF11- 10 8E 00
FF14- 00       2410 REPLAY LDY   #0     START OF FILE.
FF15- A6 A0    2420 NEXT   LDA   ,Y+    GET DATA AND POINT Y TO NEXT FILE.
FF17- B7 50 00 2430        STA   LATCH  DISPLAY.
FF1A- 27 0F    2440        BEQ   BACK   0 MEANS END OF SHORTENED FILE.
FF1C- BD FF 09 2450        JSR   DELAY
FF1F- BD FE CB 2460        JSR   TGEN   MAKE CHARACTERISTIC TONE.
FF22- BD FF 09 2470        JSR   DELAY
FF25- 10 8C 00
FF28- 0F       2480        CMPY  #$F    END OF FILE (Y < $F) ?
FF29- 25 EA    2490        BLO   NEXT    NO: SHOW NEXT KEY.
FF2B- 3B       2500 BACK   RTI           YES: BACK TO MAIN PROGRAM.
               2510 ****************************************************
               2520 * A FIRQ INTERRUPT WHICH STOPS LOADING AND FILLS $00
               2530 *  BYTES TO END OF FILE. STOPS GAME SHORT OF 15 KEYS.
               2540 *
FF2C- 10 9F 16 2550 RESTAR STY   LONG   RESET FILE LENGTH TO PRESENT VALUE.
FF2F- 6F A0    2560 EMPTY  CLR   ,Y+    CLEAR FILE FROM PRESENT TO END.
FF31- 10 8C 00
FF34- 10       2570        CMPY  #$10
```

```
FF35- 25 F8    2580        BLO   EMPTY
FF37- 86 33    2590        LDA   #$33   CODE BYTE $22 IDENTIFIES FIRQ DONE.
FF39- BD FE CB 2600        JSR   TGEN   MAKE TONE & WAIT TO AVOID 2ND FIRQ.
FF3C- 3B       2610        RTI
               2620 ***************************************************
               2630        .OR   $FFF6
               2640        .TA   $41F6
FFF6- FF 2C    2650        .DA   RESTAR   FIRQ
FFF8- FF 11    2660        .DA   REPLAY   IRQ
FFFA- FF 11    2670        .DA   REPLAY   SWI
FFFC- FF 11    2680        .DA   REPLAY   NMI
FFFE- FE 00    2690        .DA   START    RESTART
               2700        .EN
FE61- 86 0F    1550 FINISH LDA   #$0F    YES: LIGHT ALL LEDS AND
FE63- B7 50 00 1560        STA   LATCH    MAKE A TONE TO SHOW THAT
FE66- 86 30    1570        LDA   #$30     IT'S TIME FOR PLAYER TO
FE68- BD FE CB 1580        JSR   TGEN     REPEAT KEY SERIES.
FE6B- 1C EF    1590 PLAY   ANDCC #$EF   CLEAR I MASK TO ALLOW REPLAY.
FE6D- 10 8E 00
FE70- 00       1600        LDY   #$0    POINT TO START OF FILE AGAIN.
FE71- B6 30 00 1610 FILL   LDA   INPUT  READ KEYS.
FE74- 43       1620        COMA          PRESSED = 1-BIT IN LO NIBBLE.
FE75- 84 0F    1630        ANDA  #$0F
FE77- 4D       1640        TSTA         ANY KEY PRESSED?
FE78- 27 F7    1650        BEQ   FILL    NO: KEEP LOOKING.
FE7A- B7 50 00 1660        STA   LATCH   YES: DISPLAY KEY.
FE7D- BD FF 09 1670        JSR   DELAY    FOR A SEC.
FE80- A1 A0    1680        CMPA  ,Y+    IS KEY SAME AS ENTERED DURING
FE82- 27 15    1690        BEQ   GOOD    ENTER ROUTINE?
FE84- 86 E0    1700 LOSE   LDA   #$E0    NO: SET "BRONX CHEER" TONE.
FE86- BD FE CB 1710        JSR   TGEN
FE89- 86 FF    1720        LDA   #$FF    REALLY RUB IT IN.
FE8B- BD FE CB 1730        JSR   TGEN
FE8E- 31 3F    1740        LEAY  -1,Y   IF Y = 1 YOU GOT 0 RIGHT; SO DEC Y.
FE90- 10 BF 4F
FE93- FF       1750        STY   LATCH-1 DISPLAY NUMBER OF CORRECT KEYS.
FE94- BD FF 09 1760        JSR   DELAY
FE97- 20 D2    1770        BRA   PLAY   YOU GET TO TRY AGAIN FROM START.
               1780 *
FE99- BD FE CB 1790 GOOD   JSR   TGEN   GOOD KEY: MAKE ITS TONE
FE9C- 10 9C 16 1800        CMPY  LONG   DID YOU GET THEM ALL RIGHT?
FE9F- 25 D0    1810        BLO   FILL    NO: GET NEXT KEY.
FEA1- 86 4F    1820 WIN    LDA   #$4F
FEA3- B7 50 00 1830        STA   LATCH  MAKE FIREWORKS FOR WINNER.
FEA6- BD FE CB 1840        JSR   TGEN
FEA9- 86 40    1850        LDA   #$40   HIGH NIBBLE SETS TONE
FEAB- B7 50 00 1860        STA   LATCH   (WITH A LITTLE HELP FROM
FEAE- BD FE CB 1870        JSR   TGEN      FROM THE LOW NIBBLE)
FEB1- 86 35    1880        LDA   #$35
FEB3- B7 50 00 1890        STA   LATCH
FEB6- BD FE CB 1900        JSR   TGEN   LOW NIBBLE LIGHTS LEDS
FEB9- 86 2A    1910        LDA   #$2A    WHERE 1 BITS ARE.
FEBB- B7 50 00 1920        STA   LATCH
FEBE- BD FE CB 1930        JSR   TGEN
FEC1- 86 2F    1940        LDA   #$2F
FEC3- B7 50 00 1950        STA   LATCH
FEC6- BD FE CB 1960        JSR   TGEN
FEC9- 20 A0    1970        BRA   PLAY   LET SOMEONE ELSE TRY.
               1980 ***********************************************
               1990 * TONE GENERATOR SUBROUTINE
               2000 *
FECB- 81 01    2010 TGEN   CMPA  #1     IS KEY INPUT A "1"?
FECD- 27 0E    2020        BEQ   TONE1
FECF- 81 02    2030        CMPA  #2     A "2"?
FED1- 27 0E    2040        BEQ   TONE2
```

```
FED3- 81 04     2050         CMPA  #4       A "3"?
FED5- 27 0E     2060         BEQ   TONE3
FED7- 81 08     2070         CMPA  #8       IS IT A "4" KEY THEN?
FED9- 27 0E     2080         BEQ   TONE4
FEDB- 26 0E     2090         BNE   SOUND    IF NONE, MAKE TONE SET BY "A".
FEDD- 86 78     2100 TONE1   LDA   #$78
FEDF- 20 0A     2110         BRA   SOUND
FEE1- 86 68     2120 TONE2   LDA   #$68
FEE3- 20 06     2130         BRA   SOUND
```

Answers to Chapter 19 Review Questions.

1. They are identical. **2.** S is used to stack machine registers during interrupts and subroutines. U is a data pointer available for use exclusively by the programmer. **3.** D = $AB34 **4.** It adds 256 to X, pointing it to the same word on the next page. **5.** COMA; COMB. **6.** It loads DP with $C2 without destroying the contents of A. **7.** The 6809 address lines can assume a high-impedance state when the processor is not active. **8.** It asserts address $FFFF.
9. None **10.** FIRQ stacks only 3 bytes; IRQ stacks 12. **11.** ORCC #$10
12. ANDCC #$3E **13.** LEAY −4,Y **14.** LDS #$01FF
15. PULS B **16.** $FFA7 **17.** NEG forms the 2's complement by complementing all bits and then adding 1 to the *low-order bit only.* **18.** ANDB #$55; ANDA #$55 **19.** D = $566D, Z = 0, C = 0 **20.** Shift $AB38 right, yielding $559C. Divide this by $95, yielding $93. Shift $93 left, yielding $126. **21.** You can't do it directly, since A is 8 bits and X is 16 bits.
22. $1F, $52 **23.** PSHU A,B,X,CC **24.** $36, $17 **25.** $6FDA
26. Yes. Use operand $E000, which is −$2000. **27.** Neither is valid.
28. Generally both address and data. **29.** AVMA **30.** 11 cycles

CHAPTER 19 QUESTIONS AND PROBLEMS

Digits after the decimal point refer to section numbers in Chapter 19.

Basic Level

1.1 How many bytes of programmable registers are there in the 6809? (Exclude the program counter.)

2.1 What kind of data does the 6809's DP register normally hold?

3.2 A 6809 uses a 4-MHz crystal. What are the frequencies at the 6809 pins *E* and *Q*?

4.2 How should the MRDY, $\overline{\text{DMA}}$, and $\overline{\text{FIRQ}}$ inputs of the 6809 be connected in a minimum-size system?

5.3 Write a 6809 source-code instruction that will clear the N, Z, V, and C flags without changing E, F, H, or I.

6.3 Write a 6809 source-code instruction that will increment the X register by two.

7.3 What is the difference between the 6809 instructions PULS B and PULU B?

8.4 How can the D register of the 6809 be incremented?

9.4 Give all the options for placement of the multiplier, multiplicand, and product in the 6809's MUL instruction.

10.5 Write the 6809 source code to push registers D, X, and the status register on the user stack.

11.5 What is the complete object code for a Long Branch Always to address $D135? The first byte of the LBRA instruction resides at address $F4B9.

12.6 What is the name of the circuit comprised of the two flip-flops in Figure 19.9? (Review Chapter 0 if it is not immediately apparent.)

13.6 In Figure 19.8, what is the minimum guaranteed time during which write data from the CPU will be available to be read by a RAM?

Advanced Level

14.1 What is the D register in the 6809, and what can it be used for?

15.1 Explain the difference in operation of the 6809 stack pointer, compared to the 6800 stack pointer.

16.2 What is the relationship between the 6809's Q and E signals.

17.2 Explain how the 6809's BA and BS output signals are to be interpreted.

18.3 What is being loaded with the data from Y in the instruction LEAX 0,Y? If it is not an address, why is the instruction called Load Effective Address?

19.3 Write the 6809 source code for a routine to read a byte of data from BUFR to A, read an offset value to B, and store the data to a file at an address X + B.

20.4 The X register contains $A045 and B contains $F0. Compare the results obtained from the two instructions

ABX versus LEAX B,X

21.4 Make a set of binary displays to show how the DIVIDE routine of Section 19.4 would handle $1B4 divided by $25.

22.5 What is the difference between the instructions

TFR A,DP and EXG A,DP?

23.5 List three types of 6809 instructions that use a postbyte, and explain what information the postbyte contains in each case.

24.6 List and explain two major differences between the 6809 and the 6809E.

25.6 Draw a logic diagram using E OR Q logic to select a 2764 EPROM uniquely at address range $C000 through $DFFF. On a 1-MHz clock, how much time is available for the EPROM to respond?

26.7 Calculate the slowest acceptable EPROM response time for the system of Figure 19.12.

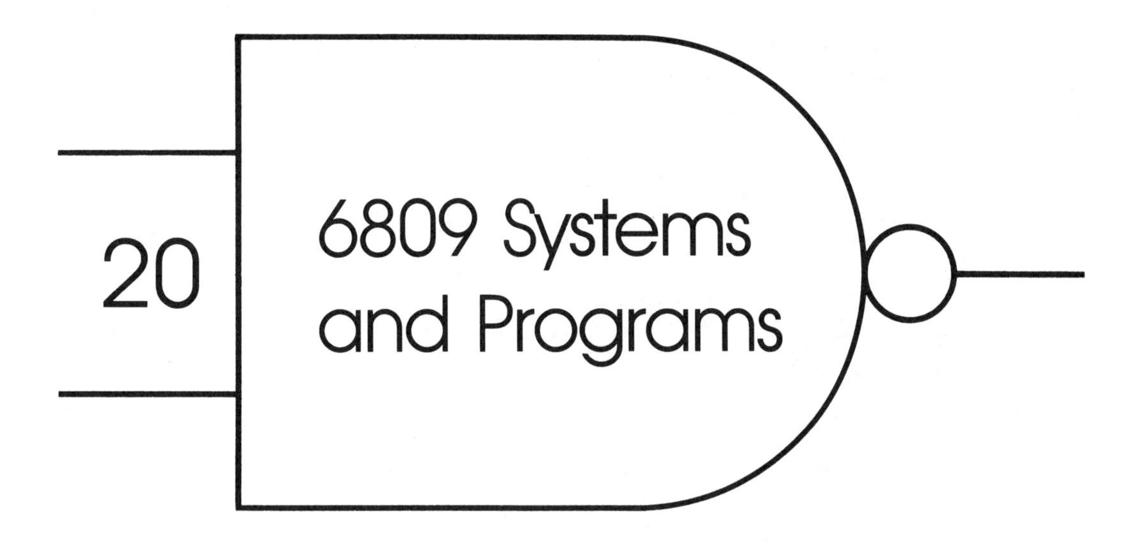

20.1 ADDRESSING MODES OF THE 6809

Section Overview

Most of the 6809 instruction modes and assembly-language forms are similar enough to those of the 6802 that they can be picked up easily. The numerous "indexed" addressing modes, however, are likely to be confusing. The assembly forms for three of these modes are given below as an orientation, using as an example the *Store A* instruction:

Extended Indirect	**STA ($1234)**	**Effective address is contained in locations $1234:1235.**
Program Counter Relative	**STA 100,PCR**	**Effective address is 100 bytes after next-instruction address.**
P C Relative, Indirect	**STA (100,PCR)**	**Effective address is contained at locations 100 bytes and 101 bytes after next-instruction address.**

The 6809 was born in the middle of an advertising war in which the contenders were all trying to claim, "My processor has more addressing modes than your processor." This is at least part of the reason why the 6809 has, by one way of counting, 37 addressing modes. The amount of fear and confusion that this causes for newcomers is quite unnecessary, for the following reasons:

1. Most of these addressing modes result from dividing instructions according to their machine-language structure, when they are perfectly identical from the point of view of the programmer or user. For the programmer, the number of different modes should be counted as 10.
2. You can start using the 6809 with just the 6 addressing modes of the 6802. You will find that the machine's new registers and instructions make it a far superior machine all by themselves. Then, when you begin to feel comfortable with the basic structure of the 6809, you can begin to learn and employ the new addressing modes one by one.

We will now review the 6809 addressing modes, starting with the original 6 and introducing 3 new ones. Section 20.2 considers the remaining addressing modes of the 6809.

The inherent mode may be considered to include those instructions that do not reference memory, and instructions that reference stack memory only, such as PSH, PUL, JSR, and RTI. Here are some examples:

- MUL RTS NOP SEX — One-byte inherent mode.
- INCA NEGB TSTA — Sometimes called *accumulator* mode.
- TFR CC,A EXG DP,B — Also called *register* mode. Uses post byte.
- PSHS A,DP,CC PULU Y — Stack operations; use post byte.

The immediate mode has the operand data in the 1 or 2 program bytes following the instruction op-code (1 byte for 8-bit registers such as B and 2 bytes for 16-bit registers, such as S.) The op-codes for LDS and LDY are 2 bytes long, as are the CMP op codes for D, S, U, and Y. The first byte in these cases is the *prebyte* ($10 or $11), made necessary by the simple fact that there are only 256 op-codes possible with 1 byte. The X register is sometimes preferred to the Y because, for loads, stores, and compares it uses no prebyte, and is 1 byte and 1 cycle shorter. Here are some samples of immediate-mode assembly forms.

```
ANDCC #$FE    LDD #5280    LDY #FILE
```

The relative mode is used chiefly for conditional branch instructions, although the Branch Always instruction provides a relative-mode Jump. They are the same as their 6802 counterparts; the first byte is the op-code, and the second byte gives the 2's complement displacement from the next-instruction address if the branch is taken.

The long-relative instructions require 4 bytes and 5 cycles (if the branch is not taken) or 6 cycles (if the branch is taken). They use the same op-codes as the short branches, with the addition of prebyte $10. Of course, they require a 2-byte operand to span the range −32 768 to +32 767. Exceptions are long-relatives LBRA, which does not require a prebyte (3 bytes, 5 cycles) and LBSR, which requires 3 bytes and 9 cycles.

Direct-page addressing may be chosen by the assembler automatically if the operand is within the page currently specified by DP, or this mode may be specified by the programmer. It is the programmer's responsibility to advise the assembler of where the DP register should be pointed with a directive such as

```
SETDP   $F0
```

The program must agree with this directive:

```
LDA   #$F0
TFR   A, DP
```

The opportunity for confusion as program segments are patched and modified is so great that it is often best to point the DP register at the most-used RAM buffer or table page and leave it there. If it becomes necessary to squeeze every last byte and every last microsecond out of a program, it may be less trouble to buy a 68B09 (2-MHz version) and some faster ROMs than to try to save the situation by jumping the DP register around.

Extended addressing is the "normal" addressing mode of the 6809 as well as the 6802 and is widely used. The only difference is that operations involving S or Y (and compares involving D or U) use a prebyte, making the instruction 4 bytes long instead of 3.

Indexed addressing for the 6809 is not a single addressing mode but a whole subset of modes, some of which are not indexing modes at all. Each op-code listed in the *indexed* column of the *6809 Programmer's Reference Chart* must be followed by a postbyte, which defines the particular addressing mode under the *indexed mode* umbrella, as illustrated in Figure 20.1. Motorola chose this expansion of the 6800's indexed-mode instructions with postbytes as a way of enlarging the 6809's instruction set while maintaining a high degree of compatibility with the earlier 6800/6802/6801 processors. Unfortunately, they had figuratively painted themselves into a corner when it came to naming the new modes. Here is what we mean.

About half of the new modes under the *indexed* umbrella are indirect-addressing modes. This means that the address specified does not contain the data; rather, this address and the subsequent address contain 2 bytes, which form the effective address (EA) where the operand data is found. The intermediate address specified by the instruction operand is called an indirect address (IA). These instruction modes are

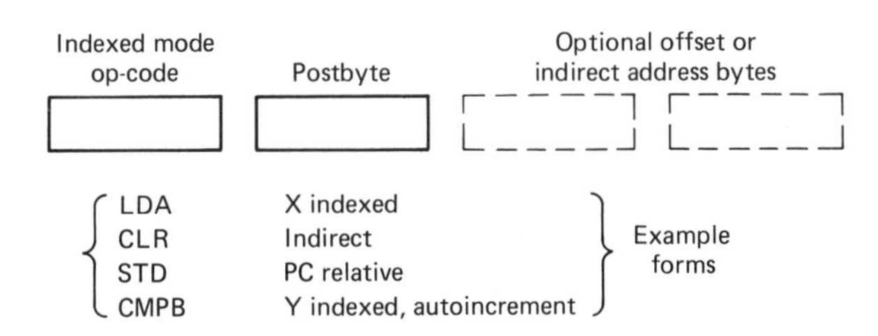

Figure 20.1 The postbyte added to an indexed-mode op-code specifies one of several quite-different addressing modes.

called, quite properly, *indirect*. The remaining new modes should logically be called *direct*, but that name was already taken, so they were dubbed in doubletalk—nonindirect. Of course, the *direct* mode should have been called *zero-page* in the 6800 and perhaps *one-page* in the 6809, but nobody foresaw in 1974 what verbal contortions the name *direct* was to cause.

To confuse the nomenclature further, three of the new addressing modes (generated by postbyte expansion of the original *indexed* instructions) are *called* indexed modes, but they are *not* indexed modes, since they do not involve X, Y, S, or U. We will explain these three modes now, saving the true indexed modes for Section 20.2.

Indirect addressing, extended mode is specified to the 6809 by an indexed op-code with postbyte $9F. The assembly-language instruction forms are such as these.

Object Code	Source Code	Comments
A6 9F 00 5A	LDA ($5A)	Indirect Address (IA) is at $005A:$005B.
6F 9F 00 4D	CLR (POINT)	Clear address pointed to by bytes at POINT:POINT+1, where POINT = $004D.

The parentheses are the assembler cues for *indirect address*. Figure 20.2 shows how a table address can be set up in the indirect RAM address, allowing access to any byte in the table from a ROM-based program. This is possible with the 6800 (See

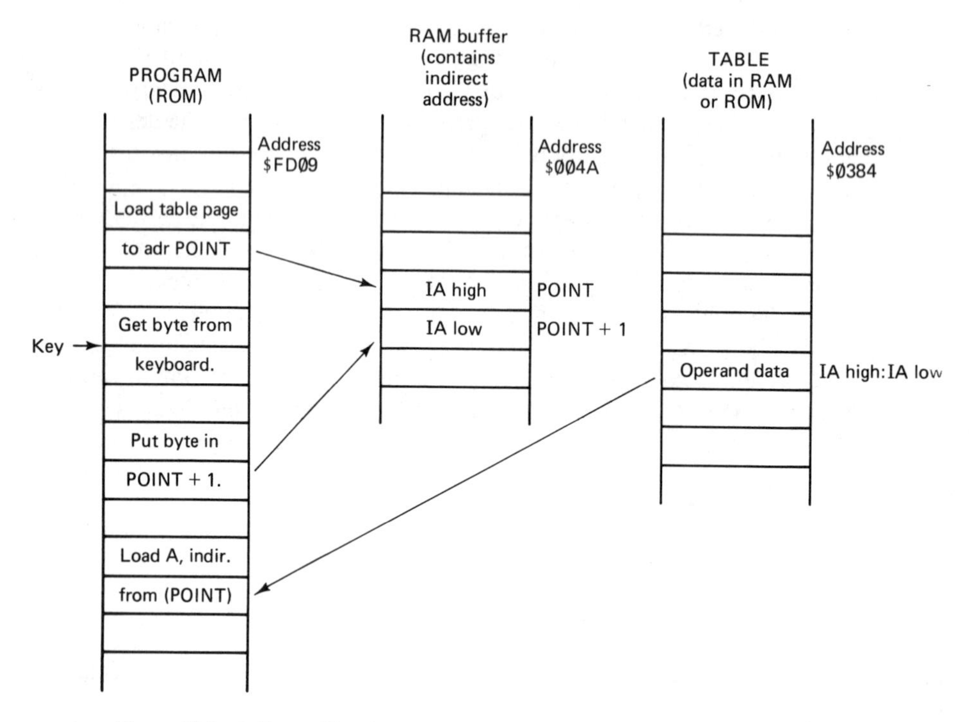

Figure 20.2 Indirect addressing gets the final effective address of the operand from a pair of RAM bytes identified by the instruction.

Section 7.6), but the 6809 does it more simply, without involving the index register. Indirect addressing is used less frequently than nonindirect addressing in 6809 programming.

Program-counter relative addressing simply extends the relative addressing mode to the entire set of memory-reference instructions. The 6802 limited this mode to conditional branches plus BRA (a relative *jump*) and BSR (a relative *jump to subroutine*.) The mode is implemented by adding the postbyte $8C to the *indexed* op-code byte and a third byte, which gives the 2's complement offset from the next-instruction address.

If an offset beyond the 8-bit range (−128 to +127) is required, a 16-bit-offset relative mode is available. The postbyte here is $8D, and *two* offset bytes follow. These 4-byte relative instructions have a range from −32 768 to +32 767.

The choice between 8-bit and 16-bit offset is made by the assembler, so the two modes appear identical to the programmer. Here are some sample assembly-language forms. The letters PCR after the operand cue the assembler to use the indexed relative mode.

Object Code	Source Code	Comments
E6 8C 20	LDB CONST,PCR	CONST is an address label 32 bytes from next-instruction address.
6D 8D 01 00	TST #$100,PCR	Test byte is 256 addresses from next-instruction address.

Program-counter relative addressing facilitates the writing of position-independent code. For example, a program may be loaded from a disk into the target-system RAM wherever there is room. The program may contain a data table that is in a fixed position relative to the program, but the absolute address of the table will vary each time the program is loaded. Relative addressing enables the table to be accessed regardless of its position in the target system. In dedicated ROM-based systems it finds little application.

There is an *indirect* program-counter-relative addressing mode in which two intermediate-address bytes are specified at a fixed distance from the current program-counter address. These bytes concatenate to form the effective address. The assembler form is

```
LDA (POINT,PCR)
```

This mode is not one that a novice 6809 programmer would expect to use frequently, so we will give it no more attention here.

20.2 INDEXED ADDRESSING MODES

Section Overview

The 6809 indexed-mode instructions can use X, Y, S, or U as the pointer register. As a convenience for accessing files, the pointer registers

can be automatically postincremented or predecremented (by 1 or by 2) each time they are used. Example forms are

```
LDA ,X+      Single postincrement
STD ,--Y     Double predecrement
```

The address in the pointer register can also be offset by a 2's complement value, either listed as a constant or held in accumulator A, B, or D. Example assembly statements are:

```
CMPB 3,U     Operand is 3 bytes beyond address in U.
COM  A,X     Operand is at address A + X.
```

The autoincrement and offset modes cannot be combined in one instruction. Indirect modes are available in each case.

Autoincrement/decrement: Data tables are often accessed sequentially by loading the data register and incrementing the pointer register. In 6800 code:

```
LDAA 0,X
INX
```

The 6809 allows these operations to be combined:

```
LDA ,X+
```

The plus sign indicates that X is to be incremented after the load. No offset is allowed in this mode. Two increments can be made, as required for tables containing 16-bit data in two concatenated bytes:

```
LDD ,Y++
```

In working backward through a table, the pointer register is decremented *before* the operation to maintain compatibility with the predecrementing character of the *pull from stack* operations:

```
LSR ,-X
```

The equivalent 6800 instructions are

```
DEX
LSR 0,X
```

Double decrement is also possible:

```
STY ,--U
```

The autoincrement/decrement modes are very widely used in 6809 programming. Figure 20.3 shows the postbytes used to implement the various versions of this mode.

Offsets to the pointer addresses can be specified as either constant values (from −32 768 to +32 767) or as the 2's complement value in one of the accumulators A, B, or D. The assembler forms are the same for all versions of this mode. Here are some example statements:

LDA, X	No offset. X points to operand address.
DEC −2,Y	Decrement data at address Y − 2.
CLR $400,S	Clear data at address 4 pages beyond system-stack-pointer address.
STB A,X	Store B at address X + A. A is 2's complement; range −128 to +127.
TST D,U	Set flags as if AND #Ø with data at address U + D. D is 2's complement.

These modes use the basic *indexed* op-code with a postbyte as specified in Figure 20.4.

The accumulator-offset modes are particularly useful for accessing data in two-dimensional arrays. A hypothetical example, illustrated in Figure 20.5, has ignition-coil dwell-angle data stored in a table as a function of throttle position (64 increments)

POSTBYTE	X	Y	U	S
Auto increment by 1	8Ø	AØ	CØ	EØ
Auto increment by 2	81	A1	C1	E1
Auto decrement by 1	82	A2	C2	E2
Auto decrement by 2	83	A3	C3	E3

Figure 20.3 The indexed-mode postbyte can specify any one of four 16-bit registers as the pointer and can cause that register to be automatically incremented or decremented to point to the next byte (or next double byte) in the table.

Postbytes for indexed with offset

	X	Y	U	S	
No offset	84	A4	C4	E4	
Constant offset, 0 to 15	Øp	2p	4p	6p	*
Constant offset, −1 to −16	1n	3n	5n	7n	*
Constant offset, −128 to +127	88	A8	C8	E8	+1 byte
Constant offset, −32768 to +32767	89	A9	C9	E9	+2 bytes
A offset, 2's complement	86	A6	C6	E6	
B offset, 2's complement	85	A5	C5	E5	
D offset, 2's complement	8B	AB	CB	EB	

*p = Ø to F for 0 to 15 offset
n = F to Ø for −1 to −16 offset

Figure 20.4 The indexed-mode postbyte can specify a fixed offset or an offset contained in an accumulator to be added to the normal address in the pointer register.

and engine temperature (16 positions). The following routine picks up the listed dwell after reading the throttle position and temperature.

```
LDX TABSTR     Point X to start of table.
LDB THROT      Get throttle position in bits Ø to 5.
LDA TEMP       Get temp in bits Ø to 3.
LSLB           Take out two blank
LSLB             bits B6 and B7.
LSRA           Shift D
RORB             right.
LSRA             twice, to fill
RORB             bits DØ–D9.
ANDA #3          Clear D15 to D1Ø.
LDAA D,X       Pick up byte D steps into table.
```

The offset-indexed modes are used relatively often in 6809 programming.

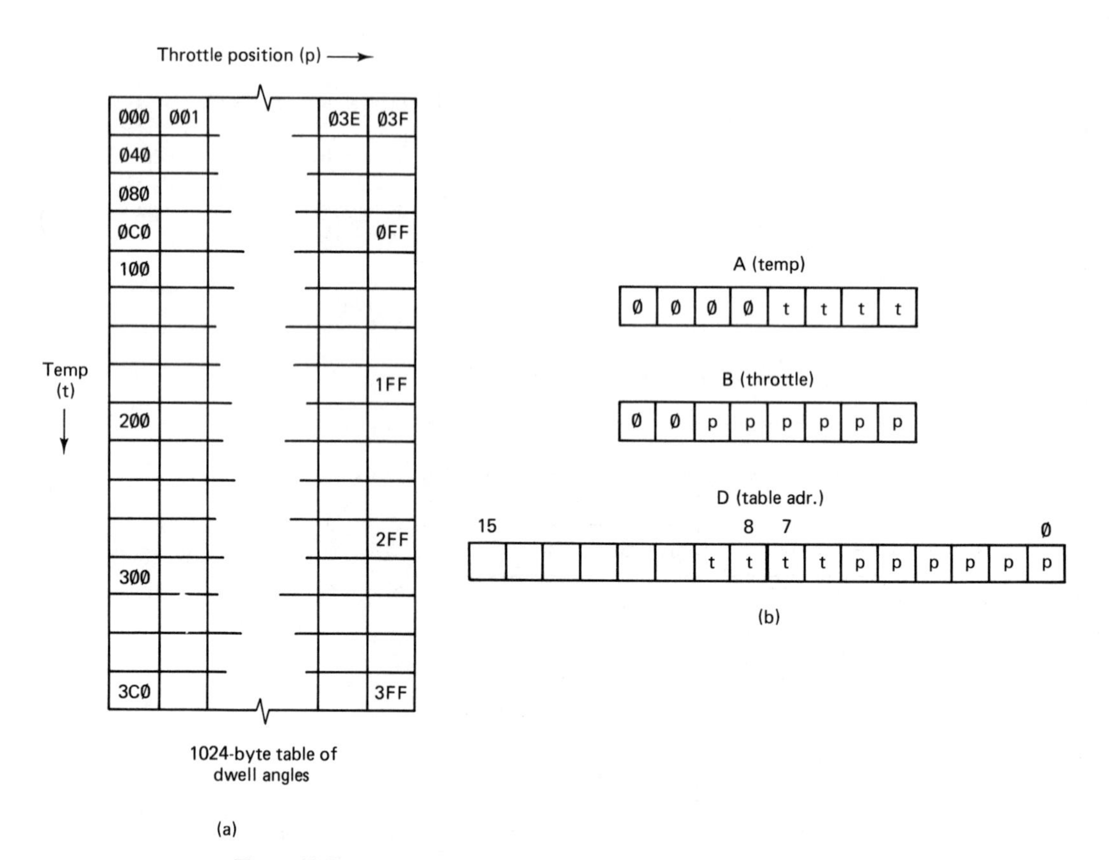

Figure 20.5 (a) A two-dimensional data array giving optimum dwell angle for all combinations of throttle position and temperature. (b) Combining the two input data words to produce a single table-address word.

There are *indirect addressing* versions of the autoincrement/decrement modes and of the offset-indexed modes, but they are less frequently used. We will not go into them in this text.

REVIEW OF SECTIONS 20.1 and 20.2

Answers appear at the end of Chapter 20.

1. Name four 6809 inherent-mode instructions that require a postbyte.
2. Fast search of a file of data is required to find any bytes greater than $B0. Which pointer register should be used to access the file?
3. Here is the object code for a Long Branch Always instruction. What is the address of the next instruction to be executed?

```
7B59   16 E3 92
```

4. How does the 6809 assembler program know the current contents of the direct-page register?
5. How would you instruct the 6809 to go to addresses $4132 : 4133 to find the 2 bytes of an address, which address it then fills with the data from accumulator B?
6. What does the following instruction do?

```
F580        EC 8D 03 2A     LDD FILE, PCR
```

7. Where does the data loaded to A come from in the sequence

```
LDX #$ABCD
LDA ,X+
```

8. Where are the contents of register D stored in the sequence

```
LDX #$ABCD
STD ,--X
```

9. How many bytes does this instruction take?

```
INC  -31,U
```

10. If A contains $F0 and X contains $12A0, what address is accessed by the instruction

```
ANDB A,X
```

20.3 INTERRUPT FEATURES OF THE 6809

Section Overview

The 6809 has several new interrupt and DMA features.

- **An FIRQ (fast-interrupt) input stacks only PC and CC. It can be masked by the F bit of the CCR. The CCR's E (entire) bit goes up if an IRQ, NMI, or SWI caused the *entire* set of machine registers to be stacked. E goes to 0 if the last interrupt was an FIRQ, which does not stack the entire register set.**
- **There are three software interrupts: SWI, SWI2, and SWI3.**
- **A SYNC instruction causes the 6809 to enter a *wait* state until a hardware interrupt input is received. If the input is an IRQ or FIRQ and the I or F bit, respectively, is *set*, the processor will resume execution of the main program.**
- **Priority Interrupt Controller chips, such as the 6828, automatically generate vector addresses for a number of interrupt sources.**

The $\overline{\text{FIRQ}}$ pin of the 6809 is a negative-level-sensitive *Fast-Interrupt-Request* input. This interrupt is acknowledged only if the F flag (bit 6 of the CCR) is cleared. The F flag is *set* (as is the I flag) by a RESET, or an acknowledgement of FIRQ, NMI, or SWI. It is not affected by IRQ or the new software-interrupt instructions SWI2 and SWI3. Of course, F can be set by ORCC #$40 and cleared by ANDCC #0BF. The FIRQ is fast because it stacks only three bytes (PCLO, PCHI, and CC) compared to the 12 bytes stacked by all other hardware and software interrupts. An acknowledgment of FIRQ causes the processor to jump to an interrupt routine whose address is stored at FIRQ vector locations $FFF6 and FFF7.

The E flag (bit 7 of the CCR) is *set* when the last interrupt caused the *entire* register set of the 6809 to be stacked, which is the case for any interrupt except FIRQ. The *E* flag is cleared if the last interrupt was an FIRQ. The same return-from-interrupt instruction (RTI) is used for all 6809 interrupts. The RTI senses the condition of the E flag to determine whether to unstack three registers or 12.

Three software interrupts are available in the 6809. They are SWI, SWI2, and SWI3. The first is called SWI rather than SWI1 to maintain compatibility with the 6800/6802/6801. SWI2 and SWI3 use a prebyte plus the original op-code and are thus 2 bytes long instead of 1. Figure 20.6 shows the 6809 vector locations and the stacking order for interrupts.

Vectors		Bytes stacked
FFF0, 1	Reserved	—
FFF2, 3	SWI 3	12
FFF4, 5	SWI 2	12
FFF6, 7	FIRQ	3
FFF8, 9	IRQ	12
FFFA, B	SWI	12
FFFC, D	NMI	12
FFFE, F	RESET	—

(a)

CC	Lo Adr
A	
B	↑
DP	Push
X Hi	
X Lo	
Y Hi	Pull
Y Lo	↓
U/S Hi	
U/S Lo	
PC Hi	
PC Lo	Hi Adr

(b)

Figure 20.6 Vector addresses and interrupt stacking order for the 6809.

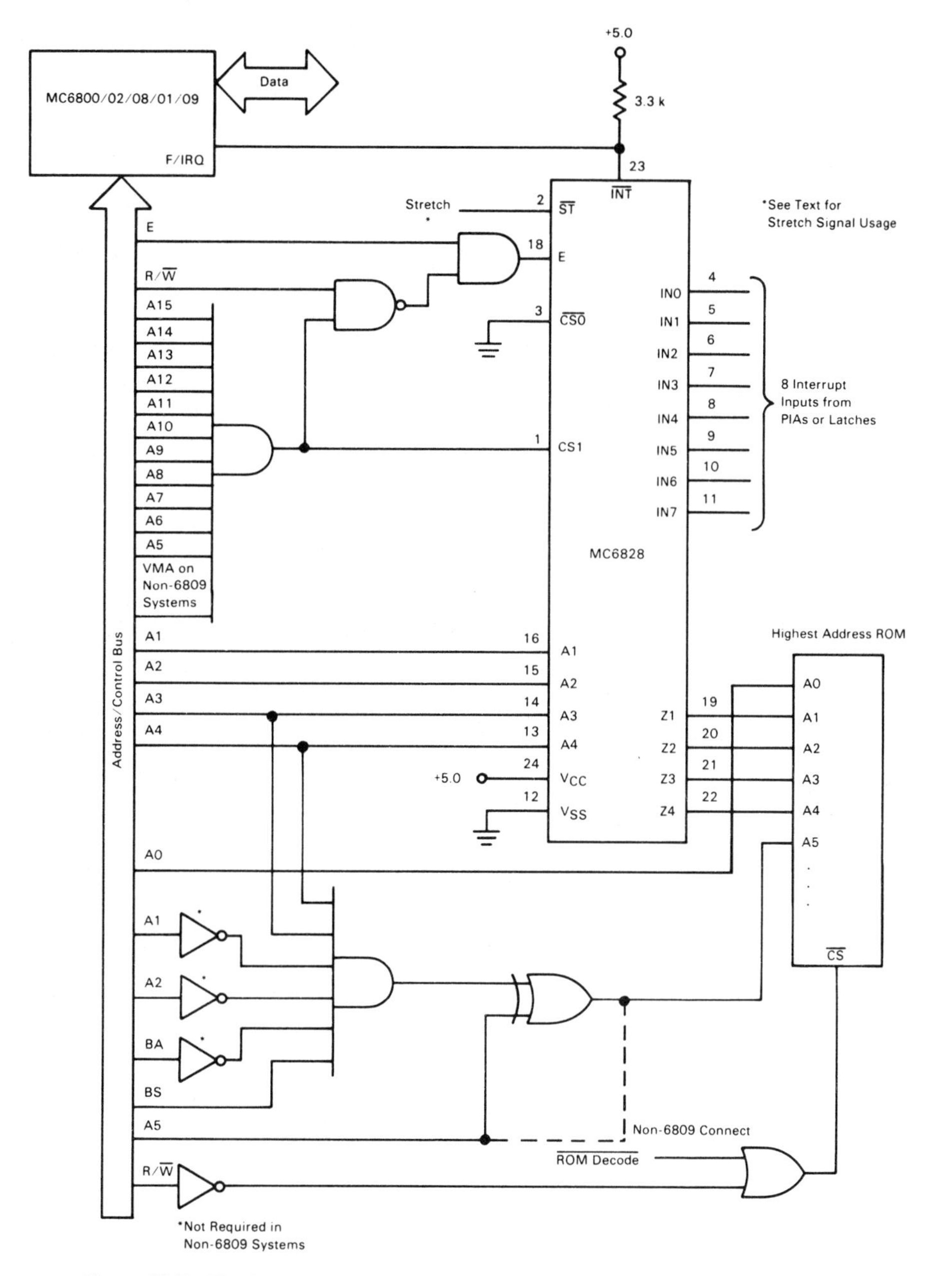

Figure 20.7 The 6828 Priority Interrupt Controller provides 8 interrupt inputs and assigns a separate vector service address to each one. Courtesy, Motorola Semiconductor Products, Inc.

The SYNC instruction causes the 6809 to enter a *wait* state, with the address and data buses and R/$\overline{W}$ in a floating state but the clocks running and BA and BS asserted *high* and *low*, respectively. The processor remains in this state until an interrupt is received on one of the three hardware-interrupt inputs. If the pertinent interrupt mask bit was cleared or if the interrupt is an NMI, the processor will leave the *wait* state, process the interrupt, and then continue with the main program. If an IRQ or FIRQ is received and the corresponding mask bit is set, the 6809 will not process the interrupt but will leave the *wait* state and resume processing the main program. SYNC is a 1-byte inherent instruction.

CWAI (clear and wait) is a 2-byte immediate-mode instruction. It ANDs the CCR with the immediate operand, allowing any bit to be cleared. It then stacks all the machine registers (except S) on the system stack and halts until an interrupt is received. When the interrupt arrives, it can be serviced more quickly, since the stacking is already done. CWAI does not tristate the 6809 buses, so it cannot be used to permit DMA or second-processor memory access. It can, however, be useful in debugging to permit static readout of the address bus and examination of the machine registers in the RAM stack.

A priority interrupt controller: As system complexity increases, the number of devices that may request service from the processor by interrupting it is likely to grow. Since the 6809 has only three interrupt lines, some scheme will have to be devised to determine which of several devices pulled the $\overline{IRQ}$ input *low*. A routine may be written to poll (read) all possible interrupting devices to identify the active one, but this is troublesome and it delays the processor's response to the devices near the end of the polling routine.

The Motorola 6828 is a *priority interrupt controller* (PIC) which

- Provides 8 interrupt inputs for one interrupt output to the processor.
- Modifies the processor's IRQ-vector fetch by expanding it to 8 vector addresses—a unique address for each of the 8 inputs.
- Provides programmable masking by priority level, from *any of the 8 accepted* to *none of the 8 accepted*.

Figure 20.7 shows the PIC connection to a 6809 system. The ROM is selected at certain *mask select* addresses if R/$\overline{W}$ is *high*, but the PIC priority level is programmed by accessing these addresses with R/$\overline{W}$ *low*.

20.4 A PROGRAMMER'S REFERENCE CHART FOR THE 6809

Section Overview

This section consists entirely of an abbreviated programmer's reference chart for the 6809. Hand assembly of 6809 programs is never attempted by serious programmers, but this chart should assist you in writing short 6809 test programs without an assembler, interpreting existing sections of

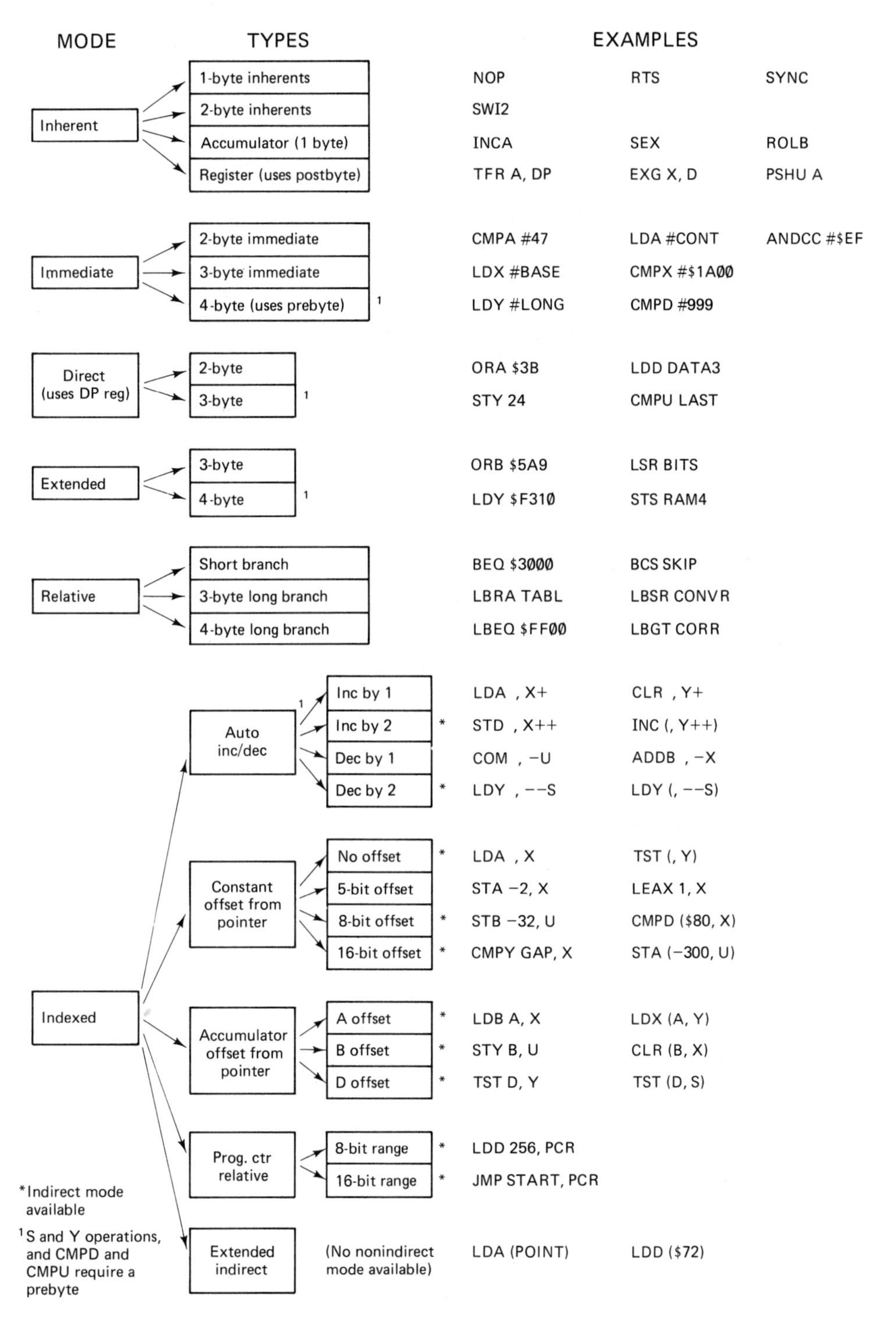

Figure 20.8 A summary of the 6809 addressing modes. Depending on how you count them there are 6, 10, or 37 different modes.

6809 code while troubleshooting and, most importantly, fitting the 6809 instruction modes and assembler forms into your brain.

The chart presented here is intended as a learning aid and is not complete. The machine cycle times have been left out, and the less frequently used *indirect-indexed* addressing modes have been omitted. The *Inherent* and *Immediate* columns have been merged, as have the *Direct* and *Relative* columns, in order to make the chart easier to read. Ease of understanding has also prompted the listing of postbytes in hex rather than binary, although this has in some cases prevented the listing of all instruction options. Finally, typical assembly-language statements are given with new instruction types to illustrate their proper forms. Figure 20.8 provides an overview of 6809 instruction modes. The Programmer's Reference Chart begins on page 503.

REVIEW OF SECTIONS 20.3 AND 20.4

Answers appear at the end of Chapter 20.

11. Which bytes are saved and where by an FIRQ?
12. What effect is produced by SWI that is not produced by SWI2 or SWI3?
13. Which of the software interrupts can be inhibited by the I-mask or F-mask bits?
14. A SYNC instruction is encountered and a WAIT state is entered. Under what conditions will the 6809 resume processing at the address where it left off?
15. How can the 6809 be *halted* with its address lines *not* floating?
16. The 6809's S register contains \$Ø1F3 when an IRQ is recognized. At what addresses the X register saved?
17. How many bytes are required for the instruction

```
LDY $12,X
```

18. If A contains \$2Ø and X contains \$41ØØ, what is the effective address containing the operand data in the instruction

```
LDB (A,X)
```

19. Give the assembly form and complete hex op-code for the 6809 instruction equivalent to DECY.
20. LOBLOK beginning at \$2ØØØ is to be compared to HIBLOK beginning at \$8ØØØ. Give the complete object code for the third instruction in the following partial routine.

```
LDX     #HIBLOK
LDD     #$AØØØ
CMPY    D,X
```

ABRIDGED 6809 PROGRAMMER'S REFERENCE

ASSEM' FORM	INHR IMMD	REL[1] DIR	EXT'D	INDX[2] (base)	OPERATION (M = Operand Data)	CCR BITS $\overline{C}$ = cleared; c = affected
ABX	3A				B + X → X (unsigned)	
ADCA	89	99	B9	A9	A + M + C → A	h n z v c
ADCB	C9	D9	F9	E9	B + M + C → B	h n z v c
ADDA	8B	9B	BB	AB	A + M → A	h n z v c
ADDB	CB	DB	FB	EB	B + M → A	h n z v c
ADDD	C3	D3	F3	E3	D + M : M + 1 → D	n z v c
ANDA	84	94	B4	A4	A AND M → A	n z $\overline{V}$
ANDB	C4	D4	F4	E4	B AND M → B	n z $\overline{V}$
ANDCC	1C				CC AND IMM BYTE → CC	Clear bits with 0s
ASLA	48				A }	3 n z v c
ASLB	58				B } C	3 n z v c
ASL		08	78	68	M }	3 n z v c
ASRA	47				A }	3 n z c
ASRB	57				B }	3 n z c
ASR		07	77	67	M }	3 n z c
BITA	85	95	B5	A5	A AND M } Sets CCR	n z $\overline{V}$
BITB	C5	D5	F5	E5	B AND M } bits only	n z $\overline{V}$
BCC[1]		24			C = 0? Carry clear? }	Borrow on subtr
BCS		25			C = 1? Carry set? }	sets C flags
BEQ		27			Z = 1? Zero result?	
BGE		2C			N XOR V = 0? Accum ≥ M? (signed)	
BGT		2E			Z OR (N XOR V) = 0? Accum > M? (signed)	
BHI		22			C AND Z = 0? Accum > M? (unsigned)	
BHS		24			Same as BCC. Accum ≥ M (unsigned)	
BLE		2F			Z OR (N XOR V) = 1? Accum ≤ M (signed)	
BLO		25			Same as BCS. Accum < M (unsigned)	
BLS		23			C + Z = 1? Accum ≤ M? (unsigned)	
BLT		2D			N XOR V = 1? Accum < M? (signed)	
BMI		2B			N = 1? Result negative? (signed)	
BNE		26			Z = 0? Accum ≠ 0?	
BPL		2A			N = 0? Result positive? (signed)	
BRA		20			No test. Branch Always.	
LBRA		16			2 signed displacement bytes. No prebyte.	
BRN		21			No test. Branch Never Taken	
BSR		8D			Unconditional Branch to Subroutine	
LBSR		17			2 signed displacement bytes. No prebyte	
BVC		28			V = 0? } V is set if the next highest bit.	
BVS		29			V = 1? } overflows into the highest (sign) bit.	
CLRA	4F				00 → A	$\overline{N}$ Z $\overline{V}$ $\overline{C}$
CLRB	5F				00 → B	$\overline{N}$ Z $\overline{V}$ $\overline{C}$
CLR		0F	7F	6F	00 → M	$\overline{N}$ Z $\overline{V}$ $\overline{C}$
CMPA	81	91	B1	A1	A − M }	3 n z v c
CMPB	C1	D1	F1	E1	B − M } Compares	3 n z v c
CMPD	10 83	10 93	10 B3	10 A3	D − M:M+1 } set flag	n z v c
CMPS	11 8C	11 9C	11 BC	11 AC	S − M:M+1 } bits only.	n z v c
CMPU	11 83	11 93	11 B3	11 A3	U − M:M+1 } Registers	n z v c
CMPX	8C	9C	BC	AC	X − M:M+1 } unchanged	n z v c
CMPY	10 8C	10 9C	10 BC	10 AC	Y − M:M+1 }	n z v c
COMA	43				$\overline{A}$ → A	n z $\overline{V}$ C
COMB	53				$\overline{B}$ → B	n z $\overline{V}$ C
COM		03	73	63	$\overline{M}$ → M	n z $\overline{V}$ C
CWAI	3C				CC AND IMMBYTE → CC; Stack. Clear 0 bits	
DAA	19				Use after ADDA, ADCA in BCD	n z $\overline{V}$ c
DECA	4A				A − 1 → A	n z v
DECB	5A				B − 1 → B	n z v
DEC		0A	7A	6A	M − 1 → M	n z v
EORA	88	98	B8	A8	A XOR M → A	n z $\overline{V}$
EORB	C8	D8	F8	E8	B XOR M → B	n z $\overline{V}$
EXG[4]	1E				Swap 8-bit OR 16-bit pair: EXG A,B	
INCA	4C				A + 1 → A	n z v
INCB	5C				B + 1 → B	n z v
INC		0C	7C	6C	M + 1 → M	n z v

ASSEM' FORM	INHR IMMD	REL[1] DIR	EXT'D	INDX[2] (base)	OPERATION (M = Operand Data)	CCR BITS $\overline{C}$ = cleared; c = affected
JMP		0E	7E	6E	Operand H,L → PC	
JSR		9D	BD	AD	Stack PCL, PCH; Operand → PC	
LDA	86	96	B6	A6	M → A	n z $\overline{V}$
LDB	C6	D6	F6	E6	M → B	n z $\overline{V}$
LDD	CC	DC	FC	EC	M:M+1 → D	n z $\overline{V}$
LDS	10 CE	10 DE	10 FE	10 EE	M:M+1 → S	n z $\overline{V}$
LDU	CE	DE	FE	EE	M:M+1 → U	n z $\overline{V}$
LDX	8E	9E	BE	AE	M:M+1 → X	n z $\overline{V}$
LDY	10 8E	10 9E	10 BE	10 AE	M:M+1 → Y	n z $\overline{V}$
LEAS				32	EA → S } See indexed postbytes below.	
LEAU				33	EA → U } Examp: LEAX 1,X = INX	
LEAX				30	EA → X } LEAY D,Y = D+Y → Y	z
LEAY				31	EA → Y } LEAU −9,X = X−9 → U	z
LSLA	48				A }	n z v c
LSLB	58				B }	n z v c
LSL		08	78	68	M }	n z v c
LSRA	44				A }	$\overline{N}$ z c
LSRB	54				B }	$\overline{N}$ z c
LSR		04	74	64	M }	$\overline{N}$ z c
MUL	3D				A × B → D (unsigned)	z 5
NEGA	40				$\overline{A}$ + 1 → A } (Forms 2's	3 n z v c
NEGB	50				$\overline{B}$ + 1 → B } complement.	3 n z v c
NEG		00	70	60	$\overline{M}$ + 1 → M }	3 n z v c
NOP	12				No Operation. 1 byte, 2 cycles	
ORA	8A	9A	BA	AA	A OR M → A.	
ORB	CA	DA	FA	EA	B OR M → B	
ORCC	1A				CC OR IMMBYTE → CC Set bits with 1s.	
PSHS[6]	34				} Push/Pull listed registers on stack.	
PSHU	36				} PUL → CC A B DP X Y U/S PC ← PSH	
PULS	35				} Examp: PSHS A, CC, U, Y	
PULU	37				} PULU X, S, B, DP	
ROLA	49				A }	n z v c
ROLB	59				B }	n z v c
ROL		09	79	69	M }	n z v c
RORA	46				A }	n z c
RORB	56				B }	n z c
ROR		06	76	66	M }	n z c
RTI	3B				Pull 12 bytes from S stack (3 if E = 0),	All Flags.
RTS	39				Pull PCH:PCL from stack.	No Flags.
SBCA	82	92	B2	A2	A − M − C → A	3 n z v c
SBCB	C2	D2	F2	E2	B − M − C → B	3 n z v c
SEX	1D				FF → A if B7 = 1. 00 → A if B7 = 0	n z $\overline{V}$
STA		97	B7	A7	A → M	n z $\overline{V}$
STB		D7	F7	E7	B → M	n z $\overline{V}$
STD		DD	FD	ED	D → M:M+1	n z $\overline{V}$
STS		10 DF	10 FF	10 EF	S → M:M+1	n z $\overline{V}$
STU		DF	FF	EF	U → M:M+1	n z $\overline{V}$
STX		9F	BF	AF	X → M:M+1	n z $\overline{V}$
STY		10 9F	10 BF	10 AF	Y → M:M+1	n z $\overline{V}$
SUBA	80	90	B0	A0	A − M → A	3 n z v c
SUBB	C0	D0	F0	E0	B − M → B	3 n z v c
SUBD	83	93	B3	A3	D − M:M+1 → D	3 n z v c
SWI	3F				} Stack 12 bytes. FFFA, B	I F
SWI2	10 3F				} Go to vector FFF4, 5	None
SWI3	11 3F				} address. FFF2, 3	None
SYNC	13				Halt, Float bus until Interrupt	None
TFR[4]	1F				Copy first register into second.	None
TSTA	4D				A − 00 } Sets CCR	n z $\overline{V}$
TSTB	5D				B − 00 } bits only.	n z $\overline{V}$
TST		0D	7D	6D	M − 00 }	n z $\overline{V}$

Notes:
1. For long branch relative, add prebyte $10; use two operand bytes.
2. All indexed instructions require a postbyte. See below.
3. State of *H* flag is undefined.
4. See EXG/TFR postbyte structure below.
5. MUL sets C = B7. Round *D* to 8 bits of *A* by ADCA #0.
6. See table below for PSH/PUL postbyte. *S* not savable on *S* stack. *U* not savable on *U* stack.

2. COMMON INDEXED-MODE POSTBYTES[7]

Operand Offset from Pointer Reg.			Auto Increment/Decrement		
Type	Byte	Operand Form	Type	Byte	Operand Form
Zero offset, X	84	,X	Post Increment, X	80	,X+
Zero offset, Y	A4	,Y	Post Inc by 2, X	81	,X++
0 to +15 offset,X[8]	00-0F	2, X	Pre Decrement, X	82	,−X
−16 to −1 offset, X	10-1F	−7,X	Pre Dec, by 2,X	83	,−XX
0 to +15 offset, Y	20-2F	13,Y	Post Increment, Y	A0	,Y+
−16 to −1 offset, Y	30-3F	−1,Y	Post Inc by 2, Y	A1	,Y++
A(signed) offset, X	86	A,X	Pre Decrement, Y	A2	,−Y
B(signed) offset, X	85	B,X	Pre Dec. by 2, Y	A3	,−−Y
D(signed) offset, X	8B	D,X	**Non Indexing types**		
A(signed) offset, Y	A6	A,Y	PC Relative, −128 to +127	8C (+1)	ADR, PCR
B(signed) offset, Y	A5	B,Y	PCR, −3276B to +32767	8D (+2)	ADR, PCR
D(signed) offset, Y	AB	D,Y	Extended Indirect[9]	9F	(ADR)

Notes: 7. Indexing by *S* and *U* is available but is not shown here.
8. Offsets range from −32,768 to +32,767 with extra bytes; not shown.
9. Most modes have indirect forms; not shown.

[4]TFR/EXG Postbytes
2 nibbles:
Source/Destination
Examp: $89 = A → B

16 bit	8 bit
0 = D	8 = A
1 = X	9 = B
2 = Y	A = CC
3 = U	B = DP
4 = S	
5 = PC	

Invalid to mix 16-with 8-bit.

[6]PSH/PUL Postbytes
Commonly used forms

01 = CC	06 = A and B
02 = A	07 = A, B, and CC
04 = B	30 = X and Y
08 = DP	3F = All but PC & S or U
10 = X	0F = All 8-bit reg.
20 = Y	F0 = All 16-bit reg.
40 = S or U	FF = All registers but stacking register
80 = PC	

20.5 CYCLE-BY-CYCLE OPERATION OF THE 6809

Section Overview

The cycle-by-cycle operation chart for the 6802 (Appendix B) provides a detailed prediction of the address, data, VMA, and R/$\overline{W}$ states as it executes each of its 192 instructions. The 6809 has several times as many instructions and instruction types as the 6802, so a similar chart would be inconveniently long in its case. Instead, a cycle-by-cycle chart is provided. The data on the buses during each cycle can be predicted for any 6809 instruction by following the chart. This section consists entirely of the chart and some notes and examples to assist you in interpreting it.

Knowing the exact state of each bus and control line as a processor executes an instruction is sometimes crucial to hardware troubleshooting and system refinement. Figures 20.9 through 20.13 give the flowchart for constructing a table of bus signals for any 6809 instruction. The entire table requires five pages. **A** is the entry point after the op code has been read. **B** is the exit point to the next instruction. **C** and **D** are entry points to the last page of the chart, which shows the action taken on the operand for memory-reference instructions other than stack operations. Here are some hints for interpreting the chart:

- Each double box in the flow lines represents a machine cycle. The state of the address bus is given in the lower box and the state of the data bus appears in the upper box for each cycle.
- The first diamond at the top of Figure 20.9 determines if an extra cycle is required. If so the address is incremented a second time to read the 2-byte op-codes when prebytes are required. This applies to the following instructions:

CMPD	LDS	SWI2
CMPS	LDY	SWI3
CMPU	STS	ALL LBXX, except LBRA and LBSR
CMPY	STY	

- The structure NNNN +1 means that the address bus outputs the next higher address.
- FFFF on the address bus indicates a VMA-not cycle.
- *High* and *low* refer to the high-order and low-order bytes, respectively, of a 16-bit register or address.

In building a data-flow list for a particular instruction, you should first go through the chart to determine the number of clock cycles required. These should be numbered to form a left-hand column. Initial addresses, data bytes, and machine-register contents should then be specified. The chart can then be traced through again to fill in a second column for address-bus contents on each cycle, a third column for data-bus contents, a fourth for R/$\overline{W}$ status, and a final column for comments.

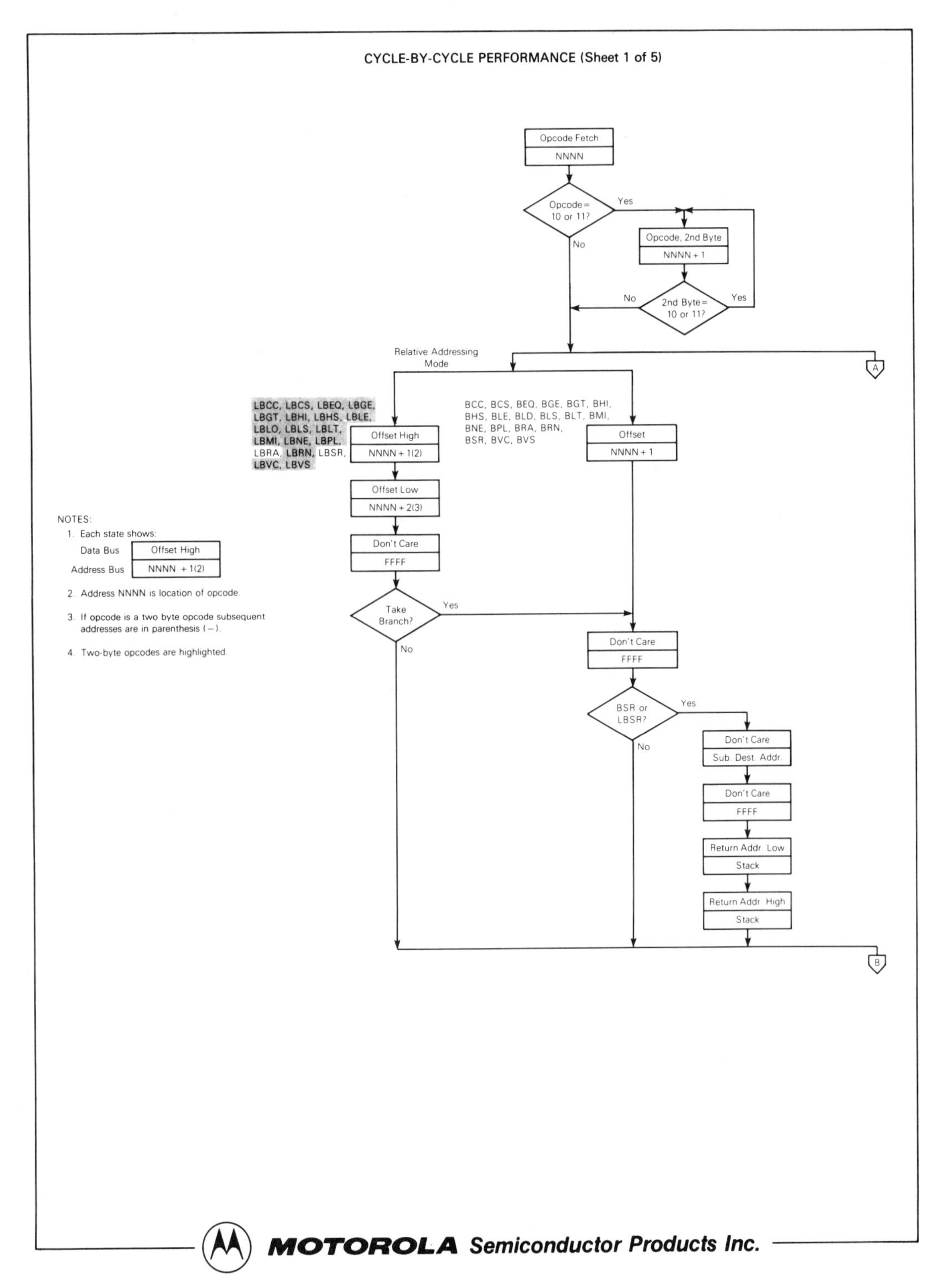

Figure 20.9 Cycle-by-cycle operation of 6809 instructions (Chart continues through Figure 20.13).

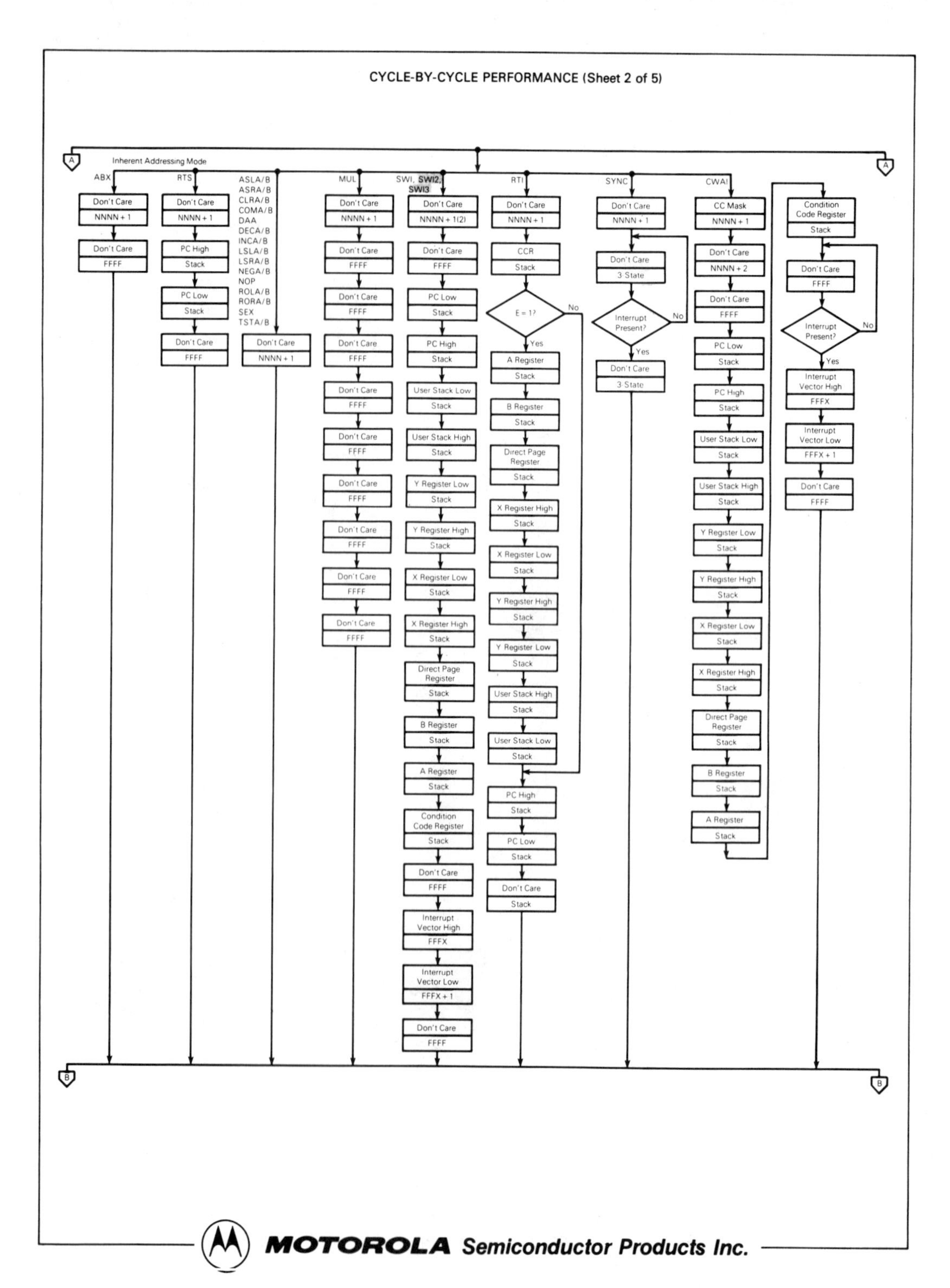

Figure 20.10 6809 cycle-by-cycle operation (continued).

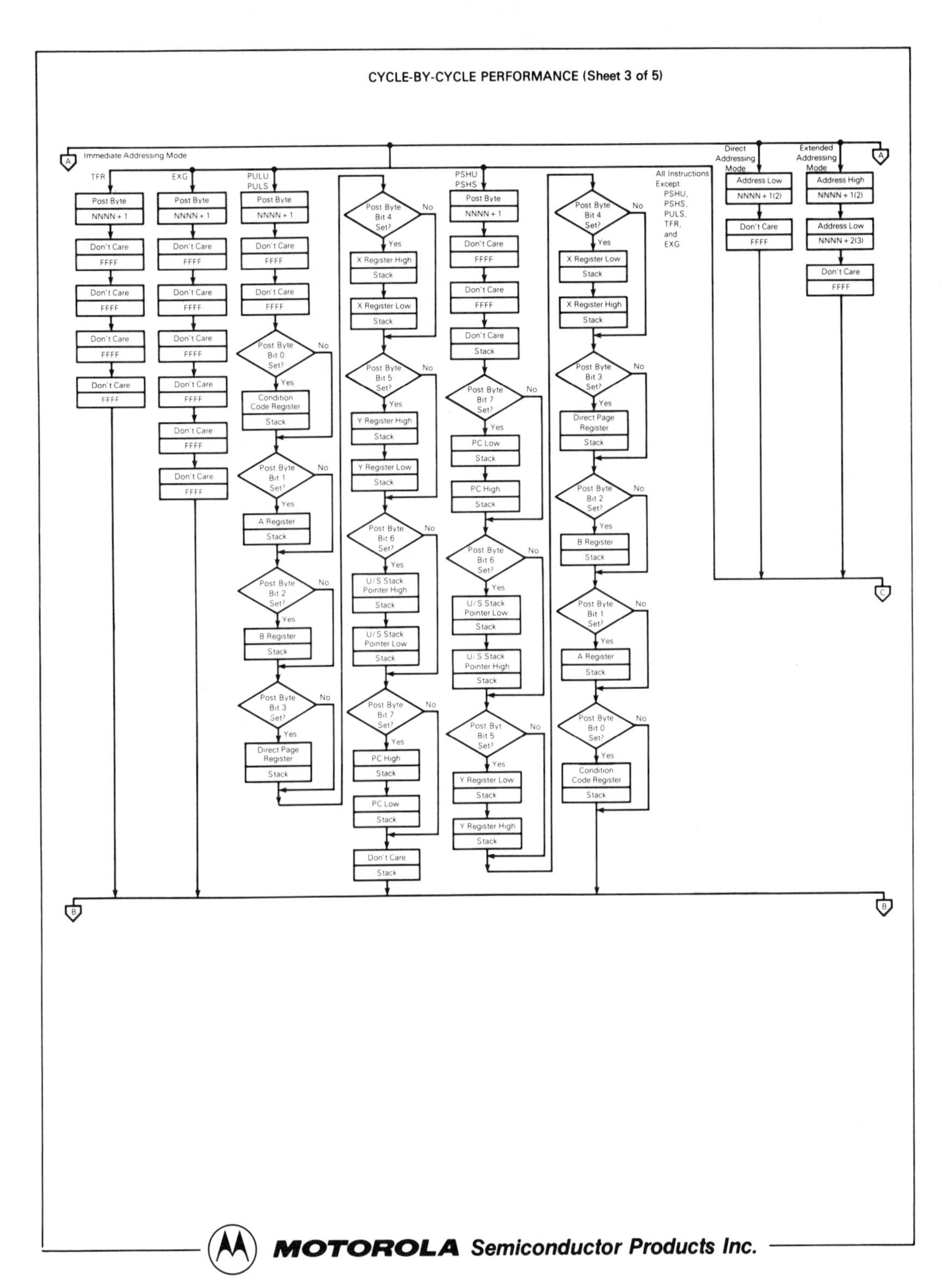

Figure 20.11 6809 cycle-by-cycle operation (continued).

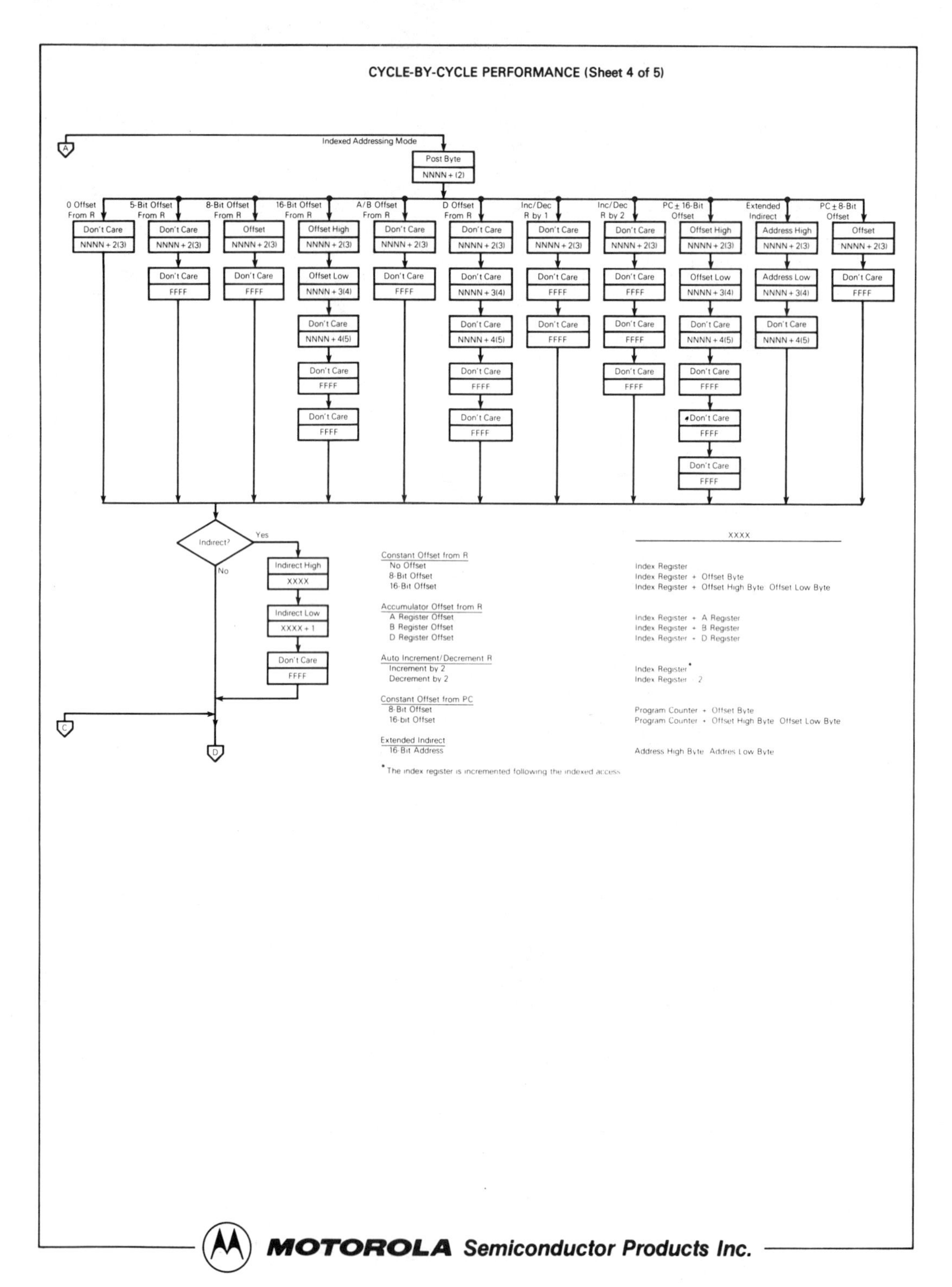

Figure 20.12 6809 cycle-by-cycle operation (continued).

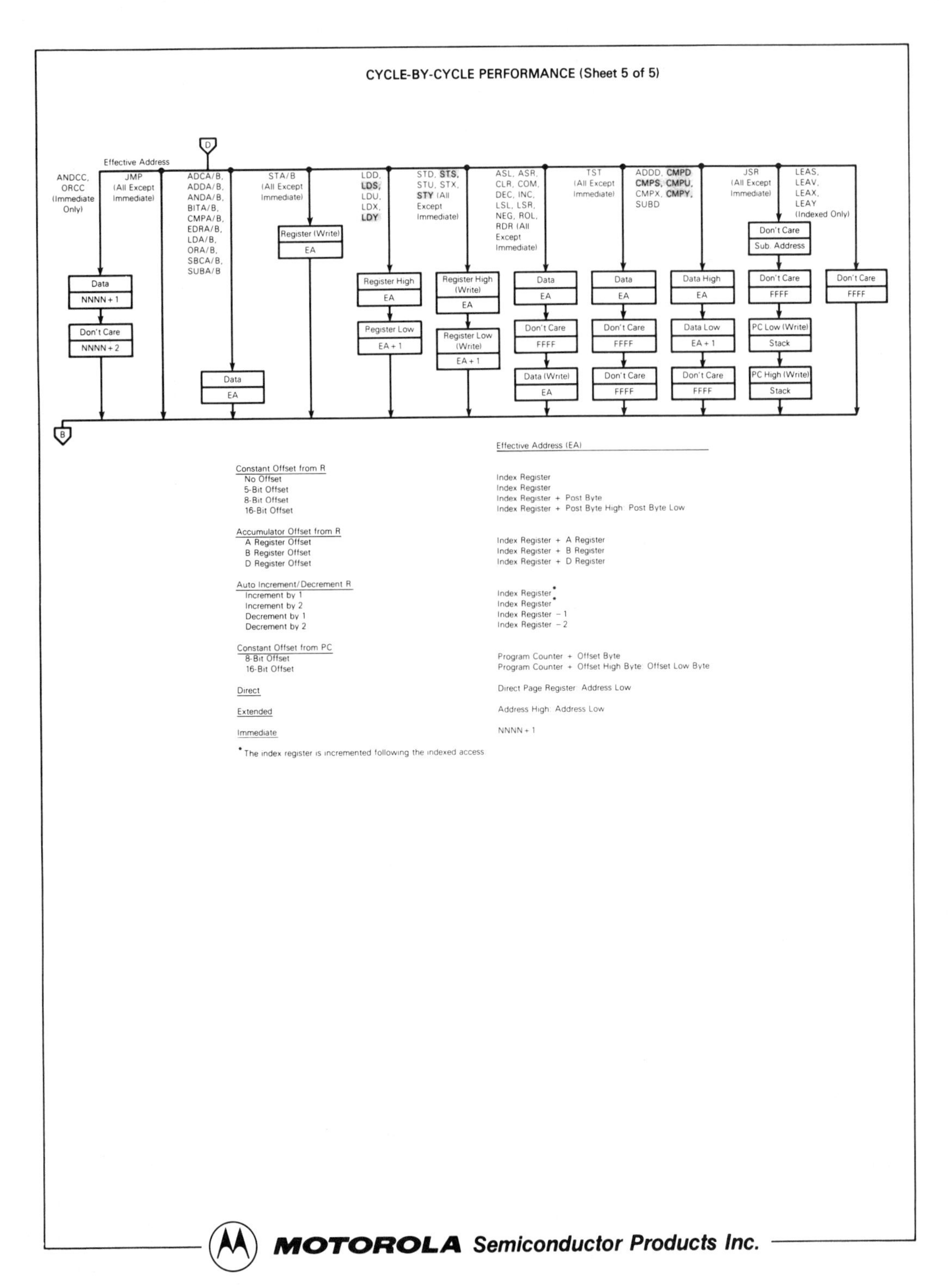

Figure 20.13 6809 cycle-by-cycle operation (continued).

Example 20.1 LDA $1234 Extended mode

Definitions: Op-code address = $A000
Data at $1234 = $BC

Cycle	Address	Data	R/$\overline{W}$	Comments
1	A000	B6	1	Op-code read.
2	A001	12	1	Operand Adr MSB read.
3	A002	34	1	Operand Adr LSB read.
4	FFFF	—	1	$\overline{VMA}$ cycle.
5	1234	BC	1	Operand data read.

One cycle is expended on the op-code read, Figure 20.9. Point *A* leads to Figure 20.11, the last column of which explains machine cycles 2 through 4. Point *C* leads, via Figure 20.12, to Figure 20.13, where the third column explains machine cycle 5.

Example 20.2 LEAX 1,X Indexed mode

Definition: Op-code address = $4567

Cycle	Address	Data	R/$\overline{W}$	Comments
1	4567	30	1	Op-code read.
2	4568	01	1	Postbyte read.
3	4569	—	1	Next-instruction op-code.
4	FFFF	—	1	$\overline{VMA}$ cycle.
5	FFFF	—	1	$\overline{VMA}$ cycle.

The first cycle appears in Figure 20.9. Cycle 2 is a postbyte read and cycles 3 and 4 are found in the second column of Figure 20.12 (5-bit offset column). Cycle 5 is found in the last column of Figure 20.13.

Example 20.3 CMPD ,X+ + Indexed, Postincrement by 2

Definitions: Op-code address = $F980
X = $4050; Data at $4050, 4051 = $AB,CD

Cycle	Address	Data	R/$\overline{W}$	Comments	
1	F980	10	1	Prebyte read, Figure 20.9.	
2	F981	A3	1	Op-code read, Figure 20.9.	
3	F982	81	1	Postbyte read, Figure 20.12.	
4	F983	—	1	Next instruction read, Figure 20.12.	
5	FFFF	—	1	$\overline{VMA}$ cycle.	Figure 20.12;
6	FFFF	—	1	$\overline{VMA}$ cycle.	INC/DEC R
7	FFFF	—	1	$\overline{VMA}$ cycle.	by 2 column.
8	4050	AB	1	Operand adr MSB read.	Figure 20.13;
9	4051	CD	1	Operand adr LSB read.	third-to-last
10	FFFF	—	1	$\overline{VMA}$ cycle.	column.

Example 20.4 (STB 2030) Indexed; extended indirect

Definition: Op-code address = \$F987; B = \$BC
Data at \$2030, 1 = \$01,45

Cycle	Address	Data	R/$\bar{W}$	Comments	
1	F987	E7	1	Op-code read.	Figure 20.9.
2	F988	9F	1	Postbyte read,	Figure 20.12.
3	F989	20	1	Operand MSB read.	Figure 20.12,
4	F98A	30	1	Operand LSB read.	second-last
5	F98B	—	1	Next instruction.	column.
6	2030	01	1	Indirect address read.	Figure 20.12,
7	2031	45	1	IA + 1 read.	indirect address
8	FFFF	—	1	$\overline{\text{VMA}}$ cycle.	
9	0145	BC	0	Effective-address write.	Figure 20.13.

20.6 THE VIDEO LINE-DRAWER IMPLEMENTED

Section Overview

This section breaks with our established tradition of putting the hardware-project description in the last section of each chapter. We place this chapter's hardware project in the second-to-last section to save a section for a brief review of another widely used 6800 descendent—the 6502. This section presents complete hardware diagrams and program listings for a 6809/6847 video line-drawing system. With a little imagination you may see that a line-drawing program could become the key subroutine in a complete graphics or CAD (computer assisted drafting) program.

The system hardware is shown in Figure 20.14. The program is ROM-resident, since it is inconveniently long to load into RAM each time you power-up. The data points, however, are stored in RAM, and are loaded from a hex keypad, two keystrokes per byte. The top endpoint is loaded first with two strokes for the *X* coordinate and two strokes for *Y*. The bottom endpoint of the line is then entered with four strokes for *X* and *Y*. Another eight strokes can then be entered to make another line.

The program follows the basic flowchart of Figure 18.16. A novel divide routine is employed, which uses a "try a bit and multiply back to see if it's right" approach. Negative-slope recognition is included, and the slope is maxed out at 255 if a vertical line produces a divide-by-zero. Figure 20.15 shows a sketch made with the completed line drawer. Here is the program listing.

```
1000 ****************************************************************
1010 *                                                              *
1020 *  LIST 20-1   ** VIDEO LINE DRAWER   ** FIG 20.14 HARDWARE *
1030 *          TESTED AND WORKING 17 AUG, 87 BY D. METZGER         *
1040 *                                                              *
1050 ****************************************************************
1060         .OR    $FE00  * UPPER RIGHT = 7F,00
```

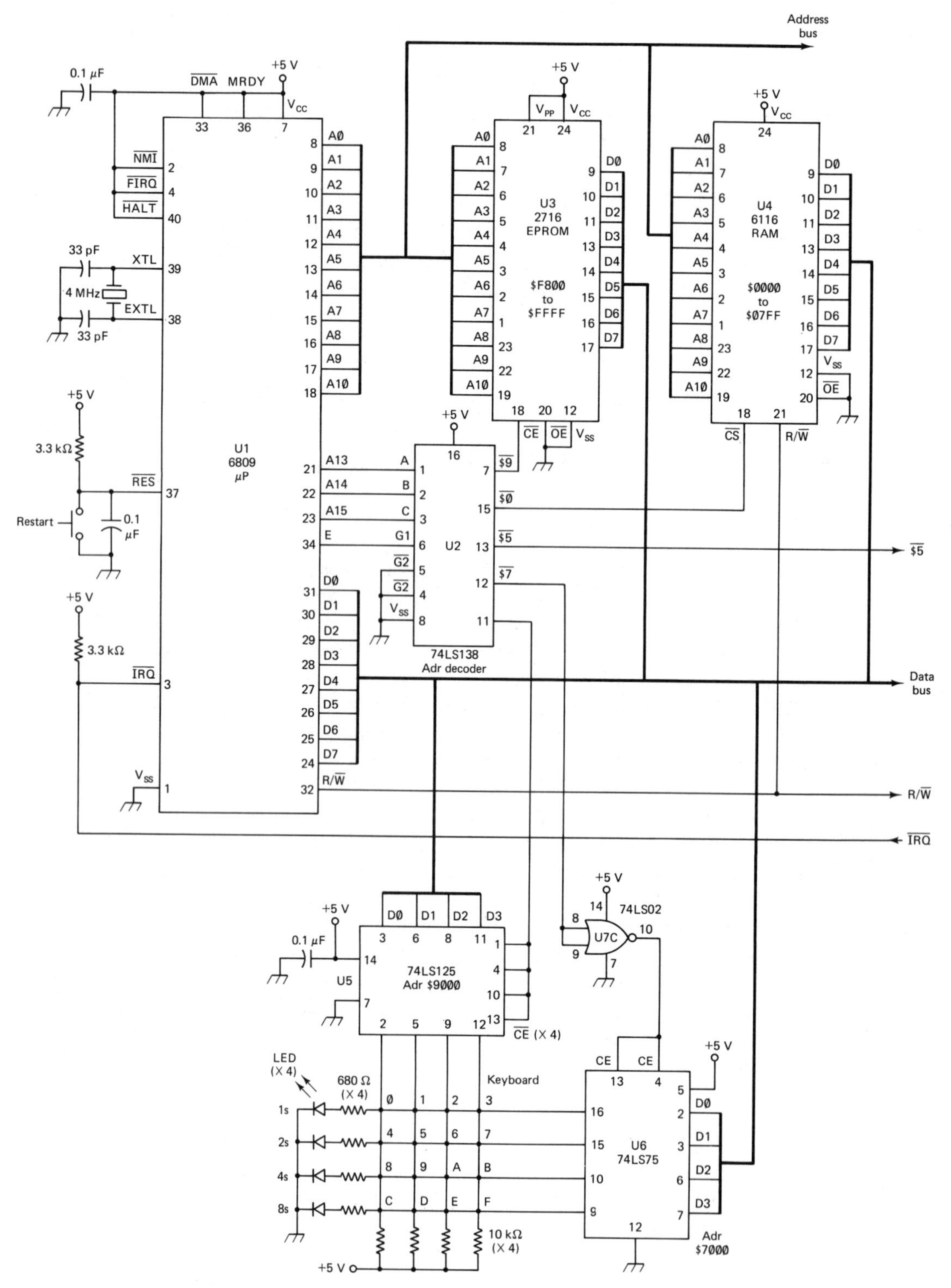

Figure 20.14 Hardware diagram for the video line-drawer project using a 6809 processor.

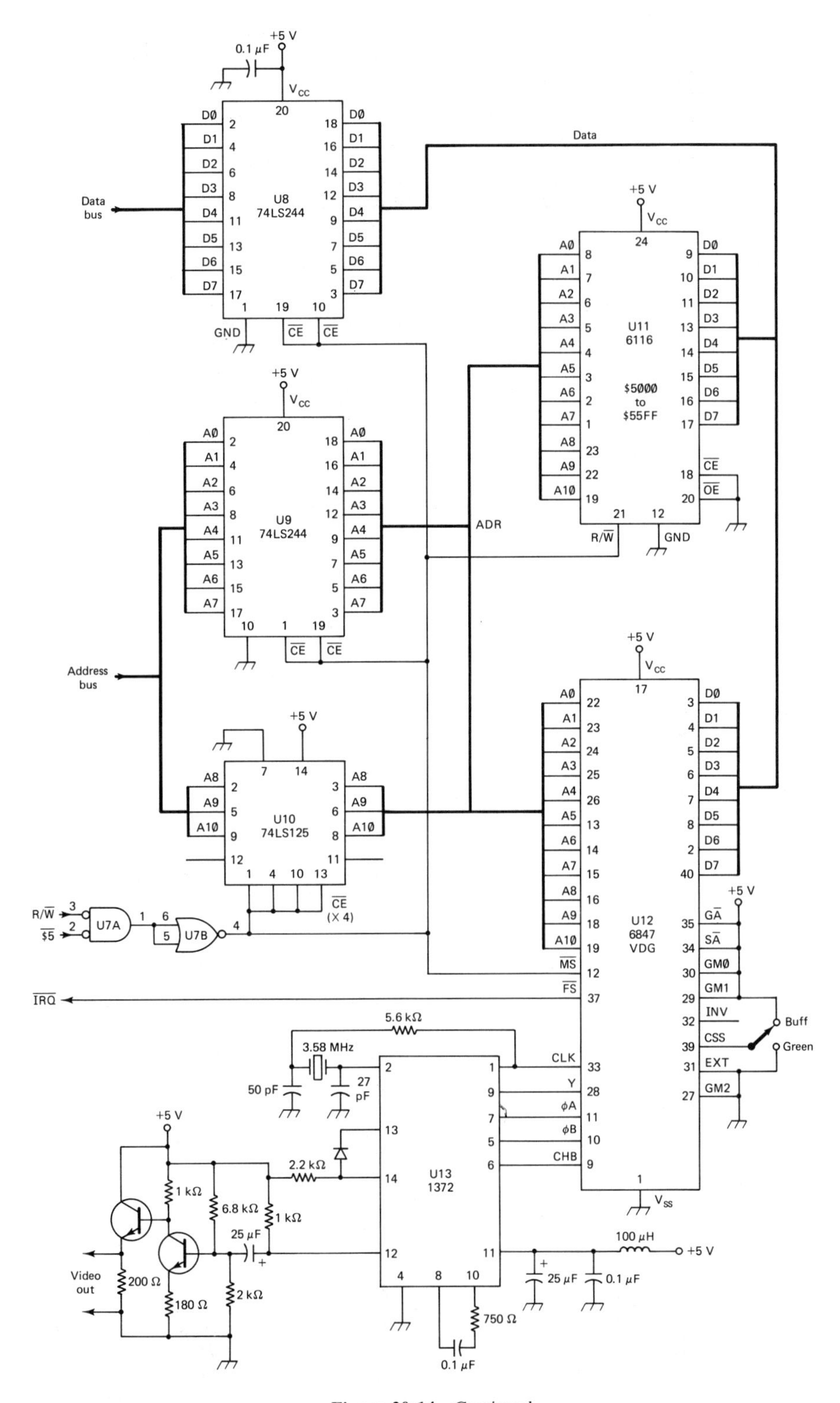

Figure 20.14 Continued

```
                   1070          .TA    $4000  * LOWER LEFT = 00,5F
0000-              1080 X1       .EQ    0      FIRST LOAD UPPER END POINTS IN HEX:
0001-              1090 Y1       .EQ    1       ($00,00 - $7F,5F)
0002-              1100 X2       .EQ    2      THEN LOAD LOWER X,Y END POINTS.
0003-              1110 Y2       .EQ    3       ($00,00 - $7F,5F)
0004-              1120 X0       .EQ    4      CURRENT X VALUE.
0005-              1130 TRY      .EQ    5      TRIAL DIVIDEND.
0006-              1140 OLDY     .EQ    6      LAST VALUE OF Y.
0007-              1150 P        .EQ    7      PRODUCT (X-X1)(Y2-Y1)
0009-              1160 COUNT    .EQ    9      FOR KEY-READ ROUTINE.
000A-              1170 DATA     .EQ    10     HOLDS FIRST KEYSTROKE.
000B-              1180 OLDAD    .EQ    11     PREVIOUS SCREEN ADDRESS AND DATA
000D-              1190 OLDDA    .EQ    13      TO AVOID OVERWRITES.
000E-              1200 FLAG     .EQ    14     FLAG = 00 IF SLOPE UPWARD.
7000-              1210 LATCH    .EQ    $7000  OUTPUT TO KEYPAD ROWS.
9000-              1220 BUFR     .EQ    $9000  INPUT FORM KEYPAD COLUMNS.
                   1230 *
FE00- CE 07 FF     1240 START    LDU    #$07FF INITIALIZE STACK POINTERS.
FE03- 10 CE 06
FE06- FF           1250          LDS    #$06FF THIS ONE FOR JSR AND IRQ'S.
FE07- BD FE 99     1260          JSR    CLEAR  CLEAR SCREEN.
FE0A- BD FE A5     1270 INPUT    JSR    BYTE   LOAD FOUR X, Y BUFFERS WITH LINE
FE0D- 97 00        1280          STA    X1      END POINTS (X1,Y1) AND (X2,Y2).
FE0F- BD FE A5     1290          JSR    BYTE    (8 KEY PRESSES; 2 FOR EACH POINT)
FE12- 97 01        1300          STA    Y1     REQUIRED: X2 > X1 AND Y2 > Y1.
FE14- 97 06        1310          STA    OLDY    (START OLDY VALUE AT Y1)
FE16- BD FE A5     1320          JSR    BYTE
FE19- 97 02        1330          STA    X2     ALSO REQUIRED: X < $80
FE1B- BD FE A5     1340          JSR    BYTE                  Y < $60
FE1E- 97 03        1350          STA    Y2
FE20- 1F 89        1360          TFR    A,B    IN CASE X2 = X1, LEADING TO DIV BY 0.
FE22- 86 FF        1370          LDA    #$FF   INIT FLAG FOR DOWN SLOPE.
FE24- 97 0E        1380          STA    FLAG
FE26- 96 00        1390          LDA    X1     LET X (CALLED X0)
FE28- 97 04        1400          STA    X0        = X1 TO START.
FE2A- 91 02        1410          CMPA   X2     CHECK SLOPE BY CMP X2 AND X1:
FE2C- 27 5E        1420          BEQ    FILIN   VERT SLOPE? SKIP MUL & DIV.
FE2E- 25 02        1430          BLO    NXTDOT  DOWN SLOPE? KEEP FLAG = FF.
FE30- 0F 0E        1440          CLR    FLAG    UP SLOPE? CLEAR FLAG.
FE32- 96 04        1450 NXTDOT   LDA    X0
FE34- 90 00        1460          SUBA   X1     A CONTAINS  X - X1
FE36- 0D 0E        1470          TST    FLAG   IF UPWARD SLOPE MAKE NEG
FE38- 26 01        1480          BNE    KEEP    RESULT POSITIVE.
FE3A- 40           1490          NEGA
FE3B- D6 03        1500 KEEP     LDB    Y2
FE3D- D0 01        1510          SUBB   Y1     B CONTAINS  Y2 - Y1.
FE3F- 3D           1520          MUL           D CONTAINS VALUE LABELED P IN
FE40- DD 07        1530          STD    P       FLOWCHART; SAVE PRODUCT IN RAM.
FE42- 96 02        1540          LDA    X2
FE44- 90 00        1550          SUBA   X1     A CONTAINS  X2 - X1
FE46- 24 01        1560          BCC    DOWN   IF UPWARD SLOPE MAKE NEG
FE48- 40           1570          NEGA           SUBTRAHEND POSITIVE
FE49- C6 80        1580 DOWN     LDB    #$80   DIVISION BY TRIAL MUL BACK:
FE4B- D7 05        1590          STB    TRY     TRY $80 * (X2 - X1) FIRST.
FE4D- DA 05        1600 DIV      ORB    TRY    TRY A 1 IN MULTIPLIER BIT POSITIONS
FE4F- 36 02        1610          PSHU   A       7 THROUGH 0
FE51- 36 04        1620          PSHU   B
FE53- 3D           1630          MUL
FE54- 10 93 07     1640          CMPD   P      PRODUCT TOO BIG?
FE57- 37 04        1650          PULU   B
FE59- 23 02        1660          BLS    OKBIT   NO: KEEP BIT IN REGISTER B.
FE5B- D0 05        1670          SUBB   TRY     YES: REMOVE BIT.
FE5D- 04 05        1680 OKBIT    LSR    TRY    MOVE TRIAL BIT LOWER.
FE5F- 37 02        1690          PULU   A      ADD BIT TO DIVIDEND & TRY 8 TIMES.
FE61- 24 EA        1700          BCC    DIV    B ENDS UP WITH DIVIDEND OR $FF IF > FF.
FE63- DB 01        1710          ADDB   Y1     NEWY = DIVIDEND + Y1.
FE65- 24 02        1720          BCC    FILL   MAX OUT NEWY IN B IF IT GOES
FE67- C6 FF        1730          LDB    #$FF    GREATER THAN $FF.
```

```
FE69- 0C 06     1740 FILL    INC    OLDY    DOES THE CALCULATED Y VALUE SKIP
FE6B- D1 06     1750         CMPB   OLDY     A VERT SPACE, MAKING A
FE6D- 22 1D     1760         BHI    FILIN    BROKEN LINE?
FE6F- 96 04     1770         LDA    X0         (GET X VALUE IN A)
FE71- BD FE E5  1780         JSR    WRITE    NO: MAKE THE PIXEL.
FE74- 96 04     1790         LDA    X0
FE76- 0D 0E     1800         TST    FLAG    IF UP SLOPE, DEC X INSTEAD
FE78- 26 09     1810         BNE    RIGHT    OF INCREMENTING IT TO NEXT VALUE.
FE7A- 4A        1820         DECA
FE7B- 97 04     1830         STA    X0
FE7D- 91 02     1840         CMPA   X2      DONE TO X2 WORKING LEFT?
FE7F- 25 89     1850         BLO    INPUT    YES: GET NEW UPPER AND LOWER END PTS.
FE81- 20 AF     1860         BRA    NXTDOT   NO: CALCULATE NEXT PIXEL POSITION.
FE83- 4C        1870 RIGHT   INCA                ADVANCE TO NEXT HORIZ POSITION.
FE84- 97 04     1880         STA    X0
FE86- 91 02     1890         CMPA   X2           DONE WORKING RIGHT TO HORIZ END?
FE88- 22 80     1900         BHI    INPUT          YES: GET 4 MORE COORDINATES.
FE8A- 20 A6     1910         BRA    NXTDOT         NO: CALCULATE NEXT Y POSITION.
FE8C- 36 04     1920 FILIN   PSHU   B          SAVE TARGET Y VALUE.
FE8E- D6 06     1930         LDB    OLDY     YES: B = Y + 1 (MOVE DOWN 1)
FE90- 96 04     1940         LDA    X0           A HOLDS X VALUE.
FE92- BD FE E5  1950         JSR    WRITE      FILL IN A Y PIXEL AT SAME X.
FE95- 37 04     1960         PULU   B
FE97- 20 D0     1970         BRA    FILL     KEEP FILLING UNTIL NEWY REACHED.
                1980 ********************************************************
                1990 * SUBROUTINE TO CLEAR SCREEN:
FE99- 8E 50 00  2000 CLEAR   LDX    #$5000 CLEAR ONLY DURING VERT RETRACE.
FE9C- 13        2010 LOOP    SYNC           WAIT FOR VERT RETRACE.
FE9D- 6F 80     2020         CLR    ,X+     CLEAR AND INCREMENT X.
FE9F- 8C 56 00  2030         CMPX   #$5600 END OF DISPLAY RAM?
FEA2- 26 F8     2040         BNE    LOOP     NO: CLEAR MORE.
FEA4- 39        2050         RTS             YES: BACK TO MAIN.
                2060 *********************************************************
                2070 * SUBROUTINE TO ACCEPT 2 KEYSTROKES & PUT VALUE IN "A".
FEA5- BD FE B4  2080 BYTE    JSR    KEY     GET HIGH DIGIT IN LO NIBBLE OF "A".
FEA8- 48        2090         ASLA
FEA9- 48        2100         ASLA           MOVE TO HIGH NIBBLE IN RAM BUFFER.
FEAA- 48        2110         ASLA
FEAB- 48        2120         ASLA
FEAC- 97 0A     2130         STA    DATA
FEAE- BD FE B4  2140         JSR    KEY     GET LO DIGIT IN LO NIBBLE OF "A".
FEB1- 9A 0A     2150         ORA    DATA    COMBINE NIBBLES.
FEB3- 39        2160         RTS
                2170 * SUBROUTINE TO ACCEPT 1 KEY & PUT 0 - F IN "A" LO NIB.
FEB4- 0F 09     2180 KEY     CLR    COUNT   START AT KEY COUNT = 0.
FEB6- C6 1E     2190         LDB    #$1E    PULL FIRST ROW LOW TO START.
FEB8- F7 70 00  2200 PULLO   STB    LATCH   PULL A ROW LOW.
FEBB- B6 90 00  2210         LDA    BUFR    READ COLUMNS 0 - 3.
FEBE- 8A F0     2220         ORA    #$F0     ANY KEY PRESSED MEANS A 0 BIT.
FEC0- 81 FF     2230         CMPA   #$FF    ALL HIGH?
FEC2- 26 0D     2240         BNE    PREST    NO: WHICH ONE IS PRESSED?
FEC4- 96 09     2250         LDA    COUNT    YES: ADD 4 TO COUNT.
FEC6- 8B 04     2260         ADDA   #4
FEC8- 97 09     2270         STA    COUNT
FECA- 1A 01     2280         ORCC   #$01    SET CARRY FLAG.
FECC- 59        2290         ROLB           PUT LOW IN NEXT HIGHER BIT.
FECD- 24 E9     2300         BCC    PULLO   GO BACK FOR 2ND, 3RD, & 4TH ROWS.
FECF- 20 E3     2310 NOKEY   BRA    KEY     IF NO KEY YET, GO BACK & KEEP LOOKING.
FED1- 46        2320 PREST   RORA           YOU FOUND THE ROW; NOW FIND COLUMN.
FED2- 24 04     2330         BCC    FOUND   SHIFT BITS RIGHT TOWARD CARRY
FED4- 0C 09     2340         INC    COUNT    UNTIL LOW BIT IS FOUND.
FED6- 20 F9     2350         BRA    PREST
FED8- 96 09     2360 FOUND   LDA    COUNT
FEDA- B7 70 00  2370         STA    LATCH   DISPLAY KEY IN BINARY.
FEDD- 8E FF FF  2380         LDX    #$FFFF   WAIT A SEC BEFORE ACCEPTING
FEE0- 30 1F     2390 WAIT    LEAX   -1,X       NEXT KEY.
FEE2- 26 FC     2400         BNE    WAIT
FEE4- 39        2410         RTS
```

```
                   2420 ***********************************************************
                   2430 * WRITE SUBROUTINE CONVERTS X (IN A) , Y (IN B) TO BIT ADR.
FEE5- 36 06        2440 WRITE   PSHU  D       SAVE X VALUE IN A AND Y VALUE IN B.
FEE7- D7 06        2450         STB   OLDY    LEAVE A RECORD OF LAST Y VALUE WRITTEN.
FEE9- 48           2460         ASLA          GET X BITS 6,5,4, & 3 INTO BITS
FEEA- 1E 89        2470         EXG   A,B      7,6,5,4 OF LOW-BYTE, D ACCUM.
FEEC- 44           2480         LSRA
FEED- 56           2490         RORB          FOUR SHIFTS RIGHT
FEEE- 44           2500         LSRA           OF ACCUM D WILL FILL
FEEF- 56           2510         RORB           BITS 11 THROUGH 0
FEF0- 44           2520         LSRA           OF SCREEN ADDRESS IN D.
FEF1- 56           2530         RORB
FEF2- 44           2540         LSRA
FEF3- 56           2550         RORB
FEF4- 8B 50        2560         ADDA  #$50    SET BITS 15 THRU 12 OF SCREEN ADR.
FEF6- 1F 01        2570         TFR   D,X     X POINTS TO SCREEN ADDRESS.
FEF8- 37 06        2580         PULU  D       RETRIEVE A & B.
FEFA- 84 07        2590         ANDA  #$07    SAVE BITS 2,1, AND 0 ONLY AS COUNTER.
FEFC- C6 80        2600         LDB   #$80    SET TO LIGHT LEFTMOST OF 8 PIXELS.
FEFE- 4D           2610 PIXEL   TSTA          IS COUNTER = 0?
FEFF- 27 04        2620         BEQ   STORE    YES: THAT'S THE BIT.
FF01- 54           2630         LSRB           NO: TRY NEXT BIT RIGHT.
FF02- 4A           2640         DECA               COUNT DOWN MAX 7 SHIFTS.
FF03- 20 F9        2650         BRA   PIXEL
FF05- 9C 0B        2660 STORE   CMPX  OLDAD   SAME ADDRESS AS LAST TIME?
FF07- 26 02        2670         BNE   DIFF     NO: NO PROBLEM.
FF09- DA 0D        2680         ORB   OLDDA    YES: KEEP OLD 1 BITS.
FF0B- 9F 0B        2690 DIFF    STX   OLDAD   REMEMBER ADDRESS AND DATA FOR
FF0D- D7 0D        2700         STB   OLDDA    NEXT TIME.
FF0F- 13           2710         SYNC          WAIT FOR IRQ FROM VDG RETRACE.
FF10- E7 84        2720         STB   0,X     LIGHT SELECTED PIXEL AT ADDRESS IN X.
FF12- 39           2730         RTS
                   2740 ************************************************************
                   2750         .OR   $FFF8
                   2760         .TA   $41F8
FFF8- FE 00        2770         .HS   FE00      * RESTART VECTOR
```

REVIEW OF SECTION 20.5 AND 20.6

21. Use the cycle-by-cycle chart to determine the number of cycles taken by a LBEQ instruction if the branch is not taken.

22. Which, if any, of the cycles in Question 21 are $\overline{\text{VMA}}$ cycles?

23. Use the cycle-by-cycle chart to determine the number of cycles for a STD ,Y instruction.

24. On which cycle(s) does a write occur?

25. How many cycles are required by an EXG X,Y instruction?

26. Referring to Question 25, are any of these *write* or $\overline{\text{VMA}}$ cycles?

27. How many cycles are required by PSHU CC,A,X?

28. Which of these are $\overline{\text{VMA}}$ and which are write cycles?

29. How many machine cycles does a *divide* take in the program for the line drawer?

30. In the line-drawing program, give the line number where a negative slope is identified.

Figure 20.15 An image produced with the video line drawer.

20.7 A PEEK AT THE 6502 PROCESSOR

Section Overview

The 6502 is another descendent of the 6800 processor. It gained immense popularity as the key chip in many early personal computers, including those from Apple, Atari, and Commodore. There are several enhanced versions of the 6502 now available, including the 16-bit 65C816. The major differences of the 6502 compared to the 6800 are as follows:

- **Reverse sequence of 16-bit data. It is always stored in memory in low-order/high-order sequence.**
- **Only one accumulator, A.**
- **Two index registers, X and Y, each 8 bits long.**
- **An 8-bit stack pointer, which comes up from *reset* fixed at address $Ø1FF.**
- **Carry flag *cleared* if a borrow occurs in a subtract operation; *set* if no borrow.**

Rated PG: This section contains potentially confusing material about a *third* processor. Professors may wish to consider whether it should be read by students who are already confused by the first two. The section is logically placed in the chapter with the other major 6800 enhancement, but it may be more profitable to tackle it *after* the 6802 and 6809 features have had time to settle and become clear in your mind.

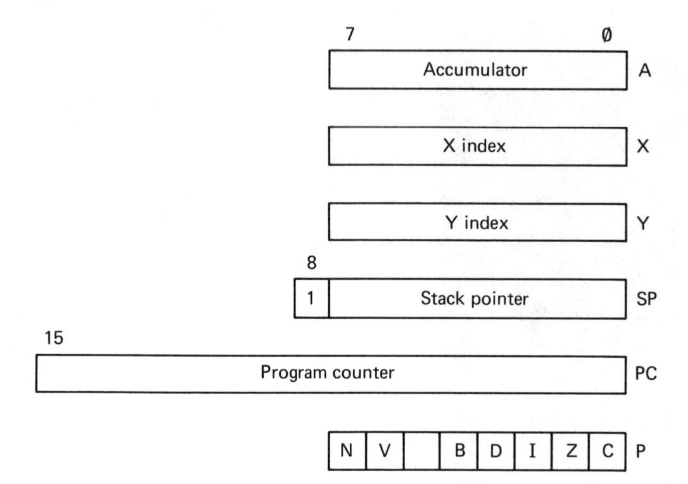

Figure 20.16 Programmer's model for the 6502 microprocessor. The stack pointer is always on page Ø1.

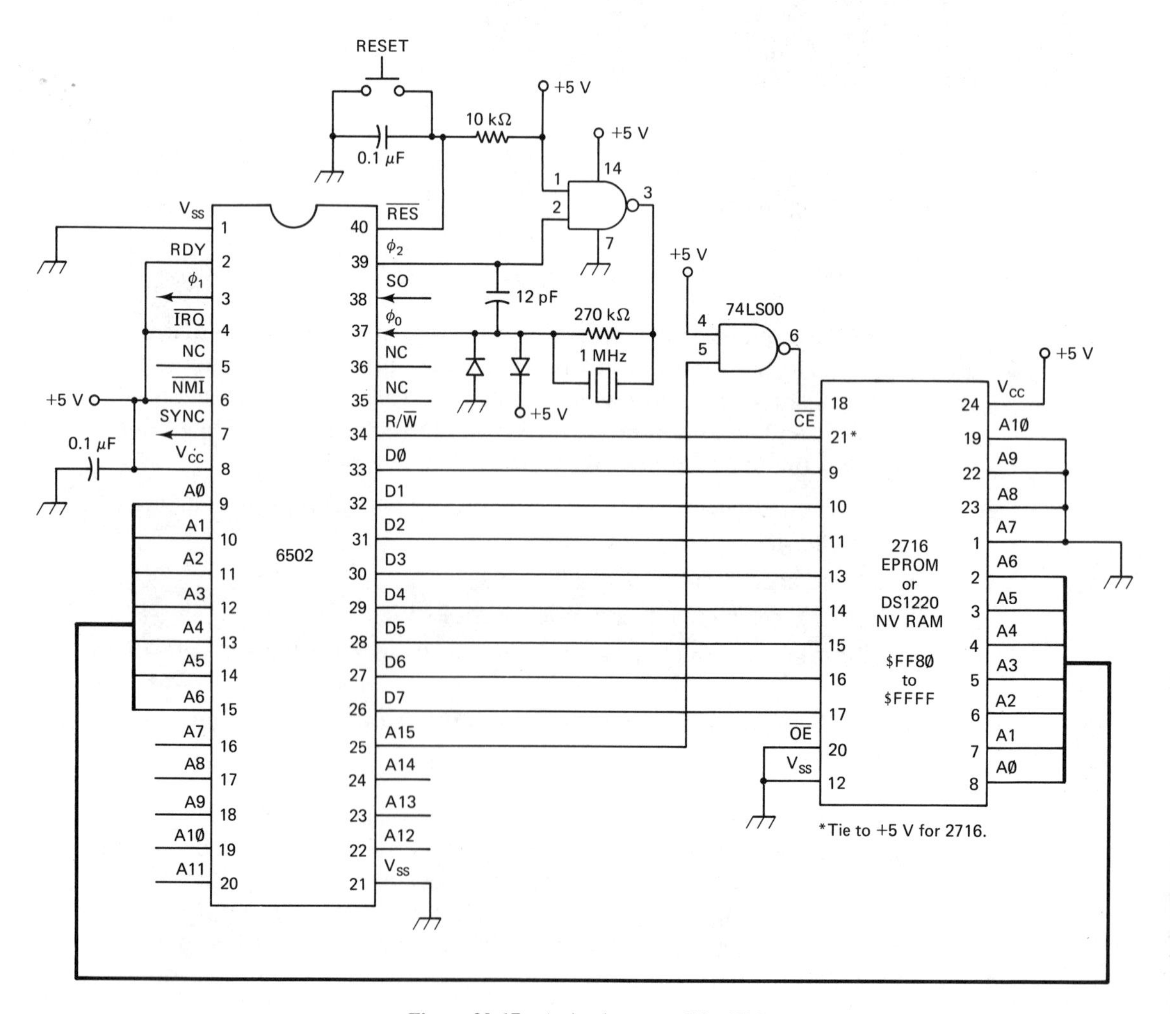

Figure 20.17 A simple-as-possible 6502 test system.

The programmer's model for the 6502 appears in Figure 20.16. The register names should all be familiar to you from the 6800 and 6809 processors. The stack pointer is automatically set to address $Ø1FF upon *reset* and covers only page 1. Most of the status-register bits are the same as in the 6802. When the D flag is *set*, the 6502 is in the BCD mode, and all additions and subtractions are automatically corrected for BCD arithmetic. The SED and CLD instructions set and clear this flag. The B flag is set by the processor when it encounters a *break* instruction. The C flag is set by a carry in an addition, but it is *cleared* by a borrow in a subtract.

A minimum test configuration for the 6502 is shown in Figure 20.17. Notice that the clock circuit requires an external gate. The crystal frequency is the same as the clock frequency, generally 1 MHz. There are two clock signals, Φ1 and Φ2. These signals have nonoverlapping transitions, as shown in Figure 20.18. The Φ2 signal is essentially the same as the *E* signal of the 6802 and 6809.

The RDY signal input is used to stretch the clock period for slow memories, as are MR in the 6802 and MRDY in the 6809.

The SYNC output goes *high* on the first cycle of each instruction. The LIC output of the 6809E is similar.

The SO input sets the overflow flag (bit 6 of the CCR) on a *high*-to-*low* level transition.

Instruction modes for the 6502 are largely the same as those of the 6800/6802. Several of them are named differently. Here is the comparison.

6800/6802	6502	Comment for 6502
Inherent	Implied	Same. Includes *accumulator* mode.
Immediate	Immediate	Same. No 2-byte operands.
Direct	Zero Page	Same. Operand on page $ØØ.
Extended	Absolute	Operand address in LO/HI order.
Relative	Relative	Same.

The 6502 indexed modes are all 3 bytes long. The base address is supplied in the second and third bytes (LO/HI order), and the 8-bit value in X or Y is added in unsigned binary to this base to produce the operand address. The range of indexing is thus only 256 bytes.

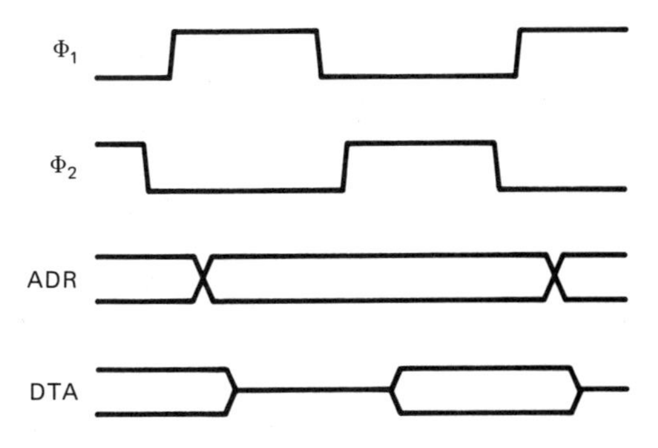

Figure 20.18 The 6502 generates a two-phase, nonoverlapping clock signal. Signal Φ2 is similar to *E* of the 6802.

There are two indirect indexing modes. The first is called *Indexed Indirect X*. A single-byte operand is added to the single byte contained in X. This produces an indirect address (IA) on page ØØ, presumably a RAM area. The final operand address is contained in IA (low-order byte) and IA + 1 (high-order byte).

The second indirect mode is named for confusion: *Indirect Indexed Y*. A single-byte operand points to an indirect address (IA) on page Ø. The processor picks up the low-order byte from IA and the high-order byte from IA + 1, concatenates them, and *adds the value of* Y to produce the address of the operand data. This is a most useful mode because the base address of a file can be changed by reloading IA + 1 on page zero, and the file can be stepped through by incrementing Y. Figure 20.19 illustrates the two indexed/indirect modes.

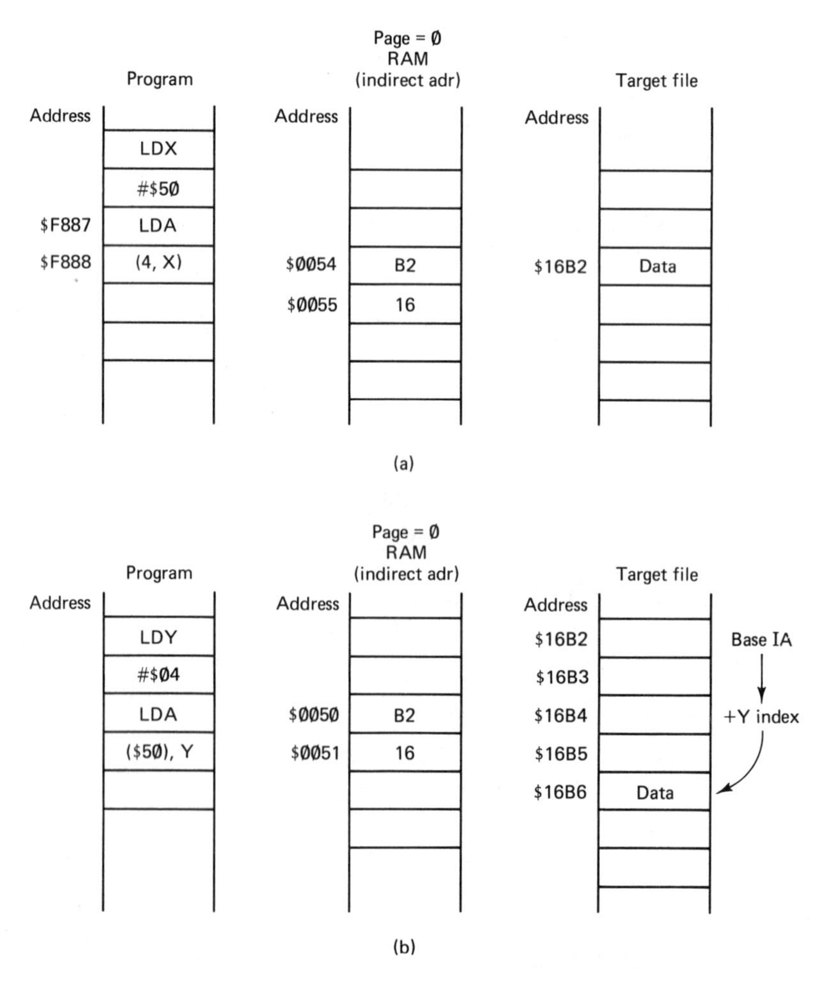

Figure 20.19 (a) *Indexed indirect X* addressing for the 6502. (b) The more useful *indirect indexed Y* addressing mode.

More 6502 notes: The 6502 has three vector addresses at the high end of its memory:

IRQ, negative-level active	\$FFFA (LO), \$FFFB (HI)
RESET, negative-level active	\$FFFC (LO), \$FFFD (HI)
NMI, negative-transition active	\$FFFE (LO), \$FFFF (HI)

There is no software-interrupt instruction, so the 6502 has one less vector address than the 6802. The software instruction BRK also accesses the \$FFFA/B vector.

All 6502 instruction mnemonics consist of three letters, making their correct forms easier to remember than the 6809's (three, four, or five letters). Here are a few examples of the assembly-language forms for each 6502 instruction mode.

Implied: CLC; INX; PHA (push A); TAX
LSR (Shift Right *Accumulator*, since no operand given)

Immediate: LDA #\$9A; CPY #END ORA #\$01
AND #\$F0 (A AND \$F0 → A)

Zero page: STA KEYBUF; LSR \$4F; ADC \$A0

Absolute: LSR DISPLAY; LSR \$3456; LDX BASE

Relative: BEQ PUT; BNE STAY; BCS SMALL

Indexed: LDA \$0500,X CMP BASE,Y

Indexed indirect *X*: SBC (\$A0,X); STA (POINT,X)

Indirect indexed *Y*: SBC (\$A0),Y; STA (POINT),Y

List 20.2 gives a 6502 test-tone program which will run on the system of Figure 20.17.

```
                 1000 * LIST 20-2.  TEST-TONE GENERATOR FOR 6502.
                 1005 * FOR FIG 20.16 HARDWARE.  326 HZ ON ADR LINE 4.
                 1010 * TESTED & RUNNING 28 JULY, 87 BY D. METZGER.
                 1020        .OR $FF8D
                 1030        .TA $408D
                 1040 *
FF8D- CA         1050 LOA4   DEX          2 US  FIRST LOOP KEEPS A4 LO FOR
FF8E- D0 FD      1060        BNE LOA4     3 US  5 US * 256 = 1.28 MS.
                 1070 *
FF90- 69 01      1080 HIA4   ADC #1       2 US  SECOND LOOP COUNTS UP ACCUM
FF92- F0 F9      1090        BEQ LOA4     2 US  JUST TO BE DIFFERENT.
FF94- 4C 90 FF   1100        JMP HIA4     3 US  7 US * 256 = 1.79 MS.
                 1110 *                         F = 1/.00307 = 326 HZ.
                 1120        .OR $FFFC
                 1130        .TA $40FC    RESET VECTOR
FFFC- 8D FF      1140        .HS 8DFF     2 BYTES IN LOBYTE/HIBYTE ORDER
                 1150        .EN
```

Answers to Chapter 20 Review Questions

1. TFR, EXG, PSH, PUL **2.** X **3.** \$5EEE

4. The programmer must supply an assembler directive, such as SETDP \$F0, agreeing with the program's setting of the DP value. **5.** STB (\$4132)

6. It loads D from addresses \$F8AE : F8AF. **7.** Address \$ABCD **8.** Address \$ABCB (high-order byte) and address \$ABCC (low-order byte). **9.** 3 bytes

10. \$1290 **11.** PCHI, PCLO, and CC are saved on the *S* stack. **12.** The I and F flags are set. **13.** None of them. **14.** An $\overline{\text{IRQ}}$ input must be received if I was set or an $\overline{\text{FIRQ}}$ if F was set. **15.** Have it encounter a CWAI instruction. **16.** Addresses \$01EB (high) and \$01EC (low). **17.** 3 bytes

18. The effective address is obtained by concatenating the 2 bytes from locations \$4120 and \$4121. What those bytes are was not given in the question.

19. LEAY −1,Y; op-code \$31 3F. **20.** 10 AC 8B **21.** 5 cycles

22. The fifth cycle is $\overline{\text{VMA}}$. **23.** 5 cycles **24.** Cycles 4 and 5 are writes.

25. 8 cycles **26.** The last six are $\overline{\text{VMA}}$ cycles. **27.** 9 cycles

28. 4 and 5 $\overline{\text{VMA}}$; 6, 7, 8, and 9 are *write*.

29. About 0.5 ms (lines 1600 through 1700, eight passes.) **30.** Line 1430.

CHAPTER 20 QUESTIONS AND PROBLEMS

Digits after the decimal point refer to section numbers in Chapter 20.

Basic Level

1.1 How many bytes are required for a 6809 load-X, immediate instruction? How many for Load-Y, immediate?

2.1 What is a 68B09, as compared with a 6809?

3.1 Why might we want to replace an extended-mode instruction with a program-counter-relative version of that instruction?

4.2 Write the 6809 source code that will store the double accumulator to the 2 bytes pointed to by X and X + 1, and then point the X register to the next available byte, X + 2.

5.2 Can the autoincrement/decrement feature of 6809 indexed addressing be combined with the accumulator offset feature?

6.3 Do the 6809 interrupts save the machine registers to RAM addresses pointed to by the system stack pointer S or the user stack pointer U?

7.3 Can the FIRQ interrupt be masked out? Why is FIRQ faster than IRQ?

8.4 What flags are affected by the 6809 TFR instruction?

9.4 What is the object code for the 6809 instruction to load A from the address pointed to by X, with no offset?

10.5 Use the 6809 cycle-by-cycle chart to determine the number of cycles taken by the instruction ADDA B,Y. Also tell which, if any, of these cycles are VMA-not cycles.

11.5 The 6802 relative- and inherent-mode instructions use instruction prefetch; that is, they read the address of the next sequential instruction while they are executing the current instruction. Is this also true of 6809 short-branch and inherent instructions?

12.7 What crystal frequency is needed for a 1-MHz 6502 clock?

13.7 What is the 6502's term for *extended*-mode addressing?

14.7 If X contains data $90 and the op code for Load A, X-indexed, is $BD, what address is accesed by the 6502 object code

```
BD C4 21
```

Advanced Level

15.1 What is the addressing mode of the 6809 instruction

```
LDD (KEYPTR)
```

Explain in what sense this is and in what sense it is not an indexed addressing mode.

16.1 Explain the difference between the 6809's direct, indirect, and nonindirect addressing modes.

17.2 Why do the 6809's autoincrement/decrement indexing modes use postincrement but *pre*decrement? Wouldn't it have been simpler to keep them the same?

18.2 List the types of offsets that are available with the 6809 indexed addressing instructions, and tell whether the offsets are signed or unsigned.

19.2 Write the 6809 source code for a routine to move a file from address range $E700 through $ECFF to a RAM area starting at $58FF and working down to lower addresses.

19.3 Explain the function of the 6809's E-flag bit.

20.3 What are the functions and advantages of a Priority Interrupt Controller IC?

20.4 Use the 6809 programmer's reference to hand assemble the program written for Problem 19.2.

21.4 What, if anything, is wrong with the instruction

```
TFR B,X
```

How would you accomplish the intended operation?

22.4 Explain the function of the Sign-Extend instruction, and tell when it should be used.

22.5 Use the 6809 cycle-by-cycle chart to determine the number of machine cycles taken by the instruction

```
PSHU A,B,CC,X,Y
```

If the initial value of U is \$Ø2F3, give the addresses to which the two bytes of X will be saved.

22.6 Is the multiply-back division routine of List 20.1 more or less efficient than the eight-step divide routine of Section 19.4 (a) in terms of byte length and (b) in terms of execution speed?

22.7 Explain how the indirect indexed *Y* mode of the 6502 can be used to access files longer than 256 bytes, even though the index register is only 8 bits wide.

21.1 Z80 MACHINE REGISTERS

Section Overview

The Z80 was introduced by Zilog in 1977 as an expanded version of Intel's very popular 8080 microprocessor. It includes all the features of the 8080 and most of those of Intel's newer 8085 in addition to a host of new instructions and registers.

The 8080 contains a primary accumulator A and a primary 16-bit address pointer HL. It also contains two general-purpose register pairs BC and DE, which can be used for data transfer or as address pointers for the accumulator A. The Z80 contains two sets of the original 8080 registers plus two index registers IX and IY.

The programmer's model for the 8080 and 8085 microprocessors is shown in Figure 21.1. The Z80 is an expansion of the 8080, so we will find it easier to learn the functions of the 8080 registers as a first step to tackling the more complex Z80.

The 8080 has one accumulator, A. This is the only register that permits a full

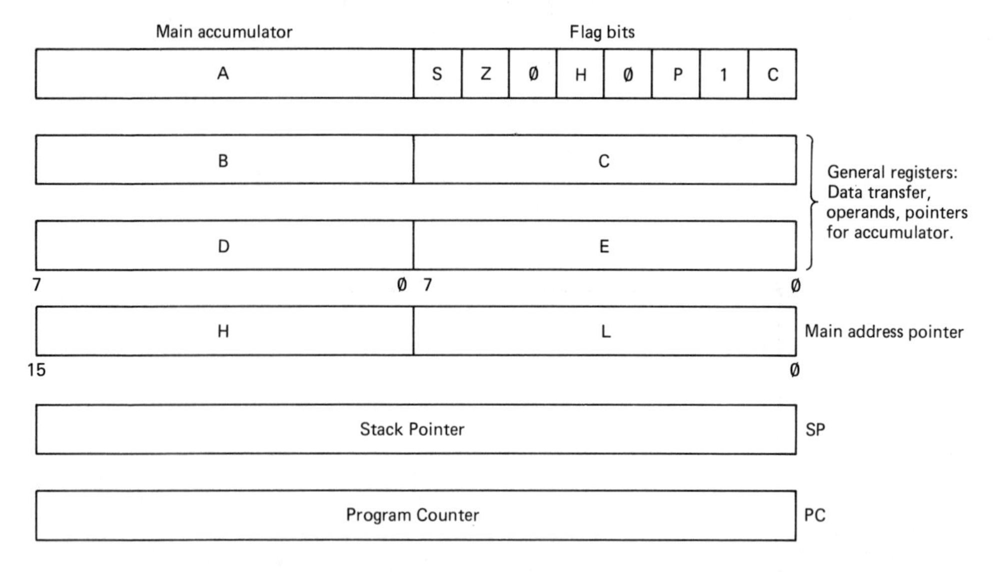

Figure 21.1 A programmer's model for the 8080 and 8085 microprocessors.

set of arithmetic and logic operations, and many instructions presume A as the active register without even mentioning it. For example, in Z80 code

AND $FØ means A *and* $FØ → A

even though the instruction mnemonic doesn't specify A.

The primary address pointer in the 8080/8085 is the 2-byte register pair, HL. This pair can be addressed as a single register or as two separate registers. For example, in Z80 code

- INC HL increments the 16-bit value in register [HL], but
- INC H increments the high-order byte of the pair, adding 256 to the contents of the total 16-bit register.

The four general-purpose registers B, C, D, and E are likewise accessed by some instructions as four separate 8-bit registers and by others as two 16-bit registers, BC and DE. However, when used separately, B, C, D, and E are capable of holding data for transfer, incrementing/decrementing, and rotating/shifting. They can hold the operand data for arithmetic and logic operations with A, but they cannot be the target or *destination* register for these operations. When used in pairs, BC and DE can serve as pointer registers for A only. This is in contrast to HL, which can point for any of the registers A, B, C, D, E, H, or L, as well as for immediate data.

The 8080/Z80 stack-pointer and program-counter registers are essentially the same as their counterparts in the 6802.

The flag register F contains the condition-code bits. C, Z, V, and H are *set* for carry, zero, overflow, and half-carry conditions, just like their counterparts in the 6802. The C flag is *set* by a borrow on a subtract operation, which is the same as the effect in the 6802. S is the *sign* flag, the same as N in the 6802 but with a different name. N in the Z80 is the *subtract* flag, set by the subtract, compare, and decrement operations. It is used by the processor in conjunction with the H flag and the DAA instruction in BCD arithmetic. *Parity* and *Overflow* share the same P/V flag. P is *set* if a logic operation produces even parity (even number of 1 bits). It is *cleared* by logic operations producing odd parity. The same flag, now called V, is *set* by arithmetic operations that produce an overflow into the sign bit.

The Z80 register set is shown in Figure 21.2. It contains the 8080 register set and adds two 16-bit index registers, IX and IY. These are convenient, but HL is still the main pointer register because instructions using IX or IY typically require twice as many bytes and twice as many cycles as those using HL.

The Interrupt-vector register is used in connection with other hardware and programming that can direct the Z80 to one of as many as 128 interrupt routines without time-consuming polling routines. The memory Refresh register is part of the Z80's built-in capability for refreshing dynamic RAMs. Both of these features are explained in Chapter 22.

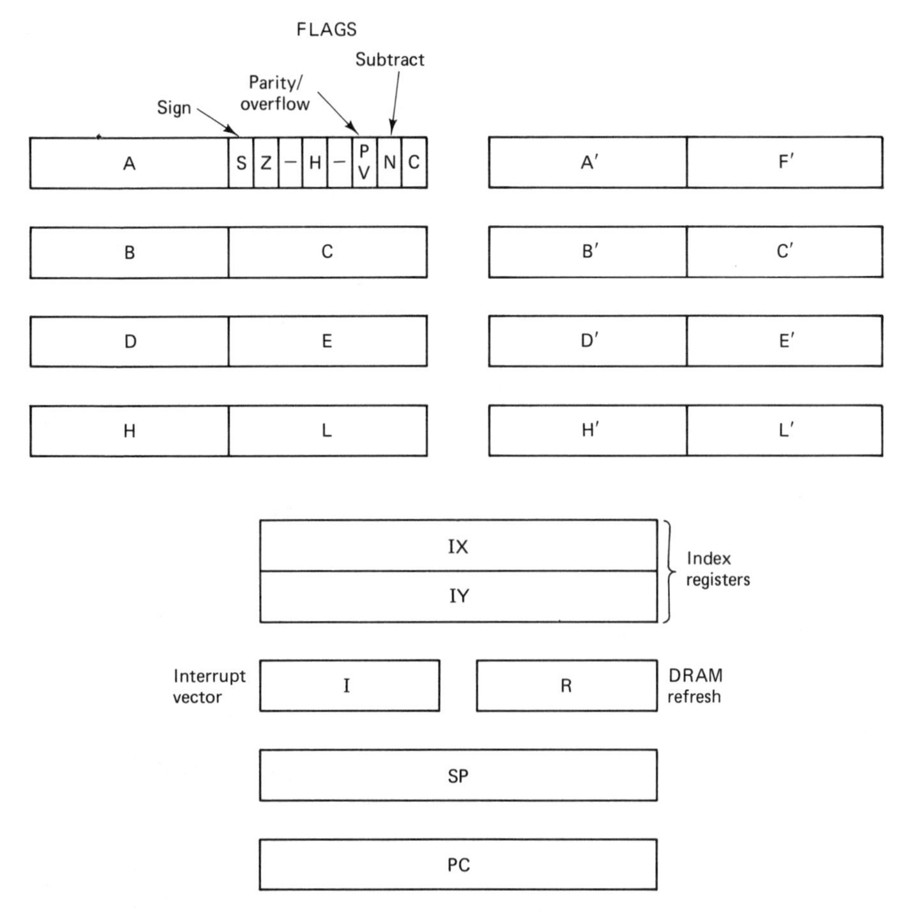

Figure 21.2 The Z80 programmer's model is an expansion on the 8080 register set.

An alternate register set is available in the Z80 for extremely fast handling of interrupts and multiple programming tasks. There is no need to stack the intermediate data of a partially completed routine to service an interrupt. The main register set can simply be left suspended while the interrupt is processed on the alternate register set. It must not be supposed that there is a second set of "primed" instructions for the primed registers; the same instructions are used for both. The processor is toggled from the main set to the alternate set and back by executing the instructions.

EX AF,AF′	to switch program references to the accumulators and flag registers
EXX	to switch program references to registers B, C, D, E, H, and L.

21.2 Z80 PIN FUNCTIONS AND TIMING

Section Overview

The Z80 comes in a 40-pin package with 16 lines devoted to address outputs and 8 lines for the data bus. The V_{CC}, *ground*, *reset*, *interrupt-request*, and $\overline{\text{NMI}}$ pins are the same as their counterparts in the 6802.

$\overline{\text{MREQ}}$ is like the 6802's VMA output. Two separate outputs, $\overline{\text{RD}}$ and $\overline{\text{WR}}$, signal a read or write to external devices; both remain *high* if external devices are not being accessed. A single-phase TTL-level clock must be supplied to the Φ input. The other Z80 pins are not used in elementary applications.

A cycle of the clock Φ is called a *T*-cycle and may have a maximum frequency of 2.5 MHz for the Z80, 4 MHz for the Z80A, 6 MHz for the Z80B, and 8 MHz for the Z80H. A machine cycle requires from 3 to 6 *T*-cycles, and an instruction may require 1 to 6 machine cycles (4 to 23 *T*-cycles).

The pin diagram for the Z80 microprocessor is given in Figure 21.3. The address and data lines have about the same drive capability as their 6802/6809 counterparts—that is, 4 LS-TTL parts plus 8 CMOS or microprocessor parts. The data and address buses are capable of a high-*Z* or floating state when the processor is not using the bus. An external device requests the Z80 to relinquish the bus by asserting the $\overline{\text{BUSRQ}}$ input. The Z80 asserts its $\overline{\text{BUSAK}}$ output to acknowledge that it has relinquished the bus.

The Z80 utilizes a single +5-V supply and requires 120 to 200 mA, essentially the same as the 6802 and 6809. A clock signal must be externally generated and applied to the Φ input pin. This signal should have a 50% duty cycle, and it must pull *high* to at least 4.4 V. A 330-Ω pullup resistor to V_{CC} will permit the clock to be driven by an ordinary LS-TTL gate. The maximum frequency of Φ is 2.5 MHz for the Z80, 4 MHz for the Z80A, 6 MHz for the Z80B, and 8 MHz for the Z80H.

The $\overline{\text{RESET}}$ input functions similarly to that of the 6802, but there is no internal Schmitt trigger, so a switch-debounce circuit should be used to drive it. There is no

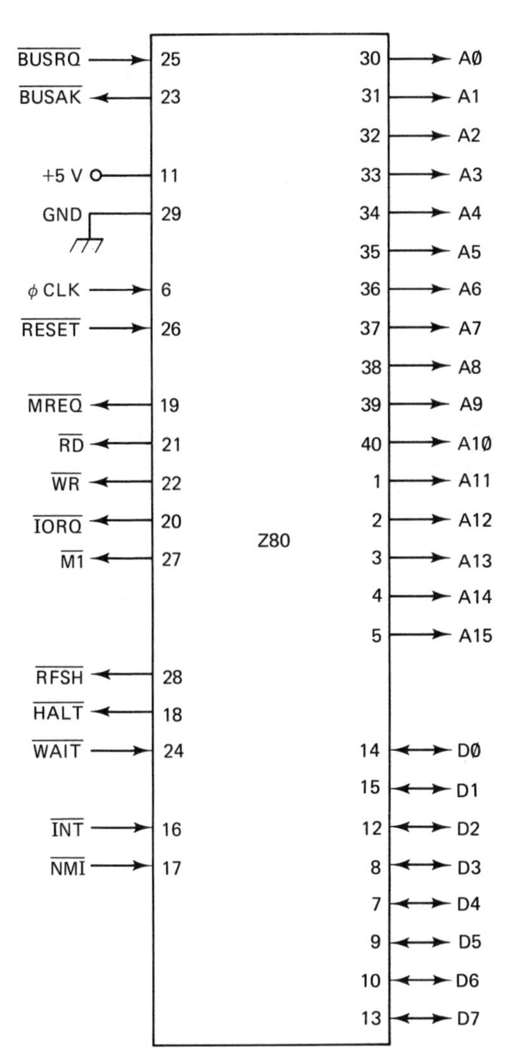

Figure 21.3 Pin functions for the Z80 microprocessor.

reset vector address in the Z80. When $\overline{\text{RESET}}$ is allowed to return to a *high* level, the processor immediately reads the data at address $\$0000$ and assumes that this is the first instruction of the main program. This has implications for the system structure. In 68XX systems we must place ROM at the high end of memory to hold the reset and interrupt vectors, and we tend to put RAM at the low end of memory to take advantage of the faster zero-page addressing. In Z80 and 808X systems we must put ROM at page 0, and we tend to put the RAM at the high end of memory.

The $\overline{\text{MREQ}}$ output is asserted only on cycles when the processor is accessing memory. It corresponds to the 6802's VMA line, except that it is *low* active. Instead of a single $R/\overline{W}$ output there are 2 lines for this function. $\overline{\text{RD}}$ goes *low* on memory or peripheral reads only and $\overline{\text{WR}}$ goes *low* on memory or peripheral writes. $\overline{\text{RD}}$ and $\overline{\text{WR}}$ contain timing information as well as data-direction, in contrast to $R/\overline{W}$ which contains only data-direction information. It is possible for $\overline{\text{RD}}$ (for example) to be asserted while $\overline{\text{MREQ}}$ is not asserted. This would mean that the processor is reading, not from memory but from an input device. $\overline{\text{IORQ}}$ (Input/Output Request) would be asserted in this case. Section 22.2 explains how the Z80 handles I/O devices differently from memory.

$\overline{\text{M1}}$ is an output that the Z80 asserts (*low*) to signal the first machine cycle of each instruction. $\overline{\text{RFSH}}$ is an output used in conjunction with the Z80's built-in ability to refresh dynamic RAMs. $\overline{\text{HALT}}$ is not an input as it is on the 6802. It is an output brought *low* when the Z80 executes a HALT instruction, and it is used to halt other parts of the system. $\overline{\text{WAIT}}$ is a cycle-stretching input used to delay the processor until slow memories can get valid data on the bus.

$\overline{\text{INT}}$ is an interrupt request input, active low, just like the $\overline{\text{IRQ}}$ on the 68XX processors. It can be masked out by disabling an internal *interrupt-enable* flip-flop. Section 22.3 covers interrupts in more detail. $\overline{\text{NMI}}$ is, of course, a nonmaskable interrupt input, and it is negative-edge sensitive, just as in the 68XX processors.

Clock timing for the Z80 is not as simple as it is for the 6802. Each clock cycle in Z80 terminology is called a *T*-state. A memory-read or memory-write *machine cycle* may consist of 3, 4, or 5 *T*-states. Figure 21.4 shows elementary *read* and *write* machine cycles. A complete instruction consists of anywhere from 1 to 6 machine cycles and may take from 4 to 23 *T*-states. The first machine cycle (the op-code fetch) takes longer than a regular memory read, and includes some extra functions.

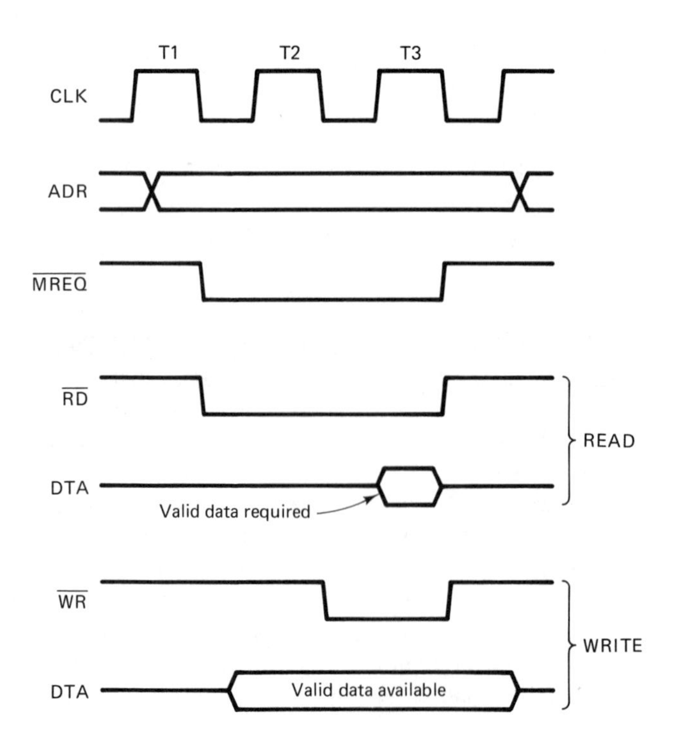

Figure 21.4 Read and write timing for a Z80 memory-access *T*-state.

REVIEW OF SECTIONS 21.1 AND 21.2

Answers appear at the end of Chapter 21.

1. NEG is a valid Z80 instruction indicating *negate* or *form 2's complement*, just NEG—nothing else specified. Where is the data that is negated?
2. Which register of the Z80 corresponds most closely to the 6802's X register?
3. True or false:
   (a) DE can serve as a pointer for A.
   (b) HL can serve as a pointer for B.
   (c) DE can serve as a pointer for B.
4. In a Z80, A contains $5B. A is then ANDed with data $95. What is the status of the P/V flag?
5. To store 2 kilobytes of data to a video-display RAM, would you prefer to use HL or IX as a pointer? Why?
6. You have just incremented the 16-bit main pointer register with the INC HL instruction. How would you now make a corresponding increment in the alternate register HL′?
7. How does the Z80 locate its starting address upon *reset*?
8. If the Z80 $\overline{\text{HALT}}$ pin is not to be used, where should it be connected?
9. What are the logic levels on the Z80 lines $\overline{\text{MREQ}}$, $\overline{\text{RD}}$, $\overline{\text{WR}}$, and $\overline{\text{IORQ}}$ when the processor writes data to RAM?
10. The instruction JP $1234 tells the Z80 to jump to address $1234. It takes 3 machine cycles or 10 *T*-states. How long will this instruction take to execute in the Z80A version of the CPU?

21.3 Z80 ASSEMBLY LANGUAGE FORMS

Section Overview

By far the most common Z80 instruction is

LD destination,source

This instruction serves all the *load*, *store*, and *transfer* functions for all of the many Z80 registers, and it has well over a hundred forms. The Z80 assumes a plain number or variable name to be immediate data; the immediate sign (#) is not used. Numbers or variable names that represent addresses must be placed in parentheses: (ADDR). In Z80 machine language when 16-bit data (such as addresses) must be stored in memory, it is stored in low-byte/high-byte order.

Register-oriented is a term often applied to the Z80 microcomputer. It has seven 8-bit registers, and each can be the source or destination for a data transfer. The two-way combination possible with A and B in the 6800 (TAB and TBA) becomes a 7-by-7 matrix of 49 instructions in the Z80. Where the 6800 has two 16-bit registers X and S, the Z80 has BC, DE, HL, IX, IY, and S, again resulting in a profusion of instruction variations. Z80 programmers tend to bring the data they need into the machine registers, process it there without reference to memory, and then send the final result out. 6800 programmers tend to be more memory-oriented, since they have so few machine registers to rely on.

The load instruction is the mainstay of the Z80 programs. It serves all the functions of the *load*, *store*, and *transfer* instructions of the 6800. It has the general form, in assembly language, of

```
LD destination,source
```

where destination may be a machine register or a memory location and source may be either of these or immediate data. In many cases the destination and/or source can be a 16-bit register, such as HL, BC, or IY.

The assembly format for most Z80 instructions follows a similar pattern. The op-code field gives the instruction mnemonic, such as LD, ADD, or INC, without any reference to a register or operand.[1] A space follows, and then the destination and source operands (if any) are given, separated by a comma. Here are two examples:

```
SBC A,E        Subtract E and Carry Flag from A; result in A.
ADD HL,BC      Add 16-bit register pairs; result in HL.
```

Often there will logically be only one entry in the operand field. Occasionally there will be none. Examples are:

```
INC BC     Increment register pair BC.
RL D       Rotate D left through carry.
CCF        Complement carry flag.
```

A special problem is that a number of instructions that obviously require a source and a destination operand may list only one operand. In such cases the listed operand is the source and the accumulator (A) is assumed as the destination. Examples are:

```
SUB B     Subtract B from A; place result in A.
OR C      Logic OR A with C; result in A.
CP D      Compare A − D to set flags. No change in data.
```

There seems to be no reason why, for example, the form for subtract without regard to carry-input is

```
SUB B
```

but the form for subtract operand and carry-input is

```
SBC A,B
```

[1]Two exceptions are DAA (decimal adjust A) and the *rotate-A* instructions (RLA, RRA, RLCA, RRCA) that reference A in the op-code field.

Such anomalies require that apprentice Z80 programmers carry and refer frequently to a programmer's reference card.

Extended and immediate addressing are indicated differently in Z80 code than they are in 68XX code. In 6800 code a simple number or variable name, unaccompanied by other symbols, refers sometimes to the data at an address and sometimes to the address itself. Here are two examples of each.

Data at Address is Referenced	Address is Referenced
LDAA $1234	JMP $1234
ADDA PLACE	BEQ PLACE

Z80 assembly language avoids this ambiguity by placing all references to addresses which point to data in parentheses. Here are some examples comparing the two assembly-language formats.

6800	Z80
LDAA $89AB	LD A,(89ABH)
STAA BUFR	LD (BUFR),A
LDX $73	LD HL,(73H)
STAB 0,X	LD (HL),B

Note that Z80 programmers tend to use the post-*H* rather than the pre-$ to indicate hex. This is just a matter of style. It is also worth noting that the Z80 stores 16-bit data in memory in reverse order. The machine code for the first Z80 instruction above, for example, is

```
3A AB 89    LD A,(89ABH)
```

Immediate data in 68XX code is always preceded by the immediate sign #. In Z80 code a plain number or variable name unaccompanied by other symbols is interpreted as immediate data. Again, some examples:

6800	Z80
LDAA #$1F	LD A,1FH
LDAB #CODE	LD B,CODE
LDX #5280	LD HL,5280

21.4 Z80 INSTRUCTION MODES AND GROUPS

Section Overview

The Z80 has most of the same addressing modes as the 6800/6802, except for the following:

- **There is no zero-page mode (except for one instruction.)**
- **The extended mode is available only for data-move (load) operations in the Z80, so it is much less heavily used than it is in the 68XX family.**
- **The mode called *register indirect* (which would be called zero-offset indexed in 6800 terminology) must be used for all arithmetic and logic operations involving memory, so it is much more heavily relied upon in Z80 programming.**
- **The *register* mode, which involves operations on or between internal data registers, is extremely important in Z80 programming. This mode is much less important in 6800 programming and would be considered part of the *inherent* mode.**
- **The mode called *bit* addressing would not be thought of as a mode at all in 6800 terms, but as a group of instructions in the *register* and *register-indirect* modes.**
- **There is a combined indexed/immediate mode in which 8-bit immediate data is transferred directly to an address contained in a 16-bit pointer register by a single instruction.**

Z80 instructions are more easily divided according to the registers involved than according to addressing modes. This is because the large number of internal registers produces huge numbers of instructions of what (in 6800 terms) would be called the inherent and indexed modes. The concept of addressing modes is taken more for granted by Z80 programmers, and in many cases it is neither particularly clear nor particularly important what the addressing mode of a given instruction is.

Implied and register addressing cover instructions that do not specify addresses outside the CPU. These would be called *inherent mode* in 6800 terms. Register instructions generally have seven forms each, to access one of the 8-bit registers A, B, C, D, E, H, or L. Three bits within the op code are reserved to specify which register is the operand. Examples of *implied* and *register* instructions are

CPL		Complement *A*.
EI		Enable interrupt (set interrupt flip-flop).
INC	IX	Increment index register *X*.
AND	L	Logic AND of *A* with *L*; result in *A*.
LD	B,H	Load *B* from *H*.
ADD	IY,DE	Add DE to IY; Result in IY.

The more common instructions of these modes require a single byte and only 4 *T*-states. The less common instructions require two bytes (both op-code bytes) and as many as 15 states.

Immediate-mode instructions have the same form and functions as their 6800 counterparts. The 8-bit *immediate* instructions require 2 bytes (an op-code and an operand), and take 7 *T*-states. Zilog calls the 16-bit *immediate* instructions *extended*

immediate mode, and these require 3 or 4 bytes and as many as 14 *T*-states. Examples of Z80 immediate instructions are

LD	H,255	Put $FF in register H.
SUB	BLOCK	Subtract value BLOCK from A.
LD	BC,1000	Put decimal 1000 in register pair BC.

Extended-mode addressing is sometimes called direct addressing in Z80 or 8080 literature. It is the same as 6800 extended addressing, but it is much less used in Z80 programming. In particular:

- Only A among the 8-bit registers can be loaded or stored by extended addressing. The others—B, C, D, E, H, and L—must use register HL, IX, or IY to point to the operand address.
- Any of the 16-bit registers BC, DE, HL, IX, IY, and SP can be loaded or stored in the extended mode by the Z80.
- Arithmetic and logic operations have no extended-mode versions in Z80 language. They can be done only on A, and the operand must be a register, immediate data, or an address pointed to by HL, IX, or IY.
- The rotate, shift, bit-test, and bit-set instructions have no extended-mode versions. They always use pointer registers to access memory.

Examples of extended-mode instructions are

LD A,(TWOBIT)	Load data from address TWOBIT into A.
LD (2345H),A	Store A to address $2345.
LD DE,(789AH)	Load DE from $789A (low-order byte) and $789B (high-order byte).

These instructions take from 13 to 20 *T*-states, and this relatively slow speed largely accounts for the limited use of extended addressing in the Z80. They generally require 3 bytes, although the loads to and from IX and IY have a 2-byte op-code and thus require 4 bytes. Remember that the extended address bytes appear in memory after the op-code in *low-byte/high-byte* order.

Register indirect addressing uses one of the pointer registers HL, BC, or DE to hold the address of the operand. This should be thought of as the primary or most-common memory addressing mode of the Z80. The word *indirect* signifies that the processor finds the operand data indirectly by first going to a pointer register to get the address where the data resides. The name was given in the early days of the 8080 and is somewhat at odds with the newer meaning of *indirect*, which denotes an address that in turn contains the final address.

HL is the main pointer register, since it can point for A, B, C, D, E, H, or L. The registers BC and DE are considered secondary pointers, since they can point only for A.

The register-indirect mode is, of course, similar to the 6800's indexed mode, except that the former has no offset capability. The Z80 has a separate *indexed addressing* mode, in which registers IX and IY serve as the address pointers. These instructions usually include a 2's complement displacement byte, which is added to IX or IY to form the final operand address. This is in contrast to the 6802's offset byte, which is added to X in straight (unsigned) binary.

The register-indirect instructions are generally preferred to the indexed instructions because they are faster. Note the comparative speeds of the following paired examples.

```
LD    A,(DE)      Load A from address in DE; 1 byte, 7 T-states.
LD    A,(IX+0)    Load A from address in IX; 3 bytes, 19 T-states.

INC   (HL)        Increment data at address of HL; 1 byte, 11 T-states.
INC   (IY+1)      Increment data at address IY + 1; 3 bytes, 23 T-states.

JP    (HL)        Jump to address given in HL; 1 byte, 4 T-states.
JP    (IX)        Jump to address given in IX; 2 bytes, 8 T-states.
```

Relative addressing in the Z80 is just like its counterpart in the 6802. The instructions consist of two bytes—the op-code and a 2's complement signed displacement. The displacement is limited to a range of -128 to $+127$ from the address of the following instruction. As with the 6802, relative instructions are used primarily for decision-making instructions. They are called conditional relative *jumps*, however. Nothing in the Z80 instruction set is called a *branch*. They require 2 bytes and 7 *T*-states if the jump is not taken, or 12 *T*-states if it is taken. Here are some example forms.

```
JR    Z,OUT       Jump relative, if result = Ø, to address OUT.
DJNZ  LOOP        Autodecrement B; jump to LOOP if result ≠ Ø.
JR    PART2       Unconditional jump, relative.
```

Bit addressing refers to a group of Z80 instructions that can access any individual bit in an 8-bit register or address and test its state, set it, or clear it. The bit is indicated by writing a number Ø through 7 as the *destination* in the assembly-language statement. The byte being accessed is indicated as the *source*. This byte may be one of the registers A, B, C, D, E, F, H, or L, or it may be a memory location pointed to by (HL), (IX + displacement), or (IY + displacement), so you may prefer to consider bit addressing to be a set of instructions under the register, register indirect, and indexed addressing modes.

There are a few combined-mode instructions. An example is *load, immediate/register-direct* in which the source data is specified as immediate data and the destination is pointed to by register HL. The form is

```
LD (HL),7EH
```

Figure 21.5 summarizes the Z80 addressing modes.

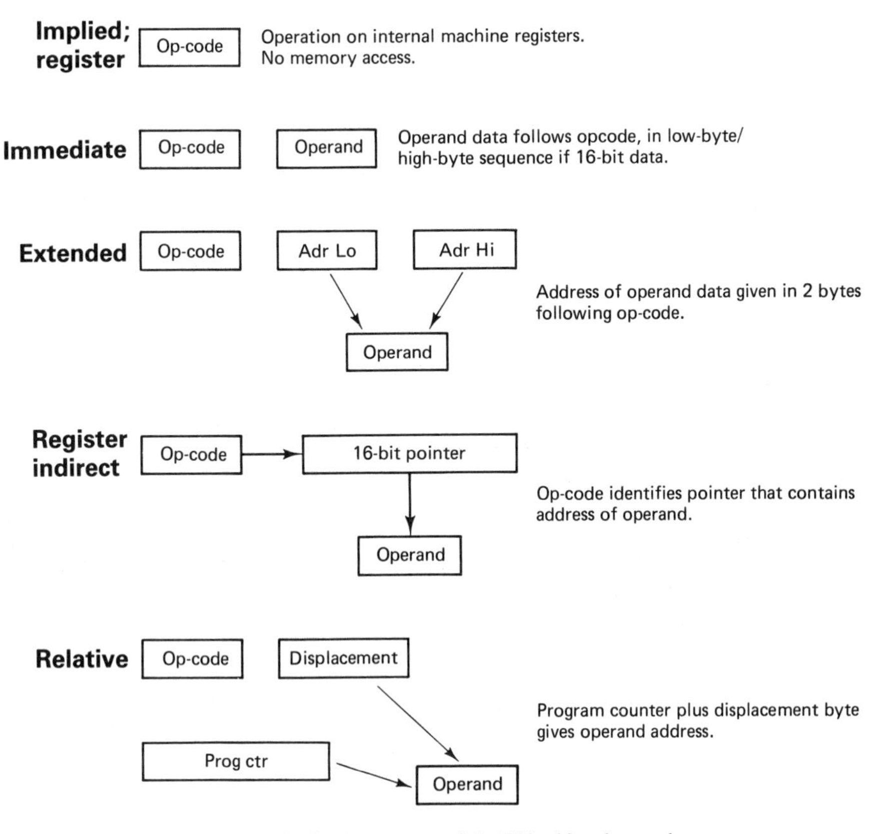

Figure 21.5 A summary of the Z80 addressing modes.

REVIEW OF SECTIONS 21.3 AND 21.4

Answers appear at the end of Chapter 21.

11. What does the Z80 instruction LD A,B do?
12. Where is the result stored in the instruction AND E?
13. What is the Z80 equivalent of the 6800 instruction ADDA #4?
14. What is the Z80 equivalent of the 6800 instruction STX $4E3?
15. In question 14 the 16-bit Z80 register contains data $ABCD. What data bytes are stored to what addresses by the Z80 instruction?
16. Two forms of one of the Z80 *register* instructions are LD A,B and LD B,C. How many forms are there of this one instruction?

17. Which of these are valid Z80 extended-mode instructions?
 a. LD B,(127) b. ADD A,(127)
 c. LD (127),A d. LD BC,(127)
18. Which of these are valid register-indirect instructions?
 a. LD A,(BC) b. LD (DE),C
 c. ADD A,(HL) d. ADD B,(HL)
19. Which is preferred, register-indirect or indexed addressing? Why?
20. What is the Z80 equivalent of the 6800 instruction BNE QUIT?

21.5 EIGHT-BIT Z80 INSTRUCTIONS

Section Overview

The Z80 instruction set is generally divided into groups by function. This section covers the 8-bit groups. The next section covers the 16-bit groups, and an abridged instruction list appears in Chapter 22.

- **The 8-bit load group consists entirely of LD instructions in the *register* (inherent), *immediate*, *extended*, *register-indirect*, and *indexed* modes.**
- **The 8-bit arithmetic and logic group contains INC and DEC instructions in the *register* and *indexed* (including register-indirect) modes, and the following instructions with A as the only allowable destination register: ADD; ADC; SUB; CP (compare); AND; OR; XOR.**
- **The Rotate group includes rotates right and left through carry (RR,RL) and "circular" rotates through 8 bits only (RRC,RLC) as well as arithmetic and logic shifts.**
- **The Bit group allows test, set, or clear of any bit in any register or memory byte.**

The 8-bit load group contains 49 register instructions of the form

LD r,r′

where r and r′ may be any of A, B, C, D, E, H, and L. These instructions are heavily used because arithmetic and logic operations always produce their results in A, and register loads are a fast way to get the results out of A to make way for the next operation.

There are seven **load-immediate** instructions of the form LD C,FFH.

There are 21 **load-indexed** (including register-indirect) instructions with a register as the destination and (HL), (IX + displacement) or (IY + displacement) specifying the source address. Remember that the displacement is an unsigned constant byte. There are 21 similar instructions with one of the 3 pointers specifying the source address and one of the 7 registers as the destination. These would be STA instructions in 6800 language.

There are 6 instructions that load **to or from A only**. The address pointer (source

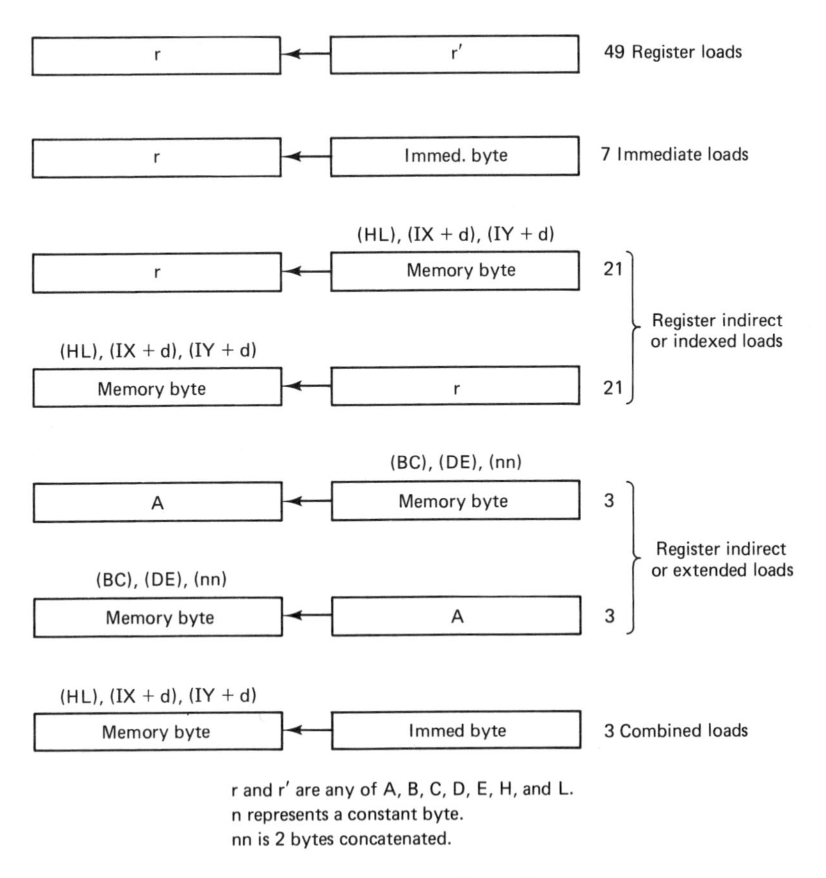

Figure 21.6 The *8-bit load* instruction group.

or destination) may be (BC), (DE), or (nn). The letters nn indicate a constant 2-byte address, extended mode.

Finally, there are 3 instructions that store an **immediate byte directly** to a memory location pointed to by (HL), (IX + d) or (IY + d), without involving any of the data registers. These have the form

```
LD (IX+8),0FH
```

Figure 21.6 illustrates the 8-bit load group of Z80 instructions. We have left out 4 instructions dealing with the interrupt and refresh registers. We will see some of these in Chapter 22.

The 8-bit arithmetic and logic group contains the following 8 instructions, where M is an 8-bit operand, however addressed.

ADD	A,M	Add byte to A; result in A.
ADC	A,M	Add byte plus carry (if set); result in A.
SUB	M	Subtract byte from A; result in A.
SBC	A,M	Subtract byte; also subtract 1 if C flag = 1.
CP	M	Compare A − M to set flag bits only.
AND	M	Logic AND of A with M; result in A.
OR	M	Logic OR of A with M; result in A.
XOR	M	Exclusive OR of A with M; result in A.

Each of these instructions can obtain the source byte from

- Data register A, B, C, D, E, H, or L (7 sources),
- An immediate byte (1 source),
- Memory via pointer (HL), (IX + d), or (IY + d) (3 sources).

There are thus 11 sources for each of 8 instructions, or 88 arithmetic and logic instructions. Notice that only A is permitted as the data destination in each case. The increment and decrement instructions can operate on any of the 7 data registers or on memory via any of the 3 pointer registers. There are thus 20 INC M and DEC M instructions.

Figure 21.7 illustrates the 8-bit arithmetic and logic instruction group.

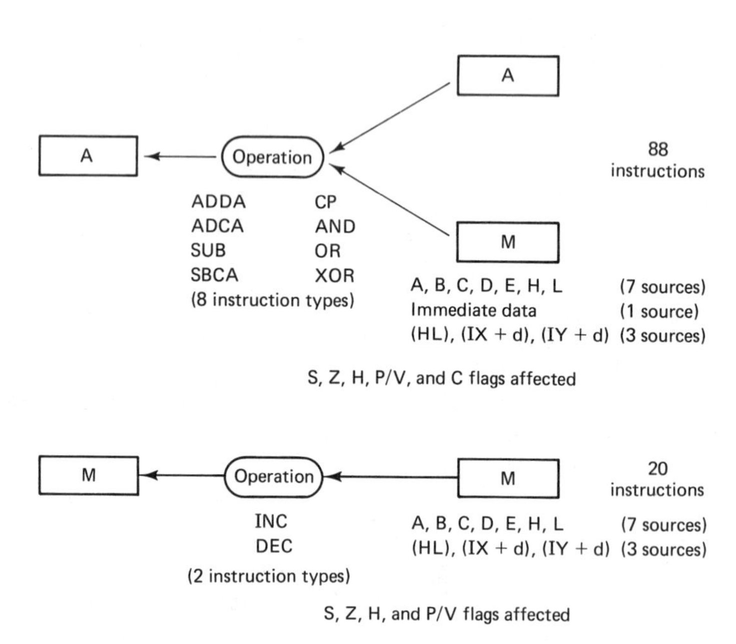

Figure 21.7 The Z80 eight-bit arithmetic and logic instruction group.

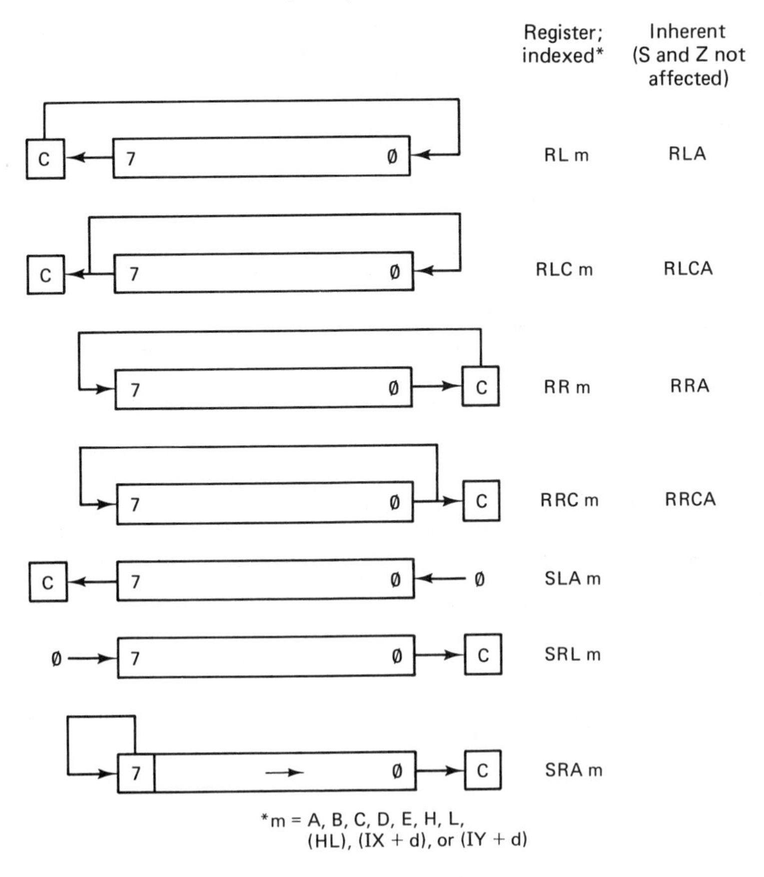

Figure 21.8 The Z80 rotates can be through the carry flag (as in the 6800) or around the 8 register bits only.

The rotate and shift group contains 4 types of rotates and 3 types of shifts, as illustrated in Figure 21.8. For each of these, the operand byte may be one of the 7 data registers or a memory byte accessed by one of the three pointers, (HL), (IX+d), and (IY+d). There are thus 6 op-codes with 10 operands each, or 60 instructions in this group.[1]

The four *rotate* instructions are duplicated in function *for the A register only*

[1]We do not treat the two special instructions RLD and RRD.

by four inherent-mode rotate instructions. These are much faster than the register-mode instructions:

RLA	Rotate left through carry, *inherent*; 1 byte, 4 *T*-states.
RL A	Rotate left through carry, *register*; 2 bytes, 8 *T*-states.

If these are counted separately, there are 64 instructions in the 8-bit rotate and shift group.

The BIT group contains three operations:

- BIT sets the Z flag to the state of the bit being tested.
- SET writes a logic 1 to the indicated bit.
- RES writes a Ø to the indicated bit.

Each of the 3 basic instructions can access data from 10 sources; the 7 registers and the memory pointed to by (HL), (IX + d), and (IY + d). The destination is considered to be one of the 8 bits. You might say that there are 3 × 10, or 30, instructions in this group. However, the destination bit is not specified by a separate byte in the machine code. It is specified by 3 bits buried in a second op-code byte along with bits specifying the source register. We are almost forced to say that there are

$$3 \text{ operations} \times 10 \text{ sources} \times 8 \text{ destinations}$$

or 240 instructions in this group. Here are some examples of the bit-group format.

BIT Ø,(HL)	Set Z flag to state of bit Ø in memory data pointed to by HL.
SET 7,H	Set bit 7 of H register to logic 1.
RES 3,(IX+Ø)	Set bit 3 of memory byte pointed to by IX register to logic Ø.

For all the instructions in this section, it should be remembered that involvement of the index registers IX or IY considerably lengthens the byte size and cycle time required.

Seven General-purpose instructions: The first three instructions listed next are available in accumulator A only:

1. CPL complements each bit in A (1's complement).
2. NEG performs a 2's complement on the data in A: $(A \leftarrow \overline{A}+1)$.
3. DAA converts data in A to BCD form following any of the 8-bit add or subtract instructions.

These four instructions do not involve any data registers:

4. SCF sets the carry flag.
5. CCF complements the carry flag. To *clear* the flag, use SCF and then CCF.
6. NOP is a place holder that takes 1 byte and 4 *T*-states.
7. HALT causes the processor to execute NOP instructions continuously at its current program-counter address.

21.6 SIXTEEN-BIT AND SUBROUTINE INSTRUCTIONS

Section Overview

The Z80 can load any of its 16-bit registers in the immediate and extended modes. It can store any of these registers in the extended mode.

Sixteen-bit additions can be done with certain 16-bit registers as the source and HL, IX, or IY as the destination. Sixteen-bit subtracts are similar but may have only HL as the destination. Any 16-bit register may be incremented and decremented, but no flags are set by these operations, making getting out of a loop a special problem.

The stack pointer (SP) can be loaded from IX, IY, or HL. Save and retrieve operations on stack are called PUSH and POP, respectively, and they can be done only on 16-bit registers or register pairs.

CALL replaces the 6800's JSR instruction, and CALLs can be conditional, depending on the state of the Z, C, P, or S flags. RET replaces the 6800's RTS, and this can also be conditional.

Jump instructions replace both the *jump* and *branches* of 6800 code, and they can be conditional or unconditional, relative or extended mode.

Sixteen-bit loads. Any 16-bit register in the Z80 can be loaded in the immediate or extended mode. Remember the *low/high* byte order when picking up 16-bit data from memory. Loading two successive memory bytes from any 16-bit register is also possible in the extended mode. Strangely enough, register-to-register transfers and memory addressing by pointer registers—two mainstays of the 8-bit load group—are entirely absent from the 16-bit load group.

Sixteen-bit additions and subtractions are possible with HL as the destination. The source registers may be BC, DE, HL, or SP. There are two *add* instructions: with carry input (ADC) and without carry input (ADD). There is only one *subtract*: SBC, which subtracts an extra 1 if the C flag was set (as by a borrow) on a previous instruction.

The index registers IX and IY may be the destination for 16-bit addition, with BC, DE, SP, or the *same* index register as the source. Register-to-register is the only mode allowed for any 16-bit addition or subtraction.

Increments and decrements are possible to any 16-bit register: BC, DE, HL, IX, IY, or SP. None of these instructions (16-bit INC or DEC) affects any of the flag bits. This and the lack of suitable alternatives (such as an *add 1, immediate*) for stepping the 16-bit registers makes it awkward to get out of a loop that uses one of these registers as a data pointer. Here are some possible ways out:

- If the file is shorter than 256 bytes, you can initialize another register (B, for example) to the proper number of counts and decrement this register along with the pointer register. DEC B affects the Z and S (sign bit 7) flags.
- You can reserve an unused byte in the data table as an end-of-file indicator (perhaps $FF, if the table contains all BCD data). Loads do not affect any flags, so it will be necessary to load to A and use CP FFH to set the Z flag on byte $FF.

- You can use one of the *block transfer* or *search* instructions described in Chapter 22.

Figure 21.9 summarizes the 16-bit instructions.

Stack operations: The Z80 stack pointer can be loaded in the immediate or extended modes, shown earlier, or by transfer from HL, IX, or IY. These instructions (LD SP,HL and LD SP,IX, and LD SP,IY) are the only 16-bit register-to-register transfers allowed in the Z80.

Z80 16-bit Instructions

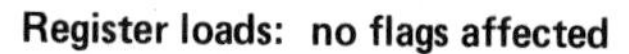

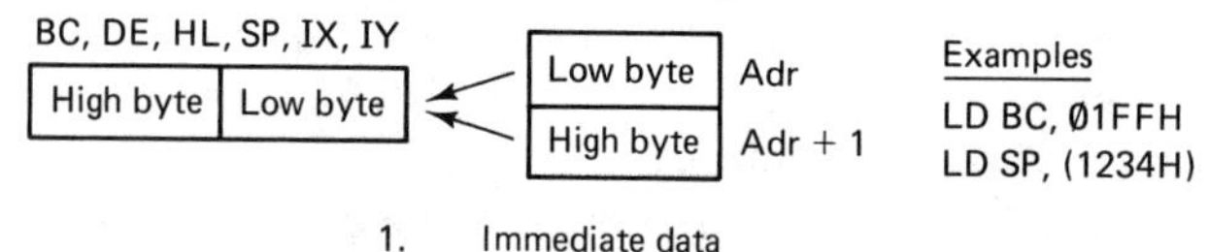

1. Immediate data
2. Data from operand memory

Register stores: no flags affected

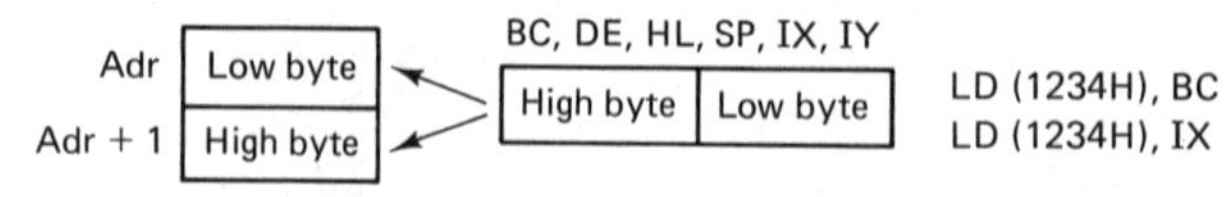

Data loaded to operand memory

Arithmetic: C flag affected

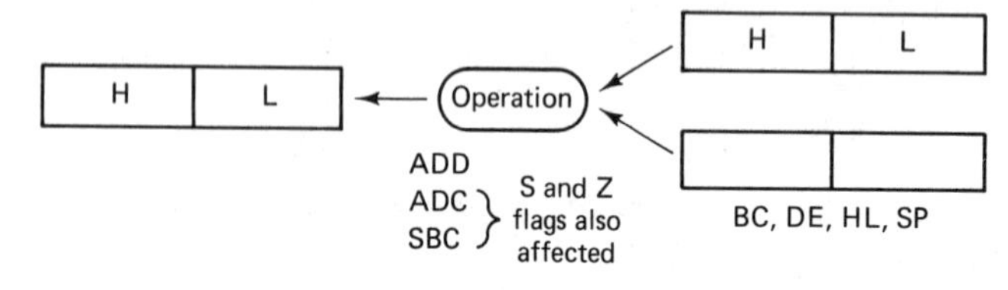

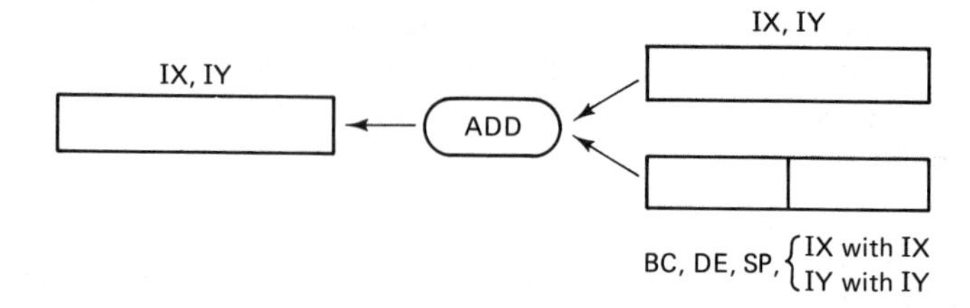

Inc/dec: no flags affected

Figure 21.9 The Z80 sixteen-bit instructions.

The PSH and PUL instructions of the 6800 become PUSH and POP, respectively, in Z80 language. Two bytes are always pushed or popped at a time, BC, DE, HL, IX, IY, and AF being the allowed pairs. F, of course, is the flag byte, paired with A to make a double byte for this instruction. Here are some example forms:

```
LD     SP,34F0H      Load stack pointer, immediate.

PUSH AF              Save A and flag byte on stack.

POP    IX            Retrieve IX from stack.
```

The JSR of the 6800 becomes the CALL instruction in Z80 language. CALL is always extended-mode, consisting of the op-code, the low-order address byte, and the high-order address byte. CALLs can be *conditional*, so that the subroutine is entered only on a specified flag state. Here are the options:

```
CALL SUBR                             Unconditional.
CALL Z,SUBR        CALL NZ,SUBR       Zero or nonzero.
CALL C,SUBR        CALL NC,SUBR       Carry or no-carry.
CALL PE,SUBR       CALL PO,SUBR       Parity even or odd.
CALL P,SUBR        CALL M,SUBR        Positive or negative.
```

The RTS instruction is RET in Z80 code, and it can also be subjected to any of the preceding conditions.

The jump instruction for the Z80 is a bit more versatile than for the 6800. There are 18 variations. There are three addressing modes for the Z80 *jump*: extended, relative, and indexed.

The extended mode is available for the unconditional *jump* instruction, and for jumps dependent upon each of the conditional tests listed with the CALLs. The relative mode is available for unconditional and for the conditions Z, NZ, C, and NC *only*. These conditional jumps serve the purpose of the *branch* instructions in the 6800. Note that the Z80 has no combinational bit tests, such as BHI or BGT.

Three unconditional indexed jumps are available, using pointers HL, IX, and IY. Figure 21.10 illustrates the Z80 *jump* group of instructions.

One relative jump is available combined with an automatic decrement of register B. For example, the following routine clears 50 bytes of memory:

```
        LD      HL,3400H      Point to start of clear area.
        LD      B,50          Load 50 counts.
        LD      C,0           Clear C to transfer 0 to memory.
LOOP    LD      (HL),C        Clear a memory location.
        INC     HL            Point to next location.
        DJNZ    LOOP          Do it 50 times, counting B to 0.
```

Z80 Jump Instructions

No flags affected

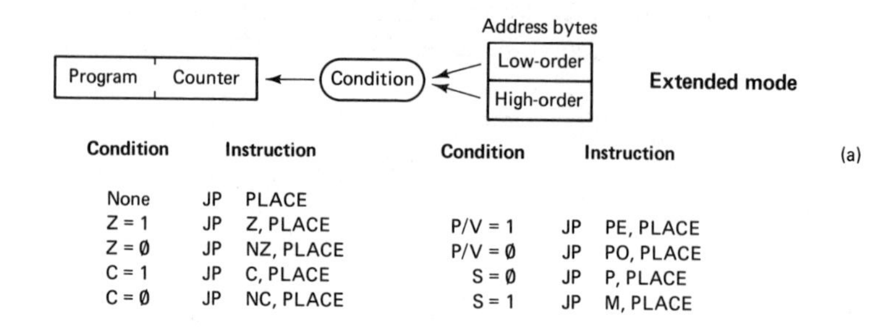

Condition	Instruction	Condition	Instruction
None	JP PLACE		
Z = 1	JP Z, PLACE	P/V = 1	JP PE, PLACE
Z = Ø	JP NZ, PLACE	P/V = Ø	JP PO, PLACE
C = 1	JP C, PLACE	S = Ø	JP P, PLACE
C = Ø	JP NC, PLACE	S = 1	JP M, PLACE

(a)

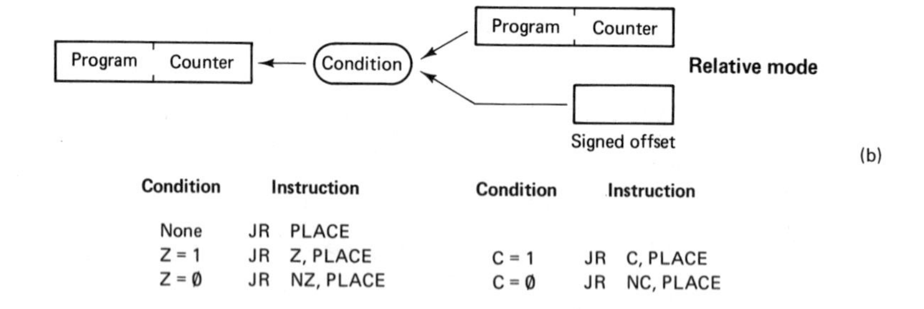

(b)

Condition	Instruction	Condition	Instruction
None	JR PLACE		
Z = 1	JR Z, PLACE	C = 1	JR C, PLACE
Z = Ø	JR NZ, PLACE	C = Ø	JR NC, PLACE

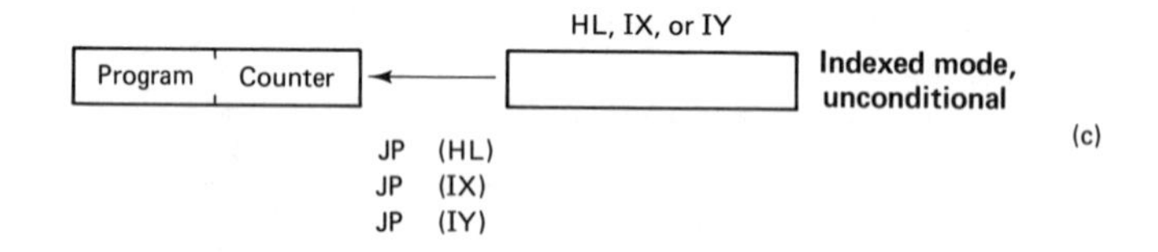

(c)

JP (HL)
JP (IX)
JP (IY)

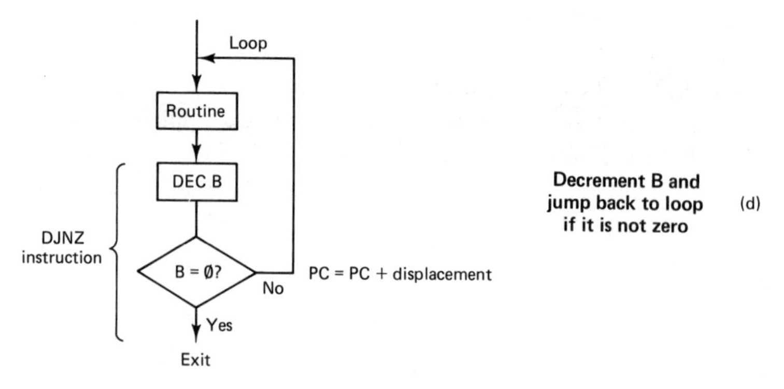

Figure 21.10 The Z80 uses conditional jumps, in the extended or relative mode, in the way that the 6800 uses branch instructions.

REVIEW OF SECTIONS 21.5 and 21.6

21. What is the effect of the instruction

LD (HL),BYTE

22. How would you add A to B, leaving the result in B?
23. For which registers can BC and DE serve as data pointers on *load* (register) instructions?
24. Which registers can be loaded in the extended mode?
25. How would you store the data from E to address 3456H without disturbing A?
26. What is the effect of the following program segment?

```
LD   HL,789AH
RES 0,(HL)
RES 1,(HL)
```

27. How can you clear the Z80 carry flag if you don't know what state it is currently in?
28. If D contains $98 and E contains $76, what data is stored to what destination by the instruction LD (ABCD),DE?
29. Give the instruction you would use to pull A from the stack.
30. Write the Z80 equivalent of this 6800 routine.

```
       LDAA   00,X     Get data from table.
       BEQ    SKIP     Don't take a zero.
       STAA   LATCH    Output nonzero bytes.
SKIP   INX             Point to next.
```

21.7 MEM GAME ON THE Z80

For our first Z80 project we return to the MEM game, first seen in Section 7.7 and implemented again on the 6809 in Section 19.7. The flowchart is the same one given in Figure 7.7, and we have tried to keep the programs as similar as possible so you can compare the Z80 and 6802 styles. Figure 21.11 shows the Z80 hardware for MEM, and the program listing follows (List 21.1).

```
                 1000 * LIST 21-1 ** MEM GAME FOR Z80. SEE LIST 7-6.
                 1010 *  FIG 21.11 HARDWARE.  FIG 7.7 FLOWCHART.
                 1020 * TESTED AND RUNNING, 31 JUL '87.   D. METZGER.
                 1030          .OR  0000H
                 1040          .TA  4000H
E000-            1050 INPUT    .EQ  0E000H        (NOTE LEADING ZERO: ASSEMBLER
C000-            1060 OUTPUT   .EQ  0C000H         WON'T TAKE A-F AS FIRST DIGIT)
                 1070 *
0000- 21 50 00   1080 LOAD     LD   HL,0050H      POINT TO START OF TABLE.
0003- 3A 00 E0   1090 KEY      LD   A,(INPUT)     GET KEY (0 = KEY PRESSED)
0006- EE FF      1100          XOR  0FFH          COMPLEMENT SO 1 = PRESSED.
0008- 32 00 C0   1110          LD   (OUTPUT),A    LIGHT LED BY PRESSED KEY.
000B- E6 0F      1120          AND  0FH           MASK OFF HIGH-ORDER 4 BITS
000D- 28 F4      1130          JR   Z,KEY         LOOK AGAIN IF NO KEY.
```

```
                 1140 *                          *REPLACE SECTION BELOW FOR "PLAY"
000F- 77         1150 SWAP    LD    (HL),A        STORE KEY VALUES IN TABLE.
0010- 00         1160         NOP                 LEAVE 2-BYTE SPACE FOR "PLAY"
0011- 00         1170         NOP                  JUMP INSTRUCTION.
                 1180 *                          *
0012- 06 FF      1190         LD    B,0FFH        SET OUTER-LOOP DELAY COUNTER
0014- 0C         1200 DELAY   INC   C             INNER-LOOP COUNTER
0015- 20 FD      1210         JR    NZ,DELAY      256 LOOPS * 16 T-STATES/LOOP
0017- 05         1220         DEC   B              * 0.5 US/T * 256 OUTER LOOPS
0018- 20 FA      1230         JR    NZ,DELAY        IS ABOUT 0.52 SEC.
001A- 23         1240         INC   HL            POINT TO NEXT TABLE ADDRESS.
001B- 7D         1250         LD    A,L           MOVE LOW HALF OF POINTER TO A
001C- FE 5F      1260         CP    5FH            TO COMPARE; REACHED 15TH ADR?
001E- 20 E3      1270         JR    NZ,KEY          NO: GET NEXT KEY.
0020- 7D         1275 DISPLY  LD    A,L           GET DATA IN A (NO EXTND-MODE LD L)
0021- 32 00 C0   1280         LD    (OUTPUT),A    LIGHT 4 LEDS AFTER 15 KEYS; OR
                 1290 *                           SHOW SCORE AFTER ERROR IN PLAY.
0024- 76         1300 STOP    HALT
                 1310 *
                 1320         .OR   SWAP
                 1330         .TA   SWAP + 4000H
                 1340 *  REPLACE LINES 1150 - 1170 WITH LINES 1360 -1380 TO
                 1350 *  CHANGE FROM "ENTER" TO "PLAY" SEQUENCE.
                 1360 *
000F- BE         1370         CP    (HL)          IS KEY VALUE SAME AS TABLE VALUE?
0010- 20 0E      1380         JR    NZ,DISPLY      NO: LO BYTE OF HL TO LATCH LEDS.
                 1390         .EN
```

Answers to Chapter 21 Review Questions

1. In accumulator A. **2.** HL **3.** (a) true, (b) true, (c) false. **4.** P/V = 1. **5.** HL, because it is faster than using IX. **6.** EXX; INC HL **7.** The Z80 always starts at $0000. **8.** It should be left unconnected.

9. $\overline{MREQ} = 0$, $\overline{RD} = 1$, $\overline{WR} = 0$, $\overline{IORQ} = 1$ **10.** 2.5 μs

11. It transfers the data from register B to register A.

12. In accumulator A. **13.** ADD A,4 **14.** LD (4E3H),HL

15. Data CD to address 4E3, and data AB to address 4E4.

16. There are 49 forms. Such forms as LD A,A are no-ops but valid. **17.** (c) and (d) are valid. **18.** (a) and (c) are valid. **19.** Register-indirect, because it typically takes half as many bytes and cycles as indexed instructions.

20. JR NZ,QUIT **21.** The value of the variable BYTE is stored to the address pointed to by register pair HL. **22.** ADD A,B then LD B,A.

23. Register A only. **24.** A and the 16-bit registers BC, DE, HL, SP, IX, and IY. **25.** LD HL,3456H then LD (HL),E **26.** The two lowest-order bits of the byte at $789A are set to 0. **27.** SCF, then CCF

28. Address $ABCD gets data $76, and address $ABCE gets data $98.

29. POP AF

30.

```
     LD A,(HL)        Get data from table
     AND A            Just to set flag bits.
     JR Z,SKIP        Don't take a zero.
     LD (LATCH),A     Output nonzero bytes.
SKIP INC HL           Point to next.
```

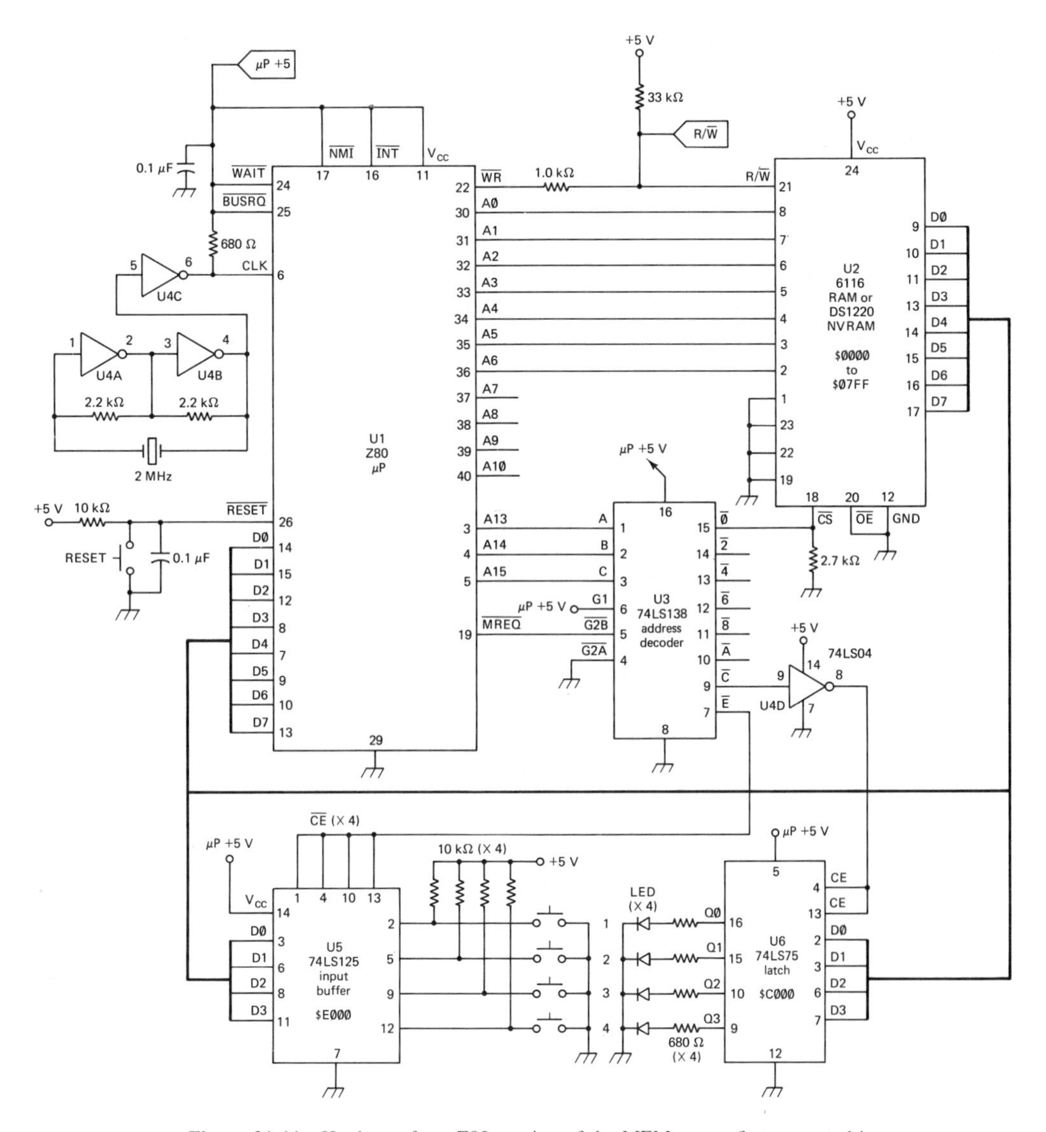

Figure 21.11 Hardware for a Z80 version of the MEM game, first presented in Section 7.7 for the 6802. The same flowchart from Chapter 7 can be used to interpret the program.

CHAPTER 21 QUESTIONS AND PROBLEMS

Digits after the decimal point refer to section numbers in Chapter 21.

Basic Level

1.1 Identify the main accumulator and the main address-pointer register in the Z80.

2.1 If register pair DE is used as an address pointer, for which of these registers can it point: A, B, or C?

3.2 The 6802 boasts an internal clock generator, a single 5-V supply, and an on-board RAM. Which of these features are also true in the Z80?

4.2 How should the $\overline{\text{WAIT}}$, $\overline{\text{INT}}$, and $\overline{\text{NMI}}$ inputs be connected in a minimum-mode Z80 test system?

5.3 What is the difference between the following instructions?

LD B,D and LD D,B

6.3 Write the Z80 source-code instructions to (a) load the accumulator with data $1C and (b) store the contents of the accumulator to address $001C.

7.4 Tell what is wrong with the following Z80 instruction, and write the two instructions that will accomplish the task intended:

```
ADD A,(4080H)
```

8.4 What 6800 addressing mode most closely resembles the Z80's *register indirect* mode?

9.5 Which address-pointer registers are used with a displacement: BC, DE, HL, IX, IY? When a displacement is used, is it one or two bytes? Signed or unsigned?

10.5 The Z80 has no CLR instructions. Devise a short Z80 routine to clear the 2 bytes at addresses 0123H and 0125H.

11.6 Which of these instructions sets the Z80's Z flag on zero result?

a. DEC B   b. DEC HL   c. LD A,H

12.6 Write the Z80 equivalent to these 6800 instructions:

a. JSR CONVR   b. PULA   c. RTS

Advanced Level

13.1 Explain how the *primed* or alternate register set of the Z80 is accessed.

14.1 When is the Z80's P/V bit a parity flag and when is it an overflow flag?

15.2 The Z80 does not have a single R/$\overline{\text{W}}$ output pin. Explain how a RAM can be signaled of the difference between a *read* and a *write* cycle.

16.2 Distinguish among machine cycles, *T*-states, and instruction cycles in the Z80. Which of them has the frequency of the CLOCK input?

17.3 Write the Z80 source code (with comments) for a routine that will subtract the data at address 34CDH from the data in the B register and store the result at the address pointed to by the HL register pair.

18.3 Explain how Z80 assembly language indicates each of the following.

a. Hex data **b.** Destination operand
c. Immediate data **d.** Address pointing to data
e. Source operand

19.4 Write source code and comments for a Z80 routine to transfer 100 bytes from a file starting at address B3ØØH to a new file starting at Ø8ØØH, both files working upward. There is no "compare pointer to end-of-file" instruction in the Z80, so you can count register C from 100 down to 0 and use JR Z to get out of the loop. Use HL and DE as source and destination pointers.

20.4 Explain the differences between the Z80's *register-indirect* and its *indexed* instructions. Which is generally preferred, and why?

21.5 Comment on these mnemonics and their significance.

a. RRC A **b.** RRCA **c.** RRC E **d.** RRCE

22.5 Make a flowchart and write the source code for a Z80 program to examine a file of bytes from address 14ØØH to address 149FH and count the number of bytes in which exactly four of their 8 bits are logic 1s. Use HL as the data-table pointer, rotate A to check the bits, count B down to check for end-of-table, and increment C to count the number of qualifying bytes. You will need to use instructions CP 4, JPZ, and JP C.

23.6 Write a Z80 routine to add the two bytes at 345ØH:3451H and 3452H:3453H, storing the results in 3454H:3455H.

24.6 Write a Z80 subroutine that will save DE on the stack and then multiply the 16-bit data in HL by 5. Use two shifts left of a copy of HL in DE to multiply by 4; then an add to the unshifted value to accomplish this. DE should be retrieved from the stack before returning from the subroutine.

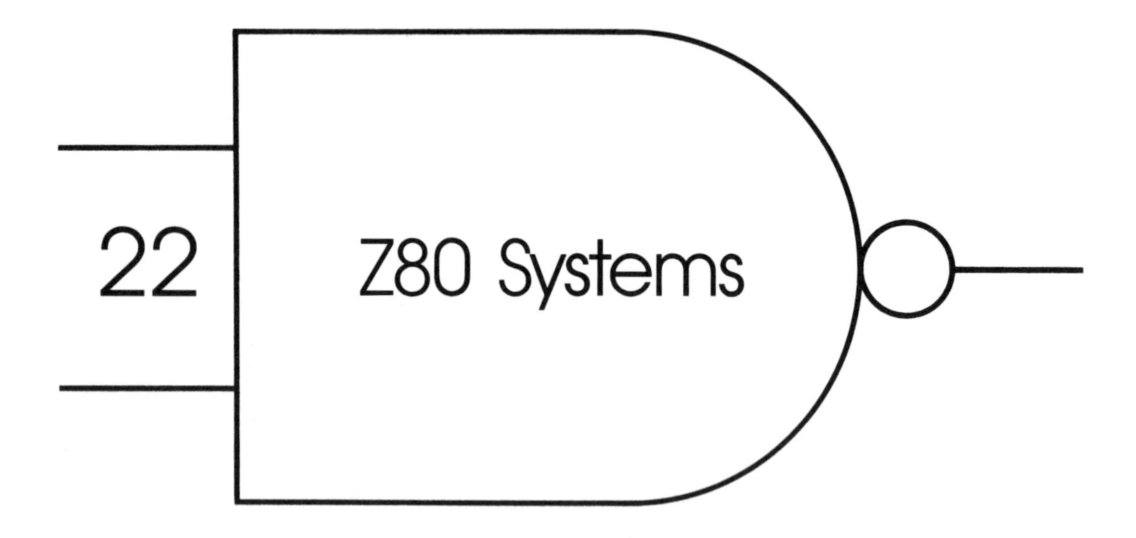

22.1 FANTASTIC Z80 INSTRUCTIONS

Section Overview

The Z80 has the capability of transferring a bock of data (up to 64 kilobytes) from any sequential memory area to any other sequential memory area with a single instruction comprising the loop. A basic 4-MHz Z80 can move an 8-kilobyte block in 33 ms with this instruction. For comparison, the basic 1-MHz 6802 requires about 164 ms and a 5-instruction loop to do the same task. In addition to this *block transfer*, there is a *block search* that allows the Z80 to examine a data block for the appearance of a given byte and report the address where that byte first appears—again with a single instruction.

The Z80 also has 6 *exchange* instructions for 16-bit registers, which are detailed in this section.

The Z80 block-transfer instruction is LDIR, which stands for *Load, Increment, and Repeat*. More specifically, this instruction causes the Z80 to do the following:

1. Load the data from the address pointed to by HL to the address pointed to by DE (think of **de**stination).
2. Increment pointer registers HL and DE and decrement loop-counter register BC (think of **b**yte **c**ounter).
3. If loop-counter BC has counted down to zero, go on to the next instruction in the program. If BC is not yet zero, return to Step 1.

A similar instruction, LDDR, causes the pointers HL (source) and DE (destination) to work backward through the file, from high address to low addresses. Figure 22.1 illustrates the block-transfer instruction. Here is a routine to transfer 8 kilobytes from a file with base address 24ØØH to a RAM with base address CØØØH.

Z80 BLOCK TRANSFER—8K FILE

```
LD   HL,2400H      Initialize source pointer,
LD   DE,C00H         destination pointer,
LD   BC,2000H        byte counter (8 kilobyte).
LDIR               Transfer; loop 8192 times.
```

The block-transfer instructions also have versions that do not automatically repeat. These are LDI and LDD, and they allow the flexibility of tests and modifications as each data byte is transferred. As an example, here is a routine that transfers 8K of data from base address 24ØØH to base address CØØØH, as before; but this time any bytes found to equal zero are left out. The destination file may thus be shorter than the source file. To understand this routine, you must realize that the P/V flag is cleared to Ø when byte counter BC reaches zero. This is a special use of the P/V flag, which allows us to exit the loop when the preset number of transfers has been

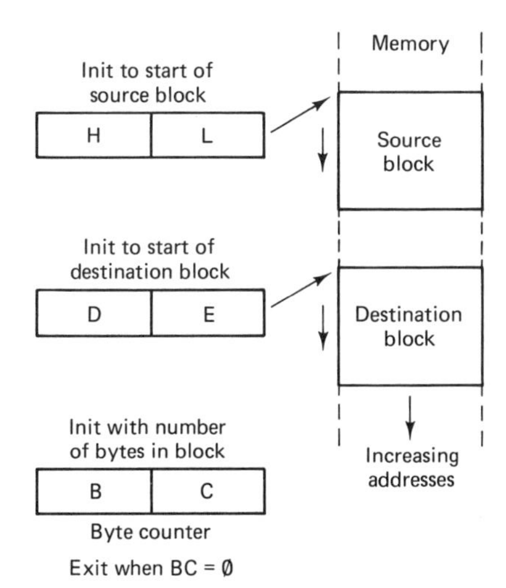

Figure 22.1 The Z80 *block transfer* allows whole files to be moved with a single instruction.

made. The destination counter DE is backed up when a zero byte is loaded, so the next byte will write over it.

```
* Z80 BLOCK TRANSFER; 8K BLOCK; DELETE 00 BYTES.
-----------------------------------------------------------------------
         LD  HL,2400H          Initialize source pointer,
         LD  DE,C000H            destination pointer,
         LD  BC,2000H            byte counter.
LOOP     LD  A,(HL)            Get the byte that will be transferred.
         LDI                   Transfer it; Increment pointers
         JP  PO,OUT            Quit if P/V flag = 0 (BC = 0).
         CP  0                 Was byte transferred equal to zero?
         JR  NZ,LOOP           No; do next byte.
         DEC  DE               Yes; back up destination counter
         JR  LOOP                and transfer next byte.
OUT      NOP                   Leave routine.
-----------------------------------------------------------------------
```

Notice that there is no relative jump conditioned on the P/V flag, so a JP (extended mode) had to be used.

The block-search instructions are similar in form. CPIR is *Compare*, *Increment*, and *Repeat*. Here are its functions:

1. Subtract the data pointed to by HL from a fixed byte in accumulator A, but use the result only to set the Z, S, and H flags.[1] Do not change A or the data.
2. Increment the pointer HL and decrement the byte counter BC.
3. If the data byte equals the byte in A or if BC = 0, go to the next program instruction. If not, go to Step 1.

The CPDR instruction is the same except that HL *decrements* through the file. CPI and CPD omit the *repeat* feature (Step 3), allowing additional tests after each byte is checked. Again, the P/V flag is cleared when byte counter BC equals zero to allow easy exit at the end of the file when using CPI or CPD. Figure 22.2 illustrates the block-search instructions.

Here is a routine to count the number of negative bytes in a table of 2's complement 8-bit numbers from address 5000H to address 57FF.

```
BLOCK SEARCH; COUNT NUMBER OF NEGATIVES IN DATA BLOCK
-----------------------------------------------------------------------
         LD    HL,5000H    Initialize start of datablock.
         LD    BC,0800H    Search 2 kilobytes.
         LD    DE,0        DE will count number of negatives.
         LD    A,FFH       FF - Neg = Pos; FF - Pos = Neg.
LOOP     CPI               Set S flag by 00 - (HL); Inc HL, Dec BC.
         JP    PO,OUT      Quit if P/V flag = 0; means BC = 0.
         JP    M,LOOP      A minus result means the byte was positive.
         INC   DE          Plus result; increment negative-counter, and
         JP    LOOP          check next byte.
OUT      NOP               Done. Answer in register DE.
-----------------------------------------------------------------------
```

[1]Strangely, the C flag is not affected.

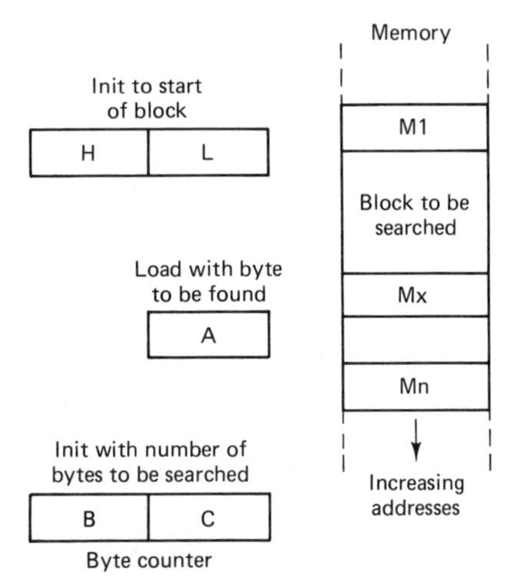

Figure 22.2 The *block-search* instruction allows a file to be examined for the appearance of a specified byte.

We will leave it to you to prove that $FF minus any negative number ($FF through $8Ø) will produce a positive result and that $FF minus any positive number will produce a negative result. This allows us to test for positive or negative numbers even though CPI does not affect the C flag.

Six register-exchange instructions are available in the Z80. Two of these extend the usefulness of the HL register by providing a place to store and retrieve it quickly, and two perform a similar service for the index registers. The final two were touched upon in Section 21.1 and provide access to the alternate register set.

- EX DE,HL switches the contents of the two named register pairs in only 4 *T*-states. Since DE cannot point for B or C, this allows HL to point for B in one function; then (after an EX DE,HL) HL can point for C in another function.
- EX (SP),HL switches the contents of HL with 2 bytes of the stack area of RAM in 19 *T*-states. Again, this allows alternating use of HL for two different functions. EX (SP),IX and EX (SP),IY perform a similar service for the index registers. A pair of exchanges must occur with no movement of the stack pointer in between if the registers are to be restored to their original states.
- EX AF,AF′ switches references to accumulator A and the flag bits to the alternate registers and back again the second time it is encountered. Since A is the only register supported by a full complement of arithmetic and logic instructions, this exchange is valuable for providing access to a second such register.
- EXX switches references to B, C, D, E, H, and L to the alternate (primed) registers and back to the main set the second time it is encountered. It is occasionally important to understand that all Z80 instructions actually operate on the main register set. The alternate set is accessed only by swapping contents

with the main set, so alternate-set data is referenced in main-set registers. For example, if C and C′ are exchanged (with EXX) but F and F′ are not exchanged, a zero flag generated by decrementing C′ will appear in F, not in F′.

22.2 THE 'STANDARD' I/O TECHNIQUE

Section Overview

The Z80 has a group of I/O instructions that cause the $\overline{IORQ}$ line to be asserted. This line is ANDed with $\overline{RD}$ for inputs or with $\overline{WR}$ for outputs to activate buffers, latches, or other I/O devices. The I/O instructions all include an 8-bit address, which appears on lines AØ through A7 to select one of 256 possible input or 256 possible output devices.

Single-instruction loops, such as OTIR (Output, Increment, and Repeat), allow files up to 256 bytes long to be loaded or dumped via peripherals at very high speeds.

The parallel-bus structure of microcomputers is taken for granted today, but try to imagine a computer in which there were no tristate outputs—only active outputs. An input device could not send its data to the CPU on the same data bus used by the memory, because the input and memory would always be fighting for the bus. There would have to be separate buses and separate instructions for memory and I/O operations. This was the situation with early computers. The idea of assigning addresses to I/O devices and treating inputs and outputs just like reads and writes to memory (memory-mapped I/O) caught on rather slowly. The 8080, as one of the early microcomputers, and the Z80, as its direct descendent, incorporated what was originally thought of as the "standard" I/O technique. Memory-mapped I/O is still possible and may be judged more desirable in some systems, even though the standard I/O system is available.

Figure 22.3 shows the hardware to implement standard I/O on a Z80 with a 74LS244 buffer and 74LS373 latch. There are three basic instruction types for input and three for output:

1. IN A,(1ØH) and OUT (2ØH),A are illustrative of the first type. Only accumulator A can transfer data in this mode, and the I/O address must be specified as an unsigned constant from Ø to 255 or as a variable with a value in this range. The data is transferred to or from accumulator A via the system data bus. The 8 low address lines (AØ through A7) assert the 8-bit I/O "address" specified in parenthesis. By using I/O address bytes with only one bit asserted, as many as 8 inputs and 8 outputs can be addressed without further decoding. With external decoding logic, as many as 256 of each can be accessed. The examples above are chosen to access the buffer and latch of Figure 22.3.
2. IN B,(C) and OUT (C),L illustrate the second mode. Data can be transferred to or from any of the 8-bit registers A, B, D, E, H, or L. The I/O address is contained in register C.

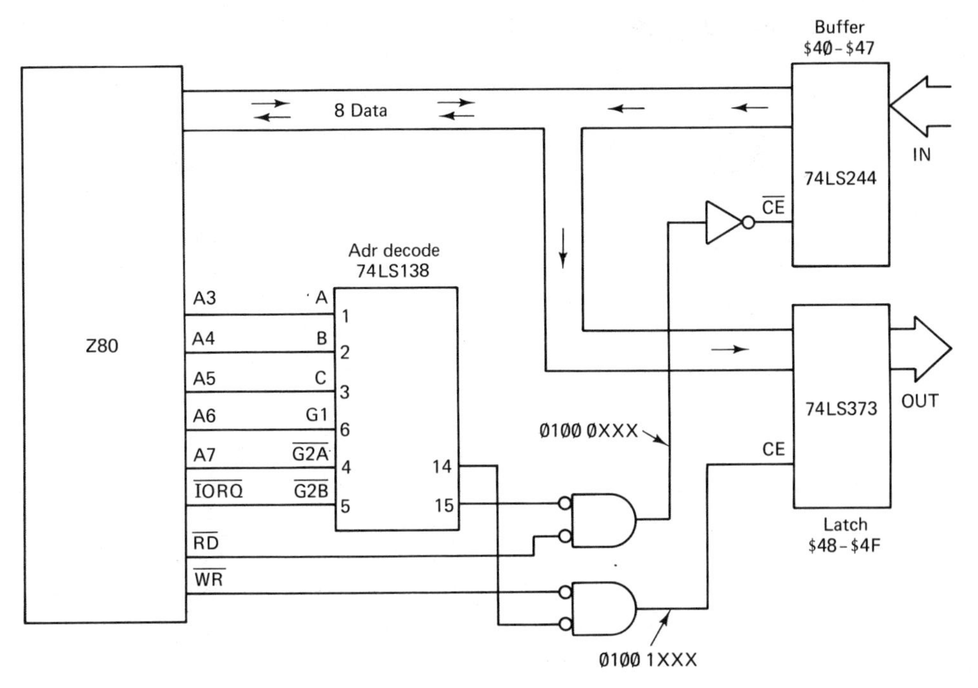

Figure 22.3 The so-called standard I/O technique uses special IN and OUT instructions to assert an I/O-enable pin of the processor.

3. INIR is a block input instruction:
   a. Data is transferred from the input device addressed by register C and immediately stored at the address pointed to by register pair HL.
   b. Pointer HL is incremented to point to the next table address to be filled, and byte counter B is decremented.
   c. If B = 0, another byte is transferred. When B = 0 the processor goes on to the next instruction.

There are a total of 8 instructions based on this idea. Here are some notes on the other 7:

- INDR is the same as before except that the HL pointer decrements through the table.
- INI and IND are nonrepeating versions of the same instructions, allowing tests to be made as each byte is input. The Z flag is set to 1 when B = 0 to allow easy exit from the loop when the preset number of bytes has been input.
- OTIR is Output, Increment, and Repeat, using HL as the data pointer, B as the byte counter, and C as the data-transfer register.
- OTDR is output, decrement through data table, and repeat.
- OUTI and OUTD are nonrepeating versions of the preceding instructions.

REVIEW OF SECTIONS 22.1 AND 22.2

Answers appear at the end of Chapter 22.

1. What is the setup procedure to cause the LDIR instruction to transfer 1000 bytes?
2. In the LDD instruction, which machine register is loaded with the data?
3. Which registers are tied up by the LDI instruction?
4. Here is a modification to the LOOP part of the Block-Search routine of page 556. What does it do?

```
LOOP    CPIR
        JP PO,OUT
        INC DE
        JP LOOP
```

5. How can you use the Z80 to execute a new routine (such as an interrupt) without losing the flag, accumulator, or pointer-register data of the present routine?
6. In a "standard" input cycle, an address appears on the address bus and $\overline{RD}$ is asserted. Why is a memory device not activated to place data on the data bus?
7. What circuit design is it that allows an input buffer, a ROM, a RAM, and the processor all to be tied in parallel on the data bus?
8. How can data be output from register D to a latch activated by a logic 1 on line A2?
9. What does this routine do?

```
        LD HL,4000H
        LD B,80H
        LD A,0
LOOP    INI
        CP C
        JP NZ,LOOP
        DEC HL
        JP LOOP
```

10. In memory-mapped I/O on a Z80, what instruction would you use to output the byte in A to a latch at base address $9000?

22.3 Z80 INTERRUPT HANDLING

Section Overview

The Z80 has two interrupt inputs, one maskable and one nonmaskable. Both stack only the 2 program-counter bytes. $\overline{NMI}$ sends the processor to address 0066H. $\overline{INT}$, similar to $\overline{IRQ}$ in the 6802, has 3 modes of operation. The simplest mode (IM1) sends the processor to address 0038H. The

other 2 modes (IMØ and IM2) permit the processor to read a vector placed on its data bus by the interrupting device. This vector sends the processor to a selected interrupt routine. Mode IMØ permits 8 different interrupts and service routines. Mode IM2 permits literally hundreds.

The nonmaskable interrupt of the Z80 is the most straightforward, so we will examine it first. A *high-to-low* transition on the $\overline{\text{NMI}}$ input causes the NMI service routine to be executed after the completion of the current instruction. The processor saves the program counter high-order and low-order bytes on the stack, decrementing the stack pointer by 2. It then jumps to address ØØ66H, reading the byte it finds there as the first instruction of the NMI service routine.

There is no vector address as there is in the 6802. The routine itself must start at address ØØ66H. If you want the NMI routine to be located elsewhere, you must place a jump instruction to the desired address at ØØ66H. The usual procedure in structuring an interrupt is first to exchange register pairs or stack the needed register pairs, and then jump to the address space chosen for the service routine. The interrupt routine must end with a second exchange or an unstacking, followed by an RET instruction.

The maskable interrupt is initiated by a *low* level on the $\overline{\text{INT}}$ input, but only if the interrupt-enable flip-flop is *set*. This flip-flop is cleared by a processor *reset* and temporarily cleared by an $\overline{\text{NMI}}$. The EI instruction enables maskable interrupts and the DI disables them. The maskable interrupt also causes only the 2 bytes of the program counter to be stacked. There are three modes of response to $\overline{\text{INT}}$, set by one of the instructions IM0, IM1, or IM2. The default mode, entered upon *reset*, is Interrupt Mode Ø, but the simplest is IM1, so we explain it first.

Interrupt mode 1 simply stacks the program counter and sends the processor to address ØØ38H upon recognition of an $\overline{\text{INT}}$ input. Again, this is the location of the first instruction of the interrupt routine, not a vector location.

Interrupt mode Ø acknowledges a valid $\overline{\text{INT}}$ input by asserting the Z80's, $\overline{\text{M1}}$, and $\overline{\text{IORQ}}$ outputs simultaneously. This condition should be used to signal the interrupting device that it must place its unique *restart* code on the data bus. The processor then reads this code, stacks PCH and PCL, and restarts at one of 8 specified addresses from ØØØØH to ØØ38H. *Restart* is also a Z80 software instruction. Here is a list of the *restart* instruction mnemonics, the data bus codes for the 8 possible interrupt vectors, and the corresponding restart addresses.

Instruction	Data Bus Code	Interrupt/Restart Address
RST 0	1100 0111	0000H
RST 8	1100 1111	0008H
RST 10H	1101 0111	0010H
RST 18H	1101 1111	0018H
RST 20H	1110 0111	0020H
RST 28H	1110 1111	0028H
RST 30H	1111 0111	0030H
RST 38H	1111 1111	0038H

The first address is seldom used, since it is used on a processor *reset*. If Interrupt Mode Ø is used, the main-program routine will have to jump to another area of memory within the first 8 bytes to avoid running into the interrupt/restart area.

Interrupt Mode 2 is designed specifically for use with the Z80 family of peripherals, such as the Z80 PIO (Parallel I/O) and Z80 SIO (Serial I/O) chips. The interrupting device must place an 8-bit code on the data bus, which becomes the low-order half of a vector address. The high-order half of the vector address must have been previously loaded into the Z80's 8-bit I register. The required instruction is LD I,A. In acknowledging the interrupt, the processor concatenates these 2 bytes to form a pointer to a selected pair of bytes in a table of vector addresses. For example, if I contains 12H and the interrupting device puts 34H on the data bus, the processor responds to INT by fetching a byte from address 1234H (call it ADRLO). It then fetches a second byte from address 1235H (call it ADRHI). It then concatenates ADRHI : ADRLO to form the address of the selected interrupt routine.

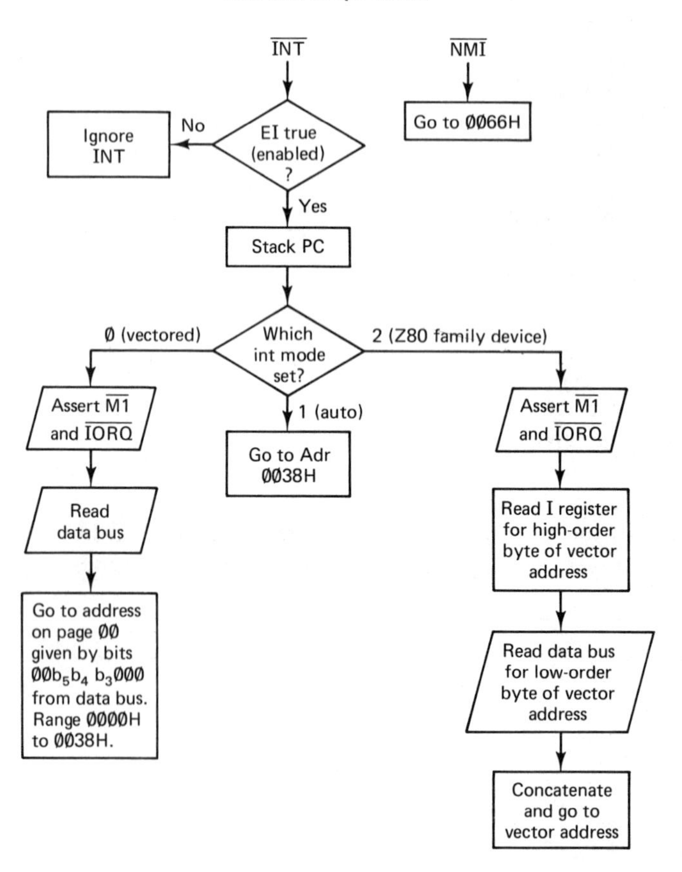

Figure 22.4 A summary of the interrupt modes of the Z80.

This system allows the external hardware to call up any of 128 routines without time-consuming polling. The number is 128 and not 256 because bit DØ (which becomes AØ) must always be a Ø to read ADRLO. It is automatically incremented to AØ = 1 in reading ADRHI. Another advantage is that the table of vector addresses and the interrupt routines can be located anywhere in memory. Finally, the same hardware request can be made to call up a different interrupt-service routine by the software operation of reloading the I register to point to a different vector table on a different page of memory.

Figure 22.4 presents a summary of the Z80 interrupt modes and their operation.

22.4 A PROGRAMMER'S REFERENCE FOR THE Z80

Section Overview

The Z80 supports such a large number of instructions, many with dozens of variations, that automatic assembly from source code is the only practical way to generate object code, even for short programs. However, a programmer's reference chart is also valuable for troubleshooting and as an aid in learning the processor's instruction set. This section presents an abridged Z80 programmer's reference chart, which has been designed as an educational tool rather than as a complete specification of the processor's instruction set.

Z80 PROGRAMMER'S REFERENCE, ABRIDGED

8-BIT LOAD GROUP

	rr →	(BC)	(DE)	(nn)
LD	A,rr	ØA	1A	3A nn
LD	(rr),A	Ø2	12	32 nn

	m→	A	B	C	D	E	H	L	(HL)	(IX + d)	(IY + d)	S, Z, V, C Flags
LD	A,m	7F	78	79	7A	7B	7C	7D	7E	DD 7E d	FD 7E d	—
LD	B,m	47	4Ø	41	42	43	44	45	46	DD 46 d	FD 46 d	—
LD	C,m	4F	48	49	4A	4B	4C	4D	4E	DD 4E d	FD 4E d	—
LD	D,m	57	5Ø	51	52	53	54	55	56	DD 56 d	FD 56 d	—
LD	E,m	5F	58	59	5A	5B	5C	5D	5E	DD 5E d	FD 5E d	—
LD	H,m	67	6Ø	61	62	63	64	65	66	DD 66 d	FD 66 d	—
LD	L,m	6F	68	69	6A	6B	6C	6D	6E	DD 6E d	FD 6E d	—
LD	(HL),m	77	7Ø	71	72	73	74	75				
LD	(IX+d),m	77^x	$7Ø^x$	71^x	72^x	73^x	74^x	75^x	(X format is DD i d)			
LD	(IY+d),m	77^y	$7Ø^y$	71^y	72^y	73^y	74^y	75^y	(Y format is FD i dØ)			
LD	m,n	3E n	Ø6 n	ØE n	16 n	1E n	26 n	2E n	36 n	DD 36 d n	FD 36 d n	

8-BIT ARITHMETIC/LOGIC GROUP

	m→	A	B	C	D	E	H	L	(HL)	(IX+d)	(IY+d)	Immed.	
ADD	A,m	87	80	81	82	83	84	85	86	86[x]	86[y]	C6 n	s z v c
ADC	A,m	8F	88	89	8A	8B	8C	8D	8E	8E[x]	8E[y]	CE n	s z v c
SUB	m	97	90	91	92	93	94	95	96	96[x]	96[y]	D6 n	s z v c
SBC	A,m	9F	98	99	9A	9B	9C	9D	9E	9E[x]	9E[y]	DE n	s z v c
AND	m	A7	A0	A1	A2	A3	A4	A5	A6	A6[x]	A6[y]	E6 n	s z p $\bar{C}$
XOR	m	AF	A8	A9	AA	AB	AC	AD	AE	AE[x]	AE[y]	EE n	s z p $\bar{C}$
OR	m	B7	B0	B1	B2	B3	B4	B5	B6	B6[x]	B6[y]	F6 n	s z p $\bar{C}$
CP	m	BF	B8	B9	BA	BB	BC	BD	BE	BE[x]	BE[y]	FE n	s z v c
INC	m	3C	04	0C	14	1C	24	2C	34	34[x]	34[y]	N/A	s z v
DEC	m	3D	05	0D	15	1D	25	2D	35	35[x]	35[y]	N/A	s z v

8-BIT ROTATE AND SHIFT GROUP

	m →	A	B	C	D	E	H	L	(HL)	(ID+d)	(IY+d)	Flags
prebyte →		CB	CB	CB	CB	CB	CB	CB	CB	DD CB	ED CB	
*optional												
RLC	m	07*	00	01	02	03	04	05	06	d 06	d 06	s z p c
RRC	m	0F*	08	09	0A	0B	0C	0D	0E	d 0E	d 0E	s z p c
RL	m	17*	10	11	12	13	14	15	16	d 16	d 16	s z p c
RR	m	1F*	18	19	1A	1B	1C	1D	1E	d 1E	d 1E	s z p c
SLA	m	27	20	21	22	23	24	25	26	d 26	d 26	s z p c
SRA	m	2F	28	29	2A	2B	2C	2D	2E	d 2E	d 2E	s z p c
SRL	m	3F	38	39	3A	3B	3C	3D	3E	d 3E	d 3E	s z p c

Notes: 1. m is an 8-bit register or byte from memory.
2. n is an immediate byte.
3. i is an instruction byte from the table.
4. d is a displacement byte.
5. (rr) is the content of the address contained in 16-bit register rr.

BIT SET AND TEST GROUP (All require prebyte CB)

BIT: affects Z
SET: no flags
RES: no flags

	m →	A	B	C	D	E	H	L	(HL)	(IX + d)	(IY + d)
BIT	0,m	47	40	41	42	43	44	45	46	DD CB d 46	FD CB d 46
BIT	1,m	4F	48	49	4A	4B	4C	4D	4E	DD CB d 4E	FD CB d 4E
BIT	2,m	57	50	51	52	53	54	55	56	DD CB d 56	FD CB d 56
BIT	3,m	5F	58	59	5A	5B	5C	5D	5E	DD CB d 5E	FD CB d 5E
BIT	4,m	67	60	61	62	63	64	65	66	DD CB d 66	FD CB d 66
BIT	5,m	6F	68	69	6A	6B	6C	6D	6E	DD CB d 6E	FD CB d 6E
BIT	6,m	77	70	71	72	73	74	75	76	DD CB d 76	FD CB d 76
BIT	7,m	7F	78	79	7A	7B	7C	7D	7E	DD CB d 7E	FD CB d 7E

SET instructions replace MSD 4, 5, 6, 7, with C, D, E, F, respectively.

RES instructions replace MSD 4, 5, 6, 7 with 8, 9, A, B, respectively.

16-BIT LOAD GROUP

rr →	BC	DE	HL	SP	IX	IY	AF	S, Z, V, C flags
LD rr,nn	01 nn	11 nn	21 nn	31 nn	DD 21 nn	FD 21 nn		—
LD rr,(nn)	ED 4B nn	ED 5B nn	2A nn	ED 7B nn	DD 2A nn	FD 2A nn		—
LD (nn),rr	ED 43 nn	ED 53 nn	22 nn	ED 73 nn	DD 22 nn	FD 22 nn		—
LD SP,rr			F9		DD F9	FD F9		—
PUSH rr	C5	D5	E5		DD E5	FD E5	F5	—
POP rr	C1	D1	E1		DD E1	FD E1	F1	—

16-BIT ARITHMETIC GROUP

	BC	DE	HL	SP	flags
ADD HL,rr	09	19	29	39	c
ADD IX,rr	DD 09	DD 19	DD 29	DD 39	c
ADD IY,rr	FD 09	FD 19	FD 29	FD 39	c
ADC HL,rr	ED 4A	ED 5A	ED 6A	ED 7A	s z v c
SBC HL,rr	ED 42	ED 52	ED 62	ED 72	s z v c
INC rr	03	13	23	33	—
DEC rr	0B	1B	2B	3B	—

JUMP GROUP AND SUBROUTINE CALLS

						parity		sign		
cc →	None	NZ	Z	NC	C	PO	PE	P	M	No flags affected
JP nn	C3									All nn operands imply
JP cc,nn		C2	CA	D2	DA	E2	EA	F2	FA	LO-ADR, HI-ADR
JR e	18									bytes after up code. All
JR cc,e		20	28	30	38	20				e operands imply one
CALL nn	CD									byte signed
CALL cc,nn		C4	CC	D4	DC	E4	EC	F4	FC	displacement byte after
RET	C9									op-code.
RET cc		C0	C8	D0	D8	E0	E8	F0	F8	

JP (HL) = E9 RETI = 4D	JP (IX) = DD E9 RETN = 45	JP (IY) = FD E9	DJNZ, e = 10 d

EXCHANGE

EX	DE,HL	EB
EX	(SP),HL	E3
EX	(SP),IX	DD E3
EX	(SP),IY	FD E3
EX	AF,AF′	08
EXX	BC,DE and HL	D9

BLOCK TRANSFER

P/V = 0 if BC = 0

LDI	ED A0	(DE) ← (HL)
LDD	ED AB	Both INC or
Load & Increment		DECR, and BC
Load & Decrement		DECR
LDIR	ED 80	
LDDR	ED B8	Same as above
Load, Inc, Repeat		but repeat until
Load, Dec, Repeat		BC = 0

BLOCK SEARCH

P/V = 0 if BC = 0
Z = 0 if CP → equal

CPI	ED A1	A − (HL), then
CPD	ED A9	INC or DEC
Compare & Inc		HL, DEC BC
Compare & Dec		
CPIR	ED B1	Same as above.
CPDR	ED B9	Repeat until
Compr, Inc, Rpt		BC = 0
Compr, Dec, Rpt		

CONTROL GROUP

DAA	27	CPL (complement A)	2F	NEG (A ← A − 0)	ED 44
NOP	00	SCF (set carry)	37	DI (disable INTR)	61
HALT	76	CCF (compl carry)	3F	EI (enable INTR)	71

INPUT/OUTPUT (greatly abridged)

			A	B	C	D	E	H	L
IN A,(n)	DB n	IN r,(C)	ED 78	ED 40	ED 48	ED 50	ED 58	ED 60	ED 68
OUT (n),A	D3 n	OUT (C),r	ED 79	ED 41	ED 49	ED 51	ED 59	ED 61	ED 69

REVIEW OF SECTIONS 22.3 AND 22.4

Answers appear at the end of Chapter 22.

11. What is the significance of Z80 address $0066?
12. What registers are stacked when the Z80 recognizes an interrupt?
13. After *reset*, how can you set up the Z80 to jump to the fixed address $0038 upon receiving a *low* level at the $\overline{\text{INT}}$ input?
14. After *reset*, how can you have the Z80 jump to address $0020 upon receiving a *low* level at the $\overline{\text{INT}}$ input? Instructions IM1 and EI have been executed.
15. The I register contains E4H. Instructions IM2 and EI have been executed. A Z80 peripheral places 50H on the data bus in response to an INT. How does the Z80 find its interrupt-service-routine address?
16. What is the machine code to store A at address $ABCD?
17. What is the machine code to store A at an address 1 byte higher than that pointed to by IX?
18. Which of these are invalid Z80 instructions?

    ADC A,H   SUB A,B   OR D   ADD 14   INC E

19. What does the instruction CPL do?
20. What assembly-language instruction will clear the lowest-order bit of the byte pointed to by register HL?

22.5 NOTES ON THE Z80 INSTRUCTION SET

Section Overview

This section presents some notes on the microprocessor instruction sets generally and comments on the relative advantages of a consistent instruction set, which manufacturers describe as *orthogonal*. It also lists 10 types of Z80 instructions and operations that might be presumed to

exist, even though they are not valid in fact. Valid and invalid forms are given for comparison.

Orthogonal: Designers of microprocessor chips must face a difficult and important decision. They often find that they can implement a very useful and impressive function in their new machine, but they don't have enough room on the chip or enough variations in op-codes to implement it for all registers. Should they implement every amazing function they can think of, each on only one register—thus producing a computationally powerful but confusing and sometimes frustrating processor? Or should they insist that no function be implemented unless it can be implemented identically on all registers and in all instruction modes—thus producing an easier-to-understand if somewhat less-impressive microprocessor?

The latter philosophy is called *orthogonal*, which, literally, means *concerning right angles*. The idea, illustrated in Figure 22.5, is that the op-codes could be placed in a matrix with instructions down the rows and registers across the columns. If it is possible to add to one register, it should be possible to add to any of them. If one register can serve as an address pointer, then any register should be able to serve as an address pointer. If there is an instruction called CMPA which affects the C flag, there should be an instruction called CMPX, and it should affect the C flag in the same way.

No processor is perfectly orthogonal, nor is any completely nonorthogonal. However, it is clear that the Intel and Zilog processors have tended toward the nonorthogonal philosophy, whereas Motorola has tended toward the orthogonal. The result is that it is quite a bit easier for a student to become proficient in 6800 assembly

Mode →	Immediate				Extended				Indexed			
Register	A	B	X	S	A	B	X	S	A	B	X	S
Operation ↓												
ADD												
ADC												
AND												
ASL												
ASR												
CLR												
↓												

Figure 22.5 An *orthogonal* instruction set would have all instructions available in all addressing modes for all machine registers.

language than in Z80 assembly language; but once proficiency is gained, Z80 programs will probably be able to do more jobs faster than 6800 programs.

General comparisons among processors are of dubious value because of the varying requirements of each project, the various talents and resources available in each company, and the continually changing state of the art. Companies with a one-person genius for a programming staff and a sell-a-million consumer item for a product may tend to prefer a nonorthogonal processor. Companies that must rely on programmers with less talent and have products that are custom programmed may tend to choose an orthogonal processor. The key phrase here is *tend to choose*. Other factors, such as processing speed, price, compatibility with existing products, and availability of desired peripheral support chips, may dominate in the decision-making process. It is a common folly among microcomputer engineers to spend time on processor comparisons when that time could be used to much better advantage in comparing assembler/editor programs and in improving program development and documentation techniques.

Z80 instructions that aren't: In 6800 programming we are often frustrated by the lack of a *push X* or confused that there is a SBA ($A - B$) but no SAB ($B - A$) instruction; but the number of registers and instructions is small enough that we soon become clear about which instructions exist and which do not exist. The task of sorting out the real from the unreal is much more formidable with the Z80. Here is some help:

1. Only A can be the destination register for any 8-bit arithmetic or logic operation, excepting increments and decrements.

No	Yes
• ADD B,18H	• ADC A,5BH
• CP C,E	• SUB (HL) *Note:* A assumed.
• AND D,A	• INC B

2. Only A and the 16-bit registers can use extended addressing.

No	Yes
• LD B,(1234H)	• LD A,(1234H)
• LD (5678H),E	• LD (5678H),BC

3. A general-purpose register pair (BC) or (DE) can point for A only, not for a general-purpose register.

No	Yes
• LD B,(DE)	• LD B,(HL)
• LD (BC),H	• LD (DE),A

4. There are no 8-bit stack operations. PUSH and POP work on register pairs or 16-bit registers only.

No	Yes
• PUSH A	• PUSH AF
• POP F	• POP BC

5. In general, the *load* operations do not set any flag bits. Increments and decrements of register pairs and 16-bit registers do not affect any flag bits. Here are two favorite tricks to get around these problems.

Set flags on load	Test HL for zero
LD B,(HL)	DEC HL
INC B	LDA H
DEC B	OR L
JP Z,OUT	JP Z,ZERO

6. In general, you cannot transfer data between 16-bit registers. An exception is that SP can be loaded from HL, IX, or IY. Here are examples and two ways to get around the problem.

No	Yes
• LD HL,DE	• LD SP,IX
• LD IX,IY	

Use the stack	A byte at a time
PUSH DE	LD H,D
POP HL	LD L,E

7. There is no *register-indirect* or *indexed* addressing mode for register pairs or for 16-bit registers.

No	Load byte by byte
• LD BC,(HL)	LD C,(HL)
• LD (IX+7),HL	INC HL
	LD B,(HL)

8. Sixteen-bit ADD, ADC, and SBC operations are available, but HL is the only allowed destination, and BC, DE, HL, or SP must be the source. In addition there is an ADD with IX as the destination and IX, SP, BC, or DE as the source. There is a similar ADD to IY with IY, SP, BC, or DE as the source. Of all these, only ADC and SBC have a full effect on the flag register: C, Z, V, and S (sign) are affected. The others affect only C.

No	Yes
• ADC HL,1234H	• ADC HL,HL
• SBC DE,HL	• SBC HL,DE
• ADD IX,IY	• ADD IX,IX

9. The conditional *jumps* (similar to 6800 *branches*) are available in the relative mode only for *carry-* and *zero-flag* tests. In the extended mode the *carry*, *zero*, *parity/overflow*, and *sign* flags can be tested.

No	Yes	
• JR P,LOOP	• JP P,LOOP	Plus sign
• JR M,LOOP	• JP M,LOOP	Minus sign
• JR PE,CHECK	• JP PE,CHECK	Parity even or overflow
• JR PO,CHECK	• JP PO,CHECK	Parity odd or no overflow

10. Unconditional jumps can be made to addresses pointed to by HL, IX or IY. Although the form is JP (HL), the program counter is loaded immediately with the contents of HL. The address reference (HL) means that the *jump* is to address HL, not that PC is loaded from an address pointed to by HL. The JP (HL) takes only 4 *T*-states, compared to 10 *T*-states for extended-mode jumps of the form JP 1234H.

No	Yes
• JP HL	• JP (HL)
• JP (DE)	• JP (IX)

22.6 A KEYBOARD MORSE CODE SENDER

Section Overview

This section presents a hardware project using the Z80 to generate the appropriate Morse code tones as letters and numbers are typed in on a keyboard. The system hardware and the keyread, file-lookup, and tone-generate routines were originally developed by student Kim Kelly as part of a term project. We deviate from our usual practice of describing a hardware project in Section 7 in order to save the last section for a quick look at the Intel 8080/8085 processors and some of the specialized Zilog and Intel peripheral chips.

The hardware diagram for the Z80 Morse Sender is given in Figure 22.6. The encoding scheme, illustrated in Figure 22.7, allows a single byte to specify up to 5 dots and dashes in any sequence, thus enabling a 36-byte file to generate all the Morse code letters and numbers. There are plenty of bit combinations left to permit

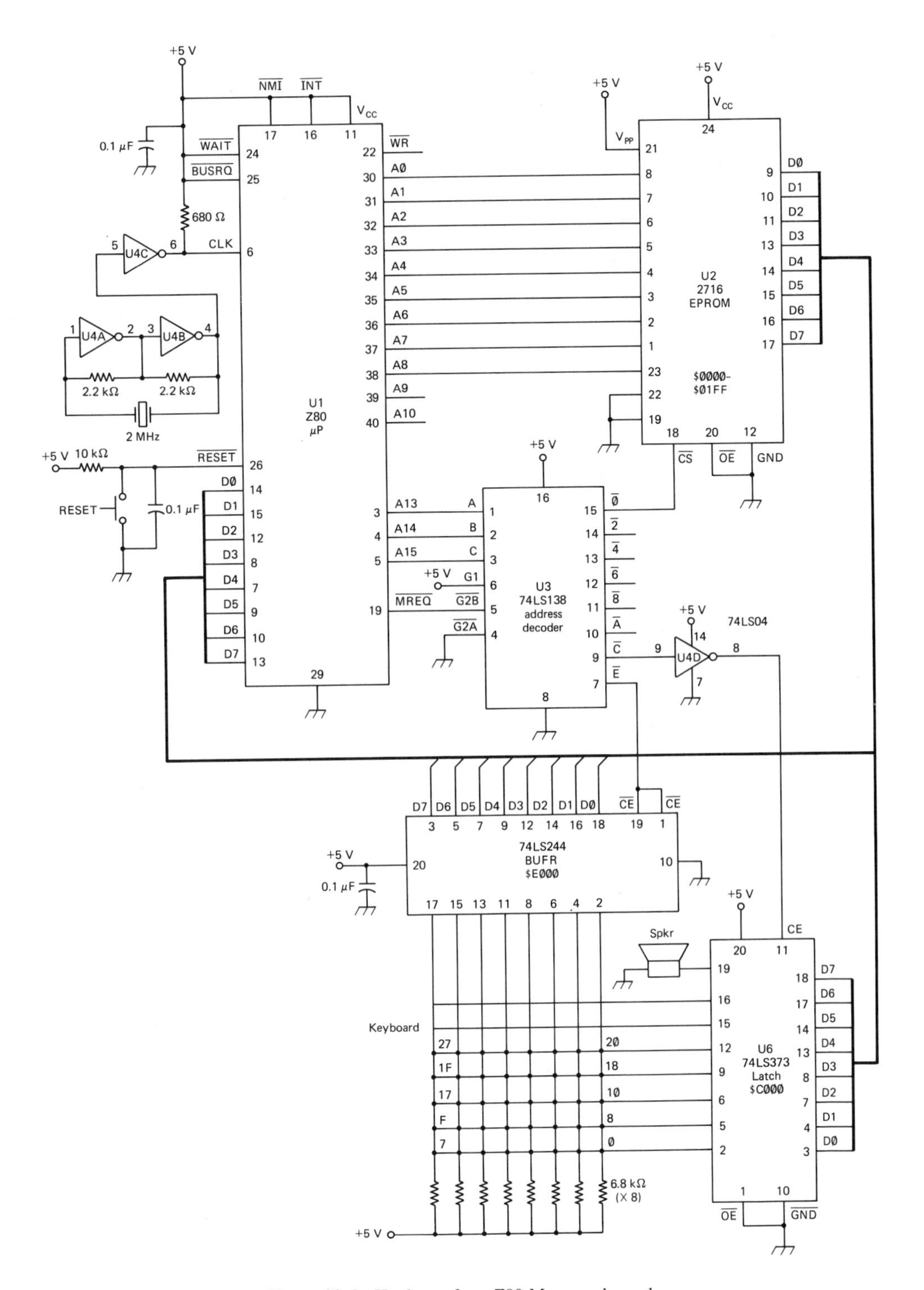

Figure 22.6 Hardware for a Z80 Morse code sender.

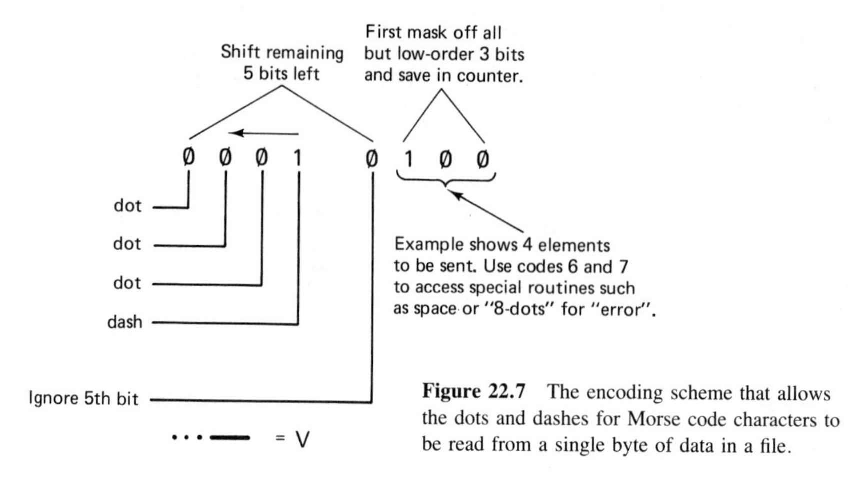

Figure 22.7 The encoding scheme that allows the dots and dashes for Morse code characters to be read from a single byte of data in a file.

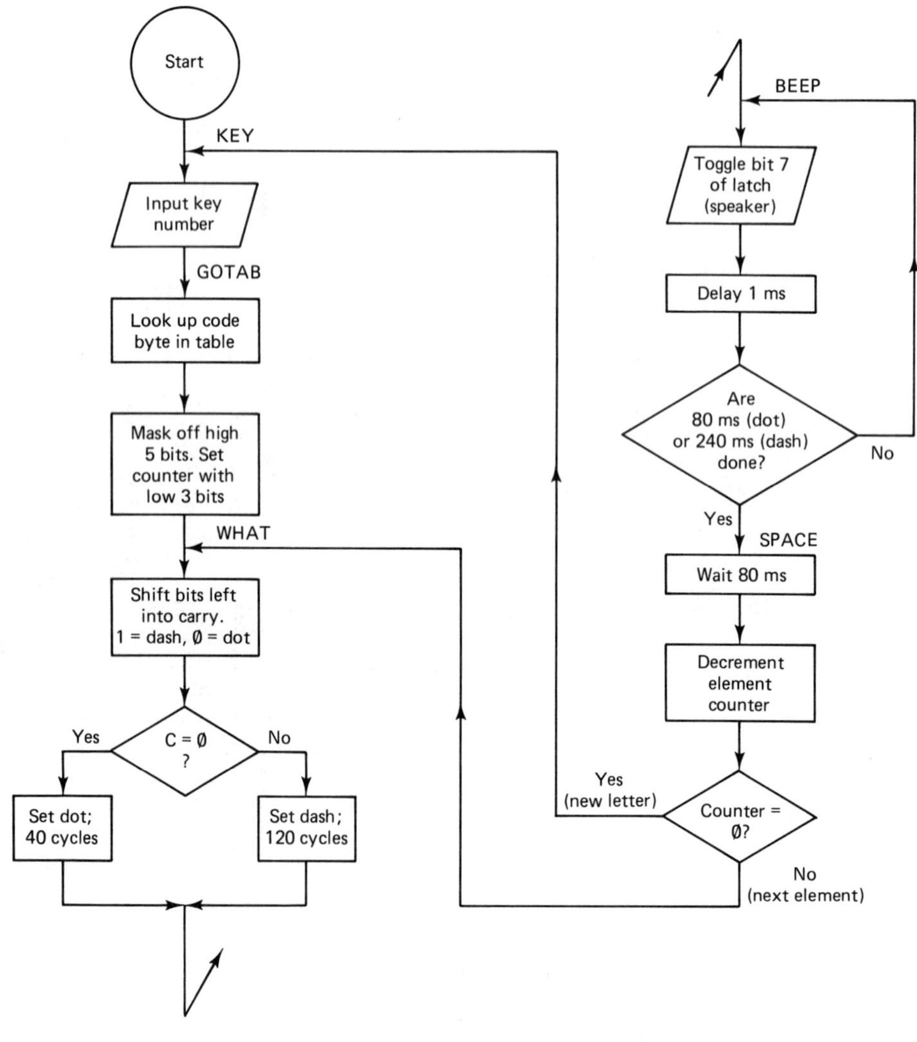

Figure 22.8 A flowchart for the Morse-sender program.

the addition of special routines to handle 6-element codes (such as periods and question marks), but we will leave the development of these routines to you. Another area for development could be a FIFO text-file generator and reader. This would allow the operator to type in a message as rapidly as skill may permit and then leave the machine to transmit the code at a slower speed until the file was emptied. Figure 22.8 gives the basic program flowchart. The program listing follows (List 22.1).

```
                   1000 * LIST 22-1   ** KEYBOARD MORSE-CODE GENERATOR FOR Z80.
                   1010 *  FIG 22.6 HARDWARE, 22.7 CONCEPT, 22.8 FLOWCHART.
                   1020 * TESTED AND RUNNING, 3 AUG '87, BY D. METZGER.
                   1030        .OR   0000H
                   1040        .TA   4000H
C000-              1050 LATCH  .EQ   0C000H  BITS 0 - 5 TO KEYBD; BIT 7 TO SPKR.
E000-              1060 BUFR   .EQ   0E000H  ALL INPUTS FROM KEYBOARD.
                   1070 *
                   1080 *READ KEYBOARD UNTIL A KEY IS PRESSED:
0000- 1E 00        1090 KEY    LD    E,0     E COUNTS KEY NUMBER.
0002- 06 FE        1100        LD    B,0FEH  B PULLS 1 ROW LOW VIA LATCH.
0004- 21 00 C0     1110        LD    HL,LATCH  POINT HL TO OUTPUT LATCH.
0007- 70           1120 ROW    LD    (HL),B  PULL A ROW LOW.
0008- 3A 00 E0     1130        LD    A,(BUFR)  READ THE ROW AND COMPLEMENT SO
000B- EE FF        1140        XOR   0FFH      A 1 MEANS A PRESSED KEY.
000D- 20 0C        1150        JR    NZ,FIND ANY KEY PRESSED?
000F- 7B           1160        LD    A,E       NO: ADD 8 TO KEY COUNT IN B;
0010- C6 08        1170        ADD   A,8         (ONLY A CAN BE ADDED TO)
0012- 5F           1180        LD    E,A
0013- FE 27        1190        CP    39      IS A INTO 6TH ROW (ALL KEYS CHECKED)?
0015- 30 E9        1200        JR    NC,KEY    YES: START KEY SEARCH AGAIN.
0017- CB 00        1210        RLC   B         NO: READY TO PULL NEXT ROW LOW.
0019- 18 EC        1220        JR    ROW
                   1230 *
                   1240 * YOU FOUND THE ROW; NOW FIND THE KEY IN THE ROW.
001B- 1F           1250 FIND   RRA           A HAS A 1 IN KEY-PRESSED POSITION.
001C- 38 03        1260        JR    C,GOTAB ROTATE 1 TOWARD CARRY, INCREMENTING E
001E- 1C           1270        INC   E         (KEY COUNTER) UNTIL 1 FALLS INTO CARRY
001F- 18 FA        1280        JR    FIND    E NOW CONTAINS KEY COUNT.
                   1290 *
                   1300 * USE KEY NUMBER TO FIND CODE BYTE IN TABLE. SPLIT INTO
                   1310 *  DOT/DASH ELEMENTS (B) AND ELEMENT COUNT (E).
0021- 16 01        1320 GOTAB  LD    D,01H   DE POINTS TO TABLE ADR HOLDING CODE.
0023- 1A           1330        LD    A,(DE)  GET CODE FROM TABLE INTO ACCUM A.
0024- 47           1340        LD    B,A     SAVE A COPY OF CODE IN B.
0025- E6 07        1350        AND   07H     KEEP CODE-ELEMENT COUNT IN C
0027- 4F           1360        LD    C,A       (MASK OFF ALL BUT LO 3 BITS)
                   1370 * DECIDE WHETHER TO MAKE A DOT OR A DASH:
0028- 16 50        1380 WHAT   LD    D,80    SET D CYCLE COUNTER FOR 80 HALF CYCLES.
002A- CB 20        1390        SLA   B       SHIFT CODE BIT INTO CARRY FLAG.
002C- 30 02        1400        JR    NC,BEEP  IF BIT WAS A 0, KEEP DOT TIMING,
002E- 16 F0        1410        LD    D,240    IF BIT WAS A 1, SET DASH TIMING.
                   1420 *
                   1430 * MAKE 80 OR 240 HALF CYCLES AT 500 HZ.
0030- 77           1440 BEEP   LD    (HL),A  OUTPUT TO LATCH (SPEAKER BIT 7)
0031- 1E 7D        1450        LD    E,125   SET E FOR 125-LOOP DELAY
0033- 1D           1460 LOOP   DEC   E        * 16 T-STATES * 0.5 US/T
0034- 20 FD        1470        JR    NZ,LOOP   = 1.0-MS DELAY (F = 500 HZ)
0036- EE 80        1480        XOR   80H     TOGGLE BIT 7 IN ACCUM A.
0038- 15           1490        DEC   D       COUNT DOWN DOT OR DASH TIMER.
0039- 20 F5        1500        JR    NZ,BEEP MAKE MORE CYCLES UNTIL DOT / DASH DONE.
```

```
                   1510 *
                   1520 * 80-MS DELAY IS SPACE BETWEEN ELEMENTS OF A LETTER.
003B- 16 50        1530 SPACE  LD   D,80      NESTED DELAY LOOP
003D- 1E 7D        1540 MILSEC LD   E,125      COUNTS 80 INTERVALS
003F- 1D           1550 WAIT   DEC  E            (OUTER LOOP)
0040- 20 FD        1560        JR   NZ,WAIT   OR 1 MS EACH
0042- 15           1570        DEC  D            (INNER LOOP)
0043- 20 F8        1580        JR   NZ,MILSEC
                   1590 *
                   1600 * DECIDE WHETHER TO MAKE ANOTHER DOT / DASH OR LOOK FOR KEY.
0045- 0D           1610        DEC  C         ELEMENT COUNTER = 0 (ALL DOT/DASH DONE)?
0046- 20 E0        1620        JR   NZ,WHAT   NO: MAKE NEXT DOT OR DASH.
0048- 18 B6        1630        JR   KEY        YES: LOOK FOR KEY.
                   1640 * NOTICE THAT NO RAM HAS BEEN REQUIRED: THE 8080 REGISTERS
                   1650 * WERE SUFFICIENT. THE ADDED Z80 REGISTERS WEREN'T EVEN USED
                   1660 *
                   1670 * TABLE OF CODE BYTES FOLLOWS. SEE FIG 17.15 KEYBOARD.
                   1680        .OR  0100H     START OF TABLE.
                   1690        .TA  4100H
0100- 23 C5 24
0103- 74 C2 0D
0106- 43 14        1700        .HS  23C52474C20D4314  ** U 7 F J M 4 R V
0108- 02 E5 83
010B- A3 8D 1D
010E- 01 A4        1710        .HS  02E583A38D1D01A4  ** I 8 D K , 3 E C
0110- E3 F5 03
0113- 4C 55 3D
0116- 63 94        1720        .HS  E3F5034C553D6394  ** O 9 S L . 2 W X
0118- B4 85 C3
011B- 04 82 05
011E- 81 84        1730        .HS  B485C30482058184  ** Y 6 G H N 5 T B
0120- 64 FD 42
0123- 8D 95 7D
0126- D4 C4        1740        .HS  64FD428D957DD4C4  ** P 0 A : / 1 Q Z
                   1750 * COMMA AND COLON SENT AS BREAK.          _..._
                   1760 * PERIOD SENT AS END MARK.                ._._.
                   1770        .EN
```

REVIEW OF SECTIONS 22.5 AND 22.6

21. Which application would be more likely to require a processor with an orthogonal instruction set; an automobile-engine controller or an industrial steel-rolling process controller?
22. On a Z80, how would you add B to C with the result in C?
23. How would you load HL from addresses 1234H (low-order) and 1235H (high-order) without using a pointer register?
24. How would you load B from address 1234H without disturbing the contents of A and without using a pointer register?
25. Which of these are valid Z80 instructions?

    LD (DE),B   LD (HL),BC   INC (BC)

26. How can you transfer the contents of IX to HL?
27. Which of these instructions set Z upon a zero result?

    LD A,(HL)   INC B   INC IX   ADD HL,DE

28. How can you subtract IX from HL with the result in HL and assurance of no extra subtract from the carry flag?
29. Using the code illustrated in Figure 22.7, what hex byte will cause the *end-of-transmission* sign to be transmitted? This sign consists of the letters AR run together:

    dot-dash-dot-dash-dot

30. A 4-MHz Z80 takes 38 *T*-states per loop and must stay in the loop for one-half cycle of a 400-Hz waveform. How many passes through the loop are necessary?

22.7 OTHER ZILOG AND INTEL 8-BIT CHIPS

Section Overview

This section presents a brief introduction to the Intel 8080 and 8085 microprocessors and to some some of the more popular peripheral support chips for the Zilog and Intel processors. It is included here as a reference because these components are so widely used that you are sure to encounter many of them in your work. It would probably be better if you did not try to read it as a contiguous part of this chapter because of the confusion that would be likely to result from similar, yet different, features of the Intel and Zilog processors.

The Intel 8080 started the microcomputer revolution in 1972. To get a minimum system operating, it requires three power supplies (+5 V, −5 V, and −12 V) and two external support chips. These are an 8224 clock generator (16-pin package) and an 8228 system controller (28-pin package). A basic 8080 system is diagrammed in Figure 22.9. We will give the 8080 no more attention than this because it has been superseded by the 8085, which is much easier to use, and retains all of the software features and most of the desirable hardware features of the 8080.

A basic 8085 system is diagrammed in Figure 22.10. Only a single +5-V supply is required for this processor, and the clock generator is on-board, requiring only a crystal to be connected externally.

The data bus is time-multiplexed with the low-order half of the address bus in the 8085. This means that the address and data signals share the same pins, which are called AD0 through AD7. Address signals are applied during the first *T*-state of a memory-reference cycle, and the ALE (Address Latch Enable) output is asserted. This signal causes an external 8-bit latch to hold the address signals until the next memory call-up. Data is then applied to the pins AD0 through AD7 (either by the processor for a *write* or by another device for a *read*) during a subsequent *T*-state. Intel chose to multiplex the buses so they could add some new control functions while maintaining the standard 40-pin package. The price to be paid is the mandatory external 8-bit latch.

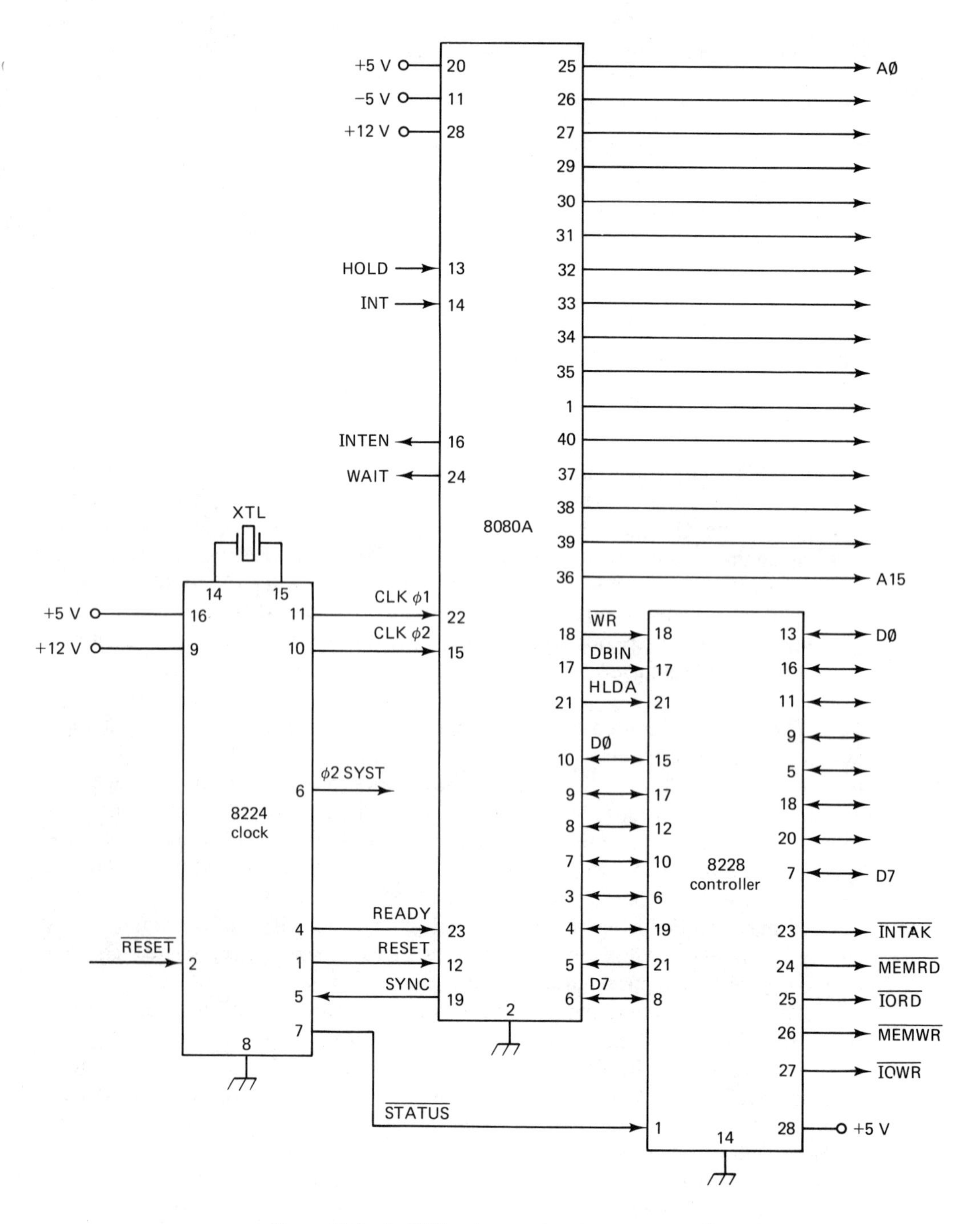

Figure 22.9 An 8080 system requires three power supplies and two system-support chips.

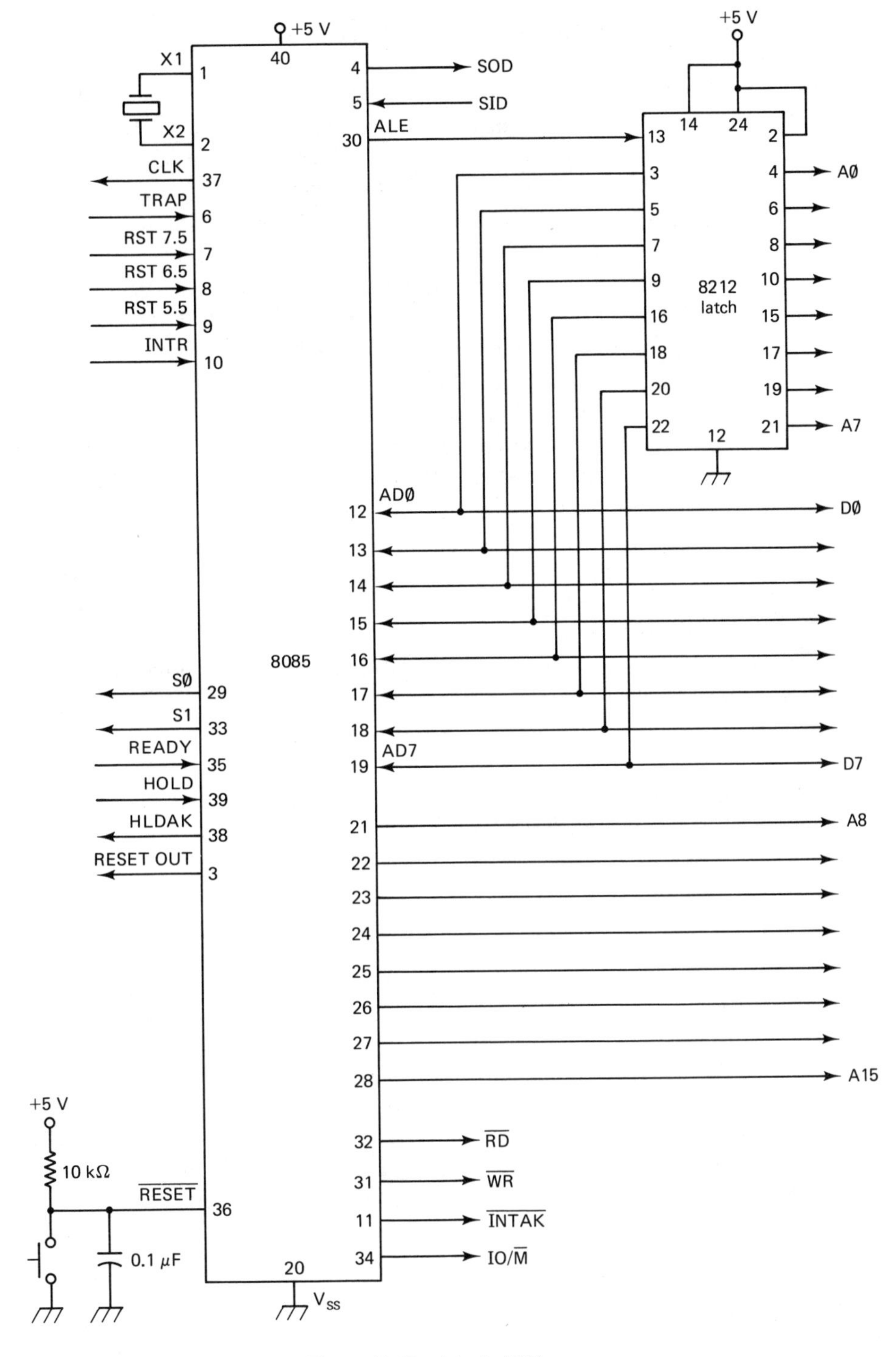

Figure 22.10 A basic 8085 system.

8085 pin description:

- $\overline{RD}$ and $\overline{WR}$ are read and write outputs, like their counterparts in the Z80.
- IO/$\overline{M}$ combines the functions of the Z80 $\overline{MREQ}$ and $\overline{IORQ}$ on one line. When it is *high*, AD0 through AD7 contain an I/O address in the first *T*-state. When it is *low*, AD0 through A15 contain a memory address.
- S0 and S1 are status outputs that are decoded with IO/$\overline{M}$ to indicate such processor states as *Memory Write*, *Op-code fetch*, and *I/O read*.
- $\overline{RESET\ IN}$ starts the processor at address 0000H when returned *high*, as does $\overline{RESET}$ in the Z80. The 8085 also has a RESET OUT (high active), which can be used to reset peripherals by the software RST instruction.
- READY is pulled *low* by slow memories to stretch the machine cycle until data is ready. It is similar to $\overline{WAIT}$ on the Z80.
- HOLD and HLDA (Hold Acknowledge) are input and output pins, respectively, used to suspend processing for DMA. They are similar to the Z80 $\overline{BUSREQ}$ and $\overline{BUSACK}$ but of the opposite logic sense.
- INTR is a maskable interrupt input, *high* active. The interrupt request, if not masked, is handled like the Z80's *Interrupt Mode 0* (see page 561). $\overline{INTA}$ acknowledges the interrupt, eliminating the need to decode two lines ($\overline{M1}$ and $\overline{IORQ}$) as is required on the Z80.
- TRAP is a nonmaskable interrupt input, like $\overline{NMI}$ on the Z80 but *high* active. It sends the processor to address 0024H.
- RST 5.5, RST 6.5, and RST 7.5 are *restart* interrupt inputs. They send the processor to addresses 002CH, 0034H, and 003CH, respectively.
- SID is a *Serial Input Data* pin. The logic level on this pin is transferred to bit 7 of accumulator A when the RIM (Read Interrupt Mask) instruction is executed. (The other 7 bits of A read bits pertaining to interrupt masks and pending interrupts.)
- SOD is a *Serial Output Data* pin. The logic level preloaded into bit A7 is asserted on this pin when the SIM (Set Interrupt Mask) instruction is executed, but only if bit A6 = 1. If A6 = 0, the SOD pin is unaffected. The lower bits of A are used to set interrupt masks.

The instruction set of the 8080 is a subset of the Z80 instruction set; at least it is at the object-code level. The instruction set of the 8085 is exactly the same as that of the 8080, except for the addition of the RIM and SIM instructions just mentioned. Only registers A, F, B, C, D, E, H, L, SP, and PC are available in the 8080/8085 (see Figure 21.1), so the Z80 instructions pertaining to IX, IY, and the alternate register set do not exist for them. Also missing from the 8080/8085 instruction set are the following:

- The relative jumps; there is no relative addressing mode.
- The block-transfer and block-search instructions.

- All the *shifts* and the *rotates*, except those operating on accumulator A. There are only four *rotates*, all of A.
- The BIT instructions for test, set, and reset.
- The I/O instructions except for two: input to A from a directly specified I/O address and a similar output from A.

The assembly-language mnemonics, or source code, for the 8080/8085 are very different from their Z80 counterparts, even though the object code (hex opcodes) are exactly the same. These differences can make the process of shifting from the Z80 to the 8085 much more confusing than it would have needed to be. Here are some general comments on the 8080/8085 assembly-language forms:

1. Specifying an address with a pointer register, which we learned to called *indexed* addressing in the 6802/6809, is called *register-indirect* in the Z80 and simply *indirect* in the 8080/8085. This is most unfortunate because indirect addressing in the 6809 is a quite different mode.
2. Specifying a 2-byte address for the operand, which we called *extended* in the 6802/6809 and the Z80, is called *direct addressing* in the 8080/8085. This same mode is called *absolute* in the 6502. Of course *direct* means *zero-page* in the 6802 and *single-page* in the 6809. Confusion reigns!
3. The general form for 8080/8085 instructions is

   MNEMONIC destination,source

   However, this pattern is followed even less rigorously than it is in Z80 instructions. Often the destination is left unmentioned and is assumed to be accumulator A. Often the source is referenced within the opcode itself.
4. Parentheses are not used to indicate "data at address." Numbers or variable names are assumed to be addresses unless the letter I at the end of the instruction mnemonic indicates *Immediate* data. Hence

   LDA 1234H = Load A from address $1234

   but

   ADI 34H = Add data $34 to A

5. The letter M as the source or destination means "the data at the memory address pointed to by HL."

6. The letter X at the end of an instruction mnemonic implies the involvement of a register pair (Think of *extended*, i.e., 16-bit). Register pairs as pointers are specified by both letters of the pair, but as data holders they are specified by only the first letter.

   LDAX BC = Load A from address pointed to by BC pair
   DCX H = Decrement register pair HL

7. The letter D is sometimes (but not always) placed at the end of a mnemonic to specify *direct* (we have called it *extended*) addressing:

LHLD 3456H = Load L from address $3456, H from $3457

8. The 8080 and 8085 use MOV instead of LD for 8-bit register-to-register and some HL-pointed transfers.

Here are examples of 8085 assembly forms, with their equivalent Z80 forms to give you a feel for reading Intel assembly-language programs.

8080/8085			Z80	
ACI	34H	Add with carry, immediate, to A	ADC	A,34H
ADD	M	Add data pointed to by HL to A	ADD	A,(HL)
ANI	MASK	AND A with byte MASK	AND	MASK
CM	MUL	Call subroutine MUL if Minus	CALL	M,MUL
CMA		Complement (2's) Accumulator	CPL	
CMC		Complement Carry flag	CCF	
CMPC		Compare A − C	CP	C
CPI	F0H	Compare A − immediate data	CP	F0H
DAD	D	Double add DE to HL	ADD	HL,DE
DCR	M	Decrement data pointed to by HL	DEC	(HL)
DCX	B	Decrement register pair BC	DEC	BC
INR	C	Increment register C	INC	C
JMP	1234H	Jump, unconditional	JP	1234H
JNZ	1234H	Jump if not zero	JP	NZ,1234H
LDA	1234H	Load A direct (extended)	LD	A,(1234H)
LDAX	BC	Load A via pointer pair BC	LD	A,(BC)
LHLD	1234H	Load HL direct (extended)	LD	HL,(1234H)
LXI	B,1234H	Load pair BC, immediate	LD	BC,1234H
MOV	A,B	Move B to A	LD	A,B
MOV	M,C	Move C to memory via pointer HL	LD	(HL),C
MVI	B,9AH	Move immediate data to B	LD	B,9AH
PCHL		Jump to (load PC with) HL	JP	(HL)
SBB	M	Subtract A − (HL) − Borrow	SBC	A,(HL)
SBI	B2H	Subtract A − $B2 − Borrow	SBC	A,B2H
SUB	D	Subtract A − D	SUB	D
SUI	BCH	Subtract A − immediate $BC	SUB	BCH
SHLD	1234H	Store HL direct to $1234,5	LD	(1234H),HL
STA	PLACE	Store A, direct address PLACE	LD	(PLACE),A
STAX	DE	Store A via pointer pair DE	LD	(DE),A
STC		Set carry flag	SCF	
XRI	4FH	Exclusive-OR, immediate, with A	XOR	4FH
XTHL		Exchange HL with top 2 stack bytes	EX	(SP),HL

Intel and Zilog peripherals are available in great number and variety. We have room here only to note the functions of some of the most popular:

- The Intel 8255 programmable peripheral interface (PPI) provides 24 programmable I/O pins in a 40-pin package, as illustrated in Figure 22.11. These are

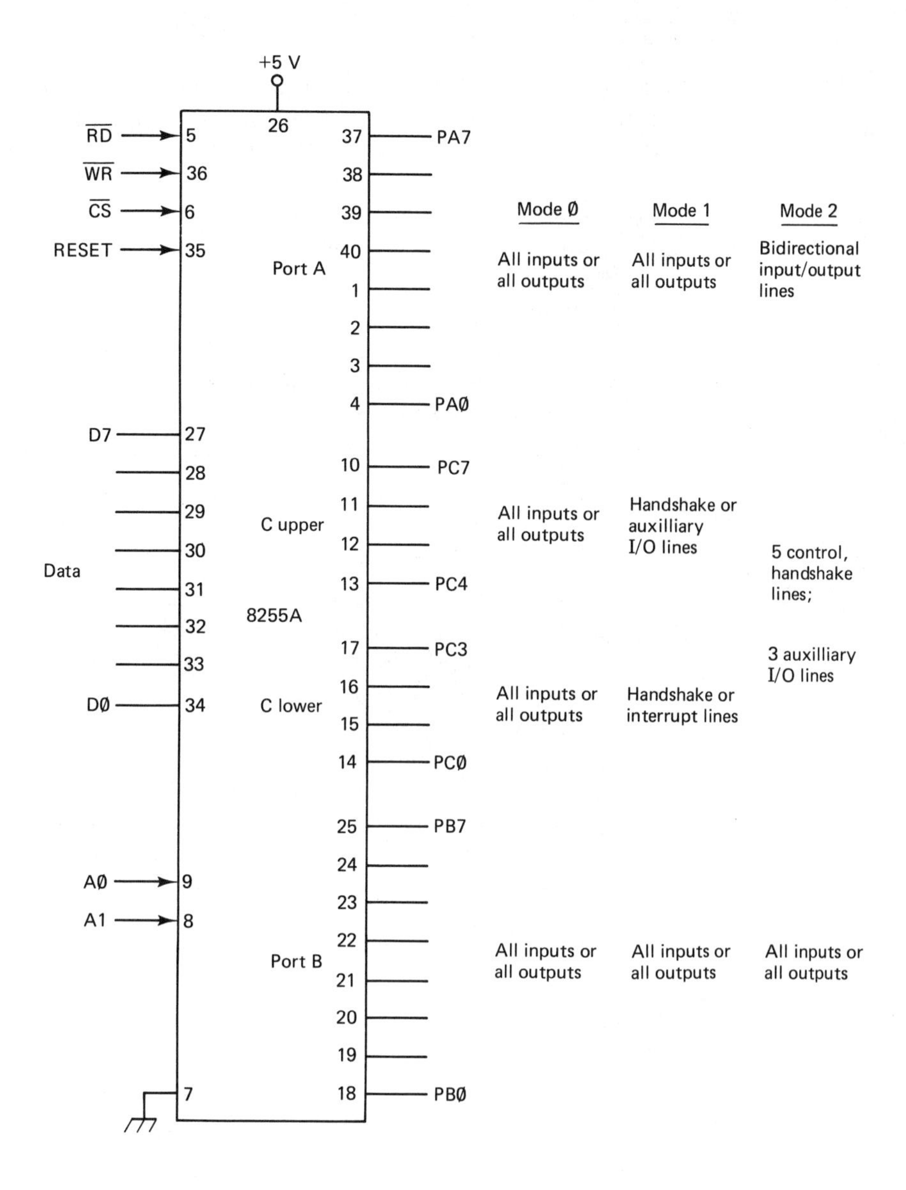

Figure 22.11 The three modes of the 8255A parallel interface chip.

similar to the two ports on the 6522 VIA of Chapter 14. There are three modes of operation:

Mode Ø Four ports: *A*, *B*, *C*-upper, and *C*-lower. Each can be configured for inputs or outputs.

Mode 1 Two ports: *A* and *B*. Each can be configured as an input or output. Port *C*-upper provides four interrupt and handshake lines for Port *A*, and Port *C*-lower provides the same for Port B.

Mode 2 Port *A* is bidirectional. Data can flow in or out under control of 5 lines from Port *C*. Port *B* is a simple input or output. The remaining three lines of Port *C* are available for auxiliary I/O.

- The Zilog Z80 PIO (Parallel Input/Output) provides two 8-bit ports with four handshake lines in a 40-pin package. There are four operating modes.

Mode 0 Eight outputs with two handshake lines. Ports *A* and *B* are set individually.
Mode 1 Eight inputs with two handshake lines.
Mode 2 Port *A* is bidirectional with four handshake and control lines. Port *B* is in Mode 3.
Mode 3 Port bits configurable bit by bit as inputs or outputs. No handshake lines.

Interrupt-enable input and output pins allow several PIOs to be connected in a series called a *daisy chain*. These inputs require a logic 1 to enable the PIO's interrupts, but the outputs do not give a logic 1 until all pending interrupts have been satisfied. Pending interrupts are thus latched in a prioritized order. Figure 22.12 illustrates.

- The Z80 CTC (Counter/Timer Circuit) provides four 8-bit counters or timers in a 28-pin package. In the timer mode the clock signal can be divided by 16 or by 256 to obtain delays of up to 65 536 clock periods. The CTC also supports the daisy-chain–interrupt prioritizing scheme.

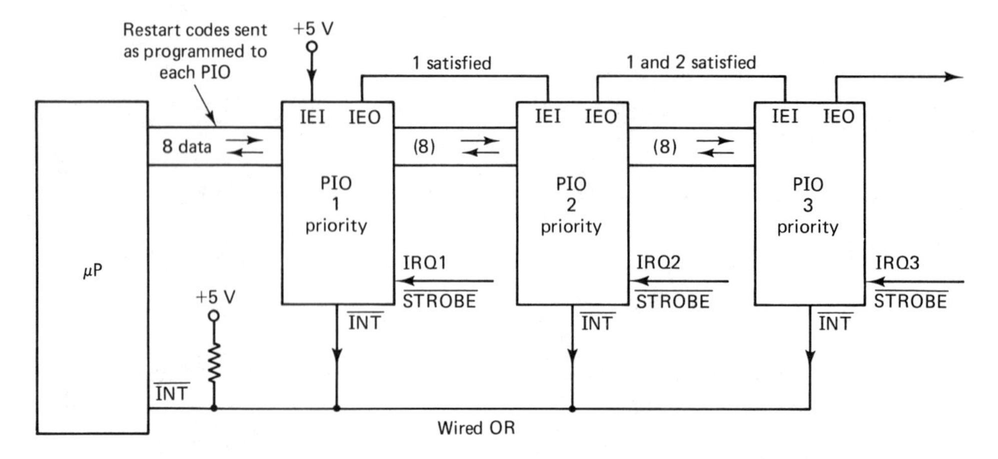

Figure 22.12 The Z80 parallel input/output chips use daisy-chain wiring to establish priorities for interrupts when there are several I/O devices in the system.

- The Z80 SIO (Serial Input/Output) chip contains two serial transmitters and receivers in a 40-pin package. There are three versions of the SIO with different functions for pins 25 and 29. There is a fourth chip called a DART (Dual Asychronous Receiver/Transmitter), whose pinout is almost (but not quite) the same as the SIO. The SIO/Ø, SIO/1, SIO/2, and DART chips are *not interchangeable*.
- DMA controllers are available for most microprocessors. Zilog's is the Z80 DMA and Intel's is the 8237, both in 40-pin packages. Because DMA timing is so complex, these devices are usually convenient to interface only with the processor for which they were designed.
- DRAM refresh chips are also quite specific, not only as to the processor but also as to the dynamic RAMs they support. The Intel 8203 is one example. It comes in a 40-pin package and supports the 2164 (64K × 1 bit) DRAM on 8080A and 8085 processors. The Z80 processor contains an internal refresh counter for support of DRAMs up to 16K. The 14 address lines of the Z80 must be broken into 7 rows and 7 columns, and *row address strobe* and *column address strobe* signals must be generated externally. This is hardly worth the effort in today's technology, since 32-kilobyte static RAMs are readily available in a single package.

Answers to Chapter 22 Review Questions

1. LD BC,3E8H **2.** No accessible machine register; data goes from (HL) to (DE), which are memory locations.
3. BC, DE, and HL **4.** It counts the number of bytes that equal FFH.
5. EX AF,AF′, then EXX **6.** $\overline{IORQ}$ is asserted but $\overline{MREQ}$ (memory request) is not. **7.** Tristate output **8.** LD C,Ø4H then OUT (C),D **9.** It reads 128 input bytes and forms a file starting at $4ØØØ, leaving out any bytes that equal zero. **10.** LD (9ØØØH),A
11. An $\overline{NMI}$ causes the processor to jump to this address.
12. Only the program counter, high- and low-order bytes.
13. Execute EI then IM1 instructions. **14.** Have the interrupting device place binary 111Ø Ø111 on the data bus in response to $\overline{M1}$ and $\overline{IORQ}$ from the Z80.
15. It picks up the low-order byte of the address from $E45Ø and the high-order byte from $E451. **16.** 32 CD AB **17.** DD 77 Ø1 **18.** SUB A,B should be SUB B and ADD 14 should be ADD A,14 **19.** It forms the 2's complement of the data in A. **20.** RES Ø, (HL)
21. The steel roller, because of the anticipated need for frequent program modification.
22. LD A,C, then ADD A,B, then LD C,A **23.** LD HL,(1234H). **24.** LD BC,(1233H). C is loaded from 1233H, but that's OK. **25.** None of them.
26. PUSH IX then POP HL **27.** Only INCB **28.** SCF, then CCF, then LD DE,IX, then SBC HL,DE **29.** The byte is $55. **30.** 132 loops are required.

CHAPTER 22 QUESTIONS AND PROBLEMS

Digits after the decimal point refer to section numbers in Chapter 22.

Basic Level

1.1 Define the functions of the BC, DE, and HL registers in the Z80's LDIR instruction.

2.1 Identify two functional differences between the instructions CPIR and CPD.

3.2 Using the Z80's "standard" I/O configuration, how many output latches are possible, assuming adequate buffering?

4.2 In standard Z80 single-byte output, what are the two ways of specifying the output address?

5.3 What is the significance of address 0038H to the Z80?

6.3 Tell whether the Z80 *interrupt* and *nonmaskable interrupt* inputs are *low* or *high* active and level or transition active.

7.4 What are the complete Z80 source and object codes to Load register IX with the immediate data 64?

8.4 Write the source and object codes for a Z80 instruction clear bit 7 of the data byte pointed to by the HL register pair.

9.5 What is an *orthogonal* instruction set?

10.5 Show how the Z80's registers HL and IY can be exchanged using the stack.

11.7 What 8085 instruction is used to read the logic level at the SIO pin, and where is this data bit stored?

12.7 Which of these Z80 instructions are available in the 8085?

RRC	LDIR	OUT A,(C)	SET 7,(HL)
JR, Z,LOOP	LD A,(IX)	EXX	

13.7 What is *direct* addressing, in 8085 terminology?

14.7 Distinguish among the following three 8085 instructions:
a. ADD M **b.** ADD 98 **c.** ADI 98

15.7 What is the significance of the letter X in each of the following 8085 instructions?
a. LXI D,0 **b.** LDAX DE

Advanced Level

16.1 Write a Z80 routine to search a file of data from addresses B000H through B7FFH. Any bytes found to equal FFH are to be written over with data byte 01H.

17.1 Write a Z80 routine to compare two files, one working upward from address 5000H and the other upward from address 5100H. Each file is 180 bytes long.

Any byte pairs that are not identical in their most-significant nibbles are to be overwritten with ØØH bytes in each file. Use the EX DE,HL instruction.

18.2 Redraw the circuit of Figure 22.3 to place the latch at I/O address BØH–B3H and the buffer at 9CH–9FH. Use 3-input negative-input NANDS and inverters as necessary.

19.2 Write a Z80 routine to input 128 bytes from I/O address KEYBD if the input is anything other than FFH. The bytes are to be stored in a file from address 3Ø8ØH to 3ØFFH.

20.3 Write a Z80 routine starting at address ØØØØH to set interrupt mode 1 and enable the interrupt. Also write a routine starting at ØØ38H to switch to the alternate register set, stack the X and Y index registers, and make an unconditioned jump to the interrupt-service routine at address 37DØH.

21.3 Draw a hardware diagram of a Z80 mode-Ø interrupt generator activated by a debounced pushbutton and set to send the processor to address ØØ1ØH. Use a 74LS244 and the necessary logic gates.

22.4 Write and hand assemble a Z80 routine to multiply 2 bytes in registers A and B, storing the results in register pair DE.

23.4 The following are all invalid Z80 instructions. For each one, devise a way to achieve the intended function.

a. ADD B,C **b.** RL HL **c.** CP HL,63ØØH
d. LD IX,HL **e.** ADD HL,A **f.** INC (1234H)

24.5 Tell what, if anything, is wrong with each of these instructions.

a. LD D,(COUNTR) **b.** SBC E,4 **c.** LD D,(BC)
d. ADD HL,IX **e.** LD (IX),B

25.5 Write the Z80 source code for a routine to convert the 4-digit BCD data in BC to its binary equivalent in register pair DE.

26.6 Make a flowchart for a program that uses two tracking address pointers to make a FIFO out of an ordinary RAM. Let the stack length be 256 words. As each new word is written in, the word written 255 entries ago is overwritten.

27.7 Explain what is meant by a *multiplexed* address/data bus, as used in the 8085 microprocessor.

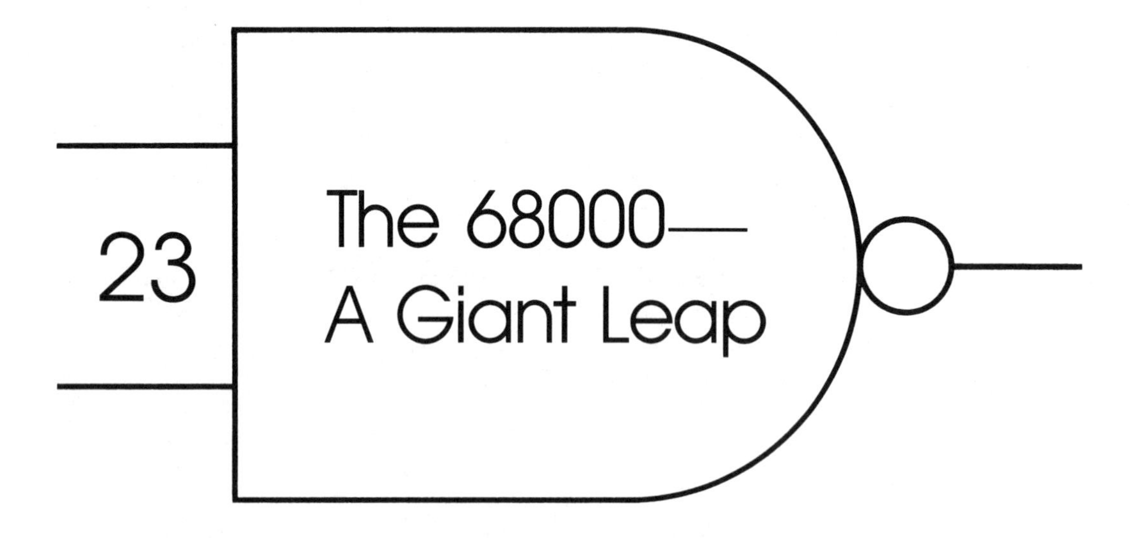

23.1 PROGRAMMER'S MODEL AND FAMILY MEMBERS

Section Overview

The 68000-family microprocessors contain 18 internal registers, each 32 bits long, plus a 16-bit status register. Eight of the 16 registers are designated as *data* holders, and seven as *address* pointers. There are 2 stack pointers and a program counter. The "normal" operand size is 16 bits, although 8-bit and 32-bit operands can be specified for most instructions. The 68000, 68010, and 68012 versions have a data bus 16 bits wide. The 68008 uses an 8-bit bus, and the 68020 matches the 32 bits from the internal data registers with 32 data pins on the device package.

The 68000 microprocessor provides the user with five major advantages over the processors we have encountered thus far:

1. It has 32-bit internal registers, allowing a data range of ±2 000 000 000. Time-consuming multiple-precision arithmetic is unnecessary in many cases.

2. It has 16 internal registers available for data and address manipulation. Time-consuming memory references are minimized because all the variables required in a calculation can often be kept right in the machine.
3. 68000-family processors are available with maximum clock rates of 8 MHz to 25 MHz. The fast versions can perform a 32-bit addition or data transfer in less than 0.5 μs.
4. The instruction set is much more orthogonal (regular) than in previous processors. For example, there are over 1000 instructions available, but there are fewer than 50 basic instruction mnemonics to learn. The same instructions apply regularly (with some exceptions) to most of the various registers and addressing modes.
5. The 23 address pins of the 68000 provide direct access to 8 million 16-bit words, in contrast to the 64 000 eight-bit words of the 8-bit processors.

The 68000 machine registers available to the programmer are shown in Figure 23.1. Data registers D0 through D7 function identically; any operation that can be performed with one of them can be performed with any of them. The same is true of address registers A0 through A6.

There are differences in the functions of the data and address registers, but there are as many similarities as differences. For example, the address registers can be either source or destination for an arithmetic or logic operation (a function usually provided only for data registers), but the data registers cannot be used as address pointers (a typical address-register function).[1] The most important similarity is that all data registers and all address registers are 32 bits long, so any address can be moved into a data register and any calculated data value can be moved into an address register with a simple register-to-register transfer.

There are two stack pointers: the user stack pointer A7 and the supervisor (or operating-system) stack pointer A7′. The processor is always in either the *user* or *supervisor* mode. In the user mode certain instructions and areas of memory are not accessible. This allows an operating system downloaded from disk to RAM to be protected against overwrites by careless or inexperienced users. The program counter contains 32 bits, allowing access of over 4×10^9 bytes of memory if all the lines are used.

The status register contains two separate bytes, one for the user, and one for the operating system. The C, V, Z, and N flags are much like their counterparts in the 6800 and 6809. The X (extend) flag is used as the carry or borrow *input* bit for additions, subtractions, and rotates. The T bit is set to put the processor in a *trace* mode for system debugging. In this mode the processor executes a single instruction and then jumps to a *trace* routine at an address specified by the system designer. This routine may cause the 68000 to display its register contents and wait for an

[1]There is a way to point with a data register, but it also ties up an address register, which must contain zero.

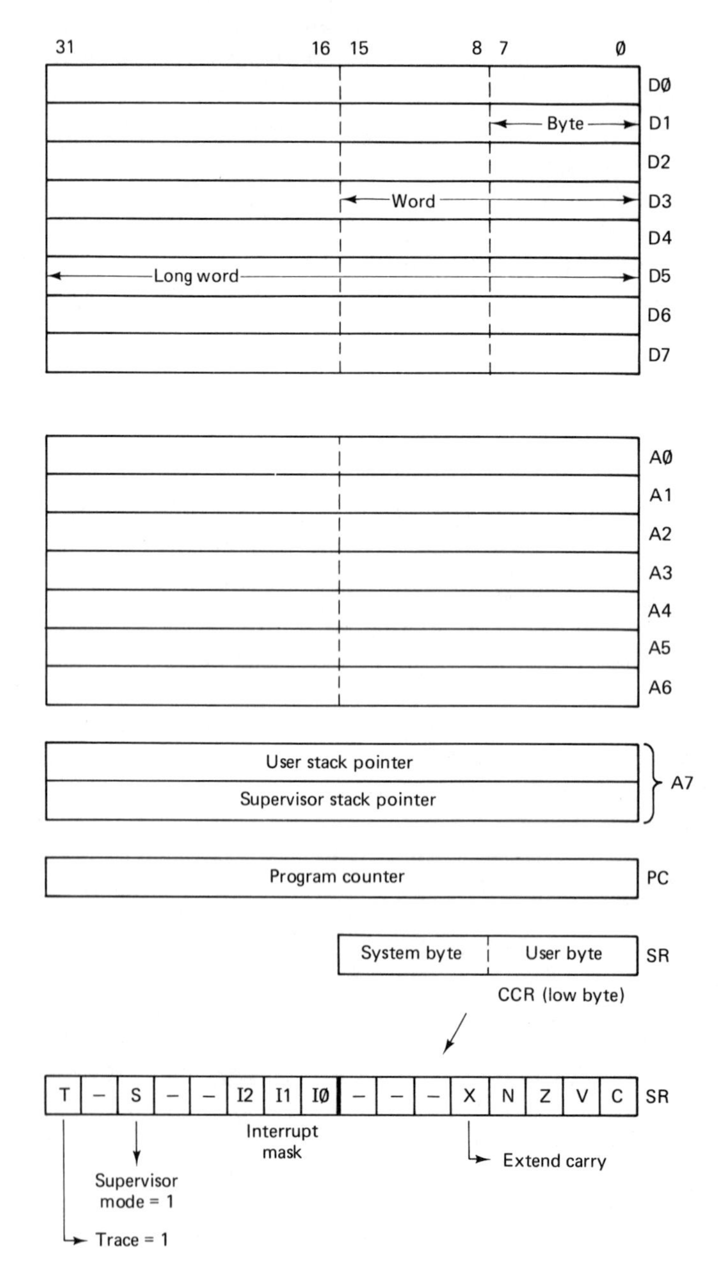

Figure 23.1 The 68000 programmer's model. All the data registers function identically, as do all of the address registers.

interrupt input before continuing, or it may test for a specified condition and tell the processor to keep returning to the main program until that condition is found.

The S bit is set to place the 68000 in the supervisor state. This will be explained in more detail in Section 24.4. The three *interrupt-mask* bits are set or cleared to form a binary number 0 through 7, which specifies to which of seven possible interrupts the processor will respond.

Bytes, words, and long words: Addresses in a 68000 system are numbered by bytes, just as they are in an 8-bit system. The range is $00 0000 through $FF FFFF (decimal 16 777 215). A *word*, in 68000 terminology, is 2 bytes, so if we tell the processor to clear address $12 3456, we actually clear a high-order byte at address $12 3456 and a low-order byte at address $12 3457. Words must always be stored in memory with the most-significant byte at an even-numbered address. If we tried to clear a word at address $12 3457, the processor would sense the error, leave the main program, and go to an address-error correction routine, assuming that the system designer has provided one. Most 68000 instructions can operate on *bytes*, 16-bit *words*, or 32-bit *long words*. Here are the assembler forms for each, using the *clear* instruction as an example. The operand size is indicated by the op-code *extensions* **.B**, **.W**, or **.L**.

CLR.B $123455	Clear a byte. Can be at even or odd address.
CLR.W $123456 (or CLR $123456)	Clear 2 bytes. Address specified is high-order byte and must be even. Note that *word* is assumed if .W is omitted.
CLR.L $123456	Clear 4 bytes, through $12 3459. Adress specified is highest-order byte and must be an even address.

Family members: The 68000 processor comes in a 64-pin wide-DIP package[2] and brings address lines A1 through A23 to the package pins, for an address capacity of 16 megabytes. The 68010 and 68012 are similar but slightly faster, since they have a faster arithmetic logic unit and the ability to prefetch and store 2 complete instructions within the processor. This means that they can perform 2-instruction loops without reading any instructions from memory, except on the first pass.

The 68020 uses the pin-grid array or "bed of nails" pin arrangement in three rows around the bottom of the package to bring out address lines A1 through A31, for a 4-gigabyte (4.29×10^9) addressing range. The 68020 is also able to store 128 prefetched instructions within the processor. This internal store (called a *cache*) significantly increases program execution speed.

The 68008 comes in a 48-pin DIP of the same width as the standard 40-pin processor ICs. This makes it easier to breadboard and wirewrap prototype systems. (Two 24-pin WW sockets can, with a little filing, be made to accept the chip.) Only address lines A0 through A19 are brought out to the pins, limiting address space to 1 megabyte. Figure 23.2 illustrates the differences among the 68000 family members.

[2]The 68000 is also available in a Pin Grid Array (PGA) package and in a plastic 64-lead chip carrier (PLCC) for surface mounting.

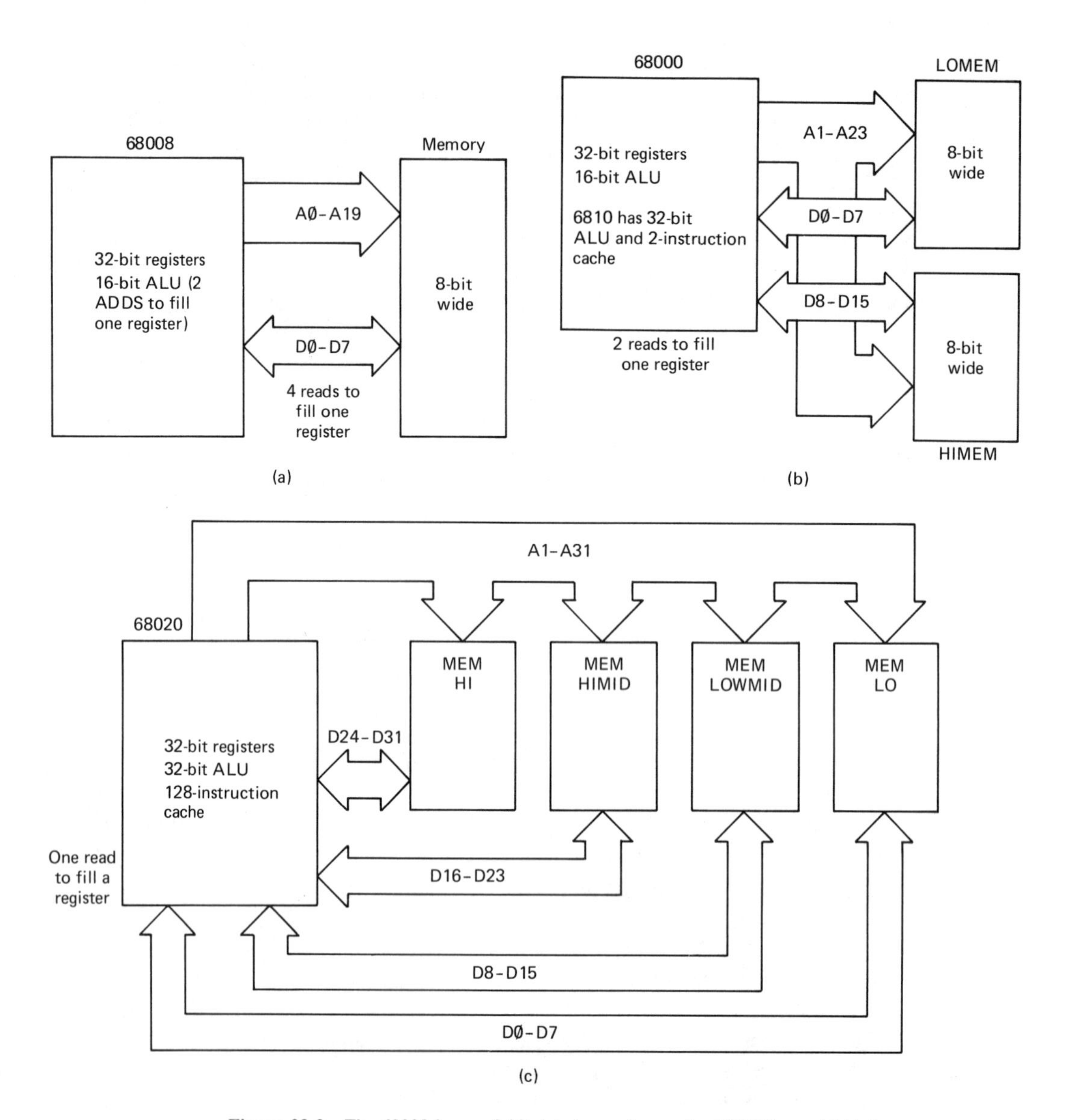

Figure 23.2 The 68008 has an 8-bit data bus, whereas the 68000 has a 16-bit bus and the 68020 has a 32-bit bus. All 68000-family processors have 32-bit internal registers.

23.2 PIN FUNCTIONS FOR THE 68000

Section Overview

This section explains the functions of the 64 pins of the 68000 and presents some simple circuits for driving the *clock* and *data-acknowledge* inputs. The 68008 pinout is then examined in terms of the changes from the more-common 68000 pinout.

The most fundamental difference in 68000 pin functions compared to the other processors we have seen is the asynchronous nature of the bus control. The processor outputs an address strobe ($\overline{\text{AS}}$) to enable an external device (say a RAM). The external device then responds with a data-transfer-acknowledge ($\overline{\text{DTACK}}$) signal to the processor when it has had its required time to read (or write) data on the bus. The processor does not go on to the next bus access until $\overline{\text{DTACK}}$ is received.

The 68000 pins can be classified by function into nine groups, as shown in Figure 23.3. We will take the classifications one at a time.

Power: The 68000 has two V_{CC} and two ground pins. The current demand is relatively heavy (300 mA), so these pins should be fed directly from a regulated 5.0-V supply with soldered No.-22 or heavier wire. The voltage between the pins should measure not less than 4.80 V. The 68000 will not tolerate a noisy supply, so two 0.1-μF capacitors should be soldered between V_{CC} and ground at the processor socket, with other bypass capacitors distributed liberally around the system supply lines.

The data lines are robust for a microprocessor. They are guaranteed to source 0.4 mA while outputting a *high* and sink 5.3 mA while holding a *low* output. (The comparable figures for a 6802 are 0.2 mA and 1.6 mA, respectively.) Still, systems that call upon a 68000 processor tend to be large fast systems, so buffering with fast TTL devices is common.

The address lines A1 through A23 are brought out. They can pull *high* with 0.4 mA and *low* with 3.2 mA, but, again, bus buffering is common. Address line A0 is not brought out because a normal word-length memory reference accesses 2 bytes at once. Thus A0 would have to be *low* (to access the high-order byte) and high (to access the low-order byte) simultaneously. Instead of A0, the 68000 has $\overline{\text{UDS}}$ and $\overline{\text{LDS}}$ (upper- and lower-byte data-strobe) outputs. For a *word* memory access, both of these will be asserted to enable a pair of 8-bit-wide memory chips together. For *byte* memory accesses, only one of $\overline{\text{UDS}}$ or $\overline{\text{LDS}}$ is asserted at a time.

The bus-control lines determine the timing and direction of data on the data bus:

- The $\overline{\text{AS}}$ (address strobe, *low*-active) output is asserted when the 68000 has a valid address on its address bus. This line, along with selected address lines, is decoded to enable the desired memory or I/O device. $\overline{\text{AS}}$ is *high* during address transitions and internal machine operations.

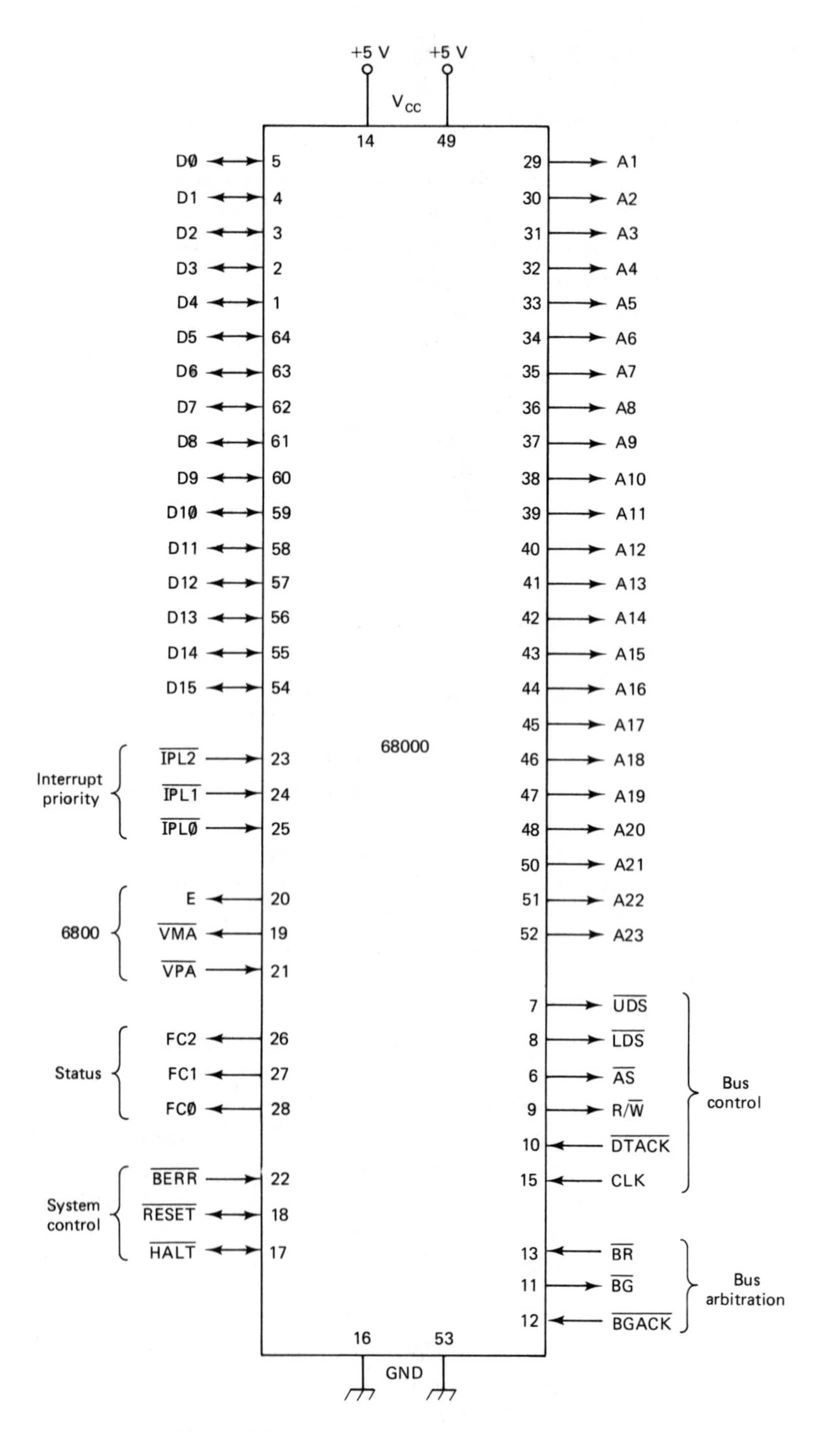

Figure 23.3 Pin functions for the 68000 microprocessor.

- The $\overline{\mathrm{DTACK}}$ (data-transfer acknowledge) input is required by the processor before it will go on to its next bus cycle. It is expected that all memory and peripheral devices in a 68000 system will assert a $\overline{\mathrm{DTACK}}$ signal to the processor when the required time for their data read or write has been completed. 68000 processors are available with internal clock periods as short as 60 ns (16.7-MHz version). This is much faster than most memory and peripheral devices, so the 68000 must generally wait for the peripherals to complete their data transfer before continuing. The $\overline{\mathrm{DTACK}}$ input requirement allows this wait time to be the minimum required for each of the various peripherals. (The wait must be an integer number of clock cycles, however.) It is said that the 68000 bus operation is *asynchronous* because peripheral devices do not have to operate at a fixed clock rate determined by the processor.

  The idea of having the processor wait for the memory is not new. The 6802 has its MR, the 6809 has its MRDY, and the Z80 has its $\overline{\mathrm{WAIT}}$ input, all of which provide the same cycle-stretching kind of function as $\overline{\mathrm{DTACK}}$. The difference is that the asynchronous stretched-cycle mode is the normal mode for the 68000.

  Figure 23.4 shows one scheme for generating DTACK signals with eight fixed delays from $\overline{\mathrm{AS}}$, depending on the area of memory being accessed. 68000-family peripherals and memory boards usually generate their own $\overline{\mathrm{DTACK}}$ signals for the processor.
- The R/$\overline{\mathrm{W}}$ output indicates whether the current bus cycle is a read or write by the processor.
- The clock input must be driven with a TTL-level signal of 50% duty cycle and a frequency not higher than the maximum specified. 68000 processors are available with frequencies of 8, 10, 12.5, and 16.7 MHz. Figure 23.5 shows a simple clock-driver circuit. Bus cycles always take an integer number of clock cycles (called *S*-states in Motorola literature), but peripheral devices need not operate on the same clock as the processor because of the asynchronous nature of the bus operation.

There are three interrupt input lines, $\overline{\mathrm{IPL0}}$, $\overline{\mathrm{IPL1}}$, and $\overline{\mathrm{IPL2}}$. These are active *low*-level inputs and must be held *high* to avoid sending the processor to an interrupt routine. They are not three independent interrupt lines; rather they form a 3-bit number to indicate the *interrupt priority level* of any pending interrupt. There are 8 interrupt priorities, from 3 *high levels* (indicating no interrupt) to 3 *low levels* (indicating nonmaskable interrupt). The 68000 interrupt-handling procedures are treated in more detail in Section 24.3. For the moment, remember that the $\overline{\mathrm{IPL}}$ inputs are negative-true, so a *low* voltage asserts a logic 1, or *condition-true* state. The $\overline{\mathrm{VPA}}$ input and the three FC outputs also have interrupt functions, as noted in the next two subsections.

Three 6800-peripheral control lines are provided to maintain compatibility between the 68000 and the many common 6800-family peripherals, such as the 6821 parallel interface, and the 6850 serial interface. $\overline{\mathrm{VPA}}$ is an input to the processor, which is asserted when a memory reference accesses an area of the memory map containing a 6800-family peripheral. When $\overline{\mathrm{VPA}}$ is asserted, the 68000 operates in

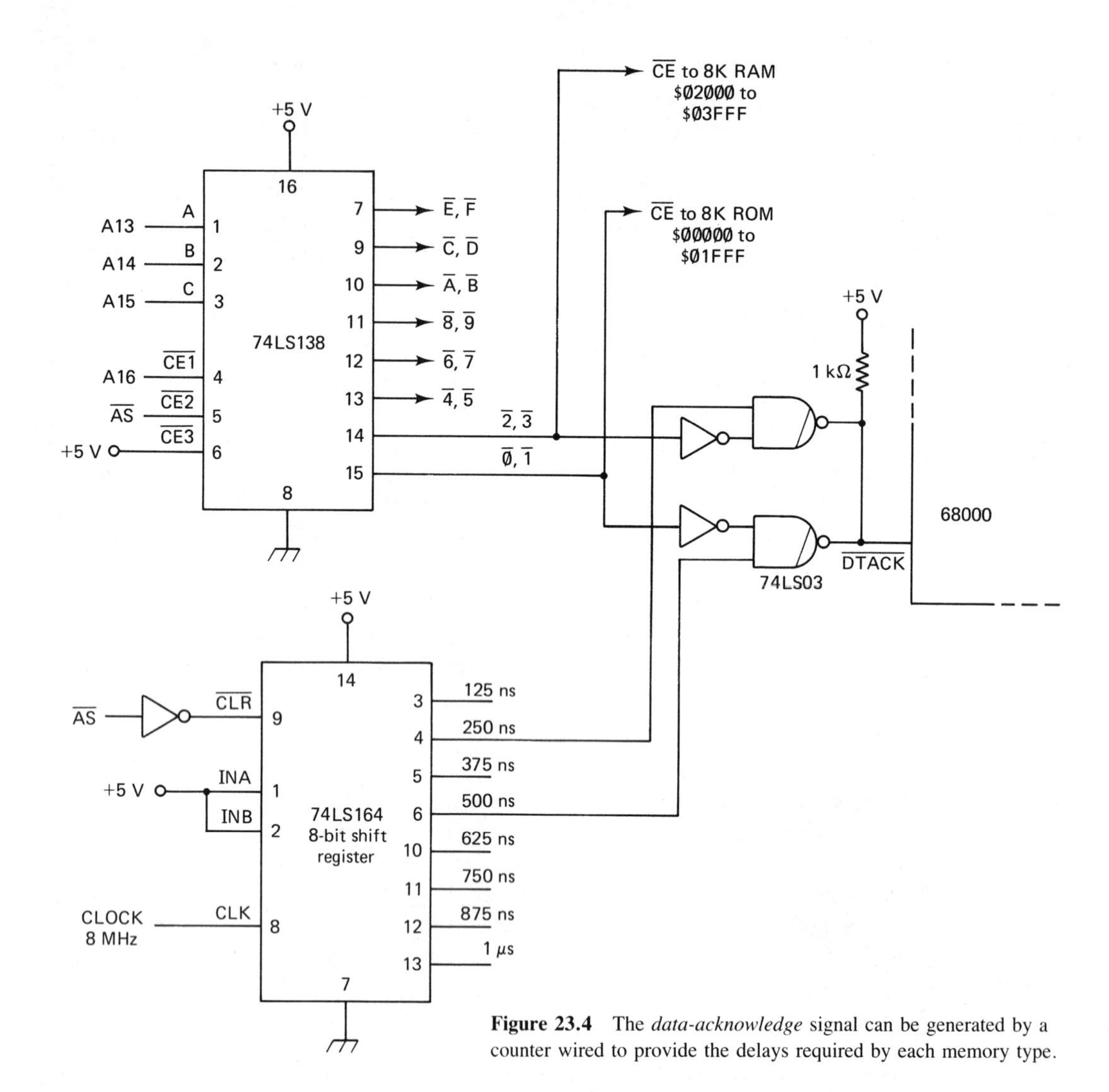

Figure 23.4 The *data-acknowledge* signal can be generated by a counter wired to provide the delays required by each memory type.

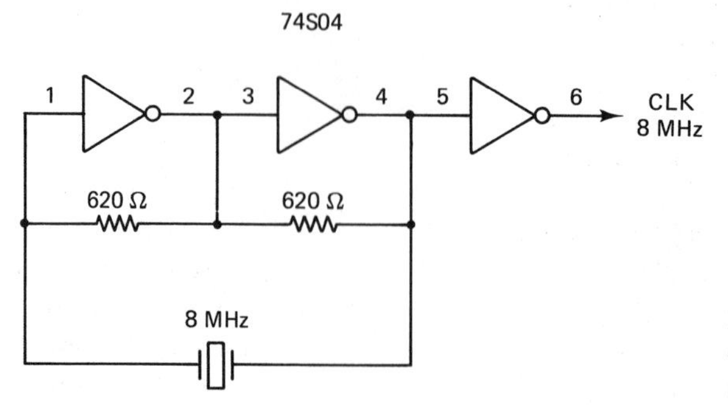

Figure 23.5 The 68000 clock generator circuit is external, but it can be very simple.

a synchronous-bus mode. It does not wait for $\overline{\text{DTACK}}$ before going on to the next bus cycle. Instead, it times the address and data signals from the *E* output signal, according to 6800-family rules. The *E* signal is *low* for six clock periods and *high* for four clock periods. An 8-MHz 68000 thus operates at a 0.8-MHz bus-cycle rate in the synchronous mode—a serious loss of speed that makes this mode relatively unpopular. $\overline{\text{VMA}}$ is an inverted version of the 6800/6802 VMA output. It is asserted only when $\overline{\text{VPA}}$ is asserted and the processor has synchronized itself to the *E* signal.

The $\overline{\text{VPA}}$ input is also used during an interrupt acknowledge to tell the processor to use the simpler *autovector* scheme to locate one of seven fixed interrupt routines. Section 24.3 covers this in more detail.

The processor status is indicated by three *function-code* outputs, FC0, FC1, and FC2. Here are the five possible states:

FC2	FC1	FC0	Processor Function
Low	Low	Low	(Undefined)
Low	Low	High	User data
Low	High	Low	User program
Low	High	High	(Undefined)
High	Low	Low	(Undefined)
High	Low	High	Supervisor data
High	High	Low	Supervisor program
High	High	High	Interrupt acknowledge

The system control lines are $\overline{\text{RESET}}$ and $\overline{\text{HALT}}$, which are bidirectional pins on the 68000, and the $\overline{\text{BERR}}$ (bus error) input. $\overline{\text{BERR}}$ may be asserted by a peripheral which is illegally accessed or is unable to respond properly. The system designer may use this input to retry the bus cycle, perform an error-handling routine, or halt the processor.

After the clock oscillator has stabilized, $\overline{\text{RESET}}$ and $\overline{\text{HALT}}$ should be brought *low* together for 100 ms with separate open-collector drivers to restart the processor. Upon *restart* the processor loads its stack pointer from addresses $0 through $3 and its program counter from $4 through $7. It then begins executing instructions at the address loaded into the program counter.

The processor itself will pull $\overline{\text{RESET}}$ *low* in response to a RESET instruction. This will reset other devices (such as a 6821 PIA, or another processor) under control of the 68000. If $\overline{\text{HALT}}$ alone is pulled *low* by an external signal, the 68000 will halt and place all its tristate lines in the high-*Z* state. Multiple bus errors will cause the 68000 to pull its $\overline{\text{HALT}}$ line *low*, indicating to external devices that the 68000 is no longer processing instructions. Figure 23.6 shows a 68000 *reset* circuit.

The bus arbitration lines are used in systems with multiple processors or DMA controllers to determine which device will control the buses:

- Bus Request ($\overline{\text{BR}}$) is an input pulled *low* by an external device that is requesting control of the bus.
- Bus Grant ($\overline{\text{BG}}$) is an output that the 68000 asserts to tell the external device that it will release control of the bus at the end of the current bus cycle.

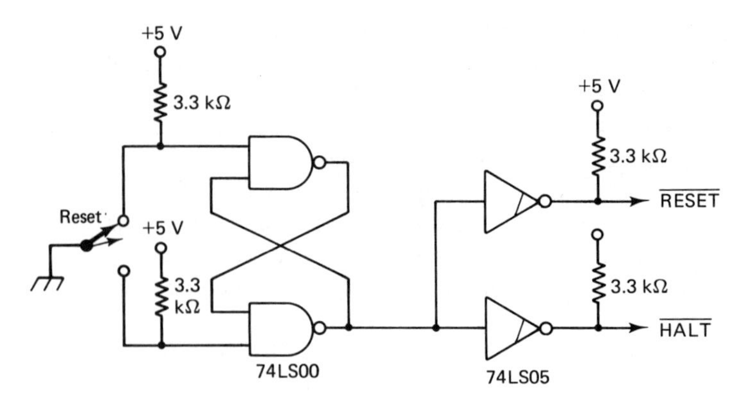

Figure 23.6 The 68000 $\overline{\text{RESET}}$ and $\overline{\text{HALT}}$ lines should be driven by open-collector devices, since they can pull active *low* as outputs themselves.

- Bus Grant Acknowledge ($\overline{\text{BGACK}}$) may be pulled *low* by the external device when $\overline{\text{BG}}$ is asserted *and* $\overline{\text{AS}}$ and $\overline{\text{DTACK}}$ are not asserted (indicating no bus activity) *and* $\overline{\text{BGACK}}$ has not already been pulled *low* by another device seeking bus mastership.

The 68008 has 16 fewer pins than the 68000. The reductions are listed next.

Pin Type	68000	68008	Pins Saved
Data	16 pins	8 pins	8
Address	23 plus $\overline{\text{UDS}}$ and $\overline{\text{LDS}}$	20 plus $\overline{\text{DS}}$	4
Interrupt	$\overline{\text{IPL0}}$, $\overline{\text{IPL1}}$, $\overline{\text{IPL2}}$	$\overline{\text{IPL0/2}}$, $\overline{\text{IPL1}}$	1
Power	2 V_{CC} pins	1 V_{CC} pin	1
VMA	$\overline{\text{VMA}}$ and $\overline{\text{BGACK}}$	No such pins	2

Only 3 interrupt levels are available in the 68008, compared to 7 in the 68000. $\overline{\text{VMA}}$ was eliminated, as the $\overline{\text{AS}}$ pin provides much the same function. The function of $\overline{\text{BGACK}}$ in assuring that only one external device takes control of the bus is left to external logic, where there is more than one such external device.

A0 is included as an address output, since the 68008 reads memory a byte at a time instead of 2 bytes at a time. The upper- and lower-byte data strobes are replaced with a single data-strobe output. A20 through A23 are eliminated, restricting the address range to 1M byte. Figure 23.7 shows the 68008 pinout.

REVIEW OF SECTIONS 23.1 AND 23.2

Answers appear at the end of Chapter 23.

1. How many address-pointer registers are available in the 68000, excluding stack pointers and program counters?

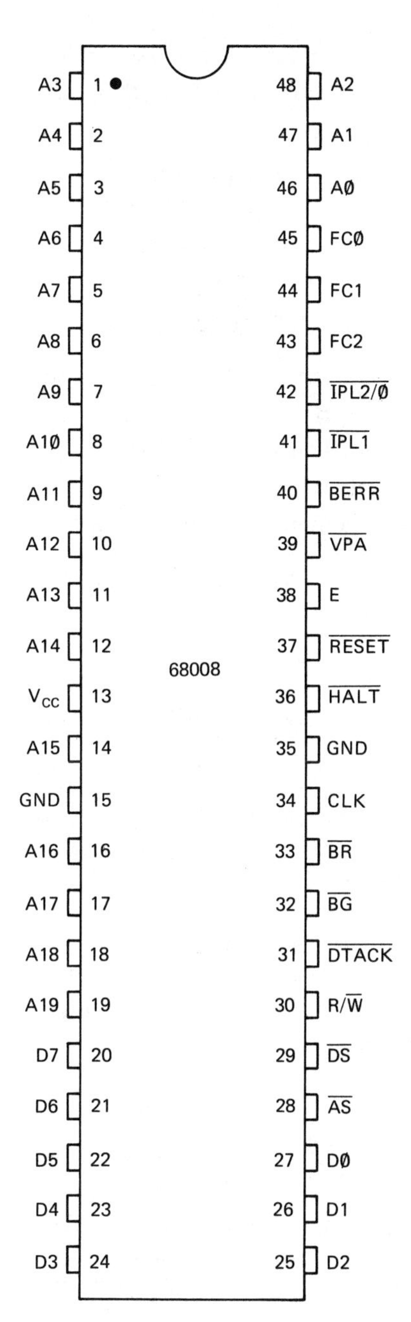

Figure 23.7 The pin diagram for the 68008 microprocessor.

2. How many registers are there that can be the destination for an addition operation?
3. How many bits are contained at address $Ø1 2ABC in a 68000 system?
4. What is wrong with the instruction

```
CLR $Ø1 23AB
```

5. Is the 68008 an 8-bit, 16-bit, or 32-bit microprocessor?
6. When the 68000 writes zeros to execute

```
CLR.W $Ø2 468A
```

   which of $\overline{UDS}$ and $\overline{LDS}$ is asserted? If both, in what order?
7. Answer Question 6 for CLR.B $Ø1 3579.
8. What is enabled to happen when $\overline{DTACK}$ is asserted?
9. How does the 68000 find its restart address?
10. Why is pin AØ absent on the 68000 but present on the 68008?

23.3 68000 ADDRESSING MODES

Section Overview

Instructions in 68000 assembly language have the basic form

OPERATION source, destination

The number sign (#) is used to indicate immediate data, and register names are enclosed in parentheses when they are used as address pointers. Operand size can, in most cases, be a *byte*, 2 bytes (a *word*), or 4 bytes (a *long* word).

There are six basic addressing modes, of which three have variations in their assembly-language forms. Here is a quick summary of the most basic form of the six modes, using *add a word to register D1* as an example instruction, where possible:

Implied	RTS	**No operand specified. It is implied by instruction.**
Register direct	ADD AØ,D1	**Source operand is an address or data register.**
Immediate	ADD #AUG,D1	**Source is a constant referenced by name or numerals.**
Absolute	ADD $89AB,D1	**Source is at address specified by label or numerals.**
Register indirect	ADD (AØ),D1	**Address register contains address of source.**
P.C. relative	ADD TABL (PCR),D1	**Source is at address given by label. Machine code uses relative addressing.**

Modes for source and destination can be mixed. In the preceding examples the destination mode is always *register direct*.

The assembly-language form for 68000 instructions is

OPERATION source, destination

The *operation* is one of about 50 mnemonics, each consisting of 2 to 5 letters. A few instructions, such as RESET, have neither source nor destination operand. Several, such as ROR (rotate right) and NOT (complement bits), have a destination operand but no source. Most have both a source and a destination operand, each of which may be addressed independently in one of five addressing modes to be described below. Notice first that the order of operands is *source*, *destination*, which is the reverse of the order used in Intel and Zilog processors.

Implied addressing is used for instructions in which neither a data source nor a data destination needs to be specified. Examples are

• NOP	No operation
• RTS	Return from subroutine
• RESET	Pull RESET line *low* to reset external devices

The form is simple, and there are no variations.

Register Direct addressing accesses data that is contained in one of the machine registers DØ through D7 or AØ through A7. Salespeople who are trying to impress a customer with the enormous variety of addressing modes the machine has may claim this as two modes—*data-register direct* and *address-register direct*—but the only difference from the programmer's point of view is that the data registers can hold byte, word, or long-word data sizes, whereas the address registers can hold only word or long-word sizes. Keeping the destination operand in the register-direct mode with the operand in D1, here are some examples of the *add* instruction with the source operand in register-direct mode.

• ADD.B DØ,D1	Add byte from register DØ to byte in D1.
• ADD.W A2,D1	Add low-order 16 bits from register A2 to low-order 16 bits in register D1.
• ADD A2,D1	Interpreted exactly the same as the preceding example, since 16 bits is the default operand word length.
• ADD.L D3,D1	Add 32 bits from register D3 to 32 bits in register D1.

Immediate addressing in the 68000 is the same as the immediate mode in the 6800. The operand data is contained in the word (or two words) immediately following the op-code word. The operand may be a decimal number, a hexadecimal number (preceded by a $ sign), or a label or variable name to which the assembler has previously assigned a value. The sign # is used to specify immediate mode. Here are some examples, again with the destination specified in the register-direct mode.

• ADD #8765,D1	Add decimal 8765 to low-order 16 bits of register D1.
• ADD.L #$F7ØB5,D1	Add hex F7ØB5 to 32 bits of D1.
• ADD.B #SPAN,D1	Add 8 bits. Value of SPAN previously defined.

There is a second version of this addressing mode called *quick immediate*. Since a 68000 machine-language op-code is 16 bits long, three instructions have saved a few bits *within the op-code* for immediate data. This permits very fast 1-byte instructions with limited immediate-data ranges. Here are three examples, one for each instruction.

• ADDQ #1,D1	Immediate data range 1 to 8. Used to increment registers.
• SUBQ #8,D1	Subtract data from contents of register D1. Data range limited to 1 through 8.
• MOVEQ # − 128,D1	Immediate data range is from − 128 through + 127 for MOVE quick.

Absolute addressing specifies the address of the operand data. This is called *extended* addressing in 6800 terminology. Actually there are two forms: absolute short, in which the address is $FFFF or less, and absolute long, in which the address is $10 000 up to the maximum as limited by the pinout ($FF FFFF for the 68000, or $F FFFF for the 68008). Absolute short is a 2-word instruction, and absolute long requires 3 words. The decision between the two is handled by the assembler and not noticed by the programmer. Here are some examples.

• ADD $F081,D1	Add operand at address $F081 (high-order byte) and $F082 (low-order byte) to D1.
• ADD.B BUFR,D1	Add byte from address labeled BUFR to D1.

Address-register-indirect addressing is an important and much-used mode in the 68000. The word *indirect* here has the meaning originally attributed to it by Intel and Zilog; that is, an internal machine register holds an address which points to the operand data. In 6800-family literature and in the 6502, this mode is called *indexed*, and the term *indirect addressing* means that pair of memory addresses are used to hold the bytes comprising the final or effective operand address.

The confusion is compounded because 68000 literature uses the term *indexed* to refer to the addition of the contents of a register to some base address to produce an effective address. In the 6809 this was called *indexed with register offset*.

There are five variations of *register-indirect* addressing in the 68000. We define and give an example for each. Notice that parentheses are used when an address register serves as a pointer to an address. However, parentheses are not used in the *absolute* mode where a number or a label name points to an address. Also, be aware that address registers *can*, but data registers *cannot*, serve as pointer registers:

1. Address-register indirect; the basic mode.

ADD (A2),D1	Add data pointed to by register A2 to contents of register D1 (16-bit operands).

2. Register indirect with postincrement of pointer register.

ADD.L (A5)+,D1	Add data pointed to by A5 to data in D1. Increment A5 by 4 to point to next long word.

3. Register indirect with predecrement of pointer register.

ADD.B −(A6),D1	First decrement pointer A6 by one. Then add byte pointed to by A6 to D1.

4. Register indirect with 16-bit signed offset.

• ADD.W −4096(A4),D1	Add the two bytes starting at address A4 − 4096 to register D1. *Example:* If A4 contains $Ø4567, the operand data is obtained from $Ø3567 and $Ø3568.
• ADD OFST(A3),D1	Similar to the preceding. The assembler must have assigned a value between −32 768 and +32 767 to the name OFST.

5. Register indirect with 8-bit signed offset and register index.

• ADD 50(A2,A3),D1	Add the 2 bytes starting at address A2+50+A3 to register D1. The index value in A3 is interpreted as a 16-bit signed number.
• ADD.L OFST(A1,D2.B),D1	Add the 4 bytes starting at address A1+OFST+D2 to register D1. The value of OFST must lie between −128 and +127. The value of the index in D2 has been restricted to −128 through +127 by the .B extension. Extensions .W or .L are also valid.

The source and destination addressing modes can always be chosen independently, but it is not possible to combine the preceding 5 modes to produce new modes. For example, you cannot form a *postincrement-with-offset* mode by combining modes 2 and 4. Also remember that the extension .B cannot be used on address registers, which must contain words or long words only.

The *register indirect with index* mode (mode 5) is useful because it allows accessing a two-dimensional array of data by loading the *row* base address into the pointer register and the *column* number into the index register. An example of this technique was given on page 496. If the offset is not used, a Ø must be typed before the parentheses.

Program-Counter-Relative addressing uses a label to point to the address of the operand. The assembler calculates the displacement of this address from the current program-counter value. This mode is useful to replace the *absolute* mode in writing position-independent code. There is a second version with indexing to allow accessing a file of data, but the displacement range is severely limited in this mode.

Assemblers differ in the way they distinguish the PC-relative mode from the absolute mode. Some use RORG in place of ORG when specifying the program origin address. This causes absolute addressing to be replaced by relative addressing whenever the instruction forms are identical and the displacement is within range. Other assemblers require the letters PC in the instruction operand.

Examples of both types are given next:

1. Program-Counter-Relative (16-bit signed displacement from PC).

ADD PROD,D1	Add two bytes starting at address PROD to register D1.
ADD PROD(PC),D1	PROD must be within −32 768 or +32 767 bytes of the next-instruction address.

2. Program-Counter-Relative with index (8-bit signed displacement).

ADD.L FILE(D5.D),D1	Add the four bytes starting at address FILE + D5 to register D1.
ADD.L FILE(PC,D5.B)D1	FILE must be within −128 through, +127 bytes of the next-instruction address. The index value in D5 is limited to −128 through, +127 by the .B extension.

Figure 23.8 summarizes the addressing modes of the 68000 microprocessor.

23.4 THE MOVE INSTRUCTION

Section Overview

The most common instruction in 68000 programs is

MOVE source, destination

This instruction can copy data from any register, memory location, or input peripheral to any register, RAM location, or output peripheral. There are three common variations of the MOVE instruction:

- **MOVEA Move to Address register; implements sign extension.**
- **MOVEQ Move Quick; immediate signed 8-bit operand.**
- **MOVEM Move Multiple registers; up to 16 registers saved or retrieved from memory in one instruction.**

Sign extension translates an 8-bit or 16-bit value to an equal 16-bit or 32-bit value by filling in the unspecified higher-order bits with Øs (positive numbers) or 1s (negative numbers). Many 68000 instructions sign-extend automatically.

There are six lesser-used versions of the MOVE instruction in 68000 assembly language.

```
                            1000 *******************************************************
                            1010 * TEST OF 68000 ADDRESSING MODES. METZGER, 12 AUG 87.
                            1020 *  TWO EXAMPLES ARE GIVEN OF EACH MODE.  DATA-REGISTER
                            1030 *  DIRECT-D0 IS USED WHERE A SECOND OPERAND IS NEEDED.
                            1040 *
000000F0-                   1050 BYTE    .EQ     $F0
0000C000-                   1060 WORD    .EQ     $C000
C0000000-                   1070 LONG    .EQ     $C0000000
                            1080 *******************************************************
00001000- 4E71              1090         NOP                     IMPLIED:
00001002- 4E75              1100         RTS
                            1110 *******************************************************
00001004- 4A43              1120         TST     D3              DATA REGISTER DIRECT:
00001006- 8240              1130         OR      D0,D1
00001008- B048              1140         CMP     A0,D0           ADDRESS REGISTER DIRECT:
0000100A- D3C0              1150         ADD.L   D0,A1
                            1160 *******************************************************
0000100C- 0200 000F         1170         AND.B   #$0F,D0         IMMEDIATE:
00001010- 0640 01F4         1180         ADD     #500,D0
                            1190 *******************************************************
00001014- 4EB9 0000
00001018- C000              1200         JSR     WORD            ABSOLUTE, WORD LENGTH:
0000101A- 0640 ABCD         1210         ADD     #$ABCD,D0
0000101E- E6F9 0001
00001022- 2345              1220         ROR     $12345          ABSOLUTE, LONG-WORD:
00001024- 9179 C000
00001028- 0000              1230         SUB     D0,LONG
                            1240 *******************************************************
0000102A- 4A12              1250         TST.B   (A2)            BASIC ADR REG INDIRECT:
0000102C- D051              1260         ADD     (A1),D0
0000102E- E2DB              1270         LSR     (A3)+           ADR REG IND, POSTINCREMNT:
00001030- 911B              1280         SUB.B   D0,(A3)+
00001032- C164              1290         AND     D0,-(A4)        ADR REG IND, PREDECREMENT:
00001034- 46A5              1300         NOT.L   -(A5)
00001036- 426E 0005         1310         CLR     5(A6)           ADR REG IND, DISPLACEMENT:
0000103A- 812A 00F0         1320         OR.B    D0,BYTE(A2)
0000103E- E7F8 0008         1330         ROL     8,(A1,D2)       ADR REG IND, DISP + INDEX:
00001042- B0B3 40FC         1340         CMP.L   -4(A3,D4),D0
                            1350 *******************************************************
00001046- 4EFA 000A         1360         JMP     THERE(PC)         PC REL, DISPLACEMENT:
0000104A- D07A 0006         1370 HERE    ADD     THERE(PC),D0
0000104E- B07B 40FA         1380         CMP     HERE(PC,D4),D0  PC REL, DISP + INDEX:
00001052- D07B 30F6         1390 THERE   ADD     HERE(PC,D3),D0
                            1400 *******************************************************
                            1410         .EN
```

Figure 23.8 A summary of the addressing modes of the 68000.

The basic MOVE instruction is used in place of the *load* and *store* instructions of the 6800 family. The source operand can be obtained by any of the 68000's addressing modes except, of course, *inherent*. The destination mode may not logically be *immediate*, nor is it valid to access the destination in either of the *program-counter-relative* modes.

The operand size may be specified as *byte* (.B), *word* (.W), or *long word* (.L), with the size defaulting to *word* if no size-extension code is used. Only the number of bits specified are affected. For example:

If D0 contains	$0123 4567
and D1 contains	$89AB CDEF,
the instruction	MOVE.B D0,D1
would leave D1 with	$89AB CD67
and the instruction	MOVE.W D0,D1
would leave D1 with	$89AB 4567.

The Z flag is set to 1 if the byte, word, or long word moved is zero; it is cleared otherwise. The N flag is set to the state of the most-significant bit moved. Here are several examples of the basic MOVE instruction.

- MOVE.B # − 1,D2 — Source: *immediate* mode; destination: *register-direct* mode. D2 gets $FF in low-order byte.
- MOVE $4000,(A0) — Source: *absolute* mode; destination: *register-indirect* mode. Two bytes from addresses $4000 : $4001 copied to locations pointed to by A0.
- MOVE.L −(A2), − 1024(A2) — Source: *register indirect with predecrement;* Destination: *register indirect with offset*. Long word pointed to by A2 is moved to location $400 bytes lower in memory. A2 automatically points to next-lower long word in table before each MOVE.
- MOVE BUFAD,(PC),0(A1,D2) — Source: *program counter relative;* destination: *register indirect with index* and zero offset. A word is read from the addresses labeled BUFAD and written to a pair of addresses at A1 + D2, where D2 is a signed 16-bit indexing number.

Sign extension is employed automatically by the three versions of the MOVE instruction that we will examine next, so we will take a moment now to explain sign extension.

Suppose that we move a byte from an input buffer to register D0. Perhaps the byte was $78, but register D0 had previously been loaded with word $ABCD. The contents of the low-order 2 bytes of D0 will now be $AB78 or, in binary,

1010 1011 0111 1000 = $AB78

If we next attempt to add D0 to the *word* in D1, we will have quite a surprise if we expect that D1 will increase by only $78! To make the 8-bit value $78 fill the 16 bits used by the next instruction operating on D0, we must wipe out the old higher-order bits and fill them with 0 bits. This is done with the instruction

EXT.W D0   Change byte length to word length

This causes the binary value in the low-order half of D0 to become

0000 0000 0111 1000 = $0078

If we need to wipe out the high 16 bits of the entire 32-bit register to preserve the word value when a long-word operation is done, we can sign-extend again.

EXT.L D0 Change word length to long-word length

The value in D0 now becomes $0000 0078, even though the first (highest-order) 4 hex digits may have previously held the digits $6789, for example.

This operation is called *sign* extension because it assumes that the smaller-sized value is a signed number. If the byte loaded from the buffer in our original example were $FE, it would be interpreted not as 254 but as −2, and an extension to word size would produce not $00FE but $FFFE—still a value of −2. In binary:

EXTEND←1111 1110
1111 1111 1111 1110

Thus negative numbers extend 1 bits to the higher-order part of the register, and positive numbers extend 0 bits. There is no byte-to-long-word extend instruction. This operation must be done in two steps: EXT.W, followed by EXT.L.

Sign extension should be used only when the data is a 2's-complement number. For other data types it is best to clear the entire register or memory location before moving word- or byte-sized data into it.

The MOVEA variation: Address registers may not contain byte-size data. Therefore, the instruction

MOVE.B D2,A3

is invalid. When an address is placed on the bus, more than 16 bits are involved in any of the 68000-family processors, so it makes little sense to move a word into the low-order half of an address register while leaving old data in the high-order half. Therefore, word-length moves to an address register are automatically sign-extended to 32 bits. For example,

MOVE #$8000,A3

causes A to point not to address $00 8000 but to address $FF 8000. If you want A3 to point to address $00 8000, you will have to use

MOVE.L #$8000,A3

Also, moves to an address register affect no flags in the condition-code register. The official mnemonic for a move to an address register is MOVEA to remind us of these differences, but most assemblers will accept MOVE with an address-register destination and treat it as MOVEA.

The MOVEQ variation is available with the source in the *immediate* mode only and the destination in the *data register direct* mode only. The source operand is embedded in the op-code word as an 8-bit signed number, so the data range is, limited to −128 through +127. The data is automatically sign-extended to 32 bits before being placed in the data register.

The N and Z flags are set by negative and zero values, respectively. The advantage of this instruction is its fast execution time and single-word length.

The MOVEM (move multiple registers) variation allows register-to-memory or memory-to-register transfer of a listed set of registers from D0 through D7 and A0 through A7, all with a single instruction. For example,

MOVEM D0/D1/D7/A0,(A1)

stores the lower 2 bytes of registers D0, D1, D7, and A0, at a series of 8 addresses ascending from the address contained in A1. As another example,

MOVE.L $4080,A3/A4/A6

loads all 32 bits of registers A3, A4, and A6 from the 12-byte series starting at $4080.

When the *predecrement* mode is selected, the instruction has the character of a push-on-stack, with the decrementing register serving as a stack pointer. When the *postincrement* mode is used, the instruction has the character of a pull-from-stack instruction. Here are two examples.

- MOVEM.L D0–D7,–(A6) — Push all data registers on a stack with A6 as stack pointer.
- MOVEM.L (A6)+,D0–D7 — Pull all data registers from A6 stack.

The MOVEM is an impressive and complex instruction variation, whose usefulness can hardly be explored fully in this two-chapter survey of the 68000. However, these few examples should provide a starting point for learning some of its many forms. MOVEM is available in word and long-word sizes only and affects no flags. It cannot be used for register-to-register transfers.

Less-frequently used variations of the MOVE instruction are introduced by the following example forms, with brief explanations.

- MOVE D0,CCR — Move source bits 4 through 0 to CCR user bits X, N, Z, V, and C, respectively.
- MOVE CCR,(A4) — Move CCR user bits to destination.
- MOVE SR,D1 — Move system and user status-register bits to destination.
- MOVE D2,SR — Move source to status register. *Privileged instruction; supervisor mode only.*
- MOVEP D1,0(A2) — Move word-sized data to tandem 8-bit peripherals at addresses A2 and A2 + 2.
- MOVE A0,USP — Move address to user stack pointer. *Privileged.*

Figure 23.9 summarizes the 68000 MOVE instruction in all its variations.

		Register direct		ADR REG INDIR					Absolute		PC REL		Immediate	Size	Comments, flags*
		Data	Adr	basic indir	post INC	pre DEC	displacement	disp + index	word	long	displacement	disp + index			
MOVE	Source	X	X	X	X	X	X	X	X	X	X	X	X	.B.W.L	MOVE n z $\overline{V}$ $\overline{C}$
	Destin.	X		X	X	X	X	X	X	X					
MOVEA	Source	X	X	X	X	X	X	X	X	X	X	X	X	.W.L	Move to address register. No flags.
	Destin.		X												
MOVEQ	Source												X	.L	Move quick to Dn, −128 to +127. n z $\overline{V}$ $\overline{C}$
	Destin.	X													
MOVEM	Reg → m			X		X	X	X	X	X				.W.L	Move multiple. No flags. An, Dn ↔ mem
	m → reg			X	X		X	X	X	X	X	X			
MOVE to	SR S	X		X	X	X	X	X	X	X	X	X	X	.W	Move to SR (2 bytes) to CCR (1 byte)
	CCR S	X		X	X	X	X	X	X	X	X	X	X		
MOVE from	SR D	X		X	X	X	X	X	X	X				.W	Move from SR
	CCR D	X		X	X	X	X	X	X	X					Move from CCR
MOVEP	Read S	X					X							.W.L	Move. No flags. Peripheral ↔ Dn
	Write D	X					X								
MOVE USP	From S		X											.L	Move USP ↔ An
	To D		X												

*C = carry set; $\overline{C}$ = carry cleared; c = carry affected.

Figure 23.9 A summary of the various forms of the 68000 MOVE instruction.

REVIEW OF SECTIONS 23.3 AND 23.4

Answers appear at the end of Chapter 23.

11. Arrange these three parts of a 68000 assembly instruction in proper order with the proper punctuation between them.

Destination; mnemonic; source

12. How many addressing modes may a single 68000 instruction have at one time?

13. How are each of these operands specified to a 68000 assembler?
a. Address \$1234 **b.** Immediate data \$1234
c. The address of register A0 **d.** The data in register A0

14. Write an instruction to add all 32 bits of D5 to A0.
15. Write an instruction to decrement register A3 by 2.
16. Write an instruction to store data $FF at the address pointed to by A4, and then automatically point A4 to the next address.
17. Write an instruction to move the word from file address A2 + D2 to a 16-bit latch at address LATCH. Assume D2 contains word-size data.
18. D0 contains $3456 789A. (a) What will it contain if we execute EXT.L D0? (b) What if instead we executed EXT.W D0?
19. If D1 contains $5678 9ABC, what does A1 contain after executing MOVEA.W D1,A1?
20. Write a single instruction to read the 4 successive long words pointed to by A3 into data registers D4 through D7.

23.5 ARITHMETIC INSTRUCTIONS FOR THE 68000

Section Overview

The 68000 arithmetic instructions include the following:

- **Binary additions and subtractions, both with and without a carry (or borrow) on the input side of the operation.**
- **BCD addition and subtraction, limited, unfortunately, to the 8-bit data size.**
- **Compare, which is a pseudosubtract for setting the N, Z, V, and C flags, and Test, which is a compare-to-zero.**
- **Multiply instructions in signed and unsigned binary with 16-bit multipliers and a 32-bit product.**
- **Divide instructions in signed and unsigned binary with a 32-bit dividend and 16-bit quotient.**

Although the 68000 is widely acclaimed as having an orthogonal instruction set, it becomes clear in this section that all addressing-mode combinations are not available for all instructions. The path from novice to accomplished programmer will require the learning of dozens of details about the 68000 instruction set.

Binary addition and subtraction operations are available in exactly the same forms, so we will cover them together, intermixing examples of each. Two rules can be remembered to define the addressing-mode combinations that are available.

1. If the destination is a data register or an address register, the source can use any addressing mode.

2. If the source is a data register or immediate data, the destination can use any addressing mode.

The restrictions can be stated more simply in the negative:

- Data in memory cannot be referenced for both source and destination in ADD or SUB operations.[3]
- If the source is an address register, the destination cannot be data in memory.

In the subtraction instruction the operation is

$$\text{Destination} - \text{source} \rightarrow \text{destination}$$

Thus to subtract D1 from D2, we write

SUB D1,D2

and read it, "Subtract D1 from D2, placing the result in D2." This will take some getting used to, because it is the opposite of the 6800 and Z80 conventions:

6800:	SUBA DADR	Subtract data at DADR from A.
Z80:	SUB A,C	Subtract C from A.

ADDA and SUBA are variations in which an address register is the destination. This variation sets no flags and can use only word or long-word operands. Words are automatically sign-extended. Most assemblers will accept the form ADD as identical with ADDA if an address register is listed as the destination.

ADDI and SUBI are variations in which the source is immediate data. Most assemblers will accept ADD as identical with ADDI if immediate data is given as the source.

ADDQ and SUBQ are quick-executing variations in which the immediate source data is limited to the values Ø through 8. The destination, once again, may use any addressing mode.

ADDX and SUBX are variations that add a carry (or subtract a borrow) generated by a previous operation. Such carry-input operations were called ADC and SBC in 6800 and 6809 language. The X reminds us that the 68000 has 2 carry flags, C and X (extend bit). Generally both are set or cleared together, but we will see some differences in Chapter 24 when we discuss the rotate instructions. The X bit is the one that is added to the source and destination operands in ADDX. These instructions are available in only 2 modes:

1. Both source and destination must be one of the data registers DØ through D7.
2. Both source and destination must be addressed in the *address-register-indirect with predecrement* mode. This mode is useful for adding or subtracting paired data in two files, from high to low addresses.

[3]Having made this rule, we will immediately break it for the ADDX variation in the next few paragraphs.

Here are examples of valid ADDX and SUBX instructions.

- SUBX.L D3,D4 — Subtract D4 – D3 – X → D4.
- ADDX.B –(A1),–(A2) — Add bytes from files pointed to by A1 and A2, with carry from last addition. Point to next most-significant bytes.

Extend, extend, and extend: We pause here to sort out the multiple meanings that have been given to the unfortunate word *extend*:

- The *extend bit* of the CCR is a carry bit used for carry and borrow inputs, and is named X.
- To *sign-extend* means to fill a larger register with data of a smaller size so that the signed value of the data is unchanged.
- A *size extension* to a 68000 instruction mnemonic is the .B, .L, or optional .W, which specifies data size.
- *Extension words* in machine code are words that follow the op-code and contain operand information.
- *Extended addressing* is a 6800-family term that corresponds to *absolute* addressing in the 68000 family.

BCD addition and subtraction are available only for byte-size operands and then only for the *data-register-direct* and *address-register-indirect with predecrement* modes. The Z, X, and C flags are affected. Here are some examples.

- ABCD D5,D6 — Add two decimal digits in D5 to D6.
- SBCD –(A3),–(A4) — Subtract BCD byte pointed to by A3 from byte pointed to by A4; decrement A3 and A4.

The compare instruction sets CCR flags N, Z, V, and C as if a subtraction had been done, but the operand data is not affected. All three data sizes are allowed, except that byte-size data cannot be held in address registers. Here is a summary of the forms that the instruction can take:

1. The source (subtrahend) can be addressed in any mode if the destination is a data or address register.
2. The destination can be addressed in any mode if the source (subtrahend) is immediate data.
3. There is a special file-comparison instruction, CMPM (compare memory), in which source and destination use the *address-register-indirect with postincrement* mode.
4. TST is, in effect, a compare to an assumed immediate value of zero. It is useful because the data tested may be accessed by any addressing mode.

Here are some examples of the compare and test instructions:

- CMP.L 0(A1,D3.B),D0 — Compare D0 minus long word starting at address 0 + A1 + D3, where D3 is a signed 8-bit index value.
- CMP.W #$FFFF, −(A5) — Decrement A5 by 2; then compare word pointed to by A5 minus data $FFFF.
- CMPM (A3)+,(A1)+ — Compare word pointed to by A1 minus word pointed to by A3. Then point to next word in each file.
- TST.B FILE(A2.L) — Compare byte at address FILE + A2 to immediate data 0. FILE must be within −128 or +127 bytes of the next instruction. The data A2 is interpreted as a 32-bit signed number.

The multiply and divide instructions should be easy to master, since there are not that many forms available. The data size must be 16 bits for multipliers, multiplicands, divisors, and quotients, and 32 bits for products and dividends. Data must be all signed or all unsigned binary. The destination must be a data register. Multiplicands and dividends start out in the destination register and are overwritten. The source (multiplier or divisor) may be addressed by any mode. The N and Z flags are affected, and the V flag is set if a division result exceeds 16 bits. Here are some examples.

- MULU D6,D7 — Multiply in straight binary, word in D6 times word in D7. Place result in 32 bits of D7.
- MULS (A4),D0 — Multiply signed 16 bit words. Place result in D0.
- DIVU $31B0,D2 — Divide 32-bit binary of D2 by 16-bit binary at address $31B0 : 31B1. Place result in D2.
- DIVS A4,D1 — Divide signed long word in D1 by signed word in A4.

Figure 23.10 summarizes the arithmetic instructions of the 68000.

23.6 BRANCH, JUMP, AND CONDITIONAL INSTRUCTIONS

Section Overview

Branch instructions for the 68000 take the same forms as their counterparts in the 6800 and 6809. The displacement range is −32 768 to +32 767 bytes, relative to the address of the next instruction. The *jump* instruction has a full range of addressing modes, including PC-relative.

There are *decrement and branch* instructions, which decrement a counter and branch back to the start of a loop until the counter runs out or a specified condition is met.

There is also an instruction to set a flag byte to $FF when a specified condition is met.

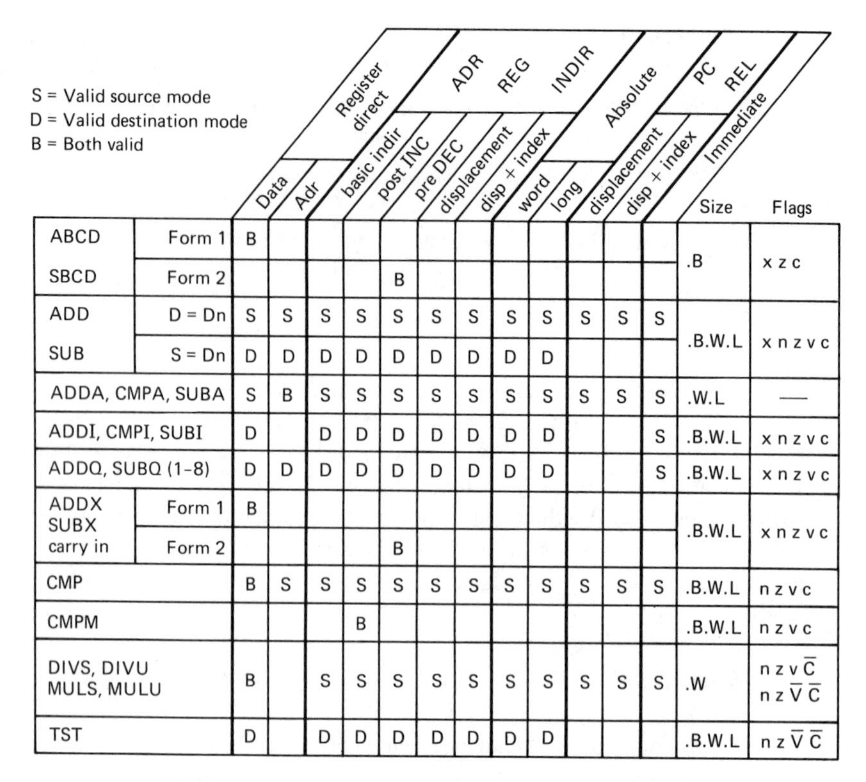

S = Valid source mode
D = Valid destination mode
B = Both valid

Instruction	Form	Register direct: Data	Register direct: Adr	ADR REG INDIR: basic indir	ADR REG INDIR: post INC	ADR REG INDIR: pre DEC	ADR REG INDIR: displacement	ADR REG INDIR: disp + index	Absolute: word	Absolute: long	PC REL: displacement	PC REL: disp + index	Immediate	Size	Flags
ABCD	Form 1	B												.B	x z c
SBCD	Form 2					B									
ADD	D = Dn	S	S	S	S	S	S	S	S	S	S	S	S	.B.W.L	x n z v c
SUB	S = Dn	D	D	D	D	D	D	D	D	D					
ADDA, CMPA, SUBA		S	B	S	S	S	S	S	S	S	S	S	S	.W.L	—
ADDI, CMPI, SUBI		D		D	D	D	D	D	D	D			S	.B.W.L	x n z v c
ADDQ, SUBQ (1-8)		D	D	D	D	D	D	D	D	D			S	.B.W.L	x n z v c
ADDX SUBX carry in	Form 1	B												.B.W.L	x n z v c
	Form 2					B									
CMP		B	S	S	S	S	S	S	S	S	S	S	S	.B.W.L	n z v c
CMPM					B									.B.W.L	n z v c
DIVS, DIVU MULS, MULU		B		S	S	S	S	S	S	S	S	S	S	.W	n z v $\overline{C}$ n z $\overline{V}$ $\overline{C}$
TST		D		D	D	D	D	D	D	D				.B.W.L	n z $\overline{V}$ $\overline{C}$

Figure 23.10 A summary of the 68000 arithmetic instructions.

MNEM	BRANCH IF
BCC	Carry clear
BCS	Carry set
BEQ	Result equal zero (Z = 1)
BGE	Destination greater than or equal to source (signed)
BGT	Destination greater than source (signed)
BHI	Destination higher than source (unsigned)
BLE	Destination less than or equal to source (signed)
BLS	Destination lower than or same as source (unsigned)
BLT	Destination less than source (signed)
BMI	Result minus (N = 1)
BNE	Result not equal to zero (Z = 0)
BPL	Result plus (N = 0)
BRA	Branch Always (unconditional)
BSR	Branch to subroutine (unconditional)
BVC	Overflow clear (no carry to sign bit)
BVS	Overflow set (carry to sign bit)

Figure 23.11 The 68000 branch instructions.

The branch instructions of the 68000 are listed in Figure 23.11. The conditions listing a source and a destination presume that the N, Z, V, and C flags were set by a subtract or compare instruction. These instructions take one word in memory if the branch address is within −128 and +127 bytes of the next instruction and two words for a displacement range of −32 768 to +32 767. The instruction forms are the same as in the 6800. Here are a few examples:

- BEQ OUT Branch to address labeled OUT if result is zero.
- BGT BIG Branch to address labeled BIG if destination was greater than source in prior compare.

The jump instruction may use any of the logically possible modes for the destination operand, which always appears in the program counter.

The decrement and branch instructions use the same condition tests as the simple branch instructions, but the function and form are quite different. They may be better described as, "Stay in the loop until the condition is true or the specified number of loops has been completed." Let us examine the following instruction as an example.

DBVS D3,FINISH

This instruction first checks the V flag in the CCR. If it is *clear*, the condition for exiting the loop has not been met, so counter register D3 is decremented. If the resulting value of D3 is not −1, the branch is taken, and the loop routine is repeated. The loop is exited, and the instruction following the DBVS is executed when

1. The specified condition V = 1 is met, or
2. The value in D3 is decremented from zero to −1.

The counter register must be one of the data registers, and only the lower 16 bits are decremented in straight binary. The range of the branch is −32 768 to +32 767 bytes.

There are two conditions (cc) of DBcc that are not listed for the Bcc instruction. These are as follows:

- DBF, sometimes listed as DBRA. This is condition *false* or *branch always*. It can be used to keep the processor in the loop for the full count in the data register.
- DBT, sometimes called DBRN. This is condition *true* or *branch never*. It can be used in program debugging to force a "fall-through" to the next instruction on the first pass.

Remember that the instruction

DBCC D1,ROUT

tells the processor to branch to ROUT *until* the carry flag is clear. This is quite different from

BCC ROUT

which tells the processor to branch to ROUT *when* the carry flag is clear.

Set flag byte according to condition causes a specified byte in memory to be loaded with $FF if the condition is true and with $00 if the condition is false. This permits several conditions to be checked and a record left of their states without any branches being taken. This is important when a branching decision must be made on the basis of a number of conditions. The operand is always byte size and may be addressed in any mode. The conditions are those listed in Figure 23.11, plus SF (always) and ST (never). Here are two example forms.

- SEQ (A5) Set byte at address of A5 if Z = 1.
- SMI $859 Set byte at address $859 if N = 1.

REVIEW OF SECTIONS 23.5 AND 23.6

21. Each of these is an invalid form. Tell what is wrong.
    a. ADD (D0),A1 b. ADD.B D2,A3 c. ADD.L A4,(A5)
22. If D3 contains $03 and D4 contains $04, what is the result of the instruction SUB.B D4, D3?
23. How does ADDX D6,D7 differ from ADD D6,D7?
24. What are the data-size and address-mode limitations on the ABCD and SBCD instructions?
25. In the instruction CMP.B #83,D5, what are the two sequential values in D5 that will cause **(a)** C = 0 and **(b)** C = 1?
26. In the instruction DIVU D1,D2 the value in D2 is $7654 0123. What is the smallest value in D1 that will *not* cause an overflow?
27. In the instruction MULS A0,D0 the word in A0 is $FFF0, and the word in D0 is $000F. What is the hex result in D0?
28. What is the addressing mode of the 68000 *branch* instructions?
29. We wish to complete the instruction DBcc D1,OVER so that the loop is exited if the Z flag is set *or* after 10 loops. What should D1 contain, and what should cc be replaced with?
30. What is the result of the instruction SVC $3456 if V = 1?

23.7 A SIMPLE-AS-POSSIBLE 68000 SYSTEM

This section presents a hardware diagram for a 68008 microprocessor interfaced with an EPROM and an 8-bit latch driving a display LED. $\overline{VPA}$ is asserted so the processor operates in the synchronous mode—10 clock cycles per bus cycle. This is painfully slow, but it eliminates the need for a $\overline{DTACK}$-generation circuit. Figure 23.12 shows the system hardware. A program for a little "chase" pattern on the LED is given in List 23.1, but the idea is for you to write and test your own programs on this simple system to begin to get used to 68000 programming and hardware.

```
                          1000 * LIST 23-1     ** CHASE LEDS AROUND: A 68008 TEST.
                          1010 *   FIG 23.12 HARDWARE.   WORKS, 6 AUG '87: METZGER.
                          1020          .OR      $0000
                          1030          .TA      $4000
00000000- 0000 07FE       1040          .HS      000007FE   STACK POINTER INIT
00000004- 0000 0400       1050          .HS      00000400   RESTART VECTOR
                          1060          .OR      $0400
                          1070          .TA      $4400
00000400- 727F            1080          MOVEQ    #$7F,D1    PULL BIT 7 LOW TO START.
00000402- 33C1 0000
00000406- 8000            1090 SHIFT    MOVE     D1,$8000     ONE BIT IN D1 BYTE IS LOW.
00000408- 343C FFFF       1100          MOVE     #$FFFF,D2  DELAY A SECOND OR SO
0000040C- 5342            1110 LOOP     SUBQ     #1,D2        BY COUNTING DOWN IN D2.
0000040E- 66FC            1120          BNE      LOOP
00000410- E219            1130          ROR.B    #1,D1      THEN PULL NEXT BIT RIGHT
00000412- 60EE            1140          BRA      SHIFT        AND WAIT AGAIN.
                          1150          .EN
```

Answers to Chapter 23 Review Questions

1. 7 **2.** 15 **3.** 8 **4.** It is a word-length instruction, but it references an odd-numbered address. **5.** A good question with no authoritative answer. It has an 8-bit data bus, a 16-bit standard operand size, and 32-bit internal data registers. Take your stand and join the fight. **6.** Both are asserted simultaneously.

7. Only $\overline{LDS}$ is asserted. **8.** The next bus cycle. **9.** The address is read, most-significant byte first, from addresses $4, $5, $6, and $7. **10.** The 68000 normally reads 2 bytes at a time; the 68008 can read only one and needs A0 to select addresses byte by byte. **11.** Mnemonic source, destination. **12.** Two; one for the source operand and one for the destination operand.

13. **(a)** $1234 **(b)** #$1234 **(c)** (A0) **(d)** A0

14. ADD.L D5,A0 **15.** SUBQ #2,A3 **16.** MOVE.B #$FF,(A4)+

17. MOVE 0(A2,D2),LATCH **18. a)** $0000 789A **b)** $3456 FF9A **19.** $FFFF 9ABC **20.** MOVEM (A3), D4-D7 **21. (a)** A data register cannot be an address pointer. **(b)** Byte data cannot be loaded to an address register. **(c)** ADD cannot have memory for a destination if an address register is the source. **22.** D3 = $FF

23. In ADDX an extra 1 is added if the X carry is set. **24.** Only 8-bit data, and only data register-direct or address-register-indirect with postincrement modes.

25. C = 0 for 83; C = 1 for 82. **26.** $7654 **27.** $FFFF FF00. **28.** Program counter relative. **29.** D1 = 9 and cc = EQ **30.** $00 is stored at address $3456.

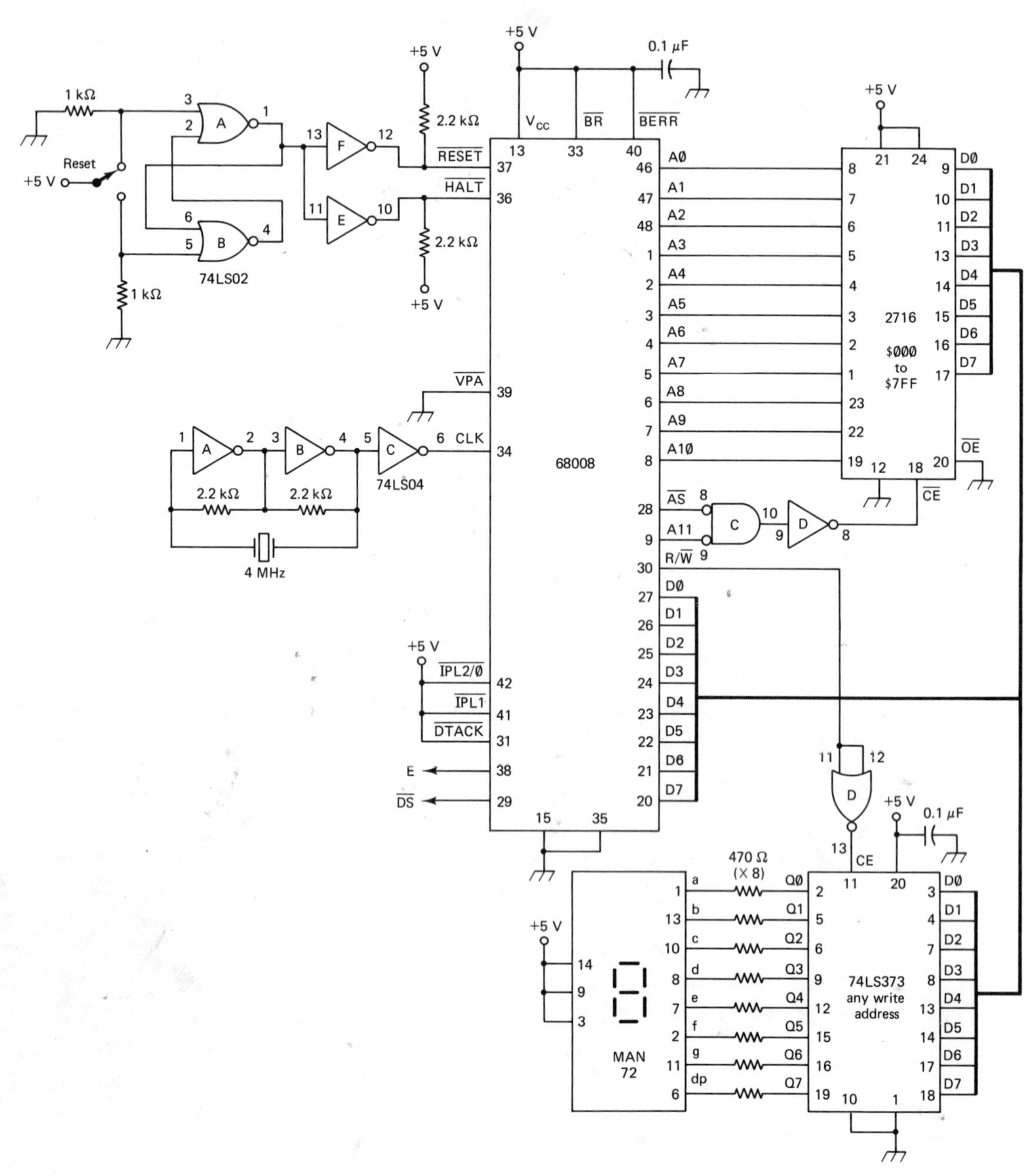

Figure 23.12 A simple 68008 system suitable for running test-loop programs. The $\overline{\text{VPA}}$ input is tied low to eliminate the need for a $\overline{\text{DTACK}}$ signal in this elementary system.

CHAPTER 23 QUESTIONS AND PROBLEMS

Digits after the decimal point refer to section numbers in Chapter 23.

Basic Level

1.1 LSR is a valid *shift* instruction in 68000 language. Write the complete source code for an instruction to shift **(a)** the byte at address $00 1234, **(b)** the 2 bytes at addresses $00 1234 and $00 1235, and **(c)** the 4 bytes at address $00 1234 through $00 1237.

2.1 What special functions are available in the D0 register that are not available in D7?

3.2 What supply voltages at approximately what currents are required by the 68000 microprocessor?

4.2 How many LS-TTL devices can be safely driven by the data bus of the 68000?

5.2 How should the 68000's three $\overline{\text{IPL}}$ lines be connected in an initial-test system that uses no interrupts?

6.3 Write 68000 assembly language instructions for the followoing.

- **a.** Add 100 to the low-order 16 bits of register D3.
- **b.** Add the 32 bits of register D4 to the 32 bits of D5.
- **c.** Increment 16-bit register D6.

7.3 Write the 68000 assembly forms for the following instructions.

- **a.** Add the byte at address $01 2345 to the low-order byte of register D0.
- **b.** Add the 32 bits from the 4 bytes starting at the address contained in register A1 to the 32 bits of D2.
- **c.** Add the 16-bit data pointed to by A2 to D7, and increment A2 by two.

8.4 Describe what the following instruction does.

```
MOVE.L D4, (A1)+
```

9.4 Write the hex contents of A0 after executing MOVE #50 000, A0.

10.4 Register D2 contains $1234 5678. What does it contain after executing

```
MOVEQ −2,D2?
```

11.5 D3 contains $0001 and D4 contains $0002. What does each contain after executing SUB D4,D3?

12.5 What major limitation applies to BCD addition in the 68000?

13.6 What addressing modes are available for the 68000 BRANCH instructions?

14.6 What is the effect of the instruction SCS (FLAG4)?

Advanced Level

15.1 What is a *cache* in computer terminology?

16.1 Explain the functions of the X, S, and T bits of the 68000 status register.

17.2 The 68000 is said to have an asynchronous data bus. Explain what this means.

18.2 Explain how the 68000 is made to operate with synchronous peripherals, such as the 6847 video display generator.

19.2 Why should the $\overline{\text{RESET}}$ and $\overline{\text{HALT}}$ inputs of the 68000 be driven by open-collector rather than active-pull-up devices?

20.3 Where is the operand data obtained from in each of these 68000 addressing modes?

a. Register direct

b. Register indirect

c. Absolute

21.3 In 68000 terminology the term *register indirect* means what we called *indexed* in 6802 terminology. The word *indexing* means something else in 68000 parlance. Explain this new use of the term *indexing*, and give examples.

22.4 Write a 68000 assembly language routine to read 2 long words from addresses $01 2ØØØ and $01 2ØØ4, add them, and store the result in the long word at an address pointed to by AØ.

23.4 Explain the effects of the instructions EXT.W D6 and EXT.L D6. Also tell when this instruction is useful and what versions of the MOVE instruction do an automatic EXT.L.

24.5 Write a 68000 routine to transfer an 8-kilobyte file of data from a block starting at address $Ø1 4ØØØ to a block starting at address $ØØ 2ØØØ. (The 68000 has a BEQ instruction which functions like the 6802's BEQ.)

25.5 Write a 68000 routine to divide an unsigned binary long word from addresses $13 1Ø3Ø through $13 1Ø33 by an unsigned word from addresses $13 1Ø34 and $13 1Ø35. The routine should check for a divide-by-zero and produce a quotient of $FFFF FFFF if such is the case. The quotient should be stored at $13 1Ø36 through $13 1Ø39.

26.6 Write a 68000 flowchart and the source code for a routine to examine a file of words from addresses $12 ØØØØ through $12 Ø3FF to identify any data greater than 2Ø ØØØ. Any such data should be reduced by 10%. *Hint:* Divide by 10 and multiply by 9.

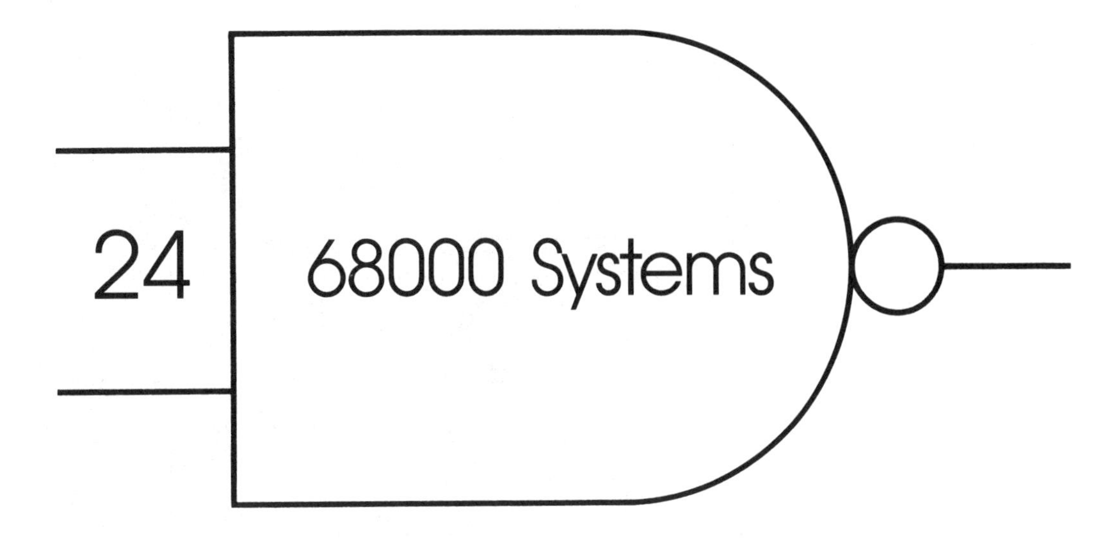

24.1 68000 INSTRUCTION WRAPUP

Section Overview

The 68000 *shift* and *rotate* instructions are similar to their 6800 counterparts, except that multiple shifts can be achieved in a single instruction. ROXL and ROXR rotate through the *extend* carry flag, but ROL and ROR rotate the specified number of register bits (8, 16, or 32) only.

Logic instructions AND, OR, EXCLUSIVE-OR, COMPLEMENT, and NEGATE (2's complement) are available.

The EXCHANGE instruction switches all 32 bits between two specified address or data registers. The SWAP instruction switches the high-order 16 bits and the low-order 16 bits of a specified data register.

LEA loads a destination address register with the effective address specified by the source operand.

The 68000 has a group of instructions which can test-for-zero, set, clear, or change (toggle) any selected bit in a register or memory location.

The rotate and shift instructions for the 68000 are summarized in Figure 24.1. For each instruction, a byte, word, or long word can be specified to be rotated or

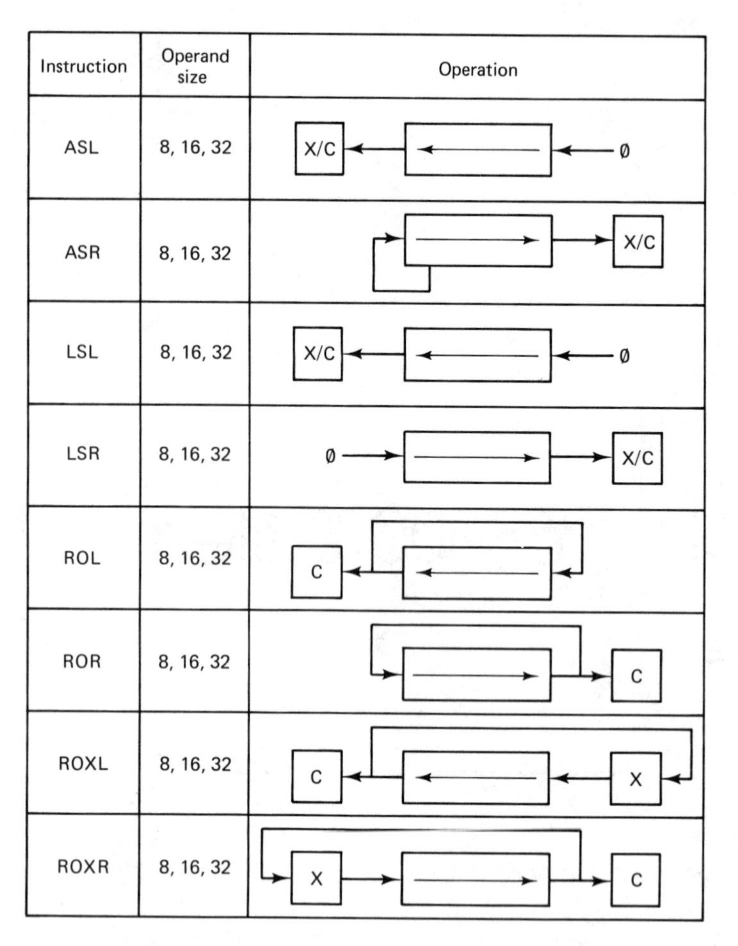

Figure 24.1 The 68000 *rotate* and *shift* instructions.

shifted by the extensions .B, .W, or .L. There are two basic forms for these instructions.

1. If a single shift or rotate is to be performed, the operand data must be in a memory location specified in any of the register-indirect modes or in the absolute mode. Here are some examples.

- ASL.B (A0) — Shift byte pointed to by A0 1 bit left. Same as LSL.
- ASL (A1)+ — Shift 16 bits pointed to by A1 1 bit left; then point A1 to next word.
- ROXL.L 0(A2,D3) — Rotate 32 bits at address A2 + D3 1 bit left through extend bit. D3 is a signed 16-bit number.
- ROR.W BITS — Rotate 16 bits at address labeled BITS 1 bit right. Bit 0 goes into bit 15.

2. If the operand data to be shifted is in one of the data registers, it is possible to specify a number of rotates or shifts to be performed with a single instruction. An immediate source operand of 1 through 8 or a data-register-direct source operand of 1 through 63 may be used to specify the number. Here are some examples.

- LSR #2,D4 — Shift 16 bits of D4 2 bits right.
- ROXR.L D5,D6 — Shift 32 bits of D6 right through X flag; number of shifts is contained in D5.

The logic operations available are generally similar to those in the 6800/6809. Logic instructions operate only on data in D0 through D7 or in memory. They cannot use the contents of A0 through A7 for either source or destination data. The familiar COM (complement) is NOT in 68000 syntax. NEG forms the 2's complement of the destination data. NEGX automatically subtracts the *extend* bit from the result, making it more negative by 1 if X was set. The destination for NOT or NEG can be a byte, word, or long word in

- A data register,

		Register direct		ADR REG INDIR					Absolute		PC REL		Immediate	Size	Flags
		Data	Adr	basic indir	post INC	pre DEC	displacement	disp + index	word	long	displacement	disp + index			
AND OR	Des = Dn	S		S	S	S	S	S	S	S	S	S	S	.B.W.L	n z $\overline{V}$ $\overline{C}$
	Sor = Dn	D		D	D	D	D	D	D	D					
ANDI, ORI, EORI	Source = immed	D		D	D	D	D	D	D	D				.B.W.L	n z $\overline{V}$ $\overline{C}$
ANDI, ORI, EORI to CCR	Destination = CCR												S	.B	x n z v c
ASL, ASR LSL, LSR		*		D	D	D	D	D	D	D				.B.W.L	x n z v c x n z $\overline{V}$ $\overline{C}$
CLR NOT		D		D	D	D	D	D	D	D				.B.W.L	$\overline{N}$ Z $\overline{V}$ $\overline{C}$ n z $\overline{V}$ $\overline{C}$
EOR	Source = Dn	D		D	D	D	D	D	D	D				.B.W.L	n z $\overline{V}$ $\overline{C}$
ROL, ROR ROXL, ROXR		*		D	D	D	D	D	D	D				.B.W.L	n z $\overline{V}$ c x n z $\overline{V}$ c

*Single-shift modes only listed. See text for multiple shifts.

Figure 24.2 The 68000 *logic* instructions.

- A memory location specified by any of the five address-register-indirect modes, or
- A memory location specified by an absolute address.

The AND and OR instructions can use any addressing mode to specify the source if a data register is the destination, or they can use any addressing mode (except immediate) to specify the destination if the source is a data register or immediate data. The EOR instruction must use a data register or immediate data as the source, but the destination can be a data register, a memory location pointed to by an address register (with auto increment/decrement, displacement, or indexing), or a memory location specified by its absolute address. The forms ANDI, ORI, and EORI indicate that the source data is immediate.

Figure 24.2 illustrates the available logic instructions and their available addressing modes. Here are some example forms.

NEG 0(A0,D1)	Negate the word at address A0 + D1.
AND (A2),D2	Logic AND 16 bits pointed to by A2 with D2; the result is in D2.
EORI.B #$F0,D3	Toggle bits 7 through 4 of D3, leaving bits 3 through 0 alone.
NOT.L RAMBIT	Complement bits at 4 bytes starting at address labeled RAMBIT.

EXG (exchange) permits data to be switched between any 2 of the 16 machine registers (A0–A7 and D0–D7). All 32 bits are always exchanged. Here is a sample instruction.

EXG D5,A6	Switch data between D5 and A6

The SWAP instruction switches data between the two 16-bit halves of a specified data register. Here is an example form.

SWAP D4	High-order 16 bits of D4 switch with low-order 16 bits.

Figure 24.3 illustrates EXG and SWAP.

Load Effective Address (LEA) loads the destination address register with the *address* that would have been used to fetch the source data. The source data itself is not touched. Here is an example comparing the MOVE and LEA instructions.

MOVE.L 8(A0,D1),A5	Load 32 bits of A5 with the data from 4 bytes of memory, starting at address A0 + D1 + 8.
LEA 8(A0,D1),A5	Load 32 bits of A5 with the address A0 + D1 + 8. The data at this address is not touched.

The Bit-Test instruction sets the Z flag of the CCR to indicate the *zero* state of an individual bit being tested. This bit may be located in any data register or in a memory location addressed in one of the register-indirect, absolute, or PC-relative modes. Similar instructions are available that *set*, *clear*, or *toggle* (change) the bit

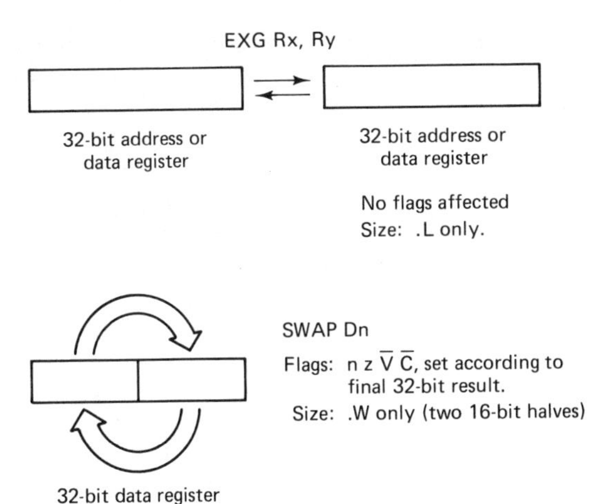

Figure 24.3 *Exchange* and *swap* instructions available in the 68000.

after testing it. The number of the bit being tested (∅ through 31) may be specified in the *source* = *immediate* or *source* = *data-register-direct* modes. Only byte and long-word data sizes are permitted. Here are some examples.

- BTST.B #∅,(A3) — Set Z = 1 if lowest-order bit of data pointed to by A3 is a ∅.
- BCLR.L #31,D2 — Set Z if D2 is positive; then clear its sign bit to ∅ (positive).
- BCHG.L D4,(A6) — Toggle (change) bit specified by D4 (data bit ∅ through 31) in long word at address pointed to by A6.

24.2 SUBROUTINES AND STACK OPERATIONS

Section Overview

Subroutines in the 68000 are similar to subroutines in the 6800. A new instruction is RTR, which automatically restores the condition-code bits saved from the main program.

The LINK and UNLINK instructions are used to set aside a stack area for data to be passed from the main program to the subroutine.

The MOVE instructions are used to push and pull data on the stack, since the stack pointer is register A7. The stack predecrements when saving data, so it is necessary to initialize the stack pointer to an even address to avoid an address-boundary error.

Subroutines in the 68000 are called by the instructions JSR or BSR. Relative addressing is always used to reach the subroutine with BSR. JSR can use relative, absolute, or address-register-indirect addressing to specify the destination. Either instruction causes only the 4-byte address of the next instruction in the main program to be saved on the stack.

To save other data on the stack the MOVE or MOVEM instructions must be used with stack pointer A7 as the destination pointer in the predecrement mode. For example:

```
         JSR      ROOTS                 Subroutine call stacks PC.
          .         .
          .         .
          .         .
ROOTS    MOVEM    D0-D7/A0-A3,-(A7)     First subroutine instruction saves all
                                        data, four address registers.
          .         .
          .         .
          .         .
         MOVEM    (A7)+,D0-D7/A0-A3     Last subroutine instruction unstacks
                                        same registers.
         RTS                            Return.
```

The new instruction RTR (return and restore) may be used in place of RTS to load the condition-code register from the stack when returning to the main program. CCR must have been stacked as the first operation in the subroutine if RTR is to be used at the end. The instruction to do this is

MOVE CCR, – (A7)

Some assemblers use – (SP) rather than – (A7) to specify *push on* stack.

The supervisor stack pointer A7′ is loaded from addresses A0, A1, A2, and A3 upon an external *reset* to the 68000. The user stack pointer may be loaded by a MOVE, immediate, to A7. The addresses loaded to the stack pointers must be *even* numbers so that *word* and *long-word* data will start on even address boundaries when stacked. If this precaution is not observed, an address-boundary error will result, which could abort the execution of the main program.

LINK is a 68000 instruction that is used to set aside an area of RAM for data to be accessed by a subroutine. For example,

LINK A6,# – 100

reserves 100 bytes of the stack for data and sets up A6 to point to the top of this data range.

REVIEW OF SECTIONS 24.1 AND 24.2

Answers appear at the end of Chapter 24.

1. If D1 contains data $A234, what do X and C contain after executing ROXL #5,D1?
2. Which of these are invalid, and why?
   a. ASR.W A3 **b.** ROL.B –3(A4) **c.** ROXL 4,(A5)
3. If D6 contains $1A8F, what does it contain after **(a)** NOT D6 and, independently, **(b)** after NEG D6?

4. In Question 3, what will D6 contain after executing SWAP D6?
5. What will A2 contain after executing LEA TABLE,A2?
6. Write an instruction to set the Z flag if bit 16 of register D3 is a logic 0.
7. What does this instruction do?

```
MOVE D5,-(A7)
```

8. How would you pull a long word from the stack and store it at an address pointed to by A2?
9. A 68000 system has ROM at address $00 0000 through $00 FFFF and RAM at addresses $01 0000 through $02 FFFF. How could the stack be positioned at the high end of the RAM?
10. What instruction would reserve a 10-byte segment of the stack for data and point A3 to the top of it?

24.3 INTERRUPTS AND EXCEPTIONS

Section Overview

The 68000 has seven levels of interrupt priority. In the autovector mode, one vector address for each level points to one of seven interrupt-service routines. The autovector mode is used if the processor's $\overline{VPA}$ input is asserted instead of $\overline{DTACK}$.

In the standard interrupt mode the interrupting device applies a vector number to the processor's D0 through D7 lines to specify one of 192 vectors pointing to one of a possible 192 interrupt routines.

Interrupts are classed as *exceptions* to the normal instruction processing of the 68000. Some other exceptions are *reset*, address-boundary error, divide-by-zero error, undefined op-code, and single-instruction *trace* mode. Each exception has a 4-byte exception vector associated with it. The processor jumps to the associated vector address upon recognizing an exception. Exception vectors take up the first 1024 bytes of the 68000's memory.

Interrupt masks and priorities: The first 3 bits of the system byte (SR bits 8, 9, and 10) are the interrupt mask bits. They establish one of 7 interrupt priority levels, from 000 (any interrupt input accepted) to 110 (only highest priority interrupt accepted). The 68000 has three *interrupt priority* input pins which are asserted *low* to request an interrupt service at 1 of 7 levels of priority, from all *high* levels (no interrupt request) to all *low* levels (nonmaskable interrupt). Figure 24.4 shows the mask bits and the lowest-priority $\overline{IPL}$ input levels that will cause an interrupt acknowledge for each. Note that mask-bit pattern 111 does not prevent a level-7 interrupt as the pattern of the other levels would suggest. This is to preserve the nonmaskable character of one of the interrupts.

	MASK BITS IN SR			LOWEST IPL INPUT LEVEL RECOGNIZED				AUTOVECTOR ADDRESS
	1Ø	9	8		$\overline{IPL2}$	$\overline{IPL1}$	$\overline{IPLØ}$	
Recognize any					high	high	high	No IRQ
	Ø	Ø	Ø	1	high	high	low	$00 0064
	Ø	Ø	1	2	high	low	high	$00 0068
	Ø	1	Ø	3	high	low	low	$00 006C
	Ø	1	1	4	low	high	high	$00 0070
	1	Ø	Ø	5	low	high	low	$00 0074
	1	Ø	1	6	low	low	high	$00 0078
	1	1	Ø	7	low	low	low	$00 007C
Recognize one only	1	1	1	7	low	low	low	$00 007C

Figure 24.4 Interrupt masking, input priority levels, and vector locations for the autovector mode.

The autovector interrupt response is the simplest, so it will be described first. When an interrupt input is acknowledged, the 68000 asserts all three of its function-code outputs *high*. To activate the autovector response mode, the $\overline{\text{VPA}}$ input must then be pulled *low*. This can be accomplished with a single NAND gate, as shown in Figure 24.5. The processor will then read 4 bytes starting at the autovector address given in Figure 24.4 and concatenate them in most-significant to least-significant order to form the address for the interrupt-service routine. Before jumping to this address, the processor will push the current program counter (4 bytes) and status register (2 bytes) on the system stack. The interrupt-service routine must save and restore any address or data registers that were used in the main program and will be reused in the interrupt routine. With 8 data registers and 7 address registers available, it is sometimes possible to dedicate different registers to main program, subroutines, and interrupts and thus avoid the need to save and restore registers.

A *return from exception* (RTE) instruction must be placed at the end of the interrupt-service routine to restore the program counter and status register as they were in the main program before the interrupt.

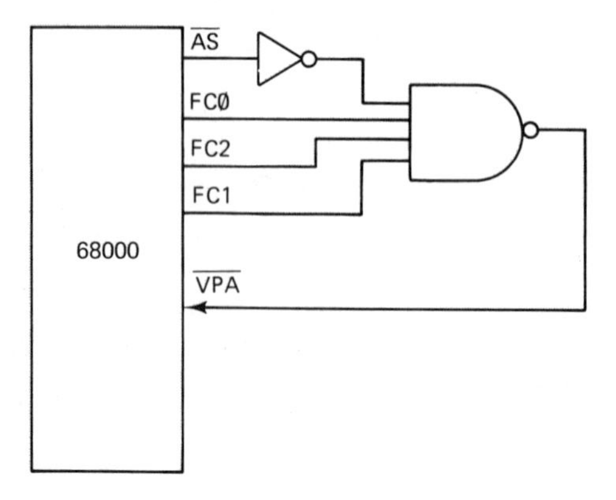

Figure 24.5 A simple logic gate responds to the interrupt-acknowledge function code (111) by putting the 68000 in the autovector mode.

Externally vectored interrupts are invoked when the interrupting device responds to the function code 111 by asserting $\overline{\text{DTACK}}$ rather than $\overline{\text{VPA}}$ to the processor. Such a device must also place a vector number from \$40 to \$FF on the data lines D7 through D0. This number will be multiplied by 4 in the processor to produce a vector address from \$00 0100 to \$00 03FC. The processor will read the 4 bytes at this address and concatenate them to form the address of the interrupt-service routine. There are 192 such addresses and routines possible. Priority levels 1 through 7 can be assigned to each interrupting device by encoding the $\overline{\text{IPL}}$ lines that are pulled *low* by an interrupt request.

Other exceptions are processed much like the interrupts. Each is assigned a vector location, and the processor picks up an address from this location upon recognizing the exception. The system designer must supply *some* address at each of the exception-vector locations. Leaving these bytes unprogrammed can result in the processor halting and leaving no indication of *why* it halted. For starters in bringing up a system, all the vectors can point to a single routine, which provides an *exception* indication on an LED and then ends in an RTE. This will at least keep the processor going so it can be followed on a logic analyzer. Later, individual routines can be written to display the particular exception encountered and possibly correct the source of the error.

A complete table of exception vectors appears in Figure 24.6. System ROM for an elementary 68000 system starts at address \$00 0000. User programs may begin at \$00 0400, since the bytes through \$00 03FF are dedicated to vector addresses.[1] The bytes from \$00 0100 through \$00 03FF could be used for user programs if externally vectored interrupts were never used, but memory is usually not in such short supply as to require this. Here are brief descriptions of the 68000 exceptions:

- *Bus error* is caused by an input to the processor's $\overline{\text{BERR}}$ input. It is up to the hardware system designer to decide what external conditions will assert $\overline{\text{BERR}}$. Although the name implies trouble, there is no reason why $\overline{\text{BERR}}$ cannot signal a normal request from a peripheral.
- *Address error* is the response when the 68000 attempts to fetch word or long-word data from an odd-numbered address.
- *Illegal instruction* results if the processor reads a word that it expects to be an instruction, but the bit pattern does not fit any defined 68000 instruction.
- *Zero divide* is quite obvious. The programmer may choose to insert a maximum positive or negative result for a division by zero.
- *The* CHK *instruction* takes the form

  ```
  CHK #300,D5
  ```

  The CHK vector will be accessed, in the preceding example, if the value in D5 exceeds 300. This vector is accessed only by a CHK instruction.

[1]More-developed systems place RAM at the beginning addresses to permit dynamic alteration of the vector table. Special circuitry is used to enable a ROM for the first four word-fetches after a *reset* to obtain the initial stack-pointer and program-counter values.

Exception Vector Assignment

Vector Number(s)	Dec	Address Hex	Space	Assignment
0	0	000	SP	Reset: initial SSP[2]
1	4	004	SP	Reset: initial PC[2]
2	8	008	SD	Bus error
3	12	00C	SD	Address error
4	16	010	SD	Illegal instruction
5	20	014	SD	Zero divide
6	24	018	SD	CHK instruction
7	28	01C	SD	TRAPV instruction
8	32	020	SD	Privilege violation
9	36	024	SD	Trace
10	40	028	SD	Line 1010 emulator
11	44	02C	SD	Line 1111 emulator
12[1]	48	030	SD	(Unassigned, reserved)
13[1]	52	034	SD	(Unassigned, reserved)
14[1]	56	038	SD	(Unassigned, reserved)
15	60	03C	SD	Uninitialized interrupt vector
16-23[1]	64	040	SD	(Unassigned, reserved)
	95	05F		—
24	96	060	SD	Spurious interrupt[3]
25	100	064	SD	Level 1 interrupt autovector
26	104	068	SD	Level 2 interrupt autovector
27	108	06C	SD	Level 3 interrupt autovector
28	112	070	SD	Level 4 interrupt autovector
29	116	074	SD	Level 5 interrupt autovector
30	120	078	SD	Level 6 interrupt autovector
31	124	07C	SD	Level 7 interrupt autovector
32-47	128	080	SD	TRAP instruction vectors[4]
	191	0BF		—
48-63[1]	192	0C0	SD	(Unassigned, reserved)
	255	0FF		—
64-255	256	100	SD	User interrupt vectors
	1023	3FF		—

Notes:

1. Vector numbers 12, 13, 14, 16 through 23, and 48 through 63 are reserved for future enhancements by Motorola. No user peripheral devices should be assigned these numbers.
2. Reset vector (0) requires four words, unlike the other vectors which only require two words, and is located in the supervisor program space.
3. The spurious interrupt vector is taken when there is a bus error indication during interrupt processing.
4. TRAP #n uses vector number 32 + n.

Figure 24.6 The complete exception-vector table for the 68000. (Courtesy Motorola SemiConductor Products, Inc.)

- TRAPV is an instruction that causes vector 7 to be accessed if the processor's V flag is set. Only the TRAPV instruction accesses this vector.
- TRAP #n is an unconditional instruction accessing vectors 32 through 37 at addresses $80 through $BF. There are 16 trap vectors, accessed by letting n equal 0 through 15, respectively, in the assembly instruction.
- *A privilege violation* occurs if the 68000 attempts to execute a supervisor-privileged instruction while in the user state.
- *Trace* is turned on by setting the T bit in the 68000 status register. In this condition the processor will save PC and SR on the supervisor stack and jump to the *trace* vector address after every instruction. The trace routine may cause the machine registers to be displayed and may halt or delay the return to the main program, thus providing a single-step or slow-step mode for system debugging. *Trace* is turned off during the trace routine and turned on again when returning from the trace routine with RTE.

24.4 LARGE-SYSTEM DEVELOPMENT

Section Overview

This section presents a collection of hints and suggestions for implementing and debugging large-scale microcomputer systems. As in another area, all the "thou shalts" and "thou shalt nots" depend on two great commandments.

1. **Implement and test your hardware and your software in small steps. Twenty sticks individually are easier to break than twenty sticks in a bundle.**
2. **Settle for nothing less than perfection—in your wiring, in your hardware diagrams, in your program flowcharting and commenting, and in your understanding of the system. Ten minutes spent avoiding an unreliable connection or a poor understanding of a program segment will save 2 hours of troubleshooting later on.**

Inputs	Outputs
Keyboard	LED display
Travel-end switches	CRT
Temperature	Printer
Pressure	Graph Plotter
Weight	Speaker
Velocity	Solenoid
Light intensity	Motor
Digital data from another system	Lamps
	Heater

Defining the problem is a frequently slighted phase of microcomputer system development. It is a truism that, "If you don't know where you're going, you'll probably have a hard time getting there." You should start with a list of the inputs and outputs available or required. Here are some common ones to give you the idea.

Next a decision should be made and recorded as to what the system output will be for every conceivable sequence and combination of the listed inputs. For example, what should happen if the light sensor says a package to be wrapped is in place, but the scale shows the weight of the package to be nearly zero?

Time spent on problem definition is often difficult to justify to management because it appears that nothing is happening except talking and doodling. Often the questions that come up seem trivial, but remember that no question is trivial to a computer; it will require specific directions for every situation it encounters. Also remember that *any* design decision is better than *no* design decision. If you aren't sure what to do if the above situation comes up, program the computer to drop the package (if there is one) on the floor and go on to the next one. If the customer doesn't like that idea, he will be sure to tell you about it, and then you can change it. The important thing for the moment is not to let yourself or the computer get stalled over an unresolved situation.

Software development is terribly expensive: Analysts estimate that the cost is on the order of $10 per line of debugged code. Programs of 100,000 lines are not unusual. That's $1 million in software-development cost. This cost may be reduced by the following means:

- Purchased software packages should be investigated. Why spend $50,000 to develop a LORAN navigation program of unproven reliability if somebody else has already developed one and is willing to license its use for $10,000? This same option is often attractive for hardware modules unless production quantities are very large. It may be less expensive to purchase CPU modules, DRAM modules, and I/O modules than to design, build, and debug your own.
- High-level languages such as PASCAL, FORTRAN, and "C" should be considered, especially where extensive mathematical calculation is required and computing speed need not be pushed to the maximum. These languages are available for almost every microprocessor, and they provide inexpensive and reliable software routines for all the basic algebraic and trigonometric functions. "C" language, in particular, is gaining wide popularity among microcomputer programmers because it retains much of the speed and bit-manipulation capabilities of assembly language while offering what the programmers call *portability*. This is the ability to compile a single version of source code on nearly any machine, regardless of what processor it uses. Routines for advanced computations, such as matrix solutions and statistics, are much easier to write in a high-level language and are even easier to buy already written in a high-level language.

Defining the register functions before you start writing the code will make 68000 programming considerably less confusing. There are 15 available registers, which the assembler references by uninformative labels such as D3 and A6. A table such as the following should be developed for each subroutine and program module and included in your program documentation:

D0	General data transfer	A0	Source-file pointer
D1	Input weight	A1	Destination-file pointer
D2	Total weight	A2	Video-memory pointer
D3	Price, this crate	A3	Price/Tax-table pointer
D4	Total price today	A4	Delay-count for timer
D5	Run-time total		

Over 20 support chips are available specifically for the 68000. These include parallel and serial I/O functions, DMA controllers, math coprocessors, and bus-arbitration, network, and terminal controllers for multiprocessor and multiuser systems. It will be a long pull, but if you are serious about learning 68000 systems you should set aside some time regularly to teach yourself these chips. The manufacturer's data sheets are the best—and often only—source of information.

System hardware testing should be carried out on one peripheral device at a time, using separate diagnostic programs to run each support chip through its paces. For systems that are targeted to run at top speed, it is a good idea to have a half-speed clock option. This will allow you to separate many timing problems from simple logic errors.

Many 68000 systems use a *watchdog timer*. This circuit, shown in Figure 24.7, asserts $\overline{\text{BERR}}$ if $\overline{\text{DTACK}}$ is not asserted once every eight E cycles (80 S states). Failure to receive a $\overline{\text{DTACK}}$ assertion can result from something as simple as calling up an unimplemented memory address. Without the watchdog, the processor would wait forever for the $\overline{\text{DTACK}}$ that never comes. The $\overline{\text{BERR}}$ exception-processing routine may attempt to recover from the error, or it may simply inform the troubleshooter of the program address that caused the failure.

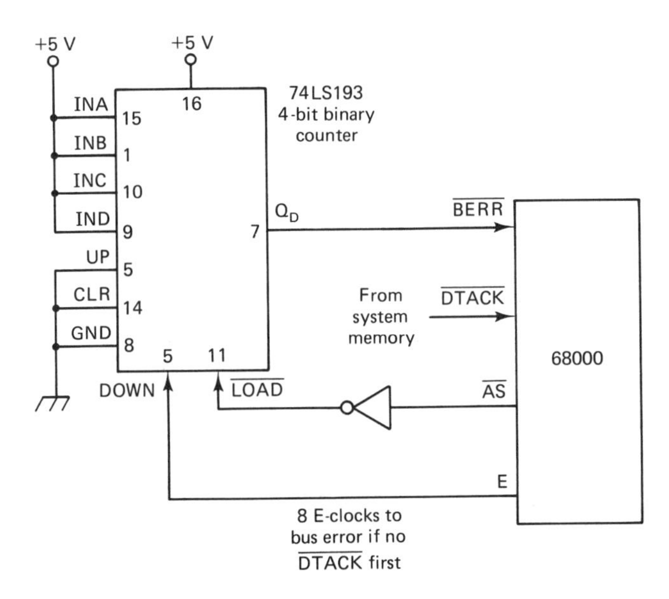

Figure 24.7 A *watchdog* timer sends the 68000 to the *bus-error* vector routine if the processor fails to get a *data acknowledge* after a predetermined time.

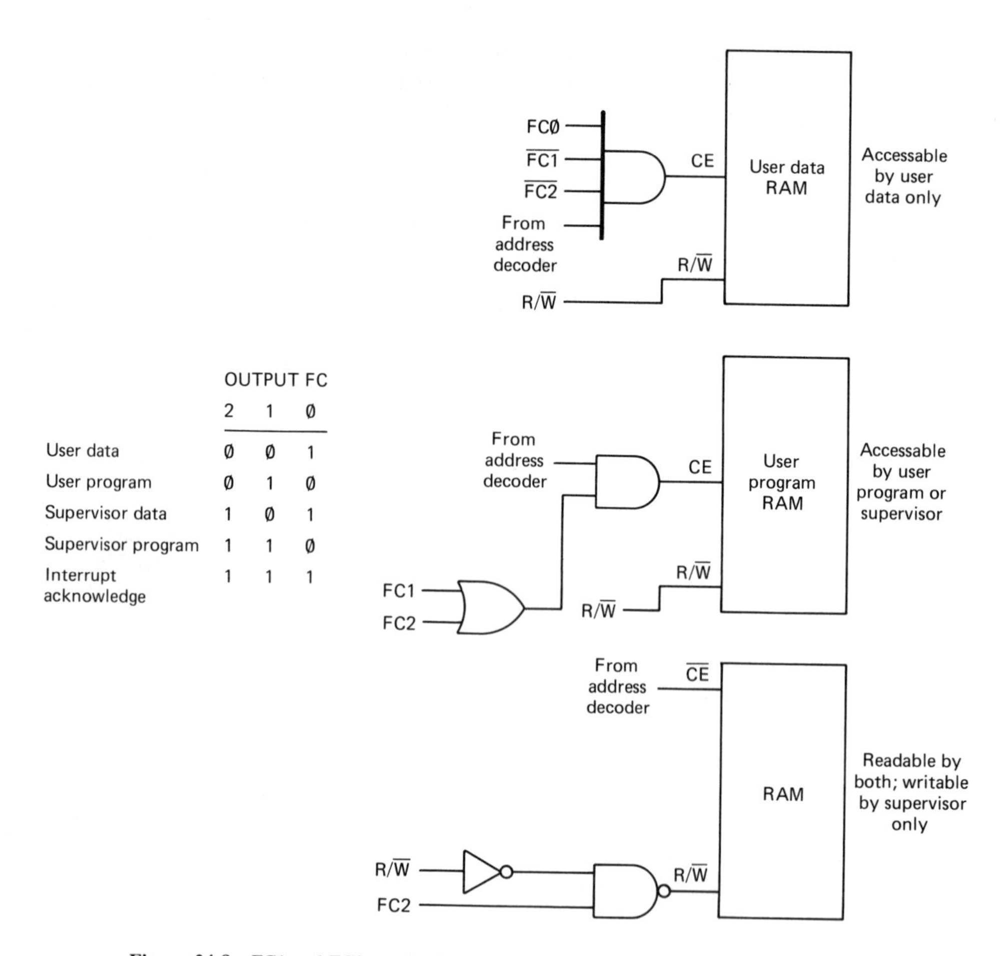

Figure 24.8 FC1 and FC2 can be decoded to enable certain memory or I/O devices only in the supervisor mode or only on program but not on data cycles.

Selected memory and I/O devices in a 68000 system can be dedicated only to the supervisor state by including FC2 in the chip-select logic, as shown in Figure 24.8. Data and program chips can be isolated with FC1, making it impossible for the user to write over a library program that was downloaded from a disk to RAM by the supervisor.

A hardware single-step circuit for the 68000 is simple to implement. It is shown in Figure 24.9. The software single step using the *trace* mode is more versatile.

System software testing should also be carried out module by module. If an untested module contains over 30 lines of code it may be a good idea to see if it can be broken into smaller sections for testing. The first tests should be done with trivial data inputs supplied artificially. Loop counters may be temporarily set to *one* to make it feasible to single step through the module. Full-speed tests can then be run with a *breakpoint* inserted at the end of the module. The breakpoint is simply a JSR or TRAP instruction, which takes the processor to a routine that displays the register contents and selected RAM locations and waits for an external key commanding it

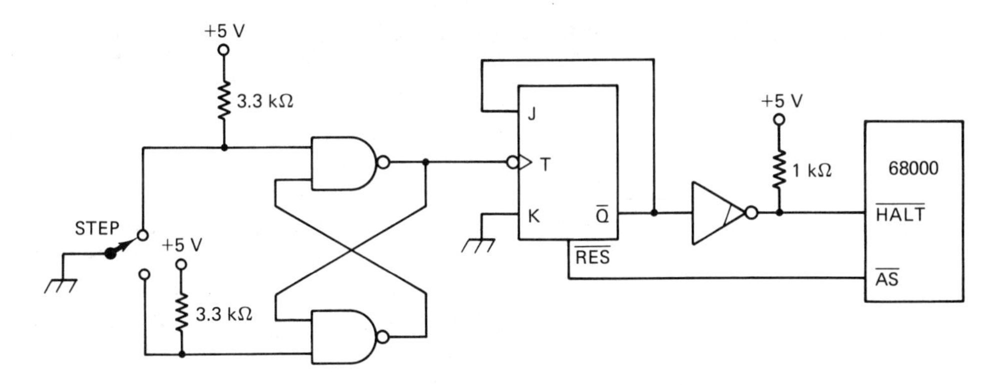

Figure 24.9 A simple single-step circuit for the 68000.

to return to the main program. If the module fails, the breakpoint can be moved closer to the beginning of the routine until a place is found, up to which things have been proceeding as anticipated.

Final testing of the module should be done with input data generated by an auxiliary routine written especially to ensure the eventual production of all possible data sequences and combinations. The auxiliary routine should save the last few hundred words of test data applied and report on any sequences that produce calls of nonexisting addresses, divisions by zero, requests for the angle whose sine is 1.5, or other oddities. Only at this point should an attempt be made to run the module in conjunction with other modules.

REVIEW OF SECTIONS 24.3 AND 24.4

Answers appear at the end of Chapter 24.

11. What levels at $\overline{\text{IPL0}}$, $\overline{\text{IPL1}}$, and $\overline{\text{IPL2}}$, respectively, will cause an interrupt with the SR bits 10, 9, and 8 at 011, but not with these bits at 100?
12. In Question 11, where is the address of the IRQ service routine found? Assume autovectoring.
13. How is autovectoring invoked?
14. What 68000 registers are stacked on interrupt acknowledge?
15. What will the instruction TRAP #2 do?
16. What will the processor do if given the following instruction?

```
MOVE.L D2,$4321
```

17. What is the estimated cost of developing a 100-line program module?
18. What is the name of the circuit that allows the 68000 to recover if it sends out an address strobe but never gets a data-acknowledge response?
19. What is a *breakpoint*?
20. What are the two great commandments for implementing microcomputer hardware and software?

24.5 CHICKEN PICKIN'—A COMPLETE 68008 SYSTEM

Section Overview

The widespread application of microcomputers is difficult to explain to laypersons, and even to electronics technicians, because few of us have the detailed technical knowledge of any given application that would allow us to appreciate the usefulness of microcomputer control. Most technical persons, for example, have some knowledge of the construction of a gasoline piston engine and of the function of the carburetor and ignition systems. Few, however, have a detailed understanding of the exact fuel-air mixture, spark advance, and dwell that will produce maximum fuel economy and minimum pollutants under all conditions of engine temperature, air pressure, throttle position, engine speed, and engine load. Those who have such understanding have no trouble appreciating the appropriateness of using a microprocessor to control an automotive engine.

Molding and inspecting glass bottles, tuning in a color television set, and preparing and analyzing bacteria cultures in a pathology laboratory are other examples of processes that are ideally suited to microcomputer control but are difficult to explain to anyone but a specialist in the field. This section presents complete hardware and software for a system using a 68008 microcomputer to solve a practical problem that anyone can understand and which is obviously not solvable by anything other than a computer. This system is based on a 6802 version presented by the author in an earlier work[2] and was developed by student Bill Harrison as a class term project.

The chicken pickin' problem: Let us say that you own a chicken ranch. You have a contract to provide a continuing supply of boxes of whole fryers to a big customer. The boxes must contain a minimum of 24 pounds of chicken and must contain whole fryers only—no parts. You receive $8 per box, whether the box weighs 24.00 lb or 24.96 lb. Obviously it is to your advantage to pack the boxes with combinations of fryers whose weights total exactly 24.00 lb.

The system, shown in Figure 24.10, weighs in 16 chickens using a self-clocked A-to-D converter to digitize the weights from a potentiometer driven from a spring scale. The weights are in 0.02-lb increments, with a maximum weight of 5.10 lb corresponding to the maximum count of 255 from the A-to-D converter.

Three 7-segment LEDs display the current chicken weight, and interrupt input $\overline{\text{IPL2/0}}$ signals the processor to accept the weight displayed. Sixteen discrete LEDs indicate the chickens "on the hanger" ready for the selection process. When the selection is completed, these LEDs indicate the chickens taken, and the display shows the total weight of the box. You may then reload to replace the chickens taken. If you load the chicken hangers with random weights in the 3- to 5-lb range, you will

[2]Daniel Metzger. *22 Microcomputer Projects to Build, Use, and Learn.* (Englewood Cliffs, N.J.: Prentice-Hall, 1985).

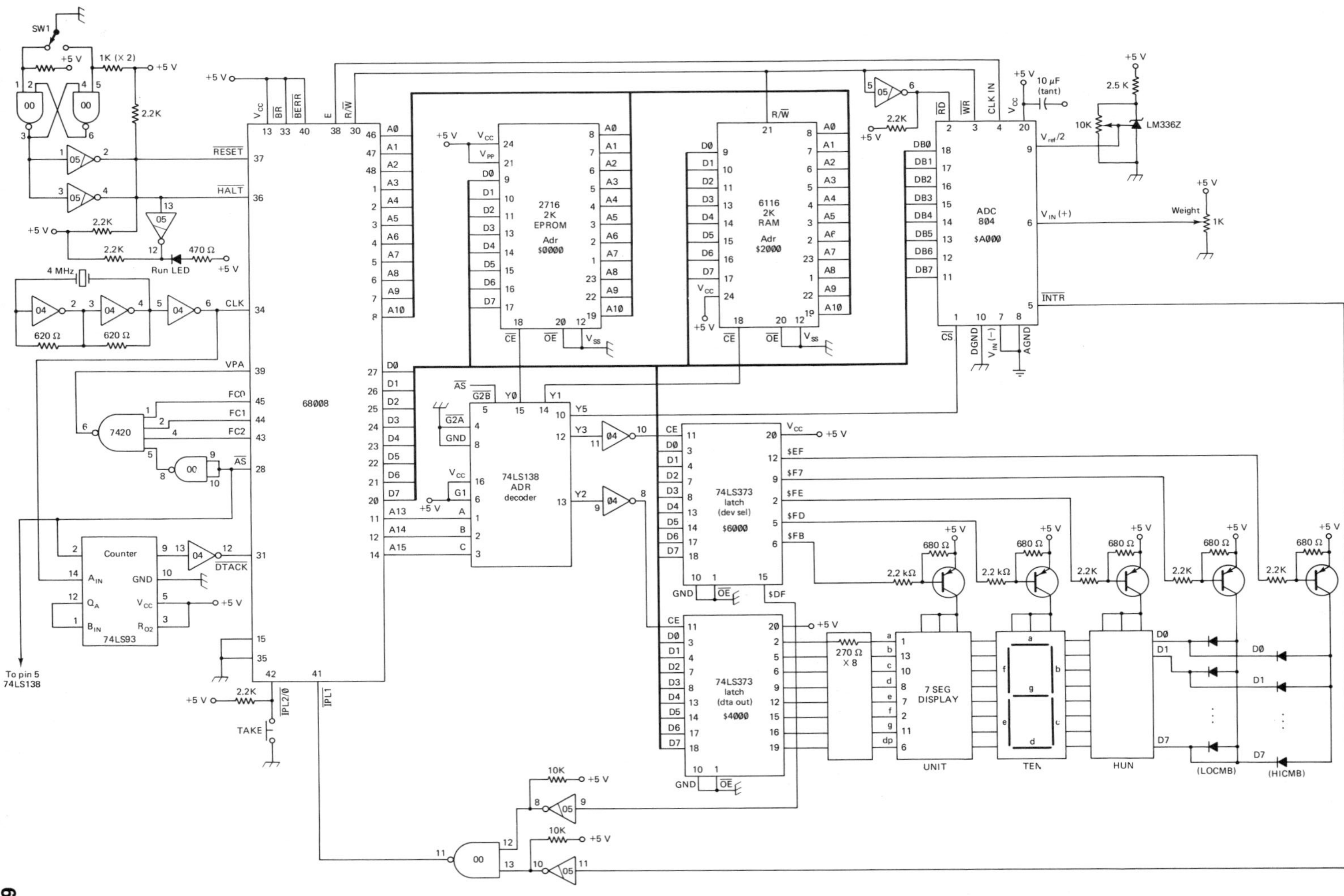

Figure 24.10 A complete 68008 hardware system for the *weigh and batch* problem called Chicken Pickin'.

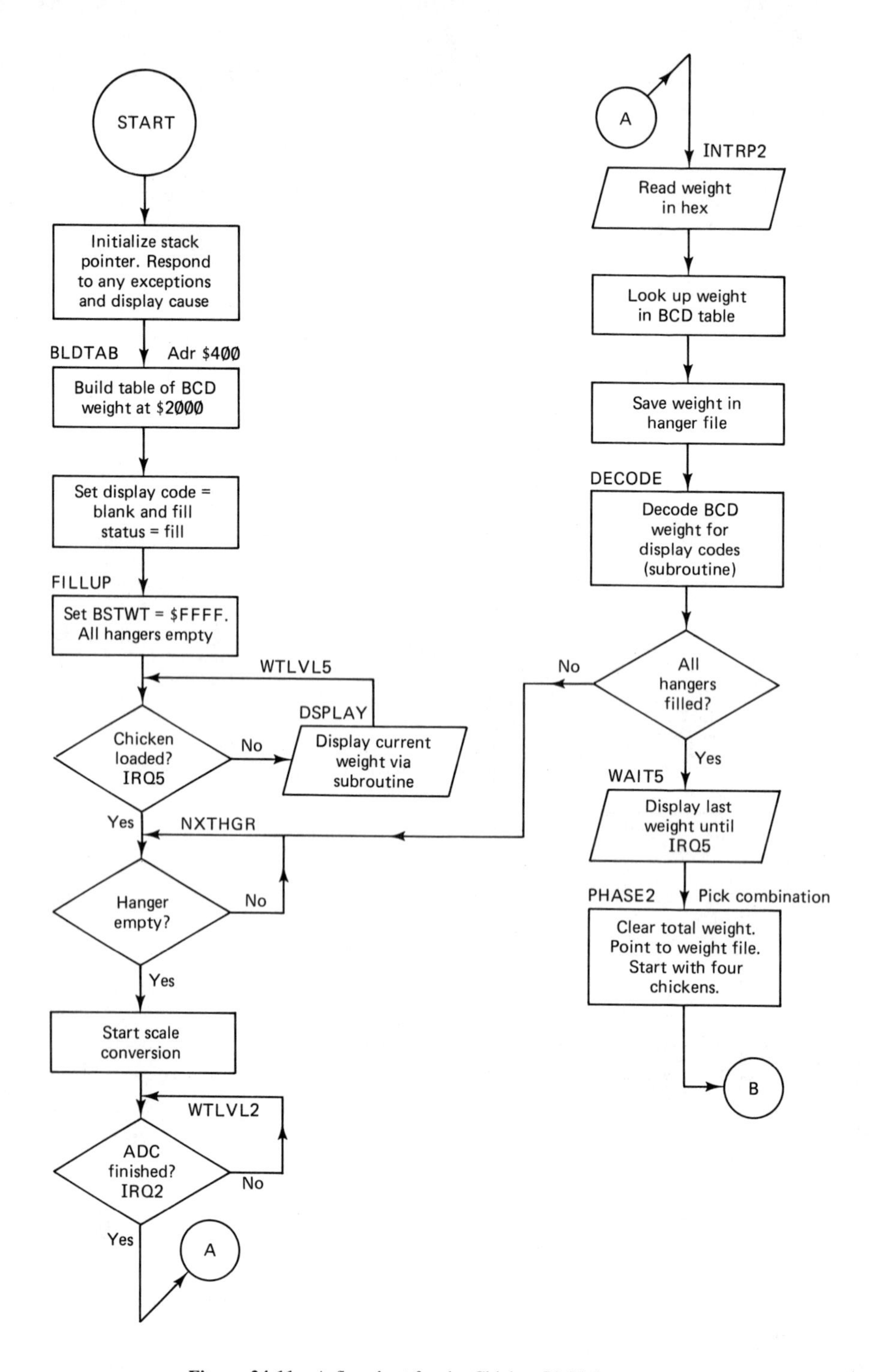

Figure 24.11 A flowchart for the Chicken Pickin' program.

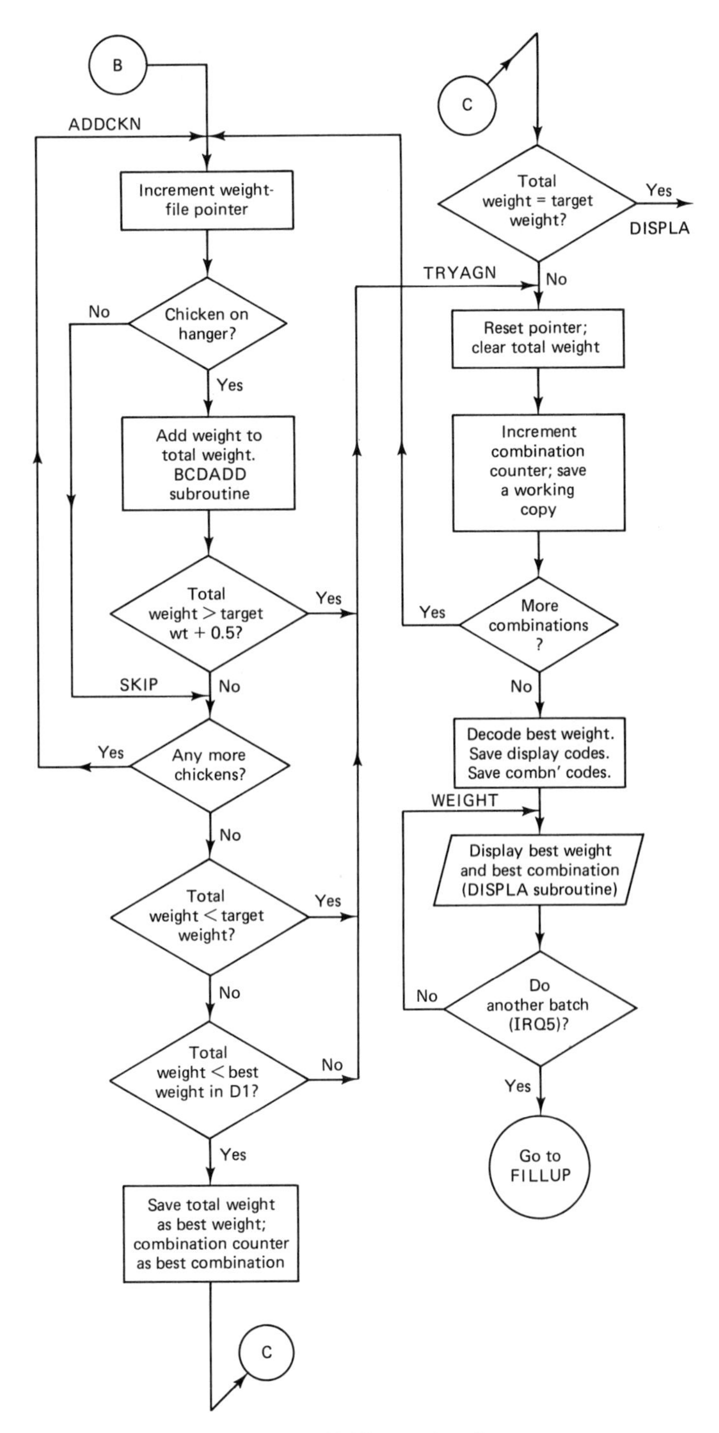

Figure 24.11 *(continued)*

probably be amazed to see how often and how quickly a combination totaling exactly 24.00 pounds is found. If you set the "scale" at 5.00 for all 16 chickens, making a 24-lb combination impossible, you will see that it takes the 68008 a full minute to try all 65 536 combinations of 16 chickens before admitting defeat.

Figure 24.11 shows the Chicken Pickin' flowchart, and the program listing follows (List 24.1). This is a first working version of the program. Quite an interesting competition can be generated if several students vie to see who can produce the fastest updated version.

```
                     1000 *----THIS IS A WORKING VERSION OF CHICKEN PICKER 68K---

                     1010 *         WRITTEN BY BILL HARRISON  APRIL 86
                     1020 *          LIST 24-1 FOR FIG 24.10 HARDWARE
                     1030 *
                     1040 *-----------EXCEPTIONS--------------------
                     1050 *
                     1060 *   WRITTEN ON 3/26/86
                     1070 *
                     1080             .OR      $0
                     1090             .TA      $4000
00000000- 0000 27FE
00000004- 0000 0400 1100             .HS      000027FE00000400
                     1110
                     1120 *
                     1130             .OR      8            THE FOLLOWING ARE
                     1140             .TA      $4008        EXCEPTION VECTORS
00000008- 0000 00C0 1150             .HS      000000C0   BUS ERROR VECTOR
0000000C- 0000 00CE 1160             .HS      000000CE   ADDRESS ERROR VECTOR
00000010- 0000 00DC 1170             .HS      000000DC   ILLEGAL INSTRUCTION
00000014- 0000 00EA 1180             .HS      000000EA   DIVIDE BY ZERO
00000018- 0000 00F8 1190             .HS      000000F8   CHECK INSTRUCTION
0000001C- 0000 00F8 1200             .HS      000000F8   TRAPV INSTRUCTION
00000020- 0000 0106 1210             .HS      00000106   PRIVILEGE VIOLATION
00000024- 0000 0114 1220             .HS      00000114   TRACE BIT ON
00000028- 0000 00F8 1230             .HS      000000F8   VECTOR #S 10-23
0000002C- 0000 00F8
00000030- 0000 00F8
00000034- 0000 00F8
00000038- 0000 00F8
0000003C- 0000 00F8 1240             .HS      000000F8000000F8000000F8000000F8000000F8
00000040- 0000 00F8
00000044- 0000 00F8
00000048- 0000 00F8
0000004C- 0000 00F8
00000050- 0000 00F8 1250             .HS      000000F8000000F8000000F8000000F8000000F8
00000054- 0000 00F8
00000058- 0000 00F8
0000005C- 0000 00F8 1260             .HS      000000F8000000F8000000F8
00000060- 0000 0122 1270             .HS      00000122   SPURIOUS INTERRUPT
00000064- 0000 0130 1280             .HS      00000130   UNUSED INT LEV 1
00000068- 0000 04A8 1290             .HS      000004A8   A/D CONV IS FINISHED
0000006C- 0000 0130 1300             .HS      00000130   UNUSED INT LEV 3
00000070- 0000 0130 1310             .HS      00000130   UNUSED INT LEV 4
00000074- 0000 047A 1320             .HS      0000047A   WEIGH OR PICK COMB
00000078- 0000 0130 1330             .HS      00000130   UNUSED INT LEV 6
0000007C- 0000 0130 1340             .HS      00000130   UNUSED NMI
00000080- 0000 00F8 1350             .HS      000000F8   UNUSED TRAP INSTR.
00000084- 0000 00F8
00000088- 0000 00F8
0000008C- 0000 00F8
00000090- 0000 00F8 1360             .HS      000000F8000000F8000000F8000000F8
```

```
00000094- 0000 00F8
00000098- 0000 00F8
0000009C- 0000 00F8
000000A0- 0000 00F8 1370        .HS      000000F8000000F8000000F8000000F8
000000A4- 0000 00F8
000000A8- 0000 00F8
000000AC- 0000 00F8
000000B0- 0000 00F8 1380        .HS      000000F8000000F8000000F8000000F8
000000B4- 0000 00F8
000000B8- 0000 00F8
000000BC- 0000 00F8 1390        .HS      000000F8000000F8000000F8
                    1400 *
                    1410        .OR      $C0        EXCEPTION ROUTINES :
                    1420        .TA      $40C0      START HERE:
                    1430 *---------DISPLAY BUS FOR BUS ERROR-------
000000C0- 183C 0092 1440        MOVE.B  #$92,D4
000000C4- 1A3C 00C1 1450        MOVE.B  #$C1,D5
000000C8- 1C3C 0083 1460        MOVE.B  #$83,D6
000000CC- 6070      1470        BRA.S   DISPLY
                    1480 *---------DISPLAY ADD FOR ADDRESS ERROR---
000000CE- 183C 00A1 1490        MOVE.B  #$A1,D4
000000D2- 1A3C 00A1 1500        MOVE.B  #$A1,D5
000000D6- 1C3C 0088 1510        MOVE.B  #$88,D6
000000DA- 6062      1520        BRA.S   DISPLY
                    1530 *---------DISPLAY ILL FOR ILLEGAL INSTR---
000000DC- 183C 00C7 1540        MOVE.B  #$C7,D4
000000E0- 1A3C 00C7 1550        MOVE.B  #$C7,D5
000000E4- 1C3C 00F9 1560        MOVE.B  #$F9,D6
000000E8- 6054      1570        BRA.S   DISPLY
                    1580 *---------DISPLAY DIV FOR ZERO DIVIDE----
000000EA- 183C 00E3 1590        MOVE.B  #$E3,D4
000000EE- 1A3C 00F9 1600        MOVE.B  #$F9,D5
000000F2- 1C3C 00A1 1610        MOVE.B  #$A1,D6
000000F6- 6046      1620        BRA.S   DISPLY
                    1630 *---------DISPLAY OTH FOR CHK,TRAPV,AND
                    1640 *        TRAP INSTRUCTIONS;  ALSO FOR
                    1650 *        VECTOR NUMBERS 10 THRU 23.
000000F8- 183C 008B 1660        MOVE.B  #$8B,D4
000000FC- 1A3C 0087 1670        MOVE.B  #$87,D5
00000100- 1C3C 00C0 1680        MOVE.B  #$C0,D6
00000104- 6038      1690        BRA.S   DISPLY
                    1700 *-----------DISPLAY FOR PRIVILEGE VIOLATION---
00000106- 183C 00E3 1710        MOVE.B  #$E3,D4
0000010A- 1A3C 00AF 1720        MOVE.B  #$AF,D5
0000010E- 1C3C 008C 1730        MOVE.B  #$8C,D6
00000112- 602A      1740        BRA.S   DISPLY
                    1750 *-----------DISPLAY FOR TRACE MODE ON----
00000114- 183C 0088 1760        MOVE.B  #$88,D4
00000118- 1A3C 00AF 1770        MOVE.B  #$AF,D5
0000011C- 1C3C 0087 1780        MOVE.B  #$87,D6
00000120- 601C      1790        BRA.S   DISPLY
                    1800 *----------DISPLAY FOR SPURIOUS INTERRUPTS
00000122- 183C 00C1 1810        MOVE.B  #$C1,D4
00000126- 1A3C 008C 1820        MOVE.B  #$8C,D5
0000012A- 1C3C 0092 1830        MOVE.B  #$92,D6
0000012E- 600E      1840        BRA.S DISPLY
                    1850 *---------DISPLAY INT FOR UNUSED INTERRUPTS
00000130- 183C 0087 1860        MOVE.B  #$87,D4
00000134- 1A3C 00AB 1870        MOVE.B  #$AB,D5
00000138- 1C3C 00F9 1880        MOVE.B  #$F9,D6
0000013C- 6000      1890        BRA.S   DISPLY
                    1900 *
0000013E- 11FC 00FE
00000142- 6000      1910 DISPLY MOVE.B  #$FE,$6000     TURN ON LOW SEG
00000144- 11C4 4000 1920        MOVE.B  D4,$4000       OUTPUT LOW LETTER
```

```
00000148- 6100 0020 1930          BSR      DELAY
0000014C- 11FC 00FD
00000150- 6000      1940          MOVE.B   #$FD,$6000      TURN ON MIDDLE SEG
00000152- 11C5 4000 1950          MOVE.B   D5,$4000        OUTPUT MIDDLE LETTER
00000156- 6100 0012 1960          BSR      DELAY
0000015A- 11FC 00FB
0000015E- 6000      1970          MOVE.B   #$FB,$6000      TURN ON HI SEG
00000160- 31C6 4000 1980          MOVE     D6,$4000        OUTPUT HI LETTER
00000164- 6100 0004 1990          BSR      DELAY
00000168- 60D4      2000          BRA.S    DISPLY
                    2010 *
0000016A- 3E3C 029B 2020 DELAY    MOVE     #$29B,D7        SET UP CNTR TO
0000016E- 57CF FFFE 2030 LOOP     DBEQ     D7,LOOP         DELAY FOR 3 MS
00000172- 4E75      2040          RTS
                    2050 *
                    2060 *----------THIS SECTION BUILDS DISPLAY-------------
                    2070 *                  CODES IN ROM
                    2080 *
                    2090          .OR      $700
                    2100          .TA      $4700
                    2110 *
00000700- C0F9 A4B0
00000704- 9992 82F8
00000708- 8090      2120          .HS      C0F9A4B0999282F88090
                    2130 *
                    2140 *-----------THIS SECTION LOADS BCD LOOKUP TABLE----
                    2150 *                     04-17-86
                    2160 *
                    2170          .OR      $0400
                    2180          .TA      $4400
                    2190 *
00002000-           2200 BOTAB    .EQ      $2000           BASE ADDRESS OF TABLE
00002220-           2210 TBLENT   .EQ      $2220           TABLE ENTRY GENERATION
                    2220 *
                    2230 *----------------REGISTER USES---------------------
                    2240 *
                    2250 *   D5--BYTE CNTR FOR BCD ADDITION
                    2260 *   A1--POINTS TO TABLE ADDRESSES
                    2270 *   A2--USED FOR SUM IN BCD ADDITION
                    2280 *   A3--HOLDS INCREMENT FOR TABLE ENTRIES
                    2290 *
00000400- 227C 0000
00000404- 2000      2300 BLDTAB MOVE.L #$2000,A1        START OF TABLE
00000406- 4278 2220 2310          CLR      TBLENT          CLEAR TBL ENT ACCUM
0000040A- 31FC 0002
0000040E- 2222      2320          MOVE     #2,$2222        WEIGHT INCREMENTS
00000410- 4240      2330          CLR      D0              FIRST WT = ZERO
00000412- 32C0      2340          MOVE     D0,(A1)+        SAVE ZERO ENTRY
00000414- 247C 0000
00000418- 2222      2350 TABLUP MOVE.L #$2222,A2        PNT PAST WT ACCUM
0000041A- 267C 0000
0000041E- 2224      2360          MOVE.L   #$2224,A3       PNT PAST INCREMENT
00000420- 3A3C 0001 2370          MOVE     #1,D5           BYTE COUNTER
00000424- 44FC 0000 2380          MOVE     #0,CCR          CLEAR EXTEND FLAG
00000428- C50B      2390 ADDLP1   ABCD     -(A3),-(A2)     ACCUM WTS IN $2620-1
0000042A- 51CD FFFC 2400          DBRA     D5,ADDLP1       ADD COMPLETE?
0000042E- 32F8 2220 2410          MOVE     TBLENT,(A1)+    SAVE BCD WT IN TABLE
00000432- B2FC 2200 2420          CMPA     #$2200,A1       TABLE COMPLETE?
00000436- 66DC      2430          BNE      TABLUP          NO;ADD AN ENTRY
                    2440 *
                    2450 *------THIS SECTION FILLS EMPTY HANGARS---------
                    2460 *        AND DISPLAYS THE WEIGHT IN BCD
                    2470 *
00000700-           2480 BOCODE   .EQ      $700            BASE ADDRESS OF CODES
00002200-           2490 BOFILE   .EQ      $2200           BASE ADDRESS OF WT FILE
```

```
000000FF-                 2500 DEVOFF .EQ     $FF           DISABLE ALL DEVICES
000000FE-                 2510 SELHUN .EQ     $FE           ENABLE HUNDRTHS SEGMENTS
000000FD-                 2520 SELTEN .EQ     $FD           ENABLE TENTHS SEGMENTS
000000FB-                 2530 SELUNT .EQ     $FB           ENABLE UNITS SEGMENTS
000000F7-                 2540 LOCMBS .EQ     $F7           ENABLE LO BYTE OF CMBNTN
000000EF-                 2550 HICMBS .EQ     $EF           ENABLE HI BYTE OF CMBNTN

000000DF-                 2560 INLV2S .EQ     $DF           ENABLE LVL 2 INTRPT GATE
00006000-                 2570 DEVSEL .EQ     $6000         DEVICE SELECTOR
00004000-                 2580 DTAOUT .EQ     $4000         DATA OUT TO DEVICES
0000A000-                 2590 DTAIN  .EQ     $A000         READ HEX BYTE FROM ADC
0000A000-                 2600 ENADC  .EQ     $A000         START A/D CONVERSION
00002224-                 2610 HUNCOD .EQ     $2224         STORE HUNDREDTHS DIGIT
00002225-                 2620 LOCODE .EQ     $2225         STORE LO BYTE OF CMBNTN
00002226-                 2630 TENCOD .EQ     $2226         STORE TENTHS DIGIT CODE
00002227-                 2640 HICODE .EQ     $2227         STORE HI BYTE OF CMBNTN
00002228-                 2650 UNTCOD .EQ     $2228         STORE UNITS DIGIT CODE
00002229-                 2660 FILLST .EQ     $2229         FF = FILL;0 = FULL
00002230-                 2670 DLYCTR .EQ     $2230         32 BIT COUNTER
00000100-                 2680 MSDLY  .EQ     $100          1 MS DLY CNT VALUE
0000FFFF-                 2690 HLFSEC .EQ     $FFFF         .5 SEC DLY VALUE
                          2700 *
                          2710 *
                          2720 *---------------REGISTER USES-------------------
                          2730 *
                          2740 *   D3--HOLDS LAST COMBINATION CHOSEN OR ALL EMPTY
                          2750 *   D4--HOLD DISPLAY CODE OFFSET; TEMP STOR UNTCOD
                          2760 *   D5--HOLDS WORKING COPY OF BCD WEIGHT
                          2770 *   D6--ACCEPTS WEIGHT FROM ADC
                          2780 *   D6--HOLDS INDEX = WEIGHT TIMES 2
                          2790 *   D6--HOLDS BCD WEIGHT FROM TABLE
                          2800 *   A4--ACCEPTS INDEX TO BCD WEIGHT DESIRED
                          2810 *   A1--POINTS TO WEIGHT TABLE BASE
                          2820 *   A0--POINTS TO WEIGHT FILE ADDRESS (16 HANGARS)
                          2830 *   A5--POINTS TO SEGMENT CODE FILE BASE ADDRESS
                          2840 *   A6--HOLD SEGMENT CODE FILE OFFSET
                          2850 *
00000438- 13FC 00FF
0000043C- 0000 A000 2860             MOVE.B  #$FF,ENADC     INIT ADC WITH WR
00000440- 31FC FFFF
00000444- 2224      2870             MOVE    #$FFFF,HUNCOD  SET DISPLAYS TO
00000446- 31FC FFFF
0000044A- 2226      2880             MOVE    #$FFFF,TENCOD  BLANKS AND FILL
0000044C- 31FC FFFF
00000450- 2228      2890             MOVE    #$FFFF,UNTCOD  STATUS TO FILLING
00000452- 363C FFFF 2900             MOVE    #$FFFF,D3      ALL HANGARS EMPTY
00000456- 227C 0000
0000045A- 2000      2910 FILLUP      MOVE.L  #BOTAB,A1      SET TABLE BASE ADD
0000045C- 207C 0000
00000460- 21FE      2920             MOVE.L  #BOFILE-2,A0   POINT TO WT FILE -2
00000462- 2A7C 0000
00000466- 0700      2930             MOVE.L  #BOCODE,A5     SET CODES BASE ADD
00000468- 9DCE      2940             SUB.L   A6,A6          CLEAR OFFSET REG
0000046A- 027C FCFF 2950             ANDI    #$FCFF,SR      ACCCPT LVL5 INTRPT
0000046E- 4EB8 052A 2960 WTLVL5      JSR     DSPLAY         DISPLAY CURRENT WT
00000472- 60FA      2970             BRA.S   WTLVL5         UNTIL HANGAR FILLS
00000474- 4A38 2229 2980 INTRP5      TST.B   FILLST         HANGRS FULL YET?
00000478- 6700 0062 2990             BEQ     PICK
0000047C- 5448      3000 NXTHGR      ADDQ    #2,A0          PNT HGR WT IN FILE
0000047E- E24B      3010             LSR     #1,D3          HANGAR EMPTY?
00000480- 64FA      3020             BCC.S   NXTHGR         YES; TRY NEXT ONE
00000482- 11FC 00FF
00000486- 6000      3030             MOVE.B  #DEVOFF,DEVSEL TURN OFF LEDS
00000488- 21FC 0000
0000048C- FFFF 2230 3040             MOVE.L  #HLFSEC,DLYCTR
```

```
00000490- 4EB8 059A 3050          JSR     DLAY            SWITCH UP YET?
00000494- 13FC 00FF
00000498- 0000 A000 3060          MOVE.B  #$FF,ENADC      START A/D CONV
0000049C- 027C F8FF 3070          ANDI    #$F8FF,SR       ACCPT ANY INT LVL
000004A0- 11FC 00DF
000004A4- 6000      3080          MOVE.B  #INLV2S,DEVSEL ENABLE INT GATE
000004A6- 60FE      3090 WTLVL2 BRA.S   WTLVL2          ADC COMPLETE?
000004A8- 1C39 0000
000004AC- A000      3100 INTRP2 MOVE.B  DTAIN,D6        YES, LOAD WEIGHT
000004AE- E34E      3110          LSL     #1,D6           X 2 FOR TBL ENTRY
000004B0- 3846      3120          MOVEA   D6,A4           INDEX FOR TBL ENTRY
000004B2- 227C 0000
000004B6- 2000      3130          MOVE.L  #BOTAB,A1       SET TABLE BASE ADD
000004B8- 3C34 9000 3140          MOVE    $0(A4,A1),D6    GET BCD WT
000004BC- 3086      3150          MOVE    D6,(A0)         STORE WT IN FILE
000004BE- 6100 0020 3160          BSR     DECODE
000004C2- 11FC 00FF
000004C6- 6000      3170          MOVE.B  #DEVOFF,DEVSEL DISABLE INTLVL2
000004C8- 027C FCFF 3180          ANDI    #$FCFF,SR        ACCPT LVL5 INTRP
000004CC- 4A43      3190          TST     D3              ALL HANGARS FULL?
000004CE- 66AC      3200          BNE     NXTHGR          NO; GO FILL THEM
000004D0- 11FC 0000
000004D4- 2229      3210          MOVE.B  #0,FILLST       STATUS = FULL
000004D6- 4EB8 052A 3220 WAIT5  JSR     DSPLAY          DISPLAY LAST WT
000004DA- 60FA      3230          BRA.S   WAIT5
000004DC- 6000 00C4 3240 PICK   BRA     PHASE2
                    3250 *
                    3260 *------------THIS ROUTINE CONVERTS THE BCD-------
                    3270 *---------------WEIGHT TO DISPLAY CODES----------
                    3280 *
000004E0- 4284      3290 DECODE CLR.L   D4
000004E2- 3A06      3300          MOVE    D6,D5           MAKE A WORK COPY
000004E4- 0245 0F0F 3310          ANDI    #$0F0F,D5       GET HUNDREDTHS
000004E8- 1805      3320          MOVE.B  D5,D4
000004EA- 3C44      3330          MOVEA   D4,A6           AND STORE FOR
000004EC- 11F5 E000
000004F0- 2224      3340          MOVE.B  $0(A5,A6),HUNCOD   DISPLAY
000004F2- E04D      3350          LSR     #8,D5
000004F4- 1805      3360          MOVE.B  D5,D4
000004F6- 3C44      3370          MOVEA   D4,A6           GET AND SAVE
000004F8- 11F5 E000
000004FC- 2228      3380          MOVE.B  $0(A5,A6),UNTCOD  UNITS
000004FE- 3A06      3390          MOVE    D6,D5
00000500- 0245 F0F0 3400          ANDI    #$F0F0,D5       NOW DO THE
00000504- E84D      3410          LSR     #4,D5           OTHER NIBBLES
00000506- 1805      3420          MOVE.B  D5,D4
00000508- 3C44      3430          MOVEA   D4,A6           GET AND SAVE
0000050A- 1836 D000 3440          MOVE.B  $0(A6,A5),D4    TENTHS DIGIT
0000050E- 0204 007F 3450          ANDI.B  #$7F,D4         ADD DECIMAL PT
00000512- 11C4 2226 3460          MOVE.B  D4,TENCOD
00000516- E84D      3470          LSR     #4,D5
00000518- 4A05      3480          TST.B   D5              IF WT > 10
0000051A- 6700 0008 3490          BEQ     NOTENS          DISPLAY DP
0000051E- 0238 007F
00000522- 2228      3500          ANDI.B  #$7F,UNTCOD     ON UNITS DIGIT
00000524- 99CC      3510 NOTENS SUB.L   A4,A4           CLEAR INDEX
00000526- 4246      3520          CLR     D6              AND OFFSET
00000528- 4E75      3530          RTS
                    3540 *
                    3550 *----------THIS ROUTINE DISPLAYS BCD----------
                    3560 *----------WEIGHTS AND COMBINATIONS-----------
                    3570 *
0000052A- 21FC 0000
0000052E- 0100 2230 3580 DSPLAY MOVE.L  #MSDLY,DLYCTR
00000532- 11F8 2224
```

```
00000536- 4000         3590         MOVE.B  HUNCOD,DTAOUT  DISPLAY HNDRDTHS
00000538- 11FC 00FE
0000053C- 6000         3600         MOVE.B  #SELHUN,DEVSEL FOR 1 MS
0000053E- 615A         3610         BSR.S   DLAY
00000540- 21FC 0000
00000544- 0100 2230 3620            MOVE.L  #MSDLY,DLYCTR
00000548- 11F8 2225
0000054C- 4000         3630         MOVE.B  LOCODE,DTAOUT  DISPLAY LO-HALF
0000054E- 11FC 00F7
00000552- 6000         3640         MOVE.B  #LOCMBS,DEVSEL OF CMB FOR 1MS
00000554- 6144         3650         BSR.S   DLAY
00000556- 21FC 0000
0000055A- 0100 2230 3660            MOVE.L  #MSDLY,DLYCTR
0000055E- 11F8 2226
00000562- 4000         3670         MOVE.B  TENCOD,DTAOUT  DISPLAY TENTHS
00000564- 11FC 00FD
00000568- 6000         3680         MOVE.B  #SELTEN,DEVSEL FOR 1MS
0000056A- 612E         3690         BSR.S   DLAY
0000056C- 21FC 0000
00000570- 0100 2230 3700            MOVE.L  #MSDLY,DLYCTR
00000574- 11F8 2227
00000578- 4000         3710         MOVE.B  HICODE,DTAOUT  DISPLAY HI-HALF
0000057A- 11FC 00EF
0000057E- 6000         3720         MOVE.B  #HICMBS,DEVSEL FOR 1 MS
00000580- 6118         3730         BSR.S   DLAY
00000582- 21FC 0000
00000586- 0100 2230 3740            MOVE.L  #MSDLY,DLYCTR
0000058A- 11F8 2228
0000058E- 4000         3750         MOVE.B  UNTCOD,DTAOUT  DISPLAY UNITS
00000590- 11FC 00FB
00000594- 6000         3760         MOVE.B  #SELUNT,DEVSEL FOR 1MS
00000596- 6102         3770         BSR.S   DLAY
00000598- 4E75         3780         RTS
                       3790 *
0000059A- 53B8 2230 3800 DLAY       SUBQ.L  #1,DLYCTR
0000059E- 66FA         3810         BNE     DLAY
000005A0- 4E75         3820         RTS

                       3840 *
                       3850 *-----------THIS PHASE PICKS THE COMBINATION--------
                       3860 *                 AND DISPLAYS IT
                       3870 *
                       3880 *
00002400-              3890 TARGET .EQ      $2400          BOXES WEIGHT 24.00 LBS
00002220-              3900 TOTWT  .EQ      $2220          ACCCUMULATOR FOR WEIGHT
00002222-              3910 BCDWT  .EQ      $2222          WT OF CURRENT CHICKEN
00002236-              3920 CMBCTR .EQ      $2236          KEEP CURR CMB HERE
00002238-              3930 BSTWT  .EQ      $2238          KEEP BEST WT SO FAR
0000223A-              3940 BSTCMB .EQ      $223A          KEEP BEST CMB HERE
                       3950 *
                       3960 *
                       3970 * D0--COMBINATION SHIFTER    A0--WEIGHT FILE POINTER
                       3980 * D1--WORKING COPY OF BEST COMBINATION
                       3990 * D6--WORKING COPY OF BEST WEIGHT TO DECODE FOR DISPLAY
                       4000 * D5--BYTE COUNT;  USED IN DECODING
                       4010 *
                       4020 *---------MODULE 0--INIT REGISTERS--------------------
                       4030 *
000005A2- 4278 2220 4040 PHASE2 CLR     TOTWT            START WITH WT = 0
000005A6- 31FC 9999
000005AA- 2238         4050         MOVE    #$9999,BSTWT     START WITH WORST WEIGHT
000005AC- 31FC 000F
000005B0- 2236         4060         MOVE    #15,CMBCTR     START WITH FOUR CHICKENS
000005B2- 203C 0000
000005B6- 000F         4070         MOVE.L  #15,D0           MAKE A COPY TO SHIFT
```

```
000005B8- 207C 0000
000005BC- 21FE      4080        MOVE.L   #BOFILE-2,A0        START WT. FILE - 2
                    4090 *
                    4100 *------MODULE 1----ADD UP WEIGHTS FOR EACH COMBINATION
                    4110 *
000005BE- 5448      4120 ADDCKN ADDQ     #2,A0             POINT TO NEXT WEIGHT
000005C0- E248      4130        LSR      #1,D0             IF HANGAR HOLDS A CHICKEN
000005C2- 640C      4140        BCC.S    SKIP              ADD THE WEIGHT TO TOTAL.
000005C4- 6100 0076 4150        BSR      BCDADD
000005C8- 0C78 2432
000005CC- 2220      4160        CMP.W    #TARGET+50,TOTWT IF TOTAL > .5 LB OVER
000005CE- 642A      4170        BCC.S    TRYAGN            THE IDEAL FORGET IT.
000005D0- 4A40      4180 SKIP   TST      D0                ANY MORE CHICKS TO ADD IN
000005D2- 66EA      4190        BNE      ADDCKN            IF SO THEN DO IT.
000005D4- 0C78 2400
000005D8- 2220      4200        CMP      #TARGET,TOTWT     IF TOTAL WT IS < TARGET
000005DA- 651E      4210        BCS.S    TRYAGN            TRY THE NEXT COMBINATION.
000005DC- 3238 2220 4220        MOVE     TOTWT,D1
000005E0- B278 2238 4230        CMP      BSTWT,D1          IF TOTWT IS LESS THAN
000005E4- 6414      4240        BCC.S    TRYAGN            BEST SO FAR THEN SAVE IT
000005E6- 31F8 2220
000005EA- 2238      4250        MOVE     TOTWT,BSTWT
000005EC- 31F8 2236
000005F0- 223A      4260        MOVE     CMBCTR,BSTCMB     AND THE COMBINATION
000005F2- 0C78 2400
000005F6- 2220      4270        CMP      #TARGET,TOTWT     IF TOTAL WEIGHT = TARGET
000005F8- 6716      4280        BEQ.S    DISPLA            DISPLAY TARGET WEIGHT
000005FA- 207C 0000
000005FE- 21FE      4290 TRYAGN MOVE.L   #BOFILE-2,A0      PNT 2 BEFORE FILE START
00000600- 5278 2236 4300        ADDQ     #1,CMBCTR         TRY NEXT COMBINATION
00000604- 670A      4310        BEQ.S    DISPLA            IF ALL COMBINATIONS HAVE
                    4320 *                                 BEEN TRIED; DISPLAY BEST
00000606- 3038 2236 4330        MOVE     CMBCTR,D0         MAKE A WORKING COPY
0000060A- 4278 2220 4340        CLR      TOTWT             RESET TOTAL WEIGHT.
0000060E- 60AE      4350        BRA.S    ADDCKN            TRY NEXT COMBINATION.
                    4360 *
                    4370 *
                    4380 *--------MODULE 2----------DISPLAY THE COMBINATION---
                    4390 *
00000610- 3238 223A 4400 DISPLA MOVE     BSTCMB,D1
00000614- 4641      4410        NOT      D1                LEDS REQUIRE A LOW
00000616- 11C1 2225 4420        MOVE.B   D1,LOCODE         SAVE LO HALF OF COMB
0000061A- E049      4430        LSR      #8,D1             MOVE HI BYTE DOWN
0000061C- 11C1 2227 4440        MOVE.B   D1,HICODE         SAVE HI HALF OF COMB
00000620- 3638 223A 4450        MOVE     BSTCMB,D3         SAVE BSTCMB FOR FILLING
00000624- 3C38 2238 4460        MOVE     BSTWT,D6          COPY TO DECODE
00000628- 11FC 00FF
0000062C- 2229      4470        MOVE.B   #$FF,FILLST       CHANGE STATUS TO FILL
0000062E- 027C FCFF 4480        ANDI     #$FCFF,SR         ACCPT LVL5 INTRPT
00000632- 6100 FEAC 4490        BSR      DECODE            DECODE WT FOR DISPLAY
00000636- 4EB8 052A 4500 WEIGHT JSR      DSPLAY            DISPLAY UNTIL INTRPT 5
0000063A- 60FA      4510        BRA.S    WEIGHT
                    4520 *
0000063C- 227C 0000
00000640- 2222      4530 BCDADD MOVE.L   #TOTWT+2,A1       1 BYT PAST TOTWT
00000642- 247C 0000
00000646- 2222      4540        MOVE.L   #BCDWT,A2         SET UP A2 FOR ADD
00000648- 34D0      4550        MOVE.W   (A0),(A2)+        PUT WT IN BCDNUM
                    4560 *                                 AND POINT PAST IT
0000064A- 3A3C 0001 4570        MOVE.W   #1,D5             ADD TWO BYTES
0000064E- 44FC 0000 4580        MOVE     #0,CCR            CLEAR EXTEND FLAG
00000652- C30A      4590 ADDLUP ABCD     -(A2),-(A1)       ADD CURR BCDWT
00000654- 51CD FFFC 4600        DBRA     D5,ADDLUP         TO TOTAL WEIGHT
00000658- 4E75      4610        RTS
                    4620 *
```

24.6 A PEEK AT THE INTEL 8086 AND 8088

Section Overview

The 8086 is a 16-bit microprocessor that has gained immense popularity, in part at least because its 8-bit-bus version (the 8088) is used in the IBM personal computer. Both processors have four 16-bit general-purpose registers and four 16-bit address-pointer registers, plus a program counter and a flag register. Both come in the standard 40-pin DIP and multiplex the data and address pins to fit all the necessary signals onto 40 pins.

The addressing modes, instruction format, and instruction set are similar (but by no means identical) to those of the Z80. An important concept, unique to the 8086/8088, is *address segmenting*. Although 1 megabyte of address space is available, only four 64-kilobyte segments are accessible at any given time—one segment each for *program code*, *operand data*, and the *stack*, plus an *extra segment* which generally is used for destination data.

The programmer's model for the 8086/8088 is shown in Figure 24.12. The general-purpose registers can be addressed as 16-bit words by specifying AX, BX, CX, or DX (think *extended* data size) or as byte-size units by specifying AH, AL, and so forth. The names given to each register imply uses in certain special instructions, but most instructions can use any of the general-purpose registers for data manipulation and any of the address pointers to specify operand data. For example, some instructions use BX as a pointer to the *base* of a file, but AX, BX, CX, or DX can be used to hold the operands in arithmetic or logic operations. As another example, some instructions use SI and DI to point to the source and destination words in a file-transfer operation, but SP, BP, SI, or DI may be used to point to operand data generally. A few instructions, such as INPUT and OUTPUT, are available for the primary accumulator, A, only.

Segmented addressing: The address pointers are only 16 bits wide, so the range is limited to 64 kilobytes for each one. The 8086 accesses a total address space of 1 megabyte (16 times 64 kilobytes) by means of four *segment-pointer* registers, which hold the base address, bits 19 through 4. Bits 3 through 0 of the base address are assumed to be 0000. The *physical address* output by the chip is obtained by adding the *logical address* specified by the program (bits 15 through 0) to the segment base (bits 19 through 4). There are four segment pointers, one each to specify the base address for

- *The program counter*, called *Instruction Pointer* (IP) in the 8086,
- *The operand data*: source and data operands in memory automatically use the *Data Segment* (DS) base,
- *The stack*: the *Stack Segment* (SS) is automatically used by PUSH and POP instructions and by subroutine calls and interrupts, and
- *The extra segment* (ES), usually accessed by the destination index register, DI.

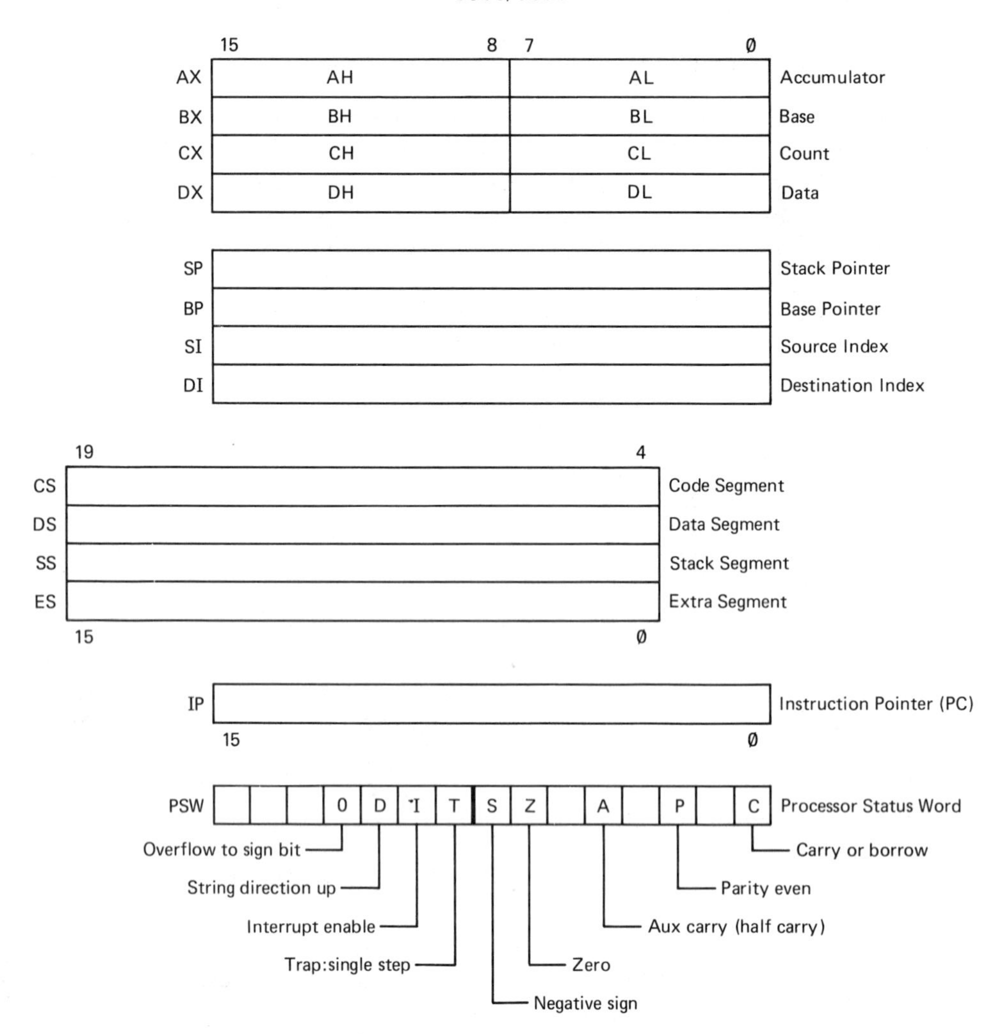

Figure 24.12 A programmer's model for the 8086 and 8088 sixteen-bit microprocessors.

Figure 24.13 illustrates how a logical address and a segment base address are combined to form a 20-bit physical address.

The address space of the 8086 is numbered by bytes, as it is for the 68000. Op-codes and data may be 1 or 2 bytes long. Two-byte words are stored in *low-byte/high-byte* order. Address line AØ appears on the 8086 pinout but is not wired directly to memory address pins in an 8086 system. Instead, it is decoded with the $\overline{\text{BHE}}$ (byte high enable) output to select one of four address-access modes, as follows:

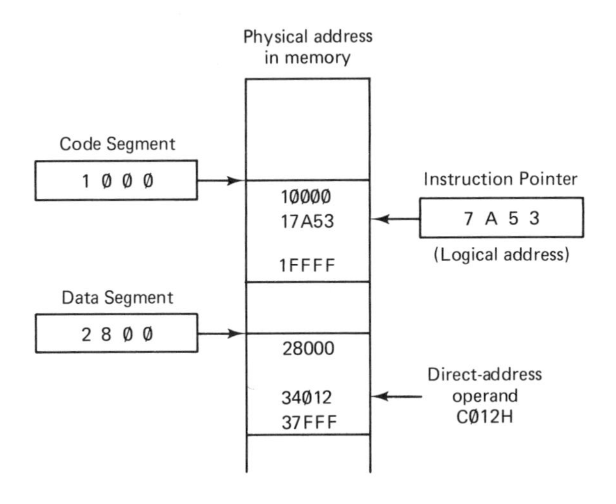

Memory reference	Segment used	Pointer used
Instruction read	CS only	IP
Operand data	DS or option[1]	Effective address[2]
String source	DS or option[1]	SI
String destination	ES only	DI
BP (base page) ops	SS or option[1]	Effective address
Stack read/write	SS only	SP

Notes:

1. One of the other segments may be specified.
2. The logical address added to the segment base may be specified by the instruction directly or relatively, depending on the instruction mode.

Figure 24.13 The 1-megabyte address space of the 8086/8088 is broken into 16 segments of 64 kilobytes each. Only four segments can be accessed without reloading the segment pointers.

$\overline{BHE}$	AØ	Data Accessed
low	*low*	Two bytes to D15–DØ
low	*high*	Odd-numbered byte to D15–8
high	*low*	Even-numbered byte to D7–DØ
high	*high*	No data accessed

The package pinout for the 8086 is shown in Figure 24.14. The pinout for the 8088 is similar, except that data is not multiplexed on address lines A8 through A15:

- Pin 33, MN/$\overline{MX}$, places the 8086/8088 in a *minimum* or maximum system configuration. The *maximum* mode redefines the functions of pins 24–31 to interface with a multiprocessor system. We will discuss only the *minimum-mode* pin functions.

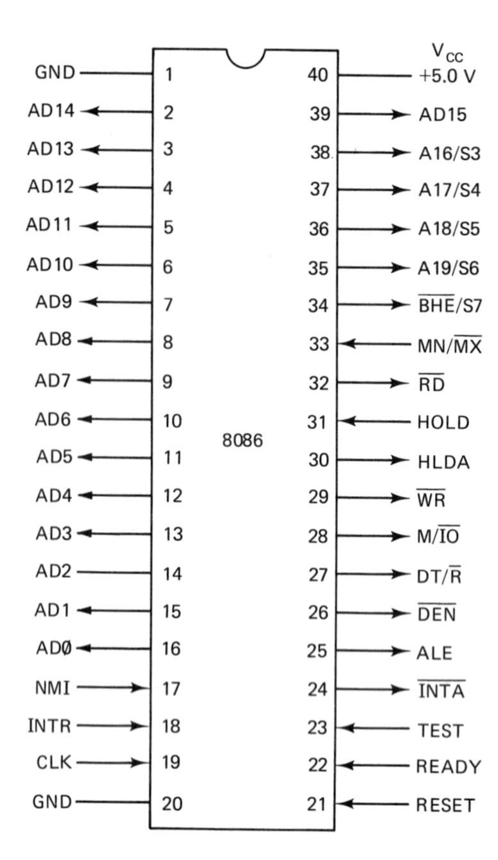

Figure 24.14 Pinout for the 8086 microprocessor.

- $\overline{\text{RD}}$ and $\overline{\text{WR}}$ are asserted during a processor *read* or *write*, respectively, of the data bus. DT/$\overline{\text{R}}$ is the *data transmit or receive* output, indicating on one line the direction of data (write/read) during the current bus cycle. $\overline{\text{DEN}}$ (data enable) signals to external devices the precise timing when they should assert data on the bus. M/$\overline{\text{IO}}$ selects access of *memory* (most instructions) or I/O *devices* (input and output instructions only).
- Address signals are asserted on A/DØ through A/D15 during the first *T*-state of each bus cycle. The ALE output is asserted to command an external address latch to hold these signals for the remainder of the bus cycle. Data is asserted on A/DØ through A/D15 for the remainder of the bus cycle, which is typically three more *T*-states.
- The high-order nibble of the physical address is asserted on pins A16 through A19 during the first *T*-state of each bus cycle. These pins contain processor status information on the remaining *T*-states.
- INTR is the interrupt-register input, acknowledged by the $\overline{\text{INTA}}$ output. NMI is the edge-sensitive nonmaskable interrupt input. $\overline{\text{TEST}}$ is an input that is asserted to cause the 8086 to resume processing after it has been placed in an idle state by execution of a WAIT instruction.

- READY can be pulled *low* by slow memories to insert *wait T*-states into a bus cycle. HOLD and HOLDA (hold acknowledge) are used by DMA devices to suspend processor operation and turn the buses over to another device.
- RESET is pulled *high* and then returned *low* to reset the processor. The *data*, *stack*, and *extra* segment pointers are set to $0000, the *code*-segment pointer is set to $FFFF, and the instruction pointer is set to $0000. An instruction is thus read from address $F FFF0, and processing continues from there.
- The clock input (typically 8 MHz) and the RESET signal are generated by the 8284 support chip.

Assembly language form for the 8086 is similar to that for the Z80. The basic form is

INSTRUCTION Destination, Source

Note that this is the opposite of the 68000 order. Data is assumed to be immediate without any # sign. Register names are enclosed in parentheses or brackets when the reference is to the data pointed to by them. The basic data-transfer instruction is MOV, similar to LD in the Z80 and MOVE in the 68000.

Addressing modes for the destination and source data may generally be chosen independently. Figure 24.15 enumerates the primary mode available and gives an example of each.

The instruction set for the 8086/8088 contains few surprises if you are already familiar with the Z80 instruction set. Here are some samples of common instructions.

- MOV AX,(BP) — Move word pointed to by BP within data segment to register A.
- MOV BL,ES:(SI) — Move byte pointed to by SI within extra segment to low-order half of B.
- XCHG CX,DI — Exchange data between registers.
- ADD AX,16 — Add decimal 16 to word data in AX.
- SUBB DX,LOSS — Subtract (with borrow input) data at address labeled LOSS from data in register DX.
- MUL CX — Multiply, unsigned, 16 bits in CX times 16 bits in AX (implied). Result in AX (low-order) and DX (high-order).
- IDIV CL — Divide, signed, 16 bits in AX by 8 bits in CL. Result in AL; remainder in AH.
- SAL DX,2 — Shift, arithmetic, left: register DX, 2 places.
- RCR AL,4 — Rotate right through carry, A low, 4 places.
- JC LOOP — Jump relative if carry set to address LOOP.
- CALL CONVR — Jump to subroutine labeled CONVR.

Note that immediate-mode operands may require a leading zero to avoid confusion with register names. For example

MOV BH,AH

means move contents of A, high-byte, to B, high-byte, whereas

MOV BH,0AH

means move hex data A (decimal 10) to B, high byte.

8086 Addressing-mode Examples

Mode	Data transfer	Source code
Register	AX ← BX	MOV AX, BX
Immediate	DH ← $3F	MOV DH, 3FH
Direct	CX ← Memory Operand; Physical address = data segment + value of DATA7	MOV CX, DATA7
Register indirect	Memory Operand ← DX; Address pointed to by register BP	MOV [BP], DX
Based with displacement	Memory; Address pointed to by register BX; 5-byte offset; AL	MOV [BX].5, AL
Based with index and displacement	Adr of BX; Index of 7 contained in source index SI; Offset of 2; AX ← Operand	MOV AX, [BX].2[SI]
Relative	Jump instructions only	JP Z, ROUT

Figure 24.15 A summary of the 8086 addressing modes.

REVIEW OF SECTIONS 24.5 and 24.6

21. Analyze Figure 24.10 to determine the location of the interrupt vector accessed by the TAKE button.
22. In Figure 24.10, is the reset switch shown in the *reset* position or the *run* position?
23. In Figure 24.10, calculate the anode current made available to each 7-segment display if the latch pulls the 2.2-kΩ base resistor down to +0.4 V, V_{BE} of the transistor is 0.7 V, and $h_{FE} = 50$.
24. If a chicken pickin' problem required 10 instructions per addition check, with the instructions averaging 10 clock cycles each on a 4-MHz clock, and the number of addition checks per combination is 16 (not much time is saved in skipping the additions), how long should it take to try all 2^{16} possible combinations?
25. How many bits are there in 8086 registers BX, BL, and BP, respectively?
26. What address is accessed by MOV (BP),0 if BP contains A987H and the data-segment register contains 4703H?
27. Are the functions of the four general-purpose registers identical?
28. What address does the 8086 go to upon RESET? Does it find there a vector address or an instruction?
29. What is the addressing mode of the source in the instruction

```
MOV (BX).4,CL
```

30. If the result of an 8086 MUL exceeds 16 bits, where is the high-order part of that result found?

24.7 A PARTING SHOT

Section Overview

This section fulfills the promise made in Section 0.2 that we would make some attempt to assess the future potentials of computers. We will divide the prophecies quite clearly into two areas:

1. **Short-term growth-area predictions, based on present economic and hardware realities and useful, perhaps, in career planning.**
2. **Long-term potentials, based on philosophical musings and useful only for café talk among friends.**

After reading this section, go to your local library and dig out some science fiction from the 1947 era. It is easy to find predictions of pocket ray guns, weekend trips to Mars, and personal flying go-carts. What is almost impossible to find is a prediction of the computer revolution that has actually occurred. The experience should help put this and other predictions in their proper perspective.

Computer growth areas: Here is a list of areas that are growing and can be expected to continue to grow. The news is not good for those of us who like the smell of rosin smoke:

1. Software generation requires more time than hardware design. A hardware problem, once solved, results in a standard board, which can be used to solve any similar hardware problem. The software, at least in part, must be written especially for each new application. Many hardware people find themselves pressured into learning and doing software. Not too many software people feel a pressure to get into hardware.
2. High-level languages, long favored in mainframe computing, will probably become increasingly important in microcomputing. High-level languages are less efficient with memory space and processing power than assembly languages, but memory chips and advanced processors are cheap. Software generation is expensive. If you don't know BASIC or FORTRAN or Pascal or "C", perhaps you had better start learning.
3. Sixteen- and 32-bit processors will probably take most of the business away from 8-bit processors. Already there is almost no cost advantage to the 8-bit machines: a 6809 costs about $5.00; a 68000 costs about $10.00 The difference melts away in the cost of the cabinet and chrome-plated knobs. Why, then, would a designer choose an 8-bit processor, which may become overtaxed as the system grows, when a 16-bit processor leaves a much more comfortable performance margin to grow into? The answer is familiarity—8-bit processors are easier to learn and easier to debug than 16-bit processors. We might estimate that the first three years of a designer's career will be spent building 8-bit systems and learning 16-bit systems. That leaves the next 30 years for building 16-bit systems.
4. Although standardized boards will probably take away most of the microcomputer design opportunities, there will be a need for someone who can connect these boards to the limit switches, thermistors, solenoids, and motor-drive controls of the real world.
5. Marketing and technical consulting will demand large numbers of trained individuals. Many persons who must purchase and use computers don't know parallel from serial, and will rely on a knowledgeable consultant to set up their systems from standard components.

The long-term potential of computing has been the subject of much reckless speculation. Perhaps we should start by laying some ground rules.

1. Linear extrapolations of current trends are not to be trusted, even if the trends have continued for decades. Figure 24.16 shows the speed of personal ground transportation in the twentieth century as an example. Standard traffic speeds of 85 mi/h in the 1980s were predicted in the 1950s, and the interstate highway system was designed with this in mind.
2. The technique of science, which has brought so much material success, has been *not* to accept an assertion unless it is supported by fact. There was no

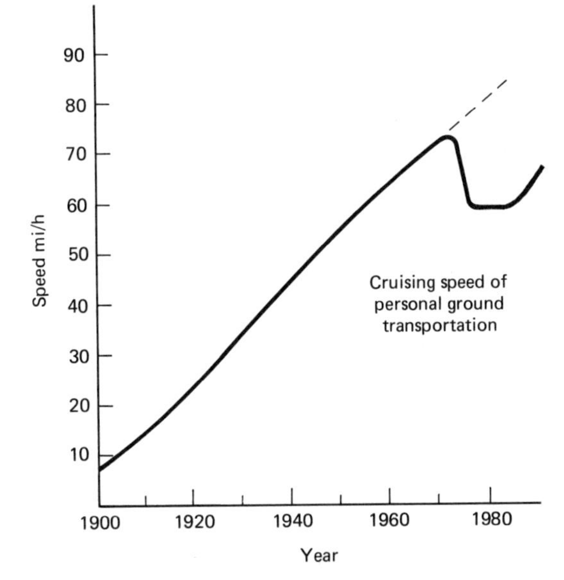

Figure 24.16 Extrapolation of long-term trends often leads to erroneous conclusions.

particular reason for Ponce De León to believe that there was a fountain of youth in Florida, and no particular reason for him to believe that there was not. He would have been well advised, however, to ignore the assertion that there was and to assume that there was not until he was shown some evidence to the contrary. Computer prognosticators often assert or imply that when the number of bits in a computer memory reaches the number of cells in a human brain, the computer will begin to think like a human brain. There is no evidence to confirm or deny this assertion, but in science the technique must be not to accept the assertion until it is proven.

The computer is often compared to a human brain because they both input and output information and they both have the ability to store and process information. This is rather like comparing meteorites to bulldozers because they both make big holes in the ground. The question is not whether a computer is like a brain, but whether a brain is at all like a computer. Having designed them, we know everything about computers. We know almost nothing about the brain.

A 68000 computer is a digital device operating in a binary system. It was once thought that the brain was also binary and that a neuron either "fired" or remained inactive. It now appears that neurons "fire" in different ways at different times and that the differences may reflect different informational contents. Thus we are unable to say with any degree of certainty that the brain is like a computer even at the elementary level of classification as digital/binary or analog. The other common questions we ask when comparing computers are totally meaningless when applied to the human brain: What's the word length? How many bits on the address bus? What's the memory cycle time? Is there a double indexed addressing mode for handling two-dimensional arrays? Is there a signed divide instruction? How many primary accumulators?

A striking ability of the human brain compared to a computer is its ability to handle analog data in the presence of deafening background noise. For example, most humans could pick Mom's face out of the 30 faces in her eighth-grade class photo. A computer would be doing well to count the number of faces correctly and probably couldn't reliably tell which were boys and which were girls. The overwhelming superiority of the brain in this area and the equally impressive superiority of the computer in mathematical and textual manipulations suggests that these two devices may be operating in fundamentally different ways.

The final question is, What are computers capable of doing, and what can they not do? You must have noticed that the 68000 is not fundamentally different from the 6800. This is true of all computers. They all execute programs composed of instructions selected from a limited instruction set. The question is thus better phrased, What can a computer *program* be written to do? This makes it clear that the answer to the first question depends on the programmers—humans—more than it depends on the computer hardware.[3] Since computers do only what humans program them to do, it may be wisest to frame the question as, What do we want to have our computers do? This makes it clear that the answers to all three questions—and the responsibility—rests with us.

Answers to Chapter 24 Review Questions

1. Both contain Ø. **2.** An address register cannot be shifted, so (a) is invalid. If multiple rotates are specified, the destination must be a data register, so (c) is invalid. **3.** (a) $E57Ø; (b) $E571 **4.** $8F1A

5. The address labeled TABLE **6.** BTST.L #16,D3

7. It pushes the low 16 bits of D5 on the stack.

8. MOVE (A7)+, A2 **9.** Load the first 4 bytes of ROM with $ØØ:02:FF:FE. The last byte is FE and not FF to avoid an odd address boundary.

10. LINK A3,#−1Ø **11.** $\overline{IPL0}$ = high, $\overline{IPL1}$ = high, $\overline{IPL2}$ = low

12. At addresses $7Ø–$73 **13.** Assert $\overline{VPA}$ when the three FC outputs all go *high* together. **14.** PC and SR

15. Send the processor to an address read from vector bytes at $84–$87 **16.** Pick up address-error vector at $ØC–$ØF. **17.** $1000 **18.** Watchdog timer

19. An instruction inserted in a program being tested, which causes processing to

[3]Computers can "program themselves" in the sense that they can help with the bookkeeping tasks in large programs. (Automatic assemblers are a low-level example.) They can also be programmed to modify their own programs to optimize results according to some specified criterion. However, the general objective of the program and the definition of *optimum* must be supplied by humans. A computer does not have—and there is no reason to expect that it will ever acquire—a will or a sense of purpose. A computer will execute a program to trigger a charge that will blow it to Hades with the same indifference that it will calculate the square root of six.

stop and data to that point to be displayed. **20.** (1) Proceed in small steps. (2) Demand perfection in what has already been done before proceeding at all.

21. Addresses $00 0074–77.

22. Reset. **23.** 37 mA **24.** 26 s

25. BX = 16, BL = 8, BP = 16 **26.** 5 19B7H

27. No **28.** It finds an instruction at address F FFF0H. **29.** Register

30. In register DX

CHAPTER 24 QUESTIONS AND PROBLEMS

Digits after the decimal point refer to section numbers in Chapter 24.

Basic Level

1.1 What is wrong with this 68000 instruction, and how should it be written to achieve the intended effect?

```
ROL.W D3
```

2.1 What is the difference between the instructions NEG and NOT?

3.2 What is the 68000 instruction that performs the function of the 6800's PHSA and PSHB instructions?

4.2 Which of these is a good stack initialization, and why?
a. MOVE #$0011FF,A7 **b.** MOVE #$0011FE,A7

5.3 Write an instruction to enable the 68000 to recognize only interrupt levels 1, 2, 3, and 4.

6.3 The 68000's three $\overline{\text{IPL}}$ inputs are connected directly to three interrupting devices. One device pulls $\overline{\text{IPL2}}$ *low*. What address does the 68000 read, assuming the interrupt level is enabled? What does it do with the data it reads?

7.4 List two phases of "defining the problem" in the process of developing a microcomputer system.

8.4 In writing code for the 68000, what should you do to keep track of what the data in each of the 15 machine registers signifies?

9.5 How many different ways are there to choose 7 chickens from among a set of 16?

10.6 What is the function of the ALE pin on the 8086 microprocessor?

11.6 Write an 8086 instruction to add immediate hex data B to the 16 bits of the A register.

12.7 What do you expect to be spending more of your time on five years from now, computer hardware or software?

Advanced Level

13.1 Explain the difference between the 68000 instructions ROXL D0 and ROL D0.

14.1 Write a 68000 routine to count the number of logic-1 bits in the 32 bits of D1 and toggle bit 31 if this number is even.

15.2 Write a 68000 instruction to save D0, D1, A1, A2, A3 and A4 on the stack and another to retrieve them.

16.3 How does the 68000 know whether to use its autovector or its external-vector mode when an interrupt request is received?

17.3 An external interrupting device places data $74 on data bus lines D0 through D7. How does the 68000 respond to this data?

18.3 What will the 68000 do if it is instructed to execute a DIV instruction when the dividend is zero?

19.4 What parts of a multiuser computer program should be reserved for the supervisor state? How can these areas be connected to prevent their access by terminal users?

20.4 Why is it usually preferable to purchase or license existing software where possible rather than to write your own for a specific application?

21.4 List some guidelines for testing microcomputer system hardware and software.

22.5 Why is it often difficult to explain to a layperson exactly why microcomputers are being employed in so many industrial products and production processes?

23.6 Explain what is meant by segmented addressing, as employed in the 8086 microprocessor? What address is accessed if the Code Segment contains $1234 and the Instruction Pointer contains $5678?

24.6 List 7 addressing modes of the 8086, and explain briefly how the operand is obtained for each.

25.7 Which is more powerful: a computer or a human brain? (You may recall that we asked this question back in Chapter 0. It may be interesting to compare your answer now with the answer you gave then.)

Appendix A Microcomputer Wiring Hints

1. Have a complete system diagram with all IC pin numbers marked before you start building.
2. The ICs come with their pins bent out at an angle to facilitate machine insertion in industrial settings. Lay the pins on a table top and carefully bend each row straight.
3. Position the ICs on your breadboard in approximately the same relative positions as they appear on the system diagram. Keep pin 1 at the top for all of them. This will make it easier to locate test points later.
4. Use the stick-on labels printed in Appendix E to make pin identification easier.
5. Use new No. 22 solid wire for breadboarding. Cut each wire about $\frac{1}{2}$ to 1 inch longer than it needs to be to reach. This will allow you to move it without pulling out one end. Strip $\frac{3}{8}$ inch of insulation from each end. If you strip too little, you run the risk of a missed connection; too much and you may get a short between two wires.
6. Use different-color wires for each function. This will be a tremendous help in troubleshooting. Using different colors for odd- and even-numbered address and data lines helps you spot wiring errors as you make them, rather than later. A suggested color code is given at the end of this appendix.

7. Wire all V_{CC} and ground pins first. Use the power strips that run vertically down the breadboards. Use an ohmmeter to check that the power strips are connected as one long bus. On some breadboards they are not. Put several 0.1-μF bypass capacitors between V_{CC} and ground.
8. Have a copy of the system diagram for notes. Brite-line every connection as you make it to avoid missing a connection.
9. Wire the address and data buses next. Route the wires in bundles such that they pass around (not over the top of) the ICs. You will want to be able to remove the ICs and get a probe on their pins.
10. Wire the clock, control, and I/O lines last. Cut resistor and capacitor leads short to avoid shorts and dangling components. Again, don't route wires over the chips.
11. Hook a 2-inch-long bare wire along the ground bus for test-instrument ground chips. Bring commonly accessed test points out to the edge of the breadboard for easy access.
12. Check that V_{CC} *at the chips* is not below +4.9 V. Long thin lines from the power supply can drop excessive voltage.
13. Don't pull chips out of their sockets with your thumb and forefinger. You'll get the leads spiked into your finger when it pulls out. Pry the chip up from underneath, or use an IC extractor tool.
14. When going from one project to the next, it is generally easier to tear the old one down and wire the new one from scratch. Attempts to modify one microprocessor system into another invariably lead to wiring errors.

MICROCOMPUTER COLOR CODE

Black	Ground
Brown	Clock, chip-select, and control
Red	V_{CC} supply; +5 V
Orange	Supply lines other than V_{CC}
Yellow	Address bus, even numbers
Green	Address bus, odd numbers
Blue	Data bus, even numbers
Violet	Data bus, odd numbers
Gray	I/O bus, even numbers
White	I/O bus, odd numbers

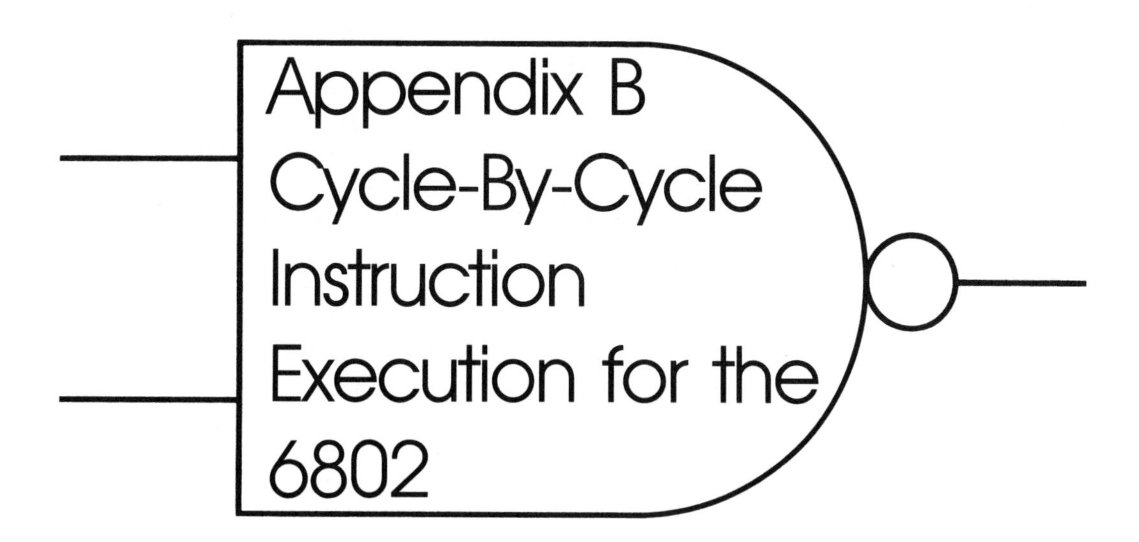

Appendix B
Cycle-By-Cycle
Instruction
Execution for the
6802

Appendix B provides a detailed description of the information present on the address bus, data bus, valid memory address line (VMA), and the read/write line (R/$\overline{W}$) during each cycle for each instruction.

This information is useful in comparing actual with expected results during debug of both software and hardware as the control program is executed. The information is categorized in groups according to addressing modes and number of cycles per instruction. (In general, instructions with the same addressing mode and number of cycles execute in the same manner; exceptions are indicated in the table.)

TABLE 8 — OPERATIONS SUMMARY

Address Mode and Instructions	Cycles	Cycle #	VMA Line	Address Bus	R/$\overline{W}$ Line	Data Bus
IMMEDIATE						
ADC EOR ADD LDA AND ORA BIT SBC CMP SUB	2	1	1	Op Code Address	1	Op Code
		2	1	Op Code Address + 1	1	Operand Data
CPX LDS LDX	3	1	1	Op Code Address	1	Op Code
		2	1	Op Code Address + 1	1	Operand Data (High Order Byte)
		3	1	Op Code Address + 2	1	Operand Data (Low Order Byte)
DIRECT						
ADC EOR ADD LDA AND ORA BIT SBC CMP SUB	3	1	1	Op Code Address	1	Op Code
		2	1	Op Code Address + 1	1	Address of Operand
		3	1	Address of Operand	1	Operand Data
CPX LDS LDX	4	1	1	Op Code Address	1	Op Code
		2	1	Op Code Address + 1	1	Address of Operand
		3	1	Address of Operand	1	Operand Data (High Order Byte)
		4	1	Operand Address + 1	1	Operand Data (Low Order Byte)
STA	4	1	1	Op Code Address	1	Op Code
		2	1	Op Code Address + 1	1	Destination Address
		3	0	Destination Address	1	Irrelevant Data (Note 1)
		4	1	Destination Address	0	Data from Accumulator
STS STX	5	1	1	Op Code Address	1	Op Code
		2	1	Op Code Address + 1	1	Address of Operand
		3	0	Address of Operand	1	Irrelevant Data (Note 1)
		4	1	Address of Operand	0	Register Data (High Order Byte)
		5	1	Address of Operand + 1	0	Register Data (Low Order Byte)
INDEXED						
JMP	4	1	1	Op Code Address	1	Op Code
		2	1	Op Code Address + 1	1	Offset
		3	0	Index Register	1	Irrelevant Data (Note 1)
		4	0	Index Register Plus Offset (w/o Carry)	1	Irrelevant Data (Note 1)
ADC EOR ADD LDA AND ORA BIT SBC CMP SUB	5	1	1	Op Code Address	1	Op Code
		2	1	Op Code Address + 1	1	Offset
		3	0	Index Register	1	Irrelevant Data (Note 1)
		4	0	Index Register Plus Offset (w/o Carry)	1	Irrelevant Data (Note 1)
		5	1	Index Register Plus Offset	1	Operand Data
CPX LDS LDX	6	1	1	Op Code Address	1	Op Code
		2	1	Op Code Address + 1	1	Offset
		3	0	Index Register	1	Irrelevant Data (Note 1)
		4	0	Index Register Plus Offset (w/o Carry)	1	Irrelevant Data (Note 1)
		5	1	Index Register Plus Offset	1	Operand Data (High Order Byte)
		6	1	Index Register Plus Offset + 1	1	Operand Data (Low Order Byte)

TABLE 8 — OPERATIONS SUMMARY (CONTINUED)

Address Mode and Instructions	Cycles	Cycle #	VMA Line	Address Bus	R/$\overline{W}$ Line	Data Bus
INDEXED (Continued)						
STA	6	1	1	Op Code Address	1	Op Code
		2	1	Op Code Address + 1	1	Offset
		3	0	Index Register	1	Irrelevant Data (Note 1)
		4	0	Index Register Plus Offset (w/o Carry)	1	Irrelevant Data (Note 1)
		5	0	Index Register Plus Offset	1	Irrelevant Data (Note 1)
		6	1	Index Register Plus Offset	0	Operand Data
ASL LSR ASR NEG CLR ROL COM ROR DEC TST INC	7	1	1	Op Code Address	1	Op Code
		2	1	Op Code Address + 1	1	Offset
		3	0	Index Register	1	Irrelevant Data (Note 1)
		4	0	Index Register Plus Offset (w/o Carry)	1	Irrelevant Data (Note 1)
		5	1	Index Register Plus Offset	1	Current Operand Data
		6	0	Index Register Plus Offset	1	Irrelevant Data (Note 1)
		7	1/0 (Note 3)	Index Register Plus Offset	0	New Operand Data (Note 3)
STS STX	7	1	1	Op Code Address	1	Op Code
		2	1	Op Code Address + 1	1	Offset
		3	0	Index Register	1	Irrelevant Data (Note 1)
		4	0	Index Register Plus Offset (w/o Carry)	1	Irrelevant Data (Note 1)
		5	0	Index Register Plus Offset	1	Irrelevant Data (Note 1)
		6	1	Index Register Plus Offset	0	Operand Data (High Order Byte)
		7	1	Index Register Plus Offset + 1	0	Operand Data (Low Order Byte)
JSR	8	1	1	Op Code Address	1	Op Code
		2	1	Op Code Address + 1	1	Offset
		3	0	Index Register	1	Irrelevant Data (Note 1)
		4	1	Stack Pointer	0	Return Address (Low Order Byte)
		5	1	Stack Pointer – 1	0	Return Address (High Order Byte)
		6	0	Stack Pointer – 2	1	Irrelevant Data (Note 1)
		7	0	Index Register	1	Irrelevant Data (Note 1)
		8	0	Index Register Plus Offset (w/o Carry)	1	Irrelevant Data (Note 1)
EXTENDED						
JMP	3	1	1	Op Code Address	1	Op Code
		2	1	Op Code Address + 1	1	Jump Address (High Order Byte)
		3	1	Op Code Address + 2	1	Jump Address (Low Order Byte)
ADC EOR ADD LDA AND ORA BIT SBC CMP SUB	4	1	1	Op Code Address	1	Op Code
		2	1	Op Code Address + 1	1	Address of Operand (High Order Byte)
		3	1	Op Code Address + 2	1	Address of Operand (Low Order Byte)
		4	1	Address of Operand	1	Operand Data
CPX LDS LDX	5	1	1	Op Code Address	1	Op Code
		2	1	Op Code Address + 1	1	Address of Operand (High Order Byte)
		3	1	Op Code Address + 2	1	Address of Operand (Low Order Byte)
		4	1	Address of Operand	1	Operand Data (High Order Byte)
		5	1	Address of Operand + 1	1	Operand Data (Low Order Byte)
STA A STA B	5	1	1	Op Code Address	1	Op Code
		2	1	Op Code Address + 1	1	Destination Address (High Order Byte)
		3	1	Op Code Address + 2	1	Destination Address (Low Order Byte)
		4	0	Operand Destination Address	1	Irrelevant Data (Note 1)
		5	1	Operand Destination Address	0	Data from Accumulator
ASL LSR ASR NEG CLR ROL COM ROR DEC TST INC	6	1	1	Op Code Address	1	Op Code
		2	1	Op Code Address + 1	1	Address of Operand (High Order Byte)
		3	1	Op Code Address + 2	1	Address of Operand (Low Order Byte)
		4	1	Address of Operand	1	Current Operand Data
		5	0	Address of Operand	1	Irrelevant Data (Note 1)
		6	1/0 (Note 3)	Address of Operand	0	New Operand Data (Note 3)

TABLE 8 — OPERATIONS SUMMARY (CONTINUED)

Address Mode and Instructions	Cycles	Cycle #	VMA Line	Address Bus	R/$\overline{W}$ Line	Data Bus
EXTENDED (Continued)						
STS STX	6	1	1	Op Code Address	1	Op Code
		2	1	Op Code Address + 1	1	Address of Operand (High Order Byte)
		3	1	Op Code Address + 2	1	Address of Operand (Low Order Byte)
		4	0	Address of Operand	1	Irrelevant Data (Note 1)
		5	1	Address of Operand	0	Operand Data (High Order Byte)
		6	1	Address of Operand + 1	0	Operand Data (Low Order Byte)
JSR	9	1	1	Op Code Address	1	Op Code
		2	1	Op Code Address + 1	1	Address of Subroutine (High Order Byte)
		3	1	Op Code Address + 2	1	Address of Subroutine (Low Order Byte)
		4	1	Subroutine Starting Address	1	Op Code of Next Instruction
		5	1	Stack Pointer	0	Return Address (Low Order Byte)
		6	1	Stack Pointer – 1	0	Return Address (High Order Byte)
		7	0	Stack Pointer – 2	1	Irrelevant Data (Note 1)
		8	0	Op Code Address + 2	1	Irrelevant Data (Note 1)
		9	1	Op Code Address + 2	1	Address of Subroutine (Low Order Byte)
INHERENT						
ABA DAA SEC ASL DEC SEI ASR INC SEV CBA LSR TAB CLC NEG TAP CLI NOP TBA CLR ROL TPA CLV ROR TST COM SBA	2	1	1	Op Code Address	1	Op Code
		2	1	Op Code Address + 1	1	Op Code of Next Instruction
DES DEX INS INX	4	1	1	Op Code Address	1	Op Code
		2	1	Op Code Address + 1	1	Op Code of Next Instruction
		3	0	Previous Register Contents	1	Irrelevant Data (Note 1)
		4	0	New Register Contents	1	Irrelevant Data (Note 1)
PSH	4	1	1	Op Code Address	1	Op Code
		2	1	Op Code Address + 1	1	Op Code of Next Instruction
		3	1	Stack Pointer	0	Accumulator Data
		4	0	Stack Pointer – 1	1	Accumulator Data
PUL	4	1	1	Op Code Address	1	Op Code
		2	1	Op Code Address + 1	1	Op Code of Next Instruction
		3	0	Stack Pointer	1	Irrelevant Data (Note 1)
		4	1	Stack Pointer + 1	1	Operand Data from Stack
TSX	4	1	1	Op Code Address	1	Op Code
		2	1	Op Code Address + 1	1	Op Code of Next Instruction
		3	0	Stack Pointer	1	Irrelevant Data (Note 1)
		4	0	New Index Register	1	Irrelevant Data (Note 1)
TXS	4	1	1	Op Code Address	1	Op Code
		2	1	Op Code Address + 1	1	Op Code of Next Instruction
		3	0	Index Register	1	Irrelevant Data
		4	0	New Stack Pointer	1	Irrelevant Data
RTS	5	1	1	Op Code Address	1	Op Code
		2	1	Op Code Address + 1	1	Irrelevant Data (Note 2)
		3	0	Stack Pointer	1	Irrelevant Data (Note 1)
		4	1	Stack Pointer + 1	1	Address of Next Instruction (High Order Byte)
		5	1	Stack Pointer + 2	1	Address of Next Instruction (Low Order Byte)

TABLE 8 — OPERATIONS SUMMARY (CONCLUDED)

Address Mode and Instructions	Cycles	Cycle #	VMA Line	Address Bus	R/$\overline{W}$ Line	Data Bus
INHERENT (Continued)						
WAI	9	1	1	Op Code Address	1	Op Code
		2	1	Op Code Address + 1	1	Op Code of Next Instruction
		3	1	Stack Pointer	0	Return Address (Low Order Byte)
		4	1	Stack Pointer − 1	0	Return Address (High Order Byte)
		5	1	Stack Pointer − 2	0	Index Register (Low Order Byte)
		6	1	Stack Pointer − 3	0	Index Register (High Order Byte)
		7	1	Stack Pointer − 4	0	Contents of Accumulator A
		8	1	Stack Pointer − 5	0	Contents of Accumulator B
		9	1	Stack Pointer − 6	1	Contents of Cond. Code Register
RTI	10	1	1	Op Code Address	1	Op Code
		2	1	Op Code Address + 1	1	Irrelevant Data (Note 2)
		3	0	Stack Pointer	1	Irrelevant Data (Note 1)
		4	1	Stack Pointer + 1	1	Contents of Cond. Code Register from Stack
		5	1	Stack Pointer + 2	1	Contents of Accumulator B from Stack
		6	1	Stack Pointer + 3	1	Contents of Accumulator A from Stack
		7	1	Stack Pointer + 4	1	Index Register from Stack (High Order Byte)
		8	1	Stack Pointer + 5	1	Index Register from Stack (Low Order Byte)
		9	1	Stack Pointer + 6	1	Next Instruction Address from Stack (High Order Byte)
		10	1	Stack Pointer + 7	1	Next Instruction Address from Stack (Low Order Byte)
SWI	12	1	1	Op Code Address	1	Op Code
		2	1	Op Code Address + 1	1	Irrelevant Data (Note 1)
		3	1	Stack Pointer	0	Return Address (Low Order Byte)
		4	1	Stack Pointer − 1	0	Return Address (High Order Byte)
		5	1	Stack Pointer − 2	0	Index Register (Low Order Byte)
		6	1	Stack Pointer − 3	0	Index Register (High Order Byte)
		7	1	Stack Pointer − 4	0	Contents of Accumulator A
		8	1	Stack Pointer − 5	0	Contents of Accumulator B
		9	1	Stack Pointer − 6	0	Contents of Cond. Code Register
		10	0	Stack Pointer − 7	1	Irrelevant Data (Note 1)
		11	1	Vector Address FFFA (Hex)	1	Address of Subroutine (High Order Byte)
		12	1	Vector Address FFFB (Hex)	1	Address of Subroutine (Low Order Byte)
RELATIVE						
BCC BHI BNE BCS BLE BPL BEQ BLS BRA BGE BLT BVC BGT BMI BVS	4	1	1	Op Code Address	1	Op Code
		2	1	Op Code Address + 1	1	Branch Offset
		3	0	Op Code Address + 2	1	Irrelevant Data (Note 1)
		4	0	Branch Address	1	Irrelevant Data (Note 1)
BSR	8	1	1	Op Code Address	1	Op Code
		2	1	Op Code Address + 1	1	Branch Offset
		3	0	Return Address of Main Program	1	Irrelevant Data (Note 1)
		4	1	Stack Pointer	0	Return Address (Low Order Byte)
		5	1	Stack Pointer − 1	0	Return Address (High Order Byte)
		6	0	Stack Pointer − 2	1	Irrelevant Data (Note 1)
		7	0	Return Address of Main Program	1	Irrelevant Data (Note 1)
		8	0	Subroutine Address (Note 4)	1	Irrelevant Data (Note 1)

NOTES:

1. If device which is addressed during this cycle uses VMA, then the Data Bus will go to the high-impedance three-state condition. Depending on bus capacitance, data from the previous cycle may be retained on the Data Bus.
2. Data is ignored by the MPU.
3. For TST, VMA = 0 and Operand data does not change.
4. MS Byte of Address Bus = MS Byte of Address of BSR instruction and LS Byte of Address Bus = LS Byte of Sub-Routine Address.

Appendix C Disassembly List for the 6800/6802

First digit \ Second digit	0	1	2	3	4	5	6	7	8	9	A	B	C	D	E	F	Mode
0		NOP					TAP	TPA	INX	DEX	CLV	SEV	CLC	SEC	CLI	SEI	Inher
1	SBA	CBA					TAB	TBA		DAA		ABA					Inher
2	BRA		BHI	BLS	BCC	BCS	BNE	BEQ	BVC	BVS	BPL	BMI	BGE	BLT	BGT	BLE	Rel
3	TSX	INS	PULA	PULB	DES	TXS	PSHA	PSHB		RTS		RTI			WAI	SWI	Inher
4	NEGA			COMA	LSRA		RORA	ASRA	ASLA	ROLA	DECA		INCA	TSTA		CLRA	Inher
5	NEGB			COMB	LSRB		RORB	ASRB	ASLB	ROLB	DECB		INCB	TSTB		CLRB	Inher
6	NEG			COM	LSR		ROR	ASR	ASL	ROL	DEC		INC	TST	JMP	CLR	Index
7	NEG			COM	LSR		ROR	ASR	ASL	ROL	DEC		INC	TST	JMP	CLR	Extnd
8	SUBA	CMPA	SBCA		ANDA	BITA	LDAA		EORA	ADCA	ORAA	ADDA	CPX	BSR Rel	LDS		Immed
9	SUBA	CMPA	SBCA		ANDA	BITA	LDAA	STAA	EORA	ADCA	ORAA	ADDA	CPX		LDS	STS	Dir
A	SUBA	CMPA	SBCA		ANDA	BITA	LDAA	STAA	EORA	ADCA	ORAA	ADDA	CPX	JSR	LDS	STS	Index
B	SUBA	CMPA	SBCA		ANDA	BITA	LDAA	STAA	EORA	ADCA	ORAA	ADDA	CPX	JSR	LDS	STS	Extnd
C	SUBB	CMPB	SBCB		ANDB	BITB	LDAB		EORB	ADCB	ORRB	ADDB			LDX		Immed
D	SUBB	CMPB	SBCB		ANDB	BITB	LDAB	STAB	EORB	ADCB	ORAB	ADDB			LDX	STX	Dir
E	SUBB	CMPB	SBCB		ANDB	BITB	LDAB	STAB	EORB	ADCB	ORAB	ADDB			LDX	STX	Index
F	SUBB	CMPB	SBCB		ANDB	BITB	LDAB	STAB	EORB	ADCB	ORAB	ADDB			LDX	STX	Extnd

D

Appendix D
An EPROM Programmer

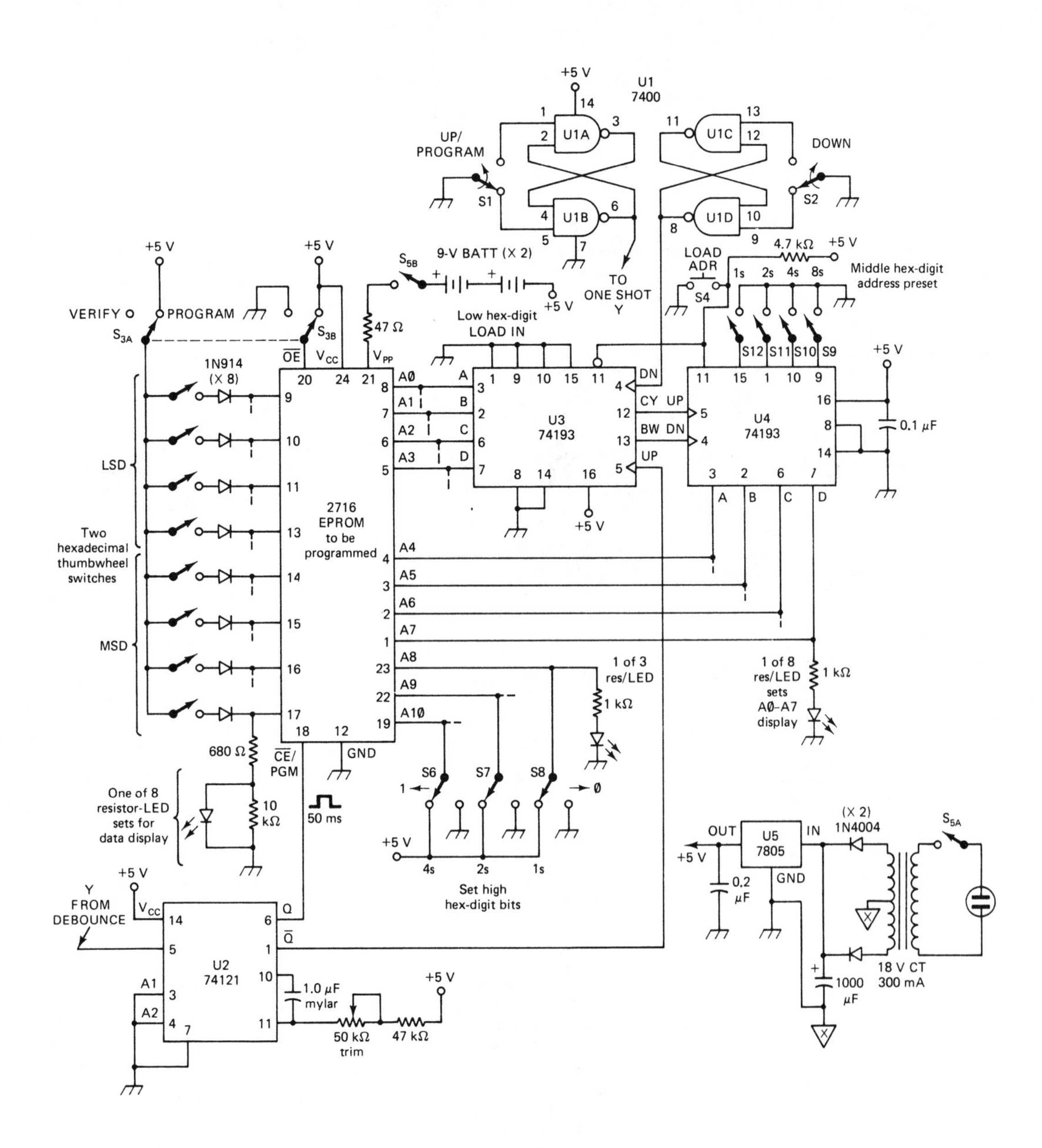
+5 V
U1
7400
U1A
U1B
U1C
U1D
UP/
PROGRAM
DOWN
S1
S2
TO
ONE SHOT
Y
LOAD
ADR
S4
4.7 kΩ
Middle hex-digit
address preset
9-V BATT (X 2)
Low hex-digit
LOAD IN
VERIFY
PROGRAM
1N914
(X 8)
LSD
Two
hexadecimal
thumbwheel
switches
MSD
2716
EPROM
to be
programmed
U3
74193
U4
74193
0.1 μF
1 of 3
res/LED
1 of 8
res/LED
sets
AØ–A7
display
1 kΩ
680 Ω
One of 8
resistor-LED
sets for
data display
10
kΩ
50 ms
Set high
hex-digit bits
Y
FROM
DEBOUNCE
U2
74121
1.0 μF
mylar
50 kΩ
trim
47 kΩ
U5
7805
0.2
μF
1000
μF
1N4004
18 V CT
300 mA

Appendix E Labels for IC Pin Functions

Left	Right
A	V_c
B	YØ
C	Y1
$\overline{CS}$	Y2
$\overline{CS}$	Y3
CS	Y4
Y7	Y5
G	Y6

74LS138

Left	Right
$\overline{CE}$	V_c
IØ	$\overline{CE}$
D7	DØ
I1	I7
D6	D1
I2	I6
D5	D2
I3	I5
D4	D3
G	I4

74LS244

Left	Right
$\overline{OE}$	V_c
QØ	Q7
DØ	D7
D1	D6
Q1	Q6
Q2	Q5
D2	D5
D3	D4
Q3	Q4
G	CE

74LS373

Left	Right
1 A3	A2 48
2 A4	A1 47
3 A5	AØ 46
4 A6	FCØ 45
5 A7	FC1 44
6 A8	FC2 43
7 A9	$\overline{I2, Ø}$ 42
8 A1Ø	$\overline{I1}$ 41
9 A11	$\overline{BER}$ 40
10 A12	$\overline{VPA}$ 39
11 A13	E → 38
12 A14	$\overline{RST}$ 37
13 V_{CC}	$\overline{HLT}$ 36
14 A15	GND 35
15 GND	CK 34
16 A16	$\overline{BR}$ 33
17 A17	$\overline{BG}$ → 32
18 A18	$\overline{DAK}$ 31
19 A19	R/$\overline{W}$ 30
20 D7	$\overline{DS}$ → 29
21 D6	$\overline{AS}$ → 28
22 D5	DØ 27
23 D4	D1 26
24 D3	D2 25

68008

Left	Right
1 V_{SS}	$\overline{CTS}$ ← 24
2 → RD	$\overline{DCD}$ ← 23
3 → RCK	DØ 22
4 → TCK	D1 21
5 ← $\overline{RTS}$	D2 20
6 ← TD	D3 19
7 ← $\overline{IRQ}$	D4 18
8 → CSØ	D5 17
9 → $\overline{CS2}$	D6 16
10 → CS1	D7 15
11 → RS	E ← 14
12 V_{CC}	R/$\overline{W}$ 13

6850

Left	Right
1 D4	D5 64
2 D3	D6 63
3 D2	D7 62
4 D1	D8 61
5 DØ	D9 60
6 ← $\overline{AS}$	D1Ø 59
7 ← $\overline{UDS}$	D11 58
8 ← $\overline{LDS}$	D12 57
9 ← R/$\overline{W}$	D13 56
10 $\overline{DTACK}$	D14 55
11 ← $\overline{BG}$	D15 54
12 $\overline{BGACK}$	GND 53
13 $\overline{BR}$	A23 52
14 V_{CC}	A22 51
15 CLK	A21 50
16 GND	V_{CC} 49
17 ↔ $\overline{HLT}$	A2Ø 48
18 ↔ $\overline{RST}$	A19 47
19 ← $\overline{VMA}$	A18 46
20 ← E	A17 45
21 $\overline{VPA}$	A16 44
22 $\overline{BERR}$	A15 43
23 $\overline{IPL2}$	A14 42
24 $\overline{IPL1}$	A13 41
25 $\overline{IPLØ}$	A12 40
26 ← FC2	A11 39
27 ← FC1	A1Ø 38
28 ← FCØ	A9 37
29 A1	A8 36
30 A2	A7 35
31 A3	A6 34
32 A4	A5 33

68000

1 A7	V_{CC} 24
2 A6	A8 23
3 A5	A9 22
4 A4	V_{PP} 21
5 A3	$\overline{OE}$ 20
6 A2	A1Ø 19
7 A1	$\overline{CE}$ 18
8 AØ	D7 17
9 DØ	D6 16
10 D1	D5 15
11 D2	D4 14
12 V_{SS}	D3 13

2716

1 V_{PP}	V_{CC} 28
2 A12	$\overline{PGM}$ 27
3 A7	NC 26
4 A6	A8 25
5 A5	A9 24
6 A4	A11 23
7 A3	$\overline{OE}$ 22
8 A2	A1Ø 21
9 A1	$\overline{CE}$ 20
10 AØ	D7 19
11 DØ	D6 18
12 D1	D5 17
13 D2	D4 16
14 GND	D3 15

2764

1 GND	V_{CC} 24
2 DØ	AØ 23
3 D1	A1 22
4 D2	A2 21
5 D3	A3 20
6 D4	A4 19
7 D5	A5 18
8 D6	A6 17
9 D7	R/$\overline{W}$ 16
10 CSØ	$\overline{CS5}$ 15
11 $\overline{CS1}$	$\overline{CS4}$ 14
12 $\overline{CS2}$	CS3 13

6810

1 V_{SS}	$\overline{RST}$ 40
2 $\overline{HLT}$	EXT 39
3 MR	XTL 38
4 $\overline{IRQ}$	E → 37
5 ← VMA	RE 36
6 $\overline{NMI}$	V_R 35
7 ← BA	R/$\overline{W}$ → 34
8 V_{CC}	DØ 33
9 AØ	D2 31
10 A1	D1 32
11 A2	D3 30
12 A3	D4 29
13 A4	D5 28
14 A5	D6 27
15 A6	D7 26
16 A7	A15 25
17 A8	A14 24
18 A9	A13 23
19 A1Ø	A12 22
20 A11	V_{SS} 21

6802

1 GND	$\overline{HLT}$ 40
2 $\overline{NMI}$	XTL 39
3 $\overline{IRQ}$	EXT 38
4 $\overline{FIR}$	$\overline{RST}$ 37
5 ← BS	MR 36
6 ← BA	Q → 35
7 V_{CC}	E → 34
8 AØ	$\overline{BRQ}$ 33
9 A1	R/$\overline{W}$ 32
10 A2	DØ 31
11 A3	D1 30
12 A4	D2 29
13 A5	D3 28
14 A6	D4 27
15 A7	D5 26
16 A8	D6 25
17 A9	D7 24
18 A1Ø	A15 23
19 A11	A14 22
20 A12	A13 21

6809

1 A11	A1Ø 40
2 A12	A9 39
3 A13	A8 38
4 A14	A7 37
5 A15	A6 36
6 CK	A5 35
7 D4	A4 34
8 D3	A3 33
9 D5	A2 32
10 D6	A1 31
11 V_{CC}	AØ 30
12 D2	GND 29
13 D7	$\overline{RF}$ → 28
14 DØ	$\overline{M1}$ → 27
15 D1	$\overline{RST}$ 26
16 $\overline{INT}$	$\overline{BR}$ → 25
17 $\overline{NMI}$	$\overline{WA}$ → 24
18 $\overline{HLT}$	$\overline{BAK}$ 23
19 ← $\overline{MR}$	$\overline{WR}$ → 22
20 ← $\overline{IO}$	$\overline{RD}$ → 21

Z80

1 GND	CA1 40
2 PAØ	CA2 39
3 PA1	RSØ 38
4 PA2	RS1 37
5 PA3	RS2 36
6 PA4	RS3 35
7 PA5	$\overline{RES}$ 34
8 PA6	DØ 33
9 PA7	D1 32
10 PBØ	D2 31
11 PB1	D3 30
12 PB2	D4 29
13 PB3	D5 28
14 PB4	D6 27
15 PB5	D7 26
16 PB6	E 25
17 PB7	CS1 24
18 CB1	$\overline{CS2}$ 23
19 CB2	R/$\overline{W}$ 22
20 V_{CC}	$\overline{IRQ}$ 21

6522

1 V_{SS}	D7 40
2 D6	CSS ← 39
3 DØ	$\overline{HS}$ → 38
4 D1	$\overline{FS}$ → 37
5 D2	$\overline{RP}$ → 36
6 D3	$\overline{AG}$ ← 35
7 D4	$\overline{AS}$ ← 34
8 D5	CLK ← 33
9 → CHB	INV ← 32
10 ← ϕB	$\overline{INT}$ ← 31
11 ← ϕA	GMØ → 30
12 → $\overline{MS}$	GM1 ← 29
13 A5	Y → 28
14 A6	GM2 ← 27
15 A7	A4 26
16 A8	A3 25
17 V_{CC}	A2 24
18 A9	A1 23
19 A1Ø	AØ 22
20 A11	A12 21

6847

1 GND	$\overline{RES}$ 40
2 RDY	ϕ_2 → 39
3 ← ϕ_1	SO 38
4 $\overline{IRQ}$	ϕ_0 37
5 NC	NC 36
6 $\overline{NMI}$	NC 35
7 SYN	R/$\overline{W}$ 34
8 V_{CC}	DØ 33
9 AØ	D1 32
10 A1	D2 31
11 A2	D3 30
12 A3	D4 29
13 A4	D5 28
14 A5	D6 27
15 A6	D7 26
16 A7	A15 25
17 A8	A14 24
18 A9	A13 23
19 A1Ø	A12 22
20 A11	GND 21

6502

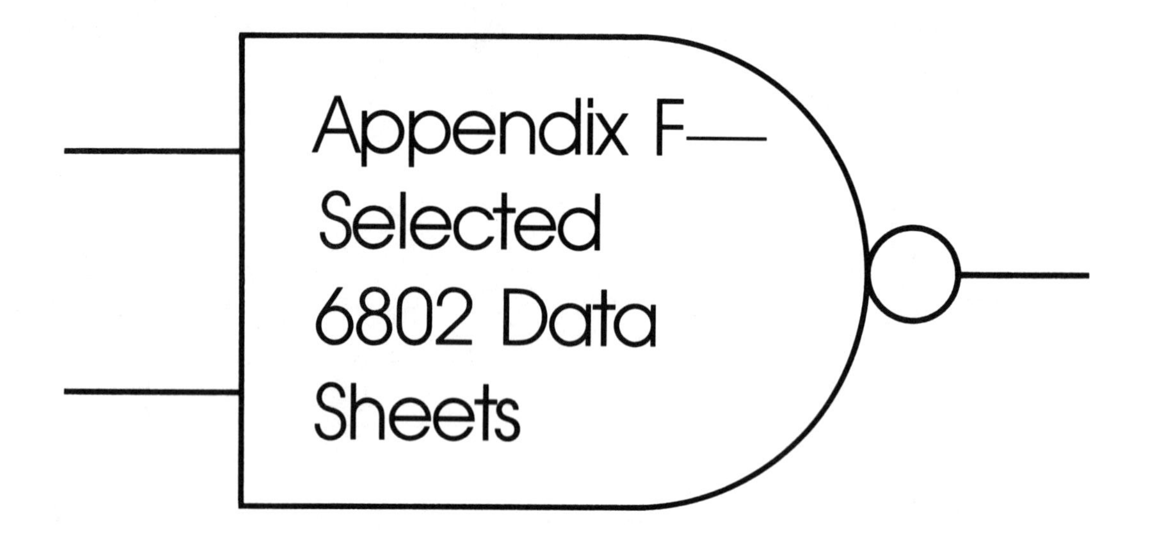
Appendix F—
Selected
6802 Data
Sheets

TYPICAL MICROCOMPUTER

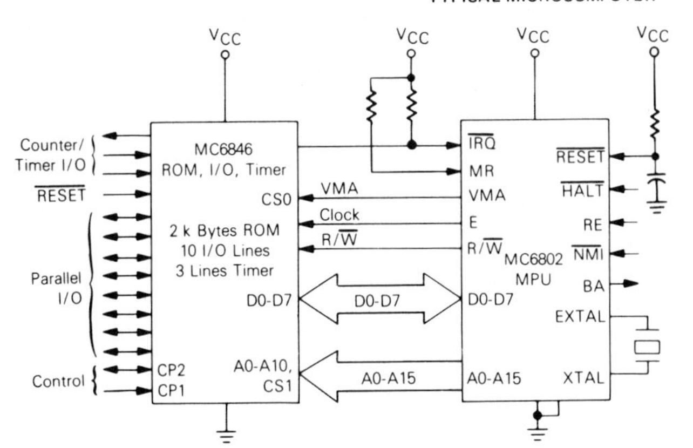

This block diagram shows a typical cost effective microcomputer. The MPU is the center of the microcomputer system and is shown in a minimum system interfacing with a ROM combination chip. It is not intended that this system be limited to this function but that it be expandable with other parts in the M6800 Microcomputer family.

MAXIMUM RATINGS

Rating	Symbol	Value	Unit
Supply Voltage	V_{CC}	−0.3 to +7.0	V
Input Voltage	V_{in}	−0.3 to +7.0	V
Operating Temperature Range MC6802, MC680A02, MC680B02 MC6802C, MC680A02C MC6802NS MC6808, MC68A08, MC68B08	T_A	 0 to +70 −40 to +85 0 to +70 0 to +70	°C
Storage Temperature Range	T_{stg}	−55 to +150	°C

This input contains circuitry to protect the inputs against damage due to high static voltages or electric fields; however, it is advised that normal precautions be taken to avoid application of any voltage higher than maximum rated voltages to this high-impedance circuit. Reliability of operation is enhanced if unused inputs are tied to an appropriate logic voltage level (e.g., either V_{SS} or V_{CC}).

THERMAL CHARACTERISTICS

Characteristic	Symbol	Value	Unit
Average Thermal Resistance (Junction to Ambient) Plastic Ceramic	θ_{JA}	 100 50	°C/W

POWER CONSIDERATIONS

The average chip-junction temperature, T_J, in °C can be obtained from:

$$T_J = T_A + (P_D \bullet \theta_{JA}) \qquad (1)$$

Where:

$T_A \equiv$ Ambient Temperature, °C

$\theta_{JA} \equiv$ Package Thermal Resistance, Junction-to-Ambient, °C/W

$P_D \equiv P_{INT} + P_{PORT}$

$P_{INT} \equiv I_{CC} \times V_{CC}$, Watts — Chip Internal Power

$P_{PORT} \equiv$ Port Power Dissipation, Watts — User Determined

For most applications $P_{PORT} \ll P_{INT}$ and can be neglected. P_{PORT} may become significant if the device is configured to drive Darlington bases or sink LED loads.

An approximate relationship between P_D and T_J (if P_{PORT} is neglected) is:

$$P_D = K \div (T_J + 273°C) \qquad (2)$$

Solving equations 1 and 2 for K gives:

$$K = P_D \bullet (T_A + 273°C) + \theta_{JA} \bullet P_D^2 \qquad (3)$$

Where K is a constant pertaining to the particular part. K can be determined from equation 3 by measuring P_D (at equilibrium) for a known T_A. Using this value of K the values of P_D and T_J can be obtained by solving equations (1) and (2) iteratively for any value of T_A.

[1]Courtesy, Motorola Semiconductor Products

DC ELECTRICAL CHARACTERISTICS (V_{CC}=5.0 Vdc ±5%, V_{SS}=0, T_A=0 to 70°C, unless otherwise noted)

Characteristic		Symbol	Min	Typ	Max	Unit
Input High Voltage	Logic, EXTAL $\overline{RESET}$	V_{IH}	V_{SS}+2.0 V_{SS}+4.0	– –	V_{CC} V_{CC}	V
Input Low Voltage	Logic, EXTAL, $\overline{RESET}$	V_{IL}	V_{SS}−0.3	–	V_{SS}+0.8	V
Input Leakage Current (V_{in}=0 to 5.25 V, V_{CC}=max)	Logic	I_{in}	–	1.0	2.5	μA
Output High Voltage (I_{Load}= −205 μA, V_{CC}=min) (I_{Load}= −145 μA, V_{CC}=min) (I_{Load}= −100 μA, V_{CC}=min)	 D0-D7 A0-A15, R/$\overline{W}$, VMA, E BA	V_{OH}	 V_{SS}+2.4 V_{SS}+2.4 V_{SS}+2.4	 – – –	 – – –	V
Output Low Voltage (I_{Load}=1.6 mA, V_{CC}=min)		V_{OL}	–	–	V_{SS}+0.4	V
Internal Power Dissipation (Measured at T_A=0°C)		P_{INT}	–	0.750	1.0	W
V_{CC} Standby	Power Down Power Up	V_{SBB} V_{SB}	4.0 4.75	– –	5.25 5.25	V*
Standby Current		I_{SBB}	–	–	8.0	mA
Capacitance # (V_{in}=0, T_A=25°C, f=1.0 MHz)	D0-D7 Logic Inputs, EXTAL	C_{in}	– –	10 6.5	12.5 10	pF
	A0-A15, R/$\overline{W}$, VMA	C_{out}	–	–	12	pF

*In power-down mode, maximum power dissipation is less than 42 mW.
#Capacitances are periodically sampled rather than 100% tested.

CONTROL TIMING (V_{CC}=5.0 V ±5%, V_{SS}=0, T_A=T_L to T_H, unless otherwise noted)

Characteristics	Symbol	MC6802 MC6802NS MC6808		MC68A02 MC68A08		MC68B02 MC68B08		Unit
		Min	Max	Min	Max	Min	Max	
Frequency of Operation	f_o	0.1	1.0	0.1	1.5	0.1	2.0	MHz
Crystal Frequency	f_{XTAL}	1.0	4.0	1.0	6.0	1.0	8.0	MHz
External Oscillator Frequency	$4xf_o$	0.4	4.0	0.4	6.0	0.4	8.0	MHz
Crystal Oscillator Start Up Time	t_{rc}	100	–	100	–	100	–	ms
Processor Controls (HALT, MR, RE, $\overline{RESET}$, $\overline{IRQ}$ $\overline{NMI}$)								
Processor Control Setup Time	t_{PCS}	200	–	140	–	110	–	ns
Processor Control Rise and Fall Time (Does Not Apply to $\overline{RESET}$)	t_{PCr}, t_{PCf}	–	100	–	100	–	100	ns

BUS TIMING CHARACTERISTICS

Ident. Number	Characteristic	Symbol	MC6802 MC6802NS MC6808		MC68A02 MC68A08		MC68B02 MC68B08		Unit
			Min	Max	Min	Max	Min	Max	
1	Cycle Time	t_{cyc}	1.0	10	0.667	10	0.5	10	μs
2	Pulse Width, E Low	PW_{EL}	450	5000	280	5000	210	5000	ns
3	Pulse Width, E High	PW_{EH}	450	9500	280	9700	220	9700	ns
4	Clock Rise and Fall Time	t_r, t_f	–	25	–	25	–	25	ns
9	Address Hold Time*	t_{AH}	20	–	20	–	20	–	ns
12	Non-Muxed Address Valid Time to E (See Note 5)	t_{AV1} t_{AV2}	160 –	– 270	100 –	– –	50 –	– –	ns
17	Read Data Setup Time	t_{DSR}	100	–	70	–	60	–	ns
18	Read Data Hold Time	t_{DHR}	10	–	10	–	10	–	ns
19	Write Data Delay Time	t_{DDW}	–	225	–	170	–	160	ns
21	Write Data Hold Time*	t_{DHW}	30	–	20	–	20	–	ns
29	Usable Access Time (See Note 4)	t_{ACC}	535	–	335	–	235	–	ns

*Address and data hold times are periodically tested rather than 100% tested.

FIGURE 2 — BUS TIMING

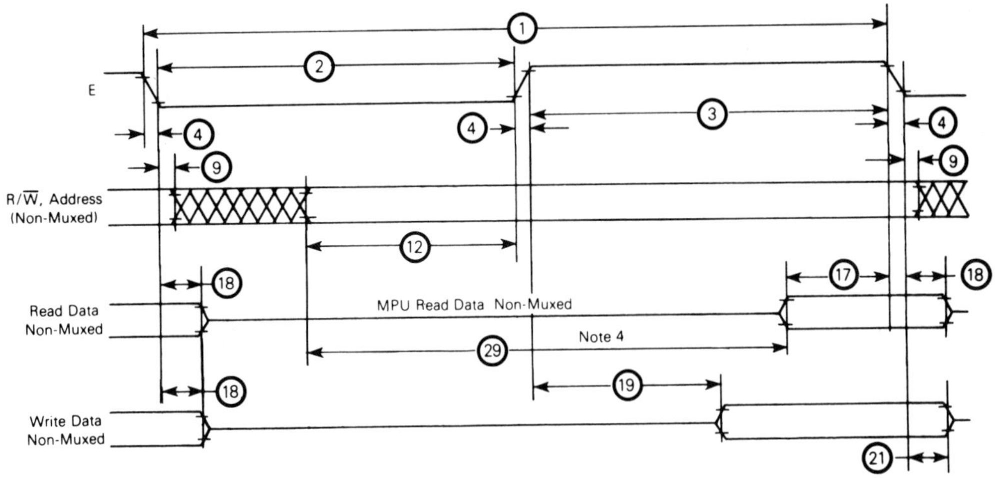

NOTES:

1. Voltage levels shown are $V_L \leq 0.4$ V, $V_H \geq 2.4$ V, unless otherwise specified.
2. Measurement points shown are 0.8 V and 2.0 V, unless otherwise noted.
3. All electricals shown for the MC6802 apply to the MC6802NS and MC6808, unless otherwise noted.
4. Usable access time is computed by: 12+3+4−17.
5. If programs are not executed from on-board RAM, TAV1 applies. If programs are to be stored and executed from on-board RAM, TAV2 applies. For normal data storage in the on-board RAM, this extended delay does not apply. Programs cannot be executed from on-board RAM when using A and B parts (MC68A02, MC68A08, MC68B02, MC68B08). On-board RAM can be used for data storage with all parts.
6. All electrical and control characteristics are referenced from: $T_L = 0°C$ minimum and $T_H = 70°C$ maximum.

FIGURE 3 — BUS TIMING TEST LOAD

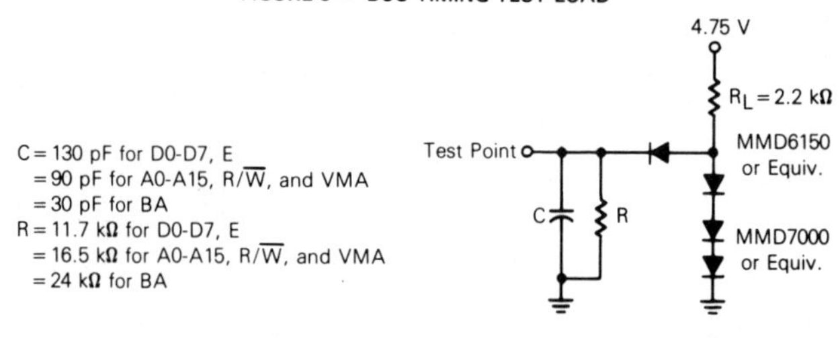

FIGURE 4 — TYPICAL DATA BUS OUTPUT DELAY versus CAPACITIVE LOADING

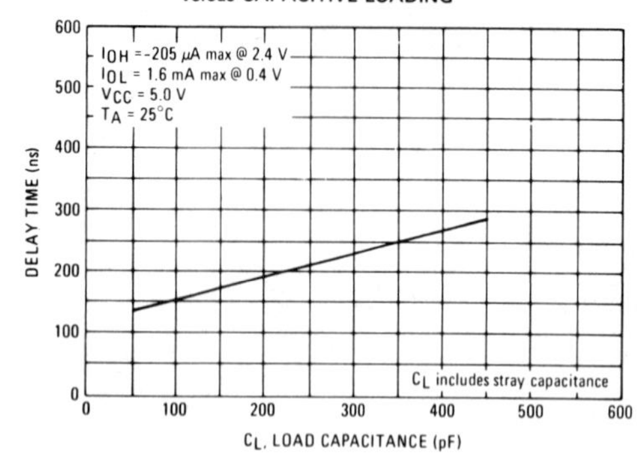

FIGURE 5 — TYPICAL READ/WRITE, VMA AND ADDRESS OUTPUT DELAY versus CAPACITIVE LOADING

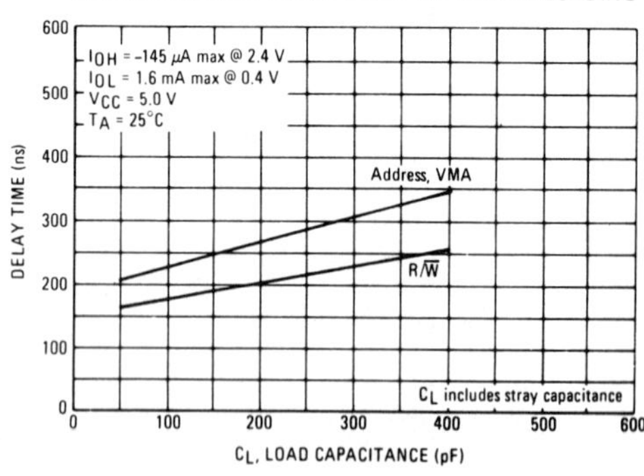

FIGURE 6 — EXPANDED BLOCK DIAGRAM

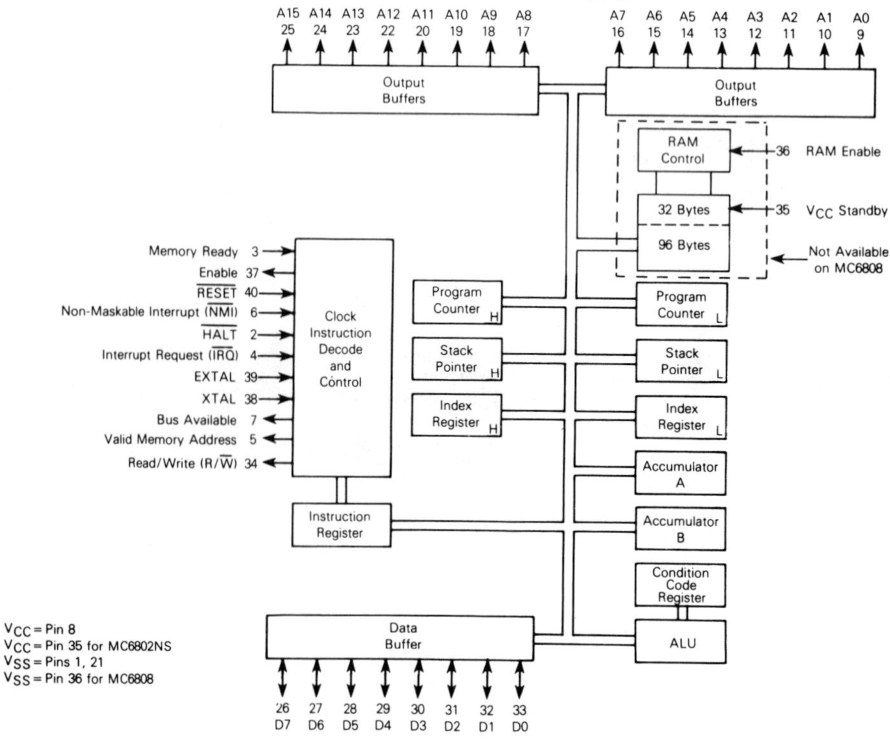

INDEX

Characters

Numbers

A

B

C

D

E

F

G

H

I

J

K

L

M

N

O

P

Q

R

S

T

U

V

W

X

Y

Z

6802 Programmer's Reference

		ADDRESSING MODES										
		IMMED INHER			DIR/REL (2 byte)		EXTND (3 byte)		INDEX (2 byte)		OPERATION (All register labels refer to contents)	AFFECTS COND. CODE REG. Example: N = set, $\overline{N}$ = cleared, n = set or cleared, n^7 = note 7 below.
OPERATIONS	MNEMON	OP	~	#	OP	~	OP	~	OP	~		
Add Acmltrs	ABA	1B	2	1							A + B → A	h n z v c
Add with Carry	ADCA	89	2	2	99	3	B9	4	A9	5	A + M + C → A	h n z v c
	ADCB	C9	2	2	D9	3	F9	4	E9	5	B + M + C → B	h n z v c
Add	ADDA	8B	2	2	9B	3	BB	4	AB	5	A + M → A	h n z v c
	ADDB	CB	2	2	DB	3	FB	4	EB	5	B + M → B	h n z v c
And	ANDA	84	2	2	94	3	B4	4	A4	5	A · M → A	n z $\overline{V}$
	ANDB	C4	2	2	D4	3	F4	4	E4	5	B · M → B	n z $\overline{V}$
Arithmetic, Shift Left	ASL						78	6	68	7	M } Shift left into	n z v^6 c
	ASLA	48	2	1							A } carry; Ø to LSB	n z v^6 c
	ASLB	58	2	1							B } [ ] ← [ ][ ][ ][ ][ ][ ][ ][ ] ← Ø (C 7 ⟵ Ø)	n z v^6 c
Arithmetic, Shift Right	ASR						77	6	67	7	M } Shift right into	n z v^6 c
	ASRA	47	2	1							A } carry; bit 7 stays	n z v^6 c
	ASRB	57	2	1							B } ↳[ ][ ][ ][ ][ ][ ][ ][ ] → [ ] (7 ⟶ Ø C)	n z v^6 c
Branch:											Test:	
If Carry Clear	BCC				24	4					C = Ø?	
If Carry Set	BCS				25	4					C = 1?	
If = Zero	BEQ				27	4					Z = 1?	
If ≥ Zero	BGE				2C	4					N ⊕ V = Ø?	
If > Zero	BGT				2E	4					Z + (N ⊕ V) = Ø?	
If Higher	BHI				22	4					C + Z = Ø?	
Bit Test	BITA	85	2	2	95	3	B5	4	A5	5	A · M } Sets CCR	n z $\overline{V}$
	BITB	C5	2	2	D5	3	F5	4	E5	5	B · M } bits only	n z $\overline{V}$
Branch:											Test:	
If ≤ Zero	BLE				2F	4					Z + (N ⊕ V) = 1?	
If Lower Or Same	BLS				23	4					C + Z = 1?	
If < Zero	BLT				2D	4					N ⊕ V = 1?	
If Minus	BMI				2B	4					N = 1?	
If Not Equal Zero	BNE				26	4					Z = Ø?	
If Plus	BPL				2A	4					N = Ø?	
Always	BRA				2Ø	4					None	
To Subroutine	BSR				8D	8					(Stack PCL, PCH)	
If Overflow Clear	BVC				28	4					V = Ø?	
If Overflow Set	BVS				29	4					V = 1?	
Compare Acmltrs	CBA	11	2	1							A − B	n z v c
Clear Carry	CLC	ØC	2	1							Ø → C	$\overline{C}$
Clear Intrpt Mask	CLI	ØE	2	1							Ø → I	$\overline{I}$
Clear	CLR						7F	6	6F	7	ØØ → M	$\overline{N}$ Z $\overline{V}$ $\overline{C}$
	CLRA	4F	2	1							ØØ → A	$\overline{N}$ Z $\overline{V}$ $\overline{C}$
	CLRB	5F	2	1							ØØ → B	$\overline{N}$ Z $\overline{V}$ $\overline{C}$
Clear Overflow	CLV	ØA	2	1							Ø → V	$\overline{V}$
Compare	CMPA	81	2	2	91	3	B1	4	A1	5	A − M } Sets CCR	n z v c
	CMPB	C1	2	2	D1	3	F1	4	E1	5	B − M } bits only	n z v c
Complement, 1's	COM						73	6	63	7	$\overline{M}$ → M	n z $\overline{V}$ C
	COMA	43	2	1							$\overline{A}$ → A	n z $\overline{V}$ C
	COMB	53	2	1							$\overline{B}$ → B	n z $\overline{V}$ C
Compare Index Reg	CPX	8C	3	3	9C	4	BC	5	AC	6	$(X_H/X_L) - (M/M+1)$	n^7 z v^8
Decimal Adjust, A	DAA	19	2	1							Converts Add. of BCD to BCD Format	n z v c^3
Decrement	DEC						7A	6	6A	7	M − 1 → M	n z v^4
	DECA	4A	2	1							A − 1 → A	n z v^4
	DECB	5A	2	1							B − 1 → B	n z v^4
Decr Stack Pntr	DES	34	4	1							SP − 1 → SP	
Decr Index Reg	DEX	Ø9	4	1							X − 1 → X	z
Exclusive OR	EORA	88	2	2	98	3	B8	4	A8	5	A ⊕ M → A	n z $\overline{V}$
	EORB	C8	2	2	D8	3	F8	4	E8	5	B ⊕ M → B	n z $\overline{V}$
Increment	INC						7C	6	6C	7	M + 1 → M	n z v^5
	INCA	4C	2	1							A + 1 → A	n z v^5
	INCB	5C	2	1							B + 1 → B	n z v^5
Incrm Stack Pntr	INS	31	4	1							SP + 1 → SP	
Incrm Index Reg	INX	Ø8	4	1							X + 1 → X	z
Jump	JMP						7E	3	6E	4	Oprnd H, L → PC	
Jump to Subrtn	JSR						BD	9	AD	8	Stack PCL, PCH	
Load Acmltr	LDAA	86	2	2	96	3	B6	4	A6	5	M → A	n z $\overline{V}$
	LDAB	C6	2	2	D6	3	F6	4	E6	5	M → B	n z $\overline{V}$
Load Stack Pntr	LDS	8E	3	3	9E	4	BE	5	AE	6	M → SP_H, (M + 1) → SP_L	n^7 z $\overline{V}$
Load Index Reg	LDX	CE	3	3	DE	4	FE	5	EE	6	M → X_H, (M + 1) → X_L	n^7 z $\overline{V}$
Logic, Shift Right	LSR						74	6	64	7	M } Shift right into	$\overline{N}$ z v^6 c
	LSRA	44	2	1							A } carry; Ø to MSB	$\overline{N}$ z v^6 c
	LSRB	54	2	1							B } Ø → [ ][ ][ ][ ][ ][ ][ ][ ] → [ ] (7 ⟶ Ø C)	$\overline{N}$ z v^6 c
Complement, 2's (Negate)	NEG						7Ø	6	6Ø	7	ØØ − M → M } Complement	n z v^1 c^2
	NEGA	4Ø	2	1							ØØ − A → A } bits and	n z v^1 c^2
	NEGB	5Ø	2	1							ØØ − B → B } add 1	n z v^1 c^2

6802 Programmer's Reference

		ADDRESSING MODES										AFFECTS COND. CODE REG. Example: N = set, $\overline{N}$ = cleared, n = set or cleared, n[7] = note 7 below.
		IMMED INHER			DIR/REL (2 byte)		EXTND (3 byte)		INDEX (2 byte)		OPERATION (All register labels refer to contents)	
OPERATIONS	MNEMON	OP	~	#	OP	~	OP	~	OP	~		
No Operation	NOP	Ø1	2	1							Inc. Prog. Cntr. Only	
Or, Inclusive	ORAA	8A	2	2	9A	3	BA	4	AA	5	A + M → A	n z $\overline{V}$
	ORAB	CA	2	2	DA	3	FA	4	EA	5	B + M → B	n z $\overline{V}$
Push Data	PSHA	36	4	1							A → M_{SP}, SP − 1 → SP	
on stack	PSHB	37	4	1							B → M_{SP}, SP − 1 → SP	
Pull Data	PULA	32	4	1							SP + 1 → SP, M_{SP} → A	
off stack	PULB	33	4	1							SP + 1 → SP, M_{SP} → B	
Rotate Left	ROL						79	6	69	7	M Rotate left	n z v[6] c
	ROLA	49	2	1							A through carry	n z v[6] c
	ROLB	59	2	1							B C 7 Ø	n z v[6] c
Rotate Right	ROR						76	6	66	7	M Rotate right	n z v[6] c
	RORA	46	2	1							A through carry	n z v[6] c
	RORB	56	2	1							B C 7 → Ø	n z v[6] c
Return From Intrpt	RTI	3B	10	1							Restore P, B. A. XH, XL, PCH, PCL from stack	h i n z v c
Return From Subrtn	RTS	39	5	1							Restore PCH, PCL from stack	
Subtract Acmltrs	SBA	1Ø	2	1							A − B → A	n z v c
Subtr. with Carry	SBCA	82	2	2	92	3	B2	4	A2	5	A − M − C → A	n z v c
	SBCB	C2	2	2	D2	3	F2	4	E2	5	B − M − C → B	n z v c
Set Carry	SEC	ØD	2	1							1 → C	C
Set Intrpt Mask	SEI	ØF	2	1							1 → I	I
Set Overflow	SEV	ØB	2	1							1 → V	V
Store Acmltr	STAA				97	4	B7	5	A7	6	A → M	n z $\overline{V}$
	STAB				D7	4	F7	5	E7	6	B → M	n z $\overline{V}$
Store Stack Pntr	STS				9F	5	BF	6	AF	7	SP_H → M, SP_L → (M + 1)	n[7] z $\overline{V}$
Store Index Reg	STX				DF	5	FF	6	EF	7	X_H → M, X_L → (M + 1)	n[7] z $\overline{V}$
Subtract	SUBA	8Ø	2	2	9Ø	3	BØ	4	AØ	5	A − M → A	n z v c
	SUBB	CØ	2	2	DØ	3	FØ	4	EØ	5	B − M → B	n z v c
Software Intrpt	SWI	3F	12	1							Stack PCL, PCH, XL, XH, A, B, P; Vector FFFA, B	I
Transfer Acmlters	TAB	16	2	1							A → B	n z $\overline{V}$
	TBA	17	2	1							B → A	n z $\overline{V}$
Acmltr A → CCR	TAP	Ø6	2	1							A → CCR	h i n z v c
CCR → Acmltr A	TPA	Ø7	2	1							CCR → A	
Test, Zero	TST						7D	6	6D	7	M − ØØ Sets CCR bits only	n z $\overline{V}$ C
or Minus	TSTA	4D	2	1							A − ØØ	n z $\overline{V}$ $\overline{C}$
	TSTB	5D	2	1							B − ØØ	n z $\overline{V}$ $\overline{C}$
Stack Pntr →Index Reg	TSX	3Ø	4	1							SP + 1 → X	
Index Reg →Stack Pntr	TXS	35	4	1							X − 1 → SP	
Wait for Intrpt	WAI	3E	9	1							Stack PC, X, A, B, CCR. Wait for NMI or valid IRQ. BA goes high.	
		IMMED INHER			DIR/REL (2 byte)		EXTND (3 byte)		INDEX (2 byte)			

Programmer's Model

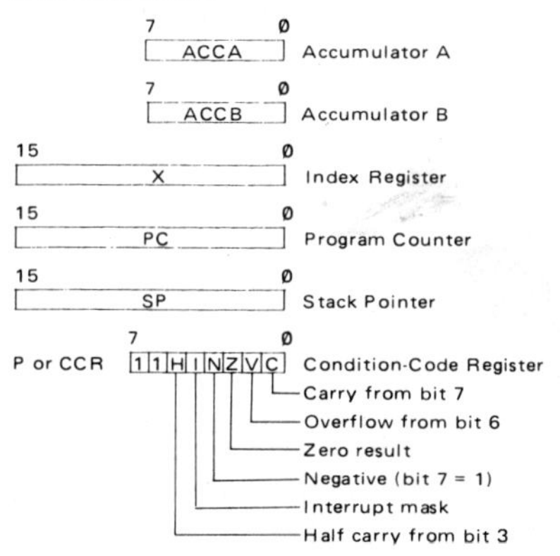

· Boolean AND
+ Boolean OR
⊕ Boolean Exclusive OR
+ Arithmetic Plus
→ Transferred into
~ Machine Cycles
\# Program Bytes

CONDITION-CODE REGISTER TESTS:
(Bit set if true and cleared otherwise)

1. (Bit V) Result = 1ØØØØØØØ?
2. (Bit C) Result = ØØØØØØØØ?
3. (Bit C) Decimal value of most significant BCD character greater than nine? (Not cleared if previously set.)
4. (Bit V) Operand = 1ØØØØØØØ prior to execution?
5. (Bit V) Operand = Ø1111111 prior to execution?
6. (Bit V) Set equal to result of N ⊕ C after shift has occurred.
7. (Bit N) MS bit of result = 1?
8. (Bit V) 2's-complement overflow from subtraction of LS bytes?